Life

The Science of Biology

Seventh Edition

Sinauer Associates, Inc.

W. H. Freeman and Company

Life

The Science of Biology SEVENTH EDITION

William K. Purves Emeritus, Harvey Mudd College • Claremont, California

David Sadava The Claremont Colleges • Claremont, California

Gordon H. Orians Emeritus, The University of Washington • Seattle, Washington

H. Craig Heller Stanford University • Stanford, California

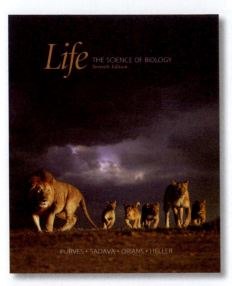

The Cover and Title Photos

The three images of African wildlife on the cover, half-title page, and frontispiece are © Steve Bloom/stevebloom.com.

Life: The Science of Biology, Seventh Edition

Address editorial correspondence to:

Sinauer Associates, Inc., 23 Plumtree Road, Sunderland, MA, 01375 U.S.A.

www.sinauer.com
email: publish@sinauer.com

Address orders to:

VHPS/W.H. Freeman & Co., Order Department, 16365 James Madison Highway, U.S. Route 15, Gordonsville, VA 22942 U.S.A.

www.whfreeman.com
Examination copy information: 1-800-446-8923
Orders: 1-888-330-8477

Library of Congress Cataloging-in-Publication Data

Life, the science of biology / William K. Purves ... [et al.].-- 7th ed.
     p.   cm.
 ISBN 0-7167-9856-5 (hardcover) – ISBN 0-7167-5808-3 (Volume 1)
 ISBN 0-7167-5809-1 (Volume 2) – ISBN 0-7167-5810-5 (Volume 3)
 ISBN 0-7167-8679-6 (Volume 4)
 1. Biology. I. Purves, William K. (William Kirkwood), 1934-

QH308.2 .L565 2003
570--dc22   2003022294

Printed in U.S.A.
First Printing January 2004
Courier Companies Inc.

To our students, especially the 25,000 we have collectively instructed
in introductory biology over the years

About the Authors

David Sadava   Bill Purves   Gordon Orians   Craig Heller

Bill Purves is Professor Emeritus of Biology as well as founder and former chair of the Department of Biology at Harvey Mudd College in Claremont, California. He received his Ph.D. from Yale University in 1959 under Arthur Galston. A fellow of the American Association for the Advancement of Science, Professor Purves has served as head of the Life Sciences Group at the University of Connecticut, Storrs, and as chair of the Department of Biological Sciences, University of California, Santa Barbara, where he won the Harold J. Plous Award for teaching excellence. His research interests focus on the chemical and physical regulation of plant growth and flowering. Professor Purves elected early retirement in 1995, after teaching introductory biology for 34 consecutive years, in order to turn his skills to writing and producing multimedia for introductory biology students.

David Sadava is the Pritzker Family Foundation Professor of Biology at the Keck Science Center of Claremont McKenna, Pitzer, and Scripps, three of The Claremont Colleges. Professor Sadava teaches and has taught courses to undergraduates in biotechnology, biochemistry, cell biology, molecular biology, plant biology, introductory biology, and cancer biology. In addition, he has taught courses in cancer to nonacademic staff members. He is a visiting professor in the Division of Biology at Caltech, and a visiting scientist in medical oncology at the City of Hope Medical Center. He is the author or coauthor of five books on cell biology and on plants, genes, and crop biotechnology. His research has resulted in over 50 papers, many coauthored with under-graduates, on topics ranging from plant biochemistry to pharmacology of narcotic analgesics to human genetic diseases. For the past decade, he and his collaborators have investigated multi-drug resistance in human small-cell lung carcinoma cells with a view to understanding and overcoming this clinical challenge.

Gordon Orians is Professor Emeritus of Biology at the University of Washington. He received his Ph.D. from the University of California, Berkeley in 1960 under Frank Pitelka. Professor Orians has been elected to the National Academy of Sciences and the American Academy of Arts and Sciences, and is a Foreign Fellow of the Royal Netherlands Academy of Arts and Sciences. He was President of the Organization for Tropical Studies, 1988–1994, and President of the Ecological Society of America, 1995–1996. He is a recipient of the Distinguished Service Award of the American Institute of Biological Sciences. Professor Orians is a leading authority in ecology, conservation biology, and evolution. His research on behavioral ecology, plant–herbivore interactions, community structure, and environmental policy has taken him to six continents. He now devotes full time to writing and to helping apply scientific information to environmental decision-making.

Craig Heller is the Lorry I. Lokey/Business Wire Professor in Biological Sciences and Human Biology at Stanford University. He earned his Ph.D. from the Department of Biology at Yale University in 1970, and then spent two years as a Postdoctoral Fellow at Scripps Institute of Oceanography studying how the brain regulates body temperature in mammals. Dr. Heller has taught at Stanford since 1972, served as Director of the Program in Human Biology, Chairman of the Biological Sciences Department, and Associate Dean of Research. Dr. Heller is a fellow of the American Association for the Advancement of Science and a recipient of the Walter J. Gores Award for excellence in teaching. His research focus is on the neurobiology of sleep and circadian rhythms, mammalian hibernation, the regulation of body temperature, and the physiology of human performance. Over the years, Dr. Heller has done research on systems ranging from sleeping college students to diving seals to hibernating bears to exercising athletes.

Preface

Like populations, textbooks evolve. Not only must the content change, but our goals must be rethought as well. We have tried, in this Seventh Edition of *Life*, to emphasize those things that will best prepare students for their future careers. The store of biological knowledge increases ever more rapidly. This requires us to seek a careful balance between thoroughness of coverage and appropriate treatment of the process, or processes, of science. We have retained and expanded the emphasis on experiment—on how things were and are learned. The emphasis remains on concepts. However, because different instructors emphasize different topics, and because a key role of the textbook is as a "place to look things up," this book is comprehensive as well. We provide sufficient detail to meet most needs without making the book too voluminous. We have enhanced our emphasis on an evolutionary theme, and have added new material on such important cutting-edge topics as evolutionary developmental biology ("evo–devo") and earth systems.

Experimental Focus

Since the First Edition of this book, we have been committed to answering the question, "How do we know?" As the book has evolved, this commitment has steadily deepened.

Obviously, we can't provide the experimental or observational evidence for every fact or theory we discuss. However, we have selected the key experiments underlying some of the most important biological principles. Some are very recent, at the cutting edge of current research; others are classics. To supplement and highlight the text discussions, we have created unique Experiment figures that show how experiments, field observations, and comparative methods help biologists formulate and test hypotheses. Other figures highlight some of the many laboratory and field methods used to do this research. In this edition there are more than 100 Experiment and Research Method figures (examples at right). In addition, we have 20 new Experiment tutorials on www.thelifewire.com, the Website/CD that was created for *Life*.

We hope that, in tandem with the frequent discussions of experimental evidence, these figures and tutorials will help students understand and appreciate the nature of biology as a vital, ongoing experimental science.

New Chapters, A New Unit, and New Essays

This edition features two new chapters that reflect current trends in biological research. Chapter 21 ("Development and Evolutionary Change") introduces students to evolutionary developmental biology, a rapidly growing field that deals with how the molecular genetics of the developing organism affects the evolution of complex morphology and biochemistry. In addition, the interaction of environment and embryogenesis on the ultimate form of an organism is covered at length.

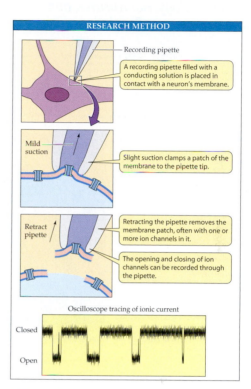

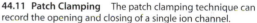

44.11 Patch Clamping The patch clamping technique can record the opening and closing of a single ion channel.

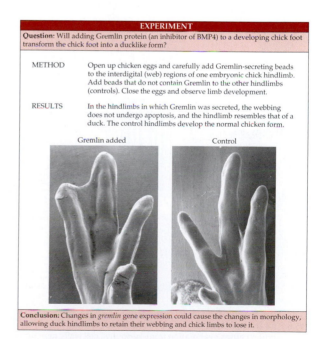

21.7 Changing the Form of an Appendage In this experiment, chick hindlimbs exposed to Gremlin-secreting beads developed ducklike webbed feet.

Part Seven · THE BIOLOGY OF ANIMALS

41 Physiology, Homeostasis, and Temperature Regulation

The Tour de France, a 3-week, 3,500-km bicycle race, is arguably the most extreme and demanding of all athletic events. Competitors are on their bikes 5 to 7 hours a day, riding at an average speed of over 41 kilometers an hour across terrain that includes the mountains of the French Alps. The Tour can be compared to running 20 marathons at world-class pace in 20 days. In 2003, Lance Armstrong won the Tour for the fifth time.

How can an athlete perform at this level, and what results in a winning performance? A number of factors are involved, including determination, skill, and physiology. It is physiology that is the subject of Part Seven of this book. **Physiology** can be simply defined as the science of how organisms work. Physiological mechanisms span the range from molecular to behavioral.

You learned in earlier parts of this book that cells oxidize glucose to produce ATP, which is then used to do biological work, such as the contraction of muscles. Performance in an event such as the Tour de France is limited ultimately by the maximum sustainable rate at which the athlete's body can convert the chemical energy of food into the mechanical energy of muscles. That rate is determined by more factors than the cellular biochemical reactions you have studied. Oxygen has to be delivered to the blood, and the blood has to be pumped to the muscles and other organs. Food has to be converted to fuel molecules by the digestive system, and those fuel molecules have to be distributed to the mitochondria of the muscle cells. The waste products of cell metabolism have to be carried away and eliminated. The temperature, ion balance, and pH of muscles and other organs have to be maintained at optimal levels. All of these tasks and more are carried out by the physiological systems we will study in this part of the book. How does Lance measure up in terms of some of his physiological characteristics?

One measure of exercise capacity is the maximum rate at which a person can take up and utilize oxygen: the V_{O_2max}. For a healthy man, a typical value is about 40 ml O_2 per kg body mass per minute. Lance's V_{O_2max} is more than twice that value. Whereas a normal, fit man might burn up to 3,500 Calories on a particularly active day, during the Tour, Lance burns about 6,500 Calories a day—10,000 Calories on peak days! Because Lance has an extremely low proportion of body fat—only 4–5 percent (20 percent is normal)—he must eat, and his body must

France on 10,000 Calories a Day Lance Armstrong is a remarkable athlete largely because of the capacity of his physiological systems.

Chapter 58 ("Earth System Science") introduces students to the new field focusing on Earth as a whole, studying great cycles of materials, inputs of solar energy, and the interactions between living organisms and the physical environment that determine how Earth as a planet functions today.

We have reorganized our treatment of developmental biology to create a new unit, Part Three, Development. Developmental biology is the subdiscipline of biology that draws from a great span of other subdisciplines, from molecular biology to ecology. Thus our new unit begins with updated chapters on "Differential Gene Expression in Development" and "Animal Development: From Genes to Organism," and concludes with the new Chapter 21. This unit leads naturally forward from Part Two, Information and Heredity, and progresses to a transition to Part Four, Evolutionary Processes.

We believe that another mission in the training of new scientists is to make them aware of the links between science and society. New for the Seventh Edition are eight essays, each of which concludes a Part of *Life*. We invited eight eminent humanists and social scientists to address a topic that bridges biology and an important ethical, moral, philosophical, or economic issue. For example, Bonnie Steinbock, SUNY Albany, examines some of the ethical implications arising from today's latest research activities in her essay, "What Are the Moral Issues Surrounding Stem Cell Therapy?" which concludes Part Three, Development.

Pedagogy

In addition to our attention to updates and enhancements of content, we have again set as a major goal making the presentation as clear and helpful as possible for the student. Here are some of the ways in which we intend to make the reader's work easier and more effective:

▶ Every chapter begins with a short story (example above) to grab the reader's interest and encourage further exploration of the chapter's content.

▶ We have increased the number of bulleted lists (like this one) that highlight key points.

▶ All second-level headings (the workhorses) are declarative sentences that describe at a glance the text and figures that follow.

▶ We have added more interim summaries and bridges between topics to keep the reader on track.

▶ All chapter summaries include references to key figures and now also provide convenient links to the appropriate tutorials and activities on the Website/CD.

▶ Responding to adopters' requests, we have put the Self-Quizzes back in the text, since some students prefer to access them directly in the chapters.

▶ Each chapter concludes with four or five "For Discussion" items that help the reader synthesize the chapter's main concepts.

▶ The balloon captions (example at left) that we introduced in the Fifth Edition are now further streamlined and positioned for maximum pedagogical effectiveness.

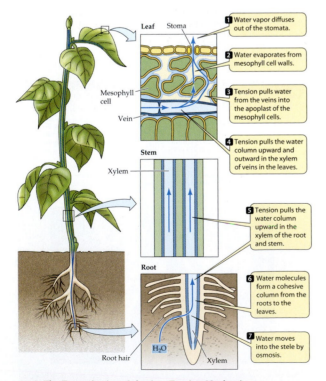

1 Water vapor diffuses out of the stomata.

2 Water evaporates from mesophyll cell walls.

3 Tension pulls water from the veins into the apoplast of the mesophyll cells.

4 Tension pulls the water column upward and outward in the xylem of veins in the leaves.

5 Tension pulls the water column upward in the xylem of the root and stem.

6 Water molecules form a cohesive column from the roots to the leaves.

7 Water moves into the stele by osmosis.

Leaf Stoma
Mesophyll cell
Vein
Stem
Xylem
Root
Root hair H_2O Xylem

36.8 The Transpiration–Cohesion–Tension Mechanism
Transpiration causes evaporation from mesophyll cell walls, generating tension on the xylem. Cohesion among water molecules in the xylem transmits the tension from the leaf to the root, causing water to move from the soil to the atmosphere.

Although our much-praised art program appears very similar to the Sixth Edition's, it has undergone a very significant pedagogical upgrade. Our new artist, Elizabeth Morales, who has been illustrating biology textbooks for more than 20 years, worked with each author to evaluate the effectiveness of every piece of line art in the book. The result is many hundreds of simplifications and improvements in clarity.

Media and New Video Collection

The Seventh Edition media and supplements are built around two main goals: (1) to provide the student with a collection of tools to help digest and truly understand the vast amount of material presented in this textbook; and (2) to provide the instructor with the richest possible collection of resources to aid in effectively teaching the course: preparing, presenting the lecture, providing course materials online, and assessing student comprehension.

Working with a dozen contributing authors and an experienced scientific multimedia studio, we have put together an outstanding package that is built specifically for this textbook. For example, the collection of over 100 in-depth animated tutorials was created using textbook art as the basis for the animations, and the introductions, conclusions, and quizzes were matched in level, terminology, and content to the Seventh Edition of *Life*.

In our continuing effort to provide instructors with outstanding visual resources for the lecture, and in response to many requests from biology professors, we are introducing a new feature for the Seventh Edition: "Seeing Life: Video Sequences in Biology." This collection of approximately 200 outstanding video segments (over two hours of footage that spans the book's coverage; example at right) can help capture the attention and imagination of your students with stunning moving images of biological phenomena. Each video segment is fully narrated.

(For a detailed description of all the media and supplements available for the Seventh Edition, please turn to "*Life*'s Supplements" on page xvi.)

The Eight Parts

Part One, The Cell, leads from basic chemistry to cell structure, membranes, and energetics. Chapters 3 ("Large Molecules") and 4 ("Cells: The Basic Units of Life") now integrate ideas on the origin and evolution of cells. The discussion of thermodynamics in Chapter 6 has been reduced and is now focused on biological applications. The art in the chapters on respiration and photosynthesis has been streamlined for clarity.

Part Two, Information and Heredity, retains the order of principles of genetics and molecular biology in the first chapters followed by applications of them in the later chapters. We have updated all of the material on genomics, and added newly emerging approaches of study, including RNA interference.

Part Three, Development, brings together and integrates topics in developmental biology to build upon the detailed treatment of genetics in Part Two and set the stage for the discussions of evolutionary processes in Part Four. We show how new insights into the ways in which genes and environment interact to yield the forms of adult organisms are providing important new perspectives on the origins of evolutionary novelties.

Part Four, Evolutionary Processes, begins with an overview of the history of life on Earth, followed by a detailed treatment of the evolutionary mechanisms and processes that are being investigated to explain those patterns. Chapter 25 ("Constructing and Using Phylogenies") has been updated to incorporate the most recent methods of inferring evolutionary relationships among organisms. Chapter 26 ("Molecular and Genomic Evolution") describes some of the exciting new information on how the genomes of organisms have evolved and how processes of genomic evolution help us understand the evolution of the diversity of life.

Part Five, The Evolution of Diversity, has been updated to reflect current views on phylogeny. We have retained the strong evolutionary thread, emphasizing lineages over some classically defined groups. The treatment of flowering plants other than monocots and eudicots has been upgraded. We have retained the organization of the chapters on animal diversity to reflect the three great lineages of animals—Lophotrochozoans, Ecdysozoans, and Deuterostomes—while incorporating new information on evolutionary relationships among animals revealed by new fossil finds and improved methods of inferring phylogenetic relationships.

In Part Six, The Biology of Flowering Plants, we have improved the explanations of bulk flow in xylem and phloem by modifying and simplifying the art and focusing the text more directly on key mechanisms. New material on clock genes, auxin carriers, RNA silencing, and other topics has been added.

Part Seven, The Biology of Animals, is about how animals work. Although we give major coverage of human physiology, we try to embed it in the background of comparative animal physiology. We have made an effort to offer a complete and broad coverage of physiology while still introducing the student to new advances. For example, the story of the molecular mechanism of the biological clock has advanced very rapidly since the last edition of this book. The genetic control of sexual behavior in fruit flies is another example where the two ends of the biological spectrum—molecular to behavioral—are coming together. As in previous editions, throughout Part Seven, we bring the student back to issues of control and regulation. These are the most central concepts in all of physiology.

Part Eight, Ecology and Biogeography, has been substantially reorganized and updated. Chapter 53 ("Behavioral Ecology") now emphasizes how the decisions that individual animals make can influence population dynamics and community structure. Chapter 54 ("Population Ecology") is organized around the key questions about populations that ecologists attempt to answer. Chapter 55 ("Communities and Ecosystems") combines and integrates material that was separated into two chapters in the Sixth Edition, providing a more integrated treatment of those topics. Chapter 57 ("Conservation Biology") now gives more emphasis on how science is used in the service of conserving Earth's biological diversity. Finally, the new chapter on "Earth System Science" (Chapter 58) introduces students to this rapidly developing field that looks at Earth as a whole.

Full Book or Paperbacks

We again provide *Life* both as the full book and as a cluster of paperbacks. For the Seventh Edition, the new Part Three is an additional fourth paperback. Thus, instructors who want to use less than the whole book, or who want their students to have more portable units, can choose from these split volumes:

▶ Volume I, The Cell and Heredity, includes: Part One, The Cell (Chapters 2–8); and Part Two, Information and Heredity (Chapters 9–18).
▶ Volume II, Evolution, Diversity, and Ecology, includes: Part Four, Evolutionary Processes (Chapters 22–26); Part Five, The Evolution of Diversity (Chapters 27–34); and Part Eight, Ecology and Biogeography (Chapters 53–58).
▶ Volume III, Plants and Animals, includes: Part Six, The Biology of Flowering Plants (Chapters 35–40); and Part Seven, The Biology of Animals (Chapters 41–52).
▶ Volume IV, Development, includes Part Three, Development (Chapters 19–21)

Note that Volumes I, II, and III include the book's front matter, Glossary, and Index plus Chapter 1.

There Are Many People To Thank

When we met in Sunderland with the key editorial and marketing people from Sinauer Associates and W. H. Freeman to plan this Seventh Edition, we determined that a central goal would be to involve and seek advice from a greater number of our teaching colleagues. This turned out to be a rich idea. We now have more than twice as many instructors to thank for their help in crafting this edition. We began the process by recruiting adopters of the Sixth Edition to report on what worked and what could be improved. With this input, we created the plan for the Seventh Edition and wrote the first drafts. Then, every chapter was reviewed by at least five introductory biology teachers. In addition to checking for accuracy and clarity, they helped us make decisions on material to cut or add. Many productive e-mail exchanges took place at this stage to the book's benefit.

After the chapters and final art were put into page proofs, we built in another round of reviews to help catch and eliminate lingering errors. This final check also provided suggestions for making the text and figures more precise. Finally, and concurrently with the manuscript and accuracy reviews, we got critical scrutiny of all of the animations, tutorials, and activities for the book's Website. We heartily thank all of the people who contributed these reviews. It's a demanding process but there is no doubt that a better book and supporting media have resulted because of it.

We wish to especially thank Scott Gilbert for providing an excellent draft of the new Chapter 21 on evolution and development.

As mentioned earlier, our new artist, Elizabeth Morales, has also made a very large contribution by assuring that *Life*'s extensive illustration program is as effective as possible. Many of the concepts that are illustrated in the book are complex. It takes an illustrator with Elizabeth's talent to render them both artistically and with maximum clarity.

The exact same team that worked with or within Sinauer Associates on the Sixth Edition on the many facets of editing the book was on board again. James Funston provided forceful and insightful developmental editing. Norma Roche contributed her elegant copy editing. Carol Wigg yet again deftly coordinated the entire editorial process and crafted many of the new captions. Jeff Johnson once again delivered the elegant interior and cover designs and coordinated the layout process. Susan McGlew orchestrated the mammoth reviewing process described above. And David McIntyre produced another dazzling array of new photographs for our selection.

W. H. Freeman's marketing and sales group has again succeeded in bringing *Life* to a wider audience. They are both effective ambassadors and skillful transmitters of information. We depend on their expertise and energy to keep us in touch with how *Life* is perceived by its users.

The constant asset we have had in our efforts to produce a better and better book to help students learn and appreciate the science of *Life* has been Andy Sinauer. For over 34 years Andy has run a company that produces the highest quality books in the biological sciences. His strategy has been to maintain a staff of talented, dedicated people and to give each book his personal attention from recruitment of authors to marketing. We feel that we have had more than our fair share of Andy's attention. He is the constant motivator to find ways to make our book and the teaching of biology more effective. He gently but firmly keeps us on track, and he is always ready to deal with the biggest crisis or the smallest detail. Andy, we are fortunate to work with you and we greatly appreciate all you have done to make *Life* a book of which we are exceedingly proud.

Bill Purves   *David Sadava*   *Gordon Orians*   *Craig Heller*

Reviewers for the Seventh Edition

Between-Editions Reviewers

Annalisa Berta, San Diego State University

Edward L. Braun, University of Florida

John Carlson, Yale University

Susan Cockayne, Brigham Young University

Craig Coleman, Brigham Young University

Karen Curto, University of Pittsburgh

Mark Decker, University of Minnesota

William Eickmeier, Vanderbilt University

William Eldred, Boston University

Ross Feldberg, Tufts University

Steven K. Fisher, University of California, Berkeley

Merrill L. Gassman, University of Illinois, Chicago

Hans-Willi Honegger, Vanderbilt University

Michele Igo, University of California, Davis

Dan Janik, University of Wisconsin, Eau Claire

Jeffrey Jensen, University of Maryland, College Park

Norman Johnson, Ohio State University

Rebecca Kimball, University of Florida

Todd Kostman, University of Wisconsin, Osh Kosh

Mary Lehman, Longwood University

Carl Luciano, Indiana University of Pennsylvania

Paula Mabee, University of South Dakota

David Magrane, Morehead State University

Charles Mallery, University of Miami

Shawn Meagher, Western Illinois University

Ken Mossman, Arizona State University

Darrel Murray, University of Illinois, Chicago

John New, Loyola University Chicago

Lou Pech, Carroll College

Lee Pike, East Tennessee State University

Don Potts, University of California, Santa Cruz

Eric Ribbens, Western Illinois University

Chris Romero, Front Range Community College, Larimer

Al Ruesink, Indiana University

Robert Savage, Williams College

Erik Scully, Towson University

Mark Storey, Texarkana College

James Traniello, Boston University

Nancy Wade, Old Dominion University

Raymond P. White, City College of San Francisco

Elizabeth Willott, University of Arizona

Manuscript Reviewers

Heather Addy, University of Calgary

Sylvester Allred, Northern Arizona University

Robert Angus, University of Alabama, Birmingham

David Armstrong, University of Colorado

Art Ayers, Albertson College of Idaho

Ellen Baker, Santa Monica College

Sharon Balchak, Notre Dame College

Monique Barakat, Stanford University

Ruth Beattie, University of Kentucky

Spencer Benson, University of Maryland, College Park

Andrew Blaustein, Oregon State University

J. Jose Bonner, Indiana University

Thomas Boyle, University of Massachusetts, Amherst

Bryan Brendley, Gannon University

George Brooks, Ohio University, Zanesville

Angela Brown, University of Idaho

James Brown, North Carolina State University

Patrick J. Bryan, Central Washington University

Matthew Buechner, University of Kansas

Art Buikema, Virginia Polytechnic Institute and State University

Warren Burggren, University of North Texas

Nancy T. Burley, University of California, Irvine

Rob Carey, Pima Community College

Clint Carter, Vanderbilt University

Clare Chatot, Ball State University

Helen C. Chuang, Southern Utah University

Elizabeth Conner, University of Massachusetts, Amherst

Ron Cooper, University of California, Los Angeles

Greg Crowther, University of Puget Sound

Sid Das, University of Texas, El Paso

Alan Day, University of Western Ontario

Mark Decker, University of Minnesota

Carmen Domingo, San Francisco State University

Ernest DuBrul, University of Toledo

Stephen Ebbs, Southern Illinois University, Carbondale

Alex Enyedi, Western Michigan University

Gordon L. Fain, University of California, Los Angeles

Paul Ferguson, University of Illinois, Urbana-Champaign

Steven K. Fisher, University of California, Berkeley

Gregory Florant, Colorado State University

Ellen Freund, National Marine Fisheries Service

Javier Gago, Glendale College

Merrill L. Gassman, University of Illinois, Chicago

Daniel Geiger, Natural History Museum of Los Angeles County

Michael Ghedotti, Regis University

Ken Gobalet, California State University, Bakersfield

Deborah Gordon, Stanford University

Dina Gould Halme, Massachusetts Institute of Technology

Dana Haine, Central Piedmont Community College

Leslie Hickok, University of Tennessee, Knoxville

Robert Hinrichsen, Indiana University of Pennsylvania

Christie Howard, University of Nevada, Reno

Andrew Jarosz, Michigan State University

Walter S. Judd, University of Florida

John Kalb, Canisius College

Larry Katz, Rutgers University, Cook College

Laura Katz, Smith College

Steve Kelso, University of Illinois, Chicago

Travis Knowles, Francis Marion University

Allen Kurta, Eastern Michigan University

Andrew Lack, Oxford Brookes University

Ralph Larson, San Francisco State University

Howard Laten, Loyola University Chicago

Carl Luciano, Indiana University of Pennsylvania

Paula Mabee, University of South Dakota

Barbara Mable, University of Guelph

Nancy Magill, Coe College

Charles Mallery, University of Miami

Jim Manser, Harvey Mudd College

Ron Markle, Northern Arizona University

Patrick H. Masson, University of Wisconsin, Madison

Paul Mayes, Muscatine Community College

Kenneth W. McCravy, Western Illinois University

Wayne Merkley, Drake University

Frank Messina, Utah State University

Darrel Murray, University of Illinois, Chicago

Bill Newcomb, Queens University

Gregory Nishiyama, College of the Canyons

Tom Oeltmann, Vanderbilt University

Laura Olsen, University of Michigan

Sanford E. Ostroy, Purdue University

Tom Owens, Cornell University

Julia Thom Oxford, Boise State University

Aparna D.N. Palmer, Mesa State College

M. Theresa Pavlovitch, Pasadena City College

Craig Peebles, University of Pittsburgh

Karen Perkins, Mount St. Mary's College

Jeff Pommerville, Maricopa Community Colleges

Ellen Porzig, Stanford University

Chris Romero, Front Range Community College

Donald Ruch, Ball State University

Al Ruesink, Indiana University

Walter Sakai, Santa Monica College

Stan Schein, University of California, Los Angeles

Daniel Scheirer, Northeastern University

Nicci Schoob, University of Illinois, Urbana-Champaign

Rodney J. Scott, Wheaton College

Neil Shay, University of Notre Dame

Jim Shinkle, Trinity University

Margaret Silliker, DePaul University

Philip Snider, University of Houston, University Park

Mitchell Sogin, Marine Biological Laboratory, Woods Hole

John Stiller, East Carolina University

Mark Sturtevant, University of Michigan, Flint

Cecil Stushnoff, Colorado State University

Kevin Swier, Chicago State University

Iain E.P. Taylor, University of British Columbia

Gerald Thrush, California State University, San Bernardino

Stephen Timme, Pittsburg State University

David Tissue, Texas Tech University

F. Daniel Vogt, Plattsburgh State University of New York

Jonathan Wenger, Concordia University, St. Paul

Lisa Werner, Pima Community College

David Wessner, Davidson College

Dave Westenberg, University of Missouri, Rolla

Mary White, Southeastern Louisiana University

Barny Whitman, University of Georgia

Elizabeth Willott, University of Arizona

Christopher Wills, University of California, San Diego

Michelle Withers, Louisiana State University

Jay Zimmerman, St. John's University, Queens

Accuracy Reviewers

Heather Addy, University of Calgary

Sylvester Allred, Northern Arizona University

Vernon Bauer, Francis Marion University

Wade Bell, Virginia Military Institute

Graeme Berlyn, Yale University

Michael Black, California Polytechnic State University

Andrew Blaustein, Oregon State University

Franklyn Bolander, University of South Carolina

Thomas Boyle, University of Massachusetts, Amherst

Patrick J. Bryan, Central Washington University

Matthew Buechner, University of Kansas

Art Buikema, Virginia Polytechnic Institute and State University

Warren Burggren, University of North Texas

Nancy Burley, University of California, Irvine

Naomi Capuccino, Carleton University

Domenic Castignetti, Loyola University Chicago

David Champlin, University of Southern Maine

Elisabeth Ciletti, Pasadena City College

Keith Clay, Indiana University

William Collins, Stony Brook State University of New York

Ronald Cooper, University of California, Los Angeles

Charles Creutz, University of Toledo

Mike Dalbey, University of California, Santa Cruz

Deborah Dardis, Southeastern Louisiana University

Gerald Deitzer, University of Maryland

Laura DiCaprio, Ohio University

Chuck Duggins, University of South Carolina

Nancy Elwess, Plattsburgh State University of New York

Ray Evert, University of Wisconsin

Marvin Friedman, Hunter College

Eve Gallman, University of Illinois, Urbana-Champaign

Charles Galt, California State University, Long Beach

Michael Ghedotti, Regis University

Diane Gorski, University of Wyoming

Brian K. Hall, Dalhousie University

Susan Han, University of Massachusetts, Amherst

Dennis C. Haney, Furman University

Mike Hart, Dalhousie University

Paul Hasegawa, Purdue University

Albert Herrera, University of Southern California

Hans-Willi Honegger, Vanderbilt University

David Jenkins, University of Alabama, Birmingham

Walter S. Judd, University of Florida

Thomas Kane, University of Cincinnati

Loren Knapp, University of South Carolina

William Kroll, Loyola University Chicago

Josephine Kurdziel, University of Michigan

Sandra F. Larson, Furman University

Howard Laten, Loyola University Chicago

Mary Lehman, Longwood University

Greg Lewis, Furman University

Min-Ken Liao, Furman University

Carol Maillet, Augustana College

Richard Malkin, University of California, Berkeley

Charles Mallery, University of Miami

Richard McCarty, The Johns Hopkins University

Jill Miller, Amherst College

Subhash Minocha, University of New Hampshire

Marty Nemeroff, Rutgers University

Seán O'Connell, Western Carolina University

Peter O'Day, University of Oregon

William H. Outlaw, Jr., Florida State University

Randall Packer, George Washington University

Craig Peebles, University of Pittsburgh

David Polcyn, California State University, San Bernardino

Ron Poole, McGill University

Leo Racich, Southern Illinois University, Edwardsville

Wendy Raymond, Williams College

Robert Reed, Southern Utah University

Eric Ribbens, Western Illinois University

Patricia Rugaber, Coastal Georgia Community College

Andy Schroeder, Harvard University

Stylianos Scordilis, Smith College

Tim Shannon, Francis Marion University

Margaret Silliker, DePaul University

Heidi Sleister, Drake University

Andrew Smith, Arizona State University

Jim Smith, Michigan State University

Philip Snider, University of Houston, University Park

Mitchell Sogin, Marine Biological Laboratory, Woods Hole

Frederick W. Spiegel, University of Arkansas, Fayetteville

Kevin Swier, Chicago State University

Doug Thrower, University of California, Santa Barbara

Briana Timmerman, University of South Carolina

Elizabeth Van Volkenburgh, University of Washington

Benjamin Weeks, Adelphi University

Ted Wilson, Winona State University

Greg Wray, Duke University

Media and Supplements Reviewers

Michael Black, California Polytechnic State University

Steve Brewer, University of Massachusetts, Amherst

Mark Browning, Purdue University

Bob Cabin, Plattsburgh State University of New York

Mark Decker, University of Minnesota

Ernest Dubrul, University of Toledo

William Eldred, Boston University

Joanne Ellzey, University of Texas, El Paso

James Franzen, University of Pittsburgh

Tejendra Gill, University of Houston

Jon Glase, Cornell University

Dominic Lannutti, El Paso Community College

Peter Lortz, North Seattle Community College

Charles Mallery, University of Miami

Philip Meneely, Haverford College

Nancy Raffetto, University of Wisconsin, Madison

Ken Robinson, Purdue University

Robert Schmidt, University of California, San Diego

Allen Shearn, The Johns Hopkins University

Thomas Terry, University of Connecticut

Lisa Werner, Pima Community College

To the Student

There are a few things you can do to help you get the most from this book and from your course. For openers, read the book actively—don't just read passively, but do things that force you to think as you read. If we pose questions, stop and think about them. Ask questions of the text as you go. Do you understand what is being said? Does it relate to something you already know? Is it supported by experimental or other evidence? Does that evidence convince you? How does this passage fit into the chapter as a whole? Annotate the book—write down comments in the margins about things you don't understand, or about how one part relates to another, or even when you find an idea particularly interesting. People remember things they think about much better than they remember things they have read passively. Highlighting is passive; copying is drudge work; questioning and commenting are active and well worthwhile.

"Read" the illustrations actively too. You will find the balloon captions in the illustrations especially useful—they are there to guide you through the complexities of some topics and to highlight the major points.

The chapter summaries will help you quickly review the high points of what you have read. They also identify particular illustrations that you should study to help organize the material in your mind. Add concepts and details to the framework by reviewing the text. Also in the summaries are keyed reminders of the tutorials and activities on the book's website. A way to review the material in slightly more detail after reading the chapter is to go back and look at the boldfaced terms. You can use the boldfaced terms to pose questions—and see if you can answer those questions. The boldfacing will probably be more useful on a second reading than on the first.

The "Self-Quiz" in each chapter is a convenient way to measure your mastery of the material. All answers are at the end of the book. Use the "For Discussion" questions that end each chapter. These questions are usually open-ended and are intended to cause you to reflect on the material.

The glossary and the index can help you a great deal. When you are uncertain of the meaning of a term, check the glossary first—it has more than 1,500 definitions. If you don't find a term in the glossary or you want a more thorough discussion of it, use the index to find where it's discussed.

The Website

The Seventh Edition Student Website (www.thelifewire.com; also available on CD at your instructor's request) is designed to help you learn the vast amount of material we are presenting in this book in a variety of ways. Throughout the book, you will see this icon on headings and figure titles. Wherever you see the icon, you will find a corresponding animated tutorial or activity on the website. They will reinforce your understanding of the key concepts presented in the book. Another important feature of the website is the extensive set of Interactive Quizzes. These quizzes incorporate figures from the book, thorough answer feedback, and links to electronic versions of book pages. There is a second quiz for each chapter, the Online Quiz, the results of which can be emailed to your instructor if he/she requests. Also on the website are key terms flashcards, a full glossary (with audio pronunciations of difficult terms), suggested readings, and two useful documents: Math for Life and Student Survival Skills. The website has been built with you in mind, we hope you find it to be an important resource in your study of introductory biology.

What If the Going Gets Tough?

Most students occasionally have difficulty in courses, including biology courses. If you find that you are slipping behind in the course, or if a particular topic is giving you an unreasonable amount of trouble, here are some useful steps you might take. First, the basics: attend class, take careful lecture notes, and read the textbook assignments. Second, note that one of the most important roles of studying is to discover what you don't know, so that you can do something about it. Use the index, the glossary, the chapter summaries, and the text itself to try to answer any questions you have and to help you organize the material. Make a habit of looking over your lecture notes within 24 hours of when you take them—find out right away what points are unclear, and get them straightened out in your mind. The website can help by providing an additional perspective.

If none of these self-help remedies does the trick, get help! Other students are often a good source of help, because they are dealing with the material at the same level as you are. Study groups can be very useful, as long as the participants are all committed to learning the material. Tutors are almost always helpful, as are faculty members. The main thing is to get help when you need it. It is not a good idea to be strong and silent and drift into a low grade.

But don't make the grade the point of this or any other course. You are in college to learn, to pursue interesting subjects, and to enjoy the subjects you are pursuing. We hope you'll enjoy the pursuit of biology.

Bill Purves   *David Sadava*   *Gordon Orians*   *Craig Heller*

Life's Supplements

For the Student

Student Website: www.thelifewire.com

(Also available as a CD optionally bundled with the book)

The *Life*, Seventh Edition Website offers the student a wealth of in-depth, self-directed review material. With the help of three contributing authors and an experienced scientific multimedia design studio, we've created a collection of resources that take advantage of the flexibility and interactivity of the electronic medium to help the student master the many complex concepts presented in the textbook. Features of the site include:

▶ **Interactive Summaries:** This is the most convenient way to review the entire chapter. The summary contains links to all the key figures from the chapter as well as all of the relevant animated tutorials and activities.

▶ **Animated Tutorials:** These 106 in-depth tutorials feature an introduction, a detailed animation, a conclusion, and a quiz on the topic covered. The clear presentation of complex topics makes these tutorials a powerful learning tool. New to the Seventh Edition are 20 **Key Experiment** tutorials that expand upon some of the most important experiments depicted in the book, and 10 additional new tutorials.

▶ **Activities:** Over 120 interactive activities help the student learn important facts through a wide range of exercises, such as labeling steps in processes or parts of structures, building a phylogenetic tree, and identifying different types of organisms.

▶ **Flashcards:** For each chapter of the book, there is at least one Flashcard activity that allows students to review and then test themselves on the key terminology from the chapter.

▶ **Interactive Quizzes:** Every question includes a figure from the book, thorough feedback on answer choices, and links to electronic versions of book pages, where the related material is highlighted for immediate reinforcement of concepts.

▶ **Online Quizzes:** An additional review tool, these quizzes can be taken online, and, at the instructor's request, students can submit their scores electronically.

▶ **Plus:** A full **Glossary** with audio pronunciations, **Suggested Readings**, and two useful study aid documents: **Math for Life** and **Student Survival Skills**.

Order ISBN 0-7167-5807-5, Seventh Edition Student CD, or ISBN 0-7167-8851-9, Text with CD

Study Guide

Edward M. Dzialowski, *University of North Texas*; Betty McGuire, *Smith College*; Lindsay Goodloe, *Cornell University*; Nancy Guild, *University of Colorado*; and Paula Mabee, *University of South Dakota*

For each chapter of the text, the Study Guide provides a detailed review of Important Concepts, a Big Picture overview, Common Problem Areas to pay particular attention to, Study Strategies, and a full set of Knowledge and Synthesis questions and Application questions, all with answers and explanations. Order ISBN 0-7167-5811-3

Lecture Notebook

This useful tool consists of all the artwork from the textbook (more than 1,000 images) presented in the order in which they appear in the text, with ample space for note-taking. Since the notebook has already done the drawing, students can focus their attention on the concepts. They will absorb the material more efficiently during class, and their notes will be clearer, more accurate, and more useful when they study from them later. Order ISBN 0-7167-5812-1

Animated Tutorial 11.1: The Meselson–Stahl Experiment

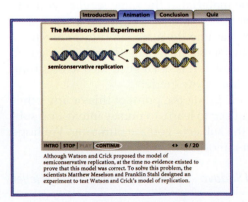

Although Watson and Crick proposed the model of semiconservative replication, at the time no evidence existed to prove that this model was correct. To solve this problem, the scientists Matthew Meselson and Franklin Stahl designed an experiment to test Watson and Crick's model of replication.

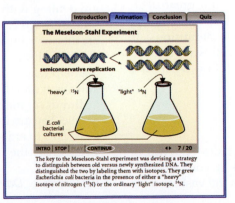

The key to the Meselson-Stahl experiment was devising a strategy to distinguish between old versus newly synthesized DNA. They distinguished the two by labeling them with isotopes. They grew *Escherichia coli* bacteria in the presence of either a "heavy" isotope of nitrogen (^{15}N) or the ordinary "light" isotope, ^{14}N.

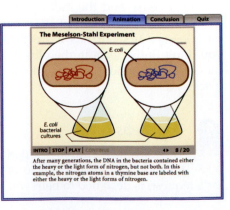

After many generations, the DNA in the bacteria contained either the heavy or the light form of nitrogen, but not both. In this example, the nitrogen atoms in a thymine base are labeled with either the heavy or the light forms of nitrogen.

For the Instructor

Instructor's Media Library

As classrooms and introductory biology courses become more and more technically sophisticated, instructors are using an ever wider array of electronic resources in planning, managing, assessing, and in presenting the lecture. In order to give instructors the widest possible range of resources to help them engage students and better communicate the material, we have assembled an unparalleled collection of media resources. The Seventh Edition of *Life* features the most expansive Instructor's Media Library yet, available on a set of CDs and DVDs. The Media Library includes:

▶ **Textbook Illustrations, Photos, and Tables:** Every image from the textbook is provided in both JPEG (high- and low-res) and PDF formats. These high-quality images have all been optimized for use in PowerPoint® or other presentation software.

▶ **Unlabeled Figures:** New for the Seventh Edition, and in response to instructor requests, every figure in the textbook is now provided in an additional, unlabeled format, in which all text labels have been removed. These are a useful resource for student quizzing and custom presentation development.

▶ **Supplemental Photos:** This collection of over 1,500 photos (in addition to those in the text) is a rich resource of visual imagery. All are organized by chapter and include explanatory captions for easy insertion into lectures.

▶ **Chapter Outlines**, **Lecture Notes**, and the complete **Test File** are all available in Microsoft Word® format for easy use in lecture and exam preparation.

▶ **An Intuitive Browser Interface** provides a quick and easy way to preview all of the content in the Media Library.

▶ **Seeing Life: Video Sequences in Biology:** New for the Seventh Edition, this collection of approximately 200 video segments covering topics across the entire textbook helps demonstrate the complexity and beauty of life. The videos are available on both DVD (for stunning image quality) and CD (for use in PowerPoint or other software).

PowerPoint Resources

For each chapter of the textbook, the Media Library offers several different types of PowerPoint presentations. These give instructors the flexibility to build presentations in the manner that best suits their needs.

▶ **Lecture PowerPoints** are designed to form the basis of a complete lecture. They combine detailed text with selected figures to create a thorough presentation.

▶ **Complete Art and Photo PowerPoints** include every piece of art, photo, table, and supplemental photo placed onto slides, ready for use in presentations.

Layered PowerPoint 37.7: A Nodule Forms

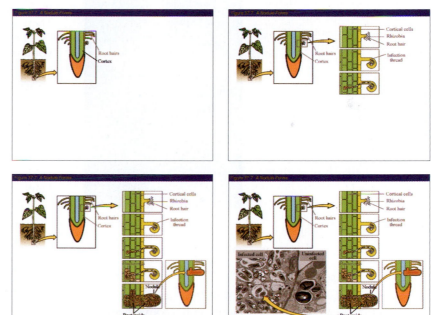

▶ **Layered Art PowerPoints** offer detailed, customizable presentations of key figures. These have been broken down into their component parts and are presented in a step-by-step, layered, and animated manner, allowing instructors to introduce elements of a figure at their own pace.

▶ **Video and Animation PowerPoints** include every video and animation in the Instructor's Media Library placed into PowerPoint, allowing instructors to simply copy the slides they want into their own presentations.

Video Sequence: The Venus Flytrap, *Dionaea* **sp.**

Instructor's Resource Kit

The Resource Kit is the central tool for lecture planning and development using the Seventh Edition of *Life*. The Kit combines several extensive instructor resources into one convenient binder:

▶ **Instructor's Manual:** The IM includes a chapter Overview, a "What's New" guide to the Seventh Edition, Key Concepts and Information, a Chapter Outline, and the list of Key Terms for each chapter.

▶ **Lecture Notes:** These detailed bulleted notes cover all the content presented in each chapter and are designed to be used as the basis for a complete lecture.

▶ **Media Guide:** New for the Seventh Edition, the Media Guide is a visual guide to all the media resources available in the Instructor's Media Library. Thumbnails and descriptions are provided for every animation, activity, video, supplemental photo, and lecture PowerPoint.

▶ **Custom Labs Information:** This is a list of all the lab separates available for instructors who choose to create a custom-published lab manual.

Supplemental Photo 20.33: Comet-tail moth (*Argema mittrei*), adult male

Overhead Transparencies

The set of transparencies includes all the line art figures from the textbook (over 1,000 images), along with a convenient binder. Labels and artwork have been resized and color-corrected for optimal projection.

Printed Test File

Catherine Ueckert, *Northern Arizona University;* Betty McGuire, *Smith College;* Chris Romero, *Front Range Community College;* Paula Mabee, *University of South Dakota;* and Erica Bergquist, *Holyoke Community College*

The Test File offers a bank of thousands of questions from which to create exams. Each chapter includes a set of fill-in-the-blank and multiple-choice questions that cover the full range of content presented in the text. To aid in selecting a range of questions, difficult and conceptual questions are indicated.

Computerized Test Bank CD

The entire printed Test File, plus the textbook end-of-chapter Self-Quizzes, the Student Website Online Quizzes and the Study Guide questions are all included in Brownstone's easy-to-use Diploma® software for an extensive collection of test questions in an easy-to-use interface.

Online Testing

Using the Computerized Test Bank and Diploma Online Testing, instructors can easily create and administer secure exams over a network and over the Internet. For more information, visit the Brownstone Research Group Website: www.brownstone.net.

Instructor's Website (www.thelifewire.com)

An extensive collection of instructor's media, as well as electronic versions of other instructor supplements, is available online for instant access anytime.

Course Management

As a service for adopters using WebCT® or Blackboard® for their courses, full electronic course packs are available. These include all the Student Website contents, plus all the Test File questions and other instructor resources. Contact your sales representative for more information.

Laboratory Manuals

Biology in the Laboratory, Third Edition
Doris R. Helms, Carl W. Helms, Robert J. Kosinski, and John C. Cummings
Order ISBN 0-7167-3146-0

Laboratory Outlines in Biology-VI
Peter Abramoff and Robert G. Thomson
Order ISBN 0-7167-2633-5

Anatomy and Dissection of the Frog, Second Edition
Warren F. Walker, Jr.
Order ISBN 0-7167-2636-X

Anatomy and Dissection of the Rat, Third Edition
Warren F. Walker, Jr. and Dominique Homberger
Order ISBN 0-7167-2635-1

Anatomy and Dissection of the Fetal Pig, Fifth Edition
Warren F. Walker, Jr and Dominique Homberger
Order ISBN 0-7167-2637-8

Custom Publishing for Laboratory Manuals

(available at www.custompub.whfreeman.com)

Instructors can build and order customized lab manuals in just minutes, choosing material from Freeman's acclaimed biology laboratory manuals—lab-tested experiments that have been used successfully by hundreds of thousands of students. Instructors determine the manual's content (including their own material) and a streamlined production process provides a quick turnaround to meet crucial deadlines.

Special Custom Lab Manual Option

The publishers of *Life*, in collaboration with Hayden-McNeil Publishing, offer enhanced customization for instructors adopting 1,000 books or more over two years. This option uses state-of-the-art customization technology to give instructors complete freedom to not only combine Freeman's biology labs with original material, but to also make editorial and page layout changes to the experiments.

Contents in Brief

Contents

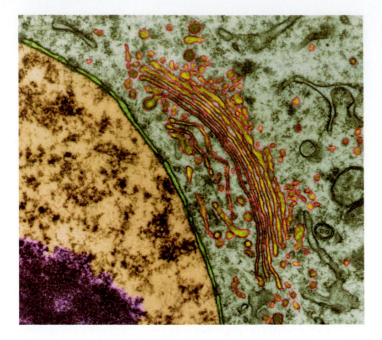

Part Two • INFORMATION AND HEREDITY

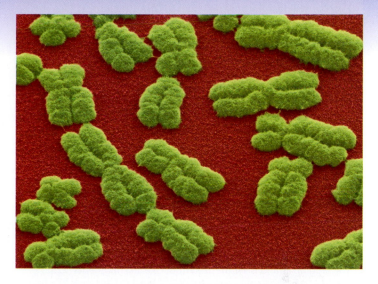

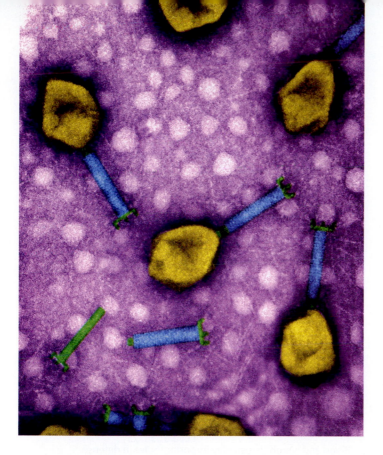

15 Cell Signaling and Communication 301

14 The Eukaryotic Genome and Its Expression 279

16 Recombinant DNA and Biotechnology 317

LIFE ESSAY: What are the ethical issues surrounding genetic modification of nature?
by Gary Comstock 389

Part Three • DEVELOPMENT

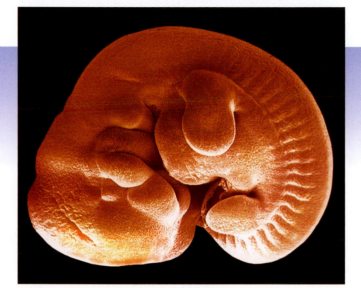

19 *Differential Gene Expression in Development* 390

20 *Animal Development: From Genes to Organism* 408

21 Development and Evolutionary Change 429

Part Four • EVOLUTIONARY PROCESSES

22 The History of Life on Earth 442

23 The Mechanisms of Evolution 460

LIFE ESSAY: How has Darwin's theory of natural selection transformed our view of humanity's place in the universe? by Daniel Dennett 523

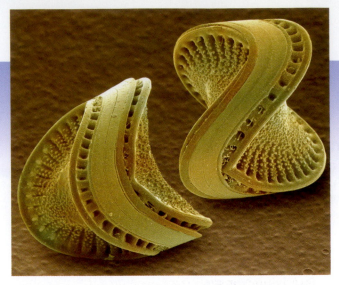

Part Five • THE EVOLUTION OF DIVERSITY

Part Six • THE BIOLOGY OF FLOWERING PLANTS

36 *Transport in Plants*   701

37 *Plant Nutrition*   716

38 *Regulation of Plant Growth*   729

LIFE ESSAY: How should we manage fire in the forest? By David E. Pesonen 779

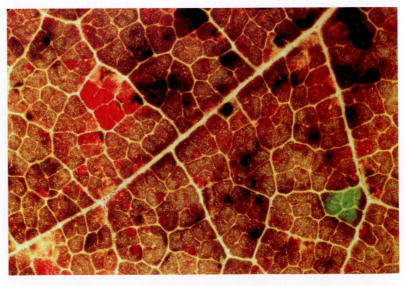

Part Seven • THE BIOLOGY OF ANIMALS

46 *The Mammalian Nervous System: Structure and Higher Functions* 885

47 *Effectors: Making Animals Move* 903

48 *Gas Exchange in Animals* 922

LIFE ESSAY: What are the ethical issues surrounding medical treatment? by Nancy S. Jecker 1023

Part Eight • ECOLOGY AND BIOGEOGRAPHY

LIFE ESSAY: Toward economic principles for sustainable ecosystems management by William E. Rees 1121

1 An Evolutionary Framework for Biology

Monster frogs—what a great topic for an undergraduate research project! That's what Stanford University sophomore Pieter Johnson thought when he was shown a jar of Pacific tree frogs with extra legs growing out of their bodies. The frogs were collected from a pond on a farm close to the old Almaden mercury mines south of San Jose, California. Scientists from all over the world were reporting alarming declines in populations of many different kinds of frogs, so perhaps these "monster" frogs would hold a clue to why frogs all over the world are in trouble. Possible causes of the deformities could have been agricultural chemicals or heavy metals leaching out of the old mines. Library research, however, suggested other possibilities to Pieter.

Pieter studied 35 ponds in the region where the deformed frogs had been found. He counted frogs in the ponds and measured chemicals in the water. Thirteen of the ponds had Pacific tree frogs, but deformed frogs were found in only four ponds. To Pieter's surprise, analysis of the water samples failed to reveal higher amounts of pesticides, industrial chemicals, or heavy metals in the ponds with deformed frogs. Also surprisingly, when he collected eggs from those ponds and hatched them in the laboratory, he always got normal frogs. The only difference he observed among the ponds he studied was that the ponds with the deformed frogs also contained freshwater snails.

Freshwater snails are hosts for many parasites. Many parasites go through complex life cycles with several stages, each of which requires a specific host animal. Pieter focused on the possibility that some parasite that used freshwater snails as intermediate hosts was infecting the frogs and causing their deformities. Pieter found a candidate with this type of life cycle: a small flatworm called *Ribeiroia*, which was present in the ponds where the deformed frogs were found.

Pieter then did an experiment. He collected frog eggs from regions where there were no records of deformed frogs or of *Ribeiroia*. He hatched the eggs in the laboratory in containers with and without the parasite. When the parasite was present in the contain-

A Monster Phenomenon As a college sophomore, Pieter Johnson studied ponds that were home to Pacific tree frogs (*Hyla regilla*), trying to discover a reason for the presence of so many deformed frogs. What appears in the inset to be a tail is an extra leg.

1.1 The Many Faces of a Life   The caterpillar, pupa, and adult are all stages in the life cycle of a monarch butterfly (*Danaeus plexippus*). The caterpillar harvests the matter and energy needed to metabolize the millions of chemical reactions that will result in its growth and transformation, first into a pupa and finally into an adult butterfly specialized for reproduction and dispersal. The transition from one stage to another is triggered by internal chemical signals.

ers, 85 percent of the frogs developed deformities. Further experiments showed why not all the frogs were deformed: The infection had to occur before a tadpole started to grow legs. When tadpoles with already developing legs were infected, they did not become deformed.

Pieter's project started with a question based on an observation in nature. He formulated several possible answers, made observations to narrow down the list of answers, and then did experiments to test what he thought was the most likely answer. His experiments enabled him to reach a conclusion: that these deformities were caused by *Ribeiroia*. Pieter's project is a good example of the application of scientific methods in biology.

Biology is the scientific study of living things. Biologists study processes from the level of molecules to the level of entire ecosystems. They study events that happen in millionths of seconds and events that occur over millions of years. Biologists ask many different kinds of questions and use a wide range of tools, but they all use the same scientific methods. Their goals are to understand how organisms (and assemblages of organisms) function, and to use that knowledge to help solve problems.

In this chapter, we will take a closer look at what biologists do. First, we will describe the characteristics of living things, the major evolutionary events that have occurred during the history of life on Earth, and the evolutionary tree of life. Then we will discuss the methods biologists use to investigate how life functions. At the end of the chapter, we will discuss how scientific knowledge is used to shape public policy.

What Is Life?

Before we probe more deeply into the study of life, we need to agree on what life is. Although we all know a living thing when we see one, it is difficult to define life unambiguously. One concise definition of **life** is: *an organized genetic unit capable of metabolism, reproduction, and evolution.* Much of this book is devoted to describing these characteristics of life and how they work together to enable organisms to survive and reproduce (Figure 1.1). The following brief overview will guide your study of these characteristics.

Metabolism involves conversions of matter and energy

Metabolism, the total chemical activity of a living organism, consists of thousands of individual chemical reactions. Chemical reactions result in the capture of matter and energy and its conversion to different forms, as we will see in Part One of this book. For an organism to function, these reactions, many of which are occurring simultaneously, must be coordinated. Genes provide that control. The nature of the genetic material that controls these lifelong events has been understood only within the last 100 years. Much of Part Two is devoted to the story of its discovery.

The external environment can change rapidly and unpredictably in ways that are beyond an organism's control. An organism can remain healthy only if its internal environment remains within a given range of physical and chemical conditions. Organisms maintain relatively constant internal environments by making metabolic adjustments to conditions such as changes in temperature, the presence or absence of sunlight, or the presence of foreign agents inside their bodies. Maintenance of a relatively stable internal condition, such as a human's constant body temperature, is called **homeostasis**.

The adjustments that organisms make to maintain homeostasis are usually not obvious, because nothing appears to change. However, at some time during their lives, many organisms respond to changing conditions not by maintaining their status, but by undergoing a major reorganization. An early form of such reorganization was the evolution of resting spores, a well protected, inactive form in which organisms survived stressful environments. A striking example that evolved much later is seen in many insects, such as butterflies. In response to internal chemical signals, a caterpillar changes into a pupa and then into an adult butterfly (see Figure 1.1).

Reproduction continues life and provides the basis for evolution

Reproduction with variation is a major characteristic of life. Without reproduction, life would quickly come to an end. The earliest single-celled organisms reproduced by duplicating their genetic material and then dividing in two. The two resulting daughter cells were identical to each other and to the parent cell, except for mutations that occurred during the process of gene duplication. Such errors, although rare, provided the raw material for biological evolution. The combination of reproduction and errors in the duplication of genetic material results in **biological evolution**, a change in the genetic composition of a population of organisms over time.

The diversification of life has been driven in part by variation in the physical environment. There are cold places and warm places, as well as places that are cold during some parts of the year and warm during other parts. Some places (oceans, lakes, rivers) are wet; others (deserts) are usually very dry. No single kind of living thing can perform well in all these environments. In addition, living things generate their own diversity. Once plants evolved, they became a source of food for other living things. Eaters of plants were, in turn, potential food for other organisms. And when living things die, they become food for still other organisms. The differences among living things that enable them to live in different kinds of environments and adopt different lifestyles are called **adaptations**. The great diversity of living things contributes to making biology a fascinating science and Earth a rich and rewarding place to live.

For a long period of time, there was no life on Earth. Then there was an extended period of only unicellular life, followed by a proliferation of multicellular life. In other words, the nature and diversity of life has changed over time. Identification of the processes that result in biological evolution

was one of the great scientific advances of the nineteenth century. These processes will be discussed in detail in Part Four of this book. Here we will briefly describe how they were discovered.

Biological Evolution: Changes over Billions of Years

Long before the mechanisms of biological evolution were understood, some people realized that organisms had changed over time and that living organisms had evolved from organisms that were no longer alive on Earth. In the 1760s, the French naturalist Count George-Louis Leclerc de Buffon (1707–1788) wrote his *Natural History of Animals*, which contained a clear statement of the possibility of evolution. Buffon observed that the limb bones of all mammals were remarkably similar in many details (Figure 1.2). He also noticed that the legs of certain mammals, such as pigs, have toes that never touch the ground and appear to be of no use. He found it difficult to explain the presence of these seemingly useless small toes by the commonly held belief that Earth and all its creatures had been divinely created in their current forms relatively recently. To explain these observations, Buffon suggested that the limb bones of mammals might all have been inherited from a common ancestor, and that pigs might have functionless toes because they inherited them from ancestors that had fully formed and functional toes.

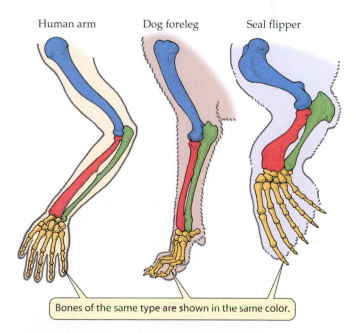

Human arm   Dog foreleg   Seal flipper

Bones of the same type are shown in the same color.

1.2 All Mammals Have Similar Limbs Mammalian forelimbs have different purposes: Humans use theirs for manipulating objects, dogs use theirs for walking, and seals use theirs for swimming. But the numbers and types of their bones are similar, indicating that they have been modified over time from the forelimbs of a common ancestor.

Buffon did not attempt to explain how such changes took place, but his student Jean-Baptiste de Lamarck (1744–1829) proposed a mechanism for such changes. Lamarck suggested that a lineage of organisms could change gradually over many generations as offspring inherited structures that had become larger and more highly developed as a result of continued use or, conversely, had become smaller and less developed as a result of lack of use. Today scientists do not believe that evolutionary changes are produced by this mechanism. But Lamarck had made an important effort to explain how living things change over time.

Darwin provided a mechanistic explanation of biological evolution

By 1858, the climate of opinion (among many biologists, at least) was receptive to a new theory of evolutionary processes proposed independently by Charles Darwin and Alfred Russel Wallace. By that time, geologists had accumulated evidence that Earth had existed and changed over millions of years, not merely a few thousand years, as most people had previously believed.

You will learn more about Darwin's theory of evolution by natural selection in Chapter 23, but its essential features are simple. You will need to be familiar with these ideas to understand the rest of this book. Darwin's theory rests on three observations and one conclusion he drew from them. The three observations are:

▶ The reproductive rates of all organisms, even slowly reproducing ones, are sufficiently high that populations would quickly become enormous if death rates were not equally high.
▶ Within each type of organism, there are differences among individuals.
▶ Offspring are similar to their parents because they inherit their parents' features.

Based on these observations (evidence), Darwin drew the following conclusion:

▶ The differences among individuals influence how well those individuals survive and reproduce. Any traits that increase the probability that their bearers will survive and reproduce are passed on to their offspring and to their offspring's offspring.

Darwin called the differential survival and reproductive success of individuals **natural selection**. He called the resulting pattern "descent with modification."

Biologists began a major conceptual shift a little more than a century ago with the acceptance of long-term evolutionary change and the gradual recognition that natural selection is the process that adapts organisms to their environments. The shift has taken a long time because it required abandoning many components of an earlier worldview. The pre-Darwinian view held that the world was young, and that organisms had been divinely created in their current forms. In the Darwinian view, the world is ancient, and both Earth and its inhabitants have changed over time. Ancestral forms were very different from the organisms that exist today. Living organisms evolved their particular features because ancestors with those features survived and reproduced more successfully than did ancestors with different features.

Major Events in the History of Life on Earth

The history of life on Earth, depicted on the scale of a 30-day calendar, is outlined in Figure 1.3. The profound changes that have occurred over the 4 billion years of this history are the result of natural processes that can be identified and studied using scientific methods. In this section, we will set the stage for the rest of this book by describing some of the most important of these changes. These six major evolutionary events will provide us with a framework for discussing both life's characteristics and how those characteristics evolved. By recognizing them, you will be able to better appreciate both the unity and diversity of life.

Life arose from nonlife via chemical evolution

The first life must have come from nonlife. All matter, living and nonliving, is made up of chemicals. The smallest chemical units are atoms, which bond together into molecules (the properties of these units are the subject of Chapter 2). The processes of **chemical evolution** that led to the appearance of life began nearly 4 billion years ago, when random inorganic chemical interactions produced molecules that had the remarkable property of acting as templates to form similar molecules. Some of the chemicals involved may have come to Earth from space, but chemical evolution continued on Earth.

The information stored in these simple molecules enabled the synthesis of larger molecules with complex but relatively stable shapes. Because they were both complex and stable, these molecules could participate in increasing numbers and kinds of chemical reactions. Certain types of large molecules are found in all living systems; the properties and functions of these complex molecules are the subject of Chapter 3.

Biological evolution began when cells formed

About 3.8 billion years ago, interacting systems of molecules came to be enclosed in compartments. Within those units—**cells**—control was exerted over the entrance, retention, and exit of molecules, as well as over the chemical reactions taking place. The origin of cells marked the beginning of bio-

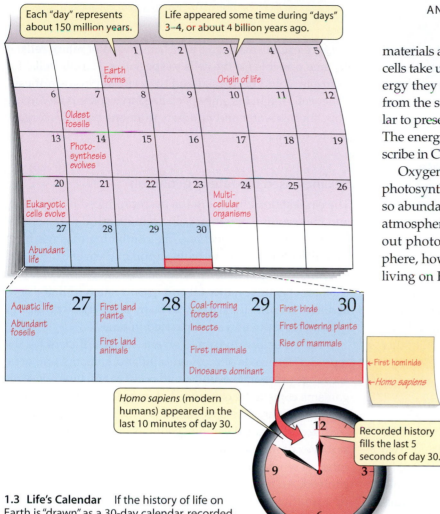

Each "day" represents about 150 million years.

Life appeared some time during "days" 3–4, or about 4 billion years ago.

Earth forms — Origin of life

Oldest fossils

Photo-synthesis evolves

Eukaryotic cells evolve

Multi-cellular organisms

Abundant life

Aquatic life **27**	First land plants **28**	Coal-forming forests **29**	First birds **30**
Abundant fossils	First land animals	Insects	First flowering plants
		First mammals	Rise of mammals
		Dinosaurs dominant	←First hominids ←Homo sapiens

Homo sapiens (modern humans) appeared in the last 10 minutes of day 30.

Recorded history fills the last 5 seconds of day 30.

1.3 Life's Calendar If the history of life on Earth is "drawn" as a 30-day calendar, recorded human history takes up only the last 5 seconds.

logical evolution. Cells and the membranes that enclose them are the subjects of Chapters 4 and 5.

Cells are so effective at capturing energy and replicating themselves—two fundamental characteristics of life—that since they evolved, cells have been the unit on which all life is built. Experiments by the French chemist and microbiologist Louis Pasteur and other scientists during the nineteenth century (described in Chapter 3) convinced most scientists that, under present conditions on Earth, cells do not arise from non-cellular material, but come only from other cells.

For 2 billion years after cells originated, all organisms were *unicellular* (had only one cell). They were confined to the oceans, where they were shielded from lethal ultraviolet light. These simple cells, called **prokaryotic cells**, had no internal membrane-enclosed compartments.

Photosynthesis changed the course of evolution

A major event that took place about 2.5 billion years ago was the evolution of **photosynthesis**: the ability to use the energy of sunlight to power metabolism. All cells must obtain raw

materials and energy to fuel their metabolism. Photosynthetic cells take up raw materials from their environment, but the energy they use to metabolize those chemicals comes directly from the sun. Early photosynthetic cells were probably similar to present-day prokaryotes called *cyanobacteria* (Figure 1.4). The energy-capturing process they used, which we will describe in Chapter 8, is the basis of nearly all life on Earth today.

Oxygen gas (O_2) is a by-product of photosynthesis. Once photosynthesis evolved, photosynthetic prokaryotes became so abundant that they released vast quantities of O_2 into the atmosphere. The O_2 we breathe today would not exist without photosynthesis. When it first appeared in the atmosphere, however, O_2 was poisonous to most organisms then living on Earth. Those prokaryotes that evolved a tolerance to O_2 were able to successfully colonize environments emptied of other organisms and proliferate in great abundance. For those prokaryotes, the presence of oxygen opened up new avenues of evolution. Metabolic reactions that use O_2, called *aerobic metabolism*, are more efficient than the anaerobic (non-oxygen-using) metabolism that earlier prokaryotes had used. Aerobic metabolism allowed cells to grow larger, and it came to be used by most organisms on Earth.

Over a much longer time frame, the vast quantities of oxygen released by photosynthesis had another effect. Formed from O_2, ozone (O_3) began to accumulate in the upper atmosphere. The ozone slowly formed a dense layer that acted as a shield, inter-

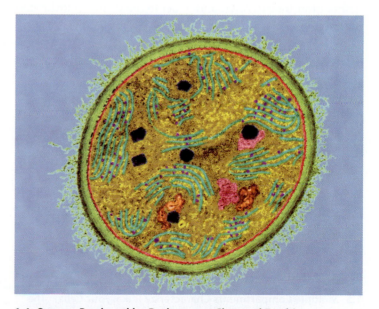

1.4 Oxygen Produced by Prokaryotes Changed Earth's Atmosphere This modern cyanobacterium may be very similar to early photosynthetic prokaryotes.

cepting much of the sun's deadly ultraviolet radiation. Eventually (although only within the last 800 million years of evolution), the presence of this shield allowed organisms to leave the protection of the ocean and establish new lifestyles on Earth's land surfaces.

Cells with complex internal compartments arose

As the ages passed, some prokaryotic cells became large enough to attack, engulf, and digest smaller prokaryotes, becoming the first predators. Usually the smaller cells were destroyed within the predators' cells, but some of these smaller cells survived and became permanently integrated into the operation of their host cells. In this manner, cells with complex internal compartments, called **eukaryotic cells**, arose. The hereditary material of eukaryotic cells is contained within a membrane-enclosed nucleus and is organized into discrete units. Other compartments are specialized for other purposes, such as photosynthesis (Figure 1.5).

Multicellularity arose and cells became specialized

Until slightly more than 1 billion years ago, only unicellular organisms (both prokaryotic and eukaryotic) existed. Two key developments made the evolution of *multicellular* organisms—organisms consisting of more than one cell—possible. One was the ability of a cell to change its structure and functioning to meet the challenges of a changing environment. This was accomplished when prokaryotes evolved the ability to transform themselves from rapidly growing cells into resting spores that could survive harsh environmental conditions. The second development allowed cells to stick together after they divided and to act together in a coordinated manner.

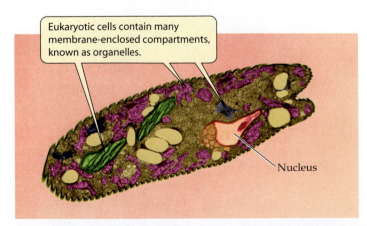

Eukaryotic cells contain many membrane-enclosed compartments, known as organelles.

Nucleus

1.5 Multiple Compartments Characterize Eukaryotic Cells The nucleus and other specialized compartments of eukaryotic cells evolved from small prokaryotes that were ingested by larger prokaryotic cells.

Once organisms began to be composed of many cells, it became possible for the cells to specialize. Certain cells, for example, could be specialized to perform photosynthesis. Other cells might become specialized to transport raw materials, such as water and nitrogen, from one part of an organism to another.

Sex increased the rate of evolution

The earliest unicellular organisms reproduced by dividing, and the resulting daughter cells were identical to the parent cell. But *sexual recombination*—the combining of genes from two different cells in one cell—appeared early during the evolution of life. Early prokaryotes engaged in **sex** (exchanges of genetic material) and reproduction (cell division) at different times. Even today in many unicellular organisms, sex and reproduction are separated in time.

Simple nuclear division—*mitosis*—was sufficient for the reproductive needs of most unicellular organisms, and gene exchange (a separate event) could occur at any time. Once organisms came to be composed of many cells, however, certain cells began to be specialized for sex. Only these specialized sex cells, called *gametes*, could exchange genes, and the sex lives of multicellular organisms became more complicated. A whole new method of nuclear division—*meiosis*—evolved. An intricate and complex process, meiosis opened up a multitude of possibilities for genetic recombination between gametes. Mitosis and meiosis are explained and compared in Chapter 9.

Sex increased the rate of evolution because an organism that exchanges genetic information with another individual produces offspring that are more genetically variable than the offspring of an organism that reproduces by mitotic division of its own cells. Some of these varied offspring are likely to survive and reproduce better than others in a variable and changing environment. It is this genetic variation that natural selection acts on.

Levels of Organization of Life

Biology can be visualized as a hierarchy of units, ordered from the smallest to the largest. These units are molecules, cells, tissues, organs, organisms, populations, communities, and the biosphere (Figure 1.6).

The organism is the central unit of study in biology; Parts Six and Seven of this book discuss organismic biology in detail. But to understand organisms, biologists study life at all its levels of organization. They study molecules, chemical reactions, and cells to understand the functioning of tissues and organs. They study organs and organ systems to determine how organisms maintain homeostasis. At higher levels in the hierarchy, biologists study how organisms interact with one

1.6 From Molecules to the Biosphere: The Hierarchy of Life

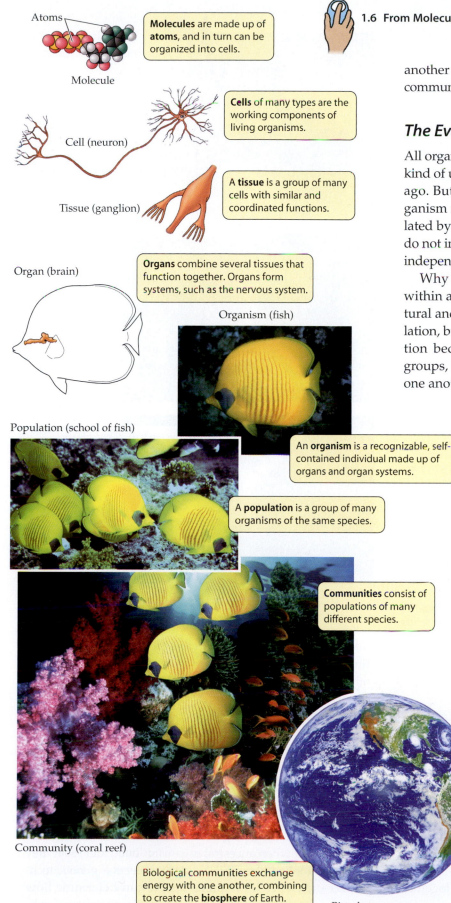

Atoms

Molecules are made up of **atoms**, and in turn can be organized into cells.

Molecule

Cells of many types are the working components of living organisms.

Cell (neuron)

A **tissue** is a group of many cells with similar and coordinated functions.

Tissue (ganglion)

Organ (brain)

Organs combine several tissues that function together. Organs form systems, such as the nervous system.

Organism (fish)

An **organism** is a recognizable, self-contained individual made up of organs and organ systems.

Population (school of fish)

A **population** is a group of many organisms of the same species.

Communities consist of populations of many different species.

Community (coral reef)

Biological communities exchange energy with one another, combining to create the **biosphere** of Earth.

Biosphere

another to form social systems, populations, and ecological communities, which are the subjects of Part Eight of this book.

The Evolutionary Tree of Life

All organisms on Earth today are the descendants of a single kind of unicellular organism that lived almost 4 billion years ago. But if that were the whole story, only one kind of organism might exist on Earth today. Instead, Earth is populated by many millions of different kinds of organisms that do not interbreed with one another. We call these genetically independent kinds **species**.

Why are there so many species? As long as individuals within a population mate at random and reproduce, structural and functional changes may evolve within that population, but only one species will exist. However, if a population becomes separated and isolated into two or more groups, individuals within each group will mate only with one another. When this happens, structural and functional differences between the groups may accumulate over time, and the groups may evolve into different species. The splitting of groups of organisms into separate species has resulted in the great diversity of life found on Earth today. The ways in which species form are explained in Chapter 24.

Sometimes humans refer to a species as "primitive" or "advanced." These and similar terms, such as "lower" and "higher," are best avoided in biology because they imply that some organisms function better than others. In fact, *all* living organisms are successfully adapted to their environments. The shape and strength of a bird's beak, or the form and dispersal mechanisms of a plant's seeds are examples of the rich array of adaptations found among living organisms (Figure 1.7). The abundance and success of prokaryotes—all of which are relatively simple organisms—readily demonstrates that they are highly functional. In this book, we use the terms *simple* and *complex* to refer to the level of complexity of a particular organism. We use the terms *ancestral* and *derived* to distinguish characteristics that appeared earlier from those that appeared later in evolution.

As many as 30 million species of organisms may live on Earth today. Many times that number lived in the past, but are now ex-

(a)

The strong, curved beak of the bald eagle is able to tear the flesh from large fish and other sizeable prey.

The curlew uses its long, curved, pointed beak to extract small crustaceans from the surface of mud, sand, and soil.

The roseate spoonbill moves its bill through the water, from which it filters food items.

(b)

The coconut seed is covered by a thick husk that protects it as it drifts across thousands of miles of ocean.

Mammals and birds eat blackberries, then disseminate the seeds when they defecate.

The seeds of milkweeds are surrounded by "kites" of fibers that carry them on wind currents.

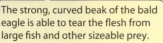

1.7 Adaptations to the Environment (a) Bird beaks are adapted to specific types of food items. (b) Plants cannot move, but their seeds have adaptations that allow them to travel varying distances from the parent plant.

tinct. This diversity is the result of millions of splits in populations, known as *speciation events*. The unfolding of these events can be expressed as an evolutionary "tree" showing the order in which populations split and eventually evolved into new species (see Figure 1.8). An evolutionary tree, with its "trunk" and its increasingly finer "branches," traces the descendants coming from ancestors that lived at different times in the past. That is, a tree shows the evolutionary relationships among species and groups of species. The organisms on any one branch share a common ancestor at the base of that branch. The most closely related groups are together on the same branch. More distantly related organisms are on

different branches. In this book, we adopt the convention that time flows from left to right, so the tree in Figure 1.8 (and other trees in this book) lies on its side, with its root—the ancestor of all life—at the left.

The U.S. National Science Foundation is sponsoring a major initiative, called Assembling the Tree of Life (ATOL). Its goal is to determine the evolutionary relationships among all species on Earth. Achieving this goal is possible today because, for the first time, biologists have the technology to assemble the complete tree of life, from microbes to mammals. Data for ATOL come from a variety of sources. *Fossils*—the preserved remains of organisms that lived in the past—tell us where and when ancestral organisms lived and what they may have looked like. With modern molecular genetic techniques such as DNA sequencing, we can determine how many genes different species share, and information tech-

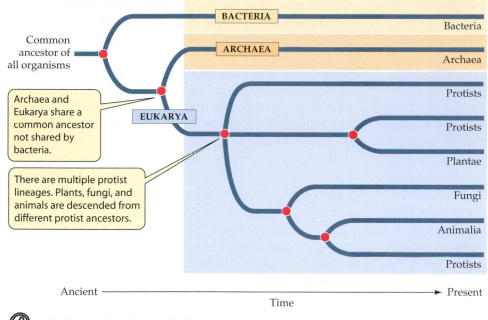

Common ancestor of all organisms

BACTERIA

Bacteria

ARCHAEA

Archaea

Archaea and Eukarya share a common ancestor not shared by bacteria.

EUKARYA

Protists

Protists

Plantae

There are multiple protist lineages. Plants, fungi, and animals are descended from different protist ancestors.

Fungi

Animalia

Protists

Ancient ——————————————————→ Present

Time

1.8 A Provisional Tree of Life The classification system used in this book divides Earth's organisms into three domains; Bacteria, Archaea, and Eukarya. Protists are descendants of multiple ancestors.

lieved to have separated into distinct evolutionary lineages very early during the evolution of life. The two prokaryotic domains are described in Chapter 27.

Members of the other domain—**Eukarya**—have eukaryotic cells. The Eukarya are divided into four groups: Protista, Plantae, Fungi, and Animalia. The Protista (protists), the subject of Chapter 28, contains mostly single-celled organisms. The other three groups, referred to as *kingdoms*, are believed to have arisen from ancestral protists.

Some bacteria, some protists, and most members of the kingdom Plantae (plants) convert light energy to chemical energy by photosynthesis. These organisms are called *autotrophs* ("self-feeders"). The biological molecules they produce are the primary food for nearly all other living organisms. The kingdom Plantae is covered in Chapters 29 and 30.

The kingdom Fungi, the subject of Chapter 31, includes molds, mushrooms, yeasts, and other similar organisms, all of which are *heterotrophs* ("other-feeders")—that is, they require a source of energy-rich molecules synthesized by other organisms. Fungi break down food molecules in their environment and then absorb the breakdown products into their cells. They are important as decomposers of the dead materials of other organisms.

Members of the kingdom Animalia (animals) are heterotrophs that ingest their food source, digest the food outside their cells, and then absorb the breakdown products. Animals eat other forms of life to obtain their raw materials and energy. This kingdom is covered in Chapters 32, 33, and 34.

We will discuss the principal levels used in today's classification scheme for living organisms in Chapter 25. But to understand some of the terms we will use in the intervening chapters, you need to know that each species of organism is identified by two Latinized names (a *binomial*). The first name identifies the **genus**—a group of species that share a recent common ancestor—of which the species is a member. The second name is the species name. To avoid confusion, no combination of two names is assigned to more than one species. For example, the scientific name of the human species is *Homo sapiens: Homo* is our genus and *sapiens* is our species. The Pacific tree frogs Pieter Johnson studied are called, in scientific nomenclature, *Hyla regilla*.

Biology is the study of all of Earth's organisms, both those living today and those that lived in the past, so even extinct species are given binomials. These unique and exact names

nology enables us to synthesize masses of genetic data. The ATOL initiative, one of the grandest projects of modern biology, is projected to take at least two decades and to involve hundreds of scientists working in a diverse array of fields. The reason it will take so long to complete is that most of Earth' species have not yet been described.

The Tree of Life will be an information framework for biology in much the same way that the periodic table of elements is an information framework for chemistry and physics. Evolution has conducted several billion years of free research and development. Every living thing carries a genetic "package" that has been tested by natural selection. Scientists can now unwrap and study these packages, learning much about the processes that produced them.

Although much remains to be accomplished, biologists know enough to have assembled a provisional tree of life, the broad outlines of which are shown in Figure 1.8. The branching patterns of this tree are based on a rich array of evidence, but no fossils are available to help us determine the earliest divisions in the lineages of life because those simple organisms had no parts that could be preserved as fossils. Therefore, molecular evidence has been used to separate all living organisms into three major **domains**. Organisms belonging to a particular domain have been evolving separately from organisms in the other domains for more than a billion years.

Organisms in the domains **Archaea** and **Bacteria** are prokaryotes. Archaea and Bacteria differ so fundamentally from one another in their metabolic processes that they are be-

illuminate the tremendous diversity of life, and are important tools for biologists because, as in all the sciences, precise and unambiguous communication of research information is critical.

Biology Is a Science

To study the rich variety of living things, biologists employ many different methods. Direct observations by unaided senses are central to many scientific investigations, but scientists also use many tools that augment the human senses. For example, to study objects that are too small to be seen with the unaided eye, scientists use microscopes. To observe and magnify remote objects, scientists use telescopes. To study events that happened thousands to millions of years ago, scientists "read" radioactive isotopes of chemical elements that decay at specific rates.

Conceptual tools guide scientific research

In addition to such technical tools, scientists use a variety of conceptual tools to help them answer questions about nature. The method that underlies most scientific research is the **hypothesis-prediction (H–P) approach**. The H–P approach allows scientists to modify their conclusions as new information becomes available. The method has five steps:

1. Making *observations*
2. Asking *questions*
3. Forming *hypotheses*, which are tentative answers to the questions
4. Making *predictions* based on the hypotheses
5. *Testing* the predictions by making additional observations or conducting experiments

If the results of the testing support the hypothesis, it is subjected to additional predictions and tests. If they continue to support it, confidence in its correctness increases, and the hypothesis comes to be considered a **theory**. If the results do not support the hypothesis, it is abandoned or modified in accordance with the new information. Then new predictions are made, and more tests are conducted.

Hypotheses are tested in two major ways

Tests of hypotheses are varied, but most are of two types: controlled experiments and the comparative method. When possible, scientists use **controlled experiments** to test predictions from hypotheses. That is what Pieter Johnson was doing when he hatched frog eggs in the laboratory. He predicted that if his hypothesis—that the parasite *Ribeiroia* caused deformities in frogs—was correct, then frogs raised

with the parasite would develop deformities and frogs raised in the absence of the parasite would not. The advantage of controlled experiments is that *all factors other than the one hypothesized to be causing the effect can be kept constant*; that is, any other factors that might influence the outcome (such as water temperature and pH in Pieter's experiment) are controlled. The most powerful experiments are those that have the ability to demonstrate that the hypothesis or the predictions made from it are wrong.

But many hypotheses cannot be tested with controlled experiments. Such hypotheses are tested by making predictions about patterns that should exist in nature if the hypothesis is correct. Data are then gathered to determine whether those patterns in fact do exist. This approach is called the **comparative method**. It is the primary approach of scientists in some fields, such as astronomy, in which experiments are rarely possible. Biologists regularly use the comparative method.

A single piece of supporting evidence rarely leads to widespread acceptance of a hypothesis. Similarly, a single contrary result rarely leads to abandonment of a hypothesis. Results that do not support the hypothesis can be obtained for many reasons, only one of which is that the hypothesis is wrong. For example, incorrect predictions can be made from a correct hypothesis. Poor experimental design, or the use of an inappropriate organism, can also lead to erroneous results.

We will now show how the H–P method was used by other researchers to investigate the larger question that concerned Pieter Johnson: Why are amphibian populations declining dramatically in many places on Earth?

STEP 1: MAKING OBSERVATIONS. Amphibian populations, like populations of most organisms, fluctuate over time. Before we decide that the current declines are different from "normal" population fluctuations, we first need to establish that they are unusual. To assess whether the current declines are unusual, an international group of scientists has been gathering worldwide data on amphibian populations. The group's data show that amphibian populations are declining seriously in some parts of the world, especially western North America, Central America, and northeastern Australia, but not others, such as the Amazon Basin. Their data also show that population declines are greater in mountains than in adjacent lowlands. These scientists also discovered that no data on population trends in amphibians are available from Africa or Asia.

STEP 2: ASKING QUESTIONS. Two questions were suggested by these observations: Why are amphibian declines greater at high elevations? Why are amphibians declining in some regions, but not in others?

STEPS 3 AND 4: FORMULATING HYPOTHESES AND MAKING PREDICTIONS. To develop hypotheses about the first question, scientists first identified the environmental factors that change with elevation. Temperatures drop and rainfall increases with elevation worldwide, and in temperate regions, summer levels of ultraviolet-B (UV-B) radiation increase about 18 percent per 1,000 meters of elevation gain. One hypothesis is that declines in the populations of some amphibian species are due to global increases in UV-B radiation resulting from reductions in atmospheric ozone concentrations. If increased levels of UV-B are adversely affecting amphibian populations, we predict that experimentally reducing UV-B over ponds where amphibian eggs are incubating and larvae are developing should improve their survival.

STEP 5: TESTING HYPOTHESES. The hypothesis that exposure to increased levels of UV-B might contribute to amphibian population decline was tested by comparing the responses of tadpoles of two species of frogs that live in Australian mountains. One species (*Litoria verreauxii*) had disappeared from high elevations; the other (*Crinia signifera*) had not. Because at higher elevations tadpoles are exposed to higher levels of UV-B radiation, experimenters predicted that *L. verreauxii* would survive less well than *C. signifera* if exposed to UV-B radiation typical of high elevations (Figure 1.9).

As predicted, when exposed to UV-B radiation, individuals of *C. signifera* survived well, but all individuals of *L. verreauxii* died within two weeks. Among control populations raised in tanks covered by filters that blocked UV transmission, individuals of both species survived well. Thus, the results supported the hypothesis.

Figure 1.9 describes one of many experiments in which the UV-B hypothesis has been tested. Some other experiments have yielded similar results, while others have shown no effects of UV-B exposure, or have shown a negative effect of UV-B exposure only when it is associated with low pH.

Several hypotheses have also been proposed to account for regional differences in amphibian population declines, including the adverse effects of habitat alteration by humans. Two obvious forms of human habitat alteration are air pollution from areas of urban and industrial growth, and the airborne pesticides used in agriculture.

A straightforward prediction from the habitat alteration hypothesis is that amphibian declines should be more noticeable in areas exposed to higher amounts of human-generated air

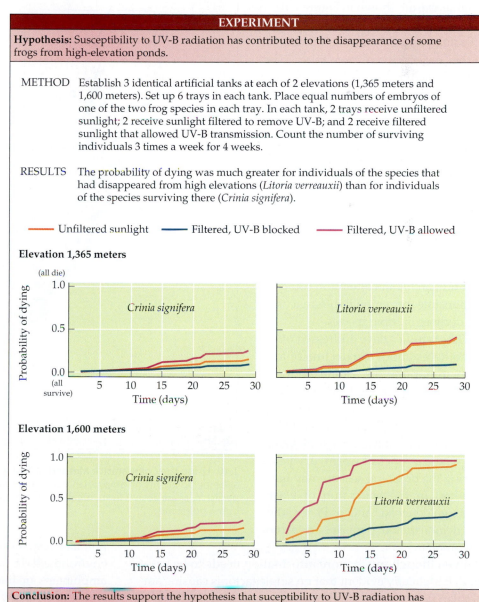

EXPERIMENT

Hypothesis: Susceptibility to UV-B radiation has contributed to the disappearance of some frogs from high-elevation ponds.

METHOD Establish 3 identical artificial tanks at each of 2 elevations (1,365 meters and 1,600 meters). Set up 6 trays in each tank. Place equal numbers of embryos of one of the two frog species in each tray. In each tank, 2 trays receive unfiltered sunlight; 2 receive sunlight filtered to remove UV-B; and 2 receive filtered sunlight that allowed UV-B transmission. Count the number of surviving individuals 3 times a week for 4 weeks.

RESULTS The probability of dying was much greater for individuals of the species that had disappeared from high elevations (*Litoria verreauxii*) than for individuals of the species surviving there (*Crinia signifera*).

Conclusion: The results support the hypothesis that susceptibility to UV-B radiation has contributed to the disappearance of *Litoria verreauxii* from high elevations.

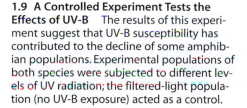

1.9 A Controlled Experiment Tests the Effects of UV-B The results of this experiment suggest that UV-B susceptibility has contributed to the decline of some amphibian populations. Experimental populations of both species were subjected to different levels of UV radiation; the filtered-light population (no UV-B exposure) acted as a control.

pollutants than in areas with less exposure. This hypothesis has been tested using the comparative method. The extensive tests compared population trends in eight species of amphibians in the state of California. The species studied included four frog species of the genus *Rana*, two species of toads, and a salamander species. The bases of the tests were simple *censuses* (surveys and counts) to determine whether populations of a given species were present or absent at each of the hundreds of study sites across the state. The census results for one of the eight species, the frog *Rana aurora*, are shown in Figure 1.10.

The map in Figure 1.10 shows a significant trend for *R. aurora*: Populations of this amphibian are more likely to be *absent* from sites downwind of large urban and agricultural areas (and thus exposed to heavy airborne pollution), and *present* in sites upwind (not heavily exposed). This type of data is the basis of the comparative method. In this particular study, meticulous tallying and comparison of similar data for all eight species showed that some species exhibited significant declines in exposed areas, but others (including the toads), did not. Therefore, we may conclude that human habitat alteration could be responsible for regional differences in the declines of some species.

Other studies have addressed other hypotheses about the decline of amphibian populations. Some evidence indicates that smoke from extensive fires also is adversely affecting amphibians. Climate change is clearly important in areas such as Central America, where a series of warm, dry years during the breeding season may have resulted in the extinction of Costa Rica's golden toad. And, as Pieter Johnson demonstrated, parasites are part of the problem.

Even though much more information needs to be gathered, it is already evident that no single factor is causing amphibian declines. This finding is not surprising, because no two places on Earth are the same, and no two species of am-

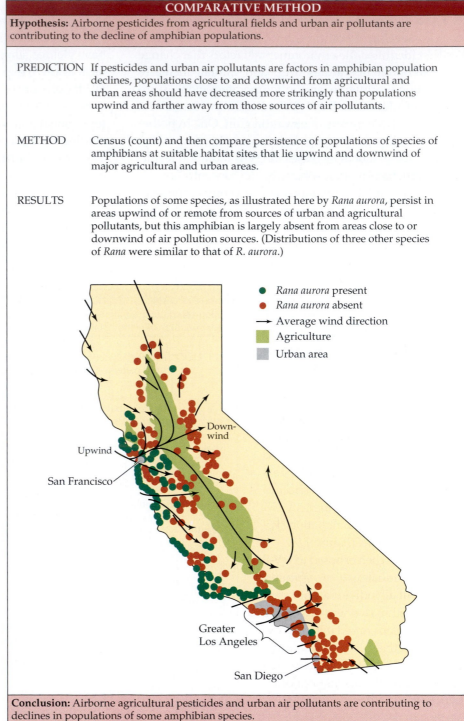

COMPARATIVE METHOD

Hypothesis: Airborne pesticides from agricultural fields and urban air pollutants are contributing to the decline of amphibian populations.

PREDICTION   If pesticides and urban air pollutants are factors in amphibian population declines, populations close to and downwind from agricultural and urban areas should have decreased more strikingly than populations upwind and farther away from those sources of air pollutants.

METHOD   Census (count) and then compare persistence of populations of species of amphibians at suitable habitat sites that lie upwind and downwind of major agricultural and urban areas.

RESULTS   Populations of some species, as illustrated here by *Rana aurora*, persist in areas upwind of or remote from sources of urban and agricultural pollutants, but this amphibian is largely absent from areas close to or downwind of air pollution sources. (Distributions of three other species of *Rana* were similar to that of *R. aurora*.)

- ● *Rana aurora* present
- ● *Rana aurora* absent
- → Average wind direction
- Agriculture
- Urban area

Conclusion: Airborne agricultural pesticides and urban air pollutants are contributing to declines in populations of some amphibian species.

1.10 Using the Comparative Method to Test a Hypothesis   The effects of human-generated airborne pollutants on amphibian populations can be assessed by determining whether species persist in, or are absent from, suitable habitats that lie upwind or downwind from sources of airborne pollutants.

phibians respond in exactly the same way to changes in the environment. In their responses to environmental changes, amphibians are like most living things. They live in complex and changing environments, and they interact with many other species.

Simple explanations that account for everything should not be expected or trusted. Its complexities make biology a difficult science, but they also make it exciting and challenging.

Not all forms of inquiry are scientific

Scientific methods are the most powerful tools that humans have developed to understand how the world works. Their strength is founded on the development of hypotheses that can be tested. The process is self-correcting because if the evidence fails to support a hypothesis, it is either abandoned or modified and subjected to further tests. In addition, because scientists publish detailed descriptions of the methods they use to test hypotheses, other scientists can—and often do—repeat those experiments. Therefore, any error or dishonesty usually is discovered. That is why, in contrast to politicians, scientists around the world usually trust one another's results.

If you understand the methods of science, you can distinguish science from non-science. Art, music, and literature, activities that contribute massively to the quality of human life, are not science. They help us understand what it means to live in a complex world. Religion is not science either. Religious beliefs give us meaning and spiritual guidance, and they form a basis for establishing values. Scientific information helps create the context in which values are discussed and established, but it cannot tell us what those values should be.

Biology has implications for public policy

The study of biology has long had major implications for human life. Agriculture and medicine are two important fields of applied biology. People have been speculating about the causes of diseases and searching for methods of combating them since ancient times. Today, with the deciphering of the genetic code and the ability to manipulate the genetic constitution of organisms, vast new possibilities exist for improvements in the control of human diseases and agricultural productivity. At the same time, these capabilities have raised important ethical and policy issues. How much and in what ways should we tinker with the genetics of people and other species? Does it matter whether organisms are changed by traditional breeding experiments or by gene transfers? How safe are genetically modified organisms in the environment and in human foods?

Another reason for studying biology is to understand the effects of the vastly increased human population on its environment. Our use of renewable and nonrenewable natural resources is putting stress on the ability of the environment to produce the goods and services upon which our society depends. Human activities are changing global climates, causing the extinction of a large number of species, and resulting in the spread of new human diseases and the resurgence of old ones. For example, the rapid spread of SARS and West Nile virus was facilitated by modern modes of transportation. Biological knowledge is vital for determining the causes of these changes, for devising wise policies to deal with them, and for drawing attention to the marvelous diversity of living organisms that provides goods and services for humankind and also enriches our lives aesthetically and spiritually.

Biologists are increasingly called upon to advise governmental agencies concerning the laws, rules, and regulations by means of which society deals with the increasing number of problems and challenges that have at least a partial biological basis. As we discuss these issues in many chapters of this book, you will see that the use of biological information is essential if wise public policies are to be established and implemented.

Throughout this book we will share with you the excitement of studying living things and illustrate the rich array of methods that biologists use to determine why the world of living things looks and functions as it does. The most important motivator of most biologists is curiosity. People are fascinated by the richness and diversity of life and want to learn more about organisms and how they interact with one another.

Humans probably evolved to be curious because individuals who were motivated to learn about their surroundings were likely to have survived and reproduced better, on average, than their less curious relatives. In other words, curiosity is adaptive! There are vast numbers of questions for which we do not yet have answers, and new discoveries usually engender questions no one thought to ask before. Perhaps *your* curiosity will lead to an important new idea.

Chapter Summary

What Is Life?

▶ Life can be defined as an organized genetic unit capable of metabolism, reproduction, and evolution.

▶ Metabolism, the total chemical activity of a living organism, is controlled by genes.

▶ Biological evolution is a change in the genetic composition of a population of organisms over time.

Biological Evolution: Changes over Billions of Years

▶ Charles Darwin's theory of natural selection rests on three simple observations and one conclusion drawn from them: Any heritable traits that increase the probability that their bearers will survive and reproduce are passed on to their offspring. **Review Figure 1.2**

Major Events in the History of Life on Earth

▶ Life arose from nonlife about 4 billion years ago by means of chemical evolution. **Review Figure 1.3**

▶ Biological evolution began about 3.8 billion years ago when interacting systems of molecules became enclosed in membranes to form cells.

▶ Photosynthetic prokaryotes released large amounts of oxygen into Earth's atmosphere, making aerobic metabolism possible.

▶ Complex eukaryotic cells evolved by incorporation of smaller cells that survived being ingested.

▶ Multicellular organisms appeared when cells evolved the ability to transform themselves and to stick together and communicate after they divided. The individual cells of multicellular organisms became modified to carry out varied functions within the organism.

▶ The evolution of sex sped up rates of biological evolution.

Levels of Organization of Life

▶ Life is organized hierarchically, from molecules to the biosphere. **Review Figure 1.6. See Web/CD Activity 1.1**

The Evolutionary Tree of Life

▶ A major effort called Assembling the Tree of Life (ATOL) is underway to determine the evolutionary relationships among all species on Earth.

▶ The hierarchy of evolutionary relationships can be represented as an evolutionary tree. **Review Figure 1.8. See Web/CD Activity 1.2**

▶ Species are grouped into three domains: Archaea, Bacteria, and Eukarya. The domains Archaea and Bacteria consist of prokaryotic cells. The domain Eukarya contains the Protists, Plantae, Fungi, and Animalia.

Biology Is a Science

▶ Biologists use a variety of technical and conceptual tools to study living things.

▶ The hypothesis-prediction (H–P) approach is used in most biological investigations. Hypotheses are tentative answers to questions. Predictions are made on the basis of a hypothesis.

The predictions are tested by experiments and comparative observations. **Review Figures 1.9 and 1.10**

▶ Science can tell us how the world works, but it does not form the basis for establishing meaning and values.

▶ Biologists are often called upon to advise governmental agencies on the solution of important problems that have a biological component.

For Discussion

1. The information Darwin used to develop his theory of evolution by natural selection was well known to his contemporaries. Why was it so difficult for people to think of such an obvious mechanism of evolutionary change?

2. According to the theory of evolution by natural selection, a species evolves certain features because they improve the chances that its members will survive and reproduce. There is no evidence, however, that evolutionary mechanisms have foresight or that organisms can anticipate future conditions. What, then, do biologists mean when they say, for example, that wings are "for flying"?

3. The first organisms appeared nearly 4 billion years ago, but multicellular organisms were slow to appear. Why did the evolution of multicellularity take so long?

4. Why is it so important in science that we design and perform tests capable of falsifying a hypothesis?

5. What features characterize questions that can be answered only by using a comparative approach?

6. Experiments show that not all amphibian declines are caused by a single factor. Does this surprise you? What kinds of environmental factors might be capable of affecting amphibian populations everywhere on Earth? What factors are likely to act only locally?

22 The History of Life on Earth

When you want to know what time it is, you probably look at your watch, or at the clock on the wall or on your computer. You could also listen for an announcement of the time on the radio or television. But suppose the electric power system failed and you lost your watch. How would you tell time then? You would use the cue that people have used during most of human history: the cycle of day and night. We are so accustomed to having time-measuring devices all around us that we forget that these devices are recent inventions. When Galileo studied the motion of a ball rolling down an inclined plane about 400 years ago, he used his pulse to mark off equal intervals of time.

The development of the science of biology is intimately linked to changing concepts of time, especially of the age of Earth. Biology as we know it could not and did not develop very far until about 150 years ago, when geologists provided evidence that Earth was ancient. Before 1850, most people believed that Earth was only a few thousand years old. Charles Darwin could not have developed his theory of evolution by natural selection if he had not read the works of Charles Lyell, who was England's leading geologist during Darwin's lifetime. Lyell suggested that existing landforms could be explained by the action, over very long time periods, of the same forces that are still acting on them today. That is, Lyell argued that it is not necessary to postulate sudden catastrophes as the reason for dramatic geological changes. As we pointed out in Chapter 1, Darwin's theory was based on the assumption that Earth was very old and that millions of years were available for life's evolution.

The goals of Part Four of this book are to document the history of life on Earth, to describe patterns of evolutionary change, and to investigate the agents that cause them. We begin this chapter by asking, How do we know that Earth is ancient? What is the evidence that life evolved early during Earth's history and has continued to evolve since then? We will first examine how events in the distant past can be dated. We will review the major changes in physical conditions on Earth during the past 4 billion years, look at how those changes have affected life, and describe some patterns in the evolution of life. In Chapter 23 we will discuss

Sunset at Stonehenge Even the earliest humans felt the need to keep track of time and seasons. The arches of Stonehenge, an astronomical "timepiece" on Salisbury Plain in England, date back to about 2000 B.C., and radioisotope dating has revealed some wooden structures unearthed here to be more than 8,000 years old.

the processes by which life evolves. In subsequent chapters, we will see how biologists determine the evolutionary histories of organisms and how the millions of species that live today (as well as many more that became extinct) were derived from a single common ancestor.

Defining Biological Evolution

Understanding evolution is important to the study of biology because the features of all organisms, including humans, are best understood in light of evolution. Furthermore, evolutionary changes are taking place all around us, some of which have powerful implications for human welfare. For example, our attempts to control populations of species we consider pests and to increase populations of those we consider desirable make humans powerful agents of evolutionary change. In addition to producing the results we desire, these efforts often cause undesirable outcomes, such as the evolution of resistance to antibiotics by pathogens and to pesticides by pests. Medicine and agriculture can respond creatively to the evolutionary changes they are causing only if their practitioners understand how and why those changes happen. But what exactly is biological evolution?

Biological evolution is a change over time in the genetic composition of a population of organisms. Many such changes happen rapidly enough to be studied directly and manipulated experimentally. Plant and animal breeding by agriculturalists and responses of organisms to environmental shifts over decades provide good examples of such short-term evolution. Other changes, such as the appearance of new species and evolutionary lineages, usually take place over much longer time frames. The fossil record is the primary source of direct evidence of those changes.

To understand the long-term patterns of evolutionary change that we will document in this chapter, we must think in time frames spanning many millions of years and imagine events and conditions very different from those we now observe. The Earth of the distant past is, to us, a foreign planet inhabited by strange organisms. The continents were not where they are today, and climates were sometimes dramatically different from those of today. One of the remarkable achievements of twentieth-century science has been the development of sophisticated techniques, using rates of decay of various radioisotopes, changes in Earth's magnetic field, and the presence or absence of certain molecules, for inferring past conditions and dating them accurately.

Determining Earth's Age

It is difficult to age rocks because any type of rock could have been formed at any time during Earth's history. It is easier to determine the ages of rocks relative to one another. The first

22.1 Young Rocks Lie on Top of Old Rocks   The oldest rocks visible in this photo of the Grand Canyon formed about 540 million years ago. The youngest rocks, at the top, are about 500 million years old.

person to recognize that this could be done was the seventeenth-century Danish physician Nicolaus Steno. Steno realized that in undisturbed sedimentary rock, the oldest layers, or *strata*, lie at the bottom, and successively higher strata are progressively younger (Figure 22.1).

Geologists subsequently combined Steno's insight with their observations of **fossils**—preserved remains of ancient organisms—contained within the rocks. They discovered that fossils of similar organisms were found in widely separated places on Earth, that certain organisms were always found in younger rocks than others, and that organisms in more recent strata were more similar to modern organisms than were those found in lower, more ancient strata. With this information, they learned much about the relative ages of sedimentary rocks and patterns in the evolution of life. But they could not tell how old the rocks were. A method of dating rocks did not become available until the discovery of radioactivity at the beginning of the twentieth century.

Radioactivity provides a way to date rocks

Radioactive isotopes decay in a predictable pattern over long time periods (see Chapter 2). During each successive time interval, an equal fraction of the remaining radioactive material of any radioisotope decays, either changing to another element or becoming the stable isotope of the same element.

For example, in 14.3 days, one-half of any sample of phosphorus-32 (^{32}P) decays to its stable isotope, phosphorus-31 (^{31}P). During the next 14.3 days, one-half of the remaining half decays, leaving one-fourth of the original ^{32}P. After 42.9 days, three *half-lives* have passed, so one-eighth (that is, $^1/_2 \times {}^1/_2 \times {}^1/_2$) of the original ^{32}P remains.

Each radioisotope has a characteristic half-life. Tritium (^3H) has a half-life of 12.3 years, and carbon-14 (^{14}C) has a half-life of about 5,700 years. The half-life of potassium-40 (^{40}K) is 1.3 billion years; that of uranium-238 (^{238}U) is about 4.5 billion years.

To use a radioisotope to date a past event, we must know or estimate the concentration of the isotope at the time of that event. In the case of carbon, we know that the production of new ^{14}C in the upper atmosphere (by the reaction of neutrons with ^{14}N) just balances the natural radioactive decay of ^{14}C. Therefore, the ratio of ^{14}C to its stable isotope, ^{12}C, is relatively constant in living organisms and their environment.

However, as soon as an organism dies, it ceases to exchange carbon compounds with its environment. Its decaying ^{14}C is no longer replenished, and the ratio of ^{14}C to ^{12}C in its remains decreases through time. The ratio of ^{14}C to ^{12}C in fossil organisms can be used to date fossils (and thus the sedimentary rocks that contain those fossils) that are less than 50,000 years old with a fair degree of certainty.

Radioisotope dating methods have been expanded and refined

Dating rocks more ancient than 50,000 years requires estimating isotope concentrations in volcanic (but not in sedimentary) rocks. Sedimentary rocks are formed from materials that existed for varying lengths of time before being transported long distances to the site of their deposition. Therefore, a sedimentary rock does not contain reliable information about the date of its formation. To age sedimentary rocks, geologists search for places where volcanic ash or lava flows have intruded into beds of those rocks. The preliminary estimate of the age of the volcanic rock determines which isotope is used. The decay of potassium-40 to argon-40 has been

22.1 Earth's Geological History

RELATIVE TIME SPAN	ERA	PERIOD	ONSET	MAJOR PHYSICAL CHANGES ON EARTH
	Cenozoic	Quaternary	1.8 mya[a]	Cold/dry climate; repeated glaciations
		Tertiary	65 mya	Continents near current positions; climate cools
	Mesozoic	Cretaceous	144 mya	Northern continents attached; Gondwana begins to drift apart; meteorite strikes Yucatán Peninsula
		Jurassic	206 mya	Two large continents form: Laurasia (north) and Gondwana (south); climate warm
		Triassic	248 mya	Pangaea slowly begins to drift apart; hot/humid climate
	Paleozoic	Permian	290 mya	Continents aggregate into Pangaea; large glaciers form; dry climates form in interior of Pangaea
		Carboniferous	354 mya	Climate cools; marked latitudinal climate gradients
		Devonian	417 mya	Continents collide at end of period; asteroid probably collides with Earth
		Silurian	443 mya	Sea levels rise; two large continents form; hot/humid climate
		Ordovician	490 mya	Gondwana moves over South Pole; massive glaciation, sea level drops 50 m
		Cambrian	543 mya	O_2 levels approach current levels
Precambrian	Precambrian		600 mya	O_2 level at >5% of current level
			1.5 bya[a]	O_2 level at >1% of current level
			3.8 bya	O_2 first appears in atmosphere
			4.5 bya	

[a]mya, million years ago; bya, billion years ago.

used to date most of the ancient events in the evolution of life. Radioisotope dating, combined with fossils, is the most powerful method of determining the ages of rocks.

But there are many places where sedimentary rocks do not contain suitable volcanic intrusions and few fossils are present. In these areas, other dating methods must be used. One method, known as *paleomagnetic dating*, is based on the fact that Earth's magnetic poles move and occasionally reverse themselves. Because both sedimentary and volcanic rocks preserve a record of Earth's magnetic field at the time they were formed, paleomagnetism helps determine the ages of those rocks. Other dating methods, which we will describe in later chapters, use continental drift, sea level changes, and molecular clocks.

Using these methods, geologists have divided the history of life into eras, which in turn are subdivided into periods (Table 22.1). The boundaries between these divisions are based on major differences in the fossil organisms contained in successive layers of rocks. The divisions were established

before the ages of the eras and periods were known. The scale at the left of Table 22.1 gives a relative sense of geological time and the vast expanse of the **Precambrian** era, during which early life evolved amid stupendous physical changes on Earth.

The Changing Face of Earth

Earth has undergone many physical changes that have influenced the evolution of life. The physical events described in this section, along with the most important milestones in the history of life, are listed in Table 22.1.

The continents have changed their positions

The maps and globes that adorn our walls, shelves, and books give an impression of a static Earth. It would be easy for us to assume that the continents have always been where they are. But we would be wrong. Earth's crust consists of a number of solid *plates* approximately 40 km thick, which float on a fluid *mantle*. The mantle fluid circulates because heat produced by radioactive decay sets up convection patterns in the fluid. The plates move because material from the mantle rises and pushes them aside. Where plates are pushed together, either they move sideways past each other, or one plate moves under the other, pushing up mountain ranges. The movement of the plates and the continents they contain is known as **continental drift**.

At times, the drifting of the plates brought the continents together; at other times, they drifted apart. The positions and sizes of the continents influence ocean circulation patterns, sea levels, and global climate patterns. Mass extinctions of species, particularly marine organisms, have usually accompanied major drops in sea level, which exposed vast areas of the continental shelves, killing the marine organisms that lived in the shallow seas that had covered them (Figure 22.2).

Earth's atmosphere has changed unidirectionally

The continents have moved irregularly over Earth's surface, but some physical changes on Earth have been unidirectional. The atmosphere of early Earth probably contained little or no free oxygen (O_2). Oxygen concentrations in the atmosphere began to increase markedly about 2.5 billion years ago, when certain bacteria evolved the ability to use water as the source of hydrogen ions for photosynthesis. By chemically splitting H_2O ($2\,H_2O \rightarrow 4\,H^+ + O_2 + 4e^-$), these bacteria generated atmospheric O_2 as a waste product; in addition, they made electrons available for reducing CO_2 to form organic compounds.

One lineage of oxygen-generating bacteria evolved into the *cyanobacteria*. These ancient photosynthesizers formed

---

MAJOR EVENTS IN THE HISTORY OF LIFE

Humans evolve; many large mammals become extinct

Diversification of birds, mammals, flowering plants, and insects

Dinosaurs continue to diversify; flowering plants and mammals diversify. **Mass Extinction** at end of period (≈76% of species disappear)
Diverse dinosaurs; radiation of ray-finned fishes

Early dinosaurs; first mammals; marine invertebrates diversify; first flowering plants; **Mass Extinction** at end of period (≈65% of species disappear)

Reptiles diversify; amphibians decline; **Mass Extinction** at end of period (≈96% of species disappear)

Extensive "fern" forests; first reptiles; insects diversify

Fishes diversify; first insects and amphibians. **Mass Extinction** at end of period (≈75% of species disappear)
Jawless fishes diversify; first ray-finned fishes; plants and animals colonize land
Mass Extinction at end of period (≈75% of species disappear)
Most animal phyla present; diverse algae

Ediacaran fauna
Eukaryotes evolve; several animal phyla appear
Origin of life; prokaryotes flourish

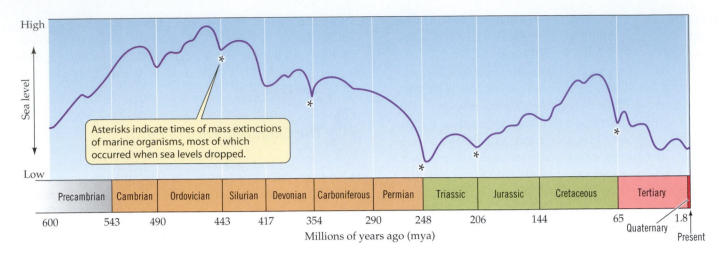

22.2 Sea Levels Have Changed Repeatedly Most mass extinctions (indicated by asterisks) of marine organisms have coincided with periods of low sea levels.

rocklike structures called *stromatolites*, which are abundantly preserved in the fossil record. Cyanobacteria are still forming stromatolites today in a few very salty places on Earth (Figure 22.3). Cyanobacteria liberated enough O_2 to open the way for the evolution of oxidation reactions as the energy source for the synthesis of ATP. Their ability to split water doubtless contributed to their extraordinary success.

The evolution of life irrevocably changed the physical nature of the planet. Living organisms not only added O_2 to Earth's atmosphere, but also removed most of the CO_2 from the atmosphere by taking it up and transferring it to ocean sediments with their remains when they died. When it first appeared, oxygen was poisonous to the anaerobic prokaryotes that inhabited Earth at the time. Those prokaryotes that evolved the ability to metabolize O_2 not only survived, but also gained a number of advantages. Aerobic (oxygen-using) metabolism proceeds at higher rates and can extract more energy from compounds than the anaerobic metabolism prevalent among living things until then (see Chapter 7). Consequently, organisms with aerobic metabolism have replaced anaerobes in most of Earth's environments.

An atmosphere rich in O_2 also made possible larger cells and more complex organisms. Small, unicellular aquatic organisms can obtain enough O_2 by simple diffusion even when O_2 concentrations are very low. Larger unicellular organisms have lower surface area-to-volume ratios (see Fig-

ure 4.3). In order to obtain enough O_2 by simple diffusion, they must live in an environment with a relatively high concentration of O_2. Bacteria can thrive on 1 percent of the current atmospheric O_2 levels, but eukaryotic cells require oxygen levels that are at least 2 to 3 percent of current atmospheric concentrations.

About 1,500 million years ago (mya), O_2 concentrations became high enough for large eukaryotic cells to flourish and diversify (Figure 22.4). Further increases in atmospheric O_2

(a)

22.3 Stromatolites (a) A vertical section through a fossil stromatolite. (b) These rocklike structures are living stromatolites that thrive in the very salty waters of Shark Bay, Western Australia. Layers of cyanobacteria are found in the uppermost parts of the structures.

(b)

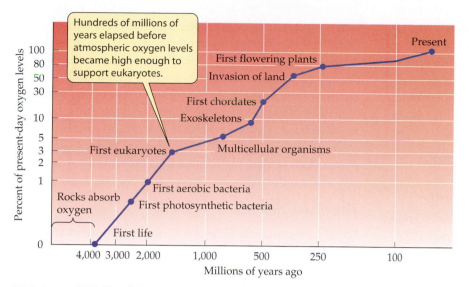

22.4 Larger Cells Need More Oxygen As oxygen concentrations in the atmosphere rose, the complexity of life increased. Although aerobic prokaryotes can flourish with less, larger eukaryotic cells with lower surface area-to-volume ratios require at least 2 to 3 percent of current atmospheric O_2 concentrations. (Both axes of the graph are on logarithmic scales.)

levels 700 to 570 mya enabled multicellular organisms to evolve. The fact that it took many millions of years for Earth to develop an oxygenated atmosphere probably explains why only unicellular prokaryotes lived on Earth for more than a billion years.

In contrast to this largely unidirectional change in atmospheric O_2 concentration, most physical conditions on Earth have oscillated in response to the planet's internal processes, such as volcanic activity and continental drift. External events, such as collisions with meteorites, have also left their mark. In some cases, these events caused **mass extinctions**, exterminating a large proportion of the species living at the time.

Earth's climate shifts between hot/humid and cold/dry conditions

Through much of its history, Earth's climate was considerably warmer than it is today, and temperatures decreased more gradually toward the poles. At other times, however, Earth was colder than it is today. Large areas were covered with glaciers during the end of the Precambrian and during the Carboniferous, Permian, and Quaternary periods, but these cold periods were separated by long periods of milder climates (Figure 22.5). Because we are living in one of the colder periods in the history of Earth, it is difficult for us to imagine the mild climates that were found at high latitudes during much of the history of life.

Weather often changes rapidly; climates usually change slowly. However, major climatic shifts have taken place over periods as short as 5,000 to 10,000 years, primarily as a result of changes in Earth's orbit around the sun. A few climatic shifts appear to have been even more rapid. For example, during one Quaternary interglacial period, the Antarctic Ocean changed from being ice-covered to being nearly ice-free in less than 100 years. Such rapid changes are usually caused by sudden shifts in ocean currents. Climates have sometimes changed rapidly enough that extinctions caused by them appear "instantaneous" in the fossil record.

Volcanoes occasionally changed the history of life

Most volcanic eruptions produce only local or short-lived effects, but a few very large volcanic eruptions have had major consequences for life. The collision of continents during

22.5 Hot/Humid and Cold/Dry Conditions Have Alternated Over Earth's History Throughout Earth's history, periods of cold climates and glaciations (white depressions) have been separated by long periods of milder climates.

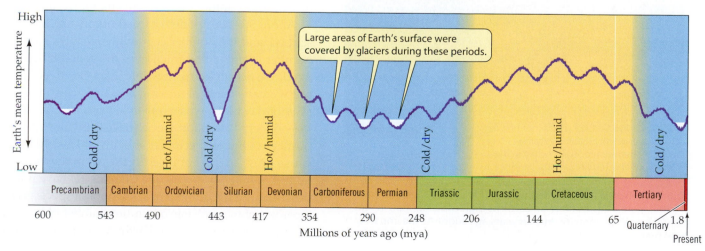

the late Permian period (about 275 mya) to form a single, gigantic land mass, called Pangaea, caused massive volcanic eruptions. The ash the volcanoes ejected into the atmosphere reduced the penetration of sunlight to Earth's surface, lowering temperatures, reducing photosynthesis, and triggering massive glaciation. Massive volcanic eruptions also occurred as the continents drifted apart during the late Triassic period and at the end of the Cretaceous.

External events have triggered changes on Earth

At least 30 meteorites between the sizes of baseballs and soccer balls hit Earth each year. Collisions with large meteorites are rare, but large meteorites have been responsible for several mass extinctions. Evidence for these collisions is found in the craters that resulted from their impact, dramatic disfigurations of rocks (microspherules and shocked quartz crystals), and within giant molecules that contain trapped helium and argon with isotopic ratios characteristic of meteorites, which are very different from the ratios found on Earth. Also, fern fossils are abundant in rocks that formed at the end of the Triassic and Cretaceous periods. Because ferns can more quickly colonize and survive in bare environments than most other plants, their abundance suggests that meteorite impacts had scoured vast areas of Earth's surface.

The first extraterrestrial impact to be documented was that of a meteorite about 10 km in diameter that caused a mass extinction at the end of the Cretaceous period, about 65 mya. The first clue that a meteorite was responsible came from the abnormally high concentrations of the element iridium in a thin layer separating rocks deposited during the Cretaceous from those deposited during the Tertiary (Figure 22.6). Iridium is abundant in some meteorites but is exceedingly rare on Earth's surface. Subsequently, a circular crater 180 km in diameter was discovered buried beneath the northern coast of the Yucatán Peninsula of Mexico. When it collided with Earth, the meteorite released the equivalent of 100 million megatons of high explosives. The force of the impact ignited massive fires, created great tidal waves, and sent up an immense dust cloud that blocked the sun, thus cooling the planet. As it settled, the dust formed the iridium-rich layer.

The Fossil Record

Fossils are a major source of information about changes on Earth during the remote past. As we saw above, pre-Darwinian geologists divided geological history into units based on their distinct fossil assemblages of animals (see Table 22.1). These divisions are marked either by mass extinctions or by dramatic increases in the diversity of major groups of organisms (called *evolutionary radiations*).

Life first evolved on Earth about 3.8 billion years ago (bya), and by about 1.5 bya, eukaryotic organisms had evolved. The fossil record of organisms that lived prior to 550 mya is fragmentary, but the available evidence suggests that the major divisions in many animal lineages predate the end of the Precambrian by more than 100 million years. The fossil record is good enough to show that the total number of species and individuals increased dramatically in late Precambrian times.

An organism is most likely to become a fossil if its dead body is deposited in an environment that lacks oxygen. However, most organisms live in aerobic environments; they decompose completely when they die. Thus, many fossil assemblages are collections of organisms that were transported by wind or water to sites that lacked oxygen. Occasionally, however, organisms, or imprints of them, are preserved where they lived. In such cases—especially if the environment in question was a cool, anaerobic swamp, where conditions for preservation were excellent—we can obtain a picture of communities of organisms that lived together.

About 300,000 species of fossil organisms have been described, and the number is growing steadily. However, this number is only a tiny fraction of the species that have ever lived. We do not know how many species lived in the past, but we have ways of making reasonable estimates. Of the present-day **biota**—that is, all living species of all kinds—approximately 1.7 million species have been named. The actual number of living species is probably at least 10 million, because most species of insects (the animal group with the largest number of species; see Chapter 33) have not yet been described. So the number of known fossil species is less than 2 percent of the probable minimum number of living species. Because life has existed on Earth for about 3.8 billion years,

A thin band rich in iridium marks the boundary between rocks deposited in the Cretaceous and Tertiary periods.

22.6 Evidence of a Meteorite Impact   Iridium is a metal common in some meteorites, but rare on Earth. Its high concentration in sediments deposited about 65 million years ago suggests the impact of a large meteorite.

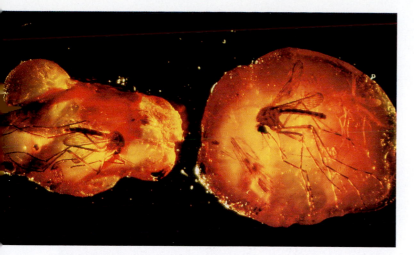

22.7 Insect Fossils These chunks of amber—fossilized tree resin—contain insects that were preserved when they were trapped in the sticky resin some 50 million years ago.

and because species last, on average, less than 10 million years, Earth's biota must have turned over many times during geological history. Thus, the total number of species that lived over evolutionary time must vastly exceed the number living today.

The number of known fossils, although it is a small fraction of the total number of extinct species, is especially large for marine animals that had hard skeletons. Among the nine major animal groups with hard-shelled members, approximately 200,000 species have been described from fossils—roughly twice the number of living marine species in these same groups. *Paleontologists* (scientists who study fossils) lean heavily on these groups in their interpretations of the evolution of life. Insects and spiders are also relatively well represented in the fossil record (Figure 22.7). The fossil record may be incomplete, but it is good enough to demonstrate clearly that organisms of particular types are found in rocks of specific ages and that new organisms appear sequentially in younger rocks. The fossil record also tells us that extinction is the eventual fate of all species.

By combining information about physical changes during Earth's history with evidence from the fossil record, scientists have composed portraits of what Earth and its inhabitants may have looked like at different times. We know in general where the continents were and how life changed over time, but many of the details are poorly known, especially for events in the more remote past. In the next section, we provide an overview of how life changed during its history on Earth. Part Five of this book will look at the evolutionary history of particular groups of organisms in more detail.

Major Patterns in the History of Life on Earth

For much of its history, life was confined to the oceans, and all organisms were small. In the Precambrian era, shallow seas teemed with life. Protists and small multicellular animals fed on floating algae. Small floating organisms, known collectively as *plankton*, were eaten by small animals that filtered them from the water. Other animals ingested sediments on the bottom of the seas and digested the remains of organisms within them. By the late Precambrian, about 650 mya, many kinds of soft-bodied invertebrates had evolved. Some of them were very different from any animals living today. They may be members of animal lineages that have no living descendants (Figure 22.8).

Life expanded rapidly during the Cambrian period

By the early **Cambrian** period (543–490 mya), at the beginning of the Paleozoic era, the O_2 concentration in the atmosphere approached its current level and the continental plates came together to form several land masses. The largest of the land masses was called Gondwana (Figure 22.9a). All of the major groups of animals that have species living today appeared during the Cambrian. This rapid diversification of life is referred to as the *Cambrian explosion*.

The most extensive fossil evidence from the Cambrian comes from the unusually well preserved animal fossils recently discovered in northeastern China (Figure 22.9b). Arthropods (crabs, shrimps and their relatives) are the most

22.8 Ediacaran Animals These fossils of soft-bodied invertebrates, excavated at Ediacara in southern Australia, formed 600 million years ago. They illustrate the diversity of life that evolved in Precambrian times.

Spriggina floundersi

Mawsonites

(a)

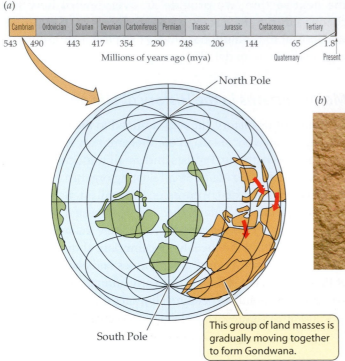

Cambrian	Ordovician	Silurian	Devonian	Carboniferous	Permian	Triassic	Jurassic	Cretaceous	Tertiary	
543	490	443	417	354	290	248	206	144	65	1.8*

Millions of years ago (mya)   Quaternary   Present

North Pole

South Pole

This group of land masses is gradually moving together to form Gondwana.

22.9 Cambrian Continents and Animals (a) Positions of the continents during mid-Cambrian times (542–490 mya). This view of Earth has been distorted so that you can see both poles. (b) Fossil beds in China have yielded well-preserved remains of Cambrian animals such as this one, called *Jianfangia*.

(b)

diverse group in this Chinese fauna; some of them were large carnivores. Trilobites, an arthropod group that was abundant and diverse during the Cambrian, suffered a major extinction at the end of the Cambrian, but they recovered and continued to be abundant during subsequent periods.

Major changes continued during the rest of the Paleozoic era

Geologists divide the remainder of the Paleozoic era into five periods: the Ordovician, Silurian, Devonian, Carboniferous, and Permian (see Table 22.1). Each period is characterized by the diversification of specific groups of organisms. Mass extinctions marked the ends of three periods: the Ordovician, Devonian, and Permian.

THE ORDOVICIAN (490–443 MYA). During the **Ordovician** period, the continents, which were located primarily in the Southern Hemisphere, were still devoid of multicellular plants. Evolutionary radiation of marine organisms was spectacular during the early Ordovician, especially among animals that filter small prey from the water, such as brachiopods and mollusks. All animals lived on the seafloor or burrowed in its sediments. At the end of the Ordovician, as massive glaciers formed over Gondwana, sea levels dropped about 50 meters and ocean temperatures dropped. About 75 percent of the animal species became extinct, probably because of these major environmental changes.

THE SILURIAN (443–417 MYA). During the **Silurian** period, the northern continents coalesced, but the general positions of the continents did not change much. Marine life rebounded from the mass extinction at the end of the Ordovician. Animals able to swim and feed above the ocean bottom appeared for the first time, but no new phyla of marine organisms evolved. The tropical sea was uninterrupted by land barriers, and most marine genera were widely distributed. On land, the first known tracheophytes (plants with true vascular tissue; see Chapter 29) appeared late in the Silurian period, about 420 mya. These plants, in the genus *Cooksonia*, were less than 50 cm tall and lacked roots and leaves (Figure 22.10). The first terrestrial arthropods—scorpions and millipedes—appeared at about the same time.

THE DEVONIAN (417–354 MYA). Rates of evolutionary change accelerated in many groups of organisms during the **Devonian** period. Both northern and southern land masses slowly moved northward (Figure 22.11a). There were great evolutionary radiations of corals and shelled squidlike cephalopods (Figure 22.11b). Fishes diversified as jawed forms replaced jawless ones, and heavy armor gave way to the less rigid outer coverings of modern fishes. All current major groups of fishes were present by the end of the period.

Terrestrial communities also changed dramatically during the Devonian. Club mosses, horsetails, and tree ferns became common toward the end of the Devonian; some attained the size of trees. Their deep roots accelerated the weathering of rocks, resulting in the development of the first forest soils. Distinct floras evolved on the two major land masses toward the end of the period, and the ancestors of gymnosperms, the

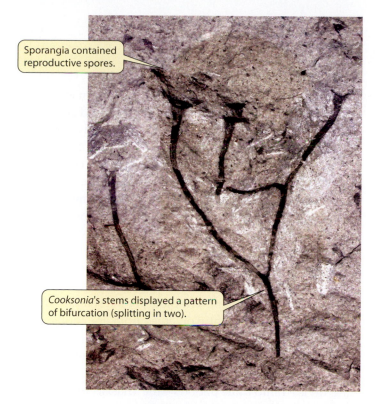

Sporangia contained reproductive spores.

Cooksonia's stems displayed a pattern of bifurcation (splitting in two).

22.10 *Cooksonia*, the Earliest Known Tracheophyte These plants were small and very simple in structure. However, they were true vascular plants (tracheophytes) with internal water-conducting cells (tracheids), well equipped to make the move from the aquatic to the terrestrial environment. This fossil of *Cooksonia pertoni* is from the Silurian (415 mya).

first plants to produce seeds, appeared in the fossil record. The first known fossils of centipedes, spiders, mites, and insects date to this period. Fishlike amphibians began to occupy the land.

An extinction of about 75 percent of all marine species marked the end of the Devonian. Paleontologists are uncertain about the cause of this mass extinction, but two large meteorites that collided with Earth at that time, one in present-day Nevada and the other in Western Australia, may have been responsible.

THE CARBONIFEROUS (354–290 MYA). Large glaciers formed over high-latitude Gondwana during the **Carboniferous** period, but extensive swamp forests grew on the tropical continents. These forests were not made up of the kinds of trees we know today, but were dominated by giant tree ferns and horsetails (see Figure 29.11). Fossilized remains of those trees formed the coal we now mine for energy.

The diversity of terrestrial animals increased greatly. Snails, scorpions, centipedes, and insects were abundant and diverse. Insects evolved wings, becoming the first animals to fly. Flight gave them access to tall plants; plant fossils from this period show evidence of chewing by insects. Amphibians became larger and better adapted to terrestrial existence. From one amphibian stock, the first reptiles evolved late in the period. In the seas, crinoids (sea lilies and feather stars) reached their greatest diversity, forming "meadows" on the seafloor (Figure 22.12)

(a)

Cambrian	Ordovician	Silurian	Devonian	Carboniferous	Permian	Triassic	Jurassic	Cretaceous	Tertiary

543 490 443 417 354 290 248 206 144 65 1.8

Millions of years ago (mya)   Quaternary   Present

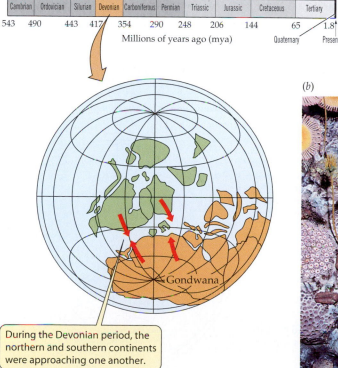

Gondwana

During the Devonian period, the northern and southern continents were approaching one another.

22.11 Devonian Continents and Marine Communities
(a) Positions of the continents during the Devonian period (417–354 mya). (b) This museum reconstruction depicts a Devonian coral reef.

(b)

Cambrian	Ordovician	Silurian	Devonian	Carboniferous	Permian	Triassic	Jurassic	Cretaceous	Tertiary

543   490   443   417   354   290   248   206   144   65   1.8↑

Millions of years ago (mya)   Quaternary   Present

22.12 A Carboniferous "Crinoid Meadow"   Crinoids, which were dominant marine animals during the Carboniferous, may have formed communities similar to this one.

THE PERMIAN (290–248 MYA).   During the **Permian** period, the continents coalesced into a supercontinent, called Pangaea. Massive volcanic eruptions resulted in outpourings of lava that covered large areas of Earth (Figure 22.13). The ash they produced blocked sunlight and cooled the climate, resulting in the largest glaciers in Earth's history.

Permian deposits contain representatives of most modern groups of insects. By the end of the period, reptiles greatly outnumbered amphibians. Late in the period, the lineage leading to mammals diverged from one reptilian lineage. In fresh waters, the Permian period was a time of extensive radiation of ray-finned fishes.

Toward the end of the Permian period, a large meteorite crashed into northwestern Australia, creating a crater about 190 km in diameter. In addition, a massive outpouring of lava flowed into the oceans, drastically reducing the oxygen content of deep ocean waters. Oceanic turnover then carried these oxygen-depleted waters toward the surface, where they released toxic concentrations of carbon dioxide and hydrogen sulfide into the surface waters and the atmosphere. These gases poisoned most of the species that had survived the impact. All in all, about 96 percent of all species on Earth became extinct at that time.

22.13 Pangaea Formed in the Permian Period   During the Permian period (290–248 mya, the interior of the "supercontinent" Pangaea experienced harsh climates. Massive lava flows spread over Earth, and the largest glaciers in Earth's history formed during this period.

Geographic differentiation increased during the Mesozoic era

The few organisms that survived the Permian mass extinction found themselves in a relatively empty world at the start of the Mesozoic era (248 mya). As Pangaea slowly separated into individual continents, the climate warmed, the glaciers melted, and the oceans rose and reflooded the continental shelves, forming huge, shallow inland seas. Life again proliferated and diversified, but different lineages came to dominate Earth. The trees that had dominated the Permian forests, for example, were replaced by new plants with seeds.

During the Mesozoic, Earth's biota, which until that time had been relatively homogeneous, became increasingly *provincialized*; that is, distinctive terrestrial floras and faunas evolved on each continent. The biotas of the shallow waters bordering the continents also diverged from one another. The provincialization that began during the Mesozoic continues to influence the geography of life today. By the end of the era, the world and its biota appeared quite modern. The Mesozoic era is divided into three periods—the Triassic, Jurassic, and Cretaceous—the first and third of which were terminated by mass extinctions, probably caused by meteorite impacts.

THE TRIASSIC (248–206 MYA).   Pangaea began to break apart during the **Triassic** period. Many invertebrate lineages became more diverse, and many burrowing forms evolved from groups living on the surfaces of bottom sediments. On land, conifers and seed ferns became the dominant trees. The first frogs and turtles appeared. A great radiation of reptiles began, which eventually gave rise to crocodilians, dinosaurs, and birds. The end of the Triassic was marked

Cambrian	Ordovician	Silurian	Devonian	Carboniferous	Permian	Triassic	Jurassic	Cretaceous	Tertiary

543   490   443   417   354   290   248   206   144   65   1.8↑

Millions of years ago (mya)   Quaternary   Present

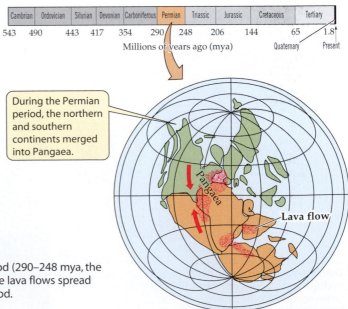

During the Permian period, the northern and southern continents merged into Pangaea.

Pangaea

Lava flow

Cambrian	Ordovician	Silurian	Devonian	Carboniferous	Permian	Triassic	Jurassic	Cretaceous	Tertiary

543   490   443   417   354   290   248   206   144   65   1.8↑
Millions of years ago (mya)   Quaternary   Present

22.14 Mesozoic Dinosaurs The dinosaurs of the Mesozoic era continue to capture our imagination. The horned animals in this artist's depiction are *Triceratops horridus*, a huge (6000 kg) herbivore of the late Cretaceous (70 mya). At the lower right is *Ornithomimus* ("bird mimic"), an ostrich-like dinosaur believed to have been able to run at high speeds.

Several groups of mammals first appeared during this time. Plant evolution continued with the likely emergence of the flowering plant lineage prevalent on Earth today (see Chapter 30).

THE CRETACEOUS (144–65 MYA). By the early **Cretaceous** period, Laurasia was completely separate from Gondwana, which was beginning to break apart. A continuous sea encircled the Tropics (Figure 22.15). Sea levels were high, and Earth was warm and humid. Life proliferated both on land and in the oceans. Marine invertebrates increased in variety and number of species. On land, dinosaurs continued to diversify. The first snakes appeared during the Cretaceous, though their lineages did not radiate until much later. Early in the Cretaceous, flowering plants began the radiation that led to their current dominance on land. Fossils of the earliest known flowering plants (Archaefructaceae), dated at 124 mya, recently were discovered in Liaoning Province in northeastern China (Figure 22.16). By the end of the period, many groups of mammals had evolved, but these mammals were generally small.

Another meteorite-caused mass extinction took place at the end of the Cretaceous period. On land, all vertebrates larger than about 25 kg in body weight, including all of the dinosaurs, apparently became extinct. Many species of insects died out, perhaps because the growth of their food

by a mass extinction that eliminated about 65 percent of the species on Earth. A large meteor that crashed into Quebec may have been responsible.

THE JURASSIC (206–144 MYA). During the **Jurassic** period, two large continents formed—Laurasia in the north, and Gondwana in the south. Diversification of many lineages proceeded. Ray-finned fishes began the great radiation that culminated in their dominance of the oceans. Salamanders and lizards first appeared. Flying reptiles (pterosaurs) evolved, and dinosaur lineages evolved into bipedal predators and large quadrupedal herbivores (Figure 22.14).

Cambrian	Ordovician	Silurian	Devonian	Carboniferous	Permian	Triassic	Jurassic	Cretaceous	Tertiary

543   490   443   417   354   290   248   206   144   65   1.8↑
Millions of years ago (mya)   Quaternary   Present

22.15 Positions of the Continents during the Cretaceous Period By the Cretaceous, Pangaea had split into two major land masses, Laurasia and Gondwana, separated by a continuous tropical sea.

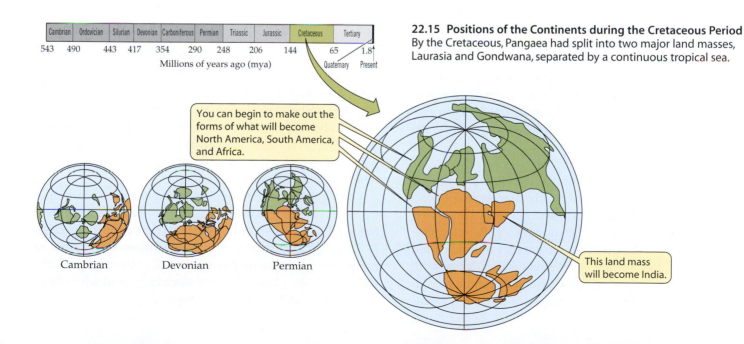

You can begin to make out the forms of what will become North America, South America, and Africa.

This land mass will become India.

Cambrian   Devonian   Permian

The "feather" pattern of its leaves indicates that *Archaefructus* lived in water.

22.16 Flowering Plants of the Cretaceous   Fossils of *Archaefructus* found in China are a minimum of 124.6 million years old. These flowering plants are early examples of the type of plants most prevalent on Earth today.

plants was greatly reduced following the impact. In the seas, many planktonic organisms and bottom-dwelling invertebrates became extinct.

The modern biota evolved during the Cenozoic era

By the early Cenozoic era (65 mya), the positions of the continents resembled those of today, but Australia was still attached to Antarctica, and the Atlantic Ocean was much narrower. The Cenozoic era was characterized by an extensive radiation of mammals, but other groups were also undergoing important changes. Flowering plants diversified extensively and came to dominate world forests, except in cool regions. The Cenozoic era is divided into two periods, the Tertiary and the Quaternary.

THE TERTIARY (65–1.8 MYA).   During the **Tertiary** period, Australia began its northward drift. By 20 mya it had nearly reached its current position. The early Tertiary was a hot/humid time, during which vegetation belts shifted latitudinally. The Tropics were probably too hot for rainforests, and were clothed in low-stature vegetation instead. In the middle of the Tertiary, however, Earth's climate became considerably drier and cooler. Many lineages of flowering plants evolved herbaceous (nonwoody) forms; grasslands spread over much of Earth.

By the beginning of the Cenozoic era, invertebrate faunas resembled those of today. It is among the vertebrates that evolutionary changes during the Tertiary period were most rapid. Living groups of reptiles, including snakes and lizards, underwent extensive radiations during this period, as did birds and mammals. Three waves of mammals dispersed from Asia to North America about 55 mya. Rodents, marsupials, primates, and hoofed mammals appeared in North America for the first time.

THE QUATERNARY (1.8 MYA TO PRESENT).   The current geological period, the **Quaternary**, is subdivided into two *epochs*, the **Pleistocene** and the **Holocene** (also known as the *Recent*). The Pleistocene epoch was a time of drastic cooling and climatic fluctuations. During four major and about 20 minor episodes of glaciation, massive glaciers spread across the continents, and animal and plant populations shifted toward the equator. The last of these glaciers retreated from temperate latitudes less than 15,000 years ago; this retreat marked the beginning of the Holocene epoch. Organisms of the Holocene are still adjusting to these changes. Many high-latitude ecological communities have occupied their current locations for no more than a few thousand years.

Interestingly, few species became extinct during these climatic fluctuations. However, the Pleistocene was the time of hominid evolution and radiation, resulting in the species *Homo sapiens*—modern humans (see Chapter 34). Many large bird and mammal species became extinct in Australia and in the Americas when *H. sapiens* arrived on these continents about 40,000 and 15,000 years ago, respectively. Human hunting may have caused these extinctions, although existing evidence does not convince all paleontologists.

Three major faunas have dominated life on Earth

The fossil record reveals three great radiations that resulted in the evolution of major new faunas (Figure 22.17). The first one, the Cambrian explosion, took place about 540 mya. The second, about 60 million years later, resulted in the Paleozoic fauna. The great Permian extinctions 300 million years later were followed by the third event, the Triassic explosion, which led to our modern fauna.

During the Cambrian explosion, organisms representing all the major body plans of present-day lineages appeared, along with a number of lineages that subsequently became extinct. The Paleozoic and Triassic explosions resulted in many new groups of organisms, but all of them had modifications of body plans that were already present when these great biological diversifications began.

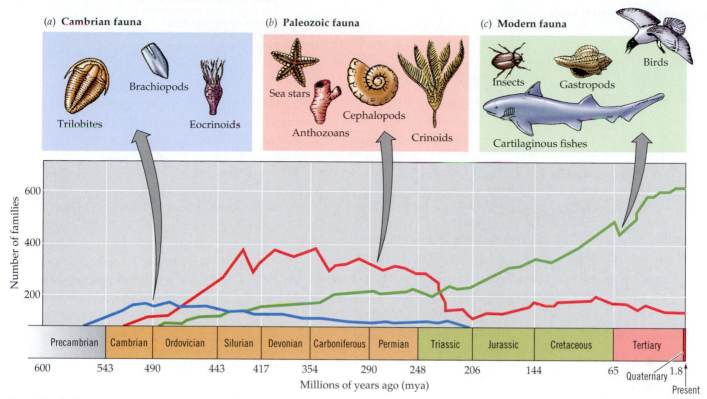

(a) Cambrian fauna

Trilobites
Brachiopods
Eocrinoids

(b) Paleozoic fauna

Sea stars
Anthozoans
Cephalopods
Crinoids

(c) Modern fauna

Insects
Gastropods
Birds
Cartilaginous fishes

Number of families

600

400

200

| Precambrian | Cambrian | Ordovician | Silurian | Devonian | Carboniferous | Permian | Triassic | Jurassic | Cretaceous | Tertiary |

600 543 490 443 417 354 290 248 206 144 65 Quaternary 1.8

Millions of years ago (mya)

Present

22.17 Evolutionary Faunas Representatives of the three great evolutionary faunas are shown, together with a graph illustrating the number of families in each fauna over time.

Rates of Evolutionary Change within Lineages

In addition to revealing the broad patterns of the evolution of life on Earth, fossils also tell us about rates of change within particular lineages of organisms. The fossil record shows that no single pattern characterizes evolutionary rates. Many species have changed very little over many millions of years. Others have changed gradually over time. Still others have undergone rapid changes over short time periods, followed by long periods of slow change. Change may incorrectly appear to be rapid if the fossil record is very incomplete, but some rapid changes are well documented by excellent temporal series of fossils. Let's look at some examples of these patterns.

have also evolved slowly. The horseshoe crabs living today are almost identical in appearance to those that lived 300 million years ago. The sandy coastlines where horseshoe crabs spawn (see Figure 33.16b) feature extremes in temperature and salt concentration that are lethal to many organisms. These harsh environments have changed relatively little over millennia. The chambered nautiluses of the late Cretaceous are indistinguishable from living species (see Figure 32.26f). Chambered nautiluses spend their days in deep, dark ocean waters, ascending to feed in food-rich surface waters only under the protective cover of darkness. Their intricate shells provide little protection against today's visually hunting fish.

Some living species closely resemble ancient ancestors

Species that have changed little over millions of years are known as "living fossils." Fossilized leaves of the genus *Ginkgo* from the Triassic, for example, are very similar to those of living trees (Figure 22.18). Animals in some marine lineages

(a)

(b)

22.18 "Living Fossils" Fossilized *Ginkgo* leaves from the Triassic (a) appear very similar to the leaves of living trees (b).

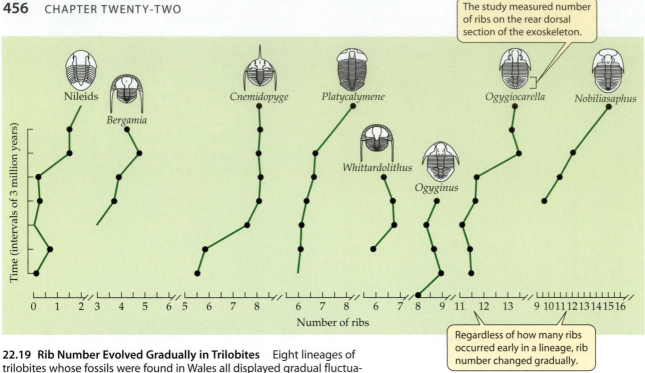

The study measured number of ribs on the rear dorsal section of the exoskeleton.

Regardless of how many ribs occurred early in a lineage, rib number changed gradually.

22.19 Rib Number Evolved Gradually in Trilobites Eight lineages of trilobites whose fossils were found in Wales all displayed gradual fluctuations in the number of rear dorsal ribs on the exoskeleton.

Evolutionary changes have been gradual in some lineages

The fossil record contains many series of fossils that demonstrate gradual change in lineages of organisms over time. A good example is the series of fossils showing changes in the number of ribs on the exoskeleton in eight lineages of trilobites during the Ordovician (Figure 22.19). Rates of change differed among the lineages, and they did not all change at the same time, but all of the changes were gradual.

Rates of evolutionary change are sometimes rapid

In the histories of some lineages, periods of gradual evolutionary change are broken by periods during which changes, either in the physical or biological environment, create conditions that favor rapid evolution of new traits. Such rapid evolutionary change is illustrated by the three-spined stickleback (*Gasterosteus aculeatus*), a widespread marine fish that

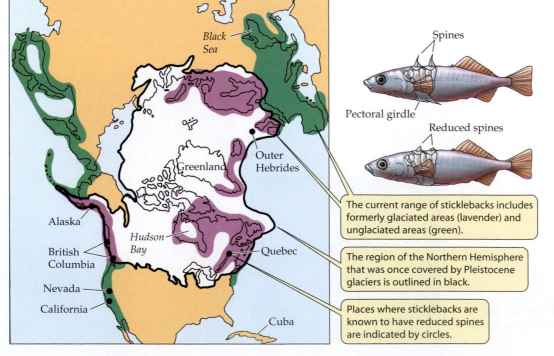

The current range of sticklebacks includes formerly glaciated areas (lavender) and unglaciated areas (green).

The region of the Northern Hemisphere that was once covered by Pleistocene glaciers is outlined in black.

Places where sticklebacks are known to have reduced spines are indicated by circles.

22.20 Natural Selection Acts on Stickleback Spines Three-spined stickleback populations with reduced spines are found principally in young lakes that were covered by ice during the most recent glacial period. These lakes lack large predatory fish, but contain predatory insects that capture the fish by grasping their spines.

Carpodacus mexicanus (male)

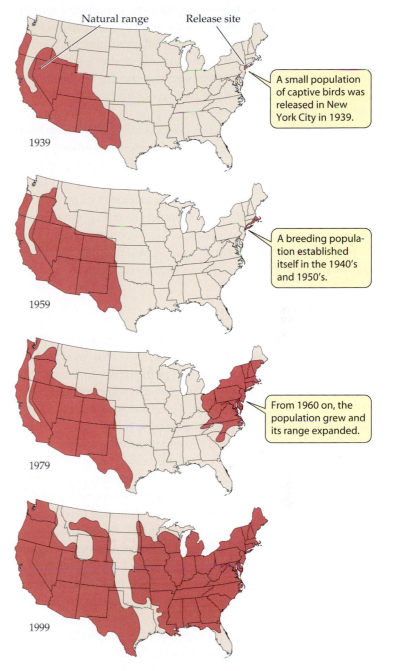

22.21 House Finches Expanded their Range in North America
The natural range of house finches in western North America has expanded somewhat during the past few decades, but the most dramatic expansion has been in the East, from a release of a few caged birds in New York City. Differences in body size also evolved rapidly during this short time period, so that northern birds are now larger than southern birds.

has repeatedly colonized new environments—freshwater lakes—during its evolutionary history.

Sticklebacks are tiny fish, usually less than 10 cm long. All marine and most freshwater populations have well-developed pectoral girdles with prominent spines that make it difficult for other fish to swallow them. However, large predatory insects can readily grasp the stickleback's spines, and they prey selectively on those stickleback individuals with the largest spines. When stickleback populations colonize freshwater habitats where predatory fish are absent but predatory insects are present, they rapidly evolve smaller spines. Populations with reduced spines are found primarily in young lakes that were covered by ice during the most recent glaciation and do not have large predatory fish (Figure 22.20). The extensive fossil record of sticklebacks shows that spine reduction evolved many times in different populations that colonized fresh water. In addition, molecular data show that each freshwater population is most closely related to an adjacent marine population, not to other freshwater populations.

At times, evolutionary change is rapid enough to be measured directly. A good example is provided by the house finch, a bird that as recently as 1939 was confined to the arid and semiarid parts of western North America. That year, some captive finches were released in New York City. Many of them survived to form a small breeding population in the immediate vicinity of the city. During the early 1960s, that population began to grow and increase its range. By the 1990s, the house finch had spread across all of the eastern United States and southern Canada (Figure 22.21). Remarkably, by 2000, birds in finch populations that had been separated for only a few decades were as different in size as birds in finch populations that had been separated for thousands of years.

Extinction rates vary over time

More than 99 percent of the species that have ever lived are extinct. Species have become extinct throughout the history of life, but extinction rates have fluctuated dramatically. Some groups had high extinction rates while others were proliferating.

A mass extinction changes the biota of the following period by selectively eliminating some types of organisms, thereby allowing others to increase in their relative abundance. For example, among the mollusks of the Atlantic coastal plain of North America, species with broad geographic ranges were less likely to become extinct during normal periods (when no mass extinctions were taking place) than were species with small geographic ranges. During the mass extinction of the late Cretaceous, however, groups of closely related mollusk species with large geographic ranges survived better than groups with small ranges, even if the individual species within the group had small ranges. Similar patterns are found in other mollusks elsewhere, suggesting that traits favoring long-term survival during normal times are often different from those that favor survival during times of mass extinctions.

At the end of the Cretaceous period, extinction rates on land were much higher among large vertebrates than among small ones. The same was true during the Pleistocene, when extinction rates were high only among large mammals and large birds. During some mass extinctions, marine organisms were heavily hit, but terrestrial organisms survived well. Other mass extinctions affected organisms living in both environments. These differences are not surprising, given that major changes on land and in the oceans did not always coincide.

The Future of Evolution

The agents of evolution are operating today just as they have been since life first appeared on Earth. However, major changes are under way as a result of the dramatic increase of Earth's human population. Until recently, human-caused extinctions affected mostly large vertebrates, but many of the species we are now exterminating are small. Humans are changing the physical and biological environment by dramatically altering Earth's vegetation, converting forests and grasslands to crops and pastures. Deliberately or inadvertently, we are moving thousands of species around the globe, reversing the relatively independent evolution of Earth's biota on different continents that began during the Mesozoic era. Humans have also taken charge of the evolution of certain species by means of artificial selection and biotechnology. As we saw in Chapter 16, modern molecular methods enable us to modify species by moving genes among even distantly related species. In short, humans have become a dominant agent of evolution. How we wield our massive influence will powerfully affect the future of life on Earth.

Chapter Summary

Defining Biological Evolution

▶ Biology is intimately linked to concepts of time. The study of biology as we know it could not have been developed until people came to understand the age of Earth.

Determining Earth's Age

▶ The relative ages of rock layers in Earth's crust were determined from their embedded fossils.

▶ Radioisotopes supplied the key for assigning absolute ages to rocks.

▶ Earth's geological history is divided into eras and periods. The boundaries between these divisions are based on differences between their fossil biotas. **Review Table 22.1**

The Changing Face of Earth

▶ Throughout Earth's history continents have drifted about, sometimes separating from one another, at other times colliding. Their collisions typically have led to periods of massive volcanic eruptions, glaciations, and major shifts in sea levels and ocean currents. **Review Figure 22.2**

▶ Earth's early atmosphere lacked free oxygen. Oxygen accumulated after prokaryotes evolved the ability to use water as their source of hydrogen ions for photosynthesis. Increasing concentrations of atmospheric oxygen made possible the evolution of eukaryotes and multicellular organisms. **Review Figure 22.4**

▶ Over Earth's history, hot/humid climatic conditions have alternated with cold/dry conditions. **Review Figure 22.5**

▶ External events, such as collisions with meteorites, also have changed conditions on Earth. Such a collision probably caused the abrupt mass extinction at the end of the Cretaceous period. **See Web/CD Tutorial 22.1**

The Fossil Record

▶ Much of what we know about the history of life on Earth comes from the study of fossils.

▶ The fossil record, although incomplete, reveals broad patterns in the evolution of life. About 300,000 fossil species have been described. The best record is that of hard-shelled animals fossilized in marine sediments.

Major Patterns in the History of Life on Earth

▶ Some lineages that evolved during Precambrian times may not have left living descendants.

▶ The diversity of life exploded during the Cambrian period. Diversification continued throughout the rest of the Paleozoic era. **Review Figures 22.9, 22.11, 22.13**

▶ Geographic differentiation of biotas increased during the Mesozoic era. **Review Figure 22.16**

▶ The modern biota evolved during the Cenozoic era.

▶ After each mass extinction, the diversity of life rebounded, but the groups of organisms that dominated the new biotas differed markedly from those characteristic of earlier biotas. **Review Figure 22.17**

Rates of Evolutionary Change within Lineages

▶ Some species, called "living fossils," closely resemble ancient ancestors.

▶ Evolutionary changes have been gradual in some lineages. **Review Figure 22.19**

▶ Rates of evolutionary change are sometimes rapid because of changes in the physical or biological environment. **Review Figures 22.20, 22.21**

The Future of Evolution

▶ The agents of evolution continue to operate today, but human intervention, both deliberate and inadvertent, now plays an unprecedented role in the history of life.

See Web/CD Activity 22.1 for a concept review of this chapter.

Self-Quiz

1. The number of species of fossil organisms that has been described is about
   a. 50,000.
   b. 100,000.
   c. 200,000.
   d. 300,000.
   e. 500,000.

2. In undisturbed strata of sedimentary rocks,
   a. the oldest rocks lie at the top.
   b. the oldest rocks lie at the bottom.
   c. the oldest rocks are in the middle.
   d. the oldest rocks are distributed among the strata of younger rocks.
   e. None of the above

3. Radioactive carbon can be used to date the ages of fossil organisms because
   a. all organisms contain many carbon compounds.
   b. radioactive carbon has a regular rate of decay to nonradioactive carbon.
   c. the ratio of radioactive to nonradioactive carbon in living organisms is always the same as that in the atmosphere.
   d. the production of new radioactive carbon in the atmosphere just balances the natural radioactive decay of ^{14}C.
   e. All of the above

4. An important unidirectional change in Earth during its history was a
   a. steady increase in volcanic activity.
   b. gradual coming together of the continents.
   c. steady increase in the oxygen content of the atmosphere.
   d. gradual warming of the climate.
   e. steady increase in Earth's precipitation.

5. The total of all species of organisms in a given region is known as the region's
   a. biota.
   b. flora.
   c. fauna.
   d. flora and fauna.
   e. diversity.

6. The coal beds we now mine for energy are the remains of
   a. trees that grew in swamps during the Carboniferous period.
   b. trees that grew in swamps during the Devonian period.
   c. trees that grew in swamps during the Permian period.
   d. small plants that grew in swamps during the Carboniferous period.
   e. None of the above

7. The cause of the mass extinction at the end of the Ordovician was probably
   a. the collision of Earth with a large meteorite.
   b. massive volcanic eruptions.
   c. massive glaciation in Gondwana.
   d. the uniting of all continents to form Pangaea.
   e. changes in Earth's orbit.

8. The cause of the mass extinction at the end of the Mesozoic era probably was
   a. continental drift.
   b. the collision of Earth with a large meteorite.
   c. changes in Earth's orbit.
   d. massive glaciation.
   e. changes in the salt concentration of the oceans.

9. The times during the history of life when many new evolutionary lineages appeared were the
   a. Precambrian, Cambrian, and Triassic.
   b. Precambrian, Cambrian, and Tertiary.
   c. Cambrian, Paleozoic, and Triassic.
   d. Cambrian, Triassic, and Devonian.
   e. Paleozoic, Triassic, and Tertiary.

10. Many scientists believe that the collision of Earth with a large meteorite was a major contributor to the mass extinction at the end of the Cretaceous period because
    a. there is an iridium-rich layer at the boundary of rocks between the Cretaceous and Cenozoic.
    b. a crater that may be the site of the collision has been found off the Yucatán Peninsula.
    c. the mass extinction at the end of the Cretaceous may have been very sudden.
    d. many planktonic organisms and bottom-dwelling invertebrates became extinct.
    e. All of the above

11. We know that organisms can evolve rapidly because
    a. the fossil record reveals periods of rapid evolutionary change.
    b. theoretical models of evolutionary change show that rapid change can be produced by natural selection.
    c. rapid evolutionary changes have been produced under artificial selection.
    d. rapid evolutionary changes have been measured in natural populations of organisms during the past century.
    e. All of the above

12. At which of the following times was there *no* mass extinction?
    a. The end of the Cretaceous period
    b. The end of the Devonian period
    c. The end of the Permian period
    d. The end of the Triassic period
    e. The end of the Silurian period

For Discussion

1. Some lineages of organisms have evolved to contain large numbers of species; other lineages have produced only a few species. Is it meaningful to consider the former more successful than the latter? What does the word "success" mean in evolution? How does your answer influence your thinking about *Homo sapiens*, the only surviving representative of the Hominidae—a family that never had many species in it?

2. Scientists date ancient events using a variety of methods, but nobody was present to witness or record those events. Accepting those dates requires us to believe in the accuracy and appropriateness of indirect measurement techniques. What other basic scientific concepts are also based on the results of indirect measurement techniques?

3. Why is it useful to be able to date past events absolutely as well as relatively?

4. If we are living during one of the cooler periods in Earth's history, why should we be concerned about human activities that are contributing to global climate warming?

5. Large meteors that collided with Earth have caused massive climatic and evolutionary changes. Should we attempt to take steps to prevent future meteorite impacts? What actions might we undertake? What adverse effects might such actions trigger?

23 *The Mechanisms of Evolution*

Newts and other salamanders can move only slowly, so they are easy prey for garter snakes. But some salamanders have evolved defensive toxic chemicals that make them less desirable as prey. The rough-skinned newt, *Taricha granulosa*, is a salamander that lives on the Pacific Coast of North America. *Taricha* sequesters in its skin a potent neurotoxin called tetrodotoxin (TTX). TTX paralyzes nerves and muscles by blocking sodium channels (see Chapter 5). Most snakes die if they eat a rough-skinned newt, but some populations of the garter snake *Thamnophis sirtalis* have evolved TTX-resistant sodium channels in their nerves and muscles. These snakes are able to eat the newts and survive—but the addition to their diet comes at a price. TTX-resistant snakes can crawl only slowly for several hours after eating a newt, and they never crawl as fast as nonresistant snakes. Thus, TTX-resistant snakes are more vulnerable to their own predators.

Pufferfish, octopuses, tunicates, and some species of frogs also use TTX as a defensive chemical. Many other species use a variety of chemicals to defend themselves against predators, and many predators have evolved resistance to those chemicals. But production of and resistance to defensive chemicals like all other adaptations, has costs as well as benefits. Such adaptations may impose a cost in the form of speed of movement, as they do on garter snakes. They may reduce the ability of the organism to function efficiently, or they may be energetically costly to develop and maintain. That is, to improve its performance in one area, the organism must accept reduced performance in some other area—a trade-off.

Biologists try to identify and measure the trade-offs that different adaptations impose because the nature and strength of these trade-offs influences how adaptations evolve. If there were no cost to TTX resistance, then snakes that live in places where toxic newts are rare would probably also be resistant to TTX—which they are not.

Charles Darwin's main contribution to biology was to propose a plausible and testable hypothesis for a mechanism that could result in the adaptation of organisms to their environments. In effect, Darwin offered a mechanistic explanation for the evolution of life on Earth, the last component of the known universe that lacked such an explanation. The mechanism that Darwin proposed can explain the evolution of all forms of life, including humans. It has been difficult for many people to accept that the same processes that determined the evolution-

An Evolutionary War Rough-skinned newts (below) evolved the ability to secrete a paralyzing neurotoxin in their skin, a trait that deters most of their predators. Some common garter snakes (above) have evolved a resistance that allows them to turn the poisonous newts into a meal.

(a)

(b)

23.1 Darwin and the Voyage of the *Beagle*   *(a)* The mission of H.M.S. *Beagle* was to chart the oceans and collect oceanographic and biological information from around the world. The map indicates the ship's path, with emphasis on the Galápagos Islands, whose organisms were an important source of Darwin's ideas on natural selection. *(b)* Charles Darwin at age 24, shortly after the *Beagle* returned to England.

ary pathways of other species also guided human evolution, but as Darwin noted, "there is grandeur in this view of life."

In this chapter we will see how Darwin developed his ideas, and then turn to the advances in our understanding of evolutionary processes since Darwin's time. We will discuss the genetic basis of evolution and show how genetic variation within populations is measured. We will describe the agents of evolution and show how biologists design studies to investigate them. Finally, we will discuss constraints on the pathways evolution can take. When you understand these processes, you will understand the mechanisms of evolution.

Charles Darwin's Theory of Evolution

As a youth, Charles Darwin was passionately interested in *natural history*—the study of how different organisms carry out their lives. He briefly studied medicine at Edin-

burgh, but he was nauseated by observing surgery conducted without anesthesia. He gave up medicine to study for a career as a clergyman of the Church of England at Cambridge University. However, he was more interested in natural history than theology, and he became a companion of scientists on the faculty, especially the botanist John Henslow. Darwin was given an unprecedented opportunity when in 1831 Henslow recommended him for a position as ship's naturalist on the H.M.S. *Beagle*, which was preparing for a survey voyage around the world (Figure 23.1).

Whenever possible during the 5-year voyage, Darwin (who was often seasick) went ashore to observe and collect specimens of plants and animals. He noticed that the species he saw in South America differed strikingly from those of Europe. He observed that the species of the temperate regions of South America (Argentina, Chile) were more similar to those of tropical South America (Brazil) than they were to European species. When he explored the Galápagos Islands, west of Ecuador, he noted that most of its animal species were found nowhere else, but were similar to those of mainland South America, 1,000 kilometers to the east. Darwin also recognized that the animals of the archipelago differed from island to island. He postulated that some animals had dispersed from mainland South America and then evolved differently on each of the islands.

When he returned to England in 1836, Darwin continued to ponder his observations. Within a decade he had developed the major features of his theory, which had two major components:

▶ Species are not immutable; they change over time. (In other words, Darwin asserted that evolution is a historical fact that can be demonstrated to have taken place.)
▶ The agent that produces these changes is natural selection.

Darwin wrote a long essay on natural selection and the origin of species in 1844, but, despite urging from his wife and colleagues, he was reluctant to publish it, preferring to assemble more evidence first.

Darwin's hand was forced in 1858 when he received a letter and manuscript from another traveling naturalist, Alfred Russel Wallace, who was studying plants and animals in the East Indies. Wallace asked Darwin to evaluate the manuscript, in which Wallace proposed a theory of natural selection almost identical to Darwin's. At first Darwin was dismayed, believing that Wallace had preempted his idea. But parts of Darwin's 1844 essay, together with Wallace's manuscript, were presented to the Linnaean Society of London on July 1, 1858, thereby giving credit for the idea to both men. Darwin then worked quickly to finish his own book, *The Origin of Species*, which was published the next year. Although both men conceived of natural selection independently, Darwin developed his ideas first, and *The Origin of Species* provided an enormous amount of evidence from many fields to support both the concept of natural selection and evolution itself, which is why these concepts are more closely associated with the name Darwin than Wallace.

The facts that Darwin used to conceive and develop his theory of evolution by natural selection were familiar to most contemporary biologists. His unique insight was to perceive the significance of relationships among them. On September 28, 1838, Darwin happened to read *An Essay on the Principle of Population* by Thomas Malthus, an economist. Malthus argued that because the rate of human population growth is greater than the rate of increase in food production, unchecked growth inevitably leads to famine. Darwin recognized that populations of all species have the potential for exponential increases in numbers. To illustrate this point, he used the following example:

> Suppose…there are eight pairs of birds, and that only four pairs of them annually… rear only four young, and that these go on rearing their young at the same rate, then at the end of seven years…there will be 2048 birds instead of the original sixteen.

Yet such rates of increase are rarely seen in nature. Therefore, Darwin reasoned that death rates in nature must also be high. Without high death rates, even the most slowly reproducing species would quickly reach enormous population sizes.

Darwin also observed that, although offspring tend to resemble their parents, the offspring of most organisms are not identical to one another or to their parents. He suggested that slight variations among individuals significantly affect the chance that a given individual will survive and reproduce. Darwin called this differential survival and he called reproduction of individuals **natural selection**.

Darwin may have used the words "natural selection" because he was familiar with the *artificial selection* of individuals with certain desirable traits by animal and plant breeders. Many of Darwin's observations on the nature of variation came from domesticated plants and animals. Darwin was a pigeon breeder, and he knew firsthand the astonishing diversity in color, size, form, and behavior that pigeon breeders could achieve (Figure 23.2). He recognized close parallels

23.2 Many Types of Pigeons Have Been Produced by Artificial Selection Charles Darwin raised pigeons as a hobby, and he saw similar forces at work in artificial and natural selection. These are just some of more than 300 varieties of pigeons that have been artificially selected by breeders to display different forms of traits such as color, size, and feather distribution.

between selection by breeders and selection in nature. As he argued in *The Origin of Species*,

> How can it be doubted, from the struggle each individual has to obtain subsistence, that any minute variation in structure, habits or instincts, adapting that individual better to the new conditions, would tell upon its vigour and health? In the struggle it would have a better chance of surviving; and those of its offspring which inherited the variation, be it ever so slight, would have a better chance.

That statement, written almost 150 years ago, still stands as a good expression of the theory of evolution by natural selection.

It is important to remember, as Darwin clearly understood, that *individuals do not evolve; populations do.* A **population** is a group of individuals of a single species that live in a particular geographic area at the same time. A major consequence of the evolution of populations is that their members become adapted to the environments in which they live.

The term **adaptation** has two meanings in evolutionary biology. The first meaning refers to the *processes* by which adaptive traits are acquired—that is, the evolutionary mechanisms that produce them. We will discuss those processes in great detail in this chapter. The second meaning refers to *traits* that enhance the survival and reproductive success of their bearers. For example, wings are adaptations for flight, and a spider's web is an adaptation for capturing flying insects.

Biologists regard an organism as being adapted to a particular environment when they can imagine—or better still, measure the performance of—a slightly different organism that reproduces and survives less well in that environment. To understand adaptation, biologists compare the performance of individuals within or among species that differ in their traits. For example, to investigate the adaptive nature of spiders' webs, we might try to determine the effectiveness of slightly different web structures in capturing insects. We might also measure changes in the webs of the same species in different environments. With these data, we could understand how variations in web structure influenced the survival and reproductive success of their builders.

When Darwin proposed his theory of evolution by natural selection, he had no examples of evolutionary agents operating in nature. Since then many studies of the action of evolutionary agents have been conducted. Similarly, many investigations have documented changes over time in the genetic composition of a population. Darwin understood the importance of heredity for his theory, but he knew nothing of the mechanisms of heredity. He devoted considerable time developing a theory of heredity, but he failed in this effort.

Fortunately, the rediscovery of Gregor Mendel's publications in the early 1900s (see Chapter 10) paved the way for the development of **population genetics**, which provides a major underpinning for Darwin's theory. Population geneticists apply Mendel's laws to entire populations of organisms. They also study variation within and among species to understand the processes that result in evolutionary changes in species through time. The perspective of population genetics given in this chapter, which emphasizes the role of variation in characteristics of adult organisms, complements the perspective of developmental biology we discussed in Chapter 21.

Genetic Variation within Populations

For a population to evolve, its members must possess heritable genetic variation, which is the raw material on which agents of evolution act. In everyday life, we do not directly observe the genetic compositions of organisms. What we do see in nature are *phenotypes*, the physical expressions of organisms' genes. The features of a phenotype are its *characters*—eye color, for example. The specific form of a character, such as brown eyes, is a *trait*. A *heritable trait* is a characteristic of an organism that is at least partly determined by its genes.

The agents of evolution generally act on phenotypes, but for the moment we will concentrate on genetic variation within populations. We will do so because genetic variation is what is passed on to offspring via gametes—eggs and sperm. The genetic constitution that governs a character is called its *genotype*. *A population evolves when individuals with different genotypes survive or reproduce at different rates.*

Recall from Chapter 10 that different forms of a gene, called *alleles*, may exist at a particular locus. A single individual has only some of the alleles found in the population to which it belongs (Figure 23.3). The sum of all copies of alleles found in the population constitutes its **gene pool**. The

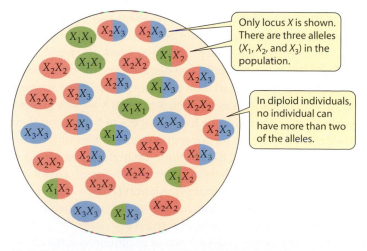

Only locus X is shown. There are three alleles (X_1, X_2, and X_3) in the population.

In diploid individuals, no individual can have more than two of the alleles.

23.3 A Gene Pool   A gene pool is the sum of all the alleles found in a population. Each of the colored circles represents an individual. The allele proportions in this gene pool for locus X are 0.20 for X_1, 0.50 for X_2, and 0.30 for X_3.

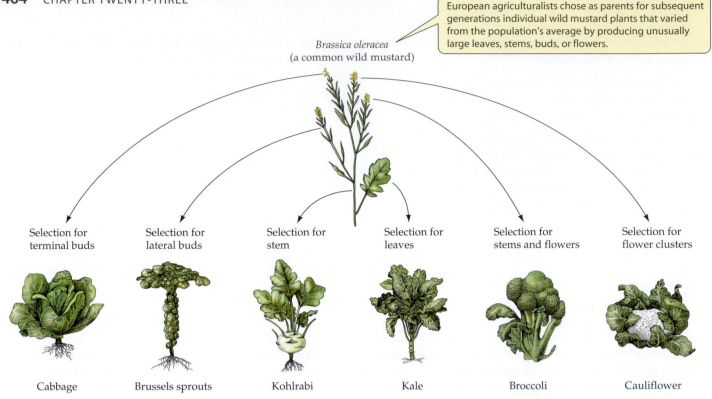

European agriculturalists chose as parents for subsequent generations individual wild mustard plants that varied from the population's average by producing unusually large leaves, stems, buds, or flowers.

Brassica oleracea (a common wild mustard)

Selection for terminal buds — Cabbage

Selection for lateral buds — Brussels sprouts

Selection for stem — Kohlrabi

Selection for leaves — Kale

Selection for stems and flowers — Broccoli

Selection for flower clusters — Cauliflower

23.4 Many Vegetables from One Species All of these crop plants have been derived from a single wild mustard species. Plant breeders produced these crops by choosing and breeding plants with unusually large buds, stems, leaves, or flowers. The results illustrate the vast amount of variation that can be present in a gene pool.

gene pool contains the variation that produces the phenotypic characters on which agents of evolution act. To understand evolution, we need to know how much genetic variation populations have, the sources of that genetic variation, and how genetic variation is maintained and expressed in populations over space and time.

Most populations are genetically variable

Nearly all populations contain some level of genetic variation for many characters. Artificial selection on different characters in a European species of wild mustard produced many important crop plants (Figure 23.4). Plant and animal breeders could achieve such results because the original population had genetic variation for the characters of interest.

Laboratory experiments also demonstrate the existence of considerable genetic variation in populations. In one such experiment, investigators chose fruit flies (*Drosophila melanogaster*) with either high or low numbers of bristles on their abdomens as parents for subsequent generations of flies. After 35 generations, all flies in both the high-bristle and low-bristle lineages had bristle numbers that fell well outside the range found in the original population (Figure 23.5). Thus, there must have been considerable variation in the original fruit fly population for selection to act on.

The study of the genetic basis of evolution is difficult because genotypes do not uniquely determine phenotypes. With dominance, for example, a particular phenotype can be produced by more than one genotype (e.g., *AA* and *Aa* individ-

uals may be phenotypically identical). Similarly, different phenotypes can be produced by a given genotype, depending on the environment encountered during development. For example, the cells of all the leaves on a tree or shrub are normally genetically identical, yet leaves on the same tree often differ in shape and size. Leaves closer to the top of an oak tree, where they receive more wind and sunlight, may be more deeply lobed than the shaded leaves growing lower down on the same tree. The same differences can be seen between the leaves of individuals growing in sunny and in shady sites.

Leaves of a white oak (*Quercus alba*)

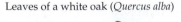

Grown in sun   Grown in shade

Thus, as we saw in Chapter 21, the phenotype of an organism is the outcome of a complex series of developmental processes that are influenced by both the environment and its genes.

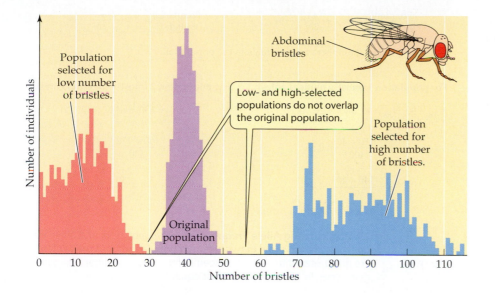

23.5 Artificial Selection Reveals Genetic Variation In artificial selection experiments with *Drosophila melanogaster*, changes in bristle number evolved rapidly. The graphs show the number of flies with different numbers of bristles after 35 generations of artificial selection.

How do we measure genetic variation?

A locally interbreeding group within a geographic population is called a **Mendelian population**. Mendelian populations are often the subjects of evolutionary studies. To measure genetic variation in a Mendelian population precisely, we would need to count every allele at every locus in every individual in it. By doing so, we could determine the relative proportions, or **frequencies**, of all alleles in the population. Fortunately, we do not need to make such complete measurements, because we can reliably estimate *allele frequencies* for a given locus by counting alleles in a sample of individuals from the population. The sum of all allele frequencies at a locus is equal to 1, so measures of allele frequency range from 0 to 1.

An allele's frequency is calculated using the following formula:

$$p = \frac{\text{number of copies of the allele in the population}}{\text{sum of alleles in the population}}$$

If only two alleles (for example, *A* and *a*) for a given locus are found among the members of a diploid population, they may combine to form three different genotypes: *AA*, *Aa*, and *aa*. Using the formula above, we can calculate the relative frequencies of alleles *A* and *a* in a population of *N* individuals as follows:

- Let N_{AA} be the number of individuals that are homozygous for the *A* allele (*AA*).
- Let N_{Aa} be the number that are heterozygous (*Aa*).
- Let N_{aa} be the number that are homozygous for the *a* allele (*aa*).

Note that $N_{AA} + N_{Aa} + N_{aa} = N$, the total number of individuals in the population, and that the total number of copies of both alleles present in the population is $2N$ because each in-

dividual is diploid. Each *AA* individual has two copies of the *A* allele, and each *Aa* individual has one copy of the *A* allele. Therefore, the total number of *A* alleles in the population is $2N_{AA} + N_{Aa}$. Similarly, the total number of *a* alleles in the population is $2N_{aa} + N_{Aa}$.

If *p* represents the frequency of *A*, and *q* represents the frequency of *a*, then

$$p = \frac{2N_{AA} + N_{Aa}}{2N}$$

and

$$q = \frac{2N_{aa} + N_{Aa}}{2N}$$

To show how this formula works, Figure 23.6 calculates allele frequencies in two populations, each containing 200 diploid individuals. Population 1 has mostly homozygotes (90 *AA*, 40 *Aa*, and 70 *aa*); population 2 has mostly heterozygotes (45 *AA*, 130 *Aa*, and 25 *aa*).

The calculations in Figure 23.6 demonstrate two important points. First, notice that for each population, $p + q = 1$. If there is only one allele in a population, its frequency is 1. If an allele is missing from a population, its frequency is 0, and the locus in that population is represented by one or more other alleles. Since $p + q = 1$, then $q = 1 - p$. So when there are only two alleles at a given locus in a population, we can calculate the frequency of one allele and then easily obtain the second allele's frequency by subtraction.

The second thing to notice is that both population 1 (consisting mostly of homozygotes) and population 2 (consisting mostly of heterozygotes) have the same allele frequencies for *A* and *a*. Therefore, they have the same gene pool for this locus. However, because the alleles in the gene pool are distributed differently, the *genotype frequencies* of the two populations differ. Genotype frequencies are calculated as the number of individuals that have the genotype divided by the total number of individuals in the population. In population 1 in Figure 23.6, the genotype frequencies are 0.45 *AA*, 0.20 *Aa*, and 0.35 *aa*.

The frequencies of different alleles at each locus and the frequencies of different genotypes in a Mendelian population describe its **genetic structure**. Allele frequencies measure the amount of genetic variation in a population; genotype frequencies show how a population's genetic variation is distributed among its members. With these measurements, it becomes possible to consider how the genetic structure of a population changes or does not change over generations.

In any population:

$$\text{Frequency of allele } A = p = \frac{2N_{AA} + N_{Aa}}{2N} \qquad \text{Frequency of allele } a = q = \frac{2N_{aa} + N_{Aa}}{2N}$$

where N is the total number of individuals in the population.

For population 1 (mostly homozygotes):

$N_{AA} = 90$, $N_{Aa} = 40$, and $N_{aa} = 70$

so

$$p = \frac{180 + 40}{400} = 0.55$$

$$q = \frac{140 + 40}{400} = 0.45$$

For population 2 (mostly heterozygotes):

$N_{AA} = 45$, $N_{Aa} = 130$, and $N_{aa} = 25$

so

$$p = \frac{90 + 130}{400} = 0.55$$

$$q = \frac{50 + 130}{400} = 0.45$$

23.6 Calculating Allele Frequencies The gene pool and allele frequencies are the same in two different populations, but the alleles are distributed differently between heterozygous and homozygous genotypes. In all cases, $p + q$ must equal 1.

The Hardy–Weinberg Equilibrium

If certain conditions are met, the genetic structure of a population may not change over time. The necessary conditions for such an equilibrium were deduced independently in 1908 by the British mathematician Godfrey Hardy and the German physician Wilhelm Weinberg. Hardy wrote his equations in response to a question posed to him by the geneticist Reginald C. Punnett (the inventor of the Punnett square) at the Cambridge University faculty club. Punnett wondered at the fact that even though the allele for short, stubby fingers (a condition called *brachydactyly*) was dominant and the allele for normal-length fingers was recessive, most people in Britain have normal-length fingers. Hardy's equations explain why dominant alleles do not necessarily replace recessive alleles in populations, as well as other features of the genetic structure of populations.

The **Hardy–Weinberg equilibrium** applies to sexually reproducing organisms. The particular example we will illustrate here assumes that the organism in question is diploid, its generations do not overlap, the gene under consideration has two alleles, and allele frequencies are identical in males and females. The Hardy–Weinberg equilibrium also applies if the gene has more than two alleles and generations overlap, but in those cases the mathematics is more complicated.

Several conditions must be met for a population to be at Hardy–Weinberg equilibrium:

▶ Mating is random
▶ Population size is very large
▶ There is no migration between populations
▶ There is no mutation
▶ Natural selection does not affect the alleles under consideration

If these conditions hold, two major consequences follow. First, the frequencies of alleles at a locus will remain constant from generation to generation. And second, after one generation of random mating, the genotype frequencies will remain in the following proportions:

Genotype	*AA*	*Aa*	*aa*
Frequency	p^2	$2pq$	q^2

Stated another way, the equation for Hardy–Weinberg equilibrium is

$$p^2 + 2pq + q^2 = 1$$

To see why, consider population 1 in Figure 23.6, in which the frequency of *A* alleles (p) is 0.55. Because we assume that individuals select mates at random, without regard to their genotype, gametes carrying *A* or *a* combine at random—that is, as predicted by the frequencies p and q. The probability that a particular sperm or egg in this example will bear an *A* allele rather than an *a* allele is 0.55. In other words, 55 out of 100 randomly sampled sperm or eggs will bear an *A* allele. Because $q = 1 - p$, the probability that a sperm or egg will bear an *a* allele is $1 - 0.55 = 0.45$.

To obtain the probability of two *A*-bearing gametes coming together at fertilization, we multiply the two independent probabilities of their occurring separately (see the discussion of probability in Chapter 10):

$$p \times p = p^2 = (0.55)^2 = 0.3025$$

Therefore, 0.3025, or 30.25 percent, of the offspring in the next generation will have the *AA* genotype. Similarly, the probability of bringing together two *a*-bearing gametes is

$$q \times q = q^2 = (0.45)^2 = 0.2025$$

Thus, 20.25 percent of the next generation will have the *aa* genotype (Figure 23.7).

Figure 23.7 also shows that there are two ways of producing a heterozygote: An *A* sperm may combine with an *a* egg, the probability of which is $p \times q$; or an *a* sperm may combine with an *A* egg, the probability of which is $q \times p$. Consequently, the overall probability of obtaining a heterozygote is $2pq$.

It is now easy to show that the allele frequencies p and q remain constant for each generation. If the frequency of *A* alleles in a randomly mating population is $p^2 + pq$, this frequency becomes $p^2 + p(1-p) = p^2 + p - p^2 = p$, the original allele frequencies are unchanged, and the population is at Hardy–Weinberg equilibrium.

If some agent, such as emigration, were to alter the allele frequencies, the genotype frequencies would automatically settle into a predictable new set in the next generation. For instance, if only *AA* and *Aa* individuals left the population, p and q would change, but there would still be *aa* individuals in the population.

Why is the Hardy–Weinberg equilibrium important?

You may already have realized that populations in nature rarely meet the stringent conditions necessary to maintain them at Hardy–Weinberg equilibrium. Why, then, is the Hardy-Weinberg equilibrium considered so important for the study of evolution? The answer is that without it, we cannot tell whether or not evolutionary agents are operating. The most important message of the Hardy–Weinberg equilibrium is that *allele frequencies remain the same from generation to generation unless some agent acts to change them.*

In order to ascertain that evolutionary agents are in play, we must estimate the actual allele or genotype frequencies present in a population and then compare them with the frequencies that would be expected at Hardy–Weinberg equilibrium. The pattern of deviation from the Hardy–Weinberg expectations tells us which assumptions are violated. Thus, we can identify the agents of evolutionary change on which we should concentrate our attention.

Evolutionary Agents and Their Effects

Evolutionary agents are forces that change the genetic structure of a population. In other words, they cause deviations from the Hardy–Weinberg equilibrium. The known evolutionary agents are mutation, gene flow, genetic drift, nonrandom mating, and natural selection. Although only natural selection results in adaptation, to understand evolutionary processes we need to discuss all of these evolutionary agents before considering natural selection in detail.

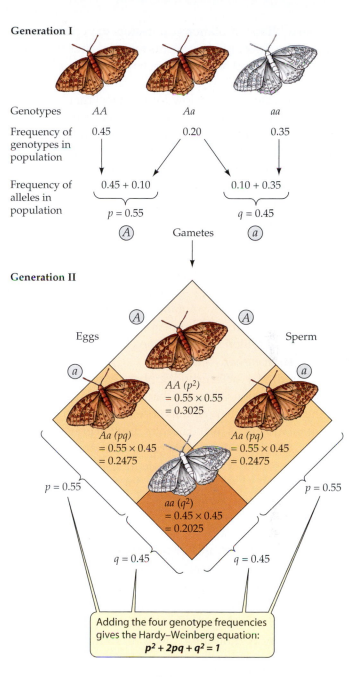

23.7 Calculating Hardy–Weinberg Genotype Frequencies   The areas within the squares are proportional to the expected frequencies of possible matings if mating is random with respect to genotype. Because there are two ways of producing a heterozygote, the probability of this event occurring is the sum of the two *Aa* squares.

Mutations are changes in the genetic material

The origin of genetic variation is mutation. A mutation, as we saw in Chapter 12, is any change in an organism's DNA. Mutations appear to be random with respect to the adaptive needs of organisms. Most mutations are harmful to their bearers or are neutral, but if environmental conditions change, previously harmful alleles may become advantageous. In addition, mutations can restore to populations alleles that other evolutionary agents remove. Thus mutations both create and help maintain genetic variation within populations.

Mutation rates are very low for most loci that have been studied. Rates as high as one mutation per locus in a thousand zygotes per generation are rare; one in a million is more typical. Nonetheless, these rates are sufficient to create considerable genetic variation because each of a large number of genes may mutate, frame-shift mutations may change many genes simultaneously, and populations often contain large numbers of individuals. For example, if the probability of a point mutation were 10^{-9} per base pair per generation, then in each human gamete, the DNA of which contains 3×10^9 base pairs, there would be an average of three new point mutations ($3 \times 10^9 \times 10^{-9} = 3$). Therefore, each zygote would carry, on average, six new mutations, and the current human population of about 8 billion people would be expected to carry about 48 billion new mutations that were not present one generation earlier.

One condition for Hardy–Weinberg equilibrium is that there be no mutation. Although this condition is never strictly met, the rate at which mutations arise at single loci is usually so low that mutations by themselves result in only very small deviations from Hardy–Weinberg expectations. If large deviations are found, it is appropriate to dismiss mutation as the cause and to look for evidence of other evolutionary agents acting on the population.

Movement of individuals or gametes, followed by reproduction, produces gene flow

Few populations are completely isolated from other populations of the same species. Migrations of individuals and movements of gametes between populations are common. If the arriving individuals or gametes reproduce in their new location, they may add new alleles to the gene pool of the population, or they may change the frequencies of alleles already present if they come from a population with different allele frequencies. For a population to be at Hardy–Weinberg equilibrium, there must be no **gene flow** from populations with different allele frequencies.

Genetic drift may cause large changes in small populations

In very small populations, **genetic drift**—the random loss of individuals and the alleles they possess—may produce large changes in allele frequencies from one generation to the next. Harmful alleles, for example, may increase in frequency because of genetic drift, and rare advantageous alleles may be lost. As we will see later, even in large populations, genetic drift can influence the frequencies of alleles that do not influence the survival and reproductive rates of their bearers.

Populations that are normally large may pass through occasional periods when only a small number of individuals survive. During these **population bottlenecks**, genetic variation can be reduced by genetic drift. How this works is illustrated in Figure 23.8, in which red and yellow beans represent two different alleles. Most of the "surviving" beans in the small sample taken from the bean population are, just by chance, red, so the new population has a much higher frequency of red beans than the previous generation had. In a natural population, the allele frequencies would be said to have "drifted."

Suppose we perform a cross of $Aa \times Aa$ individuals of a species of *Drosophila* to produce an F_1 population in which $p = q = 0.5$ and in which the genotype frequencies are 0.25 AA, 0.50 Aa, and 0.25 aa. If we randomly select 4 individuals (= 8 copies of the gene) from among the offspring to produce the F_2 generation, the allele frequencies in this small sample may differ markedly from $p = q = 0.5$. If, for example, we happen by chance to draw 2 AA homozygotes and 2 heterozygotes (Aa), the allele frequencies in this "surviving population" will be $p = 0.75$ (6 out of 8) and $q = 0.25$ (2 out of 8). If we replicate this sampling experiment 1,000 times, one of the two alleles

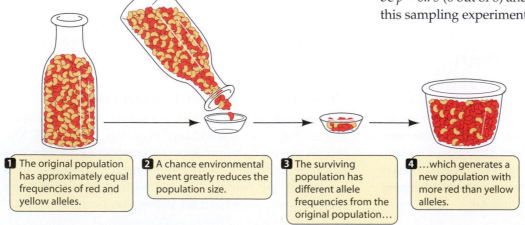

1 The original population has approximately equal frequencies of red and yellow alleles.

2 A chance environmental event greatly reduces the population size.

3 The surviving population has different allele frequencies from the original population…

4 …which generates a new population with more red than yellow alleles.

23.8 A Population Bottleneck
Population bottlenecks occur when only a few individuals survive a random event, resulting in a shift in allele frequencies within the population.

Tympanuchus cupido (male)

23.9 A Species with Low Genetic Variation   Prairie chickens in Illinois lost most of their genetic variation when the population crashed from millions to fewer than 100 individuals.

will be missing entirely from about 8 of the 1,000 "surviving populations."

These numbers show that, as it passes through a bottleneck, a population may lose much of its genetic variation. This is what happened to greater prairie chickens, millions of which lived in the prairies of North America when Europeans first arrived there. As a result of both hunting and habitat destruction, the Illinois population of prairie chickens plummeted from about 100 million birds in 1900 to fewer than 50 individuals in the 1990s (Figure 23.9). A comparison of DNA from birds collected in Illinois during the middle of the twentieth century with DNA from the surviving population in the 1990s showed that Illinois prairie chickens had lost most of their genetic diversity. As a result, both hatching success and chick survival were low. To increase the genetic diversity of Illinois prairie chickens, birds from Minnesota, Kansas, and Nebraska were introduced to Illinois. They interbred with the Illinois birds, restoring much of the genetic diversity of that population, which is now increasing in size.

When a few pioneering individuals colonize a new region, the resulting population is unlikely to have all the alleles found among members of its source population. The resulting change in genetic variation, called a **founder effect**, is equivalent to that in a large population reduced by a bottleneck. Scientists were given an opportunity to study the genetic composition of a founding population when *Drosophila subobscura*, a well-studied European species of fruit fly, was discovered near Puerto Montt, Chile, in 1978 and at Port Townsend, Washington, in 1982. In both South and North America, populations of the flies grew rapidly and expanded their ranges. Today in North America, *D. subobscura* ranges from British Columbia, Canada, to central California. In Chile it has spread across 23° of latitude, nearly as wide a range as the species has in Europe (Figure 23.10).

The *D. subobscura* founders probably reached Chile and the United States from Europe aboard the same ship, because the two populations are genetically very similar. For example, the North and South American populations have only 20 chromosomal inversions, 19 of which are the same on the two continents, whereas 80 inversions are known from European populations. North and South American populations also have lower allelic diversity at enzyme-producing genes than European populations do. Only alleles that have a frequency higher than 10 percent in European populations are present in the Americas. Thus, as expected for a small founding population, only a small part of the total genetic variation found in Europe reached the Americas. Geneticists estimate that at least ten, but no more than a hundred, flies founded the North and South American populations.

Nonrandom mating changes the frequency of homozygotes

Mating patterns may alter genotype frequencies if individuals in a population choose other individuals of certain genotypes as mates. For example, if they mate preferentially with individuals of the same genotype, then homozygous genotypes will be overrepresented, and heterozygous genotypes underrepresented, in the next generation in comparison with Hardy–Weinberg expectations. Alternatively, individuals

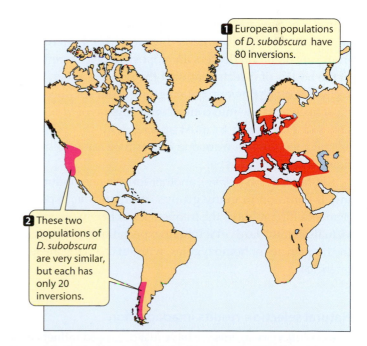

1 European populations of *D. subobscura* have 80 inversions.

2 These two populations of *D. subobscura* are very similar, but each has only 20 inversions.

23.10 A Founder Effect   Populations of the fruit fly *Drosophila subobscura* in North and South America contain less genetic variation than the European populations from which they came, as measured by the number of chromosome inversions in each population. Within two decades of arriving in the New World, the flies had increased dramatically and spread widely in spite of their reduced genetic variation.

Primula sp. (*pin* type)

23.11 Flower Structure Fosters Nonrandom Mating The structure of flowers in plant species such as the primroses ensures that pollination usually occurs between individuals of different genotypes.

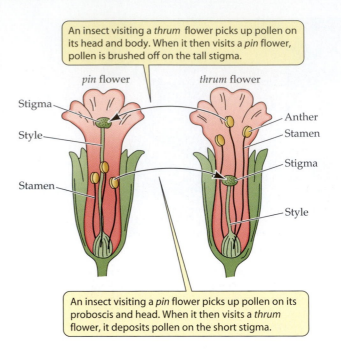

An insect visiting a *thrum* flower picks up pollen on its head and body. When it then visits a *pin* flower, pollen is brushed off on the tall stigma.

pin flower

thrum flower

Stigma

Style

Stamen

Anther

Stamen

Stigma

Style

An insect visiting a *pin* flower picks up pollen on its proboscis and head. When it then visits a *thrum* flower, it deposits pollen on the short stigma.

may mate primarily or exclusively with individuals of different genotypes.

An example of such *nonrandom mating* is provided by plant species, such as primroses (*Primula*), that bear flowers of two different types. One type, known as *pin*, has a long style (female reproductive organ) and short stamens (male reproductive organs). The other type, known as *thrum*, has a short style and long stamens (Figure 23.11). Pollen grains from *pin* and *thrum* flowers are deposited on different parts of the bodies of insects that visit the flowers. When the insects visit other flowers, pollen grains from *pin* flowers are most likely to come into contact with stigmas of *thrum* flowers, and vice versa. In most species with this reciprocal arrangement, pollen from one flower type can fertilize only flowers of the other type.

Self-fertilization (*selfing*), another form of nonrandom mating, is common in many groups of organisms, especially plants. Selfing reduces the frequencies of heterozygous individuals below Hardy–Weinberg expectations and increases the frequencies of homozygotes, without changing allele frequencies.

Natural selection results in adaptation

The evolutionary agents we have just discussed influence the frequencies of alleles and genotypes in populations. As we saw in the previous chapter, major perturbations, such as colliding continents, volcanic eruptions, and meteorite impacts, also have periodically altered the survival and reproductive rates of organisms. All of these agents dramatically affect the course of life's evolution on Earth, but none of them result in adaptations. For adaptation to occur, individuals that differ in heritable traits must survive and reproduce with different degrees of success. When some individuals contribute more offspring to the next generation than others, allele frequencies in the population change in a way that adapts individuals to the environments that influenced their success. This process is known as **natural selection**.

The reproductive contribution of a phenotype to subsequent generations relative to the contributions of other phenotypes is called its **fitness**. The word "relative" is critical: The absolute number of offspring produced by an individual does not influence the genetic structure of a population. Changes in absolute numbers of offspring are responsible for increases and decreases in the *size* of a population, but only the *relative* success of different phenotypes within a population leads to changes in allele frequencies—that is, to evolution. To contribute genes to subsequent generations, individuals must survive to reproductive age and produce offspring. The relative contribution of individuals of a particular phenotype is determined by the probability that those individuals survive multiplied by the average number of offspring they produce over their lifetimes. In other words, the *fitness of a phenotype is determined by the average rates of survival and reproduction of individuals with that phenotype.*

The Results of Natural Selection

To simplify our discussion until now, we have considered only characters influenced by alleles at a single locus. However, as we saw in Chapter 10, most characters are influenced by alleles at more than one locus. Such characters are likely to show quantitative rather than qualitative variation. For ex-

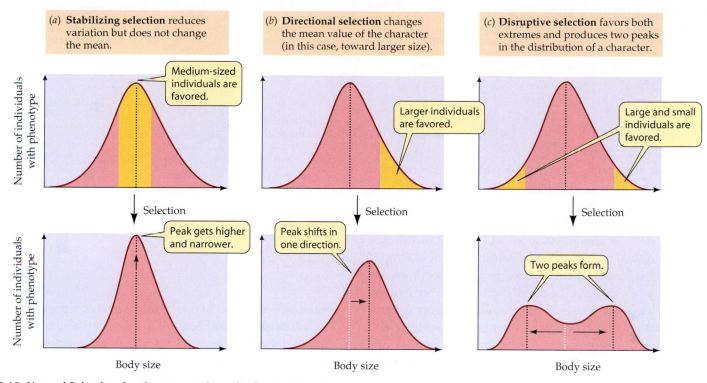

23.12 Natural Selection Can Operate on Quantitative Variation in Several Ways Each curve plots the distribution of body size in a population before selection (top) and after selection (bottom). Natural selection, by favoring the phenotype shown in yellow in the top graphs, changes the shape and position of the original curve (bottom graphs).

ample, the distribution of the sizes of individuals in a population, a character that is influenced by genes at many loci as well as by the environment, is likely to approximate the bell-shaped curves shown in the top row of Figure 23.12.

Natural selection can act on characters with quantitative variation in any one of several different ways, producing quite different results:

▸ *Stabilizing selection* preserves the average characteristics of a population by favoring average individuals.
▸ *Directional selection* changes the characteristics of a population by favoring individuals that vary in one direction from the mean of the population.
▸ *Disruptive selection* changes the characteristics of a population by favoring individuals that vary in both directions from the mean of the population.

STABILIZING SELECTION. If both the smallest and the largest individuals in a population contribute relatively fewer offspring to the next generation than those closer to the average size do, then **stabilizing selection** is operating (Figure 23.12a). Stabilizing selection reduces variation, but does not change the mean. Natural selection frequently acts in this

way, countering increases in variation brought about by genetic recombination, mutation, or migration. Rates of evolution are typically very slow because natural selection is usually stabilizing. Stabilizing selection operates, for example, on human birth weight. Babies born lighter or heavier than the population mean die at higher rates than babies whose weights are close to the mean (Figure 23.13). This was especially true before modern medical advances.

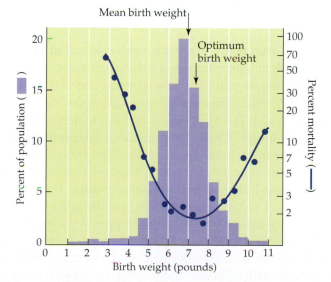

23.13 Human Birth Weight Is Influenced by Stabilizing Selection Babies that weigh more or less than average are more likely to die soon after birth than babies with weights close to the population mean.

23.14 Resistance to TTX Is Associated with the Presence of Newts Garter snakes (*Thamnophis sirtalis*) of the Pacific Coast have evolved resistance to the neurotoxin TTX produced by a prey species, the newt *Taricha granulosa*. TTX resistance evolved at least twice. The range of the newt is shown in blue.

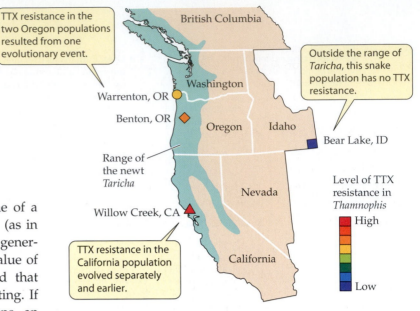

DIRECTIONAL SELECTION. If individuals at one extreme of a character distribution—the larger ones, for example (as in Figure 23.12*b*)—contribute more offspring to the next generation than other individuals do, then the average value of that character in the population will shift toward that extreme. In this case, **directional selection** is operating. If directional selection operates over many generations, an *evolutionary trend* within the population results. Such directional evolutionary trends often continue for many generations, but they may be reversed when the environment changes and different phenotypes are favored, or they may be halted when an optimum is reached, or when trade-offs oppose further change. The character then falls under stabilizing selection.

Directional selection produced the resistance to tetrodotoxin (TTX) by some garter snakes that we discussed at the beginning of this chapter. The common garter snake, *Thamnophis sirtalis*, is the only predator of the rough-skinned newt, *Taricha granulosa*, known to be resistant to TTX. Resistance to TTX has evolved independently at least twice within *T. sirtalis* populations in western North America, once in California and once in Oregon (Figure 23.14). This resistance is due to genetically based differences in the ability of sodium channels in the snake's nerves and muscles to continue functioning when exposed to variable concentrations of TTX.

DISRUPTIVE SELECTION. When **disruptive selection** operates, individuals at both extremes of a character distribution contribute more offspring to the next generation than do those close to the mean, producing two peaks in the distribution (Figure 23.12*c*). This type of selection is apparently rare.

The strikingly bimodal (two-peaked) distribution of bill sizes in the black-bellied seedcracker (*Pyrenestes ostrinus*), a West African finch (Figure 23.15), illustrates how disruptive selection can influence populations in nature. The seeds of two types of sedges (marsh plants) are the most abundant food source for these finches during part of the year. Birds with large bills can readily crack the hard seeds of the sedge *Scleria verrucosa*. Birds with small bills can crack *S. verrucosa* seeds only with difficulty, but they feed more efficiently on the soft seeds of *S. goossensii* than do birds with larger bills.

Young finches whose bills deviate markedly from the two predominant bill sizes do not survive as well as finches

whose bills are close to one of the two sizes represented by the distribution peaks. Because there are few abundant food sources in the environment, and because the seeds of the two sedges do not overlap in hardness, birds with intermediate-

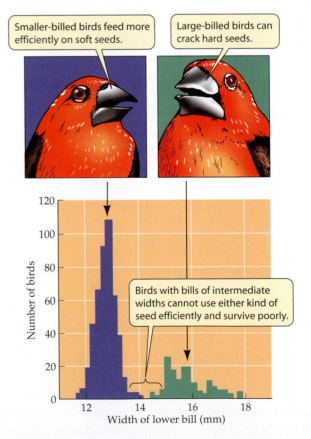

23.15 Disruptive Selection Results in a Bimodal Distribution The bimodal distribution of bill sizes in the black-bellied seedcracker of West Africa is a result of disruptive selection, which favors individuals with larger and smaller bill sizes over individuals with intermediate-sized bills.

sized bills are inefficient in using either one of the principal food sources. Disruptive selection therefore maintains a bimodal bill size distribution.

Sexual selection results in conspicuous traits

In *The Origin of Species*, Darwin devoted a few pages to sexual selection, a topic that he developed at length in another book, *The Descent of Man, and Selection in Relation to Sex*, in 1871. **Sexual selection** was Darwin's explanation for the evolution of apparently useless but conspicuous traits in males of many species, such as bright colors, long tails, horns, antlers, and elaborate courtship displays. He hypothesized that these traits either improved the ability of their bearers to compete for access to members of the other sex (*intrasexual selection*) or made their bearers more attractive to members of the other sex (*intersexual selection*). Darwin argued that female preferences for such features are also the result of sexual selection because "unornamented, or unattractive males would succeed equally in the battle for life and in leaving a numerous progeny, but for the presence of better endowed males." Sexual selection may result in species that are *sexually dimorphic*—that is, species in which males and females differ in size, shape, or color.

The concept of sexual selection was not well received by Darwin's contemporaries. However, many examples of sexual selection have been investigated in the century and a half since he first proposed the idea, and Darwin turned out to be right. For example, sexual selection is responsible for different morphological attributes of male birds that compete with each other for available females. One case in point is the remarkable tails of male African long-tailed widowbirds, which are longer than their heads and bodies combined.

To examine the role of sexual selection in the evolution of widowbird tails, a behavioral ecologist captured some male widowbirds. He shortened the tails of some males by cutting them and lengthened the tails of others by gluing on additional feathers. Male widowbirds normally select, and defend from other males, a site where they perform courtship displays to attract females. Both short-tailed and long-tailed males successfully defended their display sites, indicating that a long tail does not confer an advantage in male–male competition. However, males with artificially elongated tails attracted about four times more females than did males with shortened tails (Figure 23.16).

Why do female widowbirds prefer males with long tails? The ability to grow and maintain a costly feature such as a long tail may indicate that the male bearing it is vigorous and healthy. The hypothesis that having well-developed ornamental traits signals vigor and health has been tested experimentally with captive zebra finches. The bright red bills of male zebra finches are the result of red and yellow carotenoid

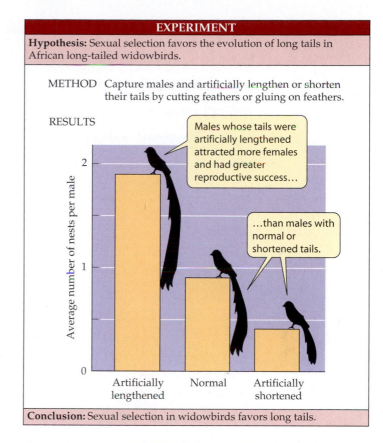

EXPERIMENT

Hypothesis: Sexual selection favors the evolution of long tails in African long-tailed widowbirds.

METHOD   Capture males and artificially lengthen or shorten their tails by cutting feathers or gluing on feathers.

RESULTS

Males whose tails were artificially lengthened attracted more females and had greater reproductive success…

…than males with normal or shortened tails.

Average number of nests per male

Artificially lengthened   Normal   Artificially shortened

Conclusion: Sexual selection in widowbirds favors long tails.

23.16 The Longer the Tail, the Better the Male   Male widowbirds with shortened tails defended their display sites successfully, but attracted fewer females (and thus fathered fewer nests of eggs) than did males with normal and lengthened tails.

pigments. Zebra finches (and most other animals) cannot synthesize carotenoids and must obtain them from their food. In addition to influencing bill color, carotenoids are antioxidants and components of the immune system. Males in good health may need to allocate fewer carotenoids to immune function than males in poorer health. If so, then females can use the brightness of his bill to assess the health of a male.

Investigators manipulated blood levels of carotenoids in male zebra finches by means of carotenoid supplements. Experimental males were given drinking water with carotenoids added; control males were given only distilled water. All the males had access to the same food. After one month, the experimental males had higher levels of carotenoids in their blood, had much brighter bills than the control males, and were preferred by female zebra finches (Figure 23.17).

Next, the investigators challenged both groups of males immunologically by injecting phytohemagglutinin (PHA) into the webs of their wings. PHA induces a response by T lymphocytes, resulting in an accumulation of white blood cells and thus a thickening of the skin. Experimental males with enhanced carotenoid levels developed thicker skins

(a)

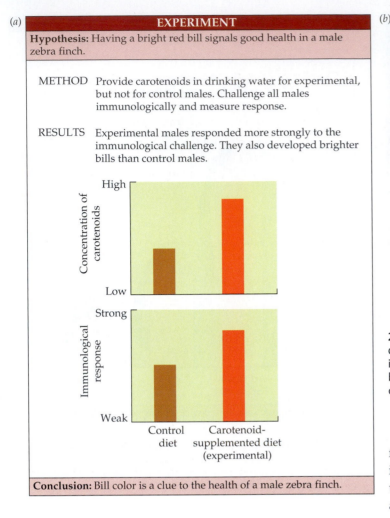

EXPERIMENT

Hypothesis: Having a bright red bill signals good health in a male zebra finch.

METHOD   Provide carotenoids in drinking water for experimental, but not for control males. Challenge all males immunologically and measure response.

RESULTS   Experimental males responded more strongly to the immunological challenge. They also developed brighter bills than control males.

Conclusion: Bill color is a clue to the health of a male zebra finch.

(b)

Taeniopygia guttata

23.17 Bright Bills Signal Good Health   (*a*) This experiment demonstrated that bright bill color in the male zebra finch does indeed indicate a healthy individual. (*b*) Female zebra finches (the bird below) preferentially choose mates with the brightest bill color—thus choosing the healthiest males.

because they responded more strongly to PHA than control males did, indicating a heightened immune system.

This experiment showed that when a female chooses the male with a bright red bill, she probably gets a mate with a healthy immune system. Such males are less likely to become infected with parasites and diseases, so they are less likely to pass on infections to their mates. Healthier males are also better able to assist with parental care than are males with duller bills.

Assessing the Costs of Adaptations

As we mentioned at the beginning of this chapter, adaptations typically impose costs as well as benefits, and the evolution of adaptations depends on the trade-off between those costs and benefits. Garter snakes in some populations, for example, can eat rough-skinned newts without being poisoned, but they pay for this ability by sacrificing crawling speed.

Determining the costs and benefits of a particular adaptation is difficult because individuals differ not only in the degree to which they possess the adaptation, but also in many other ways. How can investigators study individuals that dif-

fer only in the genetically based adaptation of interest? Such individuals can be created by recombinant DNA techniques using cloned or highly inbred populations. In plants, for example, plasmids can be used to transfer specific alleles to experimental individuals (see Figure 16.5). Control individuals also receive plasmids, but those plasmids lack the allele of interest.

Plasmid transfer techniques made it possible to measure the cost associated with resistance to the herbicide chlorosulfuron conferred by a single allele in the shale cress, *Arabidopsis thaliana*. The allele, *Csr1-1*, results in the production of an enzyme that is insensitive to chlorosulfuron. However, plants with the *Csr1-1* allele produce 34 percent fewer seeds than nonresistant plants grown under identical conditions in the absence of the herbicide (Figure 23.18).

The reason for the high cost of resistance is not fully understood, but evidence suggests that the resistance allele results in an accumulation of branched-chain amino acids that interfere with metabolism. Agriculturalists wish to alter the genotypes of plants to give them resistance to herbicides so that the herbicides applied to agricultural fields will kill the weeds, but not the crops. This experiment shows that such benefits may impose a trade-off in terms of crop yield.

We saw in the previous section that the possession of certain conspicuous features by males confers reproductive benefits. What kinds of trade-offs do these benefits impose? The cost of long tails was not measured in the experiments with

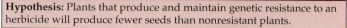

Hypothesis: Plants that produce and maintain genetic resistance to an herbicide will produce fewer seeds than nonresistant plants.

METHOD   Use a plasmid vector (see Figure 16.5) to insert an allele that confers herbicide resistance. Control plants received the same vector without the resistance gene. Measure reproductive success (seed production) of plants with and without the resistance gene.

Treated ⟶ Resistance allele   Control

RESULTS

> There is no significant difference in seed production between untreated and control plants.

Untreated plants (no resistance allele)

Control plants (empty plasmid)

Plants receiving allele that confers resistance

Seed production (thousands of seeds)
0   5   10   15   20   25

> Resistant plants produced 34% fewer seeds.

Conclusion: The cost of producing and maintaining a resistance-conferring compound is high.

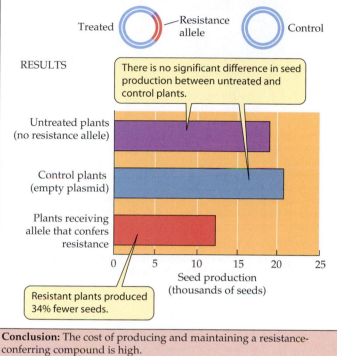

23.18 Producing and Maintaining Resistance Is Costly Possession of a gene that confers herbicide resistance greatly reduced seed production in *Arabidopsis thaliana*.

widowbirds, but related studies have been done on males of other species.

In some mammalian species, including deer, lions, and baboons, one male controls reproductive access to many females. These *polygynous* species tend to be *sexually dimorphic*—the males appear quite different from the females. Males of these species are significantly larger than females and often bear large weapons (such as horns, antlers, and large canine teeth); size and weaponry are needed to defend a male's multiple mates against other males of the species.

The costs of sexual dimorphism for males of polygynous species were assessed using the comparative method (Figure 23.19). Such males have higher parasite loads and higher mortality rates than females of their own species because maintaining a large size and bearing large weapons makes them more susceptible to parasites. In addition, when compared to parasite loads in males of closely related monogamous species (in which males and females are essentially monomorphic, appearing quite similar), the dimorphic males carried higher parasite loads in almost every case.

Maintaining Genetic Variation

Genetic drift, stabilizing selection, and directional selection all tend to reduce genetic variation within populations. Nevertheless, as we have seen, most populations have considerable genetic variation. What maintains so much genetic variation within populations? To answer this question, we will show how sexual recombination, neutral mutations, and frequency-dependent selection can maintain variation within populations, and how variation may be maintained over geographic space.

Sexual recombination amplifies the number of possible genotypes

In asexually reproducing organisms, the cells resulting from a mitotic division normally contain identical genotypes. Each new individual is genetically identical to its parent, unless there has been a mutation. When organisms exchange genetic

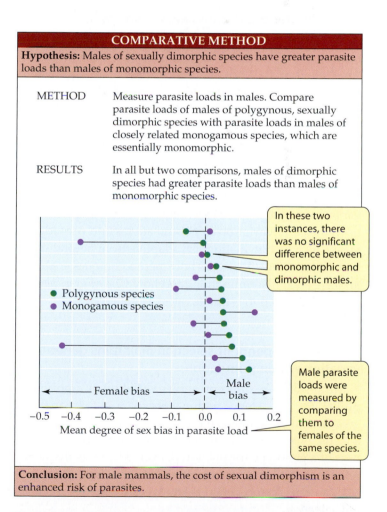

Hypothesis: Males of sexually dimorphic species have greater parasite loads than males of monomorphic species.

METHOD   Measure parasite loads in males. Compare parasite loads of males of polygynous, sexually dimorphic species with parasite loads in males of closely related monogamous species, which are essentially monomorphic.

RESULTS   In all but two comparisons, males of dimorphic species had greater parasite loads than males of monomorphic species.

> In these two instances, there was no significant difference between monomorphic and dimorphic males.

● Polygynous species
● Monogamous species

◀—— Female bias ——▶ | ◀— Male bias —▶

−0.5   −0.4   −0.3   −0.2   −0.1   0.0   0.1   0.2
Mean degree of sex bias in parasite load

> Male parasite loads were measured by comparing them to females of the same species.

Conclusion: For male mammals, the cost of sexual dimorphism is an enhanced risk of parasites.

23.19 Sexually Selected Traits Impose Costs Male mammals of sexually dimorphic, polygynous species have greater parasite loads than do males of closely related species that are not sexually dimorphic.

material during sexual reproduction, however, offspring differ from their parents because chromosomes assort randomly during meiosis, crossing-over occurs, and fertilization brings together material from two different cells (see Chapter 9).

Sexual recombination generates an endless variety of genotypic combinations that increases the evolutionary potential of populations. Because it increases the variation among the offspring produced by an individual, sexual recombination may improve the chance that at least some of those offspring will be successful in the varying and often unpredictable environments they will encounter. Sexual recombination does not influence the frequencies of alleles; rather, *sexual recombination generates new combinations of alleles on which natural selection can act.* It expands variation in a character influenced by alleles at many loci by creating new genotypes. That is why selection for bristle number in *Drosophila* (see Figure 23.5) resulted in flies with more bristles than any flies in the initial population had.

Neutral mutations accumulate within species

As we saw in Chapter 12, some mutations do not affect the functioning of the proteins encoded by the mutated genes. An allele that does not affect the fitness of an organism is called a **neutral allele**. Such alleles, untouched by natural selection, may be lost, or their frequencies may increase with time, purely by genetic drift. Therefore, neutral alleles often accumulate in a population over time, providing it with considerable genetic variation.

Much of the variation in those characters we can observe with our unaided senses is not neutral, but much molecular variation apparently is. Modern molecular techniques enable us to measure variation in neutral alleles and provide the means by which to distinguish adaptive from neutral variation. Chapter 26 will discuss how these techniques enable us to make such discriminations and how variation in neutral traits can be used to estimate rates of evolution.

Frequency-dependent selection maintains genetic variation within populations

Natural selection often preserves variation as a **polymorphism**: the coexistence within a population, at frequencies greater than mutations can produce, of two or more alleles at a locus. A polymorphism may be maintained when the fitness of a genotype (or phenotype) varies with its frequency relative to that of other genotypes (or phenotypes) in a population. This phenomenon is known as **frequency-dependent selection**.

A small fish that lives in Lake Tanganyika, in East Africa, provides an example of frequency-dependent selection. The mouth of this scale-eating fish, *Perissodus microlepis*, opens ei-

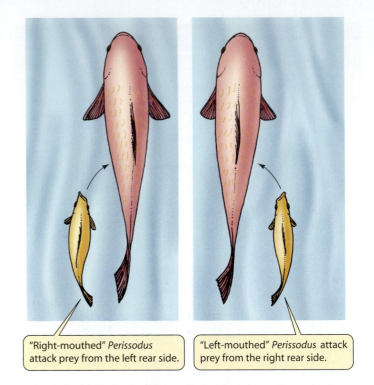

"Right-mouthed" *Perissodus* attack prey from the left rear side.

"Left-mouthed" *Perissodus* attack prey from the right rear side.

23.20 A Stable Polymorphism Frequency-dependent selection maintains equal proportions of left-mouthed and right-mouthed individuals of the scale-eating fish *Perissodus microlepis*.

ther to the right or to the left as a result of an asymmetrical jaw joint; the direction of opening is genetically determined (Figure 23.20). *P. microlepis* approaches its prey (another fish) from behind and dashes in to bite off several scales from its flank. "Right-mouthed" individuals always attack from the victim's left; "left-mouthed" individuals always attack from the victim's right. The distorted mouth enlarges the area of teeth in contact with the prey's flank, but only if the scale-eater attacks from the appropriate side.

Prey fish are alert to approaching scale-eaters, so attacks are more likely to be successful if the prey must watch both flanks. Vigilance by the prey favors equal numbers of right-mouthed and left-mouthed scale-eaters, because if one form were more common than the other, prey fish would pay more attention to potential attacks from the corresponding flank. Over an 11-year period in which the scale-eaters in Lake Tanganyika were studied, the polymorphism was found to be stable: The two forms of *P. microlepis* remained at about equal frequencies.

Genetic variation is maintained in geographically distinct subpopulations

Much of the genetic variation in large populations is preserved as differences among members in different places (subpopulations). Subpopulations often vary genetically because they are subjected to different selective pressures in different environments. Plant species, for example, may vary

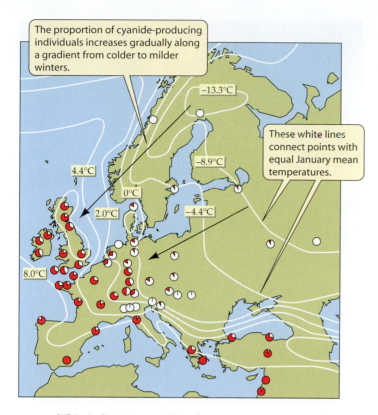

The proportion of cyanide-producing individuals increases gradually along a gradient from colder to milder winters.

−13.3°C

These white lines connect points with equal January mean temperatures.

4.4°C

−8.9°C

0°C

2.0°C

−4.4°C

8.0°C

White indicates proportion of plants not producing cyanide

Red indicates proportion of plants producing cyanide

23.21 Geographic Variation in Poisonous Clovers The frequency of cyanide-producing individuals in each population of white clover (*Trifolium repens*) is represented by the proportion of the circle that is red.

geographically in the chemicals they synthesize to defend themselves against herbivores. Some individuals of the clover *Trifolium repens* produce the poisonous chemical cyanide. Poisonous individuals are less appealing to herbivores—particularly mice and slugs—than are nonpoisonous individuals. However, clover plants that produce cyanide are more likely to be killed by frost, because freezing damages cell membranes and releases the toxic cyanide into the plant's own tissues.

In populations of *Trifolium repens*, the frequency of cyanide-producing individuals increases gradually from north to south and from east to west across Europe (Figure 23.21). Poisonous plants make up a large proportion of clover populations only in areas where winters are mild. Cyanide-producing individuals are rare where winters are cold, even though herbivores graze clovers heavily in those areas.

Constraints on Evolution

The many examples of adaptations that we have just discussed are testimony to the power of natural selection, but evolution is limited by a serious constraint: Evolutionary

changes must be based on modifications of previously existing traits, which may come to serve new functions. Engineers are able to design a completely new type of engine (jet) to power an airplane that can replace a previous type (propeller), but evolutionary changes cannot happen that way.

A striking example of such constraints on evolution is provided by the evolution of fish that spend most of their time resting on the sea bottom. One lineage, the bottom-dwelling skates and rays, is beautifully symmetrical. These fishes are descended from sharks, whose bodies were already somewhat flattened; therefore, skates and rays are able to lie on their bellies (Figure 23.22*a*).

Plaice, sole, and flounders, on the other hand, are bottom-dwelling descendants of deep-bodied, laterally flattened ancestors. Unlike sharks, these fishes cannot lie on their bellies; they must flop over on their sides. During development, the eyes of plaice and sole are grotesquely twisted around to bring both eyes to one side of the body (Figure 23.22*b*). Small shifts in the position of one eye probably helped ancestral flatfishes see better, resulting in the form found today.

(a) *Taeniura lymma*

(b) *Bothus lunatus*

23.22 Two Solutions to a Single Problem (a) Stingrays, whose ancestors were dorsally flattened, lie on their bellies. (b) Flounders, whose ancestors were laterally flattened, lie on their sides. Their eyes migrate during development so that both eyes are on the same side of the body.

Cultural Evolution

Traits can evolve by natural selection only if they are at least partly heritable. However, individuals may acquire new traits via **cultural evolution**—that is, by learning them from other individuals. Cultural evolution is most highly developed in humans, whose language and remarkable learning abilities enable new innovations to spread and be adopted at rapid rates. But the only requirement for traits to evolve via cultural evolution is that individuals have the ability to learn them. Birds, for example, copy the songs of other individuals, resulting in the evolution of song "dialects."

Many behaviors of the apes (chimpanzees, gorillas, gibbons, and orangutans) are transmitted via learning. In one study, investigators compared the behavior of four orangutan populations on the island of Borneo and two on Sumatra. The investigators identified 24 behaviors that are restricted to a single population. These behaviors are not correlated with any differences in the environments in which the populations live. Ten of the behaviors are specialized feeding techniques (Figure 23.23), including tool use. Six are alternative forms of social signals, such as kiss-squeaks. Thus, orangutan populations develop cultural distinctions as individuals copy the behavior of other individuals.

Short-Term versus Long-Term Evolution

The short-term changes in allele frequencies within populations that we have emphasized in this chapter are an important focus of study for evolutionary biologists. These changes can be observed directly, they can be manipulated experimentally, and they show us the actual processes by which evolution occurs. By themselves, however, they do not enable us to predict—or, more properly, "postdict" (because they have already happened)—the long-term evolutionary changes we described in Chapter 22.

The reason is that patterns of evolutionary change can be strongly influenced by events that occur so infrequently or so slowly that they are unlikely to be observed during short-term studies. In addition, the ways in which evolutionary agents act may change with time; even among the descendants of a single ancestral species, different lineages may evolve in different directions. Therefore, additional types of evidence, demonstrating the occurrence of rare and unusual events and trends in the fossil record, must be gathered if we wish to understand the course of evolution over billions of years.

"Postdiction" problems are not unique to evolutionary studies. For example, seismologists know the physical principles that explain how earthquakes occur, and they can pinpoint regions that are prone to earthquakes; but they cannot predict when or where an earthquake will happen.

Pongo pygmaeus

23.23 Orangutans Have Culturally Transmitted Behaviors This orangutan in Indonesia has learned to break open dead twigs and suck out the ants inside. This specialized feeding behavior is culturally transmitted from one individual to another within the individual's social group.

In subsequent chapters, we will discuss the kinds of information that biologists assemble to study long-term evolutionary changes and infer the processes that led to them.

Chapter Summary

Charles Darwin's Theory of Evolution

▶ Darwin developed his theory of evolution by natural selection by carefully observing nature, especially during his voyage around the world on the *Beagle*. **Review Figure 23.1**

▶ Darwin based this theory on well-known facts and some key inferences.

▶ Modern genetics has discovered the mechanisms of inheritance, which Darwin did not understand.

▶ Darwin had no examples of the action of natural selection, so he based his arguments on artificial selection.
See Web/CD Tutorial 23.1

Genetic Variation within Populations

▶ For a population to evolve, its members must possess heritable genetic variation, which is the raw material on which agents of evolution act.

▶ A single individual has only some of the alleles found in the population of which it is a member. **Review Figure 23.3**

▶ Considerable genetic variation characterizes most natural populations. **Review Figures 23.4, 23.5**

▶ Allele frequencies measure the amount of genetic variation in a population. Biologists estimate allele frequencies by measuring a sample of individuals from a population. The sum of all allele frequencies at a locus is equal to 1. **Review Figure 23.6**

▶ Genotype frequencies show how a population's genetic variation is distributed among its members. Populations that have the same allele frequencies may nonetheless have different genotype frequencies.

The Hardy–Weinberg Equilibrium

▶ Several conditions are required for a population to be at Hardy–Weinberg equilibrium: mating is random, the population is very large, there is no migration, there is no mutation, and natural selection is not acting on the population.

▶ In a population at Hardy–Weinberg equilibrium, allele frequencies remain the same from generation to generation. In addition, genotype frequencies will remain in the proportions $p^2 + 2pq + q^2 = 1$. **Review Figure 23.7**

▶ Biologists can determine whether an agent of evolution is acting on a population by comparing the genotype frequencies of that population with Hardy–Weinberg expectations.

See Web/CD Tutorial 23.2

Evolutionary Agents and Their Effects

▶ Changes in the genetic structure of populations are caused by several evolutionary agents: mutation, gene flow, genetic drift, nonrandom mating, and natural selection.

▶ The origin of genetic variation is mutation. Most mutations are harmful or neutral to their bearers, but some are advantageous, particularly if the environment changes.

▶ Movement of individuals or gametes from one population to another, followed by reproduction in the new location, produces gene flow. Gene flow may add new alleles to a population or may change the frequencies of alleles already present.

▶ The random loss of alleles, known as genetic drift, produces changes in allele frequencies, which may be especially dramatic in small populations. Organisms that normally have large populations may pass through occasional periods (bottlenecks) when only a small number of individuals survive. **Review Figure 23.8**

▶ New populations established by a few founding immigrants also have gene frequencies that differ from those in the parent population. **Review Figure 23.10**

▶ If individuals mate more often with other individuals of a certain genotype than would be expected on a random basis— that is, when mating is not random—genotype frequencies differ from Hardy–Weinberg expectations. **Review Figure 23.11**

▶ Self-fertilization, an extreme form of nonrandom mating, reduces the frequencies of heterozygous individuals below Hardy–Weinberg expectations without changing allele frequencies.

▶ Natural selection is the only agent of evolution that adapts populations to their environments.

▶ The reproductive contribution of a phenotype to subsequent generations relative to the contributions of other phenotypes is its fitness. The fitness of a phenotype is determined by the average rates of survival and reproduction of individuals with that phenotype.

The Results of Natural Selection

▶ Stabilizing selection reduces variation and preserves the average characteristics of a population. **Review Figures 23.12a, 23.13**

▶ Directional selection changes a character by favoring individuals that vary in one direction from the population mean. If directional selection operates over many generations, an evolutionary trend may result. **Review Figures 23.12b, 23.14**

▶ Disruptive selection changes a character by favoring individuals that vary in both directions from the population mean. **Review Figures 23.12c, 23.15**

▶ Sexually selected traits may evolve because females prefer to mate with males having those traits. **Review Figures 23.16, 23.17**

Assessing the Costs of Adaptations

▶ Possessing resistance to toxic chemicals may involve trade-offs, such as reduced reproductive output. **Review Figure 23.18. See Web/CD Tutorial 23.3**

▶ Sexually selected traits may result in higher parasite loads and mortality rates in males. **Review Figure 23.19**

Maintaining Genetic Variation

▶ Genetic drift, stabilizing selection, and directional selection all tend to reduce genetic variation, but most populations are genetically highly variable.

▶ Sexual recombination increases the evolutionary potential of populations, but it does not influence the frequencies of alleles. Rather, it generates new combinations of genetic material on which natural selection can act.

▶ Genetic variation within a population may be maintained by frequency-dependent selection. **Review Figure 23.20**

▶ Much genetic variation is maintained geographically. **Review Figures 23.21**

Constraints on Evolution

▶ Natural selection acts by modifying what already exists.

Cultural Evolution

▶ Learned traits can spread rapidly via cultural evolution.

Short-Term versus Long-Term Evolution

▶ Patterns of long-term evolutionary change can be strongly influenced by events that occur so infrequently or so slowly that they are unlikely to be observed during short-term evolutionary studies. Additional types of evidence must be gathered to understand why evolution in the long term took the particular course it did.

Self-Quiz

1. The two major components of Darwin's theory of evolution are that
   a. evolution is a fact, and mutations are the agent of evolution.
   b. evolution is a fact, and natural selection is the agent of evolution.
   c. species cannot change into other species, but natural selection can modify them.
   d. species cannot change into other species, but mutations can modify them.
   e. evolution is a hypothesis, and genetic drift is the agent of evolution.

2. To ground his theory, Charles Darwin
   a. developed a comprehensive theory of inheritance.
   b. described several evolutionary changes and identified the agents that caused them.
   c. used patterns of domestication to show how his theory differed from those patterns.
   d. assembled a broad base of supporting information from many fields.
   e. developed a mathematical model of evolutionary change.

3. The phenotype of an organism is
   a. the type specimen of its species in a museum.
   b. its genetic constitution, which governs its traits.
   c. the chronological expression of its genes.
   d. the physical expression of its genotype.
   e. the form it achieves as an adult.

4. The appropriate unit for defining and measuring genetic variation is the
   a. cell.
   b. individual.
   c. population.
   d. community.
   e. ecosystem.

5. Which statement about allele frequencies is *not* true?
   a. The sum of any set of allele frequencies is always 1.
   b. If there are two alleles at a locus and we know the frequency of one of them, we can obtain the frequency of the other by subtraction.
   c. If an allele is missing from a population, its frequency is 0.
   d. If two populations have the same gene pool for a locus, they will have the same proportion of homozygotes at that locus.
   e. If there is only one allele at a locus, its frequency is 1.

6. In a population at Hardy–Weinberg equilibrium in which the frequency of *A* alleles (*p*) is 0.3, the expected frequency of *Aa* individuals is
   a. 0.21.
   b. 0.42.
   c. 0.63.
   d. 0.18.
   e. 0.36.

7. Natural selection that preserves existing allele frequencies is called
   a. unidirectional selection.
   b. bidirectional selection.
   c. prevalent selection.
   d. stabilizing selection.
   e. preserving selection.

8. The fitness of a genotype is determined by the
   a. average rates of survival and reproduction of individuals with that genotype.
   b. individuals that have the highest rates of both survival and reproduction.
   c. individuals that have the highest rates of survival.
   d. individuals that have the highest rates of reproduction.
   e. average reproductive rate of individuals with that genotype.

9. Laboratory selection experiments with fruit flies have demonstrated that
   a. bristle number is not genetically controlled.
   b. bristle number is not genetically controlled, but changes in bristle number are caused by the environment in which the fly is raised.
   c. bristle number is genetically controlled, but there is little variation on which natural selection can act.
   d. bristle number is genetically controlled, but selection cannot result in flies having more bristles than any individual in the original population had.
   e. bristle number is genetically controlled, and selection can result in flies having more bristles than any individual in the original population had.

10. Disruptive selection maintains a bimodal distribution of bill size in the West African seedcracker because
    a. bills of intermediate shapes are difficult to form.
    b. the two major food sources of the finches differ markedly in size and hardness.
    c. males use their large bills in displays.
    d. migrants introduce different bill sizes into the population each year.
    e. older birds need larger bills than younger birds.

11. A population is said to be polymorphic for a locus if it has at least
    a. three different alleles at that locus.
    b. two different alleles at that locus.
    c. two genotypes for that locus.
    d. three genotypes for that locus.
    e. two alleles for that locus, the rarest of which is more common than expected by mutation alone.

For Discussion

1. During the past 50 years, more than 200 species of insects that attack crop plants have become highly resistant to DDT and other pesticides. Using your recently acquired knowledge of evolutionary processes, explain the rapid and widespread evolution of resistance. Propose ways of using pesticides that would slow down the rate of evolution of resistance. Now that use of DDT has been banned in the United States, what do you expect to happen to levels of resistance to DDT among insect populations? Justify your answer.

2. In what ways does artificial selection by humans differ from natural selection in nature? Was Darwin wise to base so much of his argument for natural selection on the results of artificial selection?

3. In nature, mating among individuals in a population is never truly random, immigration and emigration are common, and natural selection is seldom totally absent. Why then, does it make sense to use the Hardy–Weinberg equilibrium, which is based on assumptions known generally to be false? Can you think of other models in science that are based on false assumptions? How are such models used?

4. As far as we know, natural selection cannot adapt organisms to future events. Yet many organisms appear to respond to natural events before they happen. For example, many mammals go into hibernation while it is still quite warm. Similarly, many birds leave the temperate zone for their southern wintering grounds long before winter has arrived. How can such "anticipatory" behaviors evolve?

5. Populations of most of the thousands of species that have been introduced to areas where they were previously not found, including those that have become pests, began with a few individuals. They should therefore have begun with much less genetic variation than the parent populations have. If genetic variation is advantageous, why have so many of these species been successful in their new environments?

6. The flavors of many crop plants have been enhanced by artificial selection that has removed the bad-tasting chemicals with which they defended themselves in the wild. What problems do growing crop plants with reduced chemical defenses pose for modern agriculture?

24 Species and Their Formation

In May and June, 1993, a previously unknown disease abruptly killed 10 people in the southwestern United States. The victims experienced flu-like symptoms for several days, but then their condition deteriorated rapidly as their lungs filled with fluid. The disease agent was unknown; no cure was available, and initially 70 percent of infected people died.

Researchers from many disciplines focused on the outbreak. Within a few weeks, scientists at the U.S. Centers for Disease Control had identified the agent as a previously undescribed hantavirus—a type of virus known to be transmitted by rodents. Investigators initiated an intensive small mammal field trapping program. They quickly identified the deer mouse as the main host of the virus. The mouse sheds the virus in its feces, urine, and saliva. Humans become infected by inhaling microscopic particles of these substances that are present in the air.

Since the discovery of the hantavirus that caused the 1993 outbreak, about 25 additional hantaviruses have been described in the Western Hemisphere. People have been infected by some of these viruses in Florida, New York, Louisiana, and Texas. Other rodents, in addition to the deer mouse, harbor hantaviruses, and some of these viruses cause human diseases (Figure 24.1). Studies of the evolutionary relationships of hantaviruses and rodents indicate that rodents and hantaviruses have been evolving together for millions of years, and suggest that more kinds of hantaviruses, some of them likely to be human pathogens, have yet to be discovered.

Efforts to reduce risks to humans and develop effective cures will require an understanding of the distribution of hantavirus types among rodent species and measures of the rate at which the viruses appear to be evolving. The fact that so many new types of hantavirus have been discovered since 1993 suggests that hantaviruses may be evolving rapidly. If so, better knowledge about how new viruses form and how they have coevolved with rodents will serve important human health objectives.

All species, living and extinct, are believed to be descendants of a single ancestral species that lived more than 3 billion years ago. If speciation were a rare event, the biological world would be very different than it is today. How

A Source of Disease The deer mouse *Peromyscus maniculatus* harbors a type of hantavirus that infects people.

24.1 Hantaviruses in the New World   The map shows the many different hantaviruses found in the Western Hemisphere. Those listed in red are known to be human pathogens.

did these millions of species form? How does one species become two? These questions are the focus of this chapter. We will examine the mechanisms by which a population splits into two or more new species, and we will see how such separations are maintained. We will look at the factors that can make speciation a rapid or a very slow process. Finally, we will look at the conditions that give rise to the great diversifications called evolutionary radiations.

What Are Species?

The word **species** means, literally, "kinds." But what do we mean by "kinds?" Someone who is knowledgeable about a group of organisms, such as orchids or lizards, usually can distinguish the different species found in a particular area simply by examining their visible features. Standard field guides to birds, mammals, insects, and flowers are possible only because most species are cohesive units that change little in appearance over large geographic distances. We can easily recognize male red-winged blackbirds from New York

and from California, for example, as members of the same species (Figure 24.2*a*).

But not all members of a species look that much alike. For example, males, females, and young individuals may not resemble one another closely (Figure 24.2*b*). How do we decide whether similar but easily distinguished individuals should be assigned to different species or regarded as members of the same species?

The concept that has guided these decisions for a long time is genetic integration. If individuals of a population mate with one another, but not with individuals of other populations, they constitute a distinct group within which genes recombine; that is, they are independent evolutionary units. These independent evolutionary units are usually called species.

More than 200 years ago, the Swedish biologist Carolus Linnaeus, who originated the system of naming organisms that we use today, described hundreds of species. Because he knew nothing about the mating patterns of the organisms he was naming, Linnaeus classified them on the basis of their appearance; that is, he used a *morphological species* concept. Many of the organisms that he classified as species by their appearance are indeed independent evolutionary units. Their members look alike because they share many of the alleles that code for their body structures. In many groups of organisms for which genetic data are unavailable, species are still recognized by their morphological traits.

In 1940, Ernst Mayr proposed a definition of species that has been used by many biologists since that time. His definition, known as the *biological species concept*, says, "Species are groups of actually or potentially interbreeding natural populations which are reproductively isolated from other such groups." The words "actually or potentially" assert that, even if some members of a species live in different places and hence are unable to mate, they should not be placed in separate species if they would be likely to mate if they were together. The word "natural" is also an important part of Mayr's definition because only in nature does the exchange of genes among individuals from different populations influence evolutionary processes. The interbreeding of such individuals in captivity does not, because the resulting offspring typically spend their lives in captivity without interacting with the wild population. Since genetic integration through interbreeding maintains integrated evolutionary units, the biological species concept, although it does not apply to organisms that reproduce asexually, continues to be used by most evolutionary biologists.

Deciding whether two populations constitute different species may be difficult because speciation is often a gradual process (Figure 24.3). Once a population becomes separated into two or more populations, the daughter populations may evolve independently for a long time before they become re-

Agelaius phoeniceus (male, NY)   *Agelaius phoeniceus* (male, CA)

Agelaius phoeniceus (female)

24.2 Members of the Same Species Look Alike—or Not   (*a*) Both of these male red-winged blackbirds are obviously members of the same species, even though one is from the eastern United States and the other is from California. (*b*) Because red-winged blackbirds are sexually dimorphic (see Chapter 23), the female of the species appears quite different from the male.

productively incompatible. Alternatively, they may become reproductively incompatible before they evolve any noticeable morphological differences. But how do these differences between populations come about?

 ## How Do New Species Arise?

Speciation is the process by which one species splits into two or more daughter species, which thereafter evolve as distinct lineages. Although Charles Darwin titled his book *The Origin of Species*, he did not extensively discuss speciation, a process he called "the mystery of mysteries." He devoted most of his attention to demonstrating that species are

altered by natural selection over time. But not all evolutionary changes result in new species: A single lineage may change over time without giving rise to a new species.

The critical event in speciation is the separation of the gene pool of the ancestral species into two or more separate and isolated gene pools. Subsequently, within each isolated gene pool, allele and genotype frequencies may change as a result of the action of evolutionary agents. If two populations are isolated from each other, and sufficient differences in their genetic structure accumulate during the period of isolation, then the two populations may not be able to exchange genes when they come together again. As we will see, the amount of genetic difference that is needed to prevent gene exchange is highly variable. Gene flow among populations may be interrupted in two major ways, each of which characterizes a mode of speciation.

Allopatric speciation requires total genetic isolation

Speciation that results when a population is divided by a physical barrier is known as **allopatric** (*allo-*, "different"; *patris*, "country") or **geographic speciation** (Figure 24.4). Allopatric speciation is thought to be the dominant mode of speciation among most groups of organisms. The physical barrier that divides the range of a species may be a water body for terrestrial organisms, dry land for aquatic organisms, or a mountain range. Barriers can form when continents drift, sea levels rise, glaciers advance and retreat, and climates change. These processes continue to generate physical barriers today. The populations separated by such barriers are often large initially. They evolve differences for a variety of reasons, especially because the environments in which they live are, or become, different.

Allopatric speciation may also result when some members of a population cross an existing barrier and found a new, isolated population. Populations established in this way usually differ genetically from their parent populations because of the *founder effect*: A small group of founding individuals has

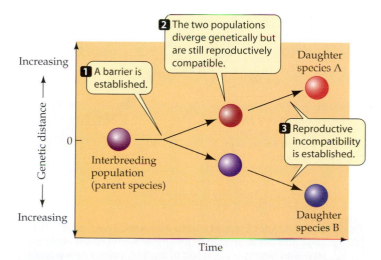

24.3 Speciation May Be a Gradual Process   In this hypothetical example, genetic divergence between two separated populations begins before reproductive incompatibility evolves.

Time

A single species is distributed over a broad range.

Sea level rises and separates two populations. Populations adapt to differing environments on opposite sides of the barrier.

If the barrier is removed, the populations may recolonize the intervening area and mingle, but do not interbreed.

Range of overlap

24.4 Allopatric Speciation Allopatric speciation may result when a population is divided into two separate populations by a physical barrier, such as rising sea levels.

Another example of allopatric speciation is found in the finches of the Galápagos archipelago, 1,000 km off the coast of Ecuador. Darwin's finches (as they are usually called, because Darwin was the first scientist to study them) arose in the Galápagos by speciation from a single South American species that colonized the islands. Today there are 14 species of Darwin's finches, all of which differ strikingly from their closest mainland relative (Figure 24.6). The islands of the Galápagos archipelago are sufficiently isolated from one another that finches seldom disperse between them. Also, environmental conditions differ among the islands. Some are relatively flat and arid; others have forested mountain slopes. Populations of finches on different islands have differentiated enough over millions of years that when occasional immigrants arrive from other islands, they either do not breed with the residents, or, if they do, the resulting offspring usually do not survive as well as those produced by pairs composed of island residents. The genetic distinctness and cohesiveness of the different species is thus maintained.

A physical barrier's effectiveness at preventing gene flow depends on the size and mobility of the species in question. What is an impenetrable barrier to a terrestrial snail may be no barrier at all to a butterfly or a bird. Popu-

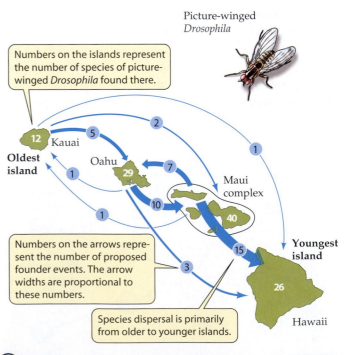

Picture-winged *Drosophila*

Numbers on the islands represent the number of species of picture-winged *Drosophila* found there.

12 Kauai
Oldest island

5

2

1

Oahu
1
29
7
Maui complex
1
10
40

Numbers on the arrows represent the number of proposed founder events. The arrow widths are proportional to these numbers.

3
15
Youngest island

26

Species dispersal is primarily from older to younger islands.

Hawaii

24.5 Founder Events Lead to Allopatric Speciation The large number of species of picture-winged *Drosophila* in the Hawaiian Islands is the result of founder events: new populations founded by individuals dispersing among the islands. The islands, which were formed in sequence as Earth's crust moved over a volcanic "hot spot," vary in age.

only an incomplete representation of the gene pool of its parent population (see Chapter 23). For example, many of the more than 800 species of the fruit fly genus *Drosophila* in the Hawaiian Islands are restricted to a single island. These species are almost certainly the descendants of new populations founded by individuals dispersing among the islands, because the closest relative of a species on one island is often a species on a neighboring island rather than a species on the same island. Biologists who have studied the chromosomes of picture-winged species of *Drosophila* believe that speciation among this group of flies has resulted from at least 45 such *founder events* (Figure 24.5).

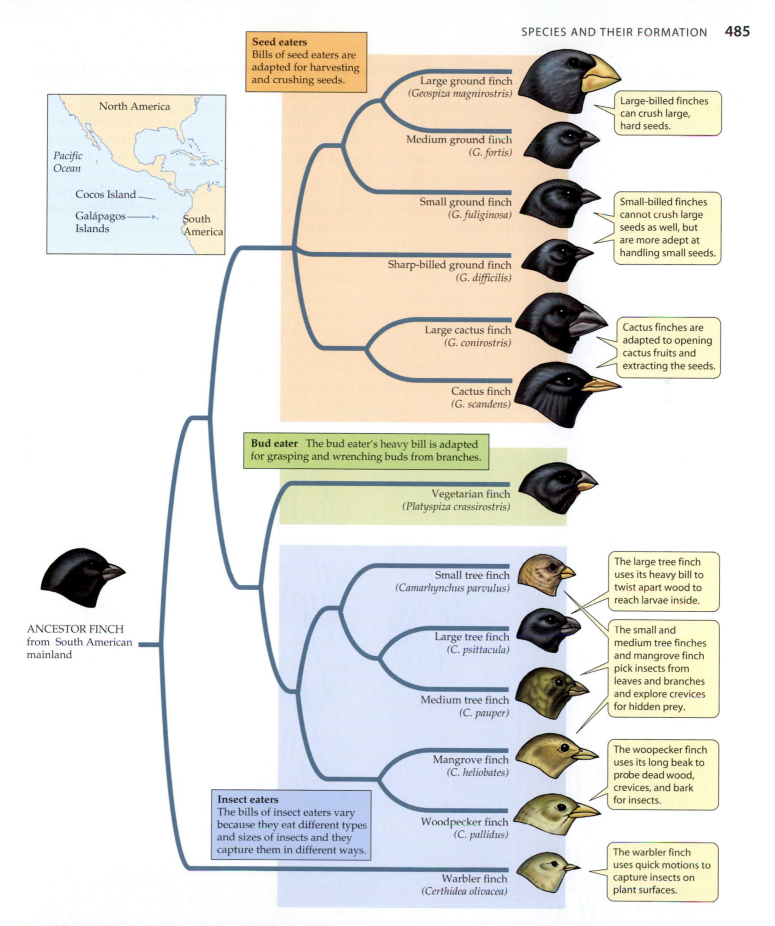

Seed eaters Bills of seed eaters are adapted for harvesting and crushing seeds.

Large ground finch
(*Geospiza magnirostris*)

Large-billed finches can crush large, hard seeds.

Medium ground finch
(*G. fortis*)

Small ground finch
(*G. fuliginosa*)

Small-billed finches cannot crush large seeds as well, but are more adept at handling small seeds.

Sharp-billed ground finch
(*G. difficilis*)

Large cactus finch
(*G. conirostris*)

Cactus finches are adapted to opening cactus fruits and extracting the seeds.

Cactus finch
(*G. scandens*)

Bud eater The bud eater's heavy bill is adapted for grasping and wrenching buds from branches.

Vegetarian finch
(*Platyspiza crassirostris*)

Small tree finch
(*Camarhynchus parvulus*)

The large tree finch uses its heavy bill to twist apart wood to reach larvae inside.

Large tree finch
(*C. psittacula*)

The small and medium tree finches and mangrove finch pick insects from leaves and branches and explore crevices for hidden prey.

Medium tree finch
(*C. pauper*)

Mangrove finch
(*C. heliobates*)

The woopecker finch uses its long beak to probe dead wood, crevices, and bark for insects.

Woodpecker finch
(*C. pallidus*)

Insect eaters
The bills of insect eaters vary because they eat different types and sizes of insects and they capture them in different ways.

The warbler finch uses quick motions to capture insects on plant surfaces.

Warbler finch
(*Certhidea olivacea*)

North America

Pacific Ocean

Cocos Island

Galápagos Islands

South America

ANCESTOR FINCH from South American mainland

24.6 Allopatric Speciation among Darwin's Finches The descendants of the ancestral finch that colonized the Galápagos archipelago several million years ago evolved into 14 different species whose members are variously adapted to feed on seeds, buds, and insects. (The fourteenth species, not pictured here, lives in Cocos Island, farther north in the Pacific Ocean.)

lations of wind-pollinated plants are isolated at the maximum distance pollen can be blown by the wind, but individual plants are effectively isolated at much shorter distances. Among animal-pollinated plants, the width of the barrier is the distance that animals can travel while carrying pollen or seeds. Even animals with great powers of dispersal are often reluctant to cross narrow strips of unsuitable habitat. For animals that cannot swim or fly, narrow water-filled gaps may be effective barriers. However, gene flow can sometimes be interrupted even in the absence of physical barriers.

Sympatric speciation occurs without physical barriers

Although physical isolation is usually required for speciation, under some circumstances speciation can occur without it. Such a partition of a gene pool is called **sympatric speciation** (*sym-*, "with"). The most common means of sympatric speciation is **polyploidy**, the production within an individual of duplicate sets of chromosomes. Polyploidy can arise either from chromosome duplication in a single species (**autopolyploidy**) or from the combining of the chromosomes of two different species (**allopolyploidy**).

An autopolyploid individual originates when (for example) cells that are normally diploid (with two sets of chromosomes) accidentally duplicate their chromosomes, resulting in a tetraploid (four sets of chromosomes) individual. Tetraploid and diploid plants of the same species are reproductively isolated because their triploid offspring are essentially sterile.

Even if triploid individuals survive to reproductive maturity, they cannot produce viable gametes because their chromosomes do not synapse correctly during meiosis (Figure 24.7). So a tetraploid plant cannot produce viable off-

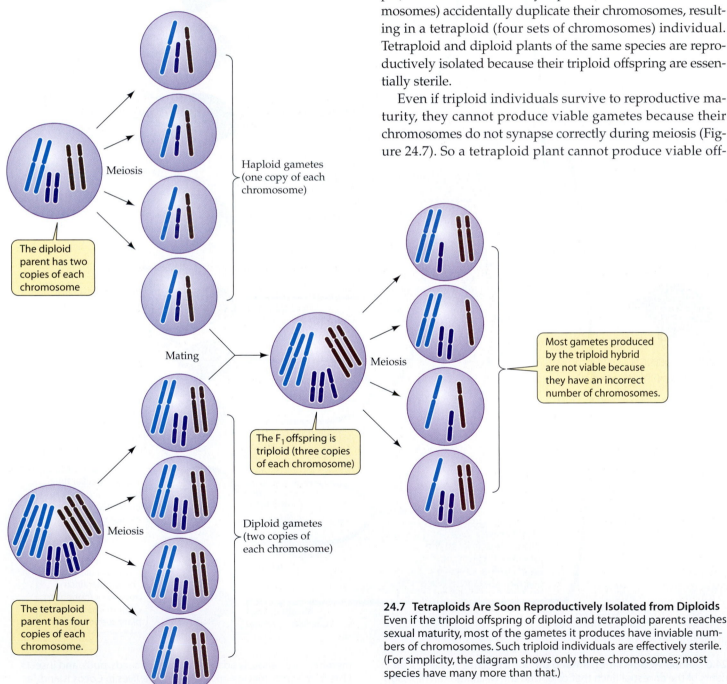

Meiosis

Haploid gametes (one copy of each chromosome)

The diploid parent has two copies of each chromosome

Mating

The tetraploid parent has four copies of each chromosome.

Meiosis

Diploid gametes (two copies of each chromosome)

The F₁ offspring is triploid (three copies of each chromosome)

Meiosis

Most gametes produced by the triploid hybrid are not viable because they have an incorrect number of chromosomes.

24.7 Tetraploids Are Soon Reproductively Isolated from Diploids Even if the triploid offspring of diploid and tetraploid parents reaches sexual maturity, most of the gametes it produces have inviable numbers of chromosomes. Such triploid individuals are effectively sterile. (For simplicity, the diagram shows only three chromosomes; most species have many more than that.)

spring by mating with a diploid individual—but it *can* do so if it self-fertilizes or mates with another tetraploid.

Allopolyploids may be produced when individuals of two different (but closely related) species interbreed, or **hybridize**. Allopolyploids are usually fertile because each of the chromosomes has a nearly identical partner with which to pair during meiosis.

New species arise by means of polyploidy much more easily among plants than among animals because plants of many species can reproduce by self-fertilization. In addition, if polyploidy arises in several offspring of a single parent, the siblings can fertilize one another. Speciation by polyploidy has been very important in the evolution of flowering plants. Botanists estimate that about 70 percent of flowering plant species and 95 percent of fern species are polyploids. Most of these arose as a result of hybridization between two species, followed by self-fertilization.

How easily allopolyploidy can produce new species is illustrated by the salsifies (*Tragopogon*), members of the sunflower family. Salsifies are weedy plants that thrive in disturbed areas around towns. People have inadvertently spread them around the world from their ancestral ranges in Eurasia. Three diploid species of salsify were introduced into North America early in the twentieth century: *T. porrifolius*, *T. pratensis*, and *T. dubius*. Two tetraploid hybrids—*T. mirus* and *T. miscellus*—between the original three diploid species were first discovered in 1950. The hybrids have spread since their discovery and today are more widespread than their diploid parents (Figure 24.8).

Studies of their genetic material have shown that both salsify hybrids have formed more than once. Some populations of *T. miscellus*—a hybrid of *T. pratensis* and *T. dubius*— have the chloroplast genome of *T. pratensis*; other populations have the chloroplast genome of *T. dubius*. Such differences among local populations of *T. miscellus* show that this allopolyploid has formed independently at least 21 times! Scientists seldom know the dates and locations of species formation so well. *T. mirus*, a hybrid of *T. porrifolius* and *T. dubius*, has formed 12 times. *T. porrifolius* prefers wet, shady places; *T. dubius* prefers dry, sunny places. *T. mirus*, however, can grow in partly shaded environments where neither parent does well. The success of these newly formed hybrid species of salsifies illustrates why so many species of flowering plants originated as polyploids.

Polyploidy, as we have just seen, can result in a new species that is completely reproductively isolated from its parent species in one generation. Allopatric speciation proceeds much more slowly, and some populations separated by a physical barrier may never acquire full reproductive isolation. Let's see how reproductive isolation may become established once two populations have been separated from each other.

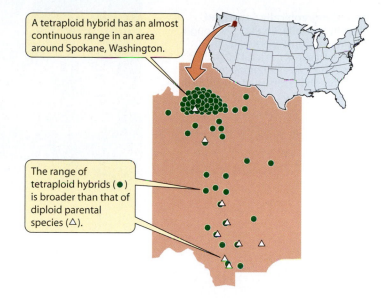

A tetraploid hybrid has an almost continuous range in an area around Spokane, Washington.

The range of tetraploid hybrids (●) is broader than that of diploid parental species (△).

24.8 Polyploids May Outperform Their Parent Species
Tragopogon species (salsifies) are members of the sunflower family. The map shows the distribution of the three diploid parent species and of the two tetraploid hybrid species of *Tragopogon* in eastern Washington and adjacent Idaho.

Completing Speciation: Reproductive Isolating Mechanisms

Once a barrier to gene flow is established, by whatever means, the separated populations may diverge genetically through the action of the evolutionary agents we described in Chapter 23. Over many generations, differences may accumulate that reduce the probability that members of the two populations could mate and produce viable offspring. In this way, reproductive isolation can evolve as an incidental by-product of genetic changes in allopatric populations.

Geographic isolation does not necessarily lead to reproductive incompatibility. For example, American sycamores and European sycamores (also known as plane trees) have been physically isolated from one another for at least 20 million years. Nevertheless, they are morphologically very similar (Figure 24.9), and they can form fertile hybrids, even though they never have an opportunity to do so in nature.

In other cases, however, genes that result in reproductive isolation between two evolving lineages spread quickly through populations as they diverge. In this section, we will examine the ways in which reproductive isolating mechanisms may arise. In the following section, we will explore what happens when reproductive isolation is incomplete.

24.9 Geographically Separated, Morphologically Similar Although they have been separated by the Atlantic Ocean for at least 20 million years, American and European sycamores have diverged very little in appearance.

(a) *Platanus occidentalis* (American sycamore)

(b) *Platanus hispanica* (European sycamore)

Prezygotic barriers operate before fertilization

Several processes that operate before fertilization—**prezygotic reproductive barriers**—may prevent individuals of different species from interbreeding:

▶ *Spatial isolation.* Individuals of different species may select different places in the environment in which to live. As a result, they may never come into contact during their respective mating periods; that is, they are reproductively isolated by location.

▶ *Temporal isolation.* Many organisms have mating periods that are as short as a few hours or days. If the mating periods of two species do not overlap, they will be reproductively isolated by time.

▶ *Mechanical isolation.* Differences in the sizes and shapes of reproductive organs may prevent the union of gametes from different species.

▶ *Gametic isolation.* Sperm of one species may not attach to the eggs of another species because the eggs do not release the appropriate attractive chemicals, or the sperm may be unable to penetrate the egg because the two gametes are chemically incompatible.

▶ *Behavioral isolation.* Individuals of a species may reject, or fail to recognize, individuals of other species as mating partners.

Two closely related species of crickets from the island of Hawaii, *Laupala paranga* and *Laupala kohalensis*, provide an example of behavioral isolation. These two species live in separate areas and do not hybridize in nature, but they will form interspecific pairs in the laboratory. The males of the two species produce genetically determined songs that differ in the number of pulses per second. The songs of hybrid males have intermediate numbers of pulses (Figure 24.10). Females are much more strongly attracted to the songs of conspecific males (males of their own species) than they are to the songs

of males of the other species. We know that this preference is genetically determined because hybrid females are most strongly attracted to the songs of hybrid males.

Sometimes the mate choice of one species is mediated by the behavior of individuals of other species. For example, whether two plant species hybridize may depend on the preferences of their pollinators. Because pollinators visit flowers to gather nutritional rewards, not to pollinate plants, pollinator behavior can be influenced only by changes in floral structures that affect the rewards the pollinators receive. Floral traits can affect reproductive isolation either by influencing pollinator behavior or by altering where pollen is deposited on the bodies of pollinators.

The evolution of floral traits that generate reproductive isolation has been studied in columbines of the genus *Aquilegia*. Columbines have undergone recent and very rapid spe-

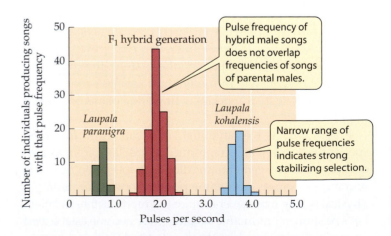

24.10 Songs of Male Crickets are Genetically Determined Hybrid males produce songs with intermediate pulse frequencies.

ciation and, at the same time, have evolved long floral nectar spurs—tubular outgrowths of petals that produce nectar at their tips. Animals pollinate these flowers while probing the spurs to collect nectar. The length of the spurs and the orientation of the flowers influence how efficiently pollinators can extract nectar. Two species, *Aquilegia formosa* and *Aquilegia pubescens*, that grow in the mountains of California can produce fertile hybrids. *A. formosa* has pendant (hanging) flowers and short spurs (Figure 24.11*a*); it is pollinated by hummingbirds. *A. pubescens* has upright flowers and long spurs (Figure 24.11*b*); it is pollinated by hawkmoths.

Investigators tested discrimination among these flowers by hawkmoths by turning *A. formosa* flowers so that they were upright. Hawkmoths still visited mostly *A. pubescens* flowers (Figure 24.11*c*), probably because the flowers of the two species differ strongly in the color of light they reflect. Genetic analyses of hybrids between the two species show that the color differences are caused by a small number of genes. Thus, although these two species are interfertile, hybrids rarely form in nature because the two species attract different pollinators.

Postzygotic barriers operate after fertilization

If individuals of two different populations overcome prezygotic reproductive barriers and interbreed, **postzygotic reproductive barriers** may still prevent gene exchange. Genetic differences that accumulated while the populations were isolated from each other may reduce the survival and reproduction of the hybrid offspring in any of several ways:

▶ *Hybrid zygote abnormality.* Hybrid zygotes may fail to mature normally, either dying during development or developing such severe abnormalities that they cannot mate as adults.

▶ *Hybrid infertility.* Hybrids may mature normally, but be infertile when they attempt to reproduce. For example, the offspring of matings between horses and donkeys—mules—are strong, but sterile; they produce no descendants.

▶ *Low hybrid viability.* Hybrid offspring may simply survive less well than offspring resulting from matings within populations.

If hybrid offspring survive poorly, more effective prezygotic barriers may evolve, because individuals that mate with individuals of the other population will leave fewer surviving offspring than individuals that mate only within their own population. This strengthening of prezygotic barriers is known as **reinforcement**. Reinforcement is difficult to detect experimentally, but it can be detected by using the comparative method (see Chapter 1). If reinforcement is occurring, then sympatric pairs of species should evolve prezygotic repro-

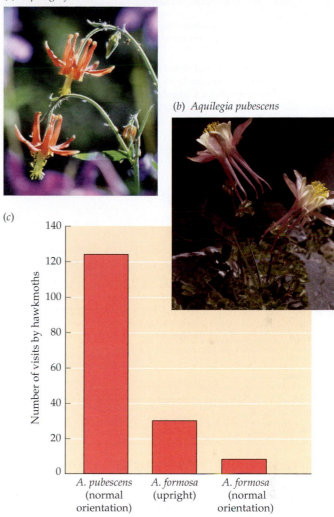

(a) Aquilegia formosa

(b) Aquilegia pubescens

(c)

24.11 Hawkmoths Favor Flowers of One Columbine Species Flowers of *Aquilegia formosa* are normally pendant (*a*), while those of *A. pubescens* are normally upright (*b*). The plants can interbreed, but the hawkmoths that pollinate *A. pubescens* distinguish between flowers of the two species, even when *A. formosa* flowers are experimentally modified to be upright (*c*).

ductive barriers more rapidly than allopatric pairs of species do. In a study of related sympatric and allopatric species of *Drosophila*, this was shown to be the case (Figure 24.12).

We can observe speciation in progress

Since speciation is a gradual process, we can find many examples of populations at different stages on the way to complete reproductive isolation. A good example of speciation in progress is found in a picture-winged fruit fly (*Rhagoletis pomenella*) in New York State. Until the mid-1800s, these fruit flies courted, mated, and deposited their eggs only on hawthorn fruits. The larvae recognized the odor of hawthorn

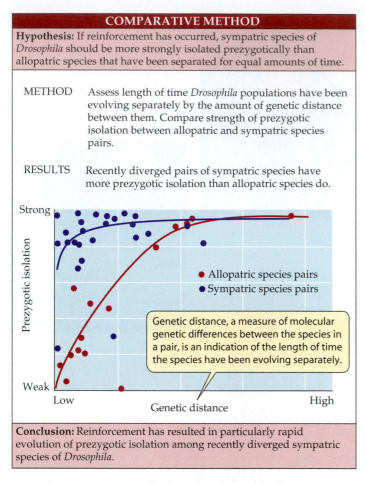

COMPARATIVE METHOD

Hypothesis: If reinforcement has occurred, sympatric species of *Drosophila* should be more strongly isolated prezygotically than allopatric species that have been separated for equal amounts of time.

METHOD   Assess length of time *Drosophila* populations have been evolving separately by the amount of genetic distance between them. Compare strength of prezygotic isolation between allopatric and sympatric species pairs.

RESULTS   Recently diverged pairs of sympatric species have more prezygotic isolation than allopatric species do.

● Allopatric species pairs
● Sympatric species pairs

Genetic distance, a measure of molecular genetic differences between the species in a pair, is an indication of the length of time the species have been evolving separately.

Conclusion: Reinforcement has resulted in particularly rapid evolution of prezygotic isolation among recently diverged sympatric species of *Drosophila*.

24.12 Prezygotic Barriers Can Evolve Rapidly   Reinforcement may occur when individuals who mate outside their own population leave few or no viable offspring. In such a situation, prezygotic isolating mechanisms can evolve extremely rapidly.

as they fed on the fruits, and when they emerged from their pupae, they used this cue to locate other hawthorn plants on which to mate and lay their eggs.

About 150 years ago, large commercial apple orchards were planted in the Hudson River valley. Apple trees are closely related to hawthorns, and a few female *Rhagoletis* laid their eggs on apples, perhaps by mistake. Their larvae did not grow as well as the larvae on hawthorn fruits, but many did survive. These larvae recognized the odor of apples, so when they emerged as adults, they sought out apple trees, where they mated with other flies reared on apples.

Today there are two types of *Rhagoletis* in the Hudson River valley that may be on the way to becoming distinct species. One feeds primarily on hawthorn fruits, the other on apples. The two incipient species are partly reproductively isolated because they mate primarily with individuals raised on the same fruit and because they emerge from their pupae at different times of the year. In addition, the apple-feeding

flies have evolved so that they now grow more rapidly on apples than they originally did.

Reproductive isolation does not develop at the same rate in all diverging populations. On the one hand, in some groups, such as Darwin's finches, there has been conspicuous morphological evolution, but many of the 14 species still interbreed and produce fertile hybrid offspring. On the other hand, reproductive isolation may develop rapidly, as it has between diverging sympatric species of *Drosophila*. Generally, reproductive isolation evolves more rapidly in species that have rapid reproductive rates (for example, most plants, fruit flies, and sea urchins) than in species that have slower reproductive rates (for example, birds and mammals).

Many species in nature form hybrids in areas where their ranges overlap, and they may continue to do so for many years. Let's examine what happens in these situations.

Hybrid Zones: Incomplete Reproductive Isolation

If contact is reestablished between formerly isolated populations before complete reproductive isolation has developed, members of the two populations may interbreed. Three outcomes of such interbreeding are possible:

▶ If hybrid offspring are as successful as those resulting from matings within each population, hybrids may spread through both populations and reproduce with other individuals. The gene pools would then combine, and no new species would result from the period of isolation.

▶ If hybrid offspring are less successful, complete reproductive isolation may evolve as reinforcement strengthens prezygotic reproductive barriers.

▶ Even if hybrid offspring are at some disadvantage, a narrow **hybrid zone** may persist if, for one or more reasons, reinforcement does not happen.

Hybrid zones are excellent natural laboratories for the study of speciation. When a hybrid zone first forms, most hybrids are offspring of crosses between purebred individuals of the two species. However, subsequent generations include a variety of individuals with different proportions of their genes derived from the original two populations. Thus, hybrid zones contain recombinant individuals resulting from many generations of hybridization. Detailed genetic studies can tell us much about why hybrid zones may be narrow and stable for long periods of time.

European toads of the genus *Bombina* have been the subject of such studies. The fire-bellied toad (*B. bombina*) lives in eastern Europe. The closely related yellow-bellied toad (*B. variegata*) lives in western and southern Europe. The ranges of the two species meet in a narrow zone stretching 4,800 km

(a) *Bombina bombina*   (b) *Bombina variegata*   (c)

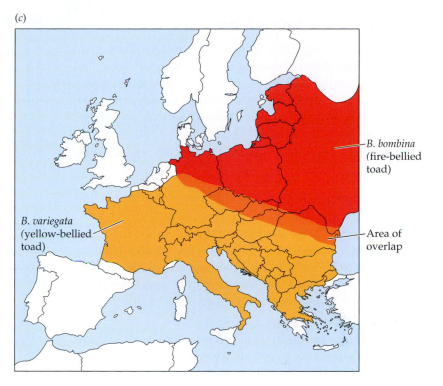

B. bombina
(fire-bellied
toad)

B. variegata
(yellow-bellied
toad)

Area of
overlap

24.13 Hybrid Zones May Be Long and Narrow   The narrow zone in Europe where fire-bellied toads (a) meet and hybridize with yellow-bellied toads (b) stretches across Europe (c). This hybrid zone has been stable for hundreds of years, but has never expanded because hybrid toads are much less fit than individuals of the parental species.

from eastern Germany to the Black Sea (Figure 24.13). Hybrids between the two species suffer from a range of defects, many of which are lethal. Those that survive often have skeletal abnormalities, such as misshapen mouths, ribs that are fused to vertebrae, and a reduced number of vertebrae.

By following the fates of thousands of toads from the hybrid zone, investigators have found that a hybrid toad is half as fit as a purebred individual. The hybrid zone is narrow because there is strong selection against hybrids, and because adult toads do not move over long distances. It has persisted for hundreds of years because individuals that move into it have not previously encountered individuals of the other species, so there has been no opportunity for reinforcement to occur.

If two species hybridize, we know that they must be similar genetically, but the absence of interbreeding tells us nothing about how dissimilar two species are. Not until modern molecular genetic techniques were developed could biologists measure genetic differences among species. These techniques show that the genetic differences that separate species are primarily differences among genes involved with reproductive isolation. The extensive data gathered on *Drosophila* indicate that fewer than ten, and often fewer than five, genes are responsible for reproductive isolation. Individuals of different species of Hawaiian *Drosophila* share nearly all of their mitochondrial DNA alleles. All of the hundreds of species of *Drosophila* that have evolved in the Hawaiian Islands during the past 32 million years, even those that have diverged morphologically, are relatively similar genetically (Figure 24.14).

Drosophila silvestris

Drosophila conspicua

Drosophila balioptera

24.14 Morphologically Different, Genetically Similar   Although these fruit flies—a small sample of the hundreds of species found only on the Hawaiian Islands—vary greatly in appearance, they are genetically similar.

Variation in Speciation Rates

Some lineages of organisms have many species; others have only a few. Hundreds of species of *Drosophila* evolved in the Hawaiian Islands, but there is only one species of horseshoe crab, even though its lineage has survived for more than 300 million years. Why do rates of speciation vary so widely among lineages?

A number of factors are known to influence speciation rates:

▶ *Species richness.* The larger the number of species in a lineage, the larger the number of opportunities for new species to form. For speciation by polyploidy, the more species in a lineage, the more species are available to hybridize with one another. For allopatric speciation, the larger the number of species living in an area, the larger the number of species whose ranges will be bisected by a given physical barrier.

▶ *Dispersal rates.* Individuals of species with poor dispersal abilities are unlikely to establish new populations by dispersing across barriers. Even narrow barriers are effective in dividing species whose members are highly sedentary.

▶ *Ecological specialization.* Populations of species restricted to habitat types that are patchy in distribution are more likely to diverge than are populations that occupy relatively continuous habitats.

▶ *Population bottlenecks.* The changes in gene pools that often occur when a population passes through a bottleneck may result in new adaptations.

▶ *Type of pollination.* Speciation rates in plants are correlated with pollination mode. Animal-pollinated plant families have, on average, 2.4 times as many species as closely related families pollinated by wind. In addition, the switch from animal to wind pollination is strongly associated with a reduction in the rate of speciation in a lineage.

▶ *Sexual selection.* Animals with complex behavior are likely to form new species at a high rate because they make sophisticated discriminations among potential mating partners. They distinguish members of their own species from members of other species, and they make subtle discriminations among members of their own species on the basis of size, shape, appearance, and behavior (see Figures 23.16 and 23.17). Such discriminations can greatly influence which individuals are most successful in producing offspring and may lead to rapid reinforcement of reproductive isolation between species.

▶ *Environmental changes.* Oscillations of climates may fragment populations of species that live in formerly continuous habitats.

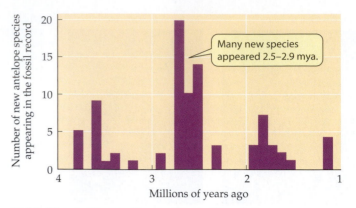

24.15 Climate Change Drove a Burst of Speciation among Antelopes   The excellent fossil record of African antelopes reveals that there was a sudden burst of speciation between 2.5 and 2.9 million years ago. At that time, the climate of Africa shifted from being consistently warm and wet to oscillating between warm and wet and cool and dry.

African antelopes provide a striking example of the influence of climate change on speciation rates. These animals experienced a burst of speciation and extinction between 2.5 and 2.9 mya. During that period, the number of known antelope species doubled, and 90 percent of all known species either first appeared or became extinct (Figure 24.15). This burst coincided with a shift in Africa from a warm, wet climate to one that oscillated between warm and wet and cooler but drier conditions. During this period, grassland and savanna environments increased and decreased, repeatedly coalescing and separating over much of Africa as the climate oscillated. The burst of speciation among antelopes resulted in many new species adapted to these environments.

Evolutionary Radiations

The fossil record reveals that, at certain times in certain lineages, speciation rates have been much higher than extinction rates. The result is the proliferation of a large number of daughter species, as happened with finches in the Galápagos archipelago (see Figure 24.6). Such an event is called an **evolutionary radiation**. What conditions cause speciation rates to be much higher than extinction rates?

Evolutionary radiations are likely when a population colonizes a new environment that contains relatively few species. As we saw in Chapter 22, evolutionary radiations occurred in many lineages on continents following mass extinctions. Evolutionary radiations occur on islands because islands lack many plant and animal groups found on the mainland. The ecological opportunities that exist on islands may stimulate rapid evolutionary changes when a new species does reach them. Water barriers also restrict gene flow among the islands in an archipelago, so populations on

Madia sativa (tarweed)

Argyroxiphium sandwicense

24.16 Rapid Evolution among Hawaiian Silverswords   The Hawaiian silverswords, three closely related genera of the sunflower family, are believed to have descended from a single common ancestor, similar to the tarweed (*Madia sativa*), that colonized Hawaii from the Pacific coast of North America. The four plants shown here are more closely related than they appear to be based on their morphology.

Dubautia menziesii

Wilkesia hobdyi

different islands may evolve adaptations to their local environments. Together, these two factors make it likely that speciation rates of newly colonizing lineages on island archipelagoes will exceed extinction rates.

Remarkable evolutionary radiations have occurred in the Hawaiian Islands, the most isolated islands in the world. The Hawaiian Islands lie 4,000 km from the nearest major land mass and 1,600 km from the nearest group of islands. The native biota of the Hawaiian Islands includes 1,000 species of flowering plants, 10,000 species of insects, 1,000 land snails, and more than 100 bird species. However, there were no amphibians, no terrestrial reptiles, and only one native mammal—a bat—on the islands until humans introduced additional species. The 10,000 known native species of insects on Hawaii are believed to have evolved from only about 400 immigrant species; only 7 immigrant species are believed to account for all the native Hawaiian land birds.

More than 90 percent of all plant species on the Hawaiian Islands are **endemic**—that is, they are found nowhere else. Several groups of flowering plants have more diverse forms and life histories on the islands, and live in a wider variety of habitats, than do their close relatives on the mainland. An outstanding example is the group of Hawaiian sunflowers called silverswords (the genera *Argyroxiphium*, *Dubautia*, and *Wilkesia*). Chloroplast DNA sequences show that these species share a relatively recent common ancestor with a species of tarweed from the Pacific coast of North America (Figure 24.16). Whereas all mainland tarweeds are small, upright, herbs (that is, nonwoody plants), the silverswords include prostrate and upright herbs, shrubs, trees, and vines. Silverword species occupy nearly all the habitats of the Hawaiian islands, from sea level to above timberline in the mountains. Despite their extraordinary morphological diversification, however, the silverswords have differentiated very little in their chloroplast genes.

The island silverswords are more diverse in size and shape than the mainland tarweeds because the original colonizers arrived on islands that had very few plant species. In particular, there were few trees and shrubs, because such large-seeded plants rarely disperse to oceanic islands. In fact, many island trees and shrubs have evolved from nonwoody ancestors. On the mainland, however, tarweeds live in ecological communities that contain tree and shrub lineages older than their own—that is, where opportunities to exploit the "tree" way of life have already been preempted.

The processes we have discussed in this chapter, operating over billions of years, have produced a world in which life is organized into millions of species, each adapted to live in a particular environment and to use environmental resources in a particular way. How these millions of species are distributed over the surface of Earth and organized into ecological communities will be a major focus of Part Eight of this book.

Chapter Summary

What Are Species?

▶ Species are independent evolutionary units. A commonly accepted definition of species is "groups of actually or potentially interbreeding natural populations which are reproductively isolated from other such groups."

▶ Because speciation is often a gradual process, it may be difficult to recognize boundaries between species. **Review Figure 24.3**

How Do New Species Arise?

▶ Not all evolutionary changes result in new species.

▶ Allopatric (geographic) speciation is the most important mode of speciation among animals and is common in other groups of organisms. **Review Figures 24.4, 24.5, 24.6.**
See Web/CD Tutorial 24.2

▶ Sympatric speciation may occur rapidly by polyploidy because polyploid offspring are sterile in crosses with members of the parent species. Polyploidy is a major factor in plant speciation but is rare among animals. **Review Figures 24.7, 24.8**
See Web/CD Tutorial 24.1

Completing Speciation: Reproductive Isolating Mechanisms

▶ Once two populations have been separated, reproductive isolating mechanisms may prevent the exchange of genes between them.

▶ Prezygotic reproductive barriers operate before fertilization. Some prezygotic barriers affect mate choice; others work by influencing pollinator behavior. **Review Figure 24.10, 24.11**

▶ Postzygotic reproductive barriers operate after fertilization by reducing the survival or fertility of hybrid offspring.

▶ If hybrid offspring survive poorly, more effective prezygotic reproductive barriers may evolve. This process is known as reinforcement. **Review Figure 24.12**

Hybrid Zones: Incomplete Reproductive Isolation

▶ Hybrid zones may develop if barriers to gene exchange fail to develop while diverging species are isolated from each other. **Review Figure 24.13**

▶ Species may differ from one another in very few genes.

Variation in Speciation Rates

▶ Rates of speciation differ greatly among lineages. Speciation rates are influenced by the number of species in a lineage, their dispersal rates, ecological specialization, experience of population bottlenecks, pollinators, and behavior, as well as by climatic changes.

Evolutionary Radiations

▶ Evolutionary radiations occur when speciation rates exceed extinction rates.

▶ High speciation rates often coincide with low extinction rates when species invade islands or other environments that contain few other species.

▶ As a result of speciation, Earth is populated with millions of species, each adapted to live in a particular environment and to use resources in a particular way.
See Web/CD Activity 24.1 for a concept review of this chapter.

Self-Quiz

1. A species is a group of
   a. actually interbreeding natural populations that are reproductively isolated from other such groups.
   b. potentially interbreeding natural populations that are reproductively isolated from other such groups.
   c. actually or potentially interbreeding natural populations that are reproductively isolated from other such groups.
   d. actually or potentially interbreeding natural populations that are reproductively connected to other such groups.
   e. actually interbreeding natural populations that are reproductively connected to other such groups.

2. Allopatric speciation may happen when
   a. continents drift apart and separate previously connected lineages.
   b. a mountain range separates formerly connected populations.
   c. different environments on two sides of a barrier cause populations to diverge.
   d. the range of a species is separated by loss of intermediate habitat.
   e. all of the above

3. Finches speciated in the Galápagos Islands because
   a. the Galápagos Islands are not far from the mainland.
   b. the Galápagos Islands are arid.
   c. the Galápagos Islands are small.
   d. the islands of the Galápagos Archipelago are sufficiently isolated from one another that there is little migration among them.
   e. the islands of the Galápagos Archipelago are close enough to one another that there is considerable migration among them.

4. Which of the following is *not* a potential prezygotic reproductive barrier?
   a. Temporal segregation of breeding seasons
   b. Differences in chemicals that attract mates
   c. Hybrid infertility
   d. Spatial segregation of mating sites
   e. Sperm cannot survive in female reproductive tracts

5. A common means of sympatric speciation is
   a. polyploidy.
   b. hybrid infertility.
   c. temporal segregation of breeding seasons.
   d. spatial segregation of mating sites.
   e. imposition of a geographic barrier.

6. Sympatric species are often similar in appearance because
   a. appearances are often of little evolutionary significance.
   b. the genetic changes accompanying speciation are often small.
   c. the genetic changes accompanying speciation are usually large.
   d. speciation usually requires major reorganization of the genome.
   e. the traits that differ among species are not the same as the traits that differ among individuals within species.

7. Narrow hybrid zones may persist for long times because
   a. hybrids are always at a disadvantage.
   b. hybrids have an advantage only in narrow zones.
   c. hybrid individuals never move far from their birthplaces.
   d. individuals that move into the zone have not previously encountered individuals of the other species, so reinforcement of isolating mechanisms has not occurred.
   e. Narrow hybrid zones are artifacts because biologists generally restrict their studies to contact zones between species.

8. Which statement about speciation is *not* true?
   a. It always takes thousands of years.
   b. It often takes thousands of years, but may happen within a single generation.
   c. Among animals, it usually requires a physical barrier.
   d. Among plants, it often happens as a result of polyploidy.
   e. It has produced the millions of species living today.

9. Speciation is often rapid within lineages in which species have complex behavior because
   a. individuals of such species make fine discriminations among potential mating partners.
   b. such species have short generation times.
   c. such species have high reproductive rates.
   d. such species have complex relationships with their environments.
   e. none of the above

10. Evolutionary radiations
    a. often happen on continents, but rarely on island archipelagos.
    b. characterize birds and plants, but not other taxonomic groups.
    c. have happened on continents as well as on islands.
    d. require major reorganizations of the genome.
    e. never happen in species-poor environments.

11. Speciation is an important component of evolution because it
    a. generates the variation upon which natural selection acts.
    b. generates the variation upon which genetic drift and mutations act.
    c. enabled Charles Darwin to perceive the mechanisms of evolution.
    d. generates the high extinction rates that drive evolutionary change.
    e. has resulted in a world with millions of species, each adapted for a particular way of life.

For Discussion

1. The snow goose of North America has two distinct color forms: blue and white. Matings between the two color forms are common. However, blue individuals pair with blue individuals and white individuals pair with white individuals much more frequently than would be expected by chance. Suppose that 75 percent of all mated pairs consisted of two individuals of the same color. What would you conclude about speciation processes in these geese? If 95 percent of pairs were the same color? If 100 percent of pairs were the same color?

2. Suppose pairs of snow geese of mixed colors were found only in a narrow zone within the broad Arctic breeding range of the geese. Would your answer to Question 1 remain the same? Would your answer change if mixed-color pairs were widely distributed across the breeding range of the geese?

3. Although many butterfly species are divided into local populations among which there is little gene flow, these species often show relatively little morphological variation among populations. Describe the studies you would conduct to determine what maintains this morphological similarity.

4. Evolutionary radiations are common and easily studied on oceanic islands, but in what types of *mainland* situations would you expect to find major evolutionary radiations? Why?

5. Fruit flies of the genus *Drosophila* are distributed worldwide, but most of the species in the genus are found on the Hawaiian Islands. What might account for this distribution pattern?

6. Evolutionary radiations take place when speciation rates exceed extinction rates. What factors can cause extinction rates to exceed speciation rates in a lineage? Name some lineages in which human activities are increasing extinction rates without increasing speciation rates.

25 Reconstructing and Using Phylogenies

Schistosomiasis is a blood infection caused by a parasitic trematode flatworm, *Schistosoma*. More than 200 million people in South America, Africa, China, Japan, and Southeast Asia have this disease. During part of its life cycle, *Schistosoma* inhabits a freshwater snail; people become infected when they come into contact with water where infected snails live. Larval *Schistosoma* swim from a snail through the water and penetrate human skin. The flatworm matures and settles in the abdominal blood vessels. The disease is progressively debilitating, causing a slow death.

Until recently, only one trematode species, *Schistosoma japonicum*, was known to infect humans, and it was believed to be transmitted by a single species of snail in the genus *Oncomelania*. Then, in the 1970s, researchers discovered a different snail species that also transmitted *Schistosoma* to humans. This discovery stimulated anatomical, genetic, and geographic research on the trematode flatworms and snails of Asia.

Investigators found that *S. japonicum* was actually a cluster of at least six different species. They also discovered that the snails that host the various *Schistosoma* parasites are closely related to one another. Of the thirteen species of *Oncomelania* in Southeast Asia, only three can host *Schistosoma*. The other species have a genetic trait that allows them to resist infection by the parasite.

This information on the evolutionary relationships among snails is of great value in efforts to combat schistosomiasis. Scientists can now quickly determine whether or not a snail is likely to be a host for *Schistosoma*, and control efforts can be directed toward only those snails that transmit *Schistosoma*.

How did investigators infer the evolutionary relationships among the *Oncomelania* snails that are hosts of *Schistosoma*? **Systematics**, the scientific study of the diversity of organisms, provides answers to such questions. In this chapter, we will describe the methods systematists use to infer evolutionary relationships among organisms, and we will show how those evolutionary relationships are incorporated into classification systems. **Taxonomy**, which is a subdivision of systematics, is the theory and practice of classifying organisms.

Asian Snails Can Transmit Schistosomiasis Workers in the rice paddies of tropical Asia are at extreme risk of contracting schistosomiasis (known in some parts of the world as bilharzia). The disease is transmitted to humans via freshwater snails that thrive in the standing water of the paddies.

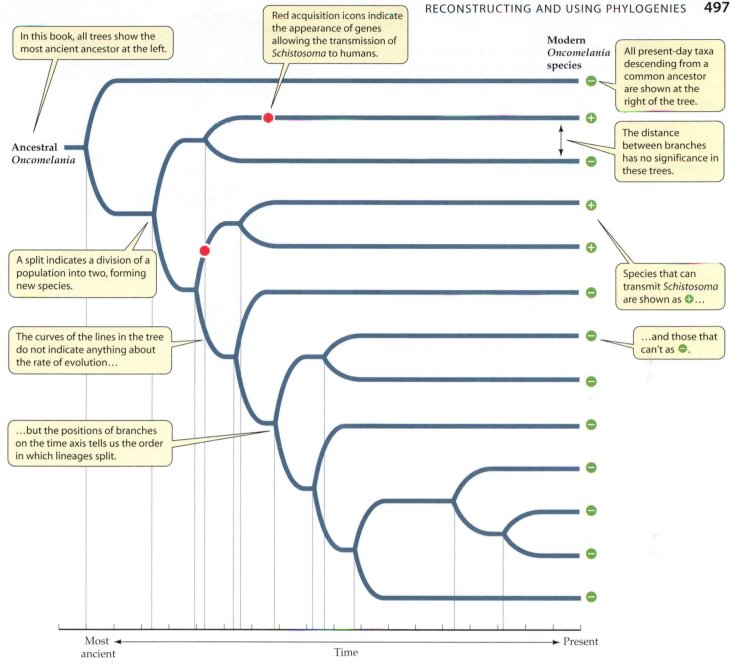

In this book, all trees show the most ancient ancestor at the left.

Red acquisition icons indicate the appearance of genes allowing the transmission of *Schistosoma* to humans.

Modern *Oncomelania* species

All present-day taxa descending from a common ancestor are shown at the right of the tree.

Ancestral *Oncomelania*

The distance between branches has no significance in these trees.

A split indicates a division of a population into two, forming new species.

The curves of the lines in the tree do not indicate anything about the rate of evolution…

Species that can transmit *Schistosoma* are shown as ⊕…

…and those that can't as ⊖.

…but the positions of branches on the time axis tells us the order in which lineages split.

Most ancient ← Time → Present

25.1 How to Read a Phylogenetic Tree A phylogenetic tree displays the order in which lineages split. This example shows the phylogeny of *Oncomelania* snails, the intermediate hosts of the human parasite *Schistosoma*. Acquisition of the ability to transmit *Schistosoma* occurred twice during the evolution of this lineage.

Phylogenetic Trees

A **phylogeny** is a hypothesis proposed by a systematist that describes the history of descent of a group of organisms from their common ancestor. A **phylogenetic tree** is a way of portraying that history. In it, a lineage is represented as a branching "tree," in which each node, or split, represents a speciation event. Thus, the tree shows the order in which lineages are hypothesized to have split. To reconstruct a phylogenetic tree, systematists analyze evolutionary changes in the traits of organisms. They are guided by Darwin's fundamental idea of descent with modification, which states that all species are descended from a common ancestor.

A phylogenetic tree may portray the evolution of all life forms; of a major evolutionary lineage such as the insects; or of a small group of organisms, such as the snail genus *Oncomelania* (Figure 25.1). The phylogenetic trees in this book depict time flowing from left (earliest) to right (most recent); it is equally common practice to draw trees with the earliest times at the bottom. Trees show the relative timing of separations between lineages of organisms, but, unless combined with other data, we do not date those separations. In the phylogenetic trees in this book, position on the horizontal axis has meaning, but vertical distance between the branches does not. Vertical distances are adjusted for legibility and clarity of presentation; they do not correlate with the degree of similarity or difference between groups.

Homologous traits are inherited from a common ancestor

The process of descent with modification means that species that share a recent common ancestor are likely to be very similar. In other words, they should share many traits, called **ancestral traits**, which they inherited from their common ancestor. Traits inherited from an ancestor in the distant past are likely to be shared by a large number of species. Traits that first appeared in a more recent ancestor should be shared by fewer species. But in all cases, the sharing of traits by a group of species indicates that they are likely to be descendants of a common ancestor.

Any features shared by two or more species that have been inherited from a common ancestor are said to be **homologous**. These features may be any heritable traits, including anatomical structures, behavior patterns, and DNA sequences. Traits that are shared by most or all of the organisms in a lineage of interest are likely to have been inherited relatively unchanged from an ancestor that lived very long ago. For example, all living vertebrates have a vertebral column, all known fossil vertebrates had a vertebral column, and all vertebrates are descended from the same common ancestor. Therefore, the vertebral column is judged to be homologous in all vertebrates.

A trait that differs from its ancestral form is called a **derived trait**. To determine how traits have changed during evolution, systematists must infer the state of the trait in an ancestor and then determine how it has been modified in the descendants. Doing so is not easy, because real evolutionary patterns are complex. Two processes generate difficulties: convergent evolution and evolutionary reversals.

▶ Independently evolved traits subjected to similar selective pressures may become superficially similar as a result of **convergent evolution**. For example, although the bones of the wings of bats and birds are homologous, having been inherited from a common ancestor, the wings of bats and birds are not homologous because they evolved independently from the forelimbs of different nonflying ancestors (Figure 25.2).

▶ A character may revert from a derived state back to an ancestral state—an **evolutionary reversal**. For example, most frogs lack teeth in the lower jaw, but the ancestor of frogs did have such teeth. One frog genus, *Amphignathodon*, has re-evolved teeth in the lower jaw.

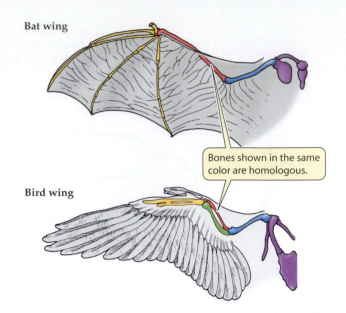

Bat wing

Bones shown in the same color are homologous.

Bird wing

25.2 The Bones Are Homologous, but the Wings Are Not The supporting bone structures of both bat wings and bird wings are derived from a common four-limbed ancestor and are thus homologous. However, the wings themselves—an adaptation for flight—evolved independently in the two groups.

Convergent evolution and evolutionary reversals generate traits that are similar for some reason other than inheritance from a common ancestor. Such traits are called **homoplastic traits** or **homoplasies**.

Depending on the group of interest, a particular trait may be ancestral or derived. For example, rats and mice (both rodents), but not dogs or other mammals, have long, continuously growing incisor teeth. Continuously growing incisors evidently developed in the common ancestor of rats and mice after their lineage separated from the one leading to dogs and other mammals, because no other mammals have that kind of incisors. Thus, if we were reconstructing a phylogeny of a group of rodents, continuously growing incisors would be an ancestral trait because all rodents have them. However, if we were reconstructing a phylogeny of all mammals, continuously growing incisors would be a derived trait unique to the rodents. The distinction between ancestral and derived traits is important in reconstructing phylogenies.

Identifying ancestral traits is sometimes difficult

Distinguishing derived traits from ancestral traits may be difficult because traits often become so dissimilar that ancestral states are unrecognizable. The leaves of plants, for example, have diverged to form many different structures. Several lines of evidence, especially details of their structure and development, indicate that protective spines, tendrils, and brightly colored structures that attract pollinators (Figure 25.3) are all modified leaves; these structures are *homologs* of one another even though they do not resemble one another closely.

The spines of the barrel cactus and the bracts of *Heliconia* are both modified leaves.

The leaves of the pitcher plant curve to hold water.

Cheiridopsis tuberculata   *Heliconia* sp.   *Sarracenia purpurea*

25.3 Homologous Structures Derived from Leaves The leaves of plants have diverged during their evolution to form many different structures, some of which bear very little resemblance to one another.

One method of distinguishing ancestral traits from derived traits is to assume that an ancestral trait should be found not only among the species of the **ingroup** (the lineage of interest), but also in outgroups. An **outgroup** is a lineage that is closely related to the ingroup, but which branches off from the ingroup before its base on the evolutionary tree. Traits found only within the ingroup, on the other hand, are likely to be derived traits.

Steps in Reconstructing Phylogenies

The first step in reconstructing a phylogeny is to select the group of organisms whose phylogeny is to be reconstructed—the ingroup—and an appropriate outgroup. The next step is to choose the characters that will be used in the analysis and to identify the possible forms of those characters (states or traits). A trait may be the presence or absence of a character, or one of the states a particular character may have, such as the number of body segments or number of appendages. The next, and usually the most difficult step, is to determine which traits are ancestral and which are derived. Finally, through phylogenetic analysis, systematists must distinguish homologous from homoplastic traits.

Because organisms differ in many ways, systematists use many characters to reconstruct phylogenies. Some of these characters, such as morphology, are readily preserved in fossils; others, such as behavior and molecular structures, rarely survive fossilization processes. Systematists use physiological, behavioral, molecular, and structural characters that can be assessed in both living and fossil organisms. The more characters that are measured, the more

likely it is that the phylogeny will reflect the actual evolutionary pattern.

Morphological and developmental traits are used in reconstructing phylogenies

An important source of information for systematists is **morphology**—that is, the sizes and shapes of body parts. Since living organisms have been studied for centuries, we have a wealth of recorded morphological data, as well as extensive museum and herbarium collections of organisms whose traits can be measured. Technological tools, such as the electron microscope and computer simulations, enable systematists to measure and analyze the structures of organisms at much finer scales, down to the level of molecules, than was formerly possible.

The early developmental stages of many organisms reveal similarities to other organisms, but those similarities may be lost during later development. For example, the larvae of marine creatures called sea squirts have a rod in the back—the *notochord*—that disappears as they develop into adults. All vertebrate animals also have a notochord at some time during their development (Figure 25.4). This shared structure is one of the reasons for believing that sea squirts are more closely related to vertebrates than would be suspected if only adult sea squirts were examined.

Fossils show us where and when organisms lived in the past and give us an idea of what they looked like. Fossils provide important evidence that helps us distinguish ancestral from derived traits. The fossil record also reveals when lineages diverged and began their independent evolutionary histories. However, few or no fossils have been found for some groups whose phylogeny we may wish to determine.

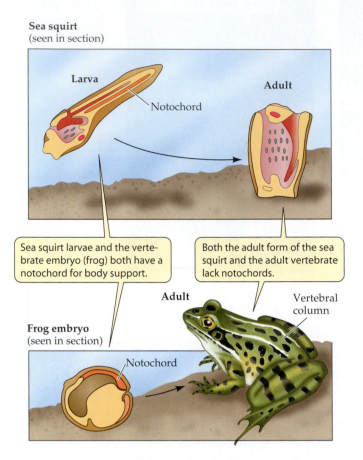

Sea squirt
(seen in section)

Larva

Adult

Notochord

Sea squirt larvae and the vertebrate embryo (frog) both have a notochord for body support.

Both the adult form of the sea squirt and the adult vertebrate lack notochords.

Adult

Vertebral column

Frog embryo
(seen in section)

Notochord

Notochord

25.4 A Larva Reveals Evolutionary Relationships Sea squirt larvae, but not adults, have a well-developed notochord (orange) that reveals their evolutionary relationship to vertebrates, all of which have a notochord at some time during their life cycle. In adult vertebrates, the vertebral column replaces the notochord as the support structure.

Molecular traits are also useful in reconstructing phylogenies

The molecules that make up organisms are also heritable traits that may diverge among lineages over evolutionary time. Molecular evolution will be discussed in detail in Chapter 26. Here we will briefly mention the molecular traits that are most useful for constructing phylogenies: the primary structures of proteins and nucleic acids (DNA and RNA).

PROTEIN PRIMARY STRUCTURE. Relatively precise information about phylogenies can be obtained by comparison of the primary structures of proteins. We can measure genetic differences between two lineages by obtaining homologous proteins from both of them and determining the number of amino acids that have changed since the lineages diverged from a common ancestor.

DNA BASE SEQUENCES. The base sequences of DNA provide excellent evidence of evolutionary relationships among organisms. The cells of eukaryotes have genes in their

mitochondria as well as in their nuclei; plant cells also have genes in their chloroplasts. The chloroplast genome (cpDNA), which is used extensively in phylogenetic studies of plants, has changed little over evolutionarily time. Mitochondrial DNA (mtDNA) has been used extensively for studies of evolutionary relationships among animals (see Figure 25.10).

Relationships among the apes were investigated by sequencing more than 10,000 base pairs making up a segment of DNA that includes a hemoglobin pseudogene (a nonfunctional DNA sequence derived early in primate evolution by duplication of a hemoglobin gene). The outgroups for the analysis were *Ateles*, the spider monkeys of tropical America, and *Macaca*, the Rhesus monkey of Asia. The DNA data strongly indicate that chimpanzees and humans share a more recent ancestor than they do with gorillas, a conclusion supported by other types of molecular data.

Reconstructing a Simple Phylogeny

To show how a phylogeny is constructed, let's consider eight vertebrate animals: the lamprey, perch, pigeon, chimpanzee, salamander, lizard, mouse, and crocodile. We will assume initially that a given derived trait evolved only once during the evolution of these animals, and that no derived traits were lost from any of the descendant groups (Table 25.1). For simplicity, we have selected traits that are either present (+) or absent (−).

As we will see in Chapter 34, a group of jawless fishes called the lampreys is thought to have separated from the lineage leading to the other vertebrates before the jaw arose. Therefore, we will choose the lamprey as the outgroup for our analysis. Derived traits are those that have been acquired by other members of the lineage since they separated from the lamprey.

We begin by noting that the chimpanzee and mouse share two unique traits: mammary glands and fur. Those traits are absent in both the outgroup and the other species of the ingroup. Therefore, we infer that mammary glands and fur are derived traits that evolved in a common ancestor of chimpanzees and mice after that lineage separated from the ones leading to the other vertebrates. In other words, we provisionally assume that mammary glands and fur evolved only once among the animals in our ingroup.

The pigeon has one unique trait: feathers. As before, we provisionally assume that feathers evolved only once, after the lineage leading to birds separated from that leading to the mouse, chimpanzee, and crocodile. By the same reasoning, we assume that keratinous scales evolved only once, after the lineage leading to crocodiles, birds, and lizards separated from the lineage leading to mammals. We assume that claws or nails evolved only once, after the lineage leading to sala-

25.1 Eight Vertebrates Ordered According to Unique Shared Derived Traits

	DERIVED TRAIT[a]							
TAXON	JAWS	LUNGS	CLAWS OR NAILS	GIZZARD	FEATHERS	FUR	MAMMARY GLANDS	KERATINOUS SCALES
Lamprey (outgroup)	–	–	–	–	–	–	–	–
Perch	+	–	–	–	–	–	–	–
Salamander	+	+	–	–	–	–	–	–
Lizard	+	+	+	–	–	–	–	+
Crocodile	+	+	+	+	–	–	–	+
Pigeon	+	+	+	+	+	–	–	+
Mouse	+	+	+	–	–	+	+	–
Chimpanzee	+	+	+	–	–	+	+	–

[a] A plus sign indicates the trait is present, a minus sign that it is absent.

manders separated from the lineage leading to those animals that have claws or nails. We make the same assumption for lungs and jaws, continuing to minimize the number of evolutionary events needed to produce the patterns of shared traits among these eight animals.

Using this information, we can reconstruct a provisional phylogeny (Figure 25.5). We assume that the animals that share unique derived traits have a common ancestor not shared with the animals lacking those traits. We assume, for example, that mice and chimpanzees, the only two animals that share fur and mammary glands, share a more recent common ancestor with each other than they do with birds

and crocodiles. Otherwise, we would need to assume that the ancestors of birds and crocodiles also had fur and mammary glands, but that they subsequently lost them—unnecessary additional assumptions.

Figure 25.5 shows a phylogeny for these eight vertebrates, based on the traits we used and the assumption that each derived trait evolved only once. This particular phylogeny was easy to construct because the animals and traits fulfilled the assumptions that derived traits appeared only once in the lineage and that they were never lost after they appeared. If we had included a snake in the group, our second assumption would have been violated, because the lizard ancestors of

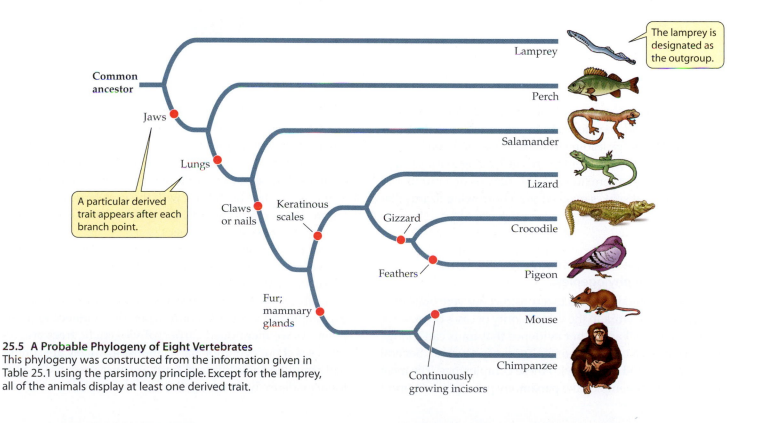

25.5 A Probable Phylogeny of Eight Vertebrates
This phylogeny was constructed from the information given in Table 25.1 using the parsimony principle. Except for the lamprey, all of the animals display at least one derived trait.

Kingdom	Plantae (plants)	
± 275,000 species		
Phylum	Angiospermae (flowering plants)	
± 250,000 species		
Class	Eudicotyledonae (true dicots)	
± 235,000 species		
Order	Rosales (roses and their allies)	
± 18,000 species		
Family	Rosaceae	
± 3,500 species		
Genus	*Rosa*	
± 500 species		
Species	*Rosa gallica*	
French rose		

snakes had limbs that were subsequently lost (along with their claws). We would need to examine additional traits to determine that the lineage leading to snakes separated from the one leading to lizards long after the lineage leading to lizards separated from the others. In fact, the analysis of a number of traits shows that snakes evolved from burrowing lizards that became adapted to a subterranean existence.

Systematists use the parsimony principle when reconstructing phylogenies

The simple method we used to reconstruct our vertebrate phylogeny does not work in the vast majority of cases because we know from fossil and other evidence that traits can change more than once or undergo evolutionary reversal. Several methods are used to deal with these complexities. The most widely used ones employ the **parsimony principle**. In its most

general form, the parsimony principle states that one should prefer the simplest hypothesis that is capable of explaining the observed data. Its application to the reconstruction of phylogenies means minimizing the number of evolutionary changes that need to be assumed over all characters in all groups in the tree (as we did with the vertebrate example in Figure 25.6). In other words, the best hypothesis is one that requires the fewest homoplasies.

Using the parsimony principle is appropriate, not because all evolutionary changes occurred parsimoniously, but because it is generally wiser to adopt the simplest explanation that can account for the observed data. More complicated explanations are accepted only when the evidence requires them. As we mentioned earlier, phylogenetic trees are hypotheses about evolutionary relationships. They are continually modified as additional traits are measured and as new fossil evidence becomes available.

Kingdom	Animalia (animals)	
> 1,500,000 species		
Phylum	Chordata (chordates)	
± 50,000 species		
Class	Aves (birds)	
8,600 species		
Order	Passeriformes (songbirds)	
5,160 species		
Family	Parulidae (wood warblers)	
125 species		
Genus	*Dendroica*	
28 species		
Species	*Dendroica fusca*	
Blackburnian warbler		

Less specific → More specific

25.6 Hierarchy in the Linnaean System The diagram shows how the hierarchical units of the Linnaean classification system are applied to the Blackburnian warbler (*Dendroica fusca*) and the moss rose (*Rosa gallica*).

Another method of reconstructing phylogenies, called the *maximum likelihood method*, is used primarily with molecular data. The computer programs employed in this method are complex. They are designed to deal with the fact that mutations commonly change nucleotide sequences (see Chapter 26).

Whatever method is employed, reconstructing the phylogeny for any group of organisms is difficult. For example, there are 34,459,425 possible phylogenetic trees for a lineage with only 11 species! Computer programs using the parsimony principle employ various search routines that calculate the shortest phylogenetic tree—the one with the fewest homoplasies—for a given data set and then compare other possible phylogenies with the shortest one. If, as is usually the case, several trees are of equal minimum length, they can be merged into a *consensus tree* that retains only those lineage splits that are found in all the shortest trees. In a consensus tree, groups whose relationships differ among the trees form

nodes with more than two branches. These nodes are considered "unresolved" because during a speciation event, a lineage typically splits into only two daughter species.

Biological Classification and Evolutionary Relationships

The biological classification system in use today was developed by the Swedish biologist Carolus Linnaeus and has been used since 1758. Linnaeus's system, referred to as *binomial nomenclature*, allows scientists throughout the world to refer unambiguously to the same organisms by the same names (Figure 25.6).

Linnaeus gave each species two names, one identifying the species itself and the other the genus to which it belongs. A **genus** (plural, genera; adjectival form, generic) is a group of closely related species. In many cases, the name of the taxonomist who first proposed the species name is added at the end. Thus, *Homo sapiens* Linnaeus is the name of the modern human species. *Homo* is the genus to which the species belongs, and *sapiens* identifies the particular species in the genus *Homo*; Linnaeus proposed the species name *Homo sapiens*. You can think of the generic name *Homo* as equivalent to your surname and the specific name *sapiens* as equivalent to your first name. The generic name is always capitalized; the name identifying the species always lowercased. Both names are always italicized, whereas common names of organisms are not. Rather than repeating a generic name when it is used several times in the same discussion, biologists often spell it out only once and abbreviate it to the initial letter thereafter (for example, *D. melanogaster* is the abbreviated form of *Drosophila melanogaster*).

Recognizing and interpreting similarities and differences among organisms is easier if the organisms are classified into groups that are ordered and ranked. Any group of organisms that is treated as a unit in a biological classification system, such as the genus *Oncomelania*, or all snails, is called a **taxon** (plural, taxa). In the Linnaean system, species and genera are further grouped into a hierarchical system of higher taxonomic categories. The taxon above the genus in the Linnaean system is the **family**. The names of animal families end in the suffix "-idae." Thus, Formicidae is the family that contains all ant species, and the family Hominidae contains humans and our recent fossil relatives, as well as chimpanzees and gorillas. Family names are based on the name of a member genus; Formicidae is based on the genus *Formica*, and Hominidae is based on *Homo*. Plant classification follows the same procedures, except that the suffix "-aceae" is used with family names instead of "-idae." Thus, Rosaceae is the family that includes the genus of roses (*Rosa*) and its close relatives.

Families, in turn, are grouped into **orders**, orders into **classes**, and classes into **phyla** (singular, phylum). The phyla of plants, fungi, and animals are grouped into the **kingdoms** Plantae, Fungi, and Animalia.

Biological classification systems and the unique names they provide for organisms are important for several reasons. They improve our ability to infer relationships among organisms. They are also an aid to memory and precise communication. It is impossible to remember the characteristics of many different organisms unless we can group them into categories based on shared characteristics. Classification systems are also useful as predictors. For example, the discovery of

biochemical precursors of the drug cortisone in certain yams of the genus *Dioscorea*, stimulated a successful search for higher concentrations of the drug in other *Dioscorea* species.

Current biological classifications reflect evolutionary relationships

Biological classification systems are designed to express relationships among organisms. The kind of relationship we wish to express influences which features we use to classify organisms. If, for instance, we were interested in a system that would help us decide what plants and animals were desirable as food, we might devise a classification based on tastiness, ease of capture, and the type of edible parts each organism possessed. Early Hindu classifications of plants were designed according to these criteria. Biologists do not use such systems today, but those systems served the needs of the people who developed them.

Most taxonomists today believe that biological classification systems should reflect the evolutionary relationships of organisms, and that taxonomic groups should be **monophyletic**. A monophyletic group (also called a **clade**) contains all the descendants of a particular ancestor and no other organisms. A group containing some members that do not share the same common ancestor is said to be **polyphyletic**. A group that contains some, but not all, of the descendants of a particular ancestor is said to be **paraphyletic** (Figure 25.7). A monophyletic group can be removed from a phylogenetic tree by a single "cut" in the tree, as shown in Figure 25.7.

Taxonomists agree that polyphyletic groups are inappropriate as taxonomic units. The classifications used today still contain many polyphyletic groups because many organisms have not been studied well enough to distinguish between characters that are homologies and those that are homoplasies. However, as soon as they detect such homoplasies,

25.7 Monophyletic, Polyphyletic, and Paraphyletic Taxa
Monophyletic groups are preferred by most taxonomists. Polyphyletic groups are considered inappropriate as taxonomic units, but taxonomists sometimes do use paraphyletic taxa.

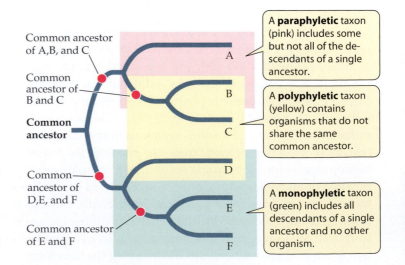

Common ancestor of A,B, and C

Common ancestor of B and C

Common ancestor

Common ancestor of D,E, and F

Common ancestor of E and F

A **paraphyletic** taxon (pink) includes some but not all of the descendants of a single ancestor.

A **polyphyletic** taxon (yellow) contains organisms that do not share the same common ancestor.

A **monophyletic** taxon (green) includes all descendants of a single ancestor and no other organism.

taxonomists change their classifications to eliminate poly-phyletic taxa.

Although most taxonomists prefer strict phylogenetic classifications, some believe that classification systems should also reflect degrees of difference among organisms. According to this view, certain paraphyletic groups that have undergone rapid evolutionary change and diversification and evolved distinctive characters should be retained. Such groups are called **grades**. This perspective can be illustrated using birds, crocodiles, and their relatives.

We have some molecular evidence that birds and crocodilians (a group that includes crocodiles and alligators) share a more recent common ancestor than crocodilians and turtles share with snakes and lizards (Figure 25.8a). Traditionally, crocodilians are grouped with snakes, lizards, and turtles in the class Reptilia. Birds are placed in a separate class, Aves (Figure 25.8b). This classification came about because crocodilians have evolved more slowly than birds since the two lineages separated. Birds rapidly evolved significant and distinctive adaptations for flight, but crocodilian traits have changed little from those of their ancestors.

As a result, crocodilians are more similar in many ancestral features to snakes and lizards than they are to birds. They look like, and are physiologically similar to, very large lizards.

Thus, the traditional class Reptilia is paraphyletic because it does not include all the descendants of its common ancestor; that is, birds are excluded (Figure 25.8b). If only monophyletic taxa were permitted, birds would be grouped with crocodilians in a single taxon separate from snakes and lizards or birds would be included within an expanded class Reptilia. Retaining birds as a separate class (that is, retaining reptiles as a paraphyletic group) emphasizes that birds have evolved unique derived traits since they separated from reptiles, and thus are a distinct grade.

Although the current preference is to change classifications to eliminate paraphyletic groups, some of the most familiar taxonomic categories—gymnosperms and reptiles, for example—are paraphyletic. Because of their familiarity and the extensive literature devoted to them, these categories are likely to remain in use even after their formal taxonomic designations have been changed to better represent their evolutionary relationships.

Phylogenetic Trees Have Many Uses

Phylogenies are usually reconstructed as part of an effort to determine evolutionary relationships among organisms. Information about these relationships is useful to scientists investigating a wide variety of biological questions. Many biological statements are really phylogenetic assertions. Any claim of an association between a trait and a group of organisms is actually a statement about when during the history of the lineage the trait first arose and about the maintenance of the trait since its first appearance. For example, the statement that the cytoskeleton is a trait possessed by all eukaryotes is an assertion that the cytoskeleton is an ancestral trait that has been maintained during the subsequent evolution of all surviving eukaryote lineages. In this section, we will illustrate how phylogenetic trees can be used to determine how many times a particular trait may have arisen during evolution, to assess when lineages split, and to explain how evolutionary radiations came about.

How many times has a trait evolved?

Most flowering plants reproduce by mating with another individual—called *outcrossing*—and have mechanisms to prevent self-fertilization. Individuals of some species, however, are *self-compatible*; that is, they can fertilize themselves with their own pollen. How can we tell how often self-compatibility has evolved in a lineage? We can do so by plot-

(a) **The evolutionary relationships**

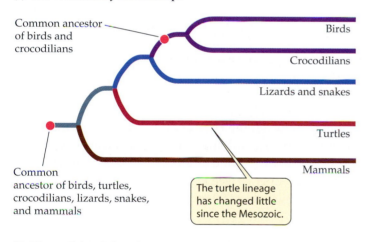

Common ancestor of birds and crocodilians

Common ancestor of birds, turtles, crocodilians, lizards, snakes, and mammals

The turtle lineage has changed little since the Mesozoic.

Birds
Crocodilians
Lizards and snakes
Turtles
Mammals

(b) **The traditional classification**

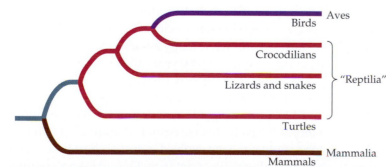

Aves
Birds
Crocodilians
Lizards and snakes
"Reptilia"
Turtles
Mammalia
Mammals

25.8 Phylogeny and Classification (a) Evolutionary relationships among mammals, reptiles, and birds. (b) The traditional classification unites crocodilians and turtles with lizards and snakes in the paraphyletic taxon "Reptilia," which excludes birds.

ting the outcrossing and selfing species on a phylogenetic tree.

The evolution of fertilization mechanisms was examined in *Linanthus* (a genus in the phlox family), a lineage of plants with a diversity of breeding systems and pollination mechanisms. The outcrossing (self-incompatible) species of *Linanthus* have long petals and are pollinated by long-tongued flies. The self-compatible species all have short petals. The investigators reconstructed a phylogeny for 12 species in the genus using a nuclear ribosomal DNA sequence (Figure 25.9). They determined whether each species was self-compatible by artificially pollinating flowers with their own pollen or pollen from other individuals and observing whether viable seeds formed.

Several lines of evidence suggest that self-incompatibility is the ancestral state in *Linanthus*. First, multiple origins of self-incompatibility have not been found in any other flowering plant family. Second, self-incompatibility depends on physiological mechanisms in both the pollen and the stigma (the female organ on which pollen lands) and requires the presence of at least three different alleles. Therefore, a change from self-incompatibility to self-compatibility would be easier than the reverse change. Third, in all self-incompatible species of *Linanthus*, the site of pollen rejection is the stigma, even though sites of pollen rejection vary greatly among other plant families.

Assuming that self-incompatibility is the ancestral state, the reconstructed phylogeny suggests that self-compatibility has evolved three times within this *Linanthus* lineage (Figure 25.9). The change to self-compatibility has been accompanied by the evolution of reduced petal size. Interestingly, the striking similarity of the flowers in the self-compatible groups led to their being classified as members of a single species. The phylogenetic analysis shows them to be members of three distinct lineages.

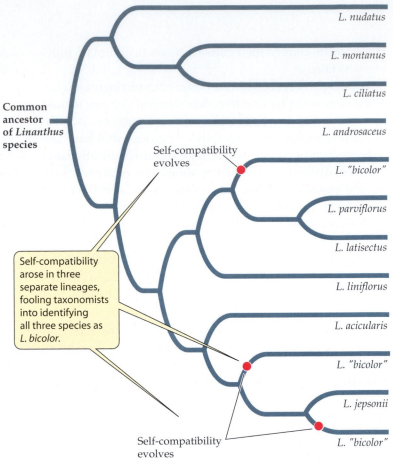

25.9 Phylogeny of a Section of the Phlox Genus *Linanthus* Self-compatibility apparently evolved three times in this lineage. Because the form of the flowers converged in the selfing lineages, taxonomists mistakenly thought that they were all members of a single species.

ferences between the rRNA of the African and the South American species in all three lineages are great enough to be consistent with a split caused by the separation of Africa from South America (see Figure 22.16), which is believed to have happened about 90 million years ago.

When did lineages split?

How phylogenetic analyses can help us determine when lineages split is illustrated by studies of characiform fishes. The approximately 1,400 species of these freshwater fishes, which are found in both South America (1,200 species) and Africa (200 species), vary greatly in size, shape, and diet. Analyses of mitochondrial DNA and anatomy suggest that the characiform fishes are monophyletic; a suggested phylogeny of part of the family is shown in Figure 25.10. Analyses using slowly evolving rRNA genes identified three lineages with closely related species on both sides of the Atlantic Ocean. Since all these species live only in fresh water, dispersal across the Atlantic Ocean is unlikely. The genetic dif-

How recently did Lake Victoria's cichlid fishes radiate?

The spectacular radiation that produced more than 500 species in one lineage of cichlid fishes in Lake Victoria, in eastern Africa, was initially assumed to have occurred over a period of about 750,000 years, the presumed age of the lake basin. However, recently discovered geological data suggests that Lake Victoria dried up completely between 15,600 and 14,700 years ago! Biologists judged that the hundreds of morphologically diverse cichlids (Figure 25.11a) in Lake Victoria could not have evolved in such a short time, so they developed alternative scenarios. One scenario assumed that the lake did not dry up completely. Another postulated that some of the fish species survived in rivers, from which they subsequently recolonized the lake. But no data were avail-

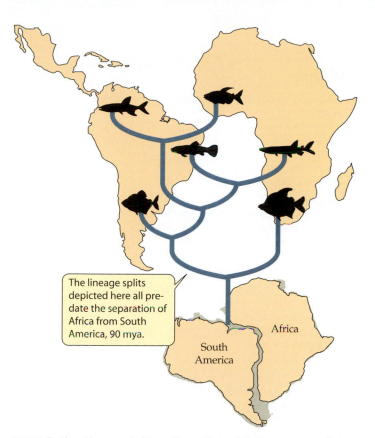

The lineage splits depicted here all predate the separation of Africa from South America, 90 mya.

Africa

South America

25.10 Dating Lineage Splits Characiform fishes are a monophyletic lineage, and species in several groups on opposite sides of the Atlantic Ocean appear similar. However, rRNA analyses indicate the African and South American lineages have been evolving separately for about 90 million years—the time at which the two continents split apart and the Atlantic Ocean became a barrier to these freshwater fishes.

able to test these possibilities. Phylogenetic analyses based on molecular data helped resolve the problem.

Using 300 mtDNA sequences, investigators reconstructed a phylogeny of the cichlid fishes of Lake Victoria and other lakes in the region. Their phylogeny suggests that the ancestors of Lake Victoria cichlids came from the geologically much older Lake Kivu. Today, Lake Kivu is home to only 15 species of cichlids, but the phylogeny suggests that fishes from Lake Kivu colonized Lake Victoria on two different occasions (Figure 25.11b). The molecular data also indicate that some of the cichlid lineages that are found only in Lake Victoria, and which therefore probably evolved there, split at least 100,000 years ago. Thus, the reconstructed phylogeny of these fishes strongly suggests that Lake Victoria did not completely dry up about 15,000 years ago, and that many fish species survived in rivers and in pockets of water that remained in the deepest part of the lake throughout the most recent dry period.

These examples illustrate how a plausible phylogeny for a lineage of organisms enables biologists to answer a variety of questions about the history of that group. They also show the value of molecular data in reconstructing phylogenies. We will consider molecules and how they are used to study evolutionary history in greater detail in the next chapter.

25.11 Origins of the Cichlid Fishes of Lake Victoria (a) The photographs show only two of the hundreds of cichlid species found in Lake Victoria. (b) Phylogenetic analysis suggests that cichlids colonized Lake Victoria by the routes shown on the map.

(a)

Harpagochromis sp.

Ptychromis sp.

(b)

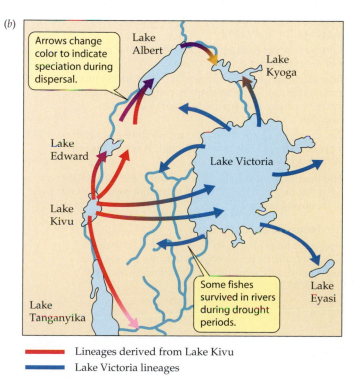

Arrows change color to indicate speciation during dispersal.

Lake Albert

Lake Kyoga

Lake Edward

Lake Victoria

Lake Kivu

Some fishes survived in rivers during drought periods.

Lake Eyasi

Lake Tanganyika

— Lineages derived from Lake Kivu
— Lake Victoria lineages

Chapter Summary

▶ Systematics is the scientific study of the diversity of organisms. Taxonomy, a subdivision of systematics, is the theory and method of classifying organisms.

Phylogenetic Trees

▶ A phylogenetic tree displays the order in which lineages split. **Review Figure 25.1**

▶ Traits inherited from a common ancestor, called ancestral traits,—are said to be homologous.

▶ A derived trait is one that differs from its form in the ancestor of a lineage.

▶ Traits that are similar as a result of convergent evolution or evolutionary reversals are said to be homoplastic. **Review Figure 25.2**

Steps in Reconstructing Phylogenies

▶ Systematists use morphological, physiological, behavioral, and molecular characters to reconstruct phylogenies.

▶ Structures in early developmental stages sometimes show evolutionary relationships that are not evident in adults. **Review Figure 25.4**

▶ Protein primary structures and the base sequences of nucleic acids are also important traits that can be used in reconstructing phylogenies.

Reconstructing a Simple Phylogeny

▶ To assess evolutionary relationships, systematists must distinguish between ancestral and derived traits within a lineage. This task is often difficult because traits can change more than once or undergo evolutionary reversal. **Review Figure 25.5**

▶ Systematists use the parsimony principle to reconstruct phylogenetic trees.
See Web/CD Activity 25.1

Biological Classification and Evolutionary Relationships

▶ Classification systems improve our ability to explain relationships among things, aid our memory, and provide unique, universally used names for organisms.

▶ Biological nomenclature assigns to each organism a unique combination of a generic and a specific name.

▶ In the universally employed Linnaean classification system, species are grouped into higher-level units called genera, families, orders, classes, phyla, and (in some cases) kingdoms. **Review Figure 25.6**

▶ Taxonomists agree that taxa should be monophyletic and that polyphyletic groups should not be recognized. **Review Figure 25.7. See Web/CD Activity 25.2**

▶ Paraphyletic taxa may be retained because of their familiarity and to highlight the fact that members of some lineages evolved unique traits. **Review Figure 25.8**

Phylogenetic Trees Have Many Uses

▶ Phylogenetic trees help biologists to determine how many times evolutionary traits have arisen, explain the geographic ranges of species, and date evolutionary radiations. **Review Figures 25.9, 25.10, 25.11**

Self-Quiz

1. Any group of organisms treated as a unit in a classification system is a
   a. species.
   b. genus.
   c. taxon.
   d. clade.
   e. phylogen.

2. A genus is a
   a. group of closely related species.
   b. group of genera.
   c. group of similar genotypes.
   d. taxonomic unit larger than a family.
   e. taxonomic unit smaller than a species.

3. A trait that is defined as one that differs from its ancestral form is called
   a. an altered trait.
   b. a homoplastic trait.
   c. a parallel trait.
   d. a derived trait.
   e. a homologous trait.

4. Identifying ancestral traits is often difficult because
   a. traits often become so dissimilar that ancient states are unrecognizable.
   b. there may be no fossils of appropriate ancestors.
   c. reversals of traits are common during evolution.
   d. traits often evolve rapidly.
   e. All of the above

5. The parsimony principle is typically used when reconstructing phylogenies because
   a. evolution is nearly always parsimonious.
   b. it is better to provisionally adopt the simplest hypothesis capable of explaining the known facts.
   c. it is easier to handle parsimonious data with computers.
   d. parsimony works well for all kinds of traits, both morphological and molecular.
   e. parsimony was used before computers were available and it continues to be used even though new methods are better.

6. Which of the following is a way of identifying ancestral traits?
   a. Determining which traits are found among fossil ancestors
   b. Using an outgroup
   c. Using a lineage that is closely related to the ingroup
   d. Examining the development of the trait
   e. All of the above

7. Traits that evolve very slowly are most useful for determining relationships at the level of
   a. phyla.
   b. genera.
   c. orders.
   d. families.
   e. species.

8. Homologous traits are
   a. similar in function.
   b. similar in structure.
   c. similar in structure but not in function.
   d. derived from a common ancestor.
   e. derived from different ancestral structures and have dissimilar structures.

9. The genes that are most extensively used to determine evolutionary relationships among plants are
   a. nuclear genes.
   b. chloroplast genes.
   c. mitochondrial genes.
   d. genes in flowers.
   e. genes in roots.

10. Which of the following is *not* a way in which phylogenies are used?
   a. To establish evolutionary relationships
   b. To determine how rapidly traits evolve
   c. To determine historical patterns of movement of organisms
   d. To help identify unknown organisms
   e. To infer evolutionary trends

11. Which of the following is *not* a major role of a classification system?
   a. To aid memory
   b. To improve predictive powers
   c. To help explain relationships among things
   d. To provide relatively stable names for things
   e. To design identification keys

For Discussion

1. Why are taxonomists concerned with identifying lineages that share a single common ancestor?

2. How are fossils used to identify ancestral and derived traits of organisms?

3. Taxonomists use the parsimony principle when reconstructing phylogenetic trees. Given that evolutionary processes are not always parsimonious, why is it used as a guiding principle?

4. A student of the evolution of frogs has proposed a strikingly new classification of frogs based on an analysis of a few mitochondrial genes from about 25 percent of frog species. Should frog taxonomists immediately accept the new classification? Why or why not?

5. Linnaeus developed his system of classification before Darwin proposed his theory of evolution by natural selection, and most classifications of organisms initially were proposed by non-evolutionists. Yet, many of these classifications are still used today, with minor modifications, by most evolutionary taxonomists. Why?

6. Classification systems summarize much information about organisms and enable us to remember the traits of many organisms. From our general knowledge, how many traits can you associate with the following names: conifer, fern, bird, mammal?

26 *Molecular and Genomic Evolution*

There are more species of insects on Earth than of all other animal groups combined. Many of those species transmit diseases to humans. Among the most dangerous are the several genera of mosquitoes that transmit malaria, dengue, lymphatic filariasis, yellow fever, and Japanese encephalitis. Worldwide, malaria is a major killer that causes more than a million deaths among children each year. How can evolutionary biology help us combat this scourge of humankind?

In 1991, a group of scientists met in Tucson, Arizona, to launch a plan to engineer transgenic mosquitoes incapable of carrying the malaria parasite. They would then test possible methods of spreading the new genotypes into wild mosquito populations and develop the basic tools needed for such an effort. In 1999, another group of experts concluded that sequencing the genome of the mosquito *Anopheles gambiae*, the most important vector of malaria in Africa, would help in achieving those goals. A sequencing initiative was launched, and within two years, in October 2002, the complete 278,000-base genome of the mosquito was published. Sequencing has also been completed for *Plasmodium falciparum*, the protist that causes the most deadly form of malaria (see Chapter 28).

Knowing the genomes of *A. gambiae* and *P. falciparum* has provided insights into the genetic makeup of insects and how they have co-evolved with the parasites they transmit to humans. But the genomes of the anopheline mosquitoes that carry malaria, such as *A. gambiae*, differ significantly from those of the culicine mosquitoes, which are the major transmitters of the other diseases mentioned above. Comparative investigations of the evolution of insect genomes, as well as the genomes of the parasites they carry, will be needed in our efforts to combat all of these debilitating diseases.

For most of its history, evolutionary biology depended on the study of obvious morphological features of organisms. During his voyage aboard the *Beagle*, Charles Darwin observed morphological differences among species found in different geographic areas. He later synthesized these observations into descriptions of how species change over time. He

A Deadly Bite The bite of the mosquito *Anopheles gambiae* transmits *Plasmodium falciparum*, a cause of malaria, into the human bloodstream. Malaria is one of the most widespread debilitating (and frequently deadly) human diseases on the planet, affecting over 600 million people.

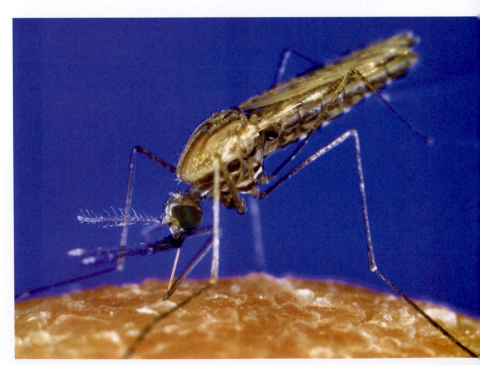

developed his theory of natural selection to explain *why* many of these morphological changes had happened, but he could not determine *how* they had occurred. Understanding of the mechanisms of evolutionary change had to await new discoveries in biochemistry more than a century later.

In this chapter we will see how molecular biologists determine the structures of nucleic acids and proteins and how they use those structures to infer both the patterns and the causes of molecular evolution. With these insights, we will explore how the functions of molecules change, where new genes come from, and how genomes change in size. Finally, we will see how knowledge of the patterns of molecular evolution can help us to solve other biological problems, such as inferring phylogenetic relationships among organisms, determining the phylogenies of genes, and combating diseases.

Genomes and Their Evolution

An organism's **genome** is the full set of genes that it contains. Most of the genes of eukaryotic organisms are found on chromosomes in the nucleus, but genes are also present in plastids (such as chloroplasts) and mitochondria. In organisms that reproduce sexually, both males and females transmit nuclear genes, but mitochondrial and chloroplast genes usually are transmitted only via the cytoplasm of eggs, as we saw in Chapter 10.

In sexually reproducing organisms, meiosis and fertilization shuffle genes in every generation, and individual gametes transmit partly randomized subsets of parental genes to offspring. For this repeated shuffling to be successful, genes must be able to operate in a wide variety of genetic backgrounds. Closely linked genes, however, are likely to remain together, which is especially important for developmental genes. Many genes that govern developmental processes are closely linked on chromosomes and are inherited together.

A gene will not be passed on to successive generations unless the individual with the gene survives and reproduces. Therefore, the capacity to cooperate with different combinations of other genes is likely to increase a gene's probability of transmission. For this reason, it is useful to view the genes of an individual as interacting members of a group, among which there are divisions of labor, but also strong interdependencies.

Students of genomic evolution look at the genome of an organism as an integrated whole, asking questions such as, How do proteins acquire new functions? Why are the genomes of different organisms so variable in size? How has enlargement of genomes been accomplished? We will return to these questions later in this chapter, but to provide some needed background, we will first describe a related field of study: molecular evolution.

The Evolution of Macromolecules

The field of **molecular evolution** investigates the evolution of macromolecules and uses the findings to reconstruct the evolutionary history of genes and the organisms that carry them. The molecules of special interest to molecular evolutionists are nucleotides, nucleic acids, amino acids, and proteins (see Chapter 3).

Nucleic acids evolve by means of nucleotide base substitutions, which in turn can result in changes in the amino acids they encode. Alterations in the structure and function of proteins result from changes in the ordering of the amino acids of which they are composed. Molecular evolutionists characterize the precise structures of these macromolecules: nucleotide sequences in nucleic acids, and primary structure in proteins. They use that information to determine how rapidly these macromolecules have changed and why they have changed.

Phylogenetic information is essential for determining the order of changes in molecular characters because knowing the order of such changes is usually the first step in inferring their causes. Conversely, knowledge of the pattern and rate of change of a given macromolecule is crucial to attempts to reconstruct the evolutionary history of groups of organisms.

Molecular evolution is driven by changes in nucleotide sequences

As we mentioned in Chapter 23, molecular evolution differs from phenotypic evolution in that random genetic drift and mutation exert important influences on the rates and direction of nucleotide changes. A *mutation*, as you will recall from Chapter 12, is any change in the genetic material. Many mutations do not alter the protein encoded by the mutated genes. The reason is that in the "universal genetic code" (see Figure 12.5), most amino acids are specified by more than one codon. Leucine, for example, is specified by six different codons: UUA, UUG, CUU, CUC, CUA, and CUG. A mutation that does not change the amino acid—UUA to UUG, for example—is known as a **synonymous** or **silent mutation** (Figure 26.1a). Synonymous mutations do not affect the functioning of a protein (and hence the organism), and are, therefore, unlikely to be influenced by natural selection.

Because they are unlikely to be influenced by natural selection, synonymous mutations can spread, at rates determined by rates of mutation and genetic drift, so that they completely replace one nucleotide base or longer sequence with another throughout an entire population or species. In this chapter we will call such replacements **substitutions** to distinguish them from a replacement of one nucleotide by another in a single individual (which is also often called a substitution). Molecular evolutionists are primarily interested in population-wide nucleotide substitutions.

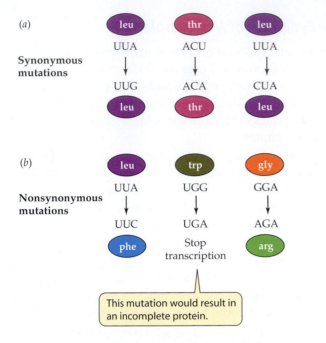

26.1 When One Base Does or Doesn't Make a Difference
(*a*) Synonymous mutations do not change the amino acid specified and do not affect protein function; such mutations are unlikely to be agents of natural selection. (*b*) Nonsynonymous mutations do change the amino acid sequence and are in general likely to have an effect (usually deleterious) on protein function and natural selection.

A mutation that *does* change the amino acid sequence encoded by a gene is known as a **nonsynonymous mutation**. For example, a change from UUA to UUC would result in phenylalanine rather than leucine in the encoded amino acid (Figure 26.1*b*). In general, nonsynonymous mutations are likely to be deleterious to the organism. But not every amino acid change alters a protein's shape (and hence its functional properties). Therefore, some nonsynonymous mutations may be selectively neutral, or nearly so.

As we saw in Chapter 23, most natural populations of organisms harbor much more genetic variation than we would expect if genetic variation were influenced primarily by natural selection. This discovery, combined with the knowledge that many mutations do not change molecular function, stimulated the development of the neutral theory of molecular evolution.

Many mutations may be selectively neutral

In 1968, Motoo Kimura proposed the *neutral theory of molecular evolution*. Kimura suggested that, at the molecular level, the majority of mutations are selectively neutral; that is, they confer neither an advantage nor a disadvantage on their bearers. Thus, the majority of evolutionary changes in macromolecules, and much of the genetic variation within species,

could result neither from directional selection of advantageous alleles nor from stabilizing selection, but rather from genetic drift.

To see why this is so, consider a population of size N and a neutral mutation rate of μ per gamete per generation at a locus. The number of new mutations would, on average, be $\mu \times 2N$, because $2N$ gene copies are available to mutate in a diploid organism. According to drift theory, the probability that a mutation will be fixed by drift alone is its frequency, p, which equals $1/(2N)$ for a newly arisen (and hence very rare) mutation. Therefore, the number of neutral mutations that arise per generation that are likely to become fixed is $2N\mu \times 1/(2N) = \mu$—which is the mutation rate!

Thus, the rate of fixation of neutral mutations is theoretically constant and is equal to the mutation rate. So if most mutations of macromolecules do not affect their functioning, macromolecules should diverge from one another at a constant rate. Indeed, in the 1960s, comparative studies of several proteins indicated that, for a particular protein, rates of amino acid substitutions were similar in all evolutionary lineages having that protein. In other words, the rate of evolution of particular proteins might be constant over time, and in effect can be a "molecular clock." We will discuss molecular clocks later in this chapter and show how, with care, they can be used to study several features of molecular evolution.

Determining and Comparing the Structure of Macromolecules

To investigate patterns of molecular evolution, biologists must first determine the precise structure of molecules. The base sequences of nucleic acids provide information about the primary structure of the proteins they encode. The invention of the polymerase chain reaction technique (PCR; see Chapter 11) allowed biologists to determine the sequences of regions of DNA not only from living tissues, but also from fossilized remains, mummified tissues, dried skins in museums, and pressed plants in herbaria, even though these objects contain only tiny amounts of DNA. DNA has been extracted and amplified from human fossils more than 30,000 years old and from plant leaf fossils 40,000 years old.

Once the sequences of molecules from different organisms have been determined, they can be compared. The purpose of comparing them is to identify the locations of deletions, insertions, and substitutions that have occurred in the molecules of interest since the organisms diverged from a common ancestor. A simple hypothetical example illustrates how this is done. In Figure 26.2 we compare two amino acid sequences (1 and 2) from homologous proteins in different organisms. The two sequences differ in the number and identity of their amino acid residues. To compare these sequences, we first observe that, although the sequences appear quite

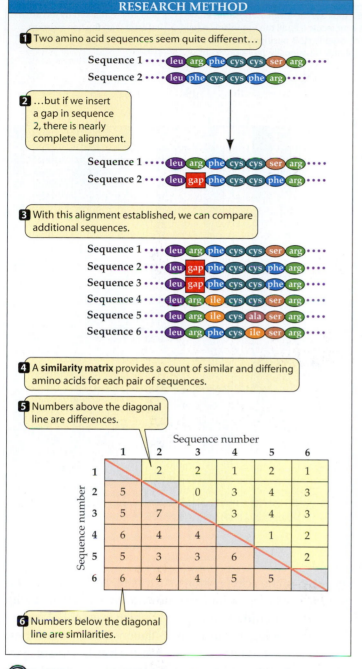

RESEARCH METHOD

1 Two amino acid sequences seem quite different...

Sequence 1 ···· leu arg phe cys cys ser arg ····
Sequence 2 ···· leu phe cys cys phe arg ····

2 ...but if we insert a gap in sequence 2, there is nearly complete alignment.

Sequence 1 ···· leu arg phe cys cys ser arg ····
Sequence 2 ···· leu gap phe cys cys phe arg ····

3 With this alignment established, we can compare additional sequences.

Sequence 1 ···· leu arg phe cys cys ser arg ····
Sequence 2 ···· leu gap phe cys cys phe arg ····
Sequence 3 ···· leu gap phe cys cys phe arg ····
Sequence 4 ···· leu arg ile cys cys ser arg ····
Sequence 5 ···· leu arg ile cys ala ser arg ····
Sequence 6 ···· leu arg phe cys ile ser arg ····

4 A **similarity matrix** provides a count of similar and differing amino acids for each pair of sequences.

5 Numbers above the diagonal line are differences.

Sequence number

	1	2	3	4	5	6
1		2	2	1	2	1
2	5		0	3	4	3
3	5	7		3	4	3
4	6	4	4		1	2
5	5	3	3	6		2
6	6	4	4	5	5	

Sequence number

6 Numbers below the diagonal line are similarities.

26.2 Amino Acid Sequence Alignment Insertion of a gap allows us to align two homologous amino acid sequences so that we can compare them. Once the alignment is established, sequences from more organisms can be added and compared. A similarity matrix sums similarities and differences between each pair of organisms. The larger the number of similarities, the more recent the presumed common ancestor of the organisms.

different, they would become similar if we were to insert a gap after the first amino acid in sequence 2 (after the leucine residue). In fact, these sequences then differ only by one amino acid at position 6 (serine or phenylalanine). Insertion of a single gap—that is, correcting a deletion—*aligns* these sequences. Longer sequences and those that have diverged more extensively require more elaborate adjustments.

After we have aligned the sequences, we can compare them by counting the number of nucleotides or amino acids that differ between them. If we add more sequences to our original example and sum the number of similar and different amino acids in each pair of sequences, we can construct a *similarity matrix* (Figure 26.2), which gives us a measure of the changes that have occurred during the divergence of the organisms.

The longer molecules have been evolving separately, the more differences they should have. Enough analyses of mammalian genes have been performed to show that the rate of nonsynonymous nucleotide substitution varies from nearly zero to about 3×10^{-9} substitutions per locus per year. Synonymous substitutions in the protein-coding regions of genes have occurred about five times more rapidly than nonsynonymous substitutions. In other words, substitution rates are highest at nucleotide positions that *do not change the amino acid being expressed* (Figure 26.3). The rate of substitution is even higher in **pseudogenes**, duplicate copies of genes that have undergone one or more mutations that eliminate their ability to be expressed. Why are these rates of substitution so dissimilar?

Rates of nucleotide substitution vary because the roles of molecules differ

The observation that rates of substitution are highest at sites and in molecules where they have no functional significance is consistent with the hypothesis that substitution rates at these sites are driven primarily by a combination of mutation and genetic drift. The much slower rates of substitution at sites that *do* affect molecular function is consistent with the

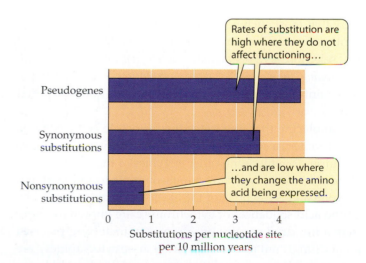

Rates of substitution are high where they do not affect functioning...

...and are low where they change the amino acid being expressed.

Pseudogenes

Synonymous substitutions

Nonsynonymous substitutions

| 0 | 1 | 2 | 3 | 4 |

Substitutions per nucleotide site per 10 million years

26.3 Rates of Base Substitution Differ Rates of nonsynonymous substitutions are much slower than rates of synonymous substitutions and substitutions in pseudogenes.

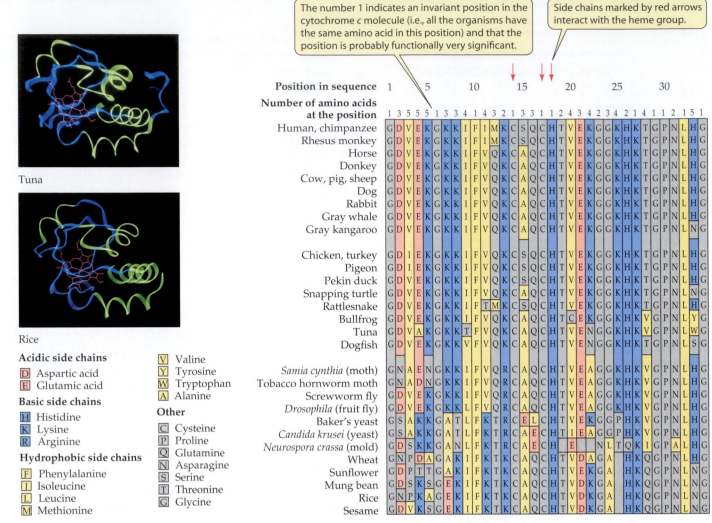

The number 1 indicates an invariant position in the cytochrome *c* molecule (i.e., all the organisms have the same amino acid in this position) and that the position is probably functionally very significant.

Side chains marked by red arrows interact with the heme group.

Acidic side chains
- D Aspartic acid
- E Glutamic acid

Basic side chains
- H Histidine
- K Lysine
- R Arginine

Hydrophobic side chains
- F Phenylalanine
- I Isoleucine
- L Leucine
- M Methionine

- V Valine
- Y Tyrosine
- W Tryptophan
- A Alanine

Other
- C Cysteine
- P Proline
- Q Glutamine
- N Asparagine
- S Serine
- T Threonine
- G Glycine

26.4 Amino Acid Sequences of Cytochrome *c* The two computer graphics show how similar the three-dimensional structures of tuna and rice cytochrome *c* are. The amino acid sequences shown in the table were obtained from analyses of cytochromes *c* from 33 species of plants, fungi, and animals.

view that most such nonsynonymous mutations are disadvantageous and are eliminated from the population by natural selection. As a result, *the more essential a molecule is for cell functioning, the slower the rate of its evolution.* These functional constraints provide part of the answer to the question posed above.

A molecule that illustrates this principle is the enzyme cytochrome *c*, a component of the respiratory chain of mitochondria. Together with other enzymes of the citric acid cycle and respiratory chain, cytochrome *c* is found in all eukaryotes and is essential for the life of the eukaryotic cell. The amino acid sequences of cytochrome *c* are known for more than a hundred species of organisms, including protists, plants, fungi, and mammals. Within these cytochromes *c* are regions that have accumulated changes relatively quickly; for example, amino acid positions 44, 89, and 100 differ among many of the organisms compared (Figure 26.4). There are also

invariant amino acid positions, such as 14, 17, 18, and 80. This particular set of invariant residues is known to interact with the iron-containing heme group, which is essential for enzyme functioning. Because any mutations that changed these amino acids would have diminished the functioning of cytochrome *c*, they would have been removed by natural selection when they arose.

Changes in macromolecules can serve as molecular clocks

Earlier in this chapter, we mentioned that the rate of evolution of some macromolecules might in effect be a *molecular clock*. To function as a molecular clock, a particular macromolecule would need to evolve at an approximately constant rate in all evolutionary lineages that possess it. But do macromolecules actually behave in this way?

Often they do. For example, if we use the fossil record to determine the time since the divergence of certain organisms, and then plot this time against the number of amino acids by which the nucleotide sequence of the organisms' cytochrome

Multiple amino acids at a position indicate a great deal of change and that the position is probably less significant.

Invariant

35   40   45   50   55   60   65   70   75   80   85   90   95   100   104

```
3 3 2 1 3 2 1 3 3 6 1 2 3 1 2 5 1 2 2 5 3 3 2 5 1 5 4 5 2 2 5 3 1 1 3 1 1 2 1 1 1 1 1 1 1 3 1 5 1 2 3 1 6 9 2 1 7 2 2 2 3 2 2 2 6 4 4 5 4

L F G R K T G Q A P G Y S Y T A A N K N K G I I W G E D T L M E Y L E N P K K Y I P G T K M I F V G I K K K E E R A D L I A Y L K K A T N E
L F G R K T G Q A P G Y S Y T A A N K N K G I I W G E D T L M E Y L E N P K K Y I P G T K M I F V G I K K K E E R A D L I A Y L K K A T N E
L F G R K T G Q A P G F T Y T D A N K N K G I T W K E E T L M E Y L E N P K K Y I P G T K M I F A G I K K K T E R E D L I A Y L K K A T N E
L F G R K T G Q A P G F S Y T D A N K N K G I T W K E E T L M E Y L E N P K K Y I P G T K M I F A G I K K K T E R E D L I A Y L K K A T N E
L F G R K T G Q A P G F S Y T D A N K N K G I T W G E E T L M E Y L E N P K K Y I P G T K M I F A G I K K K E R E D L I A Y L K K A T N E
L F G R K T G Q A V G F S Y T D A N K N K G I T W G E E T L M E Y L E N P K K Y I P G T K M I F A G I K K T E R A D L I A Y L K K A T K E
L F G R K T G Q A V G F S Y T D A N K N K G I T W G E D T L M E Y L E N P K K Y I P G T K M I F A G I K K K D E R A D L I A Y L K K A T N E
L F G R K T G Q A P G F T Y T D A N K N K G I I W G E D T L M E Y L E N P K K Y I P G T K M I F A G I K K K E R A D L I A Y L K K A T N E

L F G R K T G Q A E G F S Y T D A N K N K G I T W G E D T L M E Y L E N P K K Y I P G T K M I F A G I K K K S E R V D L I A Y L K D A T S K
L F G R K T G Q A E G F S Y T D A N K N K G I T W G E D T L M E Y L E N P K K Y I P G T K M I F A G I K K K A E R A D L I A Y L K Q A T A K
L F G R K T G Q A E G F S Y T D A N K N K G I T W G E D T L M E Y L E N P K K Y I P G T K M I F A G I K K K S E R A D L I A Y L K D A T A K
L I G R K T G Q A E G F S Y T E A N K N K G I T W G E E T L M E Y L E N P K K Y I P G T K M I F A G I K K K A E R A D L I A Y L K D A T S K
L F G R K T G Q A V G Y S Y T A A N K N K G I I W G D D T L M E Y L E N P K K Y I P G T K M V F T G L S K K K E R T N L I A Y L K E K T A A
L I G R K T G Q A A G F S Y T D A N K N K G I T W G E D T L M E Y L E N P K K Y I P G T K M I F A G I K K K E R Q D L I A Y L K S A C S K
L F G R K T G Q A E G Y S Y T D A S K N K G I V W N N D T L M E Y L E N P K K Y I P G T K M I F A G I K K K E R Q D L V A Y L K S A T S
L F G R K T G Q A Q G F S Y T D A S K N K G I T W Q Q E T L R I Y L E N P K K Y I P G T K M I F A G L K K K S E R Q D L I A Y L K K T A A S

F Y G R K T G Q A P G F S Y S N A N K A K G I T W G D D T L F E Y L E N P K K Y I P G T K M V F A G L K K A N E R A D L I A Y L K E S T K
F F G R K T G Q A P G F S Y S N A N K A K G I T W Q D D T L F E Y L E N P K K Y I P G T K M V F A G L K K A N E R A D L I A Y L K Q A T K
L F G R K T G Q A A G F A Y T N A N K A K G I T W Q D D T L F E Y L E N P K K Y I P G T K M I F A G L K K P N E R D L I A Y L K S A T K
L I G R K T G Q A A G F A Y T N A N K A K G I T W Q D D T L F E Y L E N P K K Y I P G T K M I F A G L K K P N E R D L I A Y L K S A T K
I F G R H S G Q A Q G Y S Y T D A N I K K N V L W D E N N M S E Y L T N P X K Y I P G T K M A F G L K K E K D R N D L I T Y L K K A C E
I F S R H S G Q A Q G Y S Y T D A N K R A G V E W A E P T M S D Y L E N P X K Y I P G T K M A F G L K K A K D R N D L V T Y M L E A S K
L F G R K T G Q A D G Y A Y T D A N K Q K G I T W D E N T L F E Y L E N P X K Y I P G T K M A F G L K K D K D R N D I I T F M K E A T A
L F G R Q S G S T A G Y S Y S A A N K N K A V E W E E N T L Y D Y L L N P X K Y I P G T K M V F P G L X K P Q D R A D L I A Y L K K A T S S
L F G R Q S G T T A G Y S Y S A A N K N M A V I W E E N T L Y D Y L L N P X K Y I P G T K M V F P G L X K P Q E R A D L I A Y L K T S T A
L F G R Q S G T T A G Y S Y S T A N K N M A V I W E E K T L Y D Y L L N P X K Y I P G T K M V F P G L X K P Q D R A D L I A Y L K E S T A
L F G R Q S G T T P G Y S Y S T A D K N M A V I W E E N T L Y D Y L L N P K K Y I P G T K M V F P G L K K P Q E R A D L I S Y L K E A T S
L F G R Q S G T T P G Y S Y S A A N K N M A V I W G E N T L Y D Y L L N P X K Y I P G T K M V F P G L X K P Q D R A D L I A Y L K E A T A
```

c proteins differ, we find that differences in cytochrome *c* sequences have evolved at a relatively constant rate (Figure 26.5). Many other proteins also show constancy in the rate at which they have accumulated changes over time.

Analyses of patterns of molecular evolution would be relatively easy if the rates of change were the same for all macromolecules. Unfortunately, molecular clocks tick at somewhat different rates over evolutionary time. By comparing the rates of a variety of molecular clocks, insights can be gained into why different molecules have evolved at such different rates. Such differences exist because proteins differ in the nature of the functional constraints on their evolution. For example, the rate of evolution of an enzyme would change drastically if a mutation meant that the enzyme lost its function, or if the population of the species in which the enzyme was found changed dramatically in size.

26.5 Cytochrome *c* Has Evolved at a Constant Rate Rates of substitution in cytochrome *c* are constant enough that the evolution of this molecule can be characterized by a molecular clock. The dates in this graph have been inferred from the fossil record.

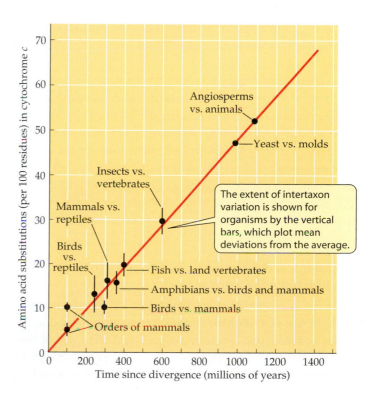

Angiosperms vs. animals

Yeast vs. molds

Insects vs. vertebrates

Mammals vs. reptiles

The extent of intertaxon variation is shown for organisms by the vertical bars, which plot mean deviations from the average.

Birds vs. reptiles

Fish vs. land vertebrates

Amphibians vs. birds and mammals

Birds vs. mammals

Orders of mammals

Amino acid substitutions (per 100 residues) in cytochrome *c*

Time since divergence (millions of years)

In addition, rates of molecular evolution are faster in organisms with short generation times than in organisms with long generation times. Short-generation organisms have more rounds of DNA replication, and thus more opportunities for errors in replication, per unit of time than long-generation organisms. For example, the rate of substitution per base per year in introns is from two to four times greater in rodents (which may reach reproductive age within 6 months and have several generations per year) than in primates (which have multi-year pre-reproductive periods and multi-year generation times).

Proteins Acquire New Functions

Evolution as we know it would not have been possible if proteins were unable to change their functional roles. The earliest forms of life must have had very few genes and proteins. Because much evidence indicates that all living organisms arose from a single ancestral lineage, the many thousands of different functional genes in modern organisms must have arisen from those few ancestral genes. How has this happened?

Proteins may acquire new functions via gene duplication

Gene duplication appears to be the most important process that enables proteins to acquire new functions. When a gene is duplicated, one copy of that gene is potentially freed from having to perform its original function; the copy is redundant if the original gene is still producing the original protein. Therefore, gene duplication may allow the evolution of entirely novel protein functions without impairing cell functioning.

Gene duplication may involve part of a gene, a single gene, a portion of a chromosome, an entire chromosome, or the whole genome. As we saw in Chapter 24, duplication of the entire genome (polyploidy) has been important in speciation, especially in plants. Autopolyploid individuals are usually viable because all of their chromosomes are duplicated. Thus, they avoid imbalances in gene expression. We will discuss gene duplication in greater detail later in this chapter, but proteins can also acquire new functions in other ways.

Physiological changes may lead to the evolution of new functions for a protein

Lysozyme is an enzyme found in almost all animals. It is produced in the tears, saliva, and milk of mammals and in the whites of bird eggs. Lysozyme digests the cell walls of bacteria, rupturing and killing them. As a result, lysozyme plays an important role as a first line of defense against invading

bacteria. All animals defend themselves against bacteria by digesting them, which is probably why most animals have lysozyme. Some animals, however, also use lysozyme in the digestion of food.

Among mammals, a mode of digestion called *foregut fermentation* has evolved twice. In animals with this mode of digestion, the foregut—the posterior esophagus and/or the stomach—has been converted into a chamber in which bacteria break down ingested plant matter by fermentation. Foregut fermenters can obtain nutrients from the otherwise indigestible cellulose that makes up a large proportion of the plant body. Foregut fermentation evolved independently in ruminants (a group of hoofed mammals that includes cows) and in certain leaf-eating monkeys, such as langurs (Figure 26.6a). We know that these evolutionary events were independent because both langurs and ruminants have close relatives that are not foregut fermenters.

In both foregut-fermenting lineages, the enzyme lysozyme has been modified to play a new, nondefensive role. Lysozyme ruptures some of the bacteria that live in the foregut, releasing nutrients that the mammal absorbs. How many changes in the lysozyme molecule allowed it to function amid the digestive enzymes and acidic conditions of the mammalian foregut? To answer this question, molecular evolutionists compared the amino acid sequences of lysozyme in foregut fermenters and in several of their nonfermenting relatives. They determined which amino acids differed and which were shared among the species (Table 26.1). Finally, they compared the pattern of these changes with the phylogenetic relationships among the species that had been established based on fossils and current morphology.

The most striking finding is that amino acid changes in lysozyme, which occurred in the absence of gene duplication, happened about twice as rapidly in the lineage leading to langurs as in any other primate lineage. This high rate of substitution shows that lysozyme went through a period of rapid change in adapting to the stomachs of langurs. The lysozymes of langurs and cows share five amino acid substitutions, all

26.1 Similarity Matrix for Lysozyme in Mammals

SPECIES	LANGUR	BABOON	HUMAN	RAT	COW	HORSE
Langur*		14	18	38	32	65
Baboon	0		14	33	39	65
Human	0	1		37	41	64
Rat	0	1	0		55	64
Cow*	5	0	0	0		71
Horse	0	0	0	0	1	

Shown above the diagonal line is the number of amino acid sequence *differences* between the two species being compared; below the line are the number of sequences uniquely *shared* by the two species. Asterisks (*) indicate foregut-fermenting species.

(a) *Presbytis entellus*

(b) *Opisthocomus hoazin*

26.6 Similar Molecular Evolution Can Take Place in Separate Lineages Foregut-fermenting mammals such as the gray langur (a) have been evolving independently from the hoatzin (b) for more than 100 million years, but each has evolved similar modifications to the enzyme lysozyme.

of which lie on the surface of the lysozyme molecule, well away from the active site (see Chapter 6). Several of the shared substitutions involve changes from arginine to lysine, which makes the proteins more resistant to attack by the pancreatic enzyme trypsin. By understanding the functional significance of amino acid substitutions, molecular evolutionists can explain the observed changes in amino acid sequences in terms of changes in the functioning of the protein.

A large body of fossil, morphological, and physiological evidence shows that langurs and cows do not share a recent common ancestor. However, langur and ruminant lysozymes share many amino acid residues that neither animal shares with the lysozymes of their own closer relatives. The lysozymes of these two animals have converged on a similar sequence despite having very different ancestry; in other words, they are homoplasies. The amino acid residues they share give these lysozymes the ability to lyse the bacteria that ferment leaves in the foregut.

An even more remarkable story emerges if we look at lysozyme in the crop of the hoatzin, a leaf-eating South American cuckoo, the only known avian foregut fermenter (Figure 26.6b). Many birds have an enlarged esophageal chamber called a *crop*. Hoatzins have a crop that contains bacteria and acts as a fermenting chamber. Many of the amino acid changes that occurred in the adaptation of hoatzin crop lysozyme are identical to the changes that evolved in ruminants and langurs. Thus, even though the hoatzin and the foregut-fermenting mammals have been evolving independently from one another for more than 100 million years, they have each evolved a similar molecule that enables them to recover nutrients from their fermenting bacteria in a highly acidic environment. The lysozyme story also illustrates why inferring phylogenies from data on single molecules can be very misleading. Identical molecular changes in this protein are not evidence of common descent.

The Evolution of Genome Size

We can now consider what determines the sizes of the genomes of different organisms. As organisms evolved to become more complex, how did the number of functional genes increase so that the organisms could carry out the greater variety of metabolic activities associated with that complexity?

Complex organisms have more DNA than do simpler ones

As we saw in Chapters 13 and 14, genome size varies tremendously among organisms. The first pattern to be detected was that genome sizes are generally correlated with organisms' complexity. The genome of *Mycoplasma genitalium*, the simplest known prokaryote, has only 470 genes. *Rickettsia prowazekii*, the prokaryote that causes typhus, has 634 genes. *Homo sapiens*, on the other hand, has about 21,000 protein-coding genes. Figure 26.7 shows the relative sizes of several prokaryotic and eukaryotic genomes.

It is not surprising that more complex genetic instructions are needed for building and maintaining a large, complex organism than a small, simple one. What is surprising is that some organisms, such as lungfishes, some salamanders, and lilies, have about 40 times as much DNA as humans do. Clearly, a lungfish or a lily is not 40 times as complex as a human. Why does genome size vary so much?

Some of the apparent variation in genome size disappears when we compare the portion of DNA that actually encodes RNAs or proteins. The size of the coding genome of organisms varies in a way that makes sense. Eukaryotes have more coding DNA than prokaryotes; plants have more coding DNA than single-celled organisms; invertebrates with wings,

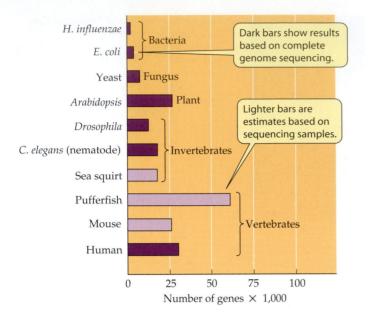

26.7 Complex Organisms Have More Genes than Simpler Organisms Genome sizes have been measured or estimated in a variety of organisms ranging from single-celled prokaryotes to vertebrates.

legs, and eyes have more coding DNA than nematodes; and vertebrates have more coding DNA than invertebrates. The organisms with the largest amount of nuclear DNA (some ferns and flowering plants) have 80,000 times as much as the simplest organisms, but no species has more than 20 times as many protein-coding genes as a bacterium. Therefore, most of the variation in genome size lies not in the number of functional genes, but in the amount of noncoding DNA (Figure 26.8).

What maintains such large quantities of noncoding DNA in the cells of most organisms? Does this noncoding DNA have a function, or is it "junk?" Most of this DNA appears to be nonfunctional. Much of it may consist of pseudogenes that are simply carried in the genome because the cost of doing so is very small. Some of it consists of parasitic transposable elements that spread through populations because they reproduce faster than the host genome. Investigators can use one type of transposable element, retrotransposons, to estimate the rates at which species lose DNA.

Retrotransposons copy themselves with the aid of RNA, as we saw in Chapter 14. The most common type of retrotransposon carries duplicated sequences at each end, called long terminal repeats (LTRs). Occasionally, LTRs join together in the host genome, at which time the DNA between them is excised. When this happens, one of the LTRs is left behind. The number of such "orphaned" LTRs in a genome is a measure of how many retrotransposons have been lost. By comparing the number of LTRs in the genomes of Hawaiian crickets of the genus *Laupala* and those of fruit flies (*Drosophila*), investigators found that *Laupala* loses DNA more

than 40 times more slowly than *Drosophila*. As a result, the genome of *Laupala* is 11 times larger than that of *Drosophila*. Why species differ so greatly in the rate at which they lose DNA is not understood.

Gene duplication can increase genome size and complexity

The identical copies of a duplicated gene can have any one of three different fates:

▶ Both copies of the gene may retain their original function, with the result that the organism produces larger quantities of the gene's RNA or protein product.
▶ One copy of the gene may be incapacitated by the accumulation of deleterious mutations and become a functionless *pseudogene*.
▶ One copy of the gene may retain its original function while the second copy accumulates enough mutations that it can perform a different function.

It is the third of the above fates that is most significant for evolution.

How often do gene duplications arise, and which of the three outcomes described above is most likely? These questions can be addressed by counting the number of synonymous nucleotide base changes in the genome of an organism. This number is then compared with the number of base changes that caused protein alterations, to see which number changed faster. Investigators have found that rates of gene duplication are fast enough for a yeast or *Drosophila* population to acquire several hundred duplicate genes over

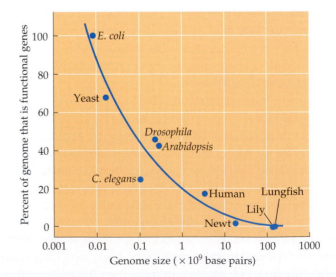

26.8 A Large Proportion of DNA Is Noncoding Most of the DNA of bacteria and yeasts encodes RNAs or proteins, but most of the DNA of more complex organisms is noncoding. Most noncoding DNA is probably nonfunctional.

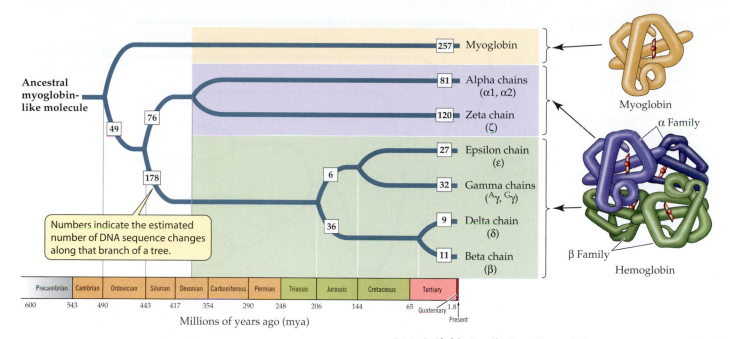

26.9 A Globin Family Gene Tree This gene tree suggests that the α-globin and β-globin gene clusters diverged about 450 mya, at about the time of the origin of the vertebrates.

the course of a million years. They also found that most of the duplicated genes in these organisms are very young. Extra genes typically are lost from a genome within 10 million years (which is rapid on an evolutionary time scale).

Although extra genes usually disappear rapidly, some duplication events lead to the evolution of genes with new functions. Several successive rounds of duplication and mutation may result in a **gene family**, a group of homologous genes with related functions, often arrayed in tandem along a chromosome. An example of this process is provided by the globin gene family (see Figure 14.7). The globins were among the first proteins to be sequenced and compared. Comparisons of their amino acid sequences strongly suggest that the different globins arose via gene duplications. How long the globins have been evolving separately can also be inferred by comparing their amino acid sequences. The greater the number of amino acid differences between two globins, the further back in time was their most recent common ancestor.

Hemoglobin, a tetramer consisting of two α-globin chains and two β-globin chains, carries oxygen in blood. Myoglobin, a monomer, is the primary oxygen storage protein in muscle. Myoglobin's affinity for O_2 is much higher than that of hemoglobin. In contrast, hemoglobin evolved to be more diversified in its role. Hemoglobin binds O_2 in the lungs or gills, where the O_2 concentration is relatively high, transports it to deep body tissues, where the O_2 concentration is low, and releases it in those areas. With its more complex tetrameric structure (see Figure 3.8), hemoglobin is able to carry four molecules of O_2, as well as hydrogen ions and carbon dioxide, in the blood.

To estimate the time of the globin gene duplication that gave rise to the α- and β-globin gene clusters, we can create a **gene tree** based on the estimated number of base substitutions necessary to account for the observed amino acid dif-

ferences between the globins. Based on this gene tree, and assuming that the rate of amino acid substitution has been relatively constant since then—about 100 substitutions per 500 million years—the two globin gene clusters are estimated to have split about 450 mya (Figure 26.9).

The Uses of Molecular Genomic Information

Information about the genomes of organisms enables biologists to investigate a variety of biological problems in new and powerful ways. Information about the rates at which molecules have evolved is used in reconstructing phylogenetic trees. Molecular genomic information can also be used to determine the phylogenies of genes, which often differ from those of organisms because genes can be transferred from one evolutionary lineage to another, a phenomenon known as *lateral gene transfer*. Recent evidence suggests that such lateral transfers may have occurred repeatedly during the evolution of plants. And, as we discussed at the beginning of this chapter, genomic information has powerful medical applications.

Molecular information is used to reconstruct phylogenies

By comparing the structures of molecules from different species, we both gain insights into how those molecules function and acquire a tool for inferring phylogenies. Molecules that have evolved slowly can be used to reconstruct relationships among organisms that diverged long ago. Molecules

26.10 Phylogeny of the engrailed Genes All of the *engrailed* genes are orthologs because they have a common ancestor. Gene duplication events have generated paralogous *engrailed* genes in the vertebrate lineages. Yellow boxes represent orthologous genes that developed after lineage splits. Green and blue boxes indicate paralogous genes created by a gene duplication.

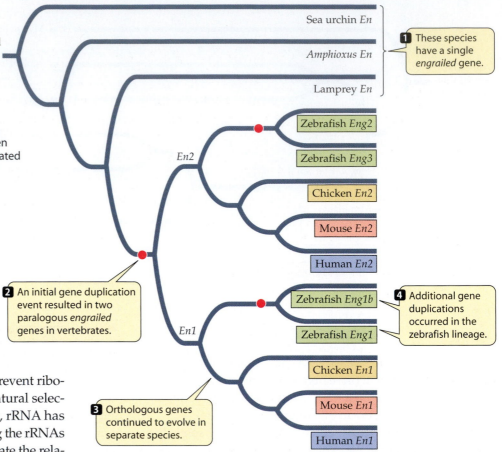

that have evolved rapidly are useful for studying organisms that share more recent common ancestors.

To investigate the evolutionary relationships of all existing organisms, we choose molecules that all organisms possess, such as ribosomal RNA. Equally important to our choice are strong functional constraints, such as those that exist for rRNA. Even minor changes in the rRNA sequence would prevent ribosomes from functioning properly, so natural selection acts to eliminate them. As a result, rRNA has evolved so slowly that differences among the rRNAs of living organisms can be used to estimate the relative timing of lineage splits that may have happened billions of years ago. No fossils exist to document the most ancient splits in the lineages of life on Earth into the three major domains—the Bacteria, the Archaea, and the Eukarya (see Figure 1.8)—so we depend on molecules such as rRNA for insight into these events.

Molecular data are also used in combination with morphological and fossil data to reconstruct phylogenies. Why do we use molecules when morphological data are available? The answer is simple: The more characters we use to reconstruct a phylogeny, the less likely we are to be misled by evolutionary reversals or convergent evolution.

Molecular data are used to determine the phylogenetic histories of genes

A gene tree shows the evolutionary relationships of members of a gene family, just as a phylogenetic tree depicts the evolutionary relationships of members of a lineage. All of the genes of a particular family have similar sequences because they have a common ancestry. Genes found in different organisms that arose from a single gene in their common ancestor are called **orthologs**. Genes that are related through gene duplication events in a single lineage are called **paralogs**.

Figure 26.10 depicts a gene tree for the members of the deuterostome *engrailed* gene family. All of the *engrailed* genes are orthologs because they have a common ancestor. Gene

duplication events have generated paralogous *engrailed* genes in some lineages. The homeotic genes of *Drosophila melanogaster* (see Chapter 19) are also paralogs.

Molecular information provides new ways to combat diseases

The effort to determine the genomes of *Anopheles* and *Plasmodium* that we described at the beginning of this chapter has already had medical benefits. Transgenic mosquitoes have been engineered to express an anti-*Plasmodium* molecule that makes them inefficient vectors of malaria in the laboratory. This advance would not have been possible without knowledge of the parasite's genome. In addition to genetic information, success in controlling malaria will require detailed knowledge of the behavior, ecology, and evolution of the many species of mosquitoes that transmit malaria in different parts of the world.

Information provided by the genomic sequence of *Treponema pallidum*, the bacterium that causes syphilis (see Figure 27.12), is being used to help develop a vaccine against this sexually transmitted disease. Syphilis may have been introduced into Europe from the Americas by members of Columbus's expedition, and the disease spread across the world during the sixteenth-century age of exploration. Effective therapies have been available since the discovery of penicillin in the mid-twentieth century, but syphilis remains a serious global health problem.

Treponema pallidum has been a difficult organism to study because it cannot be grown outside a mammalian host. The genome sequence of *T. pallidum* may provide the information necessary for the development of a culture medium. In addition, the sequence revealed that *T. pallidum* contains a family of paralogous genes (*tprA–L*) that encode proteins of the bacterial outer membrane. The identification of this gene family is guiding the search for targets for a vaccine against the disease.

Molecular data cannot solve all disease problems

Molecular data are powerful tools in our struggle with diseases, but the AIDS epidemic shows that they cannot solve all medical problems. As we saw in Chapter 13, the agent that causes AIDS is a retrovirus (HIV) that uses RNA as its genetic material. The central core of the virus contains two identical copies of the RNA genome as well as three enzymes needed to carry out the viral life cycle: reverse transcriptase, integrase, and a protease.

The highly active antiretroviral therapy (HAART) most commonly used against AIDS today employs a combination of protease inhibitors and reverse transcriptase inhibitors, each of which blocks a different stage in the viral life cycle (see Chapter 18). Unfortunately, resistant strains of HIV develop in the blood of most patients that receive HAART. HIV's mutation rate is very high, and since it has no repair enzymes, a wrong base is inserted at about one out of every 8,000 nucleotide positions. The result is that, on average, a new mutant is generated every time HIV replicates its genome—in effect, no two viruses are identical! Although biomedical scientists are now armed with detailed knowledge of the molecules that govern the different stages in the viral life cycle, HIV's extremely high mutation rate has prevented them from coming up with drugs to which the virus cannot rapidly adapt. In the absence of such drugs, reducing both the incidence and severity of HIV will require changes in human sexual behavior.

The development of molecular methods, the sequencing of the genomes of an increasing number of species, and powerful computers have ushered in a new era in the scientific study of the diversity of life. Scientific understanding of the evolutionary patterns of life on Earth and how the agents of evolution have governed these patterns is advancing more rapidly than at any prior time during the study of evolution. The range of data used by systematists is likely to continue to increase because, using modern chemical, biochemical, and microscopic methods, we will be able to measure more traits of organisms than we could previously.

By combining molecular data with information from the fossil record (which is also increasing rapidly), biologists are developing a comprehensive picture of the evolution of life on Earth. In Part Five, we will provide an extensive overview of the remarkable diversity of organisms that is the result of almost 4 billion years of evolution.

Chapter Summary

Genomes and Their Evolution
▶ A genome is the full set of genes an organism contains.
▶ The genes in an organism's genome are usefully viewed as interacting members of a group.

The Evolution of Macromolecules
▶ Molecular evolutionists characterize the structures of macromolecules and use them to determine how rapidly these macromolecules have changed and why they have changed.
▶ Mutations and genetic drift are important determinants of rates of molecular evolution. **Review Figure 26.1**

Determining and Comparing the Structure of Macromolecules
▶ Molecules are compared by aligning their sequences and counting the differences between those sequences. **Review Figure 26.2. See Web/CD Activity 26.2**
▶ Changes evolve slowly in regions of molecules that are functionally significant, but more rapidly in regions where substitutions do not affect the functioning of the molecules. **Review Figure 26.3, 26.4**
▶ Rates of substitution in some molecules are relatively constant over evolutionary time; that is, these molecules can serve as molecular clocks. **Review Figure 26.5**
See Web/CD Activity 26.1

Proteins Acquire New Functions
▶ Most new protein functions arise by gene duplication.
▶ Changes in the functions performed by proteins may also result from changes in the physiological roles of gene products.

The Evolution of Genome Size
▶ The genome sizes of organisms vary tremendously, but the amount of DNA that actually encodes RNAs or proteins varies much less. **Review Figures 26.7, 26.8**
▶ Complex organisms have more coding DNA than simpler ones.
▶ The globin family of proteins evolved via gene duplication. **Review Figure 26.9**

The Uses of Molecular and Genomic Information
▶ Molecular data can be used to infer phylogenetic relationships among organisms.
▶ Molecules that have evolved slowly are useful for determining ancient lineage splits. Molecules that have evolved rapidly are useful for determining more recent lineage splits in combination with morphological and fossil data.
▶ Molecular data are used to determine the phylogenetic histories of genes. **Review Figure 26.10**
▶ Molecular data are used to find new ways to combat diseases.
See Web/CD Activity 26.3

Self-Quiz

1. Which of the following questions do students of genomic evolution *not* try to answer?
   a. What are the forces that maintain interactions among different genes?
   b. Why are the genomes of organisms so variable in size?
   c. How has enlargement of genomes been accomplished?
   d. Why is DNA the genetic material of most organisms?
   e. How do proteins acquire new functions?

2. Molecular evolution differs from phenotypic evolution in which of the following ways?
   a. It requires changes in molecules if it is to happen.
   b. Random genetic drift and mutations usually exert greater influences on rates and directions of molecular evolution than on rates and directions of phenotypic evolution.
   c. Molecular evolution is not influenced by natural selection.
   d. Rates of molecular evolution are much slower than rates of phenotypic evolution because mutation rates typically are low.
   e. There are no important differences between molecular and phenotypic evolution.

3. Choosing the appropriate molecule for phylogenetic reconstruction does *not* require a consideration of the
   a. question being answered.
   b. rate of evolution of the molecule.
   c. phylogenetic distribution of the molecule.
   d. function of the molecule.
   e. completeness of the fossil record.

4. Ribosomal RNA sequences are useful for addressing the evolutionary relationships of lineages that diverged in ancient times because they
   a. evolve at a rapid rate.
   b. have undergone convergent evolution in many lineages.
   c. are molecules that all organisms have.
   d. consist of mainly neutral characters.
   e. are difficult to align.

5. Mitochondrial DNA sequences are useful in studying the recent evolution of closely related species because
   a. some mitochondrial genes accumulate mutations very rapidly.
   b. they are paternally inherited.
   c. they evolve only in a neutral fashion.
   d. they are highly constrained in function.
   e. they recombine every generation.

6. Issues concerning patterns of molecular evolution include
   a. evolutionary relationships among molecules.
   b. molecular clock.
   c. rate of mutation for neutral characters.
   d. the importance of gene duplication in evolution.
   e. All of the above

7. Molecules are used to reconstruct phylogenies even if a fossil record is available, because
   a. the more characters the better.
   b. molecules are more accurate characters than are fossils.
   c. molecules undergo less homoplasy than do fossil characters.
   d. molecules are less subjective characters than are fossils.
   e. molecules give us the "right" phylogeny.

8. Neutral characters
   a. are not evolving under the influence of natural selection.
   b. have a neutral pH.
   c. are not useful in reconstructing phylogenies.
   d. are subject to strong functional constraints.
   e. are not likely to evolve.

9. The concept of a molecular clock implies that
   a. many proteins show a constancy in rate of change with time.
   b. organisms evolve at a constant rate.
   c. one can date evolutionary events with molecules alone.
   d. all molecules change at the same rate in evolution.
   e. we can predict how rapidly all genes will evolve.

10. Proteins acquire new functions primarily by means of
    a. gene duplication, which frees one copy of a gene from having to perform its original function.
    b. gene duplication, which provides two copies of a gene that, working together, produce a new protein.
    c. deletions, which generate new protein shapes.
    d. deletions, which make proteins nonfunctional, thereby creating new opportunities for other proteins.
    e. None of the above

11. The lysozyme story suggests that
    a. molecules cannot change their function in evolution.
    b. selection does not act at the molecular level.
    c. molecules can help us understand the process of organismic evolution.
    d. all organisms are capable of fermenting bacteria.
    e. lysozyme has a very accurate molecular clock.

12. The actual differences in genome sizes are much less than the apparent differences because
    a. multicellular organisms are really not that much more complicated than eukaryotic protists.
    b. organisms with the largest amounts of nuclear DNA have much more noncoding DNA than organisms with smaller amounts of nuclear DNA.
    c. the sizes of many apparently large genomes have been seriously overestimated.
    d. differences in the sizes of genes account for much of the apparent difference in genome sizes.
    e. species with large genomes preferentially lose DNA by converting it to pseudogenes, which are subsequently lost.

For Discussion

1. If you were interested in reconstructing the phylogeny of a genus of fruit flies using molecular data, what kinds of molecule(s) would you choose to examine? Why? If you wanted to reconstruct the phylogeny of all vertebrates, would you use the same molecule(s)? Why or why not?

2. Discuss the relative importance of molecular characters and morphological characters in reconstructing the phylogeny of a group of organisms.

3. Existing evidence suggests that for some molecules, a molecular clock ticks at a fairly constant rate, but that rates of change differ widely among molecules. How does this variation limit how and in what ways we can use the concept of a molecular clock to help us answer questions about the evolution of both molecules and organisms?

4. One hypothesis proposed to explain the existence of large amounts of noncoding ("junk") DNA is that the cost of maintaining that DNA is so small that natural selection is too weak to reduce it. What other hypotheses might account for the existence of so much noncoding DNA?

5. If fossil evidence and molecular evidence disagree on the date of a major lineage split, which of the two kinds of evidence would you favor? Why?

6. Soon scientists will be able to produce and release into the wild genetically modified mosquitoes unable to harbor and transmit malaria parasites. What ethical issues need to be discussed before such releases are permitted?

How has Darwin's theory of natural selection transformed our view of humanity's place in the universe?

- by Daniel Dennett -

For as long as our ancestors have been making tools, it has no doubt seemed obvious that an excellent artifact can be created only by something even more excellent: a clever artificer. You never see a shoe creating a cobbler; you never see a house making a carpenter. Darwin overthrew that received wisdom. One of Darwin's earliest critics, Robert Beverley MacKenzie, could not contain his outrage:

> In the theory with which we have to deal, Absolute Ignorance is the artificer; so that we may enunciate as the fundamental principle of the whole system, that, IN ORDER TO MAKE A PERFECT AND BEAUTIFUL MACHINE, IT IS NOT REQUISITE TO KNOW HOW TO MAKE IT. This proposition will be found … to express in a few words all Mr. Darwin's meaning; who, by a strange inversion of reasoning, seems to think Absolute Ignorance fully qualified to take the place of Absolute Wisdom in all the achievements of creative skill.

This is indeed a "strange inversion of reasoning," but once the topsy-turvy perspective it implies has been accepted, most of what we have believed about who we are survives intact. We can still be in awe of the "Wisdom in all the achievements of creative skill" while attributing this wisdom not to a single Creator, but distributing it over billions of years in trillions of lineages of replicators, trying their luck in the great tournament of life, mindlessly discovering and redis-

covering the brilliant design principles that constitute the diversity of life. Tradition honors the trickle-down theory of value: what we do and think can be valuable only if it derives its value from something even more valuable—only if we are the servants, in effect, of a greater master. Darwin's "strange inversion" obliges us to rethink what could make something valuable, and then we notice that a bubble-up theory has much to recommend it. There was a time when there was no morality on this planet, and now it has evolved. Just as the air we breathe was created as a by-product of the activities of billions of years of simpler life forms, the very meaning of life on this planet has emerged from the efforts of the life forms that the atmosphere enabled.

We are animals. Are we *just* animals? The ideological tug-of-war over "human exceptionalism" can be damped, if not stopped outright, by emphasizing a few uncontroversial facts. Sight, the capacity to extract huge amounts of relevant information from a relatively safe distance, was an innovation that multiplied the opportunities of intelligent behavior: locomotion, predation, evasion, migration, and so on. Sight and flight have each evolved numerous times, but language has evolved just once, so far as we know—in our genus. (Neanderthals may have been a second talking species for a while.) Language is the key to our huge advantage in knowledge and technology. Other animal species transmit significant amounts of know-how nongenetically from parent to offspring, but without language, the lessons to be learned are rather simple pref-

Daniel Dennett is University Professor and Director of the Center for Cognitive Studies at Tufts University, Medford, Massachusetts. He is the author of many books and articles, among them *Consciousness Explained* (1991), *Darwin's Dangerous Idea* (1995), and *Freedom Evolves* (2003).

erences and prohibitions, not elaborate systems of hard-won technique and patiently gathered data.

It has taken our species thousands of years of communication and investigation to begin to find the keys to our own identities. Our newfound capacity for long-distance knowledge gives us powers that dwarf those of all the rest of the life on Earth. It has been estimated that ten thousand years ago, the human population comprised a small fraction of 1 percent of the mass of vertebrate life on land; today, we, together with our livestock and pets, make up about 98 percent of that total. We exploit an ever increasing share of the planet's resources, but we do offer something in return. Now, for the first time in its billions of years of history, our planet is protected by far-seeing sentinels, able to anticipate danger from the distant future—an asteroid on a collision course, or global warming—and devise schemes for doing something about it. The planet has finally grown its own nervous system: us. We are responsible for the future of life on the planet, in a way no other species could ever be.

Discussion Questions

1. What developments would make it possible or likely for language to evolve again, in another species?

2. Are there any good reasons, aside from tradition, for favoring a trickle-down theory of value over a bubble-up theory of value?

3. Beavers build elaborate dams; so do civil engineers. What role does language play in determining the kinds of artifacts a species can make?

4. How would you defend the hypothesis that our ancestors learned to control fire before they mastered language?

Web Links

Center for Cognitive Studies at Tufts University: ase.tufts.edu/cogstud/

Darwin Day, University of Tennessee at Knoxville: fp.bio.utk.edu/darwin/

The C. Warren Irvin, Jr., Collection of Darwiniana: www.sc.edu/library/spcoll/nathist/darwin/darwin.html

27 Bacteria and Archaea: The Prokaryotic Domains

The ancient Phoenicians called it the "river of fire." Today, Spanish astrobiologist Ricardo Amils Pibernat calls Spain's Río Tinto a possible model for the scene of the origin of the life that may have existed on Mars. The river wends its way through a huge deposit of iron pyrite—"fool's gold," or iron disulfide. Prokaryotes in the river and in the damp, acidic soil from which it arises convert the pyrite into sulfuric acid and dissolved iron. The iron gives the river its reddish brown color.

Over a period of at least 300,000 years, these prokaryotes have produced an environment seemingly hostile to life. The Río Tinto has a pH of 2 and exceptionally high concentrations of heavy metals, especially iron. The concentrations of oxygen in the river and in its source soil are extremely low. It is that soil that Amils believes resembles the kind of environment in which life could have begun on Mars. Whatever the truth of that speculation, the Río Tinto represents one of the most unusual habitats for life on Earth.

The organisms most commonly found in such extremely acidic environments belong to the two major groups of prokaryotes: Bacteria and Archaea. The bacteria live in almost every environment on Earth. The archaea are a superficially similar group of microscopic, unicellular prokaryotes. However, both the biochemistry and the genetics of bacteria differ in numerous ways from those of archaea. Not until the 1970s did biologists discover how radically different bacteria and archaea really are. And only with the sequencing of an archaeal genome in 1996 did we realize just how extensively archaea differ from both bacteria and eukaryotes.

Many biologists acknowledge the antiquity of these clades and the importance of their differences by recognizing three domains of living things: Bacteria, Archaea, and Eukarya. The domain Bacteria comprises the "true bacteria." The domain Archaea (Greek *archaios*, "ancient") comprises other prokaryotes once called (inaccurately) "ancient bacteria." The domain Eukarya includes all other living things on Earth.

Dividing the living world in this way, with two prokaryotic domains and a single domain for all the

Earth or Ancient Mars? Spain's Río Tinto owes its rusty red color—and its extreme acidity—to the action of prokaryotes on iron pyrite-rich soil.

eukaryotes, fits with the current trend toward reflecting evolutionary relationships in classification systems. In the eight chapters of Part Five, we celebrate and describe the diversity of the living world—the products of evolution. This chapter focuses on the two prokaryotic domains. Chapters 28–34 deal with the protists and the kingdoms Plantae, Fungi, and Animalia.

In this chapter, we pay close attention to the ways in which the two domains of prokaryotic organisms resemble each other as well as the ways in which they differ. We will describe the impediments to the resolution of evolutionary relationships among the prokaryotes. Then we will survey the surprising diversity of organisms within each of the two domains, relating the characteristics of the different prokaryotic groups to their roles in the biosphere and in our lives.

Why Three Domains?

What does it mean to be *different?* You and the person nearest you look very different—certainly you appear more different than the two cells shown in Figure 27.1. But the two of you are members of the same species, while these two tiny organisms that look so much alike actually are classified in entirely separate domains. Still, all three of you (you in the domain Eukarya and those two prokaryotes in the domains Bacteria and Archaea) have a lot in common. Members of all three domains

▶ conduct glycolysis;
▶ replicate their DNA semiconservatively;

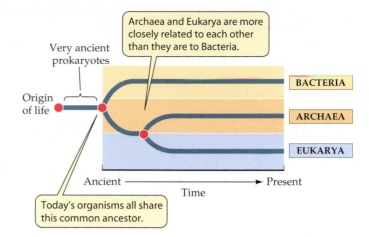

27.2 The Three Domains of the Living World   Many biologists believe that the three domains share a common prokaryotic ancestor. The relationships shown here, however, remain controversial.

▶ have DNA that encodes polypeptides;
▶ produce these polypeptides by transcription and translation and use the same genetic code;
▶ have plasma membranes and ribosomes in abundance.

There are also major differences among the domains. Members of the Eukarya have cells with nuclei, membrane-enclosed organelles, and a cytoskeleton—structures that no prokaryote has. And a glance at Table 27.1 will show you that there are also major differences (most of which cannot be seen even under the microscope) between the two prokaryotic domains. In some ways the archaea are more like us; in other ways they are more like bacteria.

Genetic studies have led many biologists to conclude that all three domains had a single common ancestor, and that the present-day archaea share a more recent common ancestor with eukaryotes than they do with bacteria (Figure 27.2). Because of the ancient time at which these three clades diverged, the major differences among the three kinds of organisms, and especially the likelihood that the archaea are more closely related to the eukaryotes than are either of those groups to the bacteria, many biologists agree that it makes sense to treat these three groups as *domains*—a higher taxonomic category than *kingdoms*. To treat all the prokaryotes as a single kingdom within a five-kingdom classification of organisms would result in a kingdom that is paraphyletic. That is, a single kingdom "Prokaryotes" would not include all the descendants of their common ancestor. (See Chapter 25, especially Figure 25.8, for a discussion of paraphyletic groups.) We will use the domain concept in this book, although it is still controversial and may have to be abandoned if new data fail to support it.

The common ancestor of all three domains was prokaryotic. Its genetic material was DNA; its machinery for tran-

Salmonella tymphimurium   *Methanospirillum hungatii*

1 µm   0.4 µm

27.1 Very Different Prokaryotes   In each image, one of the cells has nearly finished dividing. On the left are bacteria; on the right are archaea, which are more closely related to eukaryotes than they are to the bacteria.

27.1 The Three Domains of Life on Earth

CHARACTERISTIC	DOMAIN		
	BACTERIA	ARCHAEA	EUKARYA
Membrane-enclosed nucleus	Absent	Absent	*Present*
Membrane-enclosed organelles	Absent	Absent	*Present*
Peptidoglycan in cell wall	*Present*	Absent	Absent
Membrane lipids	Ester-linked	*Ether-linked*	Ester-linked
	Unbranched	*Branched*	Unbranched
Ribosomes[a]	70S	70S	*80S*
Initiator tRNA	*Formylmethionine*	Methionine	Methionine
Operons	Yes	Yes	*No*
Plasmids	Yes	Yes	*Rare*
RNA polymerases	One	One[b]	Three
Ribosomes sensitive to chloramphenicol and streptomycin	*Yes*	No	No
Ribosomes sensitive to diphtheria toxin	*No*	Yes	Yes
Some are methanogens	No	*Yes*	No
Some fix nitrogen	Yes	Yes	*No*
Some conduct chlorophyll-based photosynthesis	Yes	No	Yes

[a] 70S ribosomes are smaller than 80S ribosomes.
[b] Archaeal RNA polymerase is similar to eukaryotic polymerases.

scription and translation produced RNAs and proteins, respectively. It probably had a circular chromosome, and many of its structural genes were grouped into operons (see Chapter 13).

The Archaea, Bacteria, and Eukarya of today are all products of billions of years of natural selection and genetic drift, and they are all well adapted to present-day environments. None are "primitive." The common ancestor of the Archaea and the Eukarya probably lived more than 2 billion years ago, and the common ancestor of the Archaea, the Eukarya, and the Bacteria probably lived more than 3 billion years ago.

The earliest prokaryotic fossils date back at least 3.5 billion years, and these ancient fossils indicate that there was considerable diversity among the prokaryotes even during the earliest days of life. The prokaryotes were alone on Earth for a very long time, adapting to new environments and to changes in existing environments. They have survived to this day—and in massive numbers.

General Biology of the Prokaryotes

There are many, many prokaryotes around us—everywhere. Although most are so small that we cannot see them with the naked eye, the prokaryotes are the most successful of all creatures on Earth, if success is measured by numbers of individuals. The bacteria in one person's intestinal tract, for example, outnumber all the humans who have ever lived, and

even the total number of *human* cells in their host's body. Some of these bacteria form a thick lining along the intestinal wall. Bacteria and archaea in the oceans number more than 3×10^{28}. This stunning number is perhaps 100 million times as great as the number of stars in the visible universe.

Although small, prokaryotes play many critical roles in the biosphere, interacting in one way or another with every other living thing. In this section, we'll see that some prokaryotes perform key steps in the cycling of nitrogen, sulfur, and carbon. Other prokaryotes trap energy from the sun or from inorganic chemical sources, and some help animals digest their food. The members of the two prokaryotic domains outdo all other groups in metabolic diversity. Eukaryotes, in contrast, are much more diverse in size and shape, but their metabolism is much less diverse. In fact, much of the energy metabolism of eukaryotes is carried out in organelles—mitochondria and chloroplasts—that are descended from bacteria.

Prokaryotes are found in every conceivable habitat on the planet, from the coldest to the hottest, from the most acidic to the most alkaline, and to the saltiest. Some live where oxygen is abundant and others where there is no oxygen at all. They have established themselves at the bottom of the seas, in rocks more than 2 km into Earth's solid crust, and inside other organisms, large and small. Their effects on our environment are diverse and profound. What do these tiny but widespread organisms look like?

(a) *Enterococcus* sp.

(b) *Escherichia coli*

(c) *Leptospira interrogans*

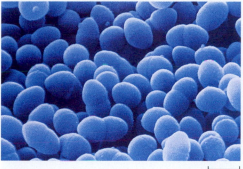
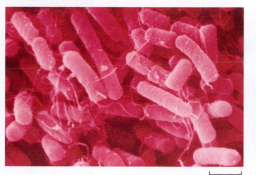
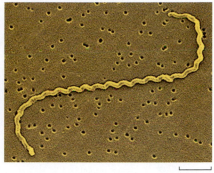

1 µm

1 µm

1 µm

Prokaryotes and their associations take a few characteristic forms

Three shapes are particularly common among the prokaryotes: spheres, rods, and curved or spiral forms (Figure 27.3). A spherical prokaryote is called a **coccus** (plural, cocci). Cocci may live singly or may associate in two- or three-dimensional arrays as chains, plates, blocks, or clusters of cells. A rod-shaped prokaryote is called a **bacillus** (plural, bacilli). Spiral forms are the third main prokaryotic shape. Bacilli and spiral forms may be single or may form chains.

Prokaryotes are almost all unicellular, although some multicellular ones are known. Associations such as chains do not signify multicellularity because each cell is fully viable and independent. These associations arise as cells adhere to one another after reproducing by fission. Associations in the form of chains are called **filaments**. Some filaments become enclosed within delicate tubular sheaths.

Prokaryotes lack nuclei, organelles, and a cytoskeleton

The architectures of prokaryotic and eukaryotic cells were compared in Chapter 4. The basic unit of archaea and bacteria is the prokaryotic cell (see Figure 4.5), which contains a full complement of genetic and protein-synthesizing systems, including DNA, RNA, and all the enzymes needed to transcribe and translate the genetic information into proteins. The prokaryotic cell also contains at least one system for generating the ATP it needs.

In what follows, bear in mind that most of what we know about the structure of prokaryotes comes from studies of bacteria. We still know relatively little about the diversity of archaea, although the pace of research on archaea is accelerating.

The prokaryotic cell differs from the eukaryotic cell in three important ways. First, the organization and replication of the genetic material differs. The DNA of the prokaryotic cell is not organized within a membrane-enclosed nucleus. DNA molecules in prokaryotes (both bacteria and archaea)

27.3 Shapes of Prokaryotic Cells (*a*) These spherical cocci of an acid-producing bacterium grow in the mammalian gut. (*b*) Rod-shaped *E. coli* are the most thoroughly studied of any bacteria—indeed, of almost any organism on Earth. (*c*) This spiral bacterium belongs to a genus of human pathogens that cause leptospirosis, an infection of the kidney and liver that is spread by contaminated water. The disease has historically been a problem for soldiers in crowded, transient campsites; this particular bacterial strain was isolated in 1915 from the blood of a soldier serving in World War I.

are usually circular; in the best-studied prokaryotes, there is a single chromosome, but there are often plasmids as well (see Chapter 13).

Second, prokaryotes have none of the membrane-enclosed cytoplasmic organelles that modern eukaryotes have—mitochondria, Golgi apparatus, and others. However, the cytoplasm of a prokaryotic cell may contain a variety of infoldings of the plasma membrane and photosynthetic membrane systems not found in eukaryotes.

Third, prokaryotic cells lack a cytoskeleton, and, without the cytoskeletal proteins, they lack mitosis. Prokaryotic cells divide by their own elaborate method, **fission**, after replicating their DNA.

Prokaryotes have distinctive modes of locomotion

Although many prokaryotes cannot move, others are *motile*. These organisms move by one of several means. Some spiral bacteria, called spirochetes, use a corkscrew-like motion made possible by modified flagella, called *axial filaments*, running along the axis of the cell beneath the outer membrane (Figure 27.4*a*). Many cyanobacteria and a few other bacteria use various poorly understood gliding mechanisms, including rolling. Various aquatic prokaryotes, including some cyanobacteria, can move slowly up and down in the water by adjusting the amount of gas in gas vesicles (Figure 27.4*b*). By far the most common type of locomotion in prokaryotes, however, is that driven by flagella.

(a)

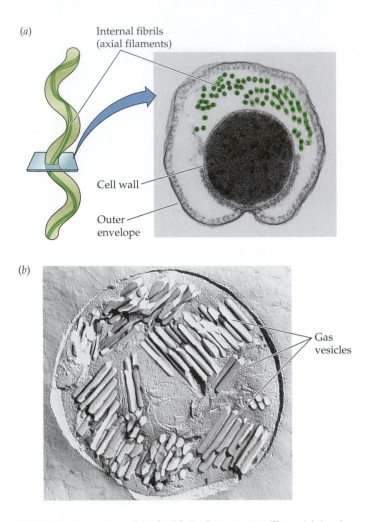

Internal fibrils
(axial filaments)

Cell wall

Outer
envelope

(b)

Gas
vesicles

27.4 Structures Associated with Prokaryote Motility (a) A spirochete from the gut of a termite, seen in cross section, shows the axial
filaments used to produce a corkscrew-like motion. (b) Gas vesicles in
a cyanobacterium, visualized by the freeze-fracture technique.

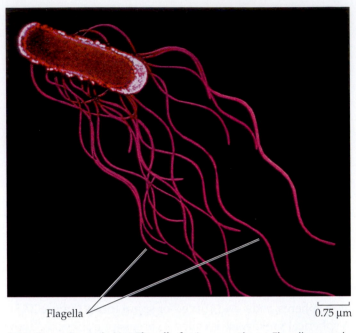

Flagella   0.75 μm

27.5 Some Bacteria Use Flagella for Locomotion Flagella propel
this rod-shaped *Salmonella*.

Bacterial flagella are slender filaments that extend singly
or in tufts from one or both ends of the cell or are randomly
distributed all around it (Figure 27.5). A bacterial flagellum
consists of a single fibril made of the protein *flagellin*, pro-
jecting from the cell surface, plus a hook and basal body responsible for motion (see Figure 4.6). In contrast, the flagellum of eukaryotes is enclosed by the plasma membrane and
usually contains a circle of nine pairs of microtubules surrounding two central microtubules, all containing the protein
tubulin, along with many other associated proteins. The
prokaryotic flagellum rotates about its base, rather than beating as a eukaryotic flagellum or cilium does.

Prokaryotes have distinctive cell walls

Most prokaryotes have a thick and relatively stiff cell wall.
This wall is quite different from the cell walls of plants and

algae, which contain cellulose and other polysaccharides, and
from those of fungi, which contain chitin. Almost all bacteria
have cell walls containing *peptidoglycan* (a polymer of amino
sugars). Archaeal cell walls are of differing types, but most
contain significant amounts of protein. One group of archaea
has pseudopeptidoglycan in its wall; as you have probably
already guessed from the prefix *pseudo-*, pseudopeptidoglycan is similar to, but distinct from, the peptidoglycan of bacteria. Peptidoglycan is a substance unique to bacteria; its absence from the walls of archaea is a key difference between
the two prokaryotic domains.

In 1884 Hans Christian Gram, a Danish physician, developed a simple staining process that has lasted into our high-
technology era as a useful tool for identifying bacteria. The
Gram stain separates most types of bacteria into two distinct
groups, Gram-positive and Gram-negative, on the basis of
their staining (Figure 27.6). A smear of cells on a microscope
slide is soaked in a violet dye and treated with iodine; it is
then washed with alcohol and counterstained with safranine
(a red dye). **Gram-positive** bacteria retain the violet dye and
appear blue to purple (Figure 27.6*a*). The alcohol washes the
violet stain out of **Gram-negative** cells; these cells then pick
up the safranine counterstain and appear pink to red (Figure
27.6*b*). Gram-staining characteristics are useful in classifying
some kinds of bacteria and are important in determining the
identity of bacteria in an unknown sample.

For many bacteria, the Gram-staining results correlate
roughly with the structure of the cell wall. Peptidoglycan
forms a thick layer outside the plasma membrane of Gram-

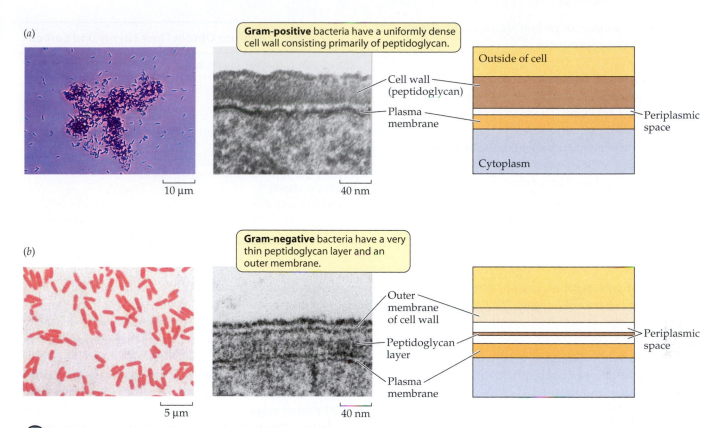

(a)

Gram-positive bacteria have a uniformly dense cell wall consisting primarily of peptidoglycan.

Outside of cell

Cell wall (peptidoglycan)

Plasma membrane

Periplasmic space

Cytoplasm

10 μm   40 nm

(b)

Gram-negative bacteria have a very thin peptidoglycan layer and an outer membrane.

Outer membrane of cell wall

Peptidoglycan layer

Plasma membrane

Periplasmic space

5 μm   40 nm

27.6 The Gram Stain and the Bacterial Cell Wall When treated with Gram stain, the cell wall components of different bacteria react in one of two ways. (a) Gram-positive bacteria have a thick peptidoglycan cell wall that retains the violet dye and appears deep blue or purple. (b) Gram-negative bacteria have a thin peptidoglycan layer that does not retain the violet dye, but picks up the counterstain and appears pink-red.

positive bacteria. The Gram-negative cell wall usually has only one-fifth as much peptidoglycan, and outside the peptidoglycan layer the cell is surrounded by a second, outer membrane quite distinct in chemical makeup from the plasma membrane (see Figure 27.6b). Between the inner (plasma) and outer membranes of Gram-negative bacteria is the *periplasmic space*. This space contains enzymes that are important in digesting some materials, transporting others, and detecting chemical gradients in the environment.

The consequences of the different features of prokaryotic cell walls are numerous and relate to the disease-causing characteristics of some prokaryotes. Indeed, the cell wall is a favorite target in medical combat against diseases that are caused by prokaryotes because it has no counterpart in eukaryotic cells. Antibiotics such as penicillin and ampicillin, as well as other agents that specifically interfere with the synthesis of peptidoglycan-containing cell walls, tend to have little, if any, effect on the cells of humans and other eukaryotes.

Prokaryotes reproduce asexually, but genetic recombination does occur

Prokaryotes reproduce by fission, an asexual process. Recall, however, that there are also processes—transformation, conjugation, and transduction—that allow the exchange of genetic information between some prokaryotes quite apart from either sex or reproduction (see Chapter 13).

Some prokaryotes multiply very rapidly. One of the fastest is the bacterium *Escherichia coli*, which under optimal conditions has a generation time of about 20 minutes. The shortest known prokaryote generation times are about 10 minutes. Generation times of 1 to 3 hours are common for others; some extend to days. Bacteria living deep in Earth's crust may suspend their growth for more than a century without dividing and then multiply for a few days before suspending growth again. What kinds of metabolism support such a diversity of growth rates?

Prokaryotes have exploited many metabolic possibilities

The long evolutionary history of the bacteria and archaea, during which they have had time to explore a wide variety of habitats, has led to the extraordinary diversity of their metabolic "lifestyles"—their use or nonuse of oxygen, their energy sources, their sources of carbon atoms, and the materials they release as waste products.

ANAEROBIC VERSUS AEROBIC METABOLISM.   Some prokaryotes can live only by anaerobic metabolism because molecular oxygen is poisonous to them. These oxygen-sensitive organisms are called **obligate anaerobes**.

Other prokaryotes can shift their metabolism between anaerobic and aerobic modes (see Chapter 7) and thus are called **facultative anaerobes**. Many facultative anaerobes alternate between anaerobic metabolism (such as fermentation) and cellular respiration as conditions dictate. **Aerotolerant anaerobes** cannot conduct cellular respiration, but are not damaged by oxygen when it is present.

At the other extreme from the obligate anaerobes, some prokaryotes are **obligate aerobes**, unable to survive for extended periods in the *absence* of oxygen. They require oxygen for cellular respiration.

NUTRITIONAL CATEGORIES.   Biologists recognize four broad nutritional categories of organisms: photoautotrophs, photoheterotrophs, chemolithotrophs, and chemoheterotrophs. Prokaryotes are represented in all four groups (Table 27.2).

Photoautotrophs perform photosynthesis. They use light as their source of energy and carbon dioxide as their source of carbon. Like the photosynthetic eukaryotes, one group of photoautotrophic bacteria, the cyanobacteria, use chlorophyll *a* as their key photosynthetic pigment and produce oxygen as a by-product of noncyclic electron transport (see Chapter 8).

By contrast, the other photosynthetic bacteria use *bacteriochlorophyll* as their key photosynthetic pigment, and they do not release oxygen gas. Some of these photosynthesizers produce particles of pure sulfur instead because hydrogen sulfide (H_2S), rather than H_2O, is their electron donor for photophosphorylation. Bacteriochlorophyll absorbs light of

27.2 How Organisms Obtain Their Energy and Carbon

NUTRITIONAL CATEGORY	ENERGY SOURCE	CARBON SOURCE
Photoautotrophs (found in all three domains)	Light	Carbon dioxide
Photoheterotrophs (some bacteria)	Light	Organic compounds
Chemolithotrophs (some bacteria, many archaea)	Inorganic substances	Carbon dioxide
Chemoheterotrophs (found in all three domains)	Organic compounds	Organic compounds

longer wavelengths than the chlorophyll used by all other photosynthesizing organisms does. As a result, bacteria using this pigment can grow in water beneath fairly dense layers of algae, using light of wavelengths that are not absorbed by the algae (Figure 27.7).

Photoheterotrophs use light as their source of energy, but must obtain their carbon atoms from organic compounds made by other organisms. They use compounds such as carbohydrates, fatty acids, and alcohols as their organic "food." The purple nonsulfur bacteria, among others, are photoheterotrophs.

Chemolithotrophs (chemoautotrophs) obtain their energy by oxidizing inorganic substances, and they use some of that energy to fix carbon dioxide. Some chemolithotrophs use reactions identical to those of the typical photosynthetic cycle (see Figure 8.3), but others use other pathways to fix carbon dioxide. Some bacteria oxidize ammonia or nitrite ions to form nitrate ions. Others oxidize hydrogen gas, hydrogen sulfide, sulfur, and other materials. Many archaea are chemolithotrophs.

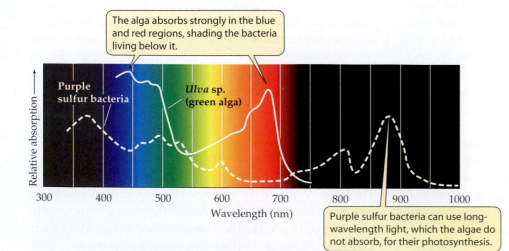

The alga absorbs strongly in the blue and red regions, shading the bacteria living below it.

Purple sulfur bacteria

Ulva sp. (green alga)

Relative absorption →

Wavelength (nm)

300   400   500   600   700   800   900   1000

Purple sulfur bacteria can use long-wavelength light, which the algae do not absorb, for their photosynthesis.

27.7 Bacteriochlorophyll Absorbs Long-Wavelength Light   The chlorophyll in *Ulva*, a green alga, absorbs no light of wavelengths longer than 750 nm. Purple sulfur bacteria, which contain bacteriochlorophyll, can conduct photosynthesis using the longer wavelengths that pass through the algae.

Deep-sea hydrothermal vent ecosystems are based on chemolithotrophic prokaryotes that are incorporated into large communities of crabs, mollusks, and giant worms, all living at a depth of 2,500 meters, below any hint of light from the sun. These bacteria obtain energy by oxidizing hydrogen sulfide and other substances released in the near-boiling water that flows from volcanic vents in the ocean floor.

Finally, **chemoheterotrophs** obtain both energy and carbon atoms from one or more complex organic compounds. Most known bacteria and archaea are chemoheterotrophs—as are all animals and fungi and many protists.

NITROGEN AND SULFUR METABOLISM. Many prokaryotes base important parts of their metabolism on reactions involving nitrogen or sulfur. For example, some bacteria carry out respiratory electron transport without using oxygen as an electron acceptor. These organisms use oxidized inorganic ions such as nitrate, nitrite, or sulfate as electron acceptors. Examples include the *denitrifiers*, bacteria that release nitrogen to the atmosphere as nitrogen gas (N_2). These normally aerobic bacteria, mostly species of the genera *Bacillus* and *Pseudomonas*, use nitrate (NO_3^-) as an electron acceptor in place of oxygen if they are kept under anaerobic conditions:

$$2\ NO_3^- + 10\ e^- + 12\ H^+ \rightarrow N_2 + 6\ H_2O$$

Nitrogen fixers convert atmospheric nitrogen gas into a chemical form usable by the nitrogen fixers themselves as well as by other organisms. They convert nitrogen gas to ammonia:

$$N_2 + 6\ H \rightarrow 2\ NH_3$$

All organisms require nitrogen for their proteins, nucleic acids, and other important compounds. The vital process of nitrogen fixation is carried out by a wide variety of archaea and bacteria, including cyanobacteria, but by no other organisms. (We'll discuss this process in detail in Chapter 37.)

Ammonia is oxidized to nitrate in the soil and in seawater by chemolithotrophic bacteria called *nitrifiers*. Bacteria of two genera, *Nitrosomonas* and *Nitrosococcus*, convert ammonia to nitrite ions (NO_2^-), and *Nitrobacter* oxidizes nitrite to nitrate (NO_3^-).

What do the nitrifiers get out of these reactions? Their chemosynthesis is powered by the energy released by the oxidation of ammonia or nitrite. For example, by passing the electrons from nitrite through an electron transport chain, *Nitrobacter* can make ATP, and using some of this ATP, it can also make NADH. With this ATP and NADH, the bacterium can convert CO_2 and H_2O to glucose.

Numerous bacteria base their metabolism on the modification of sulfur-containing ions and compounds in their environments. As examples, we have already mentioned the photoautotrophic bacteria and chemolithotrophic archaea that use H_2S as an electron donor in place of H_2O. Such uses of nitrogen and sulfur have environmental implications, as we'll see in the next section.

Prokaryotes in Their Environments

Prokaryotes live in and exploit all sorts of environments and are part of all ecosystems. In the following pages, we'll examine the roles of prokaryotes that live in soils, in water, and even in other living organisms, where they may exist in a neutral, benevolent, or parasitic relationship with their host's tissues.

Prokaryotes are important players in element cycling

Animals depend on photosynthetic plants and microorganisms for their food, directly or indirectly. But plants depend on other organisms—prokaryotes—for their own nutrition. The extent and diversity of life on Earth would not be possible without nitrogen fixation by prokaryotes. Nitrifiers are crucial to the biosphere because they convert the products of nitrogen fixation into nitrate ions, the form of nitrogen most easily used by many plants (see Figure 37.8). Plants, in turn, are the source of nitrogen compounds for animals and fungi. Denitrifiers also play a key role in keeping the nitrogen cycle going. Without denitrifiers, which convert nitrate ions back into nitrogen gas, all forms of nitrogen would leach from the soil and end up in lakes and oceans, making life on land impossible. Other prokaryotes contribute to a similar cycle of sulfur. Prokaryotes, along with fungi, return tremendous quantities of organic carbon to the atmosphere as carbon dioxide.

In the ancient past, the cyanobacteria had an equally dramatic effect on life: Their photosynthesis generated oxygen, converting Earth from an anaerobic to an aerobic environment. The result was the wholesale loss of obligate anaerobic species that could not tolerate the O_2 generated by the cyanobacteria. Only those anaerobes that were able to colonize environments that remained anaerobic survived. However, this transformation to aerobic environments made possible the evolution of cellular respiration and the subsequent explosion of eukaryotic life. What other roles do prokaryotes play in the biosphere?

Archaea help stave off global warming

A time bomb lies deep under the ocean floor. Some ten trillion tons of methane, potentially an overwhelming source of "greenhouse gas," are located there. Will this methane escape to the atmosphere, hastening global warming?

What will prevent such an escape is the presence of legions of archaea, also lying below the bottom of the seas. As methane rises from its deposits, it is metabolized by these archaea, with the result that virtually none of the methane even gets as far as the deepest waters of the ocean. Thus, these archaea play a crucial role in stabilizing the planetary environment.

Prokaryotes live on and in other organisms

Prokaryotes work together with eukaryotes in many ways. In fact, mitochondria and chloroplasts are descended from what were once free-living bacteria. Much later in evolutionary history, some plants became associated with bacteria to form cooperative nitrogen-fixing nodules on their roots (see Figure 37.5).

The tsetse fly, which transmits sleeping sickness by transferring trypanosomes (microscopic protists described in the next chapter) from one person to another, enjoys a profitable association with the bacterium *Wigglesworthia glossinidia*. Biologists who decoded the genome of *W. glossinidia* in 2002 were surprised to learn that the bacterium's tiny genome contains almost nothing but the genes needed for basic metabolism and DNA replication—and 62 genes for making ten B vitamins and other nutritional factors. Without the vitamins provided by the bacterium, the tsetse fly cannot reproduce. The bacteria, living inside the fly's cells, are in effect vitamin pills. Researchers are now trying to determine whether an attack on *W. glossinidia* may succeed in combating sleeping sickness where more obvious direct attacks on tsetse flies or the trypanosomes have failed.

Many animals, including humans, harbor a variety of bacteria and archaea in their digestive tracts. Cows depend on prokaryotes to perform important steps in digestion. Like most animals, cows cannot produce cellulase, the enzyme needed to start the digestion of the cellulose that makes up the bulk of their plant food. However, bacteria living in a special section of the gut, called the rumen, produce enough cellulase to process the cow's daily diet. Humans use some of the metabolic products—especially vitamins B_{12} and K—of bacteria living in the large intestine.

We are heavily populated, inside and out, by bacteria. Although very few of them are agents of disease, popular notions of bacteria as "germs" arouse our curiosity about those few. Let's briefly consider the roles of some bacteria as pathogens.

A small minority of bacteria are pathogens

The late nineteenth century was a productive era in the history of medicine—a time during which bacteriologists, chemists, and physicians proved that many diseases are caused by microbial agents. During this time the German physician Robert Koch laid down a set of four rules for establishing that a particular microorganism causes a particular disease:

1. The microorganism is always found in individuals with the disease.

2. The microorganism can be taken from the host and grown in pure culture.

3. A sample of the culture produces the disease when injected into a new, healthy host.

4. The newly infected host yields a new, pure culture of microorganisms identical to those obtained in the second step.

These rules, called **Koch's postulates**, were very important in a time when it was not widely accepted that microorganisms cause disease. Today medical science makes use of other, more powerful diagnostic tools. However, one important step in establishing that a coronavirus was the causal agent of SARS (Severe Acute Respiratory Syndrome), a disease that first appeared in 2003, was the satisfaction of Koch's postulates.

Only a tiny percentage of all prokaryotes are **pathogens** (disease-producing organisms), and of those that are known, all are in the domain Bacteria. For an organism to be a successful pathogen, it must overcome several hurdles:

▶ It must arrive at the body surface of a potential host.

▶ It must enter the host's body.

▶ It must evade the host's defenses.

▶ It must multiply inside the host.

▶ It must damage the host (to meet the definition of a "pathogen").

▶ It must infect a new host.

Failure to overcome any of these hurdles ends the reproductive career of a pathogenic organism. However, in spite of the many defenses available to potential hosts that we considered in Chapter 18, some bacteria are very successful pathogens.

For the host, the consequences of a bacterial infection depend on several factors. One is the **invasiveness** of the pathogen—its ability to multiply within the body of the host. Another is its **toxigenicity**—its ability to produce chemical substances (*toxins*) that are harmful to the tissues of the host. *Corynebacterium diphtheriae*, the agent that causes diphtheria, has low invasiveness and multiplies only in the throat, but its toxigenicity is so great that the entire body is affected. In contrast, *Bacillus anthracis*, which causes anthrax (a disease primarily of cattle and sheep, but also sometimes fatal in humans, as we saw in Chapter 13), has low toxigenicity but an invasiveness so great that the entire bloodstream ultimately teems with the bacteria.

There are two general types of bacterial toxins: exotoxins and endotoxins. **Endotoxins** are released when certain Gram-negative bacteria grow or lyse (burst). These toxins are lipopolysaccharides (complexes consisting of a polysaccharide and a lipid component) that form part of the outer bacterial membrane (see Figure 27.6). Endotoxins are rarely fatal; they normally cause fever, vomiting, and diarrhea.

Among the endotoxin producers are some strains of *Salmonella* and *Escherichia*.

Exotoxins are usually soluble proteins released by living, multiplying bacteria, and they may travel throughout the host's body. They are highly toxic—often fatal—to the host, but do not produce fevers. Exotoxin-induced human diseases include tetanus (from *Clostridium tetani*), botulism (from *Clostridium botulinum*), cholera (from *Vibrio cholerae*), and plague (from *Yersinia pestis*). Anthrax results from three exotoxins produced by *Bacillus anthracis*.

Remember that in spite of our frequent mention of human pathogens, only a small minority of the known prokaryotic species are pathogenic. Many more species play positive roles in our lives and in the biosphere. We make direct use of many bacteria and a few archaea in such diverse applications as cheese production, sewage treatment, and the industrial production of an amazing variety of antibiotics, vitamins, organic solvents, and other chemicals.

Pathogenic bacteria are often surprisingly difficult to combat, even with today's arsenal of antibiotics. One source of difficulty is the ability of prokaryotes to form resistant films.

Prokaryotes may form biofilms

Many unicellular microorganisms, prokaryotes in particular, tend to form dense films called **biofilms** rather than existing as clouds of individual cells. Upon contacting a solid surface, the cells lay down a gel-like polysaccharide matrix that then traps other bacteria, forming a biofilm. Once a biofilm forms, it is difficult to kill the cells. Pathogenic bacteria are hard for the immune system—and modern medicine—to combat once they form a biofilm. For example, the film may be impermeable to antibiotics. Biofilms often include a mixture of bacterial species.

The biofilm with which you are most likely to be familiar is dental plaque, the coating of bacteria and hard matrix that forms between and on your teeth unless you do a good job of flossing and brushing. Biofilms form on contact lenses, on hip replacements, and on just about any available surface. Other biofilms foul metal pipes and cause corrosion, a major problem in steam-driven electricity generation plants. Biofilms are the object of much current research. For example, some biologists are studying the chemical signals used by bacteria in biofilms to communicate with one another. By blocking the signals that lead to the production of the matrix polysaccharides, they may be able to prevent biofilms from forming.

Prokaryote Phylogeny and Diversity

The prokaryotes comprise a diverse array of microscopic organisms. To explore their diversity, let's first consider how they are classified and some of the difficulties involved in doing so.

The nucleotide sequences of prokaryotes reveal their evolutionary relationships

Why do biologists want to classify bacteria and archaea? There are three primary motivations for classification schemes: to identify unknown organisms, to reveal evolutionary relationships, and to provide universal names (see Chapter 25). Scientists and medical technologists must be able to identify bacteria quickly and accurately—when the bacteria are pathogenic, lives may depend on it.

Until recently, taxonomists based their classification schemes for the prokaryotes on readily observable phenotypic characters such as color, motility, nutritional requirements, antibiotic sensitivity, and reaction to the Gram stain. Although such schemes have facilitated the identification of prokaryotes, they have not provided insights into how these organisms evolved—a question of great interest to microbiologists and to all students of evolution. The prokaryotes and the protists (see Chapter 28) have long presented major challenges to those who attempted phylogenetic classifications. Only recently have systematists had the right tools for tackling this task.

Analyses of the nucleotide sequences of ribosomal RNA have provided us with the first apparently reliable measures of evolutionary distance among taxonomic groups. Ribosomal RNA (rRNA) is particularly useful for evolutionary studies of living organisms for several reasons:

- rRNA is evolutionarily ancient.
- No living organism lacks rRNA.
- rRNA plays the same role in translation in all organisms.
- rRNA has evolved slowly enough that sequence similarities between groups of organisms are easily found.

Let's look at just one approach to the use of rRNA for studying evolutionary relationships.

Comparisons of rRNAs from a great many organisms revealed recognizable short base sequences that are characteristic of particular taxonomic groups. These *signature sequences*, approximately 6 to 14 bases long, appear at the same approximate positions in rRNAs from related groups. For example, the signature sequence AAACUUAAAG occurs about 910 bases from one end of the small subunit of ribosomes in 100 percent of the Archaea and Eukarya tested, but in *none* of the Bacteria tested. Several signature sequences distinguish each of the three domains. Similarly, the major groups within the bacteria and archaea possess unique signature sequences.

These data sound promising, but things aren't as simple as we might wish. When biologists examined other genes and RNAs, contradictions began to appear and new questions arose. Analyses of different nucleotide sequences suggested different phylogenetic patterns. How could such a situation have arisen?

Lateral gene transfer muddied the phylogenetic waters

It is now clear that, from early in evolution to the present day, genes have been moving among prokaryotic species by **lateral gene transfer**. As we have seen, a gene from one species can become incorporated into the genome of another. Mechanisms of lateral gene transfer include transfer by plasmids and viruses and uptake of DNA by transformation. Such transfers are well documented, not just between bacterial species or archaeal species, but also across the boundaries between bacteria and archaea and between prokaryotes and eukaryotes.

A gene that has been transferred will be inherited by the recipient's progeny and in time will be recognized as part of the normal genome of the descendants. Biologists are still assessing the extent of lateral gene transfer among prokaryotes and its implications for phylogeny, especially at the early stages of evolution.

Figure 27.8 is an overview of the major clades in the domains Bacteria and Archaea that we will discuss further in this chapter. This phylogeny is based on the evidence that is currently available, but keep in mind that a new picture is likely to emerge within the next decade, based on new nucleotide sequence data and new information about the currently understudied archaea.

Mutations are a major source of prokaryotic variation

Assuming that the prokaryote groups we are about to describe do indeed represent clades, these groups are amazingly complex. A single lineage of bacteria or archaea may contain the most extraordinarily diverse species; on the other hand, a species in one group may be phenotypically almost indistinguishable from one or many species in another group. What are the sources of these phylogenetic patterns?

Although prokaryotes can acquire new alleles by transformation, transduction, or conjugation, the most important sources of genetic variation in populations of prokaryotes are probably mutation and genetic drift (see Chapter 23). Mutations, especially recessive mutations, are slow to make their presence felt in populations of humans and other diploid organisms. In contrast, a mutation in a prokaryote, which is

haploid, has immediate consequences for that organism. If it is not lethal, it will be transmitted to and expressed in the organism's daughter cells—and in their daughter cells, and so on. Thus, a beneficial mutant allele spreads rapidly.

The rapid multiplication of many prokaryotes, coupled with mutation, natural selection, and genetic drift, allows rapid phenotypic changes within their populations. Important changes, such as loss of sensitivity to an antibiotic, can occur over broad geographic areas in just a few years. Think how many significant metabolic changes could have occurred over even modest time spans, let alone over the entire history of life on Earth. When we introduce the proteobacteria, the largest group of bacteria, you will see that its different subgroups have easily and rapidly adopted and abandoned metabolic pathways under selective pressure from their environments.

The Bacteria

The best-studied prokaryotes are the bacteria. We will describe bacterial diversity using a currently popular classification scheme that enjoys considerable support from nucleotide sequence data. More than a dozen clades have been proposed under this scheme; we will describe just of a few of them here. The higher-order relationships among these groups of prokaryotes are not known. Some biologists describe them as kingdoms, some as subkingdoms, and others as phyla; here, we simply call them groups. We'll pay the closest attention to five groups: the proteobacteria, cyanobacteria, spirochetes, chlamydias, and firmicutes (see Figure 27.8). First, however, we'll mention one property that is shared by members of three other groups.

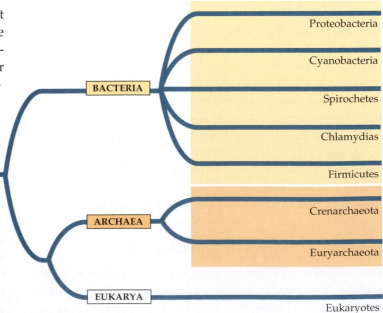

27.8 Two Domains: A Brief Overview   This abridged summary classification of the domains Bacteria and Archaea shows their relationships to each other and to the Eukarya. The relationships among the many clades of bacteria, not all of which are listed here, are unresolved at this time.

Some bacteria are heat lovers

Three of the bacterial groups that may have branched out earliest during bacterial evolution are all **thermophiles** (heat lovers), as are the most ancient of the archaea. This observation supports the hypothesis that the first living organisms were thermophiles that appeared in an environment much hotter than those that predominate today.

The Proteobacteria are a large and diverse group

By far the largest group of bacteria, in terms of numbers of described species, is the **proteobacteria**, sometimes referred to as the *purple bacteria*. Among the proteobacteria are many species of Gram-negative, bacteriochlorophyll-containing, sulfurusing photoautotrophs. How-

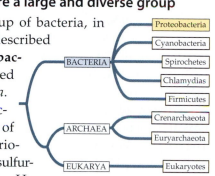

ever, the proteobacteria also include dramatically diverse bacteria that bear no resemblance to those species in phenotype. The mitochondria of eukaryotes were derived from proteobacteria by endosymbiosis.

No characteristic demonstrates the diversity of the proteobacteria more clearly than their metabolic pathways (Figure 27.9). The common ancestor of all the proteobacteria was probably a photoautotroph. Early in evolution, two groups of proteobacteria lost their ability to photosynthesize and have been chemoheterotrophs ever since. The other three groups still have photoautotrophic members, but in *each* group, some evolutionary lines have abandoned photoautotrophy and taken up other modes of nutrition. There are chemolithotrophs and chemoheterotrophs in all three groups. Why? One possibility is that each of the trends in Figure 27.9 was an evolutionary response to selective pressures encountered as these bacteria colonized new habitats that presented new challenges and opportunities.

Among the proteobacteria are some nitrogen-fixing genera, such as *Rhizobium* (see Figure 37.7), and other bacteria that contribute to the global nitrogen and sulfur cycles. *E. coli*, one of the most studied organisms on Earth, is a proteobacterium. So, too, are many of the most famous human pathogens, such as *Yersinia pestis*, *Vibrio cholerae*, and *Salmonella typhimurium*, all mentioned in our discussion of pathogens above.

Fungi cause most plant diseases, and viruses cause others, but about 200 plant diseases are of bacterial origin. *Crown gall*, with its characteristic tumors (Figure 27.10), is one of the most striking. The causal agent of crown gall is *Agrobacterium tumefaciens*, which harbors a plasmid used in recombinant DNA studies as a vehicle for inserting genes into new plant hosts (see Chapter 16).

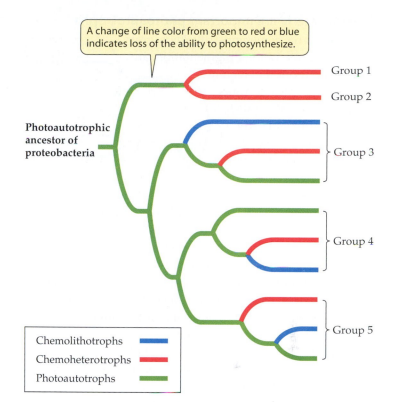

27.9 The Evolution of Metabolism in the Proteobacteria The common ancestor of all proteobacteria was probably a photoautotroph. As they encountered new environments, groups 1 and 2 lost the ability to photosynthesize; in the other three groups, some evolutionary lines became chemolithotrophs or chemoheterotrophs.

27.10 A Crown Gall This colorful tumor growing on the stem of a geranium plant is caused by the Gram-negative bacillus *Agrobacterium tumefaciens*.

Cyanobacteria are important photoautotrophs

Cyanobacteria, sometimes called *blue-green bacteria* because of their pigmentation, are photoautotrophs that require only water, nitrogen gas, oxygen, a few mineral elements, light, and carbon dioxide to survive. They use chlorophyll *a* for photosynthesis and release oxygen gas; many species also fix nitrogen. Their photosynthesis was the basis of the "oxygen revolution" that transformed Earth's atmosphere.

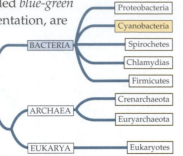

Cyanobacteria carry out the same type of photosynthesis that is characteristic of eukaryotic photosynthesizers. They contain elaborate and highly organized internal membrane systems called *photosynthetic lamellae*, or *thylakoids*. The chloroplasts of photosynthetic eukaryotes are derived from an endosymbiotic cyanobacterium.

Cyanobacteria may live free as single cells or associate in colonies. Depending on the species and on growth conditions, colonies of cyanobacteria may range from flat sheets one cell thick to filaments to spherical balls of cells.

Some filamentous colonies of cyanobacteria differentiate into three cell types: vegetative cells, spores, and heterocysts (Figure 27.11). *Vegetative cells* photosynthesize, *spores* are resting cells that can eventually develop into new filaments, and *heterocysts* are cells specialized for nitrogen fixation. All of the known cyanobacteria with heterocysts fix nitrogen. Heterocysts also have a role in reproduction: When filaments break apart to reproduce, the heterocyst may serve as a breaking point.

Spirochetes look like corkscrews

Spirochetes are Gram-negative, motile, chemoheterotrophic bacteria characterized by unique structures called axial filaments, which are modified flagella running through the periplasm (see Figure 27.4*a*). The cell body is a long cylinder coiled into a spiral (Figure 27.12). The axial filaments begin at either end of the cell and overlap in the middle, and there are typical basal bodies where they are attached to the cell wall. The basal bodies rotate, as they do in other prokaryotic flagella. Many spirochetes live in humans as parasites; a few are pathogens, including those that cause syphilis and Lyme disease. Others live free in mud or water.

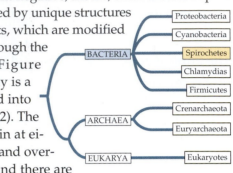

27.11 Cyanobacteria (*a*) *Anabaena* is a genus of cyanobacteria that form filamentous colonies containing three cell types. (*b*) A thin neck attaches a heterocyst to each of two vegetative cells in a filament. (*c*) Cyanobacteria appear in enormous numbers in some environments. This California pond has experienced eutrophication; phosphorus and other nutrients generated by human activity have accumulated in the pond, feeding an immense green mat (commonly referred to as "pond scum") that is made up of several species of free-living cyanobacteria.

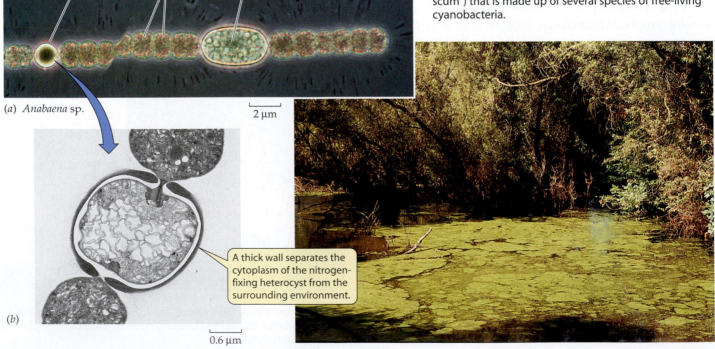

Heterocyst   Vegetative cells   Spore

(*a*) *Anabaena* sp.   2 μm

A thick wall separates the cytoplasm of the nitrogen-fixing heterocyst from the surrounding environment.

(*b*)

0.6 μm

Treponema pallidum   200 nm

27.12 A Spirochete   This corkscrew-shaped bacterium causes syphilis in humans.

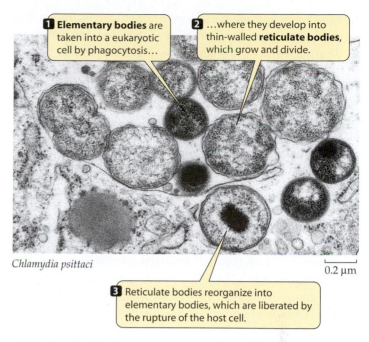

1 **Elementary bodies** are taken into a eukaryotic cell by phagocytosis...

2 ...where they develop into thin-walled **reticulate bodies**, which grow and divide.

3 Reticulate bodies reorganize into elementary bodies, which are liberated by the rupture of the host cell.

Chlamydia psittaci   0.2 μm

27.13 Chlamydias Change Form during Their Life Cycle
Elementary bodies and reticulate bodies are the two major phases of the chlamydia life cycle.

Chlamydias are extremely small

Chlamydias are among the smallest bacteria (0.2–1.5 μm in diameter). They can live only as parasites within the cells of other organisms. These tiny Gram-negative cocci are unique prokaryotes because of their complex life cycle, which involves two different forms of cells, *elementary bodies* and *reticulate bodies* (Figure 27.13). In humans, various strains of chlamydias cause eye infections (especially trachoma), sexually transmitted diseases, and some forms of pneumonia.

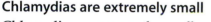

Most firmicutes are Gram-positive

The **firmicutes** are sometimes referred to as the *Gram-positive bacteria*, but some firmicutes are Gram-negative, and some have no cell wall at all. Nonetheless, the firmicutes constitute a clade.

Some firmicutes produce **endospores** (Figure 27.14)—heat-resistant resting structures—when a key nutrient such as nitrogen or carbon becomes scarce. The bacterium replicates its DNA and encapsulates one copy, along with some of its cytoplasm, in a tough cell wall heavily thickened with peptidoglycan and surrounded by a spore coat. The parent cell then breaks down, releasing the endospore. Endospore production is not a reproductive process; the endospore merely replaces the parent cell. The endospore,

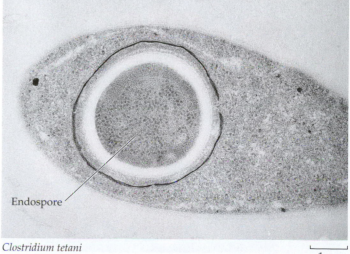

Endospore

Clostridium tetani   1 μm

27.14 The Endospore: A Structure for Waiting Out Bad Times
This firmicute, which causes tetanus, produces endospores as resistant resting structures.

however, can survive harsh environmental conditions that would kill the parent cell, such as high or low temperatures or drought, because it is *dormant*—its normal activity is suspended. Later, if it encounters favorable conditions, the endospore becomes metabolically active and divides, forming new cells like the parent. Some endospores can be reactivated after more than a thousand years of dormancy. There are credible claims of reactivation of *Bacillus* endospores after millions of years—and even one claim, of uncertain validity, of more than a billion years!

Members of this endospore-forming group of firmicutes include the many species of *Clostridium* and *Bacillus*. The toxins produced by *C. botulinum* are among the most poisonous ever discovered; the lethal dose for humans is about one-millionth of a gram (1 µg). *B. anthracis*, as noted above, is the anthrax pathogen.

The genus *Staphylococcus*—the staphylococci—includes firmicutes that are abundant on the human body surface; they are responsible for boils and many other skin problems (Figure 27.15). *S. aureus* is the best-known human pathogen in this genus; it is found in 20 to 40 percent of normal adults (and in 50 to 70 percent of hospitalized adults). It can cause respiratory, intestinal, and wound infections in addition to skin diseases.

Actinomycetes are firmicutes that develop an elaborately branched system of filaments (Figure 27.16). These bacteria closely resemble the filamentous growth habit of fungi at a reduced scale. Some actinomycetes reproduce by forming chains of spores at the tips of the filaments. In species that do not form spores, the branched, filamentous growth ceases and the structure breaks up into typical cocci or bacilli, which then reproduce by fission.

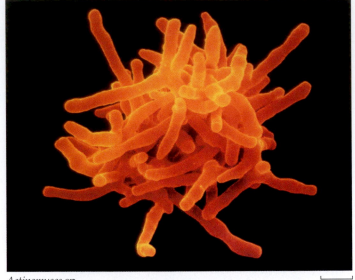

Actinomyces sp.   2 µm

27.16 Filaments of an Actinomycete   The branching filaments seen in this scanning electron micrograph are typical of actinomycetes, a medically important bacterial group.

The actinomycetes include several medically important bacteria. *Mycobacterium tuberculosis* causes tuberculosis. *Streptomyces* produces streptomycin as well as hundreds of other antibiotics. We derive most of our antibiotics from members of the actinomycetes.

Another interesting group of firmicutes, the **mycoplasmas**, lack cell walls, although some have a stiffening material outside the plasma membrane. Some of them are the smallest cellular creatures ever discovered—they are even smaller than chlamydias (Figure 27.17). The smallest mycoplasmas capable of multiplication have a diameter of about 0.2 µm. They are small in another crucial sense as well: They

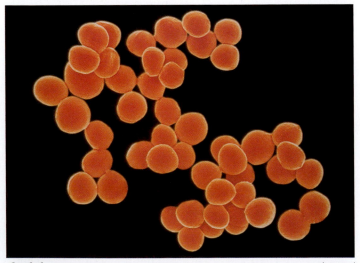

Staphylococcus aureus   1 µm

27.15 Gram-Positive Firmicutes   "Grape clusters" are the usual arrangement of Gram-positive staphylococci.

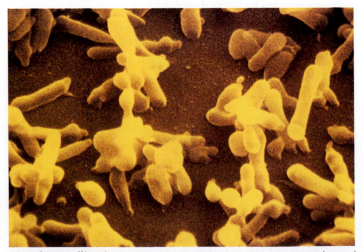

Mycoplasma gallisepticum   0.4 µm

27.17 The Tiniest Living Cells   Containing only about one-fifth as much DNA as *E. coli*, mycoplasmas are the smallest known bacteria.

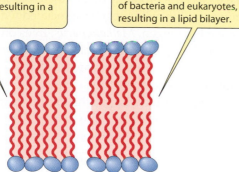

> Some archaea have long-chain hydrocarbons with glycerol at both ends, spanning the membrane and resulting in a lipid monolayer.

> Other archaeal hydrocarbons fit the same membrane template as do the fatty acids of bacteria and eukaryotes, resulting in a lipid bilayer.

27.18 Membrane Architecture in Archaea The long-chain hydrocarbons of may archaeal membranes are branched, and may have glycerol at both ends. This lipid monolayer structure (on the left) still fits into a biological membrane, however. In fact, all three domains have similar membrane structures.

have less than half as much DNA as do most other prokaryotes—but they still can grow autonomously. It has been speculated that the amount of DNA in a mycoplasma may be the minimum amount required to encode the essential properties of a living cell.

We have discussed five clades of bacteria in some detail, but other bacterial clades are well known, and there may be dozens more waiting to be discovered. This conservative estimate is based on the fact that many bacteria and archaea have never been cultured in the laboratory.

The Archaea

The domain Archaea consists mainly of prokaryotic genera that live in habitats notable for characteristics such as extreme salinity (salt content), low oxygen concentrations, high temperatures, or high or low pH. However, many archaea live in habitats that are not extreme. Perhaps the largest number of archaea live in the ocean depths.

On the face of it, the Archaea do not seem to belong together as a group. One current classification scheme divides the domain into two principal groups, **Euryarchaeota** and **Crenarchaeota**. In fact, we know relatively little about the phylogeny of archaea, in part because the study of archaea is still in its early stages. We do know that archaea share certain characteristics.

The Archaea share some unique characteristics

Two characteristics shared by all archaea are the absence of peptidoglycan in their cell walls and the presence of lipids of distinctive composition in their cell membranes (see Table 27.1). The base sequences of their ribosomal RNAs support

a close evolutionary relationship among them. Their separation from the Bacteria and Eukarya was clarified when biologists sequenced the first archaeal genome. It consisted of 1,738 genes, more than half of which were unlike any genes ever found in the other two domains.

The unusual lipids in the membranes of archaea deserve some description. They are found in all archaea, and in no bacteria or eukaryotes. Most bacterial and eukaryotic membrane lipids contain unbranched long-chain fatty acids connected to glycerol by **ester linkages**:

(Figure 27.18, right; see also Figure 3.18). In contrast, some archaeal membrane lipids contain long-chain hydrocarbons connected to glycerol by **ether linkages**:

In addition, the long-chain hydrocarbons of the archaea are branched. One class of these lipids, with hydrocarbon chains 40 carbon atoms in length, contains glycerol at *both* ends of the hydrocarbons (Figure 27.18, left). This *lipid monolayer* structure, unique to the domain Archaea, still fits in a biological membrane because the lipids are twice as long as the typical lipids in the bilayers of other membranes (see Figure 27.18). Lipid monolayers and bilayers are both found among the archaea.

In spite of this striking difference in their membrane lipids, all three domains have membranes with similar overall structures, dimensions, and functions.

Most Crenarchaeota live in hot, acidic places

Most known Crenarchaeota are both thermophilic (heat-loving) and *acidophilic* (acid-loving). Members of the genus *Sulfolobus* live in hot sulfur springs at temperatures of 70–75°C. They die of "cold" at 55°C (131°F). Hot sulfur springs are also extremely acidic. *Sulfolobus* grows best in the range from pH 2 to pH 3, but it readily tolerates pH values as low as 0.9. One species of the genus *Ferroplasma* lives at a pH near 0. Some acidophilic hyperthermophiles maintain an internal pH near 7 (neutral) in spite of their acidic environment. These

27.19 Some Would Call It Hell; Archaea Call It Home Masses of heat- and acid-loving archaea form an orange mat inside a volcanic vent on the island of Kyushu, Japan. Sulfurous residue is visible at the edges of the archaeal mat.

and other hyperthermophiles thrive where very few other organisms can even survive (Figure 27.19).

The Euryarchaeota live in many surprising places

Some species of Euryarchaeota share the property of producing methane (CH_4) by reducing carbon dioxide. All of these **methanogens** are obligate anaerobes, and methane production is the key step in their energy metabolism. Comparison of rRNA nucleotide sequences revealed a close evolutionary relationship among all these methanogens, which were previously assigned to several unrelated bacterial groups.

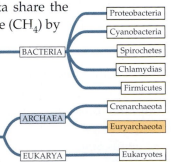

Methanogens release approximately 2 billion tons of methane gas into Earth's atmosphere each year, accounting for 80 to 90 percent of the methane in the atmosphere, including that associated with mammalian belching. Approximately a third of this methane comes from methanogens living in the guts of grazing herbivores such as cows. Methane

is increasing in Earth's atmosphere by about 1 percent per year and is a major contributor to the greenhouse effect. Most of the increase is probably due to increases in cattle and rice farming and the methanogens associated with both.

One methanogen, *Methanopyrus*, lives on the ocean bottom near blazing hydrothermal vents. *Methanopyrus* can survive and grow at 110°C. It grows best at 98°C and not at all at temperatures below 84°C.

Another group of Euryarchaeota, the *extreme halophiles* (salt lovers), lives exclusively in very salty environments. Because they contain pink carotenoids, they can be seen easily under some circumstances (Figure 27.20). Halophiles grow in the Dead Sea and in brines of all types: Pickled fish may sometimes show reddish pink spots that are colonies of halophilic archaea. Few other organisms can live in the saltiest of the homes that the extreme halophiles occupy; most would "dry" to death, losing too much water to the hypertonic environment. Extreme halophiles have been found in lakes with pH values as high as 11.5—the most alkaline environment inhabited by living organisms, and almost as alkaline as household ammonia.

Some of the extreme halophiles have a unique system for trapping light energy and using it to form ATP—without using any form of chlorophyll—when oxygen is in short supply. They use the pigment *retinal* (also found in the vertebrate eye) combined with a protein to form a light-absorbing molecule called *bacteriorhodopsin*, and they form ATP by a chemiosmotic mechanism of the sort described in Figure 7.12.

27.20 Extreme Halophiles Commercial seawater evaporating ponds, such as these in San Francisco Bay, are attractive homes for salt-loving archaea, which are easily visible because of their carotenoids.

Another member of the Euryarchaeota, *Thermoplasma*, has no cell wall. It is thermophilic and acidophilic, its metabolism is aerobic, and it lives in coal deposits. It has the smallest genome among the archaea, and perhaps the smallest (along with the mycoplasmas) of any free-living organism—1,100,000 base pairs.

In addition to these archaea that are found in amazing habitats, many Crenarchaeota and Euryarchaeota live in environments that are not extreme.

Chapter Summary

Why Three Domains?

▶ Living organisms can be divided into three domains: Bacteria, Archaea, and Eukarya. Both the Archaea and the Bacteria are prokaryotic; the Eukarya constitute the rest of the living world. The Bacteria and the Archaea are less closely related to each other than are the Archaea and the Eukarya. **Review Figure 27.2, Table 27.1**

▶ The common ancestor of all three domains lived more than 3 billion years ago, and the common ancestor of the Archaea and Eukarya at least 2 billion years ago.

See Web/CD Tutorial 27.1

General Biology of the Prokaryotes

▶ The prokaryotes are the most numerous organisms on Earth, and they occupy an enormous variety of habitats.

▶ Most prokaryotes are cocci, bacilli, or spiral forms. Some link together to form associations, but very few are truly multicellular. **Review Figure 27.3**

▶ Prokaryotes lack nuclei, membrane-enclosed organelles, and cytoskeletons. Their chromosomes are circular. They often contain plasmids. Some prokaryotes contain internal membrane systems.

▶ Many prokaryotes move by means of flagella, gas vesicles, or gliding mechanisms. Prokaryotic flagella rotate rather than beat. **Review Figures 27.4, 27.5**

▶ Prokaryotic cell walls differ from those of eukaryotes. Bacterial cell walls generally contain peptidoglycan. Differences in peptidoglycan content result in different reactions to the Gram stain. **Review Figure 27.6. See Web/CD Activity 27.1**

▶ Prokaryotes reproduce asexually by fission, but also exchange genetic information.

▶ Prokaryotes have diverse metabolic pathways and nutritional modes. They include obligate anaerobes, facultative anaerobes, and obligate aerobes. The major nutritional types are photoautotrophs, photoheterotrophs, chemolithotrophs, and chemoheterotrophs. Some prokaryotes base their energy metabolism on nitrogen- or sulfur-containing ions. **Review Figure 27.7 and Table 27.2**

Prokaryotes in Their Environments

▶ Some prokaryotes play key roles in global nitrogen and sulfur cycles. Important players in the nitrogen cycle are the nitrogen fixers, nitrifiers, and denitrifiers.

▶ Photosynthesis by cyanobacteria generated the oxygen gas that permitted the evolution of aerobic respiration and the appearance of present-day eukaryotes.

▶ Archaea lying beneath the oceans prevent large deposits of methane, a "greenhouse gas," from accumulating in the oceans and the atmosphere.

▶ Many prokaryotes live in or on other organisms, with neutral, beneficial, or harmful effects.

▶ A small minority of bacteria are pathogens. Pathogens vary with respect to their invasiveness and toxigenicity. Some produce endotoxins, which are rarely fatal to their hosts; others produce exotoxins, which tend to be highly toxic.

▶ Prokaryotes and some unicellular eukaryotes form resistant biofilms that present medical and industrial problems.

Prokaryote Phylogeny and Diversity

▶ Phylogenetic classification of prokaryotes is now based on rRNA sequences and other molecular evidence.

▶ Lateral gene transfer among prokaryotes, which has occurred throughout evolutionary history, makes it difficult to infer prokaryote phylogeny.

▶ Evolution, powered by mutation, natural selection, and genetic drift, can proceed rapidly in prokaryotes because they are haploid and can multiply rapidly.

The Bacteria

▶ There are more known bacteria than known archaea. One phylogenetic classification of the domain Bacteria groups them into more than a dozen clades. **Review Figure 27.8**

▶ The three clades that may contain the most ancient bacteria, like the most ancient archaea, are thermophiles, suggesting that life originated in a hot environment.

▶ All four nutritional types occur in the largest bacterial group, the proteobacteria. Metabolism in different groups of proteobacteria has evolved along different lines. **Review Figure 27.9**

▶ Cyanobacteria, unlike other bacteria, photosynthesize using the same pathways plants use. Many cyanobacteria fix nitrogen.

▶ Spirochetes move by means of axial filaments.

▶ Chlamydias are tiny parasites that live within the cells of other organisms.

▶ Firmicutes are diverse; some of them produce endospores as resting structures that resist harsh conditions. Actinomycetes, some of which produce important antibiotics, grow as branching filaments.

▶ Mycoplasmas, the tiniest living things, lack conventional cell walls. They have very small genomes.

The Archaea

▶ Archaea have cell walls lacking peptidoglycan, and their membrane lipids differ from those of bacteria and eukaryotes, containing branched long-chain hydrocarbons connected to glycerol by ether linkages. **Review Figure 27.18**

▶ The domain Archaea can be divided into two principal groups, Crenarchaeota and Euryarchaeota.

▶ Crenarchaeota are mostly heat-loving and often acid-loving archaea.

▶ Methanogens produce methane by reducing carbon dioxide. Some methanogens live in the guts of herbivorous animals; others occupy high-temperature environments on the ocean floor.

▶ Extreme halophiles are salt lovers that often lend a pinkish color to salty environments; some halophiles also grow in extremely alkaline environments.

▶ Archaea of the genus *Thermoplasma* lack cell walls, are thermophilic and acidophilic, and have a tiny genome (1,100,000 base pairs).

▶ Many archaea, including members of both major groups, live in environments that are not extreme.

Self-Quiz

1. Most prokaryotes
   a. are agents of disease.
   b. lack ribosomes.
   c. evolved from the most ancient eukaryotes.
   d. lack a cell wall.
   e. are chemoheterotrophs.

2. The division of the living world into three domains
   a. is strictly arbitrary.
   b. was inspired by the morphological differences between archaea and bacteria.
   c. emphasizes the greater importance of eukaryotes.
   d. was proposed by the early microscopists.
   e. is strongly supported by data on rRNA sequences.

3. Which statement about the archaeal genome is true?
   a. It is much more similar to the bacterial genome than to eukaryotic genomes.
   b. More than half of its genes are genes that are never observed in bacteria or eukaryotes.
   c. It is much smaller than the bacterial genome.
   d. It is housed in the nucleus.
   e. No archaeal genome has yet been sequenced.

4. Which statement about nitrogen metabolism is *not* true?
   a. Certain prokaryotes reduce atmospheric N_2 to ammonia.
   b. Nitrifiers are soil bacteria.
   c. Denitrifiers are strict anaerobes.
   d. Nitrifiers obtain energy by oxidizing ammonia and nitrite.
   e. Without the nitrifiers, terrestrial organisms would lack a nitrogen supply.

5. All photosynthetic bacteria
   a. use chlorophyll *a* as their photosynthetic pigment.
   b. use bacteriochlorophyll as their photosynthetic pigment.
   c. release oxygen gas.
   d. produce particles of sulfur.
   e. are photoautotrophs.

6. Gram-negative bacteria
   a. appear blue to purple following Gram staining.
   b. are the most abundant of the bacterial groups.
   c. are all either bacilli or cocci.
   d. contain no peptidoglycan in their cell walls.
   e. are all photosynthetic.

7. Endospores
   a. are produced by viruses.
   b. are reproductive structures.
   c. are very delicate and easily killed.
   d. are resting structures.
   e. lack cell walls.

8. Actinomycetes
   a. are important producers of antibiotics.
   b. belong to the kingdom Fungi.
   c. are never pathogenic to humans.
   d. are gram-negative.
   e. are the smallest known bacteria.

9. Which statement about mycoplasmas is *not* true?
   a. They lack cell walls.
   b. They are the smallest known cellular organisms.
   c. They contain the same amount of DNA as do other prokaryotes.
   d. They cannot be killed with penicillin.
   e. Some are pathogens.

10. Archaea
    a. have cytoskeletons.
    b. have distinctive lipids in their plasma membranes.
    c. survive only at moderate temperatures and near neutrality.
    d. all produce methane.
    e. have substantial amounts of peptidoglycan in their cell walls.

For Discussion

1. Why do systematic biologists find rRNA sequence data more useful than data on metabolism or cell structure for classifying prokaryotes?

2. Why does lateral gene transfer make it so difficult to arrive at agreement on prokaryote phylogeny?

3. Differentiate among the members of the following sets of related terms:
   a. prokaryotic/eukaryotic
   b. obligate anaerobe/facultative anaerobe/obligate aerobe
   c. photoautotroph/photoheterotroph/chemolithotroph/ chemoheterotroph
   d. Gram-positive/Gram-negative

4. Why are the endospores of firmicutes not considered to be reproductive structures?

5. Until fairly recently, the cyanobacteria were called blue-green algae and were not grouped with the bacteria. Suggest several reasons for this (abandoned) tendency to separate the cyanobacteria from the bacteria. Why are the cyanobacteria now grouped with the other bacteria?

6. The actinomycetes are of great commercial interest. Why?

7. Thermophiles are of great interest to molecular biologists and biochemists. Why? What practical concerns might motivate that interest?

28 Protists and the Dawn of the Eukarya

Kwame's illness began with a fever and shaking chills, followed by muscle aches, nausea, and vomiting. The child's kidneys failed, he developed seizures and went into a coma, and finally he died. Every 30 seconds, malaria kills someone somewhere—usually in sub-Saharan Africa, although malaria occurs in more than 100 countries and territories. About *600 million* people have this disease.

Mosquitoes carry the malaria pathogen from person to person. This pathogen is not a bacterium—rather, it is a tiny eukaryote, *Plasmodium falciparum*. Probably the most obvious visible difference between it and the prokaryotes is that *Plasmodium* has numerous compartments—membrane-enclosed organelles that perform specialized functions. This single-celled pathogen has a cytoskeleton, a nucleus enclosed by a nuclear envelope, and several kinds of organelles. As a member of the domain Eukarya, it differs from members of the two prokaryotic domains in other important ways as well.

The flexibility and options that arose once the eukaryotic cell had evolved resulted in a profusion of body forms and myriad specialized functions. Eukaryotic evolution has produced great diversity, especially among the multicellular clades, but even among the unicellular members of the domain. In both multicellular and unicellular forms, however, there are also many cases of convergent evolution; for example, organisms with an amoeba-like body form arose several times. These various amoebas are examples of organisms called protists.

Protists Defined

Many modern members of the Eukarya—trees, mushrooms, and dogs, not to mention ourselves—are familiar to us. We would have no problem recognizing these organisms as members of the kingdoms Plantae, Fungi, and Animalia. However, amoebas and a dazzling assortment of other eukaryotes, mostly microscopic organisms, don't fit into these three kingdoms. We call all those eukaryotes that are neither plants, animals, nor fungi **protists**. *The protists are not a clade; they are a polyphyletic group* (see Figure 1.8). Some protists are more closely related to the animals than they are to other protists. Some protists are motile, while others are stationary; some are photosynthetic, while others are heterotrophic; most are unicellular, while

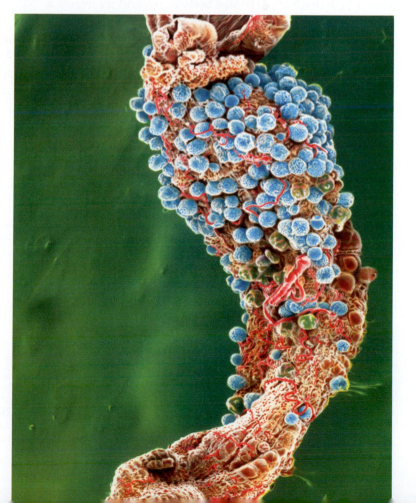

***Plasmodium falciparum*, the Malaria Parasite** This stomach wall of an *Anopheles* mosquito is covered with cells (artificially colored blue) of a particular stage of the *Plasmodium* life cycle. These *Plasmodium* cells will give rise to cells that the mosquito can transmit to humans, causing malaria.

(a) *Peridinium* sp.

(b) *Giardia* sp.

(c) *Macrocystis* sp.

15 μm   5 μm

28.1 Three Protists   (a) Most dinoflagellates are photosynthetic unicellular protists. (b) *Giardia* is a unicellular parasite of humans and other mammals. (c) Giant kelps are some of the world's longest organisms.

some giant kelps are not only multicellular but also huge, sometimes achieving lengths greater than that of a football field (Figure 28.1).

The protists include some of the most ancient eukaryotic organisms as well as the ancestors of the plants, animals, and fungi. The origin of the eukaryotic cell was one of the pivotal events in evolutionary history. In this chapter, we'll describe the origin and early diversification of the eukaryotes and the complexity achieved by some single cells. Then we'll explore some of the diversity of protist body forms and try to give a sense of developing current views of the evolutionary relationships of some of the protists.

The Origin of the Eukaryotic Cell

The eukaryotic cell differs in many ways from the prokaryotic cell. How did it originate? Given the nature of evolutionary processes, the differences cannot all have arisen simultaneously. We think we can make some reasonable inferences about the most important events, bearing in mind that the global environment underwent an enormous change—from anaerobic to aerobic—during the course of these events. As you read this chapter, keep in mind that the steps we suggest are just that: reasonable inferences. This version of the story is one of a few under current consideration. We present it as a framework for thinking about this challenging problem, not as a set of facts.

The modern eukaryotic cell arose in several steps

The essential steps in the origin of the eukaryotic cell include:

▶ The origin of a flexible cell surface
▶ The origin of a cytoskeleton
▶ The origin of a nuclear envelope

▶ The appearance of digestive vesicles
▶ The endosymbiotic acquisition of certain organelles

WHAT A FLEXIBLE CELL SURFACE ALLOWS.   Many ancient fossil prokaryotes look like rods, and we presume that they, like most present-day prokaryotic cells, had firm cell walls. The first step toward the eukaryotic condition may have been the loss of the cell wall by an ancestral prokaryotic cell. This may not seem like an obvious first step, but consider the possibilities open to a flexible cell without a wall.

First, think of cell size. As a cell grows larger, its surface area-to-volume ratio decreases (see Figure 4.3). Unless the surface area can be increased, the cell volume will reach an upper limit. If the surface is flexible, it can fold inward and elaborate itself, creating more surface area for gas and nutrient exchange (Figure 28.2). With a surface flexible enough to

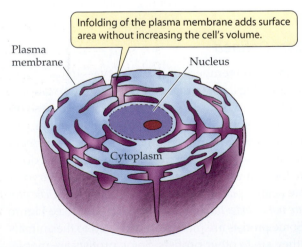

Infolding of the plasma membrane adds surface area without increasing the cell's volume.

Plasma membrane

Nucleus

Cytoplasm

28.2 Membrane Infolding   The loss of the rigid prokaryotic cell wall allowed the plasma membrane to fold inward and create more surface area.

allow infolding, the cell can exchange materials with its environment rapidly enough to sustain a larger volume and more rapid metabolism. Further, a flexible surface can pinch off bits of the environment, bringing them into the cell by endocytosis (Figure 28.3).

The chromosome of a bacterial cell is attached to a site on its plasma membrane. If that region of the plasma membrane were to fold into the cell, the first step would be taken toward the evolution of a nucleus, the key feature of the eukaryotic cell. What are some other likely early changes?

CHANGES IN CELL STRUCTURE AND FUNCTION. Other early steps in the evolution of the eukaryotic cell are likely to have included three advances: the formation of ribosome-studded internal membranes, some of which surrounded the DNA (see Figure 28.3); the appearance of a cytoskeleton; and the evolution of digestive vesicles.

A cytoskeleton made up of actin fibers and microtubules would allow the cell to manage changes in shape, to distribute daughter chromosomes, and to move materials from one part of the now much larger cell to other parts. The origin of the cytoskeleton remains a mystery, heightened by the fact that the genes that encode most of the cytoskeleton are present in neither bacteria nor archaea. An intriguing and controversial suggestion is that a fourth domain of life, now long extinct, originated these genes and transferred them laterally to an ancestor of the early eukaryotes.

From an intermediate kind of cell, the next advance was probably to a cell that we could call a *phagocyte*—a motile cell that could prey on other cells by engulfing and digesting them. The first true eukaryote possessed a cytoskeleton and a nuclear envelope. It may have had an associated endoplasmic reticulum and Golgi apparatus, and perhaps one or more flagella of the eukaryotic type.

ENDOSYMBIOSIS AND ORGANELLES. While the processes already outlined were taking place, the cyanobacteria were very busy, generating oxygen gas as a product of photosynthesis. The increasing O_2 levels in the atmosphere had disastrous consequences for most other living things because most organisms of the

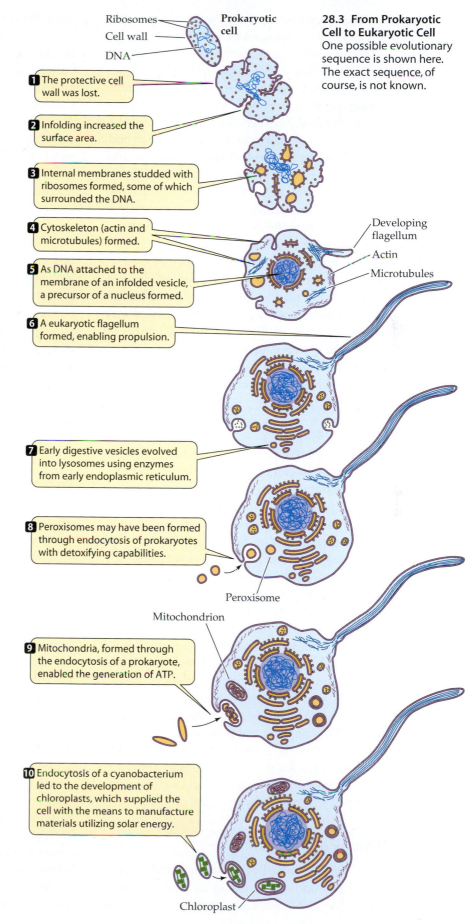

Ribosomes
Cell wall
DNA

Prokaryotic cell

28.3 From Prokaryotic Cell to Eukaryotic Cell One possible evolutionary sequence is shown here. The exact sequence, of course, is not known.

1 The protective cell wall was lost.

2 Infolding increased the surface area.

3 Internal membranes studded with ribosomes formed, some of which surrounded the DNA.

4 Cytoskeleton (actin and microtubules) formed.

5 As DNA attached to the membrane of an infolded vesicle, a precursor of a nucleus formed.

6 A eukaryotic flagellum formed, enabling propulsion.

Developing flagellum
Actin
Microtubules

7 Early digestive vesicles evolved into lysosomes using enzymes from early endoplasmic reticulum.

8 Peroxisomes may have been formed through endocytosis of prokaryotes with detoxifying capabilities.

Peroxisome

Mitochondrion

9 Mitochondria, formed through the endocytosis of a prokaryote, enabled the generation of ATP.

10 Endocytosis of a cyanobacterium led to the development of chloroplasts, which supplied the cell with the means to manufacture materials utilizing solar energy.

Chloroplast

time (archaea and bacteria) were unable to tolerate the newly aerobic, oxidizing environment. But some prokaryotes managed to cope with these changes, and—fortunately for us—so did some of the ancient phagocytes.

In Chapter 4 we introduced the concept of *endosymbiosis* (organisms living together, one inside the other; see Figure 4.18). According to one highly speculative hypothesis, the key to the survival of early phagocytes was the ingestion and incorporation of a prokaryote that took up residence within the phagocyte and evolved into the peroxisomes of today (see Figure 28.3). These organelles were able to disarm the toxic products of oxygen action, such as hydrogen peroxide. This association may have been the first important endosymbiosis in the evolution of the eukaryotic cell.

A crucial endosymbiotic event in the history of the Eukarya was the incorporation of a proteobacterium that evolved into the mitochondrion. Upon completion of this step, the basic modern eukaryotic cell was complete. Some very important eukaryotes are the result of yet another endosymbiotic step, the incorporation of a prokaryote related to today's cyanobacteria, which became the chloroplast. We'll see how this happened later in this chapter.

Many uncertainties remain

Several uncertainties cloud our current understanding of the origins of eukaryotic cells. Lateral gene transfer complicates the study of eukaryotic origins, just as it complicates the study of relationships among the prokaryote clades. At the same time, it may not have been extensive enough to account for the fact that, as genetic studies advance, more and more genes of bacterial origin are being found in eukaryotes.

An endosymbiotic origin of mitochondria and chloroplasts accounts for the presence of bacterial genes encoding enzymes for energy metabolism (respiration and photosynthesis), but it does not explain the presence of many other bacterial genes. The eukaryotic genome clearly is a mixture of genes with two distinct origins. A recent suggestion is that the Eukarya might have arisen from the mutualistic fusion (not endosymbiosis) of a Gram-negative bacterium and an archaean. There are many interesting ideas about eukaryotic origins awaiting additional data and analysis.

We can expect that these and other questions will yield to additional research. Let's leave our speculations about the origin of the eukaryotes for the moment and examine what we do know about them, beginning with the protists.

General Biology of the Protists

Most protists are aquatic. Some live in marine environments, others in fresh water, and still others in the body fluids of other organisms. The slime molds inhabit damp soil and the moist, decaying bark of rotting trees. Many other protists also live in soil water, some of them contributing to the global nitrogen cycle by preying on soil bacteria and recycling their nitrogen compounds into nitrates. Most protists are unicellular, but some are multicellular, and a few are very large.

Protists are strikingly diverse in their structure, but not so diverse in their metabolism as the prokaryotes—which is not surprising, since some of the eukaryotes' most important metabolic pathways were "borrowed" from bacteria through endosymbiosis. However, protists do display a number of nutritional modes. Some are photosynthetic autotrophs, some are heterotrophs, and some switch with ease between the autotrophic and heterotrophic modes of nutrition.

Some protists, formerly classified as animals, are sometimes referred to as **protozoans**, although biologists increasingly regard this term as inappropriate because it lumps together protist groups that are phylogenetically distant from one another. Most protozoans are ingestive heterotrophs. Similarly, there are several kinds of photosynthetic protists that some biologists still refer to as **algae** (singular, alga). Although these two terms are useful in some contexts, they do not correspond with natural phylogeny, and we generally avoid them in this book except as parts of descriptive names such as "brown algae." Let's next consider some of the other ways in which protists differ from one another.

Protists have diverse means of locomotion

Although a few protist groups consist entirely of nonmotile organisms, most groups include cells that move, either by amoeboid motion, by ciliary action, or by means of flagella.

In amoeboid motion, the cell forms **pseudopods** ("false feet") that are extensions of its constantly changing body mass. Cells such as the amoeba in Figure 28.4 simply extend a pseudopod and then flow into it. *Cilia* are tiny, hairlike organelles that beat in a coordinated fashion to move the cell forward or backward (see Figure 4.23). A eukaryotic *flagellum* moves like a whip; some flagella *push* the cell forward, others *pull* the cell forward. Cilia and eukaryotic flagella are identical in cross section; they differ only in length.

Vesicles perform a variety of functions

Unicellular organisms tend to be of microscopic size. As we noted above, an important reason that cells are small is that they need enough membrane surface area in relation to their volume to support the exchange of materials required for their existence. Many relatively large unicellular protists minimize this problem by having membrane-enclosed **vesicles** of various types that increase their effective surface area.

As we saw in Chapter 5, organisms living in fresh water are hypertonic to their environment. Many freshwater pro-

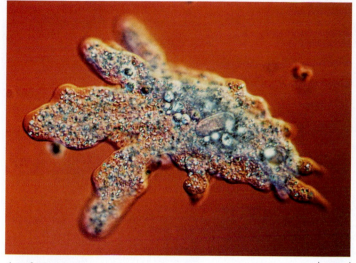

Amoeba proteus   50 μm

28.4 An Amoeba   The flowing pseudopods are constantly changing shape as the amoeba moves and feeds.

tists address this problem by means of specialized vesicles
that excrete the excess water they constantly take in by osmosis. Members of several protist groups have such **contractile vacuoles**. The excess water collects in the contractile vacuole, which then expels the water from the cell
(Figure 28.5).

A second important type of vesicle found in many protists
is the **food vacuole**. Protists such as *Paramecium* engulf solid
food by endocytosis, forming a food vacuole within which
the food is digested (Figure 28.6). Smaller vesicles containing

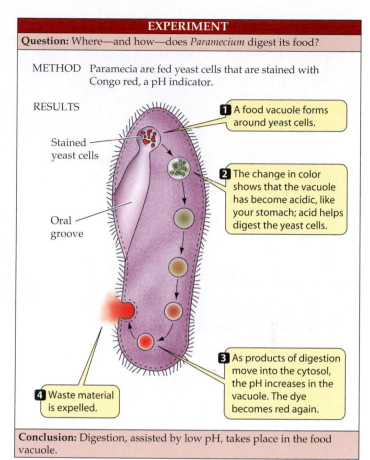

EXPERIMENT

Question: Where—and how—does *Paramecium* digest its food?

METHOD   Paramecia are fed yeast cells that are stained with
Congo red, a pH indicator.

RESULTS

Stained
yeast cells

Oral
groove

1 A food vacuole forms
around yeast cells.

2 The change in color
shows that the vacuole
has become acidic, like
your stomach; acid helps
digest the yeast cells.

3 As products of digestion
move into the cytosol,
the pH increases in the
vacuole. The dye
becomes red again.

4 Waste material
is expelled.

Conclusion: Digestion, assisted by low pH, takes place in the food
vacuole.

28.6 Food Vacuoles Handle Digestion and Excretion   An
experiment with *Paramecium* demonstrates the function of
food vacuoles. *Paramecium* ingests food by way of the oral
groove at the left. The dye Congo red turns green at acidic pH
and red at neutral or basic pH.

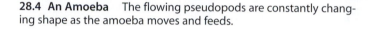

Contractile vacuole

28.5 Contractile Vacuoles Bail Out Excess Water
Water constantly enters freshwater protists by osmosis.
A pore in the cell surface allows the contractile vacuole
to expel the water it accumulates.

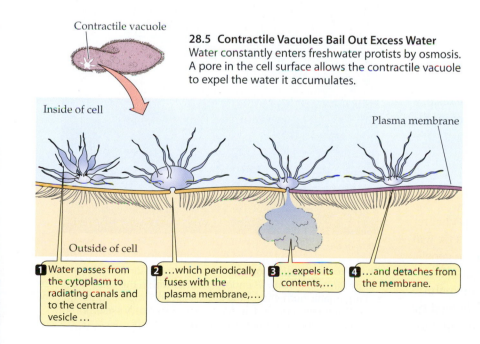

Inside of cell

Plasma membrane

Outside of cell

1 Water passes from
the cytoplasm to
radiating canals and
to the central
vesicle …

2 …which periodically
fuses with the
plasma membrane,…

3 …expels its
contents,…

4 …and detaches from
the membrane.

digested food pinch away from the food vesicle
and enter the cytoplasm. These tiny vesicles
provide a large surface area across which the
products of digestion may be absorbed by the
rest of the cell.

The cell surfaces of protists are diverse

A few protists, such as some amoebas, are surrounded by only a plasma membrane, but most
have stiffer surfaces that maintain the structural
integrity of the cell. Many protists have cell
walls, which are often complex in structure.
Other protists that lack cell walls have a variety
of ways of strengthening their surfaces. Some
have internal "shells," which the organism either produces itself, as foraminiferans do, or
makes from bits of sand and thickenings immediately beneath the plasma membrane, as
some amoebas do (Figure 28.7).

(a)

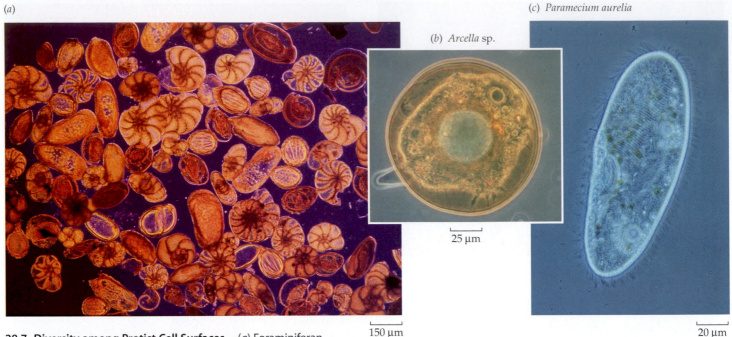

(b) Arcella sp.

(c) Paramecium aurelia

25 μm

150 μm

20 μm

28.7 Diversity among Protist Cell Surfaces (*a*) Foraminiferan shells are made of protein hardened with calcium carbonate. Several species are shown in this photograph. (*b*) This genus of shelled amoeba is commonly found in freshwater ponds and puddles. (*c*) The proteins in this *Paramecium*'s surface—known as its pellicle—make it flexible but resilient.

Many protists contain endosymbionts

Endosymbiosis is very common among the protists, and in some instances both the host and the endosymbiont are protists. Many radiolarians, for example, harbor photosynthetic protists (Figure 28.8). As a result, these radiolarians appear greenish or golden, depending on the type of endosymbiont they contain. This arrangement is beneficial to the radiolarian, for it can make use of the organic nutrients produced by its photosynthetic guest. The guest, in turn, may make use of metabolites made by the host, or it may simply receive physical protection. In other cases, the guest may be a prisoner, exploited for its photosynthetic products while receiving no benefit itself. We will take a more detailed look at the history of endosymbiosis among photosynthetic protists later in this chapter.

Both asexual and sexual reproduction occur among the protists

Although most protists practice both asexual and sexual reproduction, some groups lack sexual reproduction. As we will see, some asexually reproducing protists also engage in genetic recombination that does not directly result in reproduction.

Asexual reproductive processes in the protists include *binary fission* (splitting of the cell, with mitosis followed by cytokinesis), *multiple fission* (splitting into more than two

cells), *budding* (the outgrowth of a new cell from the surface of an old one), and the formation of *spores* (cells that are capable of developing into new organisms). Sexual reproduction also takes various forms. In some protists, as in animals, the gametes are the only haploid cells. In some other protists, by contrast, both diploid and haploid cells undergo mitosis, giving rise to alternation of generations, which will be described later in this chapter.

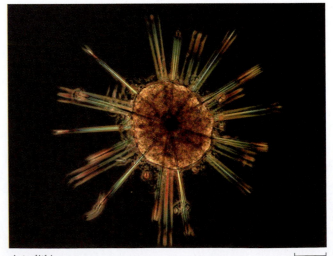

Astrolithium sp.

250 μm

28.8 Protists within Protists Photosynthetic dinoflagellates (see Figure 28.1*a*) are living as endosymbionts within this radiolarian, providing organic nutrients for the radiolarian and imparting the golden-brown pigmentation seen at the center of its glassy skeleton. Both the dinoflagellates and the radiolarian are protists.

28.1 *Major Protist Clades*

GROUP	ATTRIBUTES	EXAMPLES
Diplomonads	Unicellular, no mitochondria, two nuclei, flagella	*Giardia*
Parabasalids	Unicellular, no mitochondria, flagella and undulating membrane	*Trichomonas*
Euglenozoans	Unicellular, with flagella	
Euglenoids	Mostly photoautotrophic	*Euglena*
Kinetoplastids	Have a single large mitochondrion	*Trypanosoma*
Alveolates	Unicellular; cavities (alveoli) below cell surface	
Dinoflagellates	Pigments give golden-brown color	*Gonyaulax*
Apicomplexans	Apical complex in spores for penetration of host	*Plasmodium*
Ciliates	Cilia; two types of nuclei	*Paramecium*
Stramenopiles	Two unequal flagella, one with hairs	
Diatoms	Unicellular; photoautotrophic; two-part cell walls; no flagellum	*Thalassiosira*
Brown algae	Multicellular; marine; photoautotrophic	*Fucus, Macrocystis*
Oomycetes (water molds, powdery mildews)	Mostly coenocytic; heterotrophic	*Saprolegnia*
Red algae	No flagella; photoautrophic; phycoerythrin and phycocyanin	*Chondrus, Polysiphonia*
Chlorophytes ("Green algae"[a])	Photoautotrophic	*Ulva, Volvox*
Choanoflagellates	Resemble sponge cells; heterotrophic; with flagella	*Codosiga, Choanoeca*

[a]The green algae do not constitute a clade. The chlorophytes are a clade of green algae; a different green algal lineage gave rise to the plant kingdom.

The diversity of form, habitat, metabolism, locomotion, reproduction, and life cycles found among the protists reflects the diversity of avenues pursued during the early evolution of eukaryotes. Many of these avenues led to great success, judging from the abundance and diversity of today's protists and other eukaryotes.

Protist Diversity

The phylogeny of protists is an area of exciting, challenging research. The marvelous diversity of protist body forms and nutritional lifestyles seems reason enough for a fascination with these organisms, but questions about how the multicellular eukaryotic kingdoms originated from the protists stimulate further interest. Fortunately, the tools of molecular biology, such as rRNA sequencing, are making it possible to explore evolutionary relationships among the protists in ever greater detail and with greater confidence (see Chapters 25 and 26).

We will discuss several protist clades in this chapter, as well as a few other groups of more uncertain phylogenetic status. Some biologists refer to many of these clades as kingdoms; others refer to them as subkingdoms, and still others refer to them as phyla. This choice of words is not of immediate concern to us here, so we'll just call them "groups." We'll describe the following groups: diplomonads, parabasalids, euglenozoans, alveolates, stramenopiles, red algae, chlorophytes, and choanoflagellates (Table 28.1; Figure 28.9).

As we shall see, some of these protist clades consist of organisms with very diverse body plans. On the other hand, certain body plans, such as those of amoebas and those of slime molds, have arisen again and again during evolution, in groups only distantly related to one another. We'll begin our tour of protist clades with two of apparently ancient origin.

Diplomonads and Parabasalids

Two clades, the **diplomonads** and the **parabasalids**, appear to represent the earliest surviving branches in today's tree of eukaryotic life. It is likely that other clades diverged even earlier than the diplomonads and parabasalids. However, any such clades either were lost because of massive changes in the environment or remain hidden in rarely studied environments.

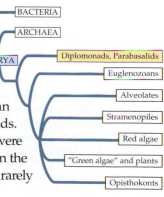

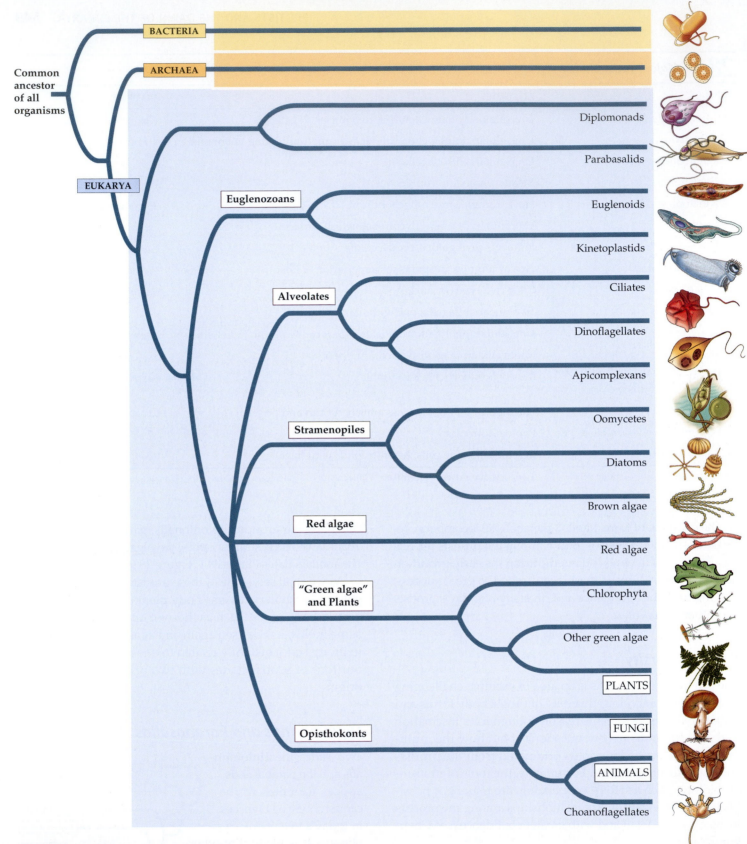

28.9 Major Protist Groups in an Evolutionary Context All of the protist groups shown here, except certain green algae, appear to be clades. The protists themselves do not constitute a clade. We also show the prokaryotic domains and the plant, fungal, and animal kingdoms to provide context. The term "opisthokont" refers to organisms that have or had (ancestrally) a flagellum in a posterior position; this group includes a protist clade as well as the fungi and animals.

Both the diplomonads and the parabasalids are unicellular organisms that lack mitochondria. This absence of mitochondria may be a derived condition: Ancestors of these organisms may have possessed mitochondria that were lost in the course of evolution. The existence of such organisms to-

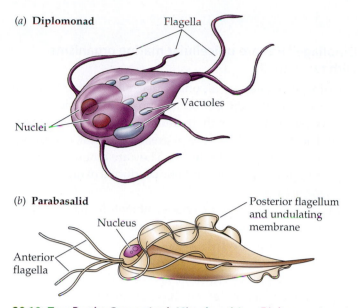

(a) **Diplomonad**

Flagella

Vacuoles

Nuclei

(b) **Parabasalid**

Nucleus

Posterior flagellum and undulating membrane

Anterior flagella

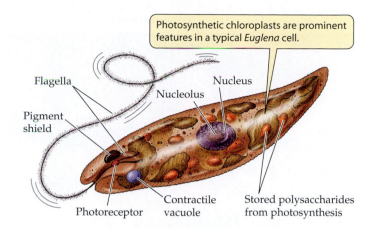

Photosynthetic chloroplasts are prominent features in a typical *Euglena* cell.

Flagella

Nucleus

Nucleolus

Pigment shield

Photoreceptor

Contractile vacuole

Stored polysaccharides from photosynthesis

28.11 A Photosynthetic Euglenoid   Several *Euglena* species are among the best-known flagellates. In this species, the second flagellum is rudimentary.

28.10 Two Protist Groups Lack Mitochondria   Diplomonads and parabasalids appear to represent the most ancient surviving branches of eukaryotic life. (a) *Giardia*, a diplomonad, has flagella and two nuclei (see also Figure 28.1b). (b) *Trichomonas*, a parabasalid, has flagella and undulating membranes. Neither of these protists possesses mitochondria.

day shows that eukaryotic life is feasible without mitochondria, and for that reason, the diplomonads and parabasalids are the focus of much attention.

Giardia lamblia, a diplomonad, is a familiar parasite that contaminates water supplies and causes the intestinal disease giardiasis (Figure 28.10a). This tiny organism has no mitochondria, chloroplasts, or other membrane-enclosed organelles, but it contains two nuclei bounded by nuclear envelopes, and it has a cytoskeleton and multiple flagella.

Trichomonas vaginalis is a parabasalid responsible for a sexually transmitted disease in humans (Figure 28.10b). Infection of the male urethra, where it may occur without symptoms, is less common than infection of the vagina. In addition to flagella, the parabasalids have undulating membranes that also contribute to the cell's locomotion.

Euglenozoans

The **euglenozoans** are a clade of *flagellates*: unicellular organisms with flagella. They reproduce asexually by binary fission. There are two subgroups of euglenozoans: euglenoids and kinetoplastids.

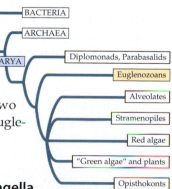

BACTERIA
ARCHAEA
EUKARYA
Diplomonads, Parabasalids
Euglenozoans
Alveolates
Stramenopiles
Red algae
"Green algae" and plants
Opisthokonts

Euglenoids have anterior flagella

The **euglenoids** possess flagella arising from a pocket at the anterior end of the cell. Many members of the group are pho-

tosynthetic. Euglenoids used to be claimed by the zoologists as animals and by the botanists as plants.

Figure 28.11 depicts a cell of the genus *Euglena*. Like most other euglenoids, this common freshwater organism has a complex cell structure. It propels itself through the water with the longer of its two flagella, which may also serve as an anchor to hold the organism in place. The flagellum provides power by means of a wavy motion that spreads from base to tip. The second flagellum is often rudimentary.

Euglena has very flexible nutritional requirements. Many species are always heterotrophic. Other species are fully autotrophic in sunlight, using chloroplasts to synthesize organic compounds through photosynthesis. The chloroplasts of euglenas are surrounded by three membranes (unlike plant chloroplasts, which have only two; we will describe the history of the third membrane later in this chapter). When kept in the dark, these euglenas lose their photosynthetic pigment and begin to feed exclusively on dead organic material floating in the water around them. Such a "bleached" *Euglena* resynthesizes its photosynthetic pigment when it is returned to the light and becomes autotrophic again. But *Euglena* cells treated with certain antibiotics or mutagens lose their photosynthetic pigment completely; neither they nor their descendants are ever autotrophs again. However, those descendants function well as heterotrophs.

Kinetoplastids have mitochondria that edit their own RNA

The **kinetoplastids** are unicellular, parasitic flagellates with a single, large mitochondrion. That mitochondrion contains a *kinetoplast*—a unique structure housing multiple, circular DNA molecules and associated proteins. Some of these DNA molecules encode "guides" that edit RNA within the mitochondrion.

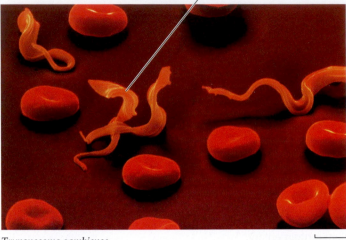

Undulating membrane

Trypanosoma gambiense   5 μm

28.12 A Parasitic Kinetoplastid   Trypanosomes, shown here among human red blood cells (round), cause sleeping sickness in mammals. A flagellum runs along one edge of the cell as part of a structure called the undulating membrane.

Some kinetoplastids are human pathogens. Sleeping sickness, one of the most dreaded diseases of Africa, is caused by the parasitic kinetoplastid *Trypanosoma* (Figure 28.12). An insect, the tsetse fly, is the *vector* (intermediate host) of *Trypanosoma*. Carrying its deadly cargo, the tsetse fly bites livestock, wild animals, and humans, infecting them with the parasite. *Trypanosoma* then multiplies in the mammalian bloodstream and produces toxins. When these parasites invade the nervous system, the neurological symptoms of sleeping sickness appear and are followed by death. Half a million people now have sleeping sickness, and 80 percent of them will die of the disease. About 3 million head of livestock die from *Trypanosoma* infections each year. Other trypanosomes cause leishmaniasis, Chagas' disease, and East Coast fever; all are major diseases in the Tropics.

Alveolates

The **alveolates** are a clade of unicellular organisms. The shared derived trait (synapomorphy) that characterizes them is the possession of cavities called *alveoli* just below their plasma membranes. They are diverse in body form. The alveolate groups we'll consider here are the dinoflagellates, apicomplexans, and ciliates.

BACTERIA
ARCHAEA
EUKARYA
Diplomonads, Parabasalids
Euglenozoans
Alveolates
Stramenopiles
Red algae
"Green algae" and plants
Opisthokonts

Dinoflagellates are unicellular marine organisms with two flagella

The **dinoflagellates** are all unicellular, and most are marine organisms. A distinctive mixture of photosynthetic and accessory pigments gives their chloroplasts a golden-brown color. The dinoflagellates are of great ecological, evolutionary, and morphological interest. They are among the most important primary photosynthetic producers of organic matter in the oceans.

Many dinoflagellates are endosymbionts living within the cells of other organisms, including various invertebrates and even other marine protists. Dinoflagellates are particularly common endosymbionts in corals, to whose growth they contribute by photosynthesis. As we will see later in this chapter, endosymbiotic events have given rise to dinoflagellates with different numbers of membranes surrounding their chloroplasts. Some dinoflagellates are nonphotosynthetic and live as parasites within other marine organisms.

Dinoflagellates have a distinctive appearance (see Figure 28.1a). They generally have two flagella, one in an equatorial groove around the cell, the other starting at the same point as the first and passing down a longitudinal groove before extending into the surrounding medium. Some dinoflagellates, notably *Pfiesteria piscida*, take on different forms, including amoeboid ones, depending on environmental conditions. It has been claimed that *P. piscida* can occur in at least two dozen distinct forms, although this claim is controversial. In any case, this remarkable dinoflagellate is highly toxic to fish and can, when present in great numbers, both stun and feed on them.

Some dinoflagellates reproduce in enormous numbers in warm and somewhat stagnant waters. The result can be a "red tide," so called because of the reddish color of the sea that results from the pigments of the dinoflagellates (Figure 28.13). During a red tide, the concentration of dinoflagellates may reach 60 million cells per liter of ocean water. *Pfiesteria* and certain other red tide species produce a potent nerve toxin that can kill tons of fish. The genus *Gonyaulax* produces a toxin that can accumulate in shellfish in amounts that, although not fatal to the shellfish, may kill a person who eats the shellfish.

Many dinoflagellates are bioluminescent. In complete darkness, cultures of these organisms emit a faint glow. If air is suddenly stirred or bubbled through the culture, the organisms each emit numerous bright flashes. A ship passing through a tropical ocean that contains a rich growth of these species produces a bow wave and wake that glow eerily as billions of these dinoflagellates discharge their light systems.

Apicomplexans are parasites with unusual spores

Exclusively parasitic organisms, the **apicomplexans** derive their name from the *apical complex*, a mass of organelles con-

The pigments of dinoflagellates make the red tide easy to see.

28.13 A Red Tide of Dinoflagellates By reproducing in astronomical numbers, the dinoflagellate *Gonyaulax tamarensis* can cause toxic red tides, such as this one along the coast of Baja California.

in parasitic protists. It appears even among parasitic dinoflagellates, a group of organisms whose nonparasitic relatives, as we have just seen, have highly distinctive, complex body forms.

Like many obligate parasites, apicomplexans have elaborate life cycles featuring asexual and sexual reproduction by a series of very dissimilar life stages. Often these stages are associated with two different types of host organisms.

Toxoplasma, a genus of apicomplexans, causes opportunistic infections in AIDS patients. There are other pathogenic apicomplexans as well. The best-known are the malarial parasites of the genus *Plasmodium*, a highly specialized group of organisms that spend part of their life cycle within human red blood cells (Figure 28.14). Although it has been almost eliminated from the United States, malaria continues to be a serious problem in many tropical countries, as we saw at the beginning of this chapter. In terms of the number of people infected, malaria is one of the world's three most serious diseases, and it kills more than a million people each year.

Female mosquitoes of the genus *Anopheles* transmit *Plasmodium* to humans. The parasite enters the human circula-

tained within the apical end of their spores. These organelles help the apicomplexan spore invade its host's tissues. Unlike many other protists, apicomplexans lack contractile vacuoles.

Apicomplexans generally have an amorphous amoeboid body form. This body form has evolved over and over again

28.14 The Life Cycle of an Apicomplexan Malaria-causing *Plasmodium* species spend part of their life cycle in humans and part in mosquitoes. The sporozoite and merozoite forms of the parasite are spores with apical complexes.

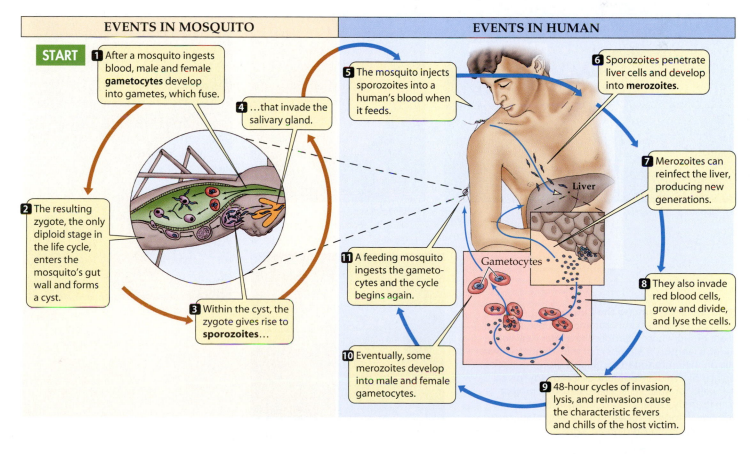

EVENTS IN MOSQUITO

EVENTS IN HUMAN

START

1 After a mosquito ingests blood, male and female **gametocytes** develop into gametes, which fuse.

4 ...that invade the salivary gland.

5 The mosquito injects sporozoites into a human's blood when it feeds.

6 Sporozoites penetrate liver cells and develop into **merozoites**.

2 The resulting zygote, the only diploid stage in the life cycle, enters the mosquito's gut wall and forms a cyst.

3 Within the cyst, the zygote gives rise to **sporozoites**...

7 Merozoites can reinfect the liver, producing new generations.

Liver

11 A feeding mosquito ingests the gametocytes and the cycle begins again.

Gametocytes

8 They also invade red blood cells, grow and divide, and lyse the cells.

10 Eventually, some merozoites develop into male and female gametocytes.

9 48-hour cycles of invasion, lysis, and reinvasion cause the characteristic fevers and chills of the host victim.

tory system when an infected *Anopheles* mosquito penetrates the human skin in search of blood. The parasites find their way to cells in the liver and the lymphatic system, change their form, multiply, and reenter the bloodstream, attacking red blood cells. The apical complex enables *Plasmodium* to enter human liver and red blood cells.

The parasites multiply inside red blood cells, which then burst, releasing new swarms of parasites. If another *Anopheles* bites the victim, the mosquito takes in *Plasmodium* cells along with blood. Some of these cells develop into gametes, which unite to form zygotes that lodge in the mosquito's gut, divide several times, and move into its salivary glands, from which they can be passed on to another human host. Thus, *Plasmodium* is an extracellular parasite in the mosquito vector and an intracellular parasite in the human host.

Plasmodium has proved to be a singularly difficult pathogen to attack. The *Plasmodium* life cycle is best broken by the removal of stagnant water, in which mosquitoes breed. The use of insecticides to reduce the *Anopheles* population can be effective, but their benefits must be weighed against the possible ecological, economic, and health risks posed by the insecticides themselves.

New hope arises from the publication, in 2002, of the genomes of both *Plasmodium falciparum* and one of its vectors, *Anopheles gambiae*, as well as a partial proteome of *P. falciparum*. These advances should lead to a better understanding of the biology of malaria and to the possible development of drugs, vaccines, or other means of dealing with this protist pathogen or its insect vectors.

Ciliates have two types of nuclei

The **ciliates** are so named because they characteristically have hairlike cilia. This group is noteworthy for its diversity and ecological importance (Figure 28.15). Almost all ciliates are heterotrophic (a few contain photosynthetic endosymbionts), and they are much more complex in body form than are most flagellates and other unicellular protists.

The definitive characteristic of ciliates is the possession of two types of nuclei, commonly a single *macronucleus* and, within the same cell, from one to several *micronuclei*. The micronuclei, which are typical eukaryotic nuclei, are essential for genetic recombination. The macronucleus is derived from micronuclei. Each macronucleus contains many copies of the genetic information, packaged in units containing very few genes each. The macronuclear DNA is transcribed and translated to regulate the life of the cell. Although we do not know how this system of macro- and micronuclei came into being,

Cilia

(a) *Paramecium bursaria*
10 μm

Cilia

(b) *Vorticella* sp.
10 μm

Tentacles

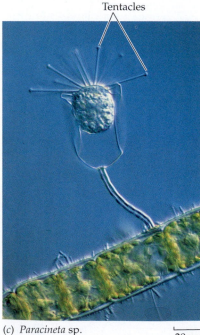

(c) *Paracineta* sp.
20 μm

Cirri

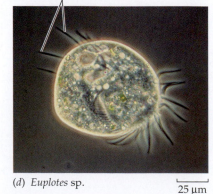

(d) *Euplotes* sp.
25 μm

28.15 Diversity among the Ciliates (*a*) A free-swimming organism, this paramecium belongs to a ciliate group whose members have many cilia of uniform length. (*b*) Members of this subgroup have cilia on their mouthparts. (*c*) In this group, tentacles replace cilia as development proceeds. (*d*) This ciliate "walks" on fused cilia, called cirri, that project from its body. Other cilia are fused into flat sheets that sweep food particles into the oral cavity; this individual has ingested green algae.

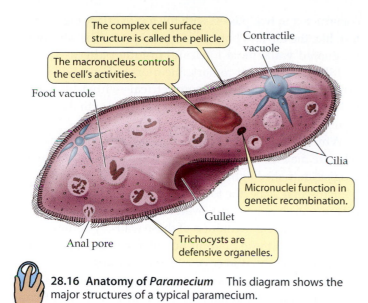

The complex cell surface structure is called the pellicle.

The macronucleus controls the cell's activities.

Food vacuole

Contractile vacuole

Cilia

Micronuclei function in genetic recombination.

Gullet

Anal pore

Trichocysts are defensive organelles.

28.16 Anatomy of *Paramecium* This diagram shows the major structures of a typical paramecium.

we do know something about the behavior of these nuclei, which we will discuss after describing the body plan of one important ciliate, *Paramecium*.

A CLOSER LOOK AT ONE CILIATE. *Paramecium*, a frequently studied ciliate genus, exemplifies the complex structure and behavior of ciliates (Figure 28.16). The slipper-shaped cell is covered by an elaborate *pellicle*, a structure composed principally of an outer membrane and an inner layer of closely packed, membrane-enclosed sacs (the alveoli) that surround the bases of the cilia. Defensive organelles called *trichocysts* are also present in the pellicle. In response to a threat, a microscopic explosion expels the trichocysts in a few milliseconds, and they emerge as sharp darts, driven forward at the tip of a long, expanding filament.

The cilia provide a form of locomotion that is generally more precise than locomotion by flagella or pseudopods. A paramecium can direct the beating of its cilia to propel itself

either forward or backward in a spiraling manner. It can also back off swiftly when it encounters a barrier or a negative stimulus. The coordination of ciliary beating is probably the result of a differential distribution of ion channels in the plasma membrane near the two ends of the cell.

REPRODUCTION WITHOUT SEX, AND SEX WITHOUT REPRODUCTION. Paramecia reproduce asexually by binary fission. The micronuclei divide mitotically. The macronuclei divide by a still unknown mechanism following a round of DNA replication.

Paramecia also have an elaborate sexual behavior called **conjugation**, in which two paramecia line up tightly against each other and fuse in the oral region of the body. Nuclear material is extensively reorganized and exchanged over the next several hours (Figure 28.17). As a result of this process, each cell ends up with two haploid micronuclei, one of its own and one from the other cell, which fuse to form a new diploid micronucleus. New macronuclei develop from the micronuclei through a series of dramatic chromosomal rearrangements. The exchange of nuclei is fully reciprocal—each of the two paramecia gives and receives an equal amount of DNA. The two organisms then separate and go their own ways, each equipped with new combinations of alleles.

Conjugation in *Paramecium* is a *sexual* process of genetic recombination, but it is not a *reproductive* process. The same two cells that begin the process are there at the end, and no new cells are created. As a rule, each asexual clone of paramecia must periodically conjugate. Experiments have shown that if some species are not permitted to conjugate, the clones can live through no more than approximately 350 cell divisions before they die out.

28.17 Paramecia Achieve Genetic Recombination by Conjugating The exchange of micronuclei by conjugating *Paramecium* individuals permits genetic recombination. After conjugation, the cells separate and continue their lives as two individuals.

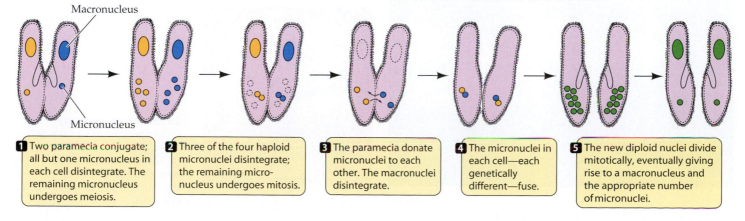

Macronucleus

Micronucleus

1 Two paramecia conjugate; all but one micronucleus in each cell disintegrate. The remaining micronucleus undergoes meiosis.

2 Three of the four haploid micronuclei disintegrate; the remaining micronucleus undergoes mitosis.

3 The paramecia donate micronuclei to each other. The macronuclei disintegrate.

4 The micronuclei in each cell—each genetically different—fuse.

5 The new diploid nuclei divide mitotically, eventually giving rise to a macronucleus and the appropriate number of micronuclei.

Stramenopiles

The shared trait that defines the **stramenopiles** is the possession of two flagella, typically unequal in length. The longer of the two bears rows of tubular hairs. Some stramenopiles lack flagella, but they are presumed to be descended from ancestors that possessed flagella. The stramenopiles include the diatoms and the brown algae, which are photosynthetic, and the oomycetes, which are not. Other stramenopile groups include some lineages that are non-photosynthetic. Most golden algae are photosynthetic, but nearly all of them become heterotrophic when light intensity is limiting or when there is a plentiful food supply; some even feed on diatoms or bacteria. Some botanists refer to the stramenopiles as the "brown plant kingdom."

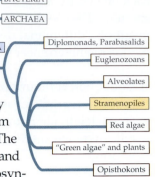

Diatoms are everywhere in the marine environment

Diatoms are single-celled organisms, although some species associate in filaments. Many have sufficient carotenoids in their chloroplasts to give them a yellow or brownish color. All make *chrysolaminarin* (a carbohydrate) and oils as photosynthetic storage products. Diatoms lack flagella.

Architectural magnificence on a microscopic scale is the hallmark of the diatoms (Figure 28.18*a*). Many diatoms deposit silicon in their cell walls. The cell wall of some species is

constructed in two pieces, with the top overlapping the bottom like the top and bottom of a petri plate. The silicon-impregnated walls have intricate, unique patterns (Figure 28.18*b*). Despite their remarkable morphological diversity, however, all diatoms are symmetrical—either bilaterally (with "right" and "left" halves) or radially (with the type of symmetry possessed by a circle).

Why are diatom cell walls so glassy and complex—and why are diatoms so abundant? A 2003 paper by German biologists may shed light on these questions. By measuring, at a microscopic scale, the forces needed to break single, living diatoms, the biologists discovered that their cell walls are exceptionally strong. Evolution of these walls by natural selection has given these diatoms an enhanced defense against predators and, thus, an edge over competitors.

Diatoms reproduce both sexually and asexually. Asexual reproduction is by binary fission and is somewhat constrained by the stiff, silica-containing cell wall. Both the top and the bottom of the "petri plate" become tops of new "plates" without changing appreciably in size; as a result, the new cell made from the former bottom is smaller than the parent cell (Figure 28.19). If this process continued indefinitely, one cell line would simply vanish, but sexual reproduction largely solves this potential problem. Gametes are formed, shed their cell walls, and fuse. The resulting zygote then increases substantially in size before a new cell wall is laid down.

Diatoms are everywhere in the marine environment and are frequently present in great numbers, making them major photosynthetic producers in coastal waters. Diatoms are also common in fresh water. Because the silicon-containing walls of dead diatom cells resist decomposition, certain sedimentary rocks are composed almost entirely of diatom skeletons that sank to the seafloor over time. Diatomaceous earth, which is obtained from such rocks, has many industrial uses, such as insulation, filtration, and metal polishing. It has also been used as an "Earth-friendly" insecticide that clogs the tracheae (breathing structures) of insects.

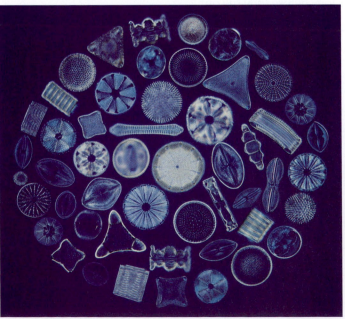

(a)

30 µm

(b)

7 µm

28.18 Diatom Diversity
(*a*) Diatoms exhibit a splendid variety of species-specific forms. (*b*) This artificially colored scanning electron micrograph shows the intricate patterning of diatom cell walls.

28.19 Diatom Reproduction   Diatoms reproduce both sexually and asexually. Half of the cells created by asexual reproduction are smaller than the parent cells. Sexual reproduction creates new parent cells with full-sized cell walls.

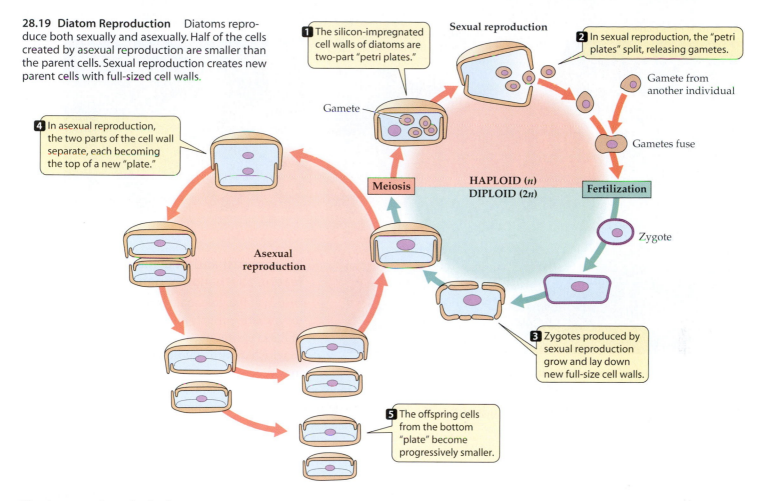

Sexual reproduction

1 The silicon-impregnated cell walls of diatoms are two-part "petri plates."

2 In sexual reproduction, the "petri plates" split, releasing gametes.

Gamete

Gamete from another individual

Gametes fuse

4 In asexual reproduction, the two parts of the cell wall separate, each becoming the top of a new "plate."

Meiosis

HAPLOID (*n*)
DIPLOID (2*n*)

Fertilization

Asexual reproduction

Zygote

3 Zygotes produced by sexual reproduction grow and lay down new full-size cell walls.

5 The offspring cells from the bottom "plate" become progressively smaller.

The brown algae include the largest protists

All the **brown algae** are multicellular. They are composed either of branched filaments (Figure 28.20) or of leaflike growths called *thalli* (singular, thallus) (Figure 28.21*a*). The brown algae obtain their namesake color from the carotenoid *fucoxanthin*, which is abundant in their chloroplasts. The combination of this yellow-orange pigment with the green of chlorophylls *a* and *c* yields a brownish tinge.

The brown algae include the largest of the protists. Giant kelps, such as those of the genus *Macrocystis*, may be up to 60 meters long (see Figure 28.1*c*). The brown algae are almost exclusively marine. Some float in the open ocean; the most famous example is the genus *Sargassum*, which forms dense mats in the Sargasso Sea in the mid-Atlantic. Most brown algae, however, are attached to rocks near the shore. A few thrive only

(a) Hormosira banksii

(b) Ectocarpus sp.

28.20 Brown Algae   (*a*) A filamentous brown alga growing in Australia. This species is sometimes called "Neptune's necklace." (*b*) A filamentous brown alga seen through a light microscope.

(a) Postelsia palmaeformis

The leaflike structures are the thalli of sea palm.

(b)

28.21 Brown Algae in a Turbulent Environment Brown algae growing in the intertidal zone on an exposed rocky shore take a tremendous pounding by the surf. (*a*) Sea palms growing along the California coast. (*b*) The tough, branched holdfast that anchors the sea palm.

where they are regularly exposed to heavy surf; a notable example is the sea palm *Postelsia palmaeformis* of the Pacific coast (Figure 28.21*a*). All of the attached forms develop a specialized structure, called a *holdfast*, that literally glues them to the rocks (Figure 28.21*b*).

Some brown algae differentiate extensively into stemlike stalks and leaflike blades, and some develop gas-filled cavities or bladders. For biochemical reasons that are only poorly understood, these bladders often contain as much as 5 percent carbon monoxide—a concentration high enough to kill a human. In addition to organ differentiation, the larger brown algae also exhibit considerable tissue differentiation. Most of the giant kelps have photosynthetic filaments only in the outermost regions of their stalks and blades. Within these photosynthetic regions lie filaments of long cells that closely resemble the nutrient-conducting tissue of plants. Called *trumpet cells* because they have flaring ends, these tubes rapidly conduct the products of photosynthesis through the body of the organism.

The cell walls of brown algae may contain as much as 25 percent *alginic acid*, a gummy polymer of sugar acids. Alginic acid cements cells and filaments together and provides good holdfast glue. It is used commercially as an emulsifier in ice cream, cosmetics, and other products.

Many protist and all plant life cycles feature alternation of generations

Brown algae, like many other multicellular photosynthetic protists and all plants, exhibit a type of life cycle known as **alternation of generations**, in which a multicellular, diploid, spore-producing organism gives rise to a multicellular, haploid, gamete-producing organism. When two haploid gametes fuse (a process called *fertilization*, or *syngamy*), a diploid organism is formed (Figure 28.22). The haploid organism, the diploid organism, or both may also reproduce asexually.

The two generations (spore-producing and gamete-producing) differ genetically (one has diploid cells and the other has haploid cells), but they may or may not differ morphologically. In **heteromorphic** alternation of generations the two generations differ morphologically; in **isomorphic** alternation of generations they do not, despite their genetic difference. We will see examples of both heteromorphic and isomorphic alternation of generations in some representative brown and green algae. In discussing the life cycles of plants and multicellular photosynthetic protists, we will use the terms **sporophyte** ("spore plant") and **gametophyte** ("gamete plant") to refer to the multicellular diploid and haploid generations, respectively.

Gametes are not produced by meiosis because the gametophyte generation is already haploid. Instead, specialized cells of the diploid sporophyte, called **sporocytes**, divide meiotically to produce four haploid spores. The spores may eventually germinate and divide mitotically to produce multicellular haploid gametophytes, which produce gametes by mitosis and cytokinesis.

Gametes, unlike spores, can produce new organisms only by fusing with other gametes. The fusion of two gametes produces a diploid zygote, which then undergoes mitotic divisions to produce a diploid organism: the sporophyte generation. The sporocytes of the sporophyte generation then undergo meiosis and produce haploid spores, starting the cycle anew.

28.22 Alternation of Generations
In many multicellular photosynthetic protists and all plants, a diploid generation that produces spores alternates with a haploid generation that produces gametes.

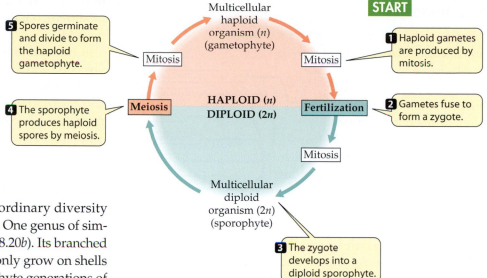

START

5 Spores germinate and divide to form the haploid gametophyte.

Multicellular haploid organism (*n*) (gametophyte)

1 Haploid gametes are produced by mitosis.

Mitosis

Mitosis

4 The sporophyte produces haploid spores by meiosis.

Meiosis

HAPLOID (*n*) DIPLOID (2*n*)

Fertilization

2 Gametes fuse to form a zygote.

Mitosis

Multicellular diploid organism (2*n*) (sporophyte)

3 The zygote develops into a diploid sporophyte.

The brown algae exemplify the extraordinary diversity found among the photosynthetic protists. One genus of simple brown algae is *Ectocarpus* (see Figure 28.20*b*). Its branched filaments, a few centimeters long, commonly grow on shells and stones. The gametophyte and sporophyte generations of *Ectocarpus* can be distinguished only by chromosome number or reproductive products (spores or gametes). Thus the generations are isomorphic.

By contrast, some kelps of the genus *Laminaria* and some other brown algae show a more complex heteromorphic alternation of generations. The larger and more obvious generation of these species is the sporophyte. Meiosis in sporocytes located on the leaflike fronds produces haploid **zoospores**—motile spores that are propelled by flagella. These spores germinate to form a tiny, filamentous gametophyte that produces either eggs or sperm. The eggs and sperm of brown algae typically have flagella.

The oomycetes include water molds and their relatives

A nonphotosynthetic stramenopile group called the **oomycetes** consists in large part of the water molds and their terrestrial relatives, such as the downy mildews. Water molds are filamentous and stationary, and they feed by absorption—that is, they secrete enzymes that digest large food molecules into smaller molecules that the water molds can absorb. If you have seen a whitish, cottony mold growing on dead fish or dead insects in water, it was probably a water mold of the common genus *Saprolegnia* (Figure 28.23).

Don't be confused by the "-mycete" in the name. That term means "fungus," and it is there because these organisms were once classified as fungi. However, we now know that the oomycetes are unrelated to the fungi.

The oomycetes are **coenocytes**: They have many nuclei enclosed in a single plasma membrane. Their filaments have no cross-walls to separate the many nuclei into discrete cells. Their cytoplasm is continuous throughout the body of the organism, and there is no single structural unit with a single nucleus, except in certain reproductive stages. A distinguishing feature of the oomycetes is their flagellated reproductive cells. Oomycetes are diploid throughout most of their life cycle and have cellulose in their cell walls.

The water molds, such as *Saprolegnia*, are all aquatic and **saprobic** (they feed on dead organic matter). Some other oomycetes are terrestrial. Although most of the terrestrial oomycetes are harmless or helpful decomposers of dead matter, a few are serious plant parasites that attack crops such as avocados, grapes, and potatoes. The water mold *Phytophthora infestans*, for example, is the causal agent of late blight of potatoes, which brought about the great Irish potato famine of 1845–1847. *P. infestans* destroyed the entire Irish potato crop in a matter of days in 1846. Among the consequences of the famine were a million deaths from starvation and the emigration of about 2 million people, mostly to the United States.

Saprolegnia sp.

28.23 An Oomycete   The filaments of a water mold radiate from the carcass of an insect.

Red Algae

Almost all **red algae** are multicellular (Figure 28.24). Some plant biologists now refer to the red algae as the "red plant kingdom." Their characteristic color is a result of the accessory photosynthetic pigment *phycoerythrin*, which is found in relatively large amounts in the chloroplasts of many species. In addition to phycoerythrin, red algae contain phycocyanin, carotenoids, and chlorophyll *a*.

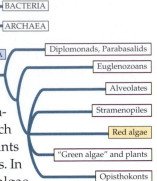

The red algae include species that grow in the shallowest tide pools as well as the algae found deepest in the ocean (as deep as 260 meters if nutrient conditions are right and the water is clear enough to permit the penetration of light). Very few red algae inhabit fresh water. Most grow attached to a substratum by a holdfast.

In a sense, the red algae, like several other groups of algae, are misnamed. They have the capacity to change the relative amounts of their various photosynthetic pigments depending on the light conditions where they are growing. Thus, the leaflike *Chondrus crispus*, a common North Atlantic red alga, may appear bright green when it is growing at or near the surface of the water and deep red when growing at greater depths. The ratio of pigments present depends to a remarkable degree on the intensity of the light that reaches the alga. In deep water, where the light is dimmest, the alga accumulates large amounts of phycoerythrin. The algae in deep water have as much chlorophyll as the green ones near the surface, but the accumulated phycoerythrin makes them look red.

In addition to being the only photosynthetic protists with phycoerythrin and phycocyanin among their pigments, the red algae have two other unique characteristics: First, they store the products of photosynthesis in the form of *floridean starch*, which is composed of very small, branched chains of approximately 15 glucose units. Second, they produce no motile, flagellated cells at any stage in their life cycle. The male gametes lack cell walls and are slightly amoeboid; the female gametes are completely immobile.

Some red algal species enhance the formation of coral reefs. Like coral animals, they possess the biochemical machinery for secreting calcium carbonate, which they deposit both in and around their cell walls. After the death of the corals and algae, the calcium carbonate persists, sometimes forming substantial rocky masses.

Some red algae produce large amounts of mucilaginous polysaccharide substances, which contain the sugar galactose with a sulfate group attached. This material readily forms solid gels and is the source of agar, a substance widely used in the laboratory for making a solid aqueous medium on which tissue cultures and many microorganisms can be grown.

Certain red algae became endosymbionts, long ago, within the cells of other, nonphotosynthetic protists, eventually giving rise to chloroplasts. They are the ancestors of the distinctive chloroplasts of the photosynthetic stramenopiles (the brown algae and the diatoms).

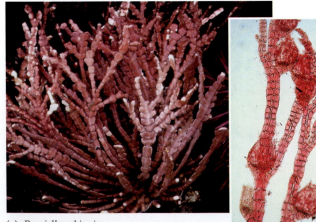

(a) *Bossiella orbigniana*

(b) *Polysiphonia* sp.

28.24 Red Algae (a) A coralline red alga grows along the coast of central Oregon. (b) Under the light microscope, both vegetative and reproductive structures can be seen in this red alga.

Chlorophytes

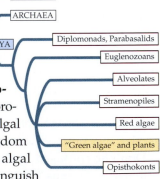

The "green algae" do not form a clade, but they include at least two multicellular clades. One major clade consists of the **chlorophytes**. A sister clade to the chlorophytes contains another green algal clade along with the plant kingdom (see Figure 28.9). The green algal clades share characters that distinguish them from other protists: Like the plants, they contain chlorophylls *a* and *b*, and their reserve of photosynthetic products is stored as starch in plastids.

The chlorophytes are the largest clade of green algae, containing more than 17,000 species. Most are aquatic—some are marine, but more are freshwater forms—but others are terrestrial, living in moist environments. The chlorophytes range in size from microscopic unicellular forms to multicellular forms many centimeters in length.

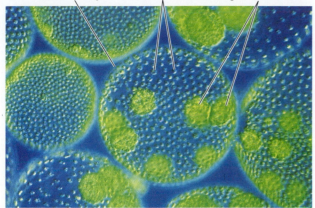

Parent colony   Somatic cells   Reproductive cells

(a) Volvox sp.

(b) Ulva lactuca

28.25 Chlorophytes (a) Volvox colonies are precisely spaced arrangements of cells. Specialized reproductive cells produce daughter colonies, which will eventually release new individuals. (b) A stand of sea lettuce submerged in a tidal pool.

While *Volvox* is colonial and spherical, *Oedogonium* is multicellular and filamentous, and each of its cells has only one nucleus. *Cladophora* is multicellular, but each cell is multinucleate. *Bryopsis* is tubular and coenocytic, forming cross-walls only when reproductive structures form. *Acetabularia* is a single, giant uninucleate cell a few centimeters long that becomes multinucleate only at the end of its reproductive stage. *Ulva lactuca* is a thin, membranous sheet a few centimeters across; its unusual appearance justifies its common name: sea lettuce (Figure 28.25b).

Chlorophyte life cycles are diverse

The life cycles of chlorophytes show great diversity. Let's examine two chlorophyte life cycles in detail, beginning with that of the sea lettuce *Ulva lactuca* (Figure 28.26). Like many chlorophytes, sea lettuce exhibits alternation of generations. The diploid sporophyte of this common seashore organism is a broad sheet only two cells thick. Some of its cells (sporocytes) differentiate and undergo meiosis and cytokinesis, producing motile haploid spores (zoospores). These swim away, each propelled by four flagella, and some eventually find a suitable place to settle. The spores then lose their flagella and begin to divide mitotically, producing a thin filament that de-

Chlorophytes vary in shape and cellular organization

We find among the chlorophytes an incredible variety in shape and construction of the algal body. *Chlamydomonas* is an example of the simplest type: unicellular and flagellated.

Surprisingly large and well-formed colonies of cells are found in such freshwater groups as the genus *Volvox* (Figure 28.25a). The cells in these colonies are not differentiated into tissues and organs, as in plants and animals, but the colonies show vividly how the preliminary step of this great evolutionary development might have been taken. In *Volvox*, the origins of cell specialization can be seen in certain cells within the colony that are specialized for reproduction.

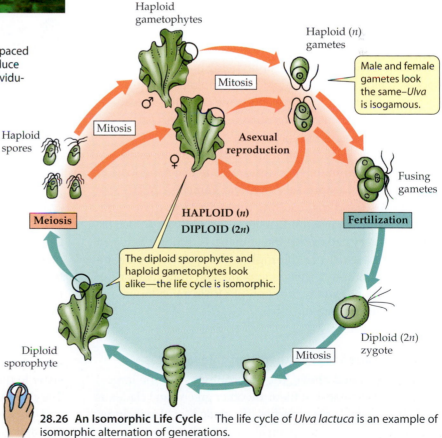

Male and female gametes look the same—*Ulva* is isogamous.

The diploid sporophytes and haploid gametophytes look alike—the life cycle is isomorphic.

Haploid gametophytes

Haploid (n) gametes

Mitosis

Asexual reproduction

Haploid spores

Mitosis

Meiosis

HAPLOID (n)
DIPLOID (2n)

Fertilization

Fusing gametes

Diploid (2n) zygote

Mitosis

Diploid sporophyte

28.26 An Isomorphic Life Cycle The life cycle of *Ulva lactuca* is an example of isomorphic alternation of generations.

velops into a broad sheet only two cells thick. The gametophyte thus produced looks just like the sporophyte—in other words, *Ulva lactuca* has an isomorphic life cycle.

In *Ulva lactuca*, an individual gametophyte can produce only male or female gametes—never both. The gametes arise mitotically within single cells (called *gametangia*), rather than within a specialized multicellular structure, as in plants. Both types of gametes bear two flagella (in contrast to the four flagella of a haploid spore) and hence are motile.

In most species of *Ulva* the female and male gametes are indistinguishable structurally, making those species **isogamous**—having gametes of identical appearance. Other chlorophytes, including some other species of *Ulva*, are **anisogamous**—having female gametes that are distinctly larger than the male gametes.

Female and male gametes come together and unite, losing their flagella as the zygote forms and settles. After resting briefly, the zygote begins mitotic division, producing a multicellular sporophyte. Any gametes that fail to find partners can settle down on a favorable substratum, lose their flagella, undergo mitosis, and produce a new gametophyte directly; in other words, the gametes can also function as zoospores. Few chlorophytes other than *Ulva* have motile gametes that can also function as zoospores.

In contrast to the isomorphic life cycle of *Ulva*, many other chlorophytes have a heteromorphic life cycle, in which sporophyte and gametophyte generations differ in structure. In one variation of the heteromorphic life cycle—the **haplontic** life cycle (Figure 28.27)—a multicellular haploid individual produces gametes that fuse to form a zygote. The zygote functions directly as a sporocyte, undergoing meiosis to produce spores, which in turn produce a new haploid individual. In the entire haplontic life cycle, only one cell—the zygote—is diploid. The filamentous organisms of the genus *Ulothrix* are examples of haplontic chlorophytes.

Some other chlorophytes have a **diplontic** life cycle like that of many animals. In a diplontic life cycle, meiosis of diploid sporocytes produces haploid gametes directly; the gametes fuse, and the resulting diploid zygote divides mitotically to form a new multicellular sporophyte. In such organisms, every cell except the gametes is diploid. Between these two extremes are chlorophytes in which the gametophyte and sporophyte generations are both multicellular, but one generation (usually the sporophyte) is much larger and more prominent than the other.

There are green algae other than chlorophytes

As we mentioned above, the chlorophytes are the largest clade of green algae, but there are other green algal clades as well. Those clades are branches of a clade that also includes the plant kingdom. The green algal clade that is sister to the

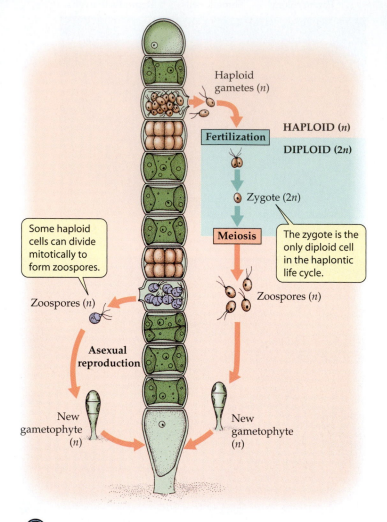

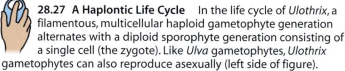

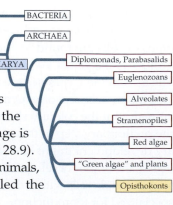

28.27 A Haplontic Life Cycle In the life cycle of *Ulothrix*, a filamentous, multicellular haploid gametophyte generation alternates with a diploid sporophyte generation consisting of a single cell (the zygote). Like *Ulva* gametophytes, *Ulothrix* gametophytes can also reproduce asexually (left side of figure).

plant kingdom, containing a group of organisms called *charophytes*, will be described in the next chapter. But now let's consider some close protist relatives of the animals.

Choanoflagellates

One group of heterotrophic protists with flagella, the **choanoflagellates**, is thought to comprise the closest relatives of the animals. The choanoflagellates are sister to the animals, and the animal–choanoflagellate lineage is sister to the fungi (see Figure 28.9). The clade consisting of fungi, animals, and choanoflagellates is called the **opisthokonts**.

(a) *Codosiga botrytis*

(b) *Choanoeca* sp.

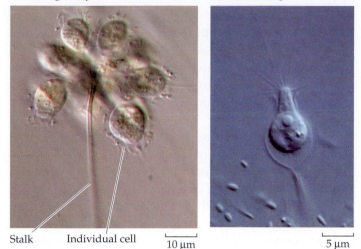

Stalk   Individual cell |10 μm| |5 μm|

28.28 A Link to the Animal Kingdom Choanoflagellates may be close relatives of the sponges, and thus represent a link between the protists and the kingdom Animalia. (a) The formation of colonies by unicellular organisms, as in this species, is one route to the evolution of multicellularity. (b) A solitary choanoflagellate illustrates the similarity of this protist group to a cell type present in the multicellular sponges (see Figure 32.4).

Choanoflagellates are colonial and are thought to be closely related to the sponges, the most ancient of the surviving phyla of animals. Choanoflagellates bear a striking resemblance to the most characteristic type of cell found in the sponges (compare Figures 28.28 and 32.4).

Sponges are also colonial rather than truly multicellular, in that they lack organized tissues and their cells can be sep-

arated and allowed to reaggregate. We turn now to a topic hinted at several times above: the history of unusual numbers of membranes surrounding the chloroplasts of some photosynthetic protists.

A History of Endosymbiosis

As we have already seen, many protists possess chloroplasts. Groups with chloroplasts appear in several distantly related protist clades. Some of these groups differ in the photosynthetic pigments their chloroplasts contain. And we've seen that not all chloroplasts have a pair of surrounding membranes—in some protists, they are surrounded by three membranes. We now understand these observations in terms of a remarkable series of endosymbioses.

All chloroplasts trace their ancestry back to the engulfment of a cyanobacterium by a larger eukaryotic cell (Figure 28.29). This event is known as *primary endosymbiosis*. The cyanobacterium, a Gram-negative bacterium, had both an inner and an outer membrane. The eukaryote's plasma membrane wrapped around the cyanobacterium as it took it up. The outer membrane and cell wall of the cyanobacterium were eventually lost. Thus, the original chloroplasts had two surrounding membranes—one from the cyanobacterium and one from the eukaryotic host cell.

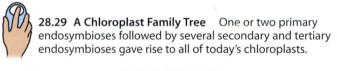
28.29 A Chloroplast Family Tree One or two primary endosymbioses followed by several secondary and tertiary endosymbioses gave rise to all of today's chloroplasts.

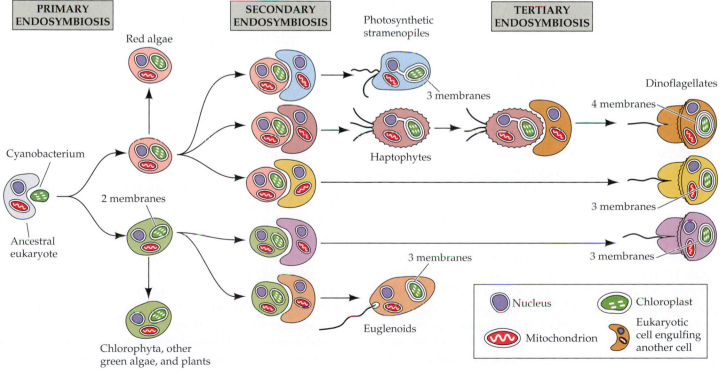

Primary endosymbiosis gave rise to the chloroplasts of the green algae and the red algae. We do not yet know whether both trace back to a single primary endosymbiosis, as is likely, with later divergence, or whether they resulted from independent occurrences of primary endosymbiosis. In either case, each line participated in further endosymbioses.

All remaining photosynthetic eukaryotes are the results of secondary or tertiary endosymbiosis. The photosynthetic euglenoids derived their chloroplasts from *secondary endosymbiosis*. Their ancestor took up a unicellular chlorophyte, retaining the endosymbiont's chloroplast and eventually losing the rest of its constituents. This history accounts for the fact that the photosynthetic euglenoids have the same photosynthetic pigments as the chlorophytes and plants. It also accounts for the third membrane of the euglenoid chloroplast, which is derived from the euglenoid's plasma membrane.

Other photosynthetic protist groups derived their chloroplasts by secondary endosymbiosis with unicellular red algae. Both the green clade and the red clade of chloroplasts appear to have been involved in more than one secondary endosymbiosis. At least one secondary endosymbiosis produced a unicellular protist that became, itself, a partner in a *tertiary endosymbiosis!* In that case, a dinoflagellate lost its plastid and took up a haptophyte protist (itself the result of secondary endosymbiosis). The result is the dinoflagellate *Karenia brevis*.

Although euglenoid chloroplasts are descendants of a chlorophyte and stramenopile chloroplasts are descendants of a red alga, this does not mean that euglenoids themselves are descendants of a chlorophyte, nor are stramenopiles themselves descendants of a red alga. The ancestors that took up green or red algae in secondary endosymbioses had their own evolutionary histories. Thus, the nuclear and chloroplast genomes of these protists have different histories. It has taken much research to piece together the clades as we now understand them.

The clades of protists that we have discussed are summarized in Table 28.1 and Figure 28.9. Now let's consider some of the body forms that have appeared repeatedly in various branches of the eukaryote family tree.

Some Recurrent Body Forms

Amoebas used to be classified together in a single protist group. We now know that the amoebas map to at least two complex assemblages that are difficult to position in phylogenetic trees. Similarly, three kinds of organisms called slime molds, once classified together, may be quite different phylogenetically. In this section, we'll look at some of the variations on these body plans found among the protists.

Amoebas form pseudopods

The pseudopods used by **amoebas** for locomotion are a hallmark of the amoeboid body plan (see Figure 28.4). This body plan has appeared by convergent evolution in various protist groups. The mechanism of amoeboid motion will be discussed in Chapter 47.

Amoebas have often been portrayed in popular writing as blobs—the simplest form of "animal" life imaginable. Superficial examination of a typical amoeba shows how such an impression might have been obtained. An amoeba consists of a single cell. It feeds on small organisms and particles of organic matter by phagocytosis, engulfing them with its pseudopods.

But amoebas are specialized protists. Many are adapted for life on the bottoms of lakes, ponds, and other bodies of water. Their creeping locomotion and their manner of engulfing food particles fit them for life close to a relatively rich supply of sedentary organisms or organic particles. Most amoebas exist as predators, parasites, or scavengers. A few are photosynthetic.

Amoebas of the free-living genus *Naegleria*, some of which can enter humans and cause a fatal disease of the nervous system, have a two-stage life cycle, one stage having amoeboid cells and the other flagellated cells. Some amoebas are shelled, living in casings of sand grains glued together (see Figure 28.7b). Others have shells secreted by the organism itself.

ACTINOPODS HAVE THIN, STIFF PSEUDOPODS. The **actinopods** are recognizable by their thin, stiff pseudopods, which are reinforced by microtubules. These pseudopods play at least four roles:

▶ They greatly increase the surface area of the cell for exchange of materials with the environment.
▶ They help the cell float in its marine or freshwater environment.
▶ They provide locomotion in some species.
▶ They are the cell's feeding organs, trapping smaller organisms and often taking them up by endocytosis.

Radiolarians, a group of actinopods found exclusively in marine environments, are perhaps the most beautiful of all microorganisms (see Figure 28.8). Almost all radiolarian species secrete glassy *endoskeletons* (internal skeletons) from which needlelike pseudopods project. Part of the skeleton is a central capsule within the cytoplasm. The skeletons of the different species are as varied as snowflakes, and many have elaborate geometric designs (Figure 28.30a). A few radiolarians are among the largest of the unicellular protists, measuring several millimeters across. Innumerable radiolarian skeletons, some as old as 700 million years, form the sediments under some tropical seas.

Heliozoans are actinopods that lack an endoskeleton (Figure 28.30b). Most heliozoans are found in fresh water. They

(a) *Podocyrtis mitra*

100 µm

(b) *Actinosphaerium eichhorni*

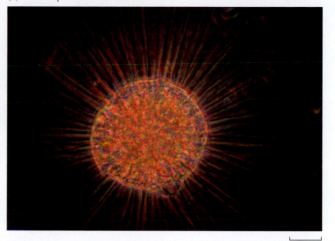

200 µm

28.30 Two Actinopods   (a) Radiolarians secrete intricate glassy skeletons such as the one shown here. A living radiolarian is shown in Figure 28.8. (b) The radiating pseudopods of a heliozoan (Greek, "sun animalcule") give it an appearance that explains its descriptive name.

roll along the substratum by shortening and elongating their pseudopods.

FORAMINIFERANS HAVE CREATED VAST LIMESTONE DEPOSITS. **Foraminiferans** are marine protists that secrete shells of calcium carbonate (see Figure 28.7a). The shells of individual foraminiferan species have distinctive shapes. Some foraminiferans live as **plankton** (free-floating microscopic organisms), and many others live at the bottom of the sea. Their long, threadlike, branched pseudopods reach out through numerous microscopic pores in the shell and interconnect to create a sticky net, which the foraminiferan uses to catch smaller plankton.

After foraminiferans reproduce (by mitosis and cytokinesis), the daughter cells abandon the parent shell and make new shells of their own. The discarded skeletons of ancient foraminiferans make up extensive limestone deposits in various parts of the world, forming a layer hundreds to thousands of meters deep over millions of square kilometers of ocean bottom. Foraminiferan skeletons also make up the sand of some beaches. A single gram of such sand may contain as many as 50,000 foraminiferan shells and shell fragments.

The shells of individual foraminiferans are easily preserved as fossils in marine sediments. Each geological period has distinctive foraminiferan species. For this reason, and because they are so abundant, the remains of foraminiferans are especially valuable as indicators in the classification and dating of sedimentary rocks, as well as in oil prospecting. They also reveal information about temperatures at the time they were alive.

Slime molds release spores from erect fruiting bodies

The three groups of **slime molds** seem so similar at first glance that they were once grouped together in a single phylum. However, the slime molds are actually so different that some biologists now classify them in different *kingdoms*. We will consider two of these groups, called acellular slime molds and cellular slime molds.

The slime molds share only general characteristics. All are motile, all ingest particulate food by endocytosis, and all form spores on erect fruiting bodies. They undergo striking changes in organization during their life cycles, and one stage consists of isolated cells that take up food particles by endocytosis. Some slime molds may cover areas of 1 meter or more in diameter while in their less aggregated stage. Such a large slime mold may weigh more than 50 grams. Slime molds of both types favor cool, moist habitats, primarily in forests. They range from colorless to brilliantly yellow and orange.

ACELLULAR SLIME MOLDS FORM MULTINUCLEATE MASSES.   If the nucleus of an amoeba began rapid mitotic division, accompanied by a tremendous increase in cytoplasm and organelles, the resulting organism might resemble the **acellular slime molds** (*myxomycetes*). During its vegetative (feeding) phase, an acellular slime mold is a wall-less mass of cytoplasm with numerous diploid nuclei. This mass streams very slowly over its substratum in a remarkable network of strands called a *plasmodium** (Figure 28.31a). The plasmodium of an acellular slime mold is another example of a coenocyte, a body in which many nuclei are enclosed in a single plasma membrane. The outer cytoplasm of the plasmodium (closest to the environment) is normally less fluid

*Do not confuse the plasmodium of an acellular slime mold with the genus *Plasmodium*, the apicomplexan that is the cause of malaria.

(a) *Physarum polycephalum*

(b) *Physarum* sp.

0.25 mm

28.31 Acellular Slime Molds (a) Plasmodia of the yellow slime mold *Physarum* cover a rock in Nova Scotia. (b) The fruiting structures—sporangiophores (yellow) and sporangia (black)—of *Physarum*.

than the interior cytoplasm and thus provides some structural rigidity.

Acellular slime molds, such as *Physarum*, provide a dramatic example of movement by *cytoplasmic streaming*. The outer cytoplasmic region of the plasmodium becomes more fluid in places, and cytoplasm rushes into those areas, stretching the plasmodium. This streaming somehow reverses its direction every few minutes as cytoplasm rushes into a new area and drains away from an older one, moving the plasmodium over its substratum. Sometimes an entire wave of plasmodium moves across the substratum, leaving strands behind. Actin filaments and a contractile protein called *myxomyosin* interact to produce the streaming movement. As it moves, the plasmodium engulfs food particles by endocytosis—predominantly bacteria, yeasts, spores of fungi, and other small organisms, as well as decaying animal and plant remains.

An acellular slime mold can grow almost indefinitely in its plasmodial stage, as long as the food supply is adequate and other conditions, such as moisture and pH, are favorable. However, one of two things can happen if conditions become unfavorable. First, the plasmodium can form an irregular mass of hardened cell-like components called a *sclerotium*. This resting structure rapidly becomes a plasmodium again when favorable conditions are restored.

Alternatively, the plasmodium can transform itself into spore-bearing fruiting structures (Figure 28.31*b*). These stalked or branched structures, called *sporangiophores*, rise from heaped masses of plasmodium. They derive their rigid-

ity from walls that form and thicken between their nuclei. The nuclei of the plasmodium are diploid, and they divide by meiosis as the sporangiophore develops. One or more knobs, called *sporangia*, develop on the end of the stalk. Within a sporangium, haploid nuclei become surrounded by walls and form spores. Eventually, as the sporangiophore dries, it sheds its spores.

The spores germinate into wall-less, flagellated, haploid cells called *swarm cells*, which can either divide mitotically to produce more haploid swarm cells or function as gametes. Swarm cells can live as separate individual cells, and can become walled and resistant resting cysts when conditions are unfavorable. When conditions improve again, the cysts release flagellated swarm cells. Two swarm cells can also fuse to form a diploid zygote, which divides by mitosis (but without a wall forming between the nuclei) and thus forms a new, coenocytic plasmodium.

CELLS RETAIN THEIR IDENTITY IN THE CELLULAR SLIME MOLDS. Whereas the plasmodium is the basic vegetative unit of the acellular slime molds, an amoeboid cell is the vegetative unit of the **cellular slime molds**. Large numbers of cells called *myxamoebas*, which have single haploid nuclei, engulf bacteria and other food particles by endocytosis and reproduce by mitosis and fission. This simple life cycle stage, consisting of swarms of independent, isolated cells, can persist indefinitely as long as food and moisture are available.

When conditions become unfavorable, however, the cellular slime molds aggregate and form fruiting structures, as do their acellular counterparts. The apparently independent myxamoebas aggregate into a mass called a *slug* or *pseudoplasmodium* (Figure 28.32). Unlike the true plasmodium of the acellular slime molds, this structure is not simply a giant sheet of cytoplasm with many nuclei; the individual myxamoebas retain their plasma membranes and, therefore, their identity.

Dictyostelium discoideum

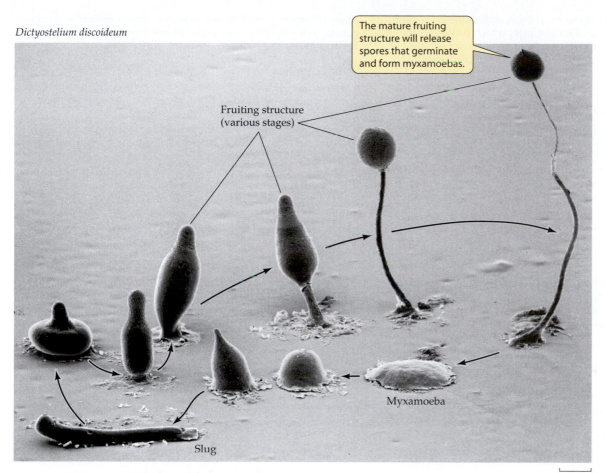

The mature fruiting structure will release spores that germinate and form myxamoebas.

Fruiting structure (various stages)

Myxamoeba

Slug

0.25 mm

28.32 A Cellular Slime Mold   The life cycle of the slime mold *Dictyostelium* is shown here in a composite micrograph.

The chemical signal that causes the myxamoebas of cellular slime molds to aggregate into a slug is 3′,5′-cyclic adenosine monophosphate (cAMP), a compound that plays many important roles as a chemical signal in animals (see Chapter 15).

A slug may migrate over its substratum for several hours before becoming motionless and reorganizing to construct a delicate, stalked fruiting structure (Figure 28.32). Cells at the top of the fruiting structure develop into thick-walled spores, which are eventually released. Later, under favorable conditions, the spores germinate, releasing myxamoebas.

The cycle from myxamoebas through slug and spores to new myxamoebas is asexual. Cellular slime molds also have a sexual cycle, in which two myxamoebas fuse. The product of this fusion develops into a spherical structure that ultimately germinates, releasing new haploid myxamoebas.

This asexual life cycle—individual cells swarming about and then aggregating to form a fruiting body—has evolved, independently, more than once. A bacterial group, the myxobacteria, also has swarming cells that aggregate and form fruiting structures. These bacteria are, of course, prokaryotes and thus have no sexual cycle.

Next we will explore the three classical kingdoms of multicellular eukaryotes. Chapters 29 and 30 deal with the kingdom Plantae (which, combined with the chlorophytes and other green algae, is called by some botanists the "green plant kingdom"). Chapter 31 presents the kingdom Fungi, and Chapters 32–34 describe the kingdom Animalia. All three of these kingdoms arose from protist ancestors.

Chapter Summary

Protists Defined

▶ In this book we define the protists simply as all eukaryotes that are not plants, fungi, or animals. The protists are a paraphyletic group, not a clade.

The Origin of the Eukaryotic Cell

▶ The modern eukaryotic cell arose from an ancestral prokaryote in several steps. Probable steps include the loss of the cell wall and infolding of the plasma membrane. **Review Figure 28.2**

▶ In subsequent steps, an infolded plasma membrane attached to the chromosome may have led to the formation of a nuclear envelope. A primitive cytoskeleton evolved. **Review Figure 28.3**

▶ The first truly eukaryotic cell was larger than its prokaryote ancestor, was probably a phagocyte, and may have possessed one or more flagella of the eukaryotic type.

▶ The incorporation of prokaryotic cells as endosymbionts gave rise to eukaryotic organelles. Peroxisomes, which protected the host cell from an oxygen-rich atmosphere, may have been the first organelles of endosymbiotic origin. Mitochondria evolved from once free-living proteobacteria, and chloroplasts evolved from once free-living cyanobacteria. **Review Figure 28.3**

General Biology of the Protists

▶ Most protists are aquatic; some live within other organisms. The great majority are unicellular and microscopic, but many are multicellular and a few are enormous.

▶ "Protozoan" is an outdated term sometimes applied to protists, mostly ingestive heterotrophs, that were once classified as animals. "Alga" is an outdated term sometimes applied to photosynthetic protists.

▶ Protists vary widely in their modes of nutrition and locomotion. Some protist cells contain contractile vacuoles, and some digest their food in food vacuoles. **Review Figures 28.5, 28.6. See Web/CD Tutorial 28.1**

▶ Protists have a variety of cell surfaces, some of them protective.

▶ Many protists contain endosymbionts. Some protists are endosymbionts in other cells, including other protists. Some endosymbiotic protists perform photosynthesis, to the advantage of their hosts.

▶ Most protists reproduce both asexually and sexually.

Protist Diversity

▶ Molecular and other techniques are enabling biologists to identify many clades of protists. **Review Table 28.1 and Figure 28.9**

Diplomonads and Parabasalids

▶ Diplomonads and parabasalids may have the most ancient roots of today's protists. Both lack mitochondria, having apparently lost them during their evolution.

▶ Diplomonads have two nuclei and multiple flagella. **Review Figure 28.10a**

▶ Parabasalids have flagella and undulating membranes. **Review Figure 28.10b**

Euglenozoans

▶ The euglenozoans are a clade of unicellular protists with flagella.

▶ Euglenoids are euglenozoans that are often photosynthetic and have anterior flagella. **Review Figure 28.11**

▶ Kinetoplastids are euglenozoans that have a single, large mitochondrion, in which RNA is edited.

Alveolates

▶ The alveolates are a clade of unicellular organisms with cavities, called alveoli, beneath their plasma membranes.

▶ Dinoflagellates are marine alveolates with a golden-brown color that results from their photosynthetic and accessory pigments. They are major contributors to world photosynthesis. Many are endosymbionts; in that role they are important contributors to coral growth. Dinoflagellates are responsible for toxic "red tides."

▶ Apicomplexans are parasitic alveolates with an amoeboid body form. Their spores, containing a mass of organelles at the apical end, are adapted to the invasion of host tissue. The apicomplexan *Plasmodium*, which causes malaria, uses two alternate hosts (humans and *Anopheles* mosquitoes). **Review Figure 28.14**

▶ Ciliates are alveolates that move rapidly by means of cilia and have two kinds of nuclei. The macronuclei control the cell by means of transcription and translation. The micronuclei are responsible for genetic recombination, accomplished by conjugation, a process that is sexual, but not reproductive. **Review Figures 28.16, 28.17. See Web/CD Activity 28.1**

Stramenopiles

▶ Stramenopiles typically have two flagella of unequal length, the longer bearing rows of tubular hairs. Some stramenopile groups are photosynthetic.

▶ Diatoms are unicellular stramenopiles, many of which have complex, two-part, glassy cell walls. They contribute extensively to world photosynthesis. **Review Figure 28.19**

▶ The brown algae are predominantly multicellular, photosynthetic stramenopiles. They include the largest of all protists, and some show considerable tissue differentiation.

▶ In many multicellular photosynthetic protists and in all plants, both haploid and diploid cells undergo mitosis, leading to an alternation of generations. The diploid sporophyte generation forms spores by meiosis, and the spores develop into haploid organisms. This haploid gametophyte generation forms gametes by mitosis, and their fusion yields zygotes that develop into the next generation of sporophytes. **Review Figure 28.22**

▶ Oomycetes are a group of nonphotosynthetic stramenopiles including water molds and downy mildews. The oomycetes are coenocytic. They are diploid for most of their life cycle.

Red Algae

▶ Red algae are multicellular, photosynthetic protists. They differ from the other photosynthetic protist groups in having a characteristic storage product (floridean starch) and lacking flagellated reproductive cells.

Chlorophytes

▶ The chlorophytes, a clade of green algae, are often multicellular. Like plants, they contain chlorophylls *a* and *b* and use starch as a storage product.

▶ The chlorophytes are sister to a clade that includes other green algae and the plant kingdom.

▶ The chlorophytes have diverse life cycles; among these are the isomorphic alternation of generations of *Ulva* and the haplontic life cycle of *Ulothrix*. **Review Figures 28.26, 28.27. See Web/CD Activities 28.2 and 28.3**

Choanoflagellates

▶ The choanoflagellates are colonial protists with flagella and a body type similar to the most characteristic type of cell found in sponges. The choanoflagellates are sister to the animal kingdom.

A History of Endosymbiosis

▶ Primary endosymbiosis of a cyanobacterium and a eukaryote gave rise to the chloroplasts of green algae, plants, and red algae. **Review Figure 28.29. See Web/CD Tutorial 28.2**

▶ Secondary endosymbioses of eukaryotes with unicellular green or red algae gave rise to the chloroplasts of euglenoids, stramenopiles, and other groups. One of those groups has given rise to another type of chloroplast by tertiary endosymbiosis.

Some Recurrent Body Forms

▶ Some similar body forms are found in several different, unrelated protist groups.

▶ Amoebas, which appear in many protist groups, move by means of pseudopods.

▶ Actinopods have thin, stiff pseudopods that serve various functions, including food capture.

▶ Foraminiferans also use pseudopods for feeding, and they secrete shells of calcium carbonate.

▶ Acellular slime molds and cellular slime molds are superficially similar, moving as slimy masses and producing stalked fruiting structures. However, they differ at the cellular level. Acellular slime molds are coenocytes with diploid nuclei. Cellular slime molds consist of individual haploid cells that aggregate into masses consisting of distinct cells.

Self-Quiz

1. Protists with flagella
   a. appear in several protist clades.
   b. are all algae.
   c. all have pseudopods.
   d. are all colonial.
   e. are never pathogenic.

2. Which statement about amoebas is *not* true?
   a. They are specialized.
   b. They use amoeboid movement.
   c. They include both naked and shelled forms.
   d. They possess pseudopods.
   e. They appeared only once in evolutionary history.

3. Apicomplexans
   a. possess flagella.
   b. possess chloroplasts.
   c. are all parasitic.
   d. are algae.
   e. include the trypanosomes that cause sleeping sickness.

4. The ciliates
   a. move by means of flagella.
   b. use amoeboid movement.
   c. include *Plasmodium*, the agent of malaria.
   d. possess both a macronucleus and micronuclei.
   e. are autotrophic.

5. The acellular slime molds
   a. include the genus *Physarum*.
   b. lack fruiting bodies.
   c. consist of large numbers of myxamoebas.
   d. consist at times of a mass called a pseudoplasmodium.
   e. possess flagella.

6. The cellular slime molds
   a. possess apical complexes.
   b. lack fruiting bodies.
   c. form a plasmodium that is a coenocyte.
   d. use cAMP as a "messenger" to signal aggregation.
   e. possess flagella.

7. The chloroplasts of photosynthetic protists
   a. are structurally identical.
   b. gave rise to mitochondria.
   c. are all descended from a once free-living cyanobacterium.
   d. all have exactly two surrounding membranes.
   e. are all descended from a once free-living red alga.

8. Which statement about the brown algae is *not* true?
   a. They are all multicellular.
   b. They use the same photosynthetic pigments as do plants.
   c. They are almost exclusively marine.
   d. A few are among the largest organisms on Earth.
   e. Some have extensive tissue differentiation.

9. The red algae
   a. are mostly unicellular.
   b. are mostly marine.
   c. owe their red color to a special form of chlorophyll.
   d. have flagella on their gametes.
   e. are all heterotrophic.

10. Which statement about the chlorophytes is *not* true?
    a. They use the same photosynthetic pigments as do plants.
    b. Some are unicellular.
    c. Some are multicellular.
    d. All are microscopic in size.
    e. They display a great diversity of life cycles.

For Discussion

1. For each type of organism below, give a single characteristic that may be used to differentiate it from the other, related organism(s) in parentheses.
   a. Foraminiferans (radiolarians)
   b. *Euglena* (*Volvox*)
   c. *Trypanosoma* (*Giardia*)
   d. Amoeba (flagellate)
   e. *Physarum* (*Dictyostelium*)

2. In what sense are sex and reproduction independent of each other in the ciliates? What does that suggest about the role of sex in biology?

3. Why are dinoflagellates and apicomplexans placed in one group of protists and brown algae and oomycetes in another?

4. Unlike many protists, apicomplexans lack contractile vacuoles. Why don't apicomplexans need a contractile vacuole?

5. Giant seaweeds (mostly brown algae) have "floats" that aid in keeping their fronds suspended at or near the surface of the water. Why is it important that the fronds be suspended in this way?

6. Why are algal pigments so much more diverse than those of plants?

7. Consider the chloroplasts of chlorophytes, euglenozoans, and red algae. For each of these groups, indicate how many membranes surround their chloroplasts, and offer a reasonable explanation in each case. Why do some dinoflagellates have more membranes around their chloroplasts than other dinoflagellates?

29 *Plants without Seeds: From Sea to Land*

 Residents of the coal-producing central Chinese city of Changsha almost never see the sun, because it is hidden behind an atmosphere dense with choking smog. Nine-tenths of the precipitation in Changsha is acid rain. China burns more coal than any other country in the world, and the resulting untreated smoke leads to disastrous conditions such as those in Changsha, the site of a major coal-fired power plant.

Coal is used for 75 percent of China's energy needs—primarily to generate electricity, but also directly for heating, smelting of metals, and other purposes. The United States produces more than half of its electricity by burning coal, and indeed has the largest coal reserves in the world. Extractable coal reserves in the U.S. exceed the total amount of oil available for pumping in all other countries combined. Where did all this coal come from?

Coal comes from the remains of seedless plants that grew in great forests hundreds of millions of years ago. (The two other "fossil fuels"—petroleum and natural gas—come from the remains of plankton that lived in ancient oceans.) Plant parts from those forests sank in swamps that were later covered by soil. Over millions of years, as the buried plant material was subjected to intense pressure and elevated temperatures, coal formed.

At the time those ancient forests flourished, the plant world also included relatives of today's mosses. These "mossy" ancestors were the first plant life on dry land. Today, mosses are among the most abundant plants on Earth, yet they seem at first glance to lack adaptations to life on land. Mosses have no advanced internal "plumbing system" to move water and nutrients within their bodies, and their leafy photosynthetic organs are only one cell thick. They require liquid water in order to reproduce, and indeed, seem at first glance to be highly dependent on external moisture. Mosses and their relatives do have effective adaptations for life in terrestrial environments, however, as is obvious from their wide distribution. Most live in moist habitats, but a few mosses even live in deserts.

An Ingredient of Coal-Based Smog
When coal burns, it produces the fly ash shown in this artificially colored image. When too much coal burns where the smoke cannot blow away, the result is disastrous smog.

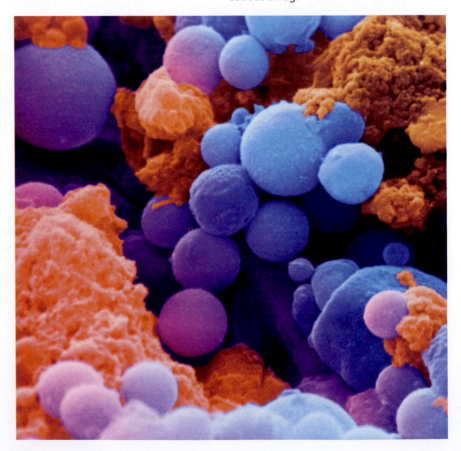

The earliest terrestrial plants invaded the land sometime during the Paleozoic era (see Table 22.1). These plants were tiny, but their metabolic activities helped convert parent rock into soil that could support the needs of their successors. Larger and larger plants evolved rapidly (in geological terms), and by the Carboniferous period (354–290 mya) great forests were widespread. However, few of the trees in those forests were like those we know today. During the tens of millions of years since the Carboniferous, those early trees have been replaced by the modern trees whose adaptations and appearance are familiar to us.

In this chapter, we will see how members of the plant kingdom invaded the land and evolved. Our descriptions here will concentrate on those plants that lack seeds. The next chapter completes our survey of the plant kingdom by considering the seed plants, which dominate the terrestrial scene today.

The Plant Kingdom

The kingdom Plantae is monophyletic—all plants descend from a single common ancestor and form a branch of the evolutionary tree of life. The shared derived trait, or synapomorphy, of the plant kingdom is development from embryos protected by tissues of the parent plant. For this reason, plants are sometimes referred to as *embryophytes*. Plants retain the derived features that they share with green algae: the use of chlorophylls *a* and *b* and the use of starch as a photosynthetic storage product. Both plants and green algae have cellulose in their cell walls.

There are other ways to define "plant" and "plant kingdom" and still come out with a monophyletic group (clade). For example, combining plants as defined above with a group of green algae called the charophytes results in a monophyletic plant kingdom with several shared derived traits, including the retention of the egg in the parent body. The addition of the chlorophytes (the remainder of the green algae) to the group just described gives another monophyletic group, with synapomorphies including the possession of chlorophyll *b*, that can be called a plant kingdom. There are no hard-and-fast criteria for defining a kingdom (or any other taxonomic rank), so these definitions of the plant kingdom are all valid.

In this book, we choose to use the first definition given above, in which the kingdom Plantae comprises only the embryophytes (Figure 29.1). Some botanists refer to a group consisting of the Plantae plus the green algae as the "green plant kingdom," to the red algae as the "red plant kingdom," and to the stramenopiles as the "brown plant kingdom."

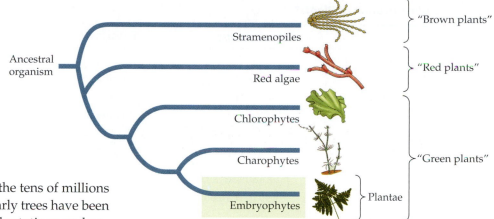

29.1 What Is a Plant?   There are three ways to define a plant kingdom, depending on which clade is chosen. In this book, we use the most restrictive definition: plants as embryophytes. Here, the two green algal clades are not considered plants.

There are ten surviving phyla of plants

The surviving members of the kingdom Plantae fall naturally into ten phyla (Table 29.1). All members of seven of those phyla possess well-developed vascular systems that transport materials throughout the plant body. We call these seven phyla, collectively, the **tracheophytes** because they all possess conducting cells called tracheids. The tracheophytes constitute a clade.

The remaining three phyla (liverworts, hornworts, and mosses), which lack tracheids, were once considered classes of a single larger phylum. In this book we use the term **nontracheophytes** to refer collectively to these three phyla. The nontracheophytes are sometimes collectively called *bryophytes*, but in this text we reserve that term for their most familiar members, the mosses. Collectively, the nontracheophytes are not a monophyletic group. They are the three basal clades of the plant kingdom.

Life cycles of plants feature alternation of generations

A universal feature of the life cycles of plants is the alternation of generations. Recall from Chapter 28 that alternation of generations has two hallmarks:

▶ The life cycle includes both multicellular diploid individuals and multicellular haploid individuals.
▶ Gametes are produced by mitosis, not by meiosis. Meiosis produces spores that develop into multicellular haploid individuals.

If we begin looking at the plant life cycle at a single-cell stage—the diploid zygote—then the first phase of the cycle

29.1 Classification of Plants[a]

PHYLUM	COMMON NAME	CHARACTERISTICS
Nontracheophytes		
Hepatophyta	Liverworts	No filamentous stage; gametophyte flat
Anthocerophyta	Hornworts	Embedded archegonia; sporophyte grows basally
Bryophyta	Mosses	Filamentous stage; sporophyte grows apically (from the tip)
Tracheophytes		
Nonseed tracheophytes		
Lycophyta	Club mosses	Microphylls in spirals; sporangia in leaf axils
Pteridophyta	Ferns and allies	Differentiation between main axis and side branches
Seed plants		
Gymnosperms		
Cycadophyta	Cycads	Compound leaves; swimming sperm; seeds on modified leaves
Ginkgophyta	Ginkgo	Deciduous; fan-shaped leaves; swimming sperm
Gnetophyta	Gnetophytes	Vessels in vascular tissue; opposite, simple leaves
Pinophyta	Conifers	Seeds in cones; needlelike or scalelike leaves
Angiosperms		
Angiospermae	Flowering plants	Endosperm; carpels; much reduced gametophytes; seeds in fruit

[a]No extinct groups are included in this classification.

features the formation, by mitosis and cytokinesis, of a multicellular embryo and eventually the mature diploid plant (Figure 29.2). This multicellular, diploid plant is the **sporophyte** ("spore plant").

Cells contained in **sporangia** (singular, sporangium, "spore vessel") on the sporophyte undergo meiosis to produce haploid, unicellular spores. By mitosis and cytokinesis, a spore forms a haploid plant. This multicellular, haploid plant is the **gametophyte** ("gamete plant") that produces haploid gametes. The fusion of two gametes (*syngamy*, or *fertilization*) results in the formation of a diploid cell—the zygote—and the cycle repeats.

The *sporophyte generation* extends from the zygote through the adult, multicellular, diploid plant; the *gametophyte generation* extends from the spore through the adult, multicellular, haploid plant to the gamete. The transitions between the generations are accomplished by fertilization and meiosis. In all plants, the sporophyte and gametophyte differ genetically: The sporophyte has diploid cells, and the gametophyte has haploid cells. In the three basal plant clades, the gametophyte generation is larger and more self-sufficient, while the sporophyte generation is dominant in those groups that appeared later in plant evolution.

Some protist life cycles also feature alternation of generations, suggesting that the plants arose from one of these protist groups. But which one?

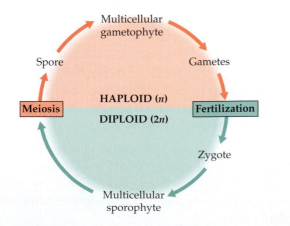

29.2 Alternation of Generations A diploid sporophyte generation that produces spores alternates with a haploid gametophyte generation that produces gametes by mitosis.

The Plantae arose from a green algal clade

Much evidence indicates that the closest living relatives of the plants are members of a clade of green algae called the **charophytes**. The charophytes, along with some other green algae and the plants, form a clade that is sister to the chlorophytes (see Figure 29.1), but we don't yet know which charophyte clade is the true sister group to the plants. Stoneworts of the genus *Chara* are charophytes that resemble plants in terms of their rRNA and DNA sequences, peroxisome con-

(a) *Chara* sp. (stonewort)

(b) *Coleochaete* sp.

29.3 The Closest Relatives of Land Plants The plant kingdom probably evolved from a common ancestor shared with the charophytes, a green algal group. (*a*) Molecular evidence seems to favor stoneworts of the genus *Chara* as sister group to the plants. (*b*) Evidence from morphology indicates that the group including this coleochaete alga may be sister to the land plants.

tents, mechanics of mitosis and cytokinesis, and chloroplast structure (Figure 29.3*a*). On the other hand, strong evidence from morphology-based cladistic analysis suggests that the sister group of the plants is a group of charophytes that includes the genus *Coleochaete* (Figure 29.3*b*). *Coleochaete*-like algae have several features found in plants, such as plasmodesmata and a tendency to protect the young sporophyte.

Whether they were more similar to stoneworts or to *Coleochaete*, the ancestors of the plants lived at the margins of ponds or marshes, ringing them with a green mat. From these marginal habitats, which were sometimes wet and sometimes dry, early plants made the transition onto land.

The Conquest of the Land

Plants, or their immediate ancestors in the green mat, first invaded the terrestrial environment between 400 and 500 million years ago. That environment differs dramatically from the aquatic environment. The most obvious difference is the availability of the water that is essential for life: It is everywhere in the aquatic environment, but hard to find and to retain in the terrestrial environment. Water provides aquatic organisms with support against gravity; a plant on land, however, must either have some other support system or sprawl unsupported on the ground. A land plant must also

use different mechanisms for dispersing its gametes and progeny than its aquatic relatives, which can simply release them into the water. How did organisms descended from aquatic ancestors adapt to such a challenging environment?

Adaptations to life on land distinguish plants from green algae

Most of the characteristics that distinguish plants from green algae are evolutionary adaptations to life on land. Several of these features probably evolved in the common ancestor of the plants:

▶ The *cuticle*, a waxy covering that retards desiccation (drying)
▶ *Gametangia*, cases that enclose plant gametes and prevent them from drying out
▶ *Embryos*, which are young sporophytes contained within a protective structure
▶ Certain *pigments* that afford protection against the mutagenic ultraviolet radiation that bathes the terrestrial environment
▶ Thick *spore walls* containing a polymer that protects the spores from desiccation and resists decay
▶ A *mutualistic* association with a fungus* that promotes nutrient uptake from the soil

Further adaptations to the terrestrial environment appeared as plants continued to evolve. One of the most important of these later adaptations was the appearance of vascular tissues.

Most present-day plants have vascular tissues

The first plants were nonvascular, lacking both water-conducting and food-conducting tissue. Although the term "nonvascular plants" is a time-honored name, it is misleading when applied to the entire nontracheophyte group, because some mosses (unlike liverworts and hornworts) do have a limited amount of simple conducting tissue. Thus the more unwieldy name "nontracheophyte" is more descriptive. The first true tracheophytes—possessing specialized conducting cells called tracheids—arose later (Figure 29.4).

The nontracheophytes (the liverworts, hornworts, and mosses) have never been large plants. Except for some of the mosses, they have no water-conducting tissue, yet some are found in dry environments. Many grow in dense masses (see Figure 29.9*a*), through which water can move by capillary action. Nontracheophytes also have leaflike structures that readily catch and hold any water that splashes onto them. These plants are small enough that minerals can be distributed throughout their bodies by diffusion.

*In a mutualistic association, both partners—here, the plant and the fungus—profit.

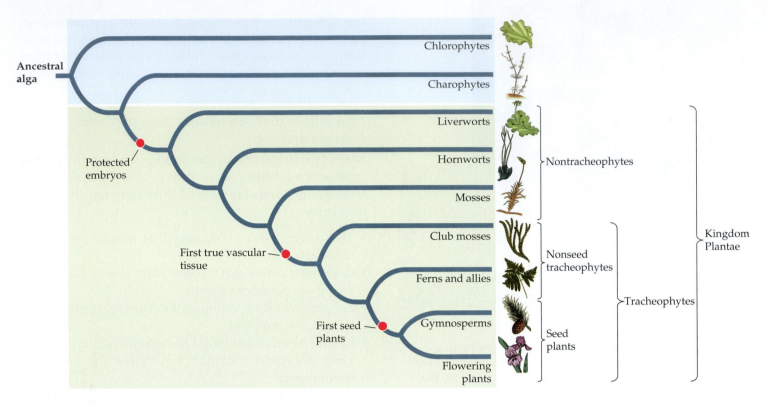

29.4 From Green Algae to Plants Three key characteristics that emerged during plant evolution—protected embryos, vascular tissues, and seeds—are all adaptations to life in a terrestrial environment. Plants with vascular tissue are called tracheophytes.

Familiar tracheophytes include the club mosses, ferns, conifers, and angiosperms (flowering plants). Tracheophytes differ from liverworts, hornworts, and mosses in crucial ways, one of which is the possession of a well-developed **vascular system** consisting of specialized tissues for the transport of materials from one part of the plant to another. One such tissue, the **phloem**, conducts the products of photosynthesis from sites where they are produced or released to sites where they are used or stored. The other vascular tissue, the **xylem**, conducts water and minerals from the soil to aerial parts of the plant; because some of its cell walls are stiffened by a substance called *lignin*, xylem also provides support in the terrestrial environment.

Nontracheophyte plants evolved tens of millions of years before the earliest tracheophytes, even though tracheophytes appear earlier in the fossil record. The oldest tracheophyte fossils date back more than 410 million years, whereas the oldest nontracheophyte fossils are only about 350 million years old, dating from a time when tracheophytes were already widely distributed. This finding simply shows that, given the differences in their structures and the chemical makeup of their cell walls, tracheophytes are more likely to form fossils than nontracheophytes are.

We will examine the adaptations of the tracheophytes later in this chapter, concentrating first on the nontracheophytes.

The Nontracheophytes: Liverworts, Hornworts, and Mosses

Most liverworts, hornworts, and mosses grow in dense mats, usually in moist habitats. The largest of these plants are only about 1 meter tall, and most are only a few centimeters tall or long. Why have the nontracheophytes not evolved to be taller? The probable answer is that they lack an efficient system for conducting water and minerals from the soil to distant parts of the plant body. To limit water loss, layers of maternal tissue protect the embryos of all nontracheophytes. All nontracheophyte clades also have a cuticle, although it is often very thin (or even absent in some species) and thus not highly effective in retarding water loss. Nontracheophytes lack the leaves, stems, and roots that characterize tracheophytes, although they have structures analogous to each.

Most nontracheophytes live on the soil or on other plants, but some grow on bare rock, dead and fallen tree trunks, and even on buildings. Nontracheophytes are widely distributed over six continents and exist very locally on the coast of the

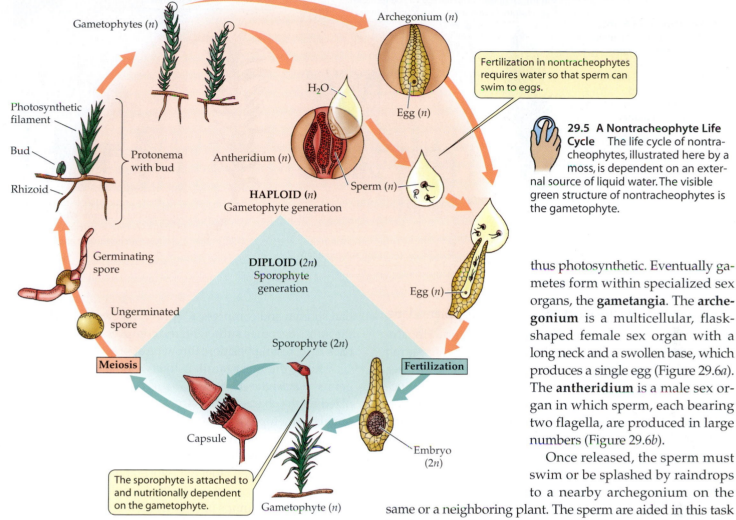

Gametophytes (*n*)

Archegonium (*n*)

Photosynthetic filament

Bud

Rhizoid

Protonema with bud

H₂O

Antheridium (*n*)

HAPLOID (*n*)
Gametophyte generation

Sperm (*n*)

Egg (*n*)

Germinating spore

Ungerminated spore

DIPLOID (2*n*)
Sporophyte generation

Egg (*n*)

Sporophyte (2*n*)

Meiosis

Fertilization

Capsule

Embryo (2*n*)

The sporophyte is attached to and nutritionally dependent on the gametophyte.

Gametophyte (*n*)

Fertilization in nontracheophytes requires water so that sperm can swim to eggs.

29.5 A Nontracheophyte Life Cycle   The life cycle of nontracheophytes, illustrated here by a moss, is dependent on an external source of liquid water. The visible green structure of nontracheophytes is the gametophyte.

thus photosynthetic. Eventually gametes form within specialized sex organs, the **gametangia**. The **archegonium** is a multicellular, flask-shaped female sex organ with a long neck and a swollen base, which produces a single egg (Figure 29.6*a*). The **antheridium** is a male sex organ in which sperm, each bearing two flagella, are produced in large numbers (Figure 29.6*b*).

Once released, the sperm must swim or be splashed by raindrops to a nearby archegonium on the same or a neighboring plant. The sperm are aided in this task by chemical attractants released by the egg or the archegonium. Before sperm can enter the archegonium, certain cells in the neck of the archegonium must break down, leaving a water-filled canal through which the sperm swim to complete their journey. Note that all of these events require liquid water.

On arrival at the egg, the nucleus of a sperm fuses with the egg nucleus to form a zygote. Mitotic divisions of the zygote produce a multicellular, diploid sporophyte embryo. The base of the archegonium grows to protect the embryo during its early development. Eventually, the developing sporophyte elongates sufficiently to break out of the archegonium, but it remains connected to the gametophyte by a "foot" that is embedded in the parent tissue and absorbs water and nutrients from it. The sporophyte remains attached to the gametophyte throughout its life. The sporophyte produces a capsule, within which meiotic divisions produce spores and thus the next gametophyte generation.

The structure and pattern of elongation of the sporophyte differ among the three nontracheophyte phyla—the liverworts (Hepatophyta), hornworts (Anthocerophyta), and mosses (Bryophyta). The probable evolutionary relationships of these three phyla and the tracheophytes can be seen in Figure 29.4.

seventh (Antarctica). They are successful plants, well adapted to their environments. Most are terrestrial. Some live in wetlands. Although a few nontracheophyte species live in fresh water, these aquatic forms are descended from terrestrial ones. There are no marine nontracheophytes.

Nontracheophyte sporophytes are dependent on gametophytes

In nontracheophytes, the conspicuous green structure visible to the naked eye is the gametophyte (Figure 29.5). In contrast, the familiar forms of tracheophytes, such as ferns and seed plants, are sporophytes. The gametophyte of nontracheophytes is photosynthetic and therefore nutritionally independent, whereas the sporophyte may or may not be photosynthetic, but is always nutritionally dependent on the gametophyte and remains permanently attached to it.

A nontracheophyte sporophyte produces unicellular, haploid spores as products of meiosis within a sporangium, or **capsule**. A spore germinates, giving rise to a multicellular, haploid gametophyte whose cells contain chloroplasts and are

29.6 Sex Organs in Plants
(a) Archegonia and (b) antheridia of the moss *Mnium* (phylum Bryophyta). The gametophytes of all plants have archegonia and antheridia, but they are much reduced in seed plants.

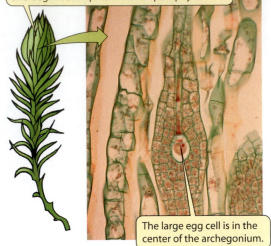

(a)

Archegonia develop at the tip of a gametophyte. In the archegonium, the egg will be fertilized and begin development into a sporophyte.

The large egg cell is in the center of the archegonium.

(b)

Antheridia are also located at the tip of a gametophyte.

These antheridia contain a large number of sperm. When released, the sperm can be carried by water to an archegonium and then swim down its neck to the egg.

Liverworts may be the most ancient surviving plant clade

The gametophytes of some **liverworts** (phylum **Hepatophyta**) are green, leaflike layers that lie flat on the ground (Figure 29.7*a*). The simplest liverwort gametophytes, however, are flat plates of cells, a centimeter or so long, that produce antheridia or archegonia on their upper surfaces and anchoring and water-absorbing filaments called **rhizoids** on their lower surfaces. Liverwort sporophytes are shorter than those of mosses and hornworts, rarely exceeding a few millimeters.

The liverwort sporophyte has a stalk that connects capsule and foot. In most species, the stalk elongates and thus raises the capsule above ground level, favoring dispersal of spores when they are released. The capsules of liverworts are simple: a globular capsule wall surrounding a mass of spores. In some species of liverworts, spores are not released by the sporophyte until the surrounding capsule wall rots. In other liverworts, however, the spores are thrown from the capsule by structures

that shorten and compress a "spring" as they dry out. When the stress becomes sufficient, the compressed spring snaps back to its resting position, throwing spores in all directions.

Among the most familiar liverworts are species of the genus *Marchantia* (Figure 29.7*a*). *Marchantia* is easily recognized by the characteristic structures on which its male and female gametophytes bear their antheridia and archegonia (Figure 29.7*b*). Like most liverworts, *Marchantia* also reproduces asexually by simple fragmentation of the gametophyte. *Marchantia* and some other liverworts and mosses also reproduce asexually by means of *gemmae* (singular, gemma), which are lens-shaped clumps of cells. In a few liverworts, the gemmae are loosely held in structures called *gemmae cups*, which promote dispersal of the gemmae by raindrops (Figure 29.7*c*).

Hornworts evolved stomata as an adaptation to terrestrial life

The phylum **Anthocerophyta** comprises the **hornworts**, so named because their sporophytes look like little horns (Fig-

29.7 Liverwort Structures Members of the phylum Hepatophyta display various characteristic structures. (*a*) Gametophytes. (*b*) Structures bearing antheridia and archegonia. (*c*) Gemmae cups.

The umbrella-like structures bear archegonia.

The disc-headed structures bear antheridia.

These cups contain gemmae—small, lens-shaped outgrowths of the plant body, each capable of developing into a new plant.

(a) *Marchantia* sp.

(b) *Marchantia* sp.

(c) *Lunularia* sp.

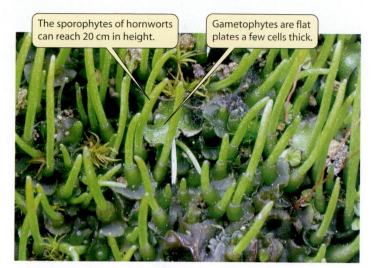

The sporophytes of hornworts can reach 20 cm in height.

Gametophytes are flat plates a few cells thick.

Anthoceros sp.

29.8 A Hornwort   The sporophytes of hornworts can resemble little horns.

ure 29.8). Hornworts appear at first glance to be liverworts with very simple gametophytes. These gametophytes consist of flat plates of cells a few cells thick.

However, the hornworts, along with the mosses and tracheophytes, share an advance over the liverwort clade in their adaptation to life on land. They have *stomata*—pores that, when open, allow the uptake of CO_2 for photosynthesis and the release of O_2. Stomata may be a shared derived trait (synapomorphy) of hornworts and all other plants except liverworts, although hornwort stomata do not close and may have evolved independently.

Hornworts have two characteristics that distinguish them from both liverworts and mosses. First, the cells of hornworts each contain a single large, platelike chloroplast, whereas the cells of other nontracheophytes contain numerous small, lens-shaped chloroplasts. Second, of all the nontracheophyte sporophytes, those of the hornworts come closest to being capable of indeterminate growth (growth without a set limit). Liverwort and moss sporophytes have a stalk that stops growing as the capsule matures, so elongation of the sporophyte is strictly limited. The hornwort sporophyte, however, has no stalk. Instead, a basal region of the capsule remains capable of indefinite cell division, continuously producing new spore-bearing tissue above. The sporophytes of some hornworts growing in mild and continuously moist conditions can become as tall as 20 centimeters. Eventually the sporophyte's growth is limited by the lack of a transport system.

To support their metabolism, the hornworts need access to nitrogen. Hornworts have internal cavities filled with mucilage; these cavities are often populated by cyanobacteria that convert atmospheric nitrogen gas into a form usable by the host plant.

We have presented the hornworts as sister to the clade consisting of mosses and tracheophytes, but this is only one pos-

sible interpretation of the current data. The exact evolutionary status of the hornworts is still unclear, and in some phylogenetic analyses they are placed as the most ancient plant clade.

Water and sugar transport mechanisms emerged in the mosses

The most familiar nontracheophytes are the **mosses** (phylum **Bryophyta**). There are more species of mosses than of liverworts and hornworts combined, and these hardy little plants are found in almost every terrestrial environment. They are often found on damp, cool ground, where they form thick mats (Figure 29.9*a*). The mosses are probably sister to the tracheophytes (see Figure 29.4).

Many mosses contain a type of cell called a *hydroid*, which dies and leaves a tiny channel through which water can

(a)

(b)

The teeth of this moss capsule help expel the thousands of spores.

29.9 The Mosses   (*a*) Dense moss forms hummocks in a valley on New Zealand's South Island. (*b*) The moss capsule, from which spores are dispersed, grows at the tip of the plant.

travel. The hydroid may be the progenitor of the tracheid, the characteristic water-conducting cell of the tracheophytes, but it lacks lignin (a waterproofing substance that also lends structural support) and the cell wall structure found in tracheids. The possession of hydroids and of a limited system for transport of sucrose by some mosses (via cells called *leptoids*) shows that the old term "nonvascular plant" is somewhat misleading when applied to mosses.

In contrast to liverworts and hornworts, the sporophytes of mosses and tracheophytes grow by **apical cell division**, in which a region at the growing tip provides an organized pattern of cell division, elongation, and differentiation. This growth pattern allows extensive and sturdy vertical growth of sporophytes. Apical cell division is a shared derived trait of mosses and tracheophytes.

The moss gametophyte that develops following spore germination is a branched, filamentous structure called a *protonema* (see Figure 29.5). Although the protonema looks a bit like a filamentous green alga, it is unique to the mosses. Some of the filaments contain chloroplasts and are photosynthetic; others, called rhizoids, are nonphotosynthetic and anchor the protonema to the substratum. After a period of linear growth, cells close to the tips of the photosynthetic filaments divide rapidly in three dimensions to form *buds*. The buds eventually differentiate a distinct tip, or apex, and produce the familiar leafy moss shoot with leaflike structures arranged spirally. These leafy shoots produce antheridia or archegonia (see Figure 29.6). The antheridia release sperm that travel through liquid water to the archegonia, where they fertilize the eggs.

Sporophyte development in most mosses follows a precise pattern, resulting ultimately in the formation of an absorptive foot anchored to the gametophyte, a stalk, and, at the tip, a swollen capsule, the sporangium. In contrast to hornworts, whose sporophytes grow from the base, the moss sporophyte stalk grows at its apical end, as tracheophytes do. Cells at the tip of the stalk divide, supporting elongation of the structure and giving rise to the capsule. For a while, the archegonial tissue grows rapidly as the stalk elongates, but eventually the archegonium is outgrown and is torn apart by the expanding sporophyte.

The lid of the capsule is shed after the completion of meiosis and spore development. In most mosses, groups of cells just below the lid form a series of toothlike structures surrounding the opening. Highly responsive to humidity, these structures dig into the mass of spores when the atmosphere is dry; then, when the atmosphere becomes moist, they fling out, scooping out the spores as they go (Figure 29.9b). The spores are thus dispersed when the surrounding air is moist—that is, when conditions favor their subsequent germination.

Mosses of the genus *Sphagnum* often grow in swampy places, where the plants begin to decompose in the water after they die. Rapidly growing upper layers compress the deeper-lying, decomposing layers. Partially decomposed plant matter is called *peat*. In some parts of the world, people derive the majority of their fuel from peat bogs. *Sphagnum*-dominated peatlands cover an area approximately half as large as the United States—more than 1 percent of Earth's surface. Long ago, continued compression of peat composed primarily of other nonseed plants gave rise to coal.

With their simple system of internal transport, the mosses are, in a sense, vascular plants. However, they are not tracheophytes because they lack true xylem and phloem.

Introducing the Tracheophytes

Although they are an extraordinarily large and diverse group, the tracheophytes can be said to have been launched by a single evolutionary event. Sometime during the Paleozoic era, probably well before the Silurian period (440 mya), the sporophyte generation of a now long-extinct plant produced a new cell type, the **tracheid** (Figure 29.10). The tracheid is the principal water-conducting element of the xylem in all tracheophytes except the angiosperms, and even in the angiosperms, tracheids persist alongside a more specialized and efficient system of vessels and fibers derived from them.

The evolution of a tissue composed of tracheids had two important consequences. First, it provided a pathway for long-distance transport of water and mineral nutrients from a source of supply to regions of need. Second, its stiff cell walls provided something almost completely lacking—and unnecessary—in the largely aquatic green algae: rigid structural support. Support is important in a terrestrial environment because plants tend to grow upward as they compete for sunlight to power photosynthesis. Thus the tracheid set the stage for the complete and permanent invasion of land by plants.

The tracheophytes feature another evolutionary novelty: a branching, independent sporophyte. A branching sporophyte can produce more spores than an unbranched body, and it can develop in complex ways. The sporophyte of a tracheophyte is nutritionally independent of the gametophyte at maturity. Among the tracheophytes, the sporophyte is the large and obvious plant that one normally notices in nature. This pattern is in contrast to the sporophyte of nontracheophytes such as mosses, which is attached to, dependent on, and usually much smaller than the gametophyte.

The present-day evolutionary descendants of the early tracheophytes belong to seven distinct phyla (see Figure 29.10). The tracheophytes have two types of life cycles, one that involves seeds and another that does not. The nonseed tracheophytes (the two basal phyla) include the club mosses and the ferns and their relatives: horsetails and whisk ferns. We will describe these phyla in detail after taking a closer

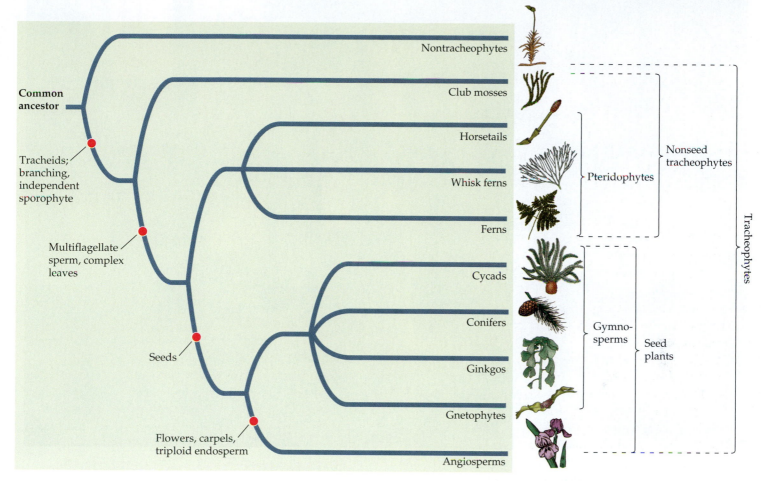

29.10 The Evolution of Today's Plants   The nine phyla of extant tracheophytes are divided between those that produce seeds and those that do not.

look at tracheophyte evolution. The five phyla of seed plants will be described in the following chapter.

Tracheophytes have been evolving for almost half a billion years

The evolution of an effective cuticle and of protective layers for the gametangia (archegonia and antheridia) helped make the first tracheophytes successful, as did the initial absence of herbivores (plant-eating animals) on land. By the late Silurian period, tracheophytes were being preserved as fossils that we can study today. Two groups of nonseed tracheophytes that still exist made their first appearances during the Devonian period (409–354 mya): the lycopods (club mosses) and the pteridophytes (including horsetails and ferns). Their proliferation made the terrestrial environment more hospitable to animals. Amphibians and insects arrived soon after the plants became established.

Trees of various kinds appeared in the Devonian period and dominated the landscape of the Carboniferous. Mighty forests of lycopods up to 40 meters tall, horsetails, and tree ferns flourished in the tropical swamps of what would become North America and Europe (Figure 29.11). The remnants of those forests are with us today as huge deposits of coal.

In the subsequent Permian period, the continents came together to form a single gigantic land mass, called Pangaea. The continental interior became warmer and drier, but late in the period glaciation was extensive. The 200-million-year reign of the lycopod–fern forests came to an end as they were replaced by forests of seed plants (gymnosperms), which dominated until other seed plants (angiosperms) became dominant less than 80 million years ago.

The earliest tracheophytes lacked roots and leaves

The earliest known tracheophytes belonged to the now-extinct phylum **Rhyniophyta**. The rhyniophytes were among the only tracheophytes in the Silurian period. The landscape at that time probably consisted of bare ground, with stands of rhyniophytes in low-lying moist areas. Early versions of the structural features of all the other tracheophyte phyla appeared in the rhyniophytes of that time. These shared features strengthen the case for the origin of all tracheophytes from a common nontracheophyte ancestor.

29.11 An Ancient Forest
This reconstruction is of a Carboniferous forest that once thrived in what is now Michigan. The dominant "trees" are lycopods of the genus *Lepidodendron*; ferns are also abundant.

In 1917, the British paleobotanists Robert Kidston and William H. Lang reported their finding of well-preserved fossils of tracheophytes embedded in Devonian rocks near Rhynie, Scotland. The preservation of these plants was remarkable, considering that the rocks were more than 395 million years old. These fossil plants had a simple vascular system of phloem and xylem. Some of the plants had flattened scales on the stems, which lacked vascular tissue and thus were not comparable to the true leaves of any other tracheophytes.

These plants also lacked roots. They were apparently anchored in the soil by horizontal portions of stem, called **rhizomes**, that bore water-absorbing rhizoids. These rhizomes also bore aerial branches, and sporangia—homologous with the nontracheophyte capsule—were found at the tips of these branches. Their branching pattern was dichotomous; that is, the shoot apex divided to produce two equivalent new branches, each pair diverging at approximately the same angle from the original stem (Figure 29.12). Scattered fragments of such plants had been found earlier, but never in such profusion or so well preserved as those discovered by Kidston and Lang.

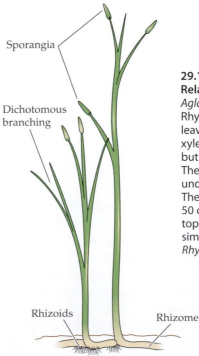

Sporangia

Dichotomous branching

Rhizoids

Rhizome

29.12 An Ancient Tracheophyte Relative This extinct plant, *Aglaophyton major* (phylum Rhyniophyta), lacked roots and leaves. It had a central column of xylem running through its stems, but true tracheids were lacking. The rhizome is a horizontal underground stem, not a root. The aerial stems were less than 50 cm tall, and some were topped by sporangia. Other very similar rhyniophytes such as *Rhynia* did have tracheids.

The presence of xylem indicated that these plants were tracheophytes. But were they sporophytes or gametophytes? Close inspection of thin sections of fossil sporangia revealed that the spores were in groups of four. In almost all living nonseed tracheophytes (with no evidence to the contrary from fossil forms), the four products of meiosis and cytokinesis remain attached to one another during their development into spores. The spores separate only when they are mature, and even after separation their walls reveal the exact geometry of how they were attached. Therefore, a group of four closely packed spores is found only immediately after meiosis, and a plant that produces such a group must be a diploid sporophyte—and so the Rhynie fossils must have been sporophytes. Gametophytes of the Rhyniophyta were also found; they, too, were branched, and depressions at the apices of the branches contained archegonia and antheridia.

Although they were apparently ancestral to the other tracheophyte phyla, the rhyniophytes themselves are long gone. None of their fossils appear anywhere after the Devonian period.

Early tracheophytes added new features

A new phylum of tracheophytes—the Lycophyta (club mosses)—also appeared in the Silurian period. Another—the Pteridophyta (ferns and fern allies)—appeared during the Devonian period. These two groups arose from rhyniophyte-like ancestors. These new groups featured specializations not found in the rhyniophytes, including one or more of the following: true roots, true leaves, and a differentiation between two types of spores.

THE ORIGIN OF ROOTS.   The rhyniophytes had only rhizoids arising from a rhizome with which to gather water and minerals. How, then, did subsequent groups of tracheophytes come to have the complex roots we see today?

It is probable that roots had their evolutionary origins as a branch, either of a rhizome or of the aboveground portion of a stem. That branch presumably penetrated the soil and branched further. The underground portion could anchor the plant firmly, and even in this primitive condition it could absorb water and minerals. The discovery of fossil plants from the Devonian period, all having horizontal stems (rhizomes) with both underground and aerial branches, supported this hypothesis.

Underground and aboveground branches, growing in sharply different environments, were subjected to very different selection pressures during the succeeding millions of years. Thus the two parts of the plant axis—the aboveground shoot system and the underground root system—diverged in structure and evolved distinct internal and external anatomies. In spite of these differences, scientists believe that the root and shoot systems of tracheophytes are homologous—that they were once part of the same organ.

THE ORIGIN OF TRUE LEAVES.   Thus far we have used the term "leaf" rather loosely. We spoke of "leafy" mosses and commented on the absence of "true leaves" in rhyniophytes. In the strictest sense, a **leaf** is a flattened photosynthetic structure emerging laterally from a main axis or stem and possessing true vascular tissue. Using this precise definition as we take a closer look at true leaves in the tracheophytes, we see that there are two different types of leaves, very likely of different evolutionary origins.

The first leaf type, the **microphyll**, is usually small and only rarely has more than a single vascular strand, at least in plants alive today. Plants in the phylum Lycophyta (club mosses), of which only a few genera survive, have such simple leaves. The evolutionary origin of microphylls is thought by some biologists to be sterile sporangia (Figure 29.13*a*). The principal characteristic of this type of leaf is that its vascular

29.13 The Evolution of Leaves   (*a*) Microphylls are thought to have evolved from sterile sporangia. (*b*) The megaphylls of pteridophytes and seed plants may have arisen as photosynthetic tissue developed between branch pairs that were "left behind" as dominant branches overtopped them.

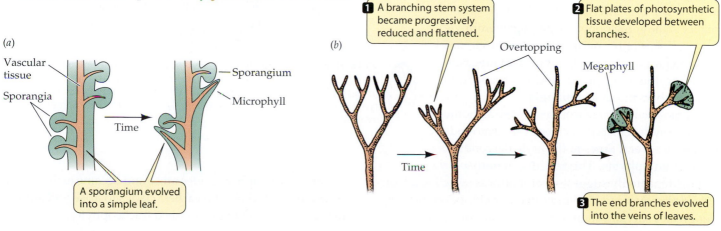

(*a*)

Vascular tissue

Sporangia

Time

Sporangium

Microphyll

A sporangium evolved into a simple leaf.

(*b*)

1 A branching stem system became progressively reduced and flattened.

2 Flat plates of photosynthetic tissue developed between branches.

Overtopping

Megaphyll

Time

3 The end branches evolved into the veins of leaves.

strand departs from the vascular system of the stem in such a way that the structure of the stem's vascular system is scarcely disturbed. This was true even in the fossil lycopod trees of the Carboniferous period, many of which had leaves many centimeters long.

The other leaf type is found in ferns and seed plants. This larger, more complex leaf is called a **megaphyll**. The megaphyll is thought to have arisen from the flattening of a dichotomously branching stem system and the development of *overtopping* (a pattern in which one branch differentiates from and grows beyond the others). This change was followed by the development of photosynthetic tissue between the members of overtopped groups of branches (Figure 29.13b). Megaphylls may have evolved more than once, in different phyla of tracheophytes showing overtopping of branches.

HOMOSPORY AND HETEROSPORY. In the most ancient of the present-day tracheophytes, both the gametophyte and the sporophyte are independent and usually photosynthetic. Spores produced by the sporophytes are of a single type, and they develop into a single type of gametophyte that bears both female and male reproductive organs. The female organ is a multicellular archegonium, typically containing a single egg. The male organ is an antheridium, containing many sperm. Such plants, which bear a single type of spore, are said to be **homosporous** (Figure 29.14a).

A different system, with two distinct types of spores, evolved somewhat later. Plants of this type are said to be **heterosporous** (Figure 29.14b). One type of spore, the **megaspore**, develops into a larger, specifically female gametophyte (a **megagametophyte**) that produces only eggs. The other type, the **microspore**, develops into a smaller, male gametophyte (a **microgametophyte**) that produces only sperm. The sporophyte produces megaspores in small numbers in **megasporangia** on the sporophyte, and microspores in large numbers in **microsporangia**.

The most ancient tracheophytes were all homosporous, but heterospory evidently evolved independently several times in the early descendants of the rhyniophytes. The fact that heterospory evolved repeatedly suggests that it affords selective advantages. Subsequent evolution in the plant kingdom featured ever greater specialization of the heterosporous condition.

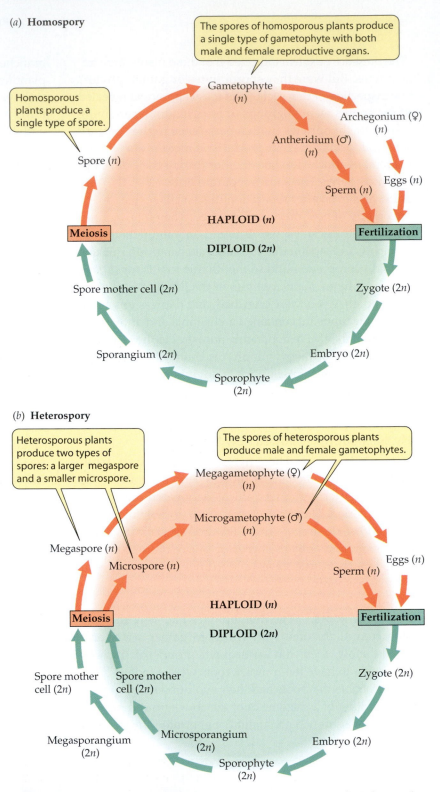

29.14 Homospory and Heterospory (a) Homosporous plants bear a single type of spore. Each gametophyte has two types of sex organs, antheridia (male) and archegonia (female). (b) Heterosporous plants, which bear two types of spores that develop into distinctly male and female gametophytes, evolved later.

Some tracheophyte clades arose and became extinct in the course of evolution. The earliest clades to arise and survive to this day belong to the nonseed tracheophytes.

The Surviving Nonseed Tracheophytes

The nonseed tracheophytes have a large, independent sporophyte and a small gametophyte that is independent of the sporophyte. The gametophytes of the surviving nonseed tracheophytes are rarely more than 1 or 2 centimeters long and are short-lived, whereas their sporophytes are often highly visible; the sporophyte of a tree fern, for example, may be 15 or 20 meters tall and may live for many years.

The most prominent resting stage in the life cycle of a nonseed tracheophyte is the single-celled spore. This feature makes their life cycle similar to those of the fungi, the green algae, and the nontracheophytes, but not, as we will see in the next chapter, to that of the seed plants. Nonseed tracheophytes must have an aqueous environment for at least one stage of their life cycle because fertilization is accomplished by a motile, flagellated sperm.

The ferns are the most abundant and diverse group of nonseed tracheophytes today, but the club mosses and horsetails were once dominant elements of Earth's vegetation. A fourth group, the whisk ferns, contains only two genera. In this section we'll look at the characteristics of these four groups and at some of the evolutionary advances that appeared in them.

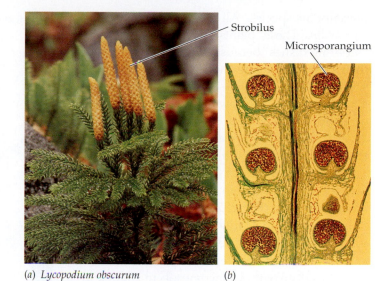

(a) *Lycopodium obscurum*　　　　　(b)

29.15 Club Mosses (a) Strobili are visible at the tips of this club moss. Club mosses have microphylls arranged spirally on their stems. (b) A thin section through a strobilus of a club moss, showing microsporangia.

The club mosses are sister to the other tracheophytes

The **club mosses** and their relatives (together called **lycopods**, phylum **Lycophyta**) diverged earlier than all other living tracheophytes—that is, the remaining tracheophytes share an ancestor that was not ancestral to the Lycophyta. There are relatively few surviving species of club mosses.

The lycopods have roots that branch dichotomously. The arrangement of vascular tissue in their stems is simpler than in the other tracheophytes. They bear only microphylls, and these simple leaves are arranged spirally on the stem. Growth in club mosses comes entirely from apical cell division, and branching is dichotomous, by a division of the apical cluster of dividing cells.

The sporangia in many club mosses are contained within conelike structures called *strobili* (singular, strobilus; Figure 29.15). A strobilus is a cluster of spore-bearing leaves inserted on an axis tucked into the upper angle between a specialized leaf and the stem. (Such an angle is called an *axil*.) Other club mosses lack strobili and bear their sporangia in the axil between a photosynthetic leaf and the stem. This placement contrasts with the apical sporangia of the rhyniophytes. There are both homosporous species and heterosporous

species of club mosses. Although only a minor element of present-day vegetation, the Lycophyta are one of two phyla that appear to have been the dominant vegetation during the Carboniferous period. One type of coal (cannel coal) is formed almost entirely from fossilized spores of the tree lycopod *Lepidodendron*—which gives us an idea of the abundance of this genus in the forests of that time (see Figure 29.11). The other major elements of Carboniferous vegetation were horsetails and ferns.

Horsetails, whisk ferns, and ferns constitute a clade

Once treated as distinct phyla, the horsetails, whisk ferns, and ferns form a clade, the phylum **Pteridophyta** (**pteridophytes**, or "ferns and fern allies"). Within that clade, the whisk ferns and the horsetails are both monophyletic; the ferns are not. However, about 97 percent of all fern species, including those with which you are most likely to be familiar, do belong to a single clade, the *leptosporangiate ferns*. In the pteridophytes—and in all seed plants—there is differentiation (overtopping) between the main axis and side branches.

HORSETAILS GROW AT THE BASES OF STEM SEGMENTS. Like the club mosses, the horsetails are represented by only a few present-day species. All are in a single genus, *Equisetum*. These plants are sometimes called "scouring rushes" because silica deposits found in their cell walls made them useful for cleaning. They have true roots that branch irregularly. Their sporangia curve back toward the stem on the ends of short stalks called *sporangiophores* (Figure 29.16*a*). Horsetails have a large sporophyte and a small gametophyte, both independent.

The small leaves of horsetails are reduced megaphylls and form in distinct whorls (circles) around the stem (Figure 29.16*b*). Growth in horsetails originates to a large extent from discs of dividing cells just above each whorl of leaves, so each segment of the stem grows from its base. Such basal growth is uncommon in plants, although it is found in the grasses, a major group of flowering plants.

PRESENT-DAY WHISK FERNS RESEMBLE THE MOST ANCIENT TRACHEOPHYTES. There once was some disagreement about whether rhyniophytes are entirely extinct. The confusion arose because of the existence today of two genera of rootless, spore-bearing plants, *Psilotum* and *Tmesipteris*, collectively called the whisk ferns. *Psilotum nudum* (Figure 29.17) has only minute scales instead of true leaves, but plants of the genus *Tmesipteris* have flattened photosynthetic organs—reduced megaphylls—with well-developed vascular tissue. Are these two genera the living relics of the rhyniophytes, or do they have more recent origins?

Psilotum and *Tmesipteris* once were thought to be evolutionarily ancient descendants of anatomically simple ancestors. That hypothesis was weakened by an enormous hole in the geological record between the rhyniophytes, which apparently became extinct more than 300 million years ago, and *Psilotum* and *Tmesipteris*, which are modern plants. DNA sequence data finally settled the question in favor of a more modern origin of the whisk ferns from fernlike ancestors. These two genera are a clade of highly specialized plants that evolved fairly recently from anatomically more complex ancestors by loss of complex leaves and true roots. Whisk fern gametophytes live below the surface of the ground and lack chlorophyll. They depend upon fungal partners for their nutrition.

Ferns evolved large, complex leaves

The sporophytes of the ferns, like those of the seed plants, have true roots, stems, and leaves. Their leaves are typically large and have branching vascular strands. Some species have small leaves as a result of evolutionary reduction, but even these small leaves have more than one vascular strand, and are thus megaphylls.

The ferns constitute a group that first appeared during the Devonian period and today consists of about 12,000 species. The ferns are not a monophyletic group, although, as already mentioned, 97 percent of the species—the leptosporangiate ferns—do constitute a monophyletic group. The leptosporangiate ferns differ from the other ferns in having sporangia with walls only one cell thick, borne on a stalk.

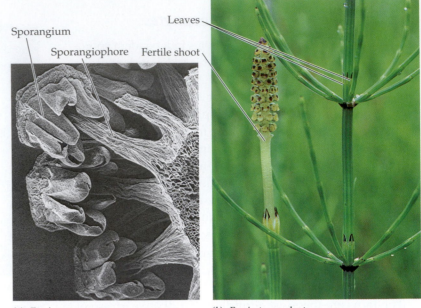

Sporangium
Sporangiophore   Fertile shoot
Leaves

(*a*) *Equisetum arvense*   (*b*) *Equisetum palustre*

29.16 Horsetails (*a*) Sporangia and sporangiophores of a horsetail. (*b*) Vegetative and fertile shoots of the marsh horsetail. Reduced megaphylls can be seen in whorls on the stem of the vegetative shoot on the right; the fertile shoot on the left is ready to disperse its spores.

Psilotum nudum

29.17 A Whisk Fern *Psilotum nudum* was once considered by some to be a surviving rhyniophyte and by others to be a fern. It is now included in the phylum Pteridophyta, and it is widespread in the Tropics and Subtropics.

(a) *Adiantum pedatum*

(b)

(c) *Marsilea mutica*

29.18 Fern Fronds Take Many Forms (a) The fronds of northern maidenhair fern form a pattern in this photograph. (b) The "fiddle-head" (developing frond) of a common forest fern; this structure will unfurl and expand to give rise to a complex adult frond such as those in (a). (c) The tiny fronds of a water fern.

Ferns are characterized by fronds (large leaves with complex vasculature; Figure 29.18a). During its development, the fern frond unfurls from a tightly coiled "fiddlehead" (Figure 29.18b). Some fern leaves become climbing organs and may grow to be as much as 30 meters long.

Because they require water for the transport of the male gametes to the female gametes, most ferns inhabit shaded, moist woodlands and swamps. Tree ferns can reach heights of 20 meters. Tree ferns are not as rigid as woody plants, and they have poorly developed root systems. Thus they do not grow in sites exposed directly to strong winds, but rather in

ravines or beneath trees in forests. The sporangia of ferns are found on the undersurfaces of the fronds, sometimes covering the whole undersurface and sometimes only at the edges. In most species the sporangia are found in clusters called *sori* (singular, sorus) (Figure 29.19).

The sporophyte generation dominates the fern life cycle

Inside the sporangia, fern spore mother cells undergo meiosis to form haploid spores. Once shed, the spores travel great distances and eventually germinate to form independent gametophytes. Old World climbing fern, *Lygodium microphyllum*, is currently spreading disastrously through the Florida Everglades, choking off the growth of other plants. This rapid spread is testimony to the effectiveness of windborne spores.

Fern gametophytes have the potential to produce both antheridia and archegonia, although not necessarily at the same time or on the same gametophyte. Sperm swim through water to archegonia—often to those on other gametophytes—where they unite with an egg. The resulting zygote develops into a new sporophyte embryo. The young sporophyte sprouts a root and can thus grow independently of the gametophyte. In the alternating generations of a fern, the gametophyte is small, delicate, and short-lived, but the sporophyte can be very large and can sometimes survive for hundreds of years (Figure 29.20).

Most ferns are homosporous. However, two groups of aquatic ferns, the Marsileaceae and Salviniaceae, are derived from a common ancestor that evolved heterospory. The megaspores and microspores of these plants (which germinate to produce female and male gametophytes, respectively) are produced in different sporangia (megasporangia and microsporangia), and the microspores are always much smaller and greater in number than the megaspores.

Dryopteris intermedia

29.19 Fern Sori Are Clusters of Sporangia Sori, each containing many spore-producing sporangia, have formed on the underside of this frond of the Midwestern fancy fern.

29.20 The Life Cycle of a Fern The most conspicuous stage in the fern life cycle is the mature, diploid sporophyte.

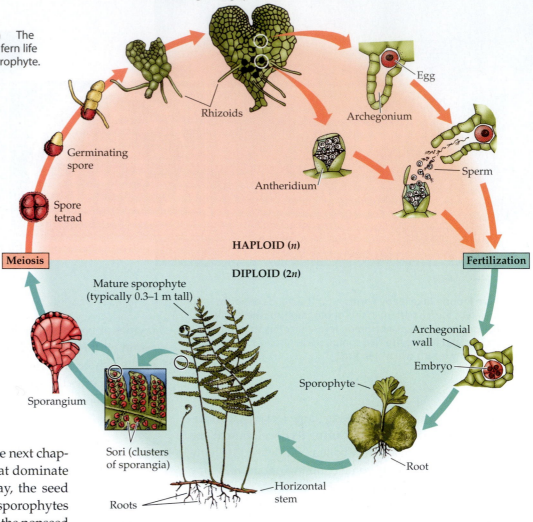

Mature gametophyte (about 0.5 cm wide)

Rhizoids

Archegonium

Egg

Antheridium

Sperm

Germinating spore

Spore tetrad

HAPLOID (n)

Meiosis

DIPLOID (2n)

Fertilization

Mature sporophyte (typically 0.3–1 m tall)

Archegonial wall

Embryo

Sporophyte

Sporangium

Sori (clusters of sporangia)

Roots

Horizontal stem

Root

A few genera of ferns produce a tuberous, fleshy gametophyte instead of the characteristic flattened, photosynthetic structure produced by most ferns. Like the gametophytes of whisk ferns, these tuberous gametophytes depend on a mutualistic fungus for nutrition; in some genera, even the sporophyte embryo must become associated with the fungus before extensive development can proceed. In Chapter 31 we will see that there are many other important plant–fungus mutalisms.

All the tracheophytes we have discussed thus far disperse themselves by spores. In the next chapter we will discuss the plants that dominate most of Earth's vegetation today, the seed plants, whose seeds afford new sporophytes protection unavailable to those of the nonseed tracheophytes.

Chapter Summary

The Plant Kingdom

▶ Plants are photosynthetic eukaryotes that develop from embryos protected by parental tissue. Like the green algae, they use chlorophylls *a* and *b* and store carbohydrates as starch. **Review Figure 29.1**

▶ Plant life cycles feature alternation of gametophyte (haploid) and sporophyte (diploid) generations. Both generations include multicellular organisms. **Review Figure 29.2**

▶ There are ten surviving phyla of plants. The three basal phyla are nontracheophytes, and the remaining seven phyla are tracheophytes. **Review Table 29.1**

▶ Plants arose from a common green algal ancestor in the charophyte clade, either a stonewort or a member of the group that includes *Coleochaete*. Descendants of this ancestral charophyte colonized the land.

The Conquest of the Land

▶ The acquisition of a cuticle, gametangia, a protected embryo, protective pigments, thick spore walls with a protective polymer, and a mutualistic association with a fungus are all defining characters of plants, and all are associated with the adaptation of plants to life on land.

▶ Tracheophytes are characterized by possession of a vascular system, consisting of water- and mineral-conducting xylem and nutrient-conducting phloem. Nontracheophytes lack a vascular system. **Review Figure 29.4**

The Nontracheophytes: Liverworts, Hornworts, and Mosses

▶ Nontracheophytes either lack vascular tissues completely or, in the case of certain mosses, have only a rudimentary system of water- and food-conducting cells.

▶ The nontracheophyte sporophyte generation is smaller than the gametophyte generation and depends on the gametophyte for water and nutrition. **Review Figures 29.5, 29.6. See Web/CD Tutorial 29.1**

▶ The nontracheophytes include the liverworts (phylum Hepatophyta), hornworts (phylum Anthocerophyta), and mosses (phylum Bryophyta).

▶ Hornwort sporophytes grow at their basal end.

▶ Hornworts, mosses, and tracheophytes have surface pores (stomata) that allow gas exchange and minimize water loss.

▶ In mosses and tracheophytes, the sporophytes grow by apical cell division.

▶ The hydroids of mosses, through which water may travel, may be ancestral to tracheids, the water-conducting cells of the tracheophytes.

Introducing the Tracheophytes

▶ The tracheophytes have vascular tissue with tracheids and other specialized cells designed to conduct water, minerals, and products of photosynthesis.

▶ Present-day tracheophytes are grouped into seven phyla. The two basal phyla are nonseed tracheophytes, and the rest are seed plants. **Review Figure 29.10**

▶ In tracheophytes, the sporophyte is larger than the gametophyte and independent of the gametophyte generation.

▶ The earliest tracheophytes, known to us only in fossil form, lacked roots and leaves. **Review Figure 29.12**

▶ Roots may have evolved from rhizomes or from branches that penetrated the ground. Microphylls are thought to have evolved from sporangia, and megaphylls may have resulted from the flattening and reduction of an overtopping, branching stem system. **Review Figure 29.13**

▶ Heterospory, the production of distinct female megaspores and male microspores, evolved on several occasions from homosporous ancestors. **Review Figure 29.14**. **See Web/CD Activities 29.1 and 29.2**

The Surviving Nonseed Tracheophytes

▶ Club mosses (phylum Lycophyta) have microphylls arranged spirally.

▶ Among the pteridophytes (phylum Pteridophyta), horsetails have reduced megaphylls in whorls. Whisk ferns lack roots; one genus has minute scales rather than leaves, and the other has reduced megaphylls with vascular tissue. Leaves with more complex vasculature are characteristic of all other phyla of tracheophytes.

▶ The ferns are not a clade, although 97 percent of fern species do constitute a clade. Ferns have megaphylls with branching vascular strands. **Review Figure 29.20**. **See Web/CD Activity 29.3**

Self-Quiz

1. Plants differ from photosynthetic protists in that only plants
   a. are photosynthetic.
   b. are multicellular.
   c. possess chloroplasts.
   d. have multicellular embryos protected by the parent.
   e. are eukaryotic.

2. Which statement about alternation of generations in plants is *not* true?
   a. It is heteromorphic.
   b. Meiosis occurs in sporangia.
   c. Gametes are always produced by meiosis.
   d. The zygote is the first cell of the sporophyte generation.
   e. The gametophyte and sporophyte differ genetically.

3. Which statement is *not* evidence for the origin of plants from the green algae?
   a. Some green algae have multicellular sporophytes and multicellular gametophytes.
   b. Both plants and green algae have cellulose in their cell walls.
   c. The two groups have the same photosynthetic and accessory pigments.
   d. Both plants and green algae produce starch as their principal storage carbohydrate.
   e. All green algae produce large, stationary eggs.

4. The nontracheophytes
   a. lack a sporophyte generation.
   b. grow in dense masses, allowing capillary movement of water.

   c. possess xylem and phloem.
   d. possess true leaves.
   e. possess true roots.

5. Which statement is *not* true of the mosses?
   a. The sporophyte is dependent on the gametophyte.
   b. Sperm are produced in archegonia.
   c. There are more species of mosses than of liverworts and hornworts combined.
   d. The sporophyte grows by apical cell division.
   e. Mosses are probably sister to the tracheophytes.

6. Megaphylls
   a. probably evolved only once.
   b. are found in all the tracheophyte phyla.
   c. probably arose from sterile sporangia.
   d. are the characteristic leaves of club mosses.
   e. are the characteristic leaves of horsetails and ferns.

7. The rhyniophytes
   a. possessed vessel elements.
   b. possessed true roots.
   c. possessed sporangia at the tips of stems.
   d. possessed leaves.
   e. lacked branching stems.

8. Club mosses and horsetails
   a. have larger gametophytes than sporophytes.
   b. possess small leaves.
   c. are represented today primarily by trees.
   d. have never been a dominant part of the vegetation.
   e. produce fruits.

9. Which statement about ferns is *not* true?
   a. The sporophyte is larger than the gametophyte.
   b. Most are heterosporous.
   c. The young sporophyte can grow independently of the gametophyte.
   d. The frond is a megaphyll.
   e. The gametophytes produce archegonia and antheridia.

10. The leptosporangiate ferns
    a. are not a monophyletic group.
    b. have sporangia with walls more than one cell thick.
    c. constitute a minority of all ferns.
    d. are pteridophytes.
    e. produce seeds.

For Discussion

1. Mosses and ferns share a common trait that makes water droplets a necessity for sexual reproduction. What is that trait?

2. Are the mosses well adapted to terrestrial life? Justify your answer.

3. Ferns display a dominant sporophyte generation (with large fronds). Describe the major advance in anatomy that enables most ferns to grow much larger than mosses.

4. What features distinguish club mosses from horsetails? What features distinguish these groups from rhyniophytes? From ferns?

5. Why did some botanists once believe that the whisk ferns should be classified together with the rhyniophytes?

6. Contrast microphylls with megaphylls in terms of structure, evolutionary origin, and occurrence among plants.

30 *The Evolution of Seed Plants*

 A violent thunderstorm moves through forested hills and valleys where summer rain has been scarce. A jagged fork of lightning strikes a tree, and it bursts into flame. Soon the flames reach dead and dry underbrush, and fire spreads to the surrounding trees. The fire rages rapidly through the forest, leaving a blackened and smoking landscape behind.

Though devastating, such fires are a natural part of the forest ecosystem. Life returns quickly following a fire in a natural grassland or forest, in part because some plants have adaptations that enable them to live with fire. One example, obvious from its common name, is fireweed. The seeds of fireweed not only survive fires, but are stimulated by high temperatures to break their dormancy and sprout. Another example is the lodgepole pine tree, which covers vast fire-prone areas in the Rocky Mountains and elsewhere. Its cones will not release their seeds unless the heat of a fire causes them to open.

Seeds are remarkable structures. They protect the plant embryo within them from environmental extremes through what may be a very long resting period. This and other properties have contributed to making seed plants the predominant plants on Earth. All of today's forests are dominated by seed plants.

In this chapter we will describe the defining characteristics of the seed plants as a group. We will survey the diversity of seed plants and describe the flowers and fruits that are characteristic of their most dominant group, the flowering plants. Finally, we will consider some of the unsolved problems in seed plant evolution.

The Seed Plants

The most recent group to appear in the evolution of the tracheophytes is the **seed plants**. The earliest fossil evidence of seed plants is found in Devonian rocks. The earliest seed plants combined characteristics of rhyniophytes and heterosporous ferns, but they had tracheids of the type found in modern seed plants. They also differed from the plants around them by having extensively thickened woody

A Forest Ablaze Fires like this one in a northern Arizona forest can pose dangers to human life and property. But they play an essential role in the life cycles of many fire-adapted seed plants.

stems, which resulted from the proliferation of xylem. This type of growth in the diameter of stems and roots is called *secondary growth*. By the Carboniferous period, new lines of seed plants had evolved, including various seed ferns, which possessed fernlike foliage but had seeds attached to their leaves.

Two clades of early seed plants are known only as fossils. These clades are basal to the surviving seed plants, which fall into two groups, the **gymnosperms** (such as pines and cycads) and the **angiosperms** (flowering plants). There are four living phyla of gymnosperms and one of angiosperms (Figure 30.1). The phylogenetic relationships among these five clades have not yet been resolved. All living gymnosperms and many angiosperms show secondary growth. The life cycles of all seed plants share major features, as we are about to see.

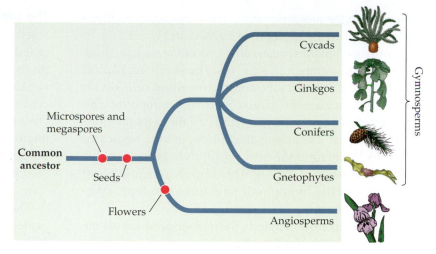

30.1 The Phyla of Living Seed Plants   There are four phyla of gymnosperms and one of angiosperms. Their exact evolutionary relationship is still uncertain.

Seed plants are heterosporous and have tiny gametophytes

In seed plants, the gametophyte generation is reduced even further than it is in the ferns (Figure 30.2). The haploid gametophyte develops partly or entirely while attached to and nutritionally dependent on the diploid sporophyte.

Among the seed plants, only the earliest types of gymnosperms (and their few survivors) had swimming sperm. All other seed plants have evolved other means of bringing eggs and sperm together. The culmination of this striking evolutionary trend was independence from the liquid water that earlier plants needed for sexual reproduction.

Seed plants are heterosporous (see Figure 29.14b). They form separate megasporangia and microsporangia on structures that are grouped on short axes, such as the cones and strobili of conifers and the flowers of angiosperms.

As in other plants, the spores of seed plants are produced by meiosis within the sporangia, but in seed plants, the megaspores are not shed. Instead, they develop into female gameto-

30.2 The Relationship between Sporophyte and Gametophyte Has Evolved   In the course of plant evolution, the gametophyte has been reduced and the sporophyte has become more prominent.

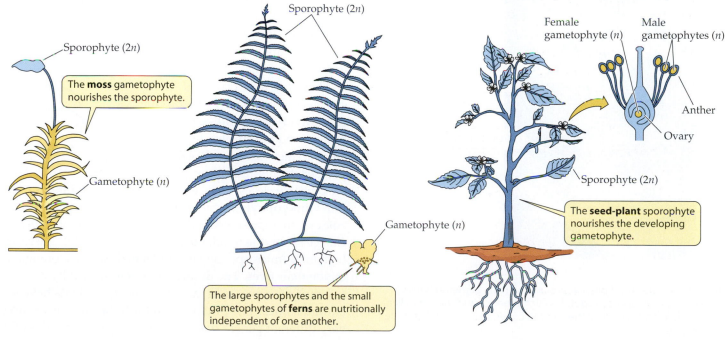

phytes within the megasporangia. These megagametophytes are dependent on the sporophyte for food and water.

In most seed plant species, only one of the meiotic products in a megasporangium survives. The surviving haploid nucleus divides mitotically, and the resulting cells divide again to produce a multicellular female gametophyte. This megagametophyte is retained within the megasporangium, where it matures. The megagametophyte, in turn, houses the early development of the next sporophyte generation following fertilization of the egg. The megasporangium is surrounded by sterile sporophytic structures that form a protective **integument**.

Within the microsporangium, the meiotic products are microspores, which divide mitotically within the spore wall one or a few times to form a male gametophyte called a **pollen grain**. Pollen grains are released from the microsporangium to be distributed by wind, an insect, a bird, or a plant breeder (Figure 30.3). A pollen grain that reaches the appropriate surface of a sporophyte of the same species develops further. It produces a slender **pollen tube** that elongates and digests its way through the sporophytic tissue toward the female gametophyte.

When the tip of the pollen tube reaches the female gametophyte, sperm are released from the tube, and fertilization occurs. The resulting diploid zygote divides repeatedly, forming a young sporophyte that develops to an embryonic stage at which growth is temporarily suspended (often referred to as a *dormant* stage). The end product at this stage is a multicellular **seed**.

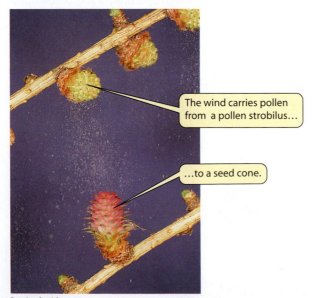

Larix decidua

The wind carries pollen from a pollen strobilus…

…to a seed cone.

30.3 Pollen Grains Pollen grains are the male gametophytes of seed plants. Conifers have strobili, which produce and release pollen. Their pollen is dispersed by the wind to cones, which contain female gametophytes.

The seed is a complex package

A seed may contain tissues from three generations. The seed coat develops from tissues of the diploid sporophyte parent (the integument). Within the megasporangium is the haploid female gametophytic tissue from the next generation, which contains a supply of nutrients for the developing embryo. (This tissue is fairly extensive in most gymnosperm seeds. In angiosperm seeds its place is taken by a tissue called endosperm, which we will describe below.) In the center of the seed is the third generation, the embryo of the new diploid sporophyte.

The seed of a gymnosperm or an angiosperm is a well-protected resting stage. The seeds of some species may remain *viable* (capable of growth and development) for many years, germinating when conditions are favorable for the growth of the sporophyte. In contrast, the embryos of non-seed plants develop directly into sporophytes, which either survive or die, depending on environmental conditions; there is no dormant stage in the life cycle.

During the dormant stage, the seed coat protects the embryo from excessive drying and may also protect it against potential predators that would otherwise eat the embryo and its nutrient reserves. Many seeds have structural adaptations that promote their dispersal by wind or, more often, by animals. When the young sporophyte resumes growth, it draws on the food reserves in the seed. The possession of seeds is a major reason for the enormous evolutionary success of the seed plants, which are the dominant life forms of Earth's modern terrestrial flora in most areas.

The Gymnosperms: Naked Seeds

The extant gymnosperms are a clade of seed plants that do not form flowers. Although there are probably fewer than 750 species of living gymnosperms, these plants are second only to the angiosperms in their dominance of the terrestrial environment.

There are four clades of living gymnosperms today. The **cycads** (phylum **Cycadophyta**) are palmlike plants of the Tropics and Subtropics, growing as tall as 20 meters (Figure 30.4*a*). Of the present-day gymnosperms, the cycads 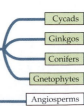 are probably closest to the earliest seed plants. **Ginkgos** (phylum **Ginkgophyta**), which were common during the Mesozoic era, are represented today by a single genus and species, *Ginkgo biloba*, the maidenhair tree (Figure 30.4*b*). There are both male (microsporangiate) and female (megasporangiate) maidenhair trees. The difference is determined by X and Y sex chromosomes, as in humans; few other plants have sex chromosomes. The phylum **Gnetophyta** consists of three very different genera that share certain characteristics with

(a) *Cycas* sp.

(b) *Ginkgo biloba*

(c) *Welwitschia mirabilis*

(d) *Sequoiadendron giganteum*

30.4 Diversity among the Gymnosperms (a) Many cycads, such as this palmlike tree, have growth forms that resemble both ferns and palms. (b) The characteristic fleshy seed coat and broad leaves of the maidenhair tree. (c) A gnetophyte growing in the Namib Desert of Africa. Two huge, straplike leaves grow throughout the life of the plant, breaking and splitting as they grow. (d) Conifers, like this giant sequoia growing in Sequoia National Park, California, dominate many modern forests.

the angiosperms. One of the gnetophytes is *Welwitschia* (Figure 30.4c), a long-lived desert plant with just two straplike leaves that sprawl on the sand and can grow as long as 3 meters. By far the most abundant of the gymnosperms are the **conifers** (phylum **Pinophyta**), cone-bearing plants such as pines and redwoods (Figure 30.4d).

All living gymnosperms except the Gnetophyta have only tracheids as water-conducting and support cells in their xylem; they lack the more specialized vessels and fibers found alongside tracheids in the angiosperms. Although this difference may make the gymnosperm water transport and support system seem less efficient than that of the angiosperms, it serves some of the largest trees known. The coast redwoods of California are the tallest gymnosperms; the largest are well over 100 m tall. Secondary xylem—wood—produced by gymnosperms is the principal resource of the timber industry.

During the Permian period, the conifers and cycads flourished. Gymnosperm forests changed over time as the gymnosperm groups evolved. Gymnosperms dominated the

Mesozoic era, during which the continents drifted apart and dinosaurs strode the Earth. They were the principal trees in all forests until less than 100 million years ago, and they still dominate many present-day forests. Let's look at the most abundant gymnosperms, the conifers, in more detail.

Conifers have cones but no motile cells

The great Douglas fir and cedar forests of the northwestern United States and the massive boreal forests of pine, fir, and spruce found in northern regions of Eurasia and North America, as well as on the upper slopes of mountain ranges everywhere, rank among the great vegetation formations of the world. All these trees belong to one phylum of gymnosperms, Pinophyta—the conifers, or cone-bearers. A **cone** is a short axis (a modified *stem*) bearing a tight cluster of scales, which are reduced *branches* specialized for reproduction (Figure 30.5*a*). A **strobilus** is a conelike cluster of scales that are modified *leaves* inserted on an axis (Figure 30.5*b*). Megaspores are produced in seed cones, and microspores are produced in pollen strobili. Seed cones are much larger than pollen strobili.

We will use the life cycle of a pine to illustrate reproduction in gymnosperms (Figure 30.6). The production of male gametophytes in the form of pollen grains frees the plant completely from its dependence on liquid water for fertilization. Instead of water, wind assists conifer pollen grains in their first stage of travel from the strobilus to the female gametophyte inside the seed cone (see Figure 30.3). The pollen tube provides the sperm with the means for the last stage of travel by elongating and digesting its way through maternal sporophytic tissue. When it reaches the female gametophyte, it releases two sperm, one of which degenerates after the other unites with an egg.

The megasporangium, in which the female gametophyte will form, is enclosed in a layer of sporophytic tissue—the integument—that will eventually develop into the seed coat. The integument, the megasporangium inside it, and the tissue attaching it to the maternal sporophyte constitute the **ovule**. The pollen grain enters through a small opening in the integument at the tip of the ovule, the **micropyle**.

Gymnosperms derive their name (which means "naked-seeded") from the fact that their ovules and seeds are not protected by ovary or fruit tissue. Most conifer ovules (which, upon fertilization, develop into seeds) are borne exposed on the upper surfaces of the modified branches that form the scales of the cone. Each cone scale lies in the angle between a modified leaf and the axis. The only protection of the ovules comes from the scales, which are tightly pressed against each other within the cone. As we have seen, some pines, such as the lodgepole pine, have such tightly closed seed cones that only fire suffices to split them open and release the seeds.

About half of the conifer species have soft, fleshy fruitlike tissues associated with their seeds; examples are the fleshy cones or "berries" of juniper and yew. Animals may eat these tissues and then disperse the seeds in their feces, often carrying them considerable distances from the parent plant. These tissues, however, are not true fruits, which are characteristic of the plant phylum that is dominant today: the angiosperms.

(a) *Pinus resinosa*   Seed cones

(b) *Pinus ponderosa*

Pollen strobili

30.5 Cones and Strobili (*a*) The scales of seed cones are modified branches. (*b*) The spore-bearing structures in pollen strobili are modified leaves.

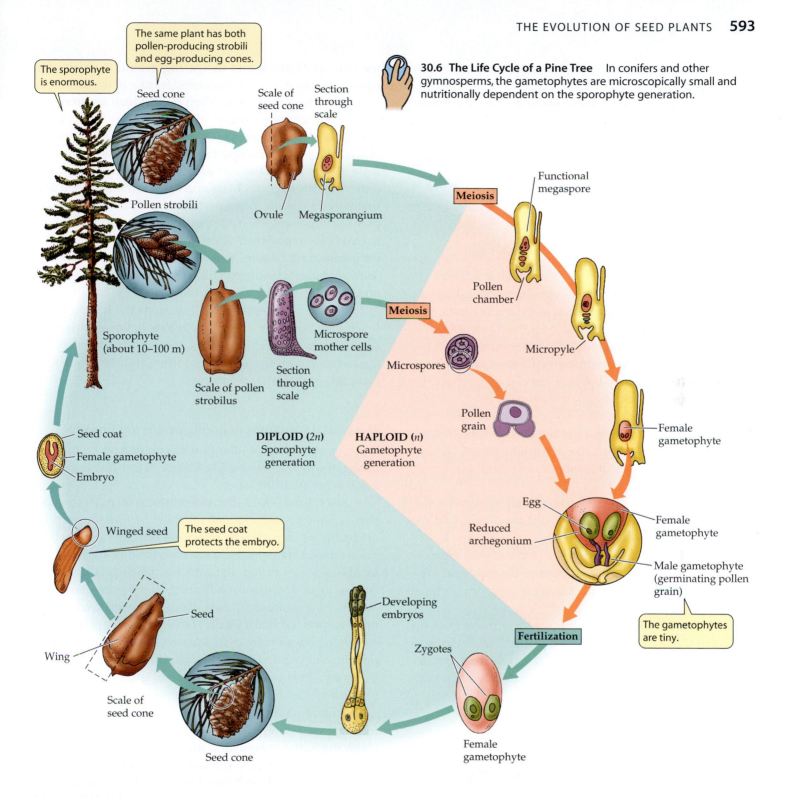

The sporophyte is enormous.

The same plant has both pollen-producing strobili and egg-producing cones.

Seed cone

Pollen strobili

Sporophyte (about 10–100 m)

Scale of seed cone

Section through scale

Ovule

Megasporangium

Scale of pollen strobilus

Section through scale

Microspore mother cells

Microspores

Pollen grain

DIPLOID (2n) Sporophyte generation

HAPLOID (n) Gametophyte generation

Seed coat

Female gametophyte

Embryo

The seed coat protects the embryo.

Winged seed

Seed

Wing

Scale of seed cone

Seed cone

Developing embryos

Zygotes

Female gametophyte

Meiosis

Functional megaspore

Pollen chamber

Meiosis

Micropyle

Female gametophyte

Egg

Reduced archegonium

Female gametophyte

Male gametophyte (germinating pollen grain)

The gametophytes are tiny.

Fertilization

30.6 The Life Cycle of a Pine Tree   In conifers and other gymnosperms, the gametophytes are microscopically small and nutritionally dependent on the sporophyte generation.

The Angiosperms: Flowering Plants

The phylum **Angiospermae** consists of the **flowering plants**, also commonly known as the **angiosperms**. This highly diverse phylum includes more than 257,000 species. The oldest evidence of angiosperms dates back to the early Cretaceous period, about 140 million years ago. The angiosperms radiated explosively and, over a period of only about 60 mil-

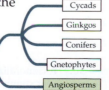

Cycads

Ginkgos

Conifers

Gnetophytes

Angiosperms

lion years, became the dominant plant life of the planet. In later chapters, when we mention "plants," we are generally referring to the angiosperms.

The female gametophyte of the angiosperms, consisting of just seven cells, is even more reduced than that of the gymnosperms. Thus, the angiosperms represent the current extreme of an evolutionary trend that runs throughout the tracheophytes: The sporophyte generation becomes larger and more independent of the gametophyte, while the gametophyte generation becomes smaller and more dependent on the sporophyte.

A number of synapomorphies (shared derived traits) characterize the angiosperms:

▶ They have double fertilization.
▶ They produce a triploid nutritive tissue called the endosperm.
▶ Their ovules and seeds are enclosed in a carpel.
▶ They have flowers.
▶ They produce fruit.
▶ Their xylem contains vessel elements and fibers.
▶ Their phloem contains companion cells.

Double fertilization was long considered the single most reliable distinguishing characteristic of the angiosperms. Two male gametes, contained within a single microgametophyte (pollen grain), participate in fertilization events within the megagametophyte of an angiosperm. One sperm combines with the egg to produce a diploid zygote, the first cell of the sporophyte generation. In most angiosperms, the other sperm nucleus combines with two other haploid nuclei of the female gametophyte to form a triploid ($3n$) nucleus. This nucleus, in turn, divides to form a triploid tissue, the **endosperm**, that nourishes the embryonic sporophyte during its early development.

Double fertilization occurs in nearly all present-day angiosperms. We are not sure when and how it evolved because there is no known fossil evidence on this point. It may have first resulted in two embryos, as it does in the three existing genera of Gnetophyta: *Ephedra*, *Gnetum*, and *Welwitschia*. Both of the fertilizations in gnetophytes produce diploid products.

The name *angiosperm* ("enclosed seed") is drawn from another distinctive character of these plants: The ovules and seeds are enclosed in a modified leaf called a **carpel**. Besides protecting the ovules and seeds, the carpel often interacts with incoming pollen to prevent self-pollination, thus favoring cross-pollination and increasing genetic diversity. Of course, the most evident diagnostic feature of angiosperms is that they have **flowers**. Production of a **fruit** is another of their unique characteristics.

Most angiosperms are also distinguished by the possession of specialized water-transporting cells called **vessel elements** in their xylem, but these cells are also found, in anatomically different form, in gnetophytes and a few ferns. A second distinctive cell type in angiosperm xylem is the **fiber**, which plays an important role in supporting the plant body. Angiosperm phloem possesses another unique cell type, called a **companion cell**. Like the gymnosperms, woody angiosperms show secondary growth, producing secondary xylem and secondary phloem and growing in diameter.

In the following sections we'll examine the structure and function of flowers, evolutionary trends in flower structure, the functions of pollen and fruits, the angiosperm life cycle, the two major groups of angiosperms, and the origin and evolution of flowering plants.

The sexual structures of angiosperms are flowers

If you examine any familiar flower, you will notice that the outer parts look somewhat like leaves. In fact, all the parts of a flower *are* modified leaves.

A generalized flower (for which there is no exact counterpart in nature) is diagrammed in Figure 30.7 for the purpose of identifying its parts. The structures bearing microsporangia are called **stamens**. Each stamen is composed of a **filament** bearing an **anther** that contains pollen-producing microsporangia. The structures bearing megasporangia are the carpels. A structure composed of one carpel or two or more fused carpels is called a **pistil**. The swollen base of the pistil, containing one or more ovules (each containing a megasporangium surrounded by its protective integument), is called the **ovary**. The apical stalk of the pistil is the **style**, and the terminal surface that receives pollen grains is the **stigma**.

In addition, a flower often has several specialized sterile (non-spore-bearing) leaves. The inner ones are called **petals** (collectively, the **corolla**) and the outer ones **sepals** (collectively, the **calyx**). The corolla and calyx, which can be quite showy, often play roles in attracting animal pollinators to the flower. The calyx more commonly protects the immature flower in bud. From base to apex, the sepals, petals, sta-

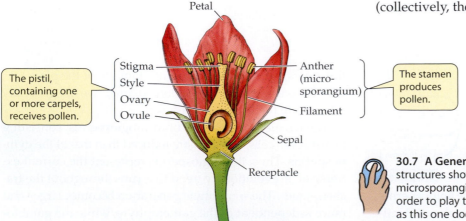

The pistil, containing one or more carpels, receives pollen.

Petal
Stigma
Style
Ovary
Ovule
Anther (micro-sporangium)
Filament
Sepal
Receptacle

The stamen produces pollen.

30.7 A Generalized Flower Not all flowers possess all the structures shown here, but they must possess a stamen (bearing microsporangia), a pistil (containing megasporangia), or both in order to play their role in reproduction. Flowers that have both, as this one does, are referred to as perfect.

(a) *Daucus carota* Compound umbel Umbels

(b) *Echinacea purpurea* Disk flowers (many) Ray flowers

Spikes

(c) *Pennisetum setaceum*

30.8 Inflorescences (a) The inflorescence of Queen Anne's lace is a compound umbel. Each umbel bears flowers on stalks that arise from a common center. (b) Coneflowers are members of the aster family; their inflorescence is a head. In a head, each of the long, petal-like structures is a ray flower; the central portion of the head consists of dozens to hundreds of disc flowers. (c) Grasses such as this fountain grass have inflorescences called spikes.

mens, and carpels (which are referred to as the floral organs; see Figure 19.12) are usually positioned in circular arrangements or whorls and attached to a central stalk called the **receptacle**.

The generalized flower shown in Figure 30.7 has both megasporangia and microsporangia; such flowers are referred to as **perfect**. Many angiosperms produce two types of flowers, one with only megasporangia and the other with only microsporangia. Consequently, either the stamens or the carpels are nonfunctional or absent in a given flower, and the flower is referred to as **imperfect**.

Species such as corn or birch, in which both megasporangiate and microsporangiate flowers occur on the same plant, are said to be **monoecious** (meaning "one-housed"—but, it must be added, one house with separate rooms). Complete separation is the rule in some other angiosperm species, such as willows and date palms; in these species, a given plant produces either flowers with stamens or flowers with pistils, but never both. Such species are said to be **dioecious** ("two-housed").

Flowers come in an astonishing variety of forms, as you will realize if you think of some of the flowers you recognize. The generalized flower shown in Figure 30.7 has distinct petals and sepals arranged in distinct whorls. In nature, however, petals and sepals sometimes are indistinguishable. Such appendages are called **tepals**. In other flowers, petals, sepals, or tepals are completely absent.

Flowers may be single, or they may be grouped together to form an **inflorescence**. Different families of flowering plants have their own, characteristic types of inflorescences, such as the compound umbels of the carrot family, the heads of the aster family, and the spikes of many grasses (Figure 30.8).

Flower structure has evolved over time

The flowers of the most basal lineages of angiosperms have a large and variable number of tepals (or sepals and petals),

carpels, and stamens (Figure 30.9a). Evolutionary change within the angiosperms has included some striking modifications of this early condition: reductions in the number of each type of floral organ to a fixed number, differentiation of petals from sepals, and changes in symmetry from radial (as in a lily or magnolia) to bilateral (as in a sweet pea or orchid), often accompanied by an extensive fusion of parts (Figure 30.9b).

According to one theory, the first carpels to evolve were modified leaves, folded but incompletely closed, and thus differing from the scales of the gymnosperms. In the groups of angiosperms that evolved later, the carpels fused and became progressively more buried in receptacle tissue (Figure 30.10a). In the flowers of the most recent groups, the other flower parts are attached at the very top of the ovary, rather than at the bottom as in Figure 30.7. The stamens of the most ancient flowers may have appeared leaflike (Figure 30.10b), little resembling those of the generalized flower in Figure 30.7.

Why do so many flowers have pistils with long styles and anthers with long filaments? Natural selection has favored length in both of these structures, probably because length increases the likelihood of successful pollination. Long filaments may bring the anthers into contact with insect bodies, or they may place the anthers in a better position to catch the wind. Similar arguments apply to long styles.

30.9 Flower Form and Evolution
(a) A magnolia flower shows the major features of early flowers: It is radially symmetrical, and the individual tepals, carpels, and stamens are separate, numerous, and attached at their bases. (b) Orchids, like this ladyslipper, have a bilaterally symmetrical structure that evolved much later. One of the three petals evolved into the complex lower "lip." Inside, the stamen and pistil are fused. There are two anthers in this species, although most orchids have only a single anther.

(a) *Magnolia grandifolia*

(b) *Cypripedium reginae*

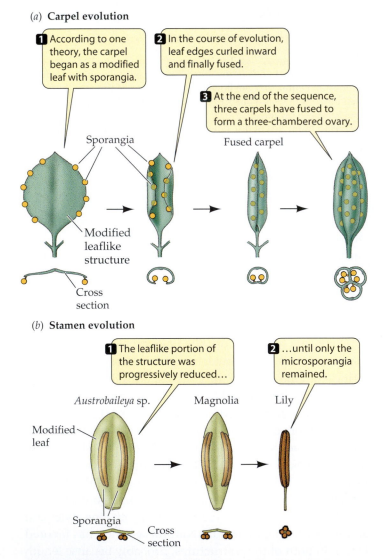

(a) **Carpel evolution**

1 According to one theory, the carpel began as a modified leaf with sporangia.

2 In the course of evolution, leaf edges curled inward and finally fused.

3 At the end of the sequence, three carpels have fused to form a three-chambered ovary.

Sporangia

Fused carpel

Modified leaflike structure

Cross section

(b) **Stamen evolution**

1 The leaflike portion of the structure was progressively reduced...

2 ...until only the microsporangia remained.

Austrobaileya sp.

Magnolia

Lily

Modified leaf

Sporangia

Cross section

30.10 Carpels and Stamens Evolved from Leaflike Structures
(a) Possible stages in the evolution of a carpel from a more leaflike structure. (b) The stamens of three modern plants show the various stages in the evolution of that organ. It is *not* implied that these species evolved one from another; they simply illustrate the structures.

A long style may serve another purpose as well. If several pollen grains land on one stigma, a pollen tube will start growing from each grain down the style toward the ovary. If there are more pollen grains than ovules, there is a "race" to fertilize the ovules. The race down the style can be viewed as "mate selection" by the plant bearing the style.

Angiosperms have coevolved with animals

Pollen has played another crucial role in the evolution of the angiosperms. Whereas many gymnosperms are wind-pollinated, most angiosperms are animal-pollinated. Animals visit flowers to obtain nectar or pollen, and in the process often carry pollen from one flower to another, or from one plant to another. Thus, in its quest for food, the animal contributes to the genetic diversity of the plant population. Insects, especially bees, are among the most important pollinators; birds and some species of bats also play major roles as pollinators.

For more than 130 million years, angiosperms and their animal pollinators have coevolved in the terrestrial environment. The animals have affected the evolution of the plants, and the plants have affected the evolution of the animals. Flower structure has become incredibly diverse under these selection pressures.

Some of the products of coevolution are highly specific; for example, some yucca species are pollinated by only one species of moth. Pollination by just one or a few animal species provides a plant species with a reliable mechanism for transferring pollen from one of its members to another.

Most plant–pollinator interactions are much less specific; that is, many different animal species pollinate the same plant species, and the same animal species pollinate many different plant species. However, even these less specific interactions have developed some specialization. Bird-pollinated flowers are often red and odorless. Many insect-pollinated flowers have characteristic odors, and bee-pollinated flowers may have conspicuous markings, or *nectar guides*, that are evident only in the ultraviolet region of the spectrum, where bees have better vision than in the red region. Coevolution and other aspects of plant–animal interactions are covered in more detail in Chapter 55.

The angiosperm life cycle features double fertilization

The life cycle of the angiosperms is summarized in Figure 30.11. The angiosperm life cycle will be considered in detail in Chapter 39, but let's look at it briefly here and compare it with the conifer life cycle in Figure 30.6.

Like all seed plants, angiosperms are heterosporous. The ovules are contained within carpels, rather than being exposed on the surfaces of scales, as in most gymnosperms. The male gametophytes, as in the gymnosperms, are pollen grains.

The ovule develops into a seed containing the products of the double fertilization that characterizes angiosperms: a diploid zygote and a triploid endosperm. The endosperm

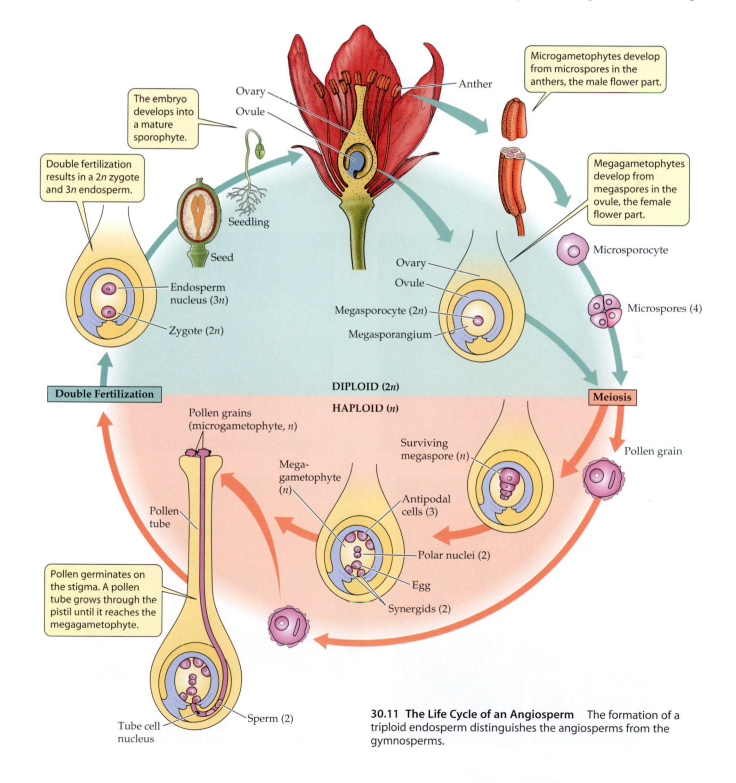

30.11 The Life Cycle of an Angiosperm The formation of a triploid endosperm distinguishes the angiosperms from the gymnosperms.

serves as storage tissue for starch or lipids, proteins, and other substances that will be needed by the developing embryo.

The zygote develops into an embryo, consisting of an embryonic axis and one or two **cotyledons**, or seed leaves. The cotyledons have different fates in different plants. In many, they serve as absorptive organs that take up and digest the endosperm. In others, they enlarge and become photosynthetic when the seed germinates. Often they play both roles.

Angiosperms produce fruits

The ovary of a flowering plant (together with the seeds it contains) develops into a fruit after fertilization. A fruit may consist only of the mature ovary and its seeds, or it may include other parts of the flower or structures associated with it. A *simple fruit*, such as a cherry (Figure 30.12*a*), is one that develops from a single carpel or several united carpels. A raspberry is an example of an *aggregate fruit* (Figure 30.12*b*)—one that develops from several separate carpels of a single flower.

Pineapples and figs are examples of *multiple fruits* (Figure 30.12*c*), formed from a cluster of flowers (an inflorescence). Fruits derived from parts in addition to the carpel and seeds are called *accessory fruits* (Figure 30.12*d*); examples are apples, pears, and strawberries. The development, ripening, and dispersal of fruits will be considered in Chapters 38 and 39.

There are several clades of angiosperms

The better-understood relationships among the angiosperm clades are shown in Figure 30.13. Two large clades include the great majority of angiosperm species: the **monocots** and the **eudicots**. The monocots are so called because they have a single embryonic cotyledon; the eudicots have two. We will describe other differences between these groups in Chapter 35.

Some familiar angiosperms belong to clades other than the monocots and eudicots (Figure 30.14). These clades include the water lilies, star anise and its relatives, and the magnoliid complex. The magnoliids are less numerous than the

30.12 Fruits Come in Many Forms and Flavors (*a*) A simple fruit (sour cherry). (*b*) An aggregate fruit (raspberry). (*c*) A multiple fruit (pineapple). (*d*) An accessory fruit (strawberry).

30.13 Evolutionary Relationships among the Angiosperms The monocots and the eudicots are the largest clades among the angiosperms. This diagram is a conservative interpretation of current data on relationships among the clades.

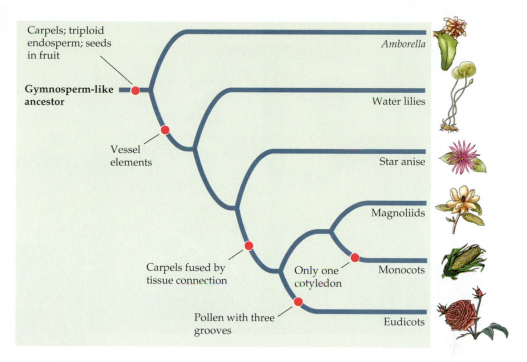

Carpels; triploid endosperm; seeds in fruit

Gymnosperm-like ancestor

Vessel elements

Carpels fused by tissue connection

Pollen with three grooves

Only one cotyledon

Amborella

Water lilies

Star anise

Magnoliids

Monocots

Eudicots

monocots and eudicots, but they include many familiar and often useful plants such as magnolias, avocados, cinnamon, and pepper.

The monocots (Figure 30.15) include grasses, cattails, lilies, orchids, and palms. The eudicots (Figure 30.16) include the vast majority of familiar seed plants, including most herbs, vines, trees, and shrubs. Among them are such diverse plants as oaks, willows, violets, snapdragons, and sunflowers.

(a) *Amborella trichopoda*

(b) *Nymphaea odorata*

(c) *Illicium floridanum*

(d) *Piper nigrum*

(e) *Aristolochia grandiflora*

(f) *Persea* sp.

30.14 Monocots and Eudicots Are Not the Only Surviving Angiosperms (a) *Amborella*, a shrub, is the closest living relative of the first angiosperms; its clade is sister to the remaining extant angiosperms. (b) The water lily clade is the next most basal clade after *Amborella*'s. (c) Star anise and its relatives belong to another basal clade. (d–f) The largest clade other than the monocots and eudicots is the magnoliid complex, represented here by (d) a black pepper, (e) Dutchman's pipe, and (f) an avocado tree. The magnolia in Figure 30.9a is another magnoliid.

(a) *Phoenix dactylifera*

(b) *Triticum* sp.

30.15 Monocots (a) Palms are among the few monocot trees. Date palms are a major food source in some areas of the world. (b) Grasses such as this cultivated wheat and the fountain grass in Figure 30.8c are monocots. (c) Monocots also include popular garden flowers such as these lilies. Many orchids (Figure 30.9b) are highly sought-after monocot flowers.

(c) *Lilium* sp.

(a) *Borzicactus samaipatanus*

(c) *Rosa rugosa*

(b) *Cornus florida*

30.16 Eudicots (a) The cactus family is a large group of eudicots, with about 1,500 species in the Americas. This cactus bears scarlet flowers for a brief period of the year. (b) The flowering dogwood is a small eudicot tree. (c) Climbing Cape Cod roses are members of the eudicot family Rosaceae, as are the familiar roses from your local florist.

Determining the oldest angiosperm clade

Which angiosperms were the earliest flowering plants was long a matter of great controversy. Two leading candidates were the magnolia family (see Figure 30.9a) and another family, the Chloranthaceae, whose flowers are much simpler than those of the magnolias. At the close of the twentieth century, however, an impressive convergence of evidence led to the conclusion that the most basal living angiosperm belongs to neither of those families, but rather to a clade that today consists of a single species of the genus *Amborella* (see Figure 30.14a). This woody shrub, with cream-colored flowers, lives only on New Caledonia, an island in the South Pacific. Its five to eight carpels are in a single whorl, and it has 30 to 100 stamens. The xylem of *Amborella* lacks vessel elements, which appeared later in angiosperm evolution. The characteristics of *Amborella* give us a good sense of what the first angiosperms might have been like. But are there extinct angiosperms that may represent still more ancient clades?

In 2002, Chinese and American botanists examined fossils of two species of a 125-million-year-old aquatic genus, *Archaefructus* (see Figure 22.16). Their studies established an extinct family, Archaefructaceae, that is posited to be the sister taxon of all other angiosperms. The flower of these plants had its ovules enclosed in carpels, as in all angiosperms. The flower had neither petals nor sepals, however, and its carpels and stamens were arranged spirally around elongated shoots. This arrangement of carpels and stamens is seen today in the magnolias.

The origin of the angiosperms remains a mystery

We have learned a lot about evolution within the angiosperm clade. But how did the angiosperms first arise? Are the angiosperms sister to any single gymnosperm phylum? A few years ago, it seemed that we were on the verge of answering these questions. But the puzzle remains as vexing today as it ever was.

Why should this be? Different phylogenetic methods, applied by different investigators, have produced apparently contradictory results. It might seem a simple matter to rectify this situation, but several questions complicate such efforts: What morphological characters should be selected as important, or should they all be treated as equally important? What algorithms should be applied to computerized analysis of data? Are all molecular differences and similarities significant, or are some of them incidental? Which fossils should be chosen for comparisons? What is the likelihood that we can find evidence of double fertilization in ancient fossils? Furthermore, it is possible that the angiosperms have no close relatives at all among living seed plants.

We are left with our original question: Where did the first angiosperm come from? Current progress in methodology gives us reason to hope that our understanding of seed plant evolution will be much improved before the present decade ends. We will see in Chapters 32–34 whether our understanding of animal evolution is any more complete.

Chapter Summary

The Seed Plants

▶ The seed plants (gymnosperms and angiosperms) are heterosporous and have greatly reduced gametophytes. **Review Figures 30.1, 30.2**

▶ Modern gymnosperms and many angiosperms have abundant xylem and extensive secondary growth.

▶ Most modern seed plants have no swimming gametes and do not require liquid water for fertilization. The male gametophyte—the pollen grain—is dispersed by wind or by animals.

▶ The seed is a well-protected resting stage that often contains nutrients that support the growth of the embryo.

The Gymnosperms: Naked Seeds

▶ The gymnosperms, once the dominant vegetation on Earth, still dominate forests in the northern parts of the Northern Hemisphere and at high elevations.

▶ The four surviving gymnosperm phyla are the Cycadophyta (perhaps the most ancient), Ginkgophyta (consisting of a single species, the maidenhair tree), Gnetophyta (which has some characters in common with the angiosperms), and Pinophyta (the familiar cone-bearing trees).

▶ Conifers have a life cycle in which naked seeds are produced on the scales of cones. Pollen is produced in strobili, which are smaller than cones. Pollen is transferred from strobili to cones by wind. **Review Figures 30.5, 30.6. See Web/CD Tutorial 30.1 and Activity 30.1**

The Angiosperms: Flowering Plants

▶ Angiosperms (phylum Angiospermae) are distinguished by double fertilization, which results in a triploid nutritive tissue, the endosperm.

▶ The ovules and seeds of angiosperms are enclosed by a carpel. Angiosperms are also characterized by the production of flowers and fruits.

▶ The vascular tissues of angiosperms contain three characteristic cell types: vessel elements, fibers, and companion cells. Woody angiosperms show secondary growth.

▶ Flowers are made up of various combinations of carpels, stamens, petals, and sepals. Perfect flowers have both carpels and stamens. **Review Figure 30.7. See Web/CD Activity 30.2**

▶ Monoecious plant species have both female and male flowers on the same plant. In dioecious species, female and male flowers are found on separate individuals.

▶ Carpels and stamens may have evolved from leaflike structures. **Review Figure 30.10**

▶ Angiosperms and the animals that pollinate them have co-evolved.

▶ The angiosperm seed contains the products of double fertilization: the diploid zygote and the triploid endosperm. **Review Figure 30.11**

▶ The largest clades of flowering plants, in terms of numbers of species, are the monocots and the eudicots. There are a few other angiosperm clades, notably the water lilies, star anise and its relatives, and the magnoliids. **Review Figure 30.13**

▶ *Amborella*, a tropical shrub, is thought to be the sole living representative of the most ancient living angiosperm clade.

▶ The evolutionary origin of the angiosperms remains a mystery.

Self-Quiz

1. Which of the following statements about seed plants is true?
   a. The phylogenetic relationships among all five phyla have been established.
   b. The sporophyte generation is more reduced than in the ferns.
   c. The gametophytes are independent of the sporophytes.
   d. All seed plant species are heterosporous.
   e. The zygote divides repeatedly to form the gametophyte.

2. The gymnosperms
   a. dominate all land masses today.
   b. have never dominated land masses.
   c. have active secondary growth.
   d. all have vessel elements.
   e. lack sporangia.

3. Conifers
   a. produce ovules in strobili and pollen in cones.
   b. depend on liquid water for fertilization.
   c. have triploid endosperm.
   d. have pollen tubes that release two sperm.
   e. have vessel elements.

4. Angiosperms
   a. have ovules and seeds enclosed in a carpel.
   b. produce triploid endosperm by the union of two eggs and one sperm.
   c. lack secondary growth.
   d. bear two kinds of cones.
   e. all have perfect flowers.

5. Which statement about flowers is *not* true?
   a. Pollen is produced in the anthers.
   b. Pollen is received on the stigma.
   c. An inflorescence is a cluster of flowers.
   d. A species having female and male flowers on the same plant is dioecious.
   e. A flower with both megasporangia and microsporangia is said to be perfect.

6. Which statement about fruits is *not* true?
   a. They develop from ovaries.
   b. They may include other parts of the flower.
   c. A multiple fruit develops from several carpels of a single flower.
   d. They are produced only by angiosperms.
   e. A cherry is a simple fruit.

7. Which statement is *not* true of angiosperm pollen?
   a. It is the male gamete.
   b. It is haploid.
   c. It produces a long tube.
   d. It interacts with the carpel.
   e. It is produced in microsporangia.

8. Which statement is *not* true of carpels?
   a. They are thought to have evolved from leaves.
   b. They bear megasporangia.
   c. They may fuse to form a pistil.
   d. They are floral organs.
   e. They were absent in *Archaefructus*.

9. *Amborella*
   a. was the first flowering plant.
   b. belongs to the first angiosperm clade.
   c. belongs to the oldest angiosperm clade still extant.
   d. is a eudicot.
   e. has vessel elements in its xylem.

10. The eudicots
    a. include many herbs, vines, shrubs, and trees.
    b. and the monocots are the only extant angiosperm clades.
    c. are not a clade.
    d. include the magnolias.
    e. include orchids and palm trees.

For Discussion

1. In most seed plant species, only one of the products of meiosis in the megasporangium survives. How might this be advantageous?

2. Suggest an explanation for the great success of the angiosperms in occupying terrestrial habitats.

3. In many locales, large gymnosperms predominate over large angiosperms. Under what conditions might gymnosperms have the advantage, and why?

4. Not all flowers possess all of the following floral organs: sepals, petals, stamens, and carpels. Which floral organ or organs do you think might be found in the flowers that have the smallest number of floral organ types? Discuss the possibilities, both for a single flower and for a species.

5. The problem of the origin of the angiosperms has long been "an abominable mystery," as Charles Darwin once put it. Scientists still do not know the nearest relatives of the angiosperms. It has often been suggested (correctly or incorrectly) that the gnetophytes are sister to the angiosperms. What pieces of evidence suggested this connection?

31 Fungi: Recyclers, Pathogens, Parasites, and Plant Partners

About 300 million Africans in 25 countries are suffering because of the invasion of crops by witchweed (*Striga*), a parasitic flowering plant. This parasite has attacked more than two-thirds of the sorghum, maize, and millet crops in sub-Saharan Africa, doing damage estimated at U.S. $7 billion each year.

In 1991 a team of Canadian scientists began a search for a solution to the *Striga* problem. By 1995 they had begun fieldwork in Mali. What was their strategy? They had isolated a strain of a fungus, the mold *Fusarium oxysporum*, that has two outstanding properties. First, it grows on *Striga*, wiping out a high percentage of the parasites. Second, it is not toxic to humans, nor does it attack the crop plants on which *Striga* is growing. Now farmers apply the fungus to their crops and are rewarded by greatly increased crop yields as *Striga* is held in check.

It may be possible to repeat this story—using a fungus to wipe out a particular type of flowering plant—in a very different context. A different strain of *F. oxysporum* preferentially attacks coca plants (the source of cocaine). There is a controversial proposal to use *F. oxysporum* to wipe out the coca plantations of Andean South America and some countries in other parts of the world.

Some other fungi attack people, not plants. Every breath we take contains large numbers of fungal spores. Some of those spores can be dangerous, and fungal diseases of humans, some of which are as yet incurable, have become a major global threat. However, other fungi are of immense commercial importance to us. Fungi are essential to plants as well. They interact with roots, greatly enhancing the roots' ability to take up water and mineral nutrients. Fungi and plants probably invaded the land together in the Paleozoic era (see Table 22.1).

Earth would be a messy place without the fungi. They are constantly at work in forests, fields, and garbage dumps, breaking down the remains of dead organisms (and even manufactured substances, such as some plastics). For almost a billion years, the ability of fungi to decompose organic substances has been essential for life on Earth, chiefly because by breaking down carbon com-

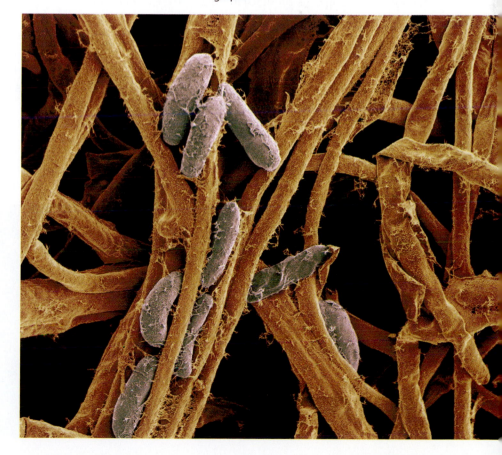

Fungus Trumps Plant The fungus *Fusarium oxysporum* is a potent pathogen of witchweed (*Striga*), a parasitic plant that attacks crops. The fungus spores are shown in blue; the fungal filaments are in tan. Both colors were added to this electron micrograph.

pounds, they return carbon and other elements to the environment, where they can be used again by other organisms.

In this chapter we will examine the general biology of the kingdom Fungi, which differs in interesting ways from the other kingdoms. We will also explore the diversity of body forms, reproductive structures, and life cycles among the four phyla of fungi, as well as the mutually beneficial associations of certain fungi with other organisms. As we begin our study, recall that the fungi and the animals are descended from a common ancestor—molds and mushrooms are more closely related to us than they are to the flowers we admired in the last chapter.

General Biology of the Fungi

The kingdom Fungi encompasses *heterotrophic organisms with absorptive nutrition and with chitin in their cell walls*. The fungi live by **absorptive nutrition**: They secrete digestive enzymes that break down large food molecules in the environment, and then absorb the breakdown products. Many fungi are *saprobes* that absorb nutrients from dead matter, others are *parasites* that absorb nutrients from living hosts (Figure 31.1), and still others are *mutualists* that live in intimate association with other organisms.

The production of **chitin**, a polysaccharide, is a synapomorphy (shared derived trait) for fungi, choanoflagellates, and animals. That is, its presence in fungi is the evidence that all fungi are more closely related to animals than any fungi are to plants. Chitin is used in the cell walls of fungi, but it is used in other ways in animals. The use of chitin in cell walls is a synapomorphy for fungi, and it allows us to distinguish between the fungi and the basal eukaryotes (protists) that resemble them. Some protists that were formerly confused with fungi include the slime molds (see Figures 28.31 and 28.32) and water molds (oomycetes; see Figure 28.23).

The alternation between gametophyte (*n*) and sporophyte (2*n*) generations that evolved in plants (see Chapter 29) is found in only the most basal group of fungi, the chytrids. The derived condition, which is found in the other three fungal clades, involves a unique state in which two haploid nuclei are present in a single cell, discussed later in this chapter. As one might expect, the chytrids, which are aquatic, possess flagellated gametes (or spores). Flagella have been lost in the terrestrial fungi.

The kingdom Fungi consists of four phyla: Chytridiomycota, Zygomycota, Ascomycota, and Basidiomycota. We distinguish the phyla on the basis of their methods and structures for sexual reproduction and, to a lesser extent, by criteria such as the presence or absence of cross-walls separating their cell-like compartments. This morphologically based phylogeny has proved largely consistent with phylogenies based on DNA sequencing. The term "fungal systematics" has an interesting anagram, "fantastic ugly mess," but we'll see that the situation isn't all that bad.

In the sections that follow, we'll consider some aspects of the general biology of the fungi, including their body structure and its intimate relationship with their environment, their nutrition, and some special aspects of their unusual sexual reproductive cycles.

Some fungi are unicellular

Unicellular forms are found in all of the fungal phyla. Unicellular members of the Zygomycota, Ascomycota, and Basidiomycota are called **yeasts**. Yeasts may reproduce by budding, by fission, or by sexual means (Figure 31.2). Their means of reproduction help us to place them in their appropriate phyla, as we will see below.

The body of a multicellular fungus is composed of hyphae

Most fungi are multicellular. The body of a multicellular fungus is called a **mycelium** (plural, mycelia). It is composed of rapidly growing individual tubular filaments called

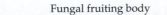

(a) Fungus

(b) Fungal fruiting body

31.1 Parasitic Fungi Attack Other Living Organisms (a) The gray masses on this ear of corn are the parasitic fungus *Ustilago maydis*, commonly called corn smut. (b) The tropical fungus whose fruiting body is growing out of the carcass of this ant has developed from a spore ingested by the ant. The spores of this fungus must be ingested by insects before they will germinate and develop. The growing fungus absorbs organic and inorganic nutrients from the ant's body, eventually killing it, after which the fruiting body produces a new crop of spores.

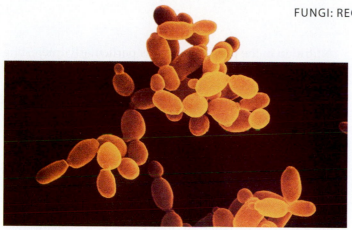

Saccharomyces sp.

31.2 Yeasts Are Unicellular Fungi Unicellular members of the fungal phyla Zygomycota, Ascomycota, and Basidiomycota are known as yeasts. Many yeasts reproduce by budding—mitosis followed by asymmetrical cell division—as those shown here are doing.

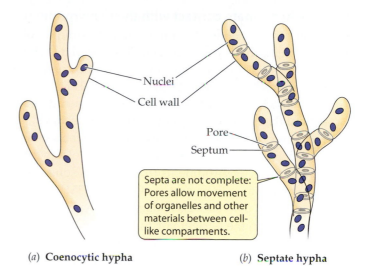

Nuclei

Cell wall

Pore

Septum

Septa are not complete: Pores allow movement of organelles and other materials between cell-like compartments.

(a) **Coenocytic hypha**   (b) **Septate hypha**

31.3 Most Hyphae Are Incompletely Divided into Separate Cells (a) Coenocytic hyphae have no septa between their nuclei. (b) Even in septate hyphae, the septa do not block the movement of organelles within the hypha.

hyphae (singular, hypha). Within hyphae of two clades, *incomplete* cross-walls called **septa** (singular, septum) divide the hypha into separate cells. Pores in the septa allow organelles—sometimes even nuclei—to move in a controlled way between cells (Figure 31.3). Other hyphae are **coenocytic** and have no septa.

Certain modified hyphae, called **rhizoids**, anchor chytrids and some other fungi to their substratum (the dead organism or other matter upon which they feed). These rhizoids are not homologous to the rhizoids of plants because they are not specialized to absorb nutrients and water. Parasitic fungi may possess modified hyphae that take up nutrients from their host.

The total hyphal growth of a mycelium (not the growth of an individual hypha) may exceed 1 km per day. The hy-

phae may be widely dispersed to forage for nutrients over a large area, or they may clump together in a cottony mass to exploit a rich nutrient source. Sometimes, when sexual spores are produced, the mycelium becomes reorganized into a *fruiting* (reproductive) *structure* such as a mushroom.

The way in which a parasitic fungus attacks a plant illustrates the absorptive role of fungal hyphae (Figure 31.4). The hyphae of a fungus invade a leaf through the stomata, through wounds, or in some cases, by direct penetration of epidermal cells. Once inside the leaf, the hyphae form a mycelium. Some hyphae produce **haustoria**, branching projections that push into the living plant cells, absorbing the nutrients within the cells. The haustoria do not break through the plant cell plasma membranes; they simply press into the cells, with the membrane fitting them like a glove. Fruiting structures may form, either within the plant body or on its surface.

Grass cells

Fungal hyphae

Spore   Stoma   Hypha

Fungal spores germinate on the surface of the leaf.

Elongating hyphae pass through stomata into the interior of the leaf.

Some hyphae penetrate cells within the leaf.

31.4 A Fungus Attacks a Leaf The white structures in the micrograph are hyphae of the fungus *Blumeria graminis*, which is growing on the dark surface of the leaf of a grass.

Fungi are in intimate contact with their environment

The filamentous hyphae of a fungus give it a unique relationship with its physical environment. The fungal mycelium has an enormous surface area-to-volume ratio compared with that of most large multicellular organisms. This large ratio is a marvelous adaptation for absorptive nutrition. Throughout the mycelium (except in fruiting structures), all the hyphae are very close to their environmental food source.

Another characteristic of some fungi is their tolerance for highly hypertonic environments (those with a solute concentration higher than their own; see Chapter 5). Many fungi are more resistant than bacteria to damage in hypertonic surroundings. Jelly in the refrigerator, for example, will not become a growth medium for bacteria because it is too hypertonic to the bacteria, but it may eventually harbor mold colonies. Their presence in the refrigerator illustrates another trait of many fungi: tolerance of temperature extremes. Many fungi tolerate temperatures as low as 5–6°C below freezing, and some tolerate temperatures as high as 50°C or more.

Fungi are absorptive heterotrophs

All fungi are heterotrophs that obtain food by direct absorption from their immediate environment. The majority are saprobes, obtaining their energy, carbon, and nitrogen directly from dead organic matter through the action of enzymes they secrete. However, as we've learned already, some are parasites, and still others form mutualistic associations with other organisms.

Saprobic fungi, along with bacteria, are the major decomposers of the biosphere, contributing to decay and thus to the recycling of the elements used by living things. In the forest, for example, the mycelia of fungi absorb nutrients from fallen trees, thus decomposing their wood. Fungi are the principal decomposers of cellulose and lignin, the main components of plant cell walls (most bacteria cannot break down these materials). Other fungi produce enzymes that decompose keratin and thus break down animal structures such as hair and nails.

Because many saprobic fungi are able to grow on artificial media, we can perform experiments to determine their exact nutritional requirements. Sugars are their favored source of carbon. Most fungi obtain nitrogen from proteins or the products of protein breakdown. Many fungi can use nitrate (NO_3^-) or ammonium (NH_4^+) ions as their sole source of nitrogen. No known fungus can get its nitrogen directly from nitrogen gas, as can some bacteria and plant–bacteria associations (see Chapter 37). Nutritional studies also reveal that most fungi are unable to synthesize their own thiamin (vitamin B_1) or biotin (another B vitamin) and must absorb these vitamins from their environment. On the other hand, fungi can synthesize some vitamins that animals cannot. Like all organisms, fungi also require some mineral elements.

Nutrition in the parasitic fungi is particularly interesting to biologists. *Facultative* parasites can attack living organisms but can also be grown by themselves on artificial media. *Obligate* parasites cannot be grown on any available medium; they can grow only on their specific living hosts, usually plants. Because their growth is limited to living hosts, they must have specialized nutritional requirements.

Some fungi have adaptations that enable them to function as active predators, trapping nearby microscopic protists or animals. The most common strategy is to secrete sticky substances from the hyphae so that passing organisms stick tightly to them. The hyphae then quickly invade the prey, growing and branching within it, spreading through its body, absorbing nutrients, and eventually killing it.

A more dramatic adaptation for predation is the constricting ring formed by some species of *Arthrobotrys*, *Dactylaria*, and *Dactylella* (Figure 31.5). All of these fungi grow in soil. When nematodes (tiny roundworms) are present in the soil, these fungi form three-celled rings with a diameter that just fits a nematode. A nematode crawling through one of these rings stimulates the fungus, causing the cells of the ring to swell and trap the worm. Fungal hyphae quickly invade and digest the unlucky victim.

Two other kinds of relationships between fungi and other organisms have nutritional consequences for the fungal partner. These relationships are highly specific, *symbiotic* (the partners live in close, permanent contact with one another), and *mutualistic* (the relationships benefit both partners). **Lichens** are associations of a fungus with a cyanobacterium, a unicellular photosynthetic protist, or both. **Mycorrhizae** (singular, mycorrhiza) are associations between fungi and the roots of plants. In these associations, the fungus obtains organic com-

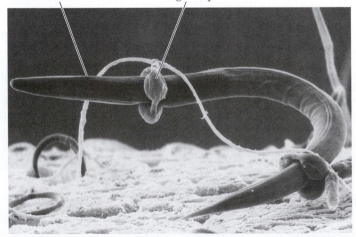

Roundworm   Fungal loop

31.5 Some Fungi Are Predators A nematode (roundworm) is trapped in sticky loops of the soil-dwelling fungus *Arthrobotrys anchonia*.

pounds from its photosynthetic partner, but provides it with minerals and water in return, so that the partner's nutrition is also promoted. In fact, many plants could not grow at all without their fungal partners. We will discuss lichens and mycorrhizae more thoroughly later in this chapter.

Most fungi reproduce both asexually and sexually

Both asexual and sexual reproduction are common among the fungi. Asexual reproduction takes several forms:

▶ The production of (usually) haploid spores within structures called **sporangia**.
▶ The production of naked spores (not enclosed in sporangia) at the tips of hyphae; such spores are called **conidia** (from the Greek *konis*, "dust").
▶ Cell division by unicellular fungi—either a relatively equal division (called *fission*) or an asymmetrical division in which a small daughter cell is produced (called *budding*).
▶ Simple breakage of the mycelium.

Asexual reproduction in fungi can be spectacular in terms of quantity. A 2.5-centimeter colony of *Penicillium* can produce as many as 400 million conidia. The air we breathe contains as many as 10,000 fungal spores per cubic meter.

Sexual reproduction in many fungi features an interesting twist. There is often no morphological distinction between female and male structures, or between female and male individuals. Rather, there is a genetically determined distinction between two *or more* **mating types**. Individuals of the same mating type cannot mate with one another, but they can mate with individuals of another mating type within the same species. This distinction prevents self-fertilization. Individuals of different mating types differ genetically from one another, but are often visually and behaviorally indistinguishable. Many protists also have mating type systems.

Fungi reproduce sexually when hyphae (or, in the chytrids, motile cells) of different mating types meet and fuse. In many fungi, the zygote nuclei formed by sexual reproduction are the only diploid nuclei in the life cycle. These nuclei undergo meiosis, producing haploid nuclei that become incorporated into spores. Haploid fungal spores, whether produced sexually in this manner or asexually, germinate, and their nuclei divide mitotically to produce hyphae. This type of life cycle, called a *haplontic* life cycle, is also characteristic of many protists (see Figure 28.27).

The presence of a dikaryon is a synapomorphy of three phyla

Certain hyphae of some Zygomycota, Ascomycota, and Basidiomycota have a nuclear configuration other than the familiar haploid or diploid states. In these fungi, sexual reproduction begins in an unusual way: The cytoplasms of two individuals of different mating types fuse (*plasmogamy*) long before their nuclei fuse (*karyogamy*), so that *two genetically different haploid nuclei coexist and divide within the same hypha*. Such a hypha is called a **dikaryon** ("two nuclei"). Because the two nuclei differ genetically, such a hypha is also called a **heterokaryon** ("different nuclei").

Eventually, specialized fruiting structures form, within which the pairs of genetically dissimilar nuclei—one from each parent—fuse, giving rise to zygotes long after the original "mating." The diploid zygote nucleus undergoes meiosis, producing four haploid nuclei. The mitotic descendants of those nuclei become spores, which give rise to the next generation of hyphae.

The reproduction of such fungi displays several unusual features. First, there are no gamete *cells*, only gamete *nuclei*. Second, there is never any true diploid tissue, although for a long period the genes of both parents are present in the dikaryon and can be expressed. In effect, the hypha is neither diploid (2*n*) nor haploid (*n*); rather, it is *dikaryotic* (*n* + *n*). A harmful recessive mutation in one nucleus may be compensated for by a normal allele on the same chromosome in the other nucleus. Dikaryosis is perhaps the most significant of the genetic peculiarities of the fungi.

Finally, although zygomycetes, ascomycetes, and basidiomycetes grow in moist places, their gamete nuclei are not motile and are not released into the environment. Therefore, liquid water is not required for fertilization.

Some fungi are pathogens

Although most human diseases are caused by bacteria or viruses, fungal pathogens are a major cause of death among people with compromised immune systems. Most people with AIDS die of fungal diseases, such as the pneumonia caused by *Pneumocystis carinii* or the incurable diarrhea caused by some other fungi. *Candida albicans* and certain other yeasts also cause severe diseases in individuals with AIDS and in individuals taking immunosuppressive drugs. Such fungal diseases are a growing international health problem. Our limited understanding of the basic biology of these fungi still hampers our ability to treat the diseases they cause. Various fungi cause other, less threatening human diseases, such as ringworm and athlete's foot.

In plants, the situation is reversed. Fungi are by far the most important plant pathogens, causing crop losses amounting to billions of dollars. Major fungal diseases of crop plants include black stem rust of wheat and other diseases of wheat, corn, and oats. Bacteria and viruses are less important as plant pathogens.

The fungus that causes root and butt rot in pine trees is an important forest pathogen with an interesting, recently dis-

31.1 Classification of Fungi

PHYLUM	COMMON NAME	FEATURES	EXAMPLES
Chytridiomycota	Chytrids	Aquatic; gametes have flagella	*Allomyces*
Zygomycota	Zygote fungi	Zygosporangium; no regularly occurring septa; usually no fleshy fruiting body	*Rhizopus*
Ascomycota	Sac fungi	Ascus; perforated septa	*Neurospora*, baker's yeast
Basidiomycota	Club fungi	Basidium; perforated septa	*Armillariella*, mushrooms

covered property. The virulence (relative ability to cause disease) of some strains of the fungus is controlled by genes in its mitochondria—even though its dikaryotic cells have two different nuclei.

Diversity in the Kingdom Fungi

In this section on fungal diversity, we'll consider four phyla—Chytridiomycota, Zygomycota, Ascomycota, and Basidiomycota (Figure 31.6; Table 31.1). The first two groups are probably not clades, but the Ascomycota and Basidiomycota are clades.

Chytrids probably resemble the ancestral fungi

The earliest-diverging fungal group is the **chytrids** (phylum **Chytridiomycota**). These aquatic microorganisms were formerly classified with the protists. However, morphological (cell walls that consist primarily of chitin) and molecular evidence support their inclusion in the kingdom Fungi as its basal

members. In this book, we use the term "chytrid" to refer to the entire phylum, but some mycologists reserve the term to apply to one of the major clades in the phylum.

Like their sister taxon, the animals, the chytrids possess flagellated gametes. The retention of this character reflects the aquatic environment in which fungi first evolved. Chytrids are the only fungi that have flagella at any life cycle stage.

Chytrids are either parasitic (on organisms such as algae, mosquito larvae, and nematodes) or saprobic, obtaining nutrients by breaking down dead organic matter. Chytrids in the compound stomachs of foregut-fermenting animals such as cows may be an exception, living in a mutualistic association with their hosts. Most chytrids live in freshwater habitats or in moist soil, but some are marine. Some chytrids are unicellular; others have mycelia made up of branching, coenocytic hyphae. Chytrids reproduce both sexually and asexually, but they do not have a dikaryon stage.

Allomyces, a well-studied genus of chytrids, displays alternation of generations. A haploid *zoospore* (a spore with flagella) comes to rest on dead plant or animal material in water and germinates to form a small, multicellular haploid mycelium. That mycelium produces female and male *gametangia* (gamete cases) (Figure 31.7). *Mitosis* in the gametangia results in the formation of haploid gametes, each with a single nucleus.

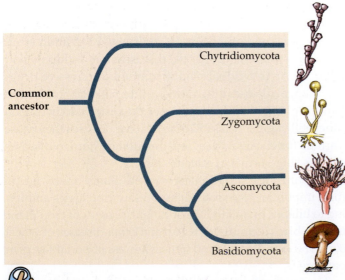

31.6 Phylogeny of the Fungi Four phyla are recognized among the fungi.

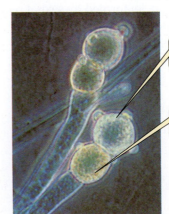

Allomyces sp.

The female gametangium contains female gametes.

The male gametangium contains male gametes.

31.7 Reproductive Structures of a Chytrid The haploid gametes produced in these gametangia will fuse with other gametes to form diploid mycelia. The male gametangia are smaller than the female gametangia and possess a light orange pigment.

Both female and male gametes have flagella. The motile female gamete produces a *pheromone*, a chemical that attracts the swimming male gamete. The two gametes fuse, and then their nuclei fuse to form a diploid zygote. Mitosis and cytokinesis in the zygote gives rise to a small, multicellular diploid organism, which produces numerous diploid flagellate zoospores. These diploid zoospores disperse and germinate to form more diploid organisms. Eventually, the diploid organism produces thick-walled resting sporangia that can survive unfavorable conditions such as dry weather or freezing. Nuclei in the resting sporangia eventually undergo meiosis, giving rise to haploid zoospores that are released into the water and begin the cycle anew.

The presence of flagellated gametes is a distinguishing feature of the chytrids. The loss of flagella is a synapomorphy that unites the remaining three fungal lineages.

Zygomycetes reproduce sexually by fusion of two gametangia

Most **zygomycetes** ("zygote fungi," phylum **Zygomycota**) have coenocytic hyphae. They produce no motile cells, and only one diploid cell— the zygote—appears in the entire life cycle. The mycelium of a zygomycete spreads over its substratum, growing forward by means of vegetative hyphae. Most zygomycetes do not form a fleshy fruiting structure; rather, the hyphae spread in an apparently random fashion, with occasional stalked **sporangiophores** reaching up into the air (Figure 31.8). These reproductive structures may bear one or many sporangia.

Pilobolus sp.

31.8 A Zygomycete This small forest of filamentous structures is made up of sporangiophores. The stalks end in tiny, rounded sporangia.

Almost 900 species of zygomycetes have been described. A very important group of zygomycetes serve as the fungal partners in the most common type of mycorrhizal association with plant roots. A zygomycete that you may be more familiar with is *Rhizopus stolonifer*, the black bread mold. *Rhizopus* reproduces asexually by producing many stalked sporangiophores, each bearing a single sporangium containing hundreds of minute spores (Figure 31.9a). As in other filamentous fungi, the spore-forming structure is separated from the rest of the hypha by a wall.

Zygomycetes reproduce sexually when adjacent hyphae of two different mating types release pheromones, which cause them to grow toward each other. These hyphae produce gametangia, which fuse to form a **zygosporangium**. Sometime later, the gamete nuclei now contained within the zygosporangium fuse to form a single multinucleate **zygospore** (Figure 31.9b). The zygosporangium develops a thick, multilayered wall that protects the zygospore. The highly resistant zygospore may remain dormant for months before its nuclei undergo meiosis and a sporangiophore sprouts. The sporangium contains the products of meiosis: haploid nuclei that are incorporated into spores. These spores disperse and germinate to form a new generation of haploid hyphae.

The next two fungal lineages that we'll discuss are related groups with many similarities, including a dikaryon stage and hyphae with septa. A key feature distinguishing between them is whether the sexual spores are borne inside a sac (in the ascomycetes) or on a pedestal (in the basidiomycetes).

The sexual reproductive structure of ascomycetes is an ascus

The **ascomycetes** ("sac fungi," phylum **Ascomycota**) are a large and diverse group of fungi distinguished by the production of sacs called **asci** (singular, ascus), which contain sexually produced **ascospores** (Figure 31.10). The ascus is the characteristic sexual reproductive structure of the ascomycetes. Ascomycete hyphae are segmented by more or less regularly spaced septa. A pore in each septum permits extensive movement of cytoplasm and organelles (including the nuclei) from one segment to the next.

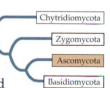

The approximately 30,000 known species of ascomycetes can be divided into two broad groups, depending on whether the asci are contained within a specialized fruiting structure. Species that have this fruiting structure, the **ascocarp**, are collectively called **euascomycetes** ("true ascomycetes"); those without ascocarps are called **hemiascomycetes** ("half ascomycetes").

(a)

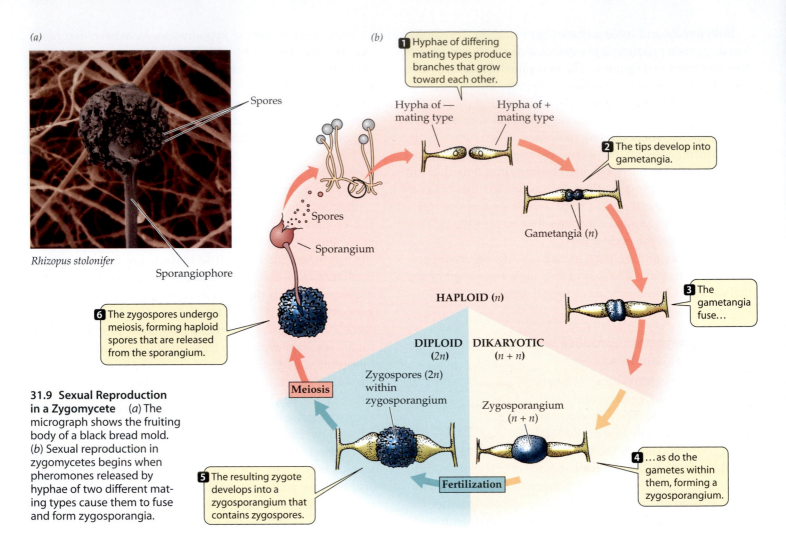

Rhizopus stolonifer

Spores

Sporangium

Sporangiophore

(b)

1 Hyphae of differing mating types produce branches that grow toward each other.

Hypha of — mating type

Hypha of + mating type

2 The tips develop into gametangia.

Gametangia (*n*)

HAPLOID (*n*)

3 The gametangia fuse...

6 The zygospores undergo meiosis, forming haploid spores that are released from the sporangium.

Meiosis

DIPLOID (*2n*)

DIKARYOTIC (*n + n*)

Zygospores (*2n*) within zygosporangium

Zygosporangium (*n + n*)

5 The resulting zygote develops into a zygosporangium that contains zygospores.

Fertilization

4 ...as do the gametes within them, forming a zygosporangium.

31.9 Sexual Reproduction in a Zygomycete *(a)* The micrograph shows the fruiting body of a black bread mold. *(b)* Sexual reproduction in zygomycetes begins when pheromones released by hyphae of two different mating types cause them to fuse and form zygosporangia.

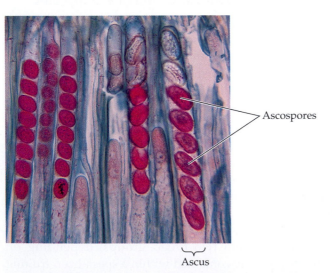

Ascospores

Ascus

31.10 Asci and Ascospores The ascomycetes are characterized by the production of ascospores within sacs called asci. Ascospores are the products of meiosis followed by a single mitotic division. Ascospores and asci do not mature all at once, and they may abort, so not every ascus in this micrograph contains eight mature ascospores.

HEMIASCOMYCETES. Most hemiascomycetes are microscopic, and many species are unicellular. Perhaps the best known are the ascomycete yeasts, especially baker's or brewer's yeast (*Saccharomyces cerevisiae*; see Figure 31.2). These yeasts are among the most important domesticated fungi. *S. cerevisiae* metabolizes glucose obtained from its environment to ethanol and carbon dioxide by fermentation. It forms carbon dioxide bubbles in bread dough and gives baked bread its light texture. Although they are baked away in bread making, the ethanol and carbon dioxide are both retained when yeast ferments grain into beer. Other yeasts live on fruits such as figs and grapes and play an important role in the making of wine.

Hemiascomycete yeasts reproduce asexually either by fission (splitting in half after mitosis) or by budding (an asymmetrical cell division in which a small daughter cell is produced; see Figure 31.2). Sexual reproduction takes place when two adjacent haploid cells of opposite mating types fuse. (We discussed the genetics of yeast mating types in Chapter 14.) In some species, the resulting zygote buds to form a diploid cell population; in others, the zygote nucleus undergoes meiosis immediately. When this diploid nucleus undergoes meiosis, the entire cell becomes an ascus. Depending on whether the products of meiosis then undergo mitosis, a yeast ascus usually contains either eight or four ascospores (see Figure 31.10). The ascospores germinate to become haploid cells. Hemiascomycetes have no dikaryon stage.

Yeasts, especially *Saccharomyces cerevisiae*, are frequently used in molecular biological research. Just as *E. coli* is the best-studied prokaryote, *S. cerevisiae* is the most completely studied eukaryote.

EUASCOMYCETES. The euascomycetes include many of the filamentous fungi known as molds. Among them are several common pink molds, one of which (*Neurospora*) Beadle and Tatum used in their pioneering genetic studies (see Figure 12.1). Many euascomycetes are parasites on flowering plants. Chestnut blight and Dutch elm disease are both caused by euascomycetes. The powdery mildews are euascomycetes that infect cereal grains, lilacs, and roses, among many other plants. They can be a serious problem to grape growers, and a great deal of research has focused on ways to control these agricultural pests.

The euascomycetes also include the cup fungi (Figure 31.11*a*,*b*). In most of these organisms the ascocarps are cup-shaped and can be as large as several centimeters across. The inner surfaces of the cups are covered with a mixture of vegetative hyphae and asci, and they produce huge numbers of spores. Although these fleshy structures appear to be composed of distinct tissue layers, microscopic examination shows that their basic organization is still filamentous—a tightly woven mycelium.

Two particularly delicious euascomycetes ascocarps are morels (Figure 31.11*a*) and truffles. Truffles grow underground in a mutualistic association with the roots of some species of oaks. Europeans traditionally used pigs to find truffles because some truffles secrete a substance that has an odor similar to a pig's sex pheromone. Unfortunately, pigs also eat truffles, so dogs are now the usual truffle hunters.

Penicillium is a genus of green molds, of which some species produce the antibiotic penicillin, presumably for defense against competing bacteria. Two species, *P. camembertii* and *P. roquefortii*, are the organisms responsible for the characteristic flavors of Camembert and Roquefort cheeses, respectively.

Brown molds of the genus *Aspergillus* are important in some human diets. *A. tamarii* acts on soybeans in the production of soy sauce, and *A. oryzae* is used in brewing the Japanese alcoholic beverage sake. Some species of *Aspergillus* that grow on nuts such as peanuts and pecans produce extremely

(a) *Morchella esculenta*

(b) *Sarcoscypha coccinea*

31.11 Two Cup Fungi (*a*) Morels, which have a spongelike ascocarp and a subtle flavor, are considered a delicacy by humans. (*b*) These brilliant red cups are the ascocarps of another cup fungus.

carcinogenic (cancer-inducing) compounds called aflatoxins. In the United States, moldy grain infected with *Aspergillus* is thrown out. In Africa, where food is scarcer, the grain gets eaten, moldy or not, and causes severe health problems.

The euascomycetes reproduce asexually by means of conidia that form at the tips of specialized hyphae (Figure 31.12). Small chains of conidia are produced by the millions and can survive for weeks in nature. The conidia are what give molds their characteristic colors.

The sexual reproductive cycle of euascomycetes includes the formation of a dikaryon. Most euascomycetes form mat-

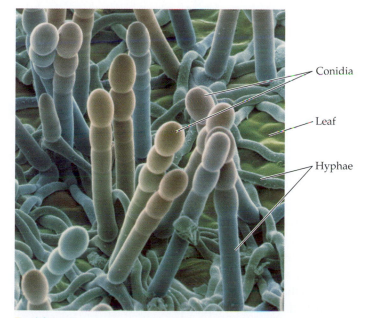

Conidia

Leaf

Hyphae

Erysiphe sp.

31.12 Conidia Chains of conidia are developing at the tips of specialized hyphae arising from this powdery mildew growing on a leaf.

31.13 The Life Cycle of a Euascomycete
This cup fungus is so named because of its cup-shaped ascocarp.

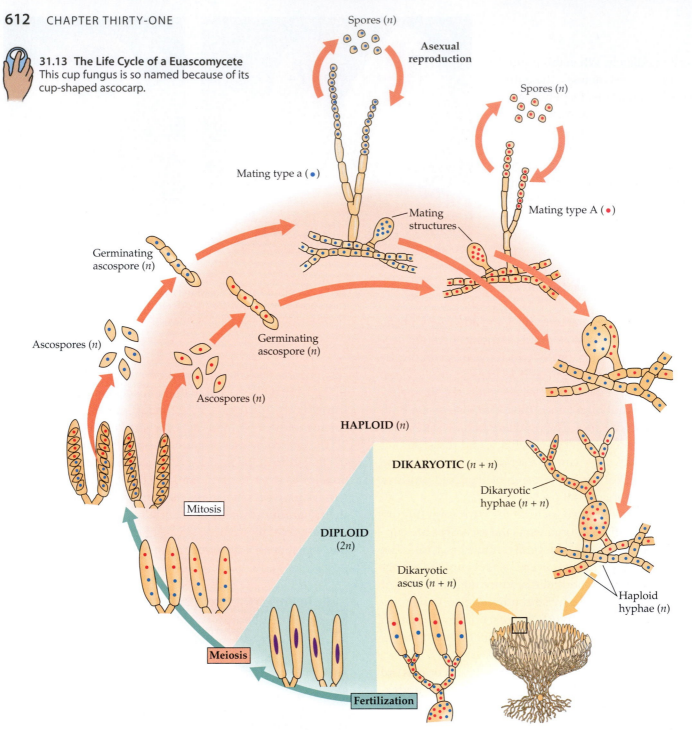

Ascocarp

ing structures, some "female" and some "male" (Figure 31.13). Nuclei from a male structure on one hypha enter a female mating structure on a hypha of a compatible mating type. Dikaryotic *ascogenous* (ascus-forming) hyphae develop from the now dikaryotic female mating structure. The introduced nuclei divide simultaneously with the host nuclei. Eventually asci form at the tips of the ascogenous hyphae. Only with the formation of asci do the nuclei finally fuse. Both nuclear fusion and the subsequent meiosis of the resulting diploid nucleus take place within individual asci. The meiotic products are incorporated into ascospores that are ultimately shed by the ascus to begin the new haploid generation.

The sexual reproductive structure of basidiomycetes is a basidium

About 25,000 species of **basidiomycetes** ("club fungi," phylum **Basidiomycota**) have been described. Basidiomycetes produce some of the most spectacular fruiting structures found anywhere among the fungi. These fruiting structures, called **basidiocarps**, include puffballs (which may be more than half a meter in diameter), mushrooms of all kinds, and the

Chytridiomycota
Zygomycota
Ascomycota
Basidiomycota

(a) *Lycoperdon perlatum*

(b) *Amanita muscaria*

(c) *Laetiporus sulphureus*

31.14 Basidiomycete Fruiting Structures The basidiocarps of the basidiomycetes are probably the most familiar structures produced by fungi. (a) When raindrops hit them, these puffballs will release clouds of spores for dispersal. (b) These mushrooms were produced by a member of a highly poisonous genus, *Amanita*, that forms mycorrhizal relationships with trees. (c) This edible bracket fungus is parasitizing a tree.

giant bracket fungi often encountered on trees and fallen logs in a damp forest (Figure 31.14). There are more than 3,250 species of mushrooms, including the familiar *Agaricus bisporus* you may enjoy on your pizza, as well as poisonous species, such as members of the genus *Amanita*. Bracket fungi do great damage to cut lumber and stands of timber. Some of the most damaging plant pathogens are basidiomycetes, including the rust fungi and the smut fungi (see Figure 31.1a) that parasitize cereal grains. In contrast, other basidiomycetes contribute to the survival of plants as fungal partners in mycorrhizae.

Some of the largest organisms on Earth are basidiomycetes. One such fungus, a member of the genus *Armillariella* growing in Michigan, covers an area of 37 acres. Its effect on plants is evident from the air, but from ground level, it is difficult to realize how large the fungus is. At the surface, only seemingly isolated clumps of mushrooms are visible. The vast body of the fungus, which weighs approximately the same as a blue whale, grows underground and consists almost entirely of microscopic hyphal filaments. Molecular studies indicate that this giant fungus is or was a single individual that arose from a single spore. It is possible that fragmentation over time may have broken it into a few separate—but still gigantic—individuals. Another, larger fungus

of the same genus, growing in the state of Washington, occupies parts of three counties.

Basidiomycete hyphae characteristically have septa with small, distinctive pores. The **basidium** (plural, basidia), a swollen cell at the tip of a hypha, is the characteristic sexual reproductive structure of the basidiomycetes. It is the site of nuclear fusion and meiosis. Thus, the basidium plays the same role in the basidiomycetes as the ascus does in the ascomycetes and the zygosporangium does in the zygomycetes.

The life cycle of the basidiomycetes is shown in Figure 31.15. After nuclei fuse in the basidium, the resulting diploid nucleus undergoes meiosis, and the four resulting haploid nuclei are incorporated into haploid **basidiospores**, which form on tiny stalks on the outside of the basidium. These basidiospores typically are forcibly discharged from their basidia and then germinate, giving rise to haploid hyphae. As these hyphae grow, haploid hyphae of different mating types meet and fuse, forming dikaryotic hyphae, each cell of which contains two nuclei, one from each parent hypha. The dikaryotic mycelium grows and eventually, when triggered by rain or another environmental cue, produces a basidiocarp. The dikaryon stage may persist for years—some basidiomycetes live for decades or even centuries. This pattern contrasts with the life cycle of the ascomycetes, in which the dikaryon is found only in the stages leading up to formation of the asci.

The elaborate basidiocarp of some fleshy basidiomycetes, such as the mushroom shown in Figure 31.15, is topped by a

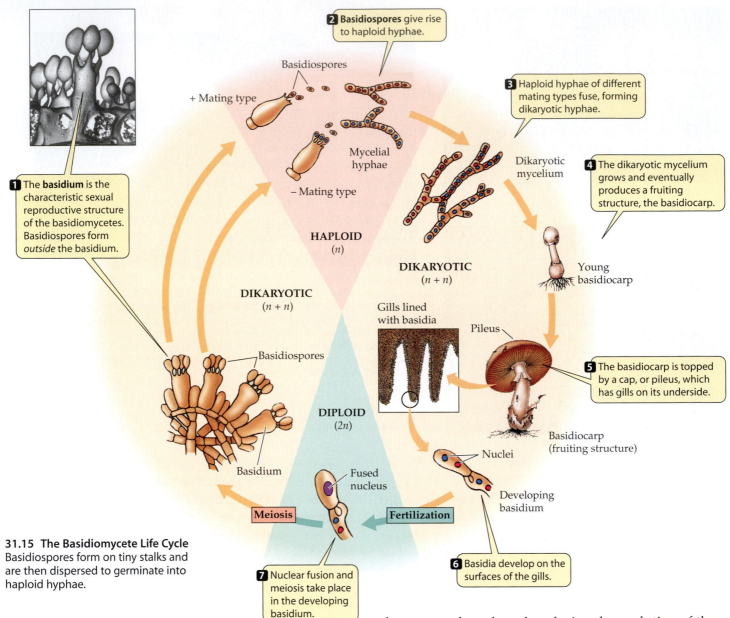

2 Basidiospores give rise to haploid hyphae.

3 Haploid hyphae of different mating types fuse, forming dikaryotic hyphae.

4 The dikaryotic mycelium grows and eventually produces a fruiting structure, the basidiocarp.

5 The basidiocarp is topped by a cap, or pileus, which has gills on its underside.

6 Basidia develop on the surfaces of the gills.

7 Nuclear fusion and meiosis take place in the developing basidium.

1 The **basidium** is the characteristic sexual reproductive structure of the basidiomycetes. Basidiospores form *outside* the basidium.

Basidiospores

+ Mating type

– Mating type

Mycelial hyphae

HAPLOID (*n*)

DIKARYOTIC (*n + n*)

Dikaryotic mycelium

Young basidiocarp

Gills lined with basidia

Pileus

Basidiocarp (fruiting structure)

Basidiospores

Basidium

DIPLOID (2*n*)

Fused nucleus

Nuclei

Developing basidium

Meiosis

Fertilization

31.15 The Basidiomycete Life Cycle
Basidiospores form on tiny stalks and are then dispersed to germinate into haploid hyphae.

cap, or *pileus*, which has structures called *gills* on its underside. Enormous numbers of basidia develop on the surfaces of the gills. The basidia discharge their basidiospores into the air spaces between adjacent gills, and the spores sift down into air currents for dispersal and germination as new haploid mycelia. A single basidiocarp of the common bracket fungus *Ganoderma applanatum* can produce as many as 4.5 *trillion* basidiospores in one growing season.

Imperfect fungi lack a sexual stage

As we have just seen, mechanisms of sexual reproduction readily distinguish members of the four phyla of fungi from one another. But many fungi, including both saprobes and parasites, appear to lack sexual stages entirely; presumably these stages have been lost during the evolution of these species or have not yet been observed. Classifying these fungi used to be difficult, but biologists now can assign most such fungi to one of the four phyla on the basis of their DNA sequences.

Fungi that have not yet been placed in any of the existing phyla are pooled together in a polyphyletic group called **deuteromycetes**, informally known as "imperfect fungi." Thus, the deuteromycete group is a holding area for species whose status is yet to be resolved. At present, about 25,000 species are classified as imperfect fungi.

If sexual structures are found on a fungus classified as a deuteromycete, that fungus is reassigned to the appropriate phylum. That happened, for example, to a fungus that produces plant growth hormones called gibberellins (see Chapter 38). Originally classified as the deuteromycete *Fusarium moniliforme*, this fungus was later found to produce asci, whereupon it was renamed *Gibberella fujikuroi* and transferred to the phylum Ascomycota.

Fungal Associations

Earlier in this chapter we mentioned mycorrhizae and lichens, two kinds of symbiotic, mutualistic associations between fungi and other organisms. Now that we have learned a bit about fungal diversity, let's consider mycorrhizae and lichens in greater detail.

Mycorrhizae are essential to many plants

Almost all tracheophytes require a symbiotic association with fungi. Unassisted, the root hairs of such plants do not absorb enough water or minerals to sustain growth. However, their roots usually do become infected with fungi, forming an association called a mycorrhiza.

In *ectomycorrhizae*, the fungus (usually a basidiomycete) wraps around the root, and its mass is often as great as that of the root itself (Figure 31.16*a*). The fungal hyphae do not penetrate the root cells. An extensive web of hyphae penetrates the soil in the area around the root, so that up to 25 percent of the soil volume near the root may be fungal hyphae. The hyphae of the fungi attached to the root increase the surface area for the absorption of water and minerals, and the mass of the mycorrhiza, like a sponge, holds water efficiently in the neighborhood of the root. Infected roots characteristically branch extensively and become swollen and club-shaped, and they lack root hairs.

In *endomycorrhizae*, the fungal (zygomycete) hyphae enter the root and penetrate the root cells, forming tree-like structures inside the cells, which become the primary site of exchange between plant and fungus (Figure 31.16*b*). As with the ectomycorrhizae, the fungus forms a vast web of hyphae leading from the root surface into the surrounding soil.

The mycorrhizal association is important to both partners. The fungus obtains important organic compounds, such as sugars and amino acids, from the plant. In return, the fungus, because of its very high surface area-to-volume ratio and ability to penetrate the fine structure of the soil, greatly increases the plant's ability to absorb water and minerals (especially phosphorus). The fungus may also provide the plant with certain growth hormones and may protect it against attack by microorganisms. Plants that have active endomycorrhizae typically are a deeper green and may resist drought and temperature extremes better than plants of the same species that have little mycorrhizal development. Attempts to introduce some plant species to new areas have failed until a bit of soil from the native area (presumably containing the fungus necessary to establish mycorrhizae) was provided. Trees without ectomycorrhizae normally will not grow at all, so the health of our forests depends on the presence of ectomycorrhizal fungi.

The partnership between plant and fungus results in a plant that is better adapted for life on land. It has been suggested that the evolution of mycorrhizae was the single most important step leading to the colonization of the terrestrial environment by living things. Fossils of mycorrhizal structures more than 300 million years old have been found, and some rocks dating back 460 million years contain structures that appear to be fossilized fungal spores. Some liverworts, which are among the most ancient terrestrial plants (see Chapter 29), form mycorrhizae.

Certain plants that live in nitrogen-poor habitats, such as cranberry bushes and orchids, invariably have mycorrhizae. Orchid seeds will not germinate in nature unless they are already infected by the fungus that will form their mycorrhizae. Plants that lack chlorophyll always have mycorrhizae, which they often share with the

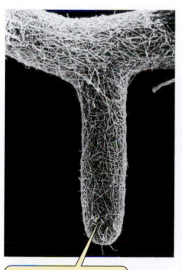

(a)

Hyphae of the fungus *Pisolithus tinctorius* cover a eucalyptus root.

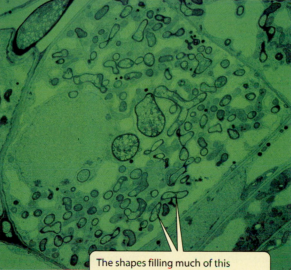

(b)

The shapes filling much of this soybean root cell are sections through the hyphae of the endomycorrhizal fungus *Glomus caledonium*.

31.16 Mycorrhizal Associations *(a)* Ectomycorrhizal fungi wrap themselves around the plant root, increasing the area available for absorption of water and nutrients. *(b)* Endomycorrhizae infect the root internally and penetrate the root cells.

(a) Foliose   Crustose   *(b)*

31.17 Lichen Body Forms   Lichens fall into three principal classes based on their body form. (*a*) These foliose and crustose lichens are growing on otherwise bare rock. (*b*) A miniature jungle of fruticose lichens.

roots of green, photosynthetic plants. In effect, these plants without chlorophyll are feeding on nearby green plants, using the fungus as a bridge.

Lichens can grow where plants cannot

A lichen is not a single organism, but rather a meshwork of two radically different organisms: a fungus and a photosynthetic microorganism. Together the organisms constituting a lichen can survive some of the harshest environments on Earth. The biota of Antarctica, for example, features more than 100 times as many species of lichens as of plants.

In spite of this hardiness, lichens are very sensitive to air pollution because they are unable to excrete toxic substances that they absorb. Hence they are not common in industrialized cities. Because of their sensitivity, lichens are good biological indicators of air pollution.

The fungal components of most lichens are ascomycetes, but a few are basidiomycetes or imperfect fungi. The photosynthetic component is most often a unicellular green alga but may be a cyanobacterium, or may include both. Relatively little experimental work has focused on lichens, perhaps because they grow so slowly—typically less than 1 centimeter per year.

There are about 13,500 "species" of lichens. Their fungal components may constitute as many as 20 percent of all fungal species, but none of these species are able to grow independently without a photosynthetic partner. Lichens are found in all sorts of exposed habitats: on tree bark, open soil, and bare rock. Reindeer "moss" (actually not a moss at all, but the lichen *Cladonia subtenuis*) covers vast areas in arctic, subarctic, and boreal regions, where it is an important part of the diets of reindeer and other large mammals. Lichens come in various forms and colors. *Crustose* (crustlike) lichens look

like colored powder dusted over their substratum (Figure 31.17*a*); *foliose* (leafy) and *fruticose* (shrubby) lichens may have complex forms (Figure 31.17*b*).

The most widely held interpretation of the lichen relationship is that it is a mutually beneficial symbiosis. The hyphae of the fungal mycelium are tightly pressed against the algae or cyanobacteria and sometimes even invade them. The bacterial or algal cells not only survive these indignities, but continue their growth and photosynthesis. In fact, the algal cells in a lichen "leak" photosynthetic products at a greater rate than do similar cells growing on their own. On the other hand, photosynthetic cells from lichens grow more rapidly on their own than when associated with a fungus. On this basis, we could consider lichen fungi as parasitic on their photosynthetic partners.

Lichens can reproduce simply by fragmentation of the vegetative body, which is called the *thallus*, or by means of specialized structures called **soredia** (singular, soredium). Soredia consist of one or a few photosynthetic cells surrounded by fungal hyphae (Figure 31.18*a*). The soredia become detached, are dispersed by air currents, and upon arriving at a favorable location, develop into a new lichen. Alternatively, if the fungal partner is an ascomycete or a basidiomycete, it may go through its sexual cycle, producing either ascospores or basidiospores. When these spores are discharged, however, they disperse alone, unaccompanied by the photosynthetic partner, and thus may not be capable of reestablishing the lichen association, or even of surviving on their own. Nevertheless, many lichens produce characteristic fruiting structures containing asci or basidia.

31.18 Lichen Anatomy (a) Soredia of a fruticose lichen. (b) Cross section showing the layers of a foliose lichen.

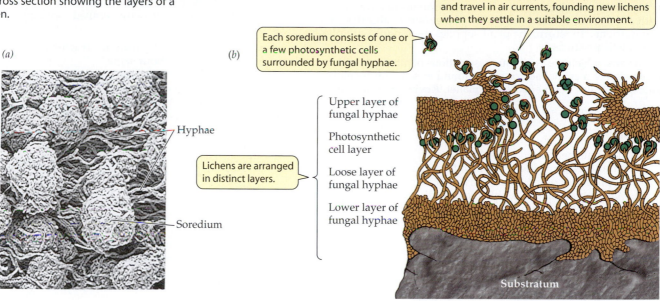

Soredia detach readily from the parent lichen and travel in air currents, founding new lichens when they settle in a suitable environment.

Each soredium consists of one or a few photosynthetic cells surrounded by fungal hyphae.

(a)

(b)

Hyphae

Soredium

Lichens are arranged in distinct layers.

Upper layer of fungal hyphae

Photosynthetic cell layer

Loose layer of fungal hyphae

Lower layer of fungal hyphae

Substratum

Visible in a cross section of a typical foliose lichen are a tight upper region of fungal hyphae, a layer of cyanobacteria or algae, a looser hyphal layer, and finally hyphal rhizoids that attach the whole structure to its substratum (Figure 31.18b). The meshwork of fungal hyphae takes up some nutrients needed by the photosynthetic cells and provides a suitably moist environment for them by holding water tenaciously. The fungi derive fixed carbon from the photosynthesis of the algal or cyanobacterial cells.

Lichens are often the first colonists on new areas of bare rock. They satisfy most of their nutritional needs from the air and rainwater, augmented by minerals absorbed from dust. A lichen begins to grow shortly after a rain, as it begins to dry. As it grows, the lichen acidifies its environment slightly, and this acid contributes to the slow breakdown of rocks, an early step in soil formation. After further drying, the lichen's photosynthesis ceases. The water content of the lichen may drop to less than 10 percent of its dry weight, at which point it becomes highly insensitive to extremes of temperature.

Whether living on their own or in symbiotic associations, fungi have spread successfully over much of Earth since their origin from a protist ancestor. That ancestor also gave rise to the choanoflagellates and the animal kingdom, as we will see in Chapter 32.

Chapter Summary

General Biology of the Fungi

▶ Fungi are heterotrophic eukaryotes with absorptive nutrition and with chitin in their cell walls. They may be saprobes, parasites, or mutualists.

▶ These four fungal phyla differ in their reproductive structures, mechanisms of spore formation, and less importantly, the presence and form of septa in their hyphae.

▶ The yeasts are unicellular fungi.

▶ The bodies of multicellular fungi are composed of multinucleate hyphae, often massed to form a mycelium. The hyphae usually have incomplete partitions (septa) that allow the movement of organelles between cells. They give fungi a large surface area-to-volume ratio, enhancing their ability to absorb nutrients. **Review Figures 31.3, 31.4**

▶ Fungi reproduce asexually by means of spores formed within sporangia, by conidia formed at the tips of hyphae, by fission or budding, or by fragmentation.

▶ Fungi reproduce sexually when hyphae of different mating types meet and fuse.

▶ In addition to the haploid and diploid states, many fungi demonstrate a third nuclear condition: the dikaryotic, or $n + n$, state.

Diversity in the Kingdom Fungi

▶ The kingdom Fungi consists of four phyla: Chytridiomycota, Zygomycota, Ascomycota, and Basidiomycota. **Review Figure 31.6, Table 31.1. See Web/CD Activity 31.1**

▶ The chytrids, with their flagellated zoospores and gametes, probably resemble the ancestral fungi.

▶ The zygomycetes reproduce sexually by fusion of gametangia. **Review Figure 31.9**

▶ The sexual reproductive structure of ascomycetes is an ascus containing ascospores. The ascomycetes are divided into two groups, euascomycetes and hemiascomycetes, on the basis of whether they have an ascocarp, or fruiting structure. **Review Figure 31.13. See Web/CD Activity 31.2**

▶ The sexual reproductive structure of basidiomycetes is a basidium, a swollen cell bearing basidiospores. **Review Figure 31.15**

▶ Imperfect fungi (deuteromycetes) lack sexual structures, but DNA sequencing can sometimes identify the phylum to which they belong.

See Web/CD Tutorial 31.1

Fungal Associations

▶ Mycorrhizae, which are symbiotic associations of a fungus with plant roots, enhance the ability of the roots to absorb water and nutrients. In return, the plant supplies the fungus with photosynthetic products.

▶ Lichens, which are symbiotic associations of a fungus with a green alga or a cyanobacterium, are found in some of the most inhospitable environments on the planet. **Review Figure 31.18**

Self-Quiz

1. Which statement about fungi is *not* true?
   a. A multicellular fungus has a body called a mycelium.
   b. Hyphae are composed of individual mycelia.
   c. Many fungi tolerate highly hypertonic environments.
   d. Many fungi tolerate low temperatures.
   e. Some fungi are anchored to their substrate by rhizoids.

2. The absorptive nutrition of fungi is aided by
   a. dikaryon formation.
   b. spore formation.
   c. the fact that they are all parasites.
   d. their large surface area-to-volume ratio.
   e. their possession of chloroplasts.

3. Which statement about fungal nutrition is *not* true?
   a. Some fungi are active predators.
   b. Some fungi form mutualistic associations with other organisms.
   c. All fungi require mineral nutrients.
   d. Fungi can make some of the compounds that are vitamins for animals.
   e. Facultative parasites can grow only on their specific hosts.

4. Which statement about dikaryosis is *not* true?
   a. The cytoplasm of two cells fuses before their nuclei fuse.
   b. The two haploid nuclei are genetically different.
   c. The two nuclei are of the same mating type.
   d. The dikaryon stage ends when the two nuclei fuse.
   e. Not all fungi have a dikaryon stage.

5. Reproductive structures consisting of one or more photosynthetic cells surrounded by fungal hyphae are called
   a. ascospores.
   b. basidiospores.
   c. conidia.
   d. soredia.
   e. gametes.

6. The zygomycetes
   a. have hyphae without regularly occurring septa.
   b. produce motile gametes.
   c. form fleshy fruiting bodies.
   d. are haploid throughout their life cycle.
   e. have sexual reproductive structures similar to those of the ascomycetes.

7. Which statement about ascomycetes is *not* true?
   a. They include yeasts.
   b. They form reproductive structures called asci.
   c. Their hyphae are segmented by septa.
   d. Many of their species have a dikaryotic state.
   e. All have fruiting structures called ascocarps.

8. The basidiomycetes
   a. often produce fleshy fruiting structures.
   b. have hyphae without septa.
   c. have no sexual stage.
   d. produce basidia within basidiospores.
   e. form diploid basidiospores.

9. The deuteromycetes
   a. have distinctive sexual stages.
   b. are all parasitic.
   c. have "lost" some members to other fungal groups.
   d. include the ascomycetes.
   e. are never components of lichens.

10. Which statement about lichens is *not* true?
    a. They can reproduce by fragmentation of the vegetative body.
    b. They are often the first colonists in a new area.
    c. They render their environment more basic (alkaline).
    d. They contribute to soil formation.
    e. They may contain less than 10 percent water by weight.

For Discussion

1. You are shown an object that looks superficially like a pale green mushroom. Describe at least three criteria (including anatomical and chemical traits) that would enable you to tell whether the object is a piece of a plant or a piece of a fungus.

2. Differentiate among the members of the following pairs of related terms:
   a. hypha/mycelium
   b. euascomycete/hemiascomycete
   c. ascus/basidium
   d. ectomycorrhiza/endomycorrhiza

3. For each type of organism listed below, give a single characteristic that may be used to differentiate it from the other, related organism(s) in parentheses.
   a. Zygomycota (Ascomycota)
   b. Basidiomycota (deuteromycetes)
   c. Ascomycota (Basidiomycota)
   d. baker's yeast (*Neurospora crassa*)

4. Many fungi are dikaryotic during part of their life cycle. Why are dikaryons described as $n + n$ instead of $2n$?

5. If all the fungi on Earth were suddenly to die, how would the surviving organisms be affected? Be thorough and specific in your answer.

6. How might the first mycorrhizae have arisen?

7. What might account for the ability of lichens to withstand the intensely cold environment of Antarctica? Be specific in your answer.

32 Animal Origins and the Evolution of Body Plans

In 1822, nearly forty years before Darwin wrote *The Origin of Species*, a French naturalist, Étienne Geoffroy Saint-Hilaire, was examining a lobster. He noticed that when he turned the lobster upside down and viewed it with its ventral surface up, its central nervous system was located above its digestive tract, which in turn was located above its heart—the same relative positions these systems have in mammals when viewed *dorsally*. His observations led Geoffroy to conclude that the differences between arthropods (such as lobsters) and vertebrates (such as mammals) could be explained if the embryos of one of those groups were inverted during development.

Geoffroy's suggestion was regarded as preposterous at the time and was largely dismissed until recently. However, the discovery of two genes that influence a system of extracellular signals involved in development has lent new support to Geoffroy's seemingly outrageous hypothesis.

A vertebrate gene called *chordin* helps to establish cells on one side of the embryo as dorsal and on the other as ventral. A probably homologous gene in fruit flies, called *sog*, acts in a similar manner, but has the opposite effect. Fly cells where *sog* is active become ventral, whereas vertebrate cells where *chordin* is active become dorsal. However, when *sog* mRNA is injected into an embryo of the frog *Xenopus*, a vertebrate, it causes dorsal development. *Chordin* mRNA injected into fruit flies promotes ventral development. In both cases, injection of the mRNA promotes the development of the portion of the embryo that contains the central nervous system!

Chordin and *sog* are among many genes that regulate similar functions in very different organisms. Such genes are providing evolutionary biologists with information that can help them understand relationships among animal lineages that separated from one another in ancient times. As we saw in Chapter 25, new knowledge about gene functions and gene sequences is increasingly being used to infer evolutionary relationships.

In this chapter, we will apply the methods described in Chapter 25 to infer evolutionary relationships among the animals. First, we will review the defining characteristics of the animal way of life. Then we will describe several lineages of simple animals. Finally, we will describe the lophotro-

Genes that Control Development A human and a lobster carry similar genes that control the development of the body axis, but these genes position their body systems inversely. A lobster's nervous system runs up its ventral (belly) surface, whereas a vertebrate's runs down its dorsal (back) surface.

chozoans, one of the three great evolutionary lineages of animals. In the next two chapters, we will discuss the other two great animal lineages, the ecdysozoans and the deuterostomes.

Animals: Descendants of a Common Ancestor

Biologists have long debated whether animals arose once or several times from protist ancestors, but enough molecular and morphological evidence has now been assembled to indicate that, with the possible exception of sponges (Porifera), the Kingdom Animalia is a monophyletic group—that is, all animals are descendants of a single ancestral lineage. This conclusion is supported by the fact that all animals share several derived traits:

▶ Similarities in their small-subunit ribosomal RNAs (see Chapter 26)
▶ Similarities in their Hox genes (see Chapter 20)
▶ Special types of cell–cell junctions: tight junctions, desmosomes, and gap junctions (see Figure 5.6)
▶ A common set of extracellular matrix molecules, including collagen (see Figure 4.26)

Animals evolved from colonial flagellated protists as a result of division of labor among their aggregated cells. Within the ancestral colonies of cells—perhaps analogous to those still existing in the chlorophyte *Volvox* or some colonial choanoflagellates (see Figures 28.25*a* and 28.28)—some cells became specialized for movement, others became specialized for nutrition, and still others differentiated into gametes. Once this specialization by function had begun, working groups of cells continued to differentiate while improving their coordination with other groups of cells. Such coordinated groups of cells evolved into the larger and more complex organisms that we now call **animals**.

Animals are multicellular heterotrophs

What traits characterize the animals? In contrast to the Bacteria, Archaea, and most protists, all animals are *multicellular*. Unlike plants, animals must take in pre-formed organic molecules because they cannot synthesize them from inorganic chemicals. They acquire these organic molecules by ingesting other organisms or their products, either living or dead, and digesting them inside their bodies; thus animals are *heterotrophs*. Most animals have *circulatory systems* that take up O_2, get rid of CO_2, and carry nutrients from their guts to other body tissues.

To acquire food, animals must expend energy either to move through the environment and position themselves where food will pass close to them, or to move the environment and the food it contains to them. The foods animals ingest include most other members of the animal kingdom as well as members of all other kingdoms. Much of the diversity of animal sizes and shapes evolved as animals acquired the ability to capture and eat many different kinds of food and to avoid becoming food for other animals. The need to locate food has favored the evolution of sensory structures to provide animals with detailed information about their environment and nervous systems to receive and coordinate that information.

The accounts in this chapter and the following two chapters serve as an orientation to the major groups of animals, their similarities and differences, and the evolutionary pathways that resulted in the current richness of animal lineages and species. But how do biologists infer evolutionary relationships among animals?

Several traits show evolutionary relationships among animals

Biologists use a variety of traits in their efforts to infer animal phylogenies. Clues to these relationships are found in the fossil record, in patterns of embryonic development, in the comparative morphology and physiology of living and fossil animals, and in the structure of animal molecules, especially small subunit rRNAs and mitochondrial genes (see Chapters 25 and 26).

Using this wide variety of comparative data, zoologists concluded that sponges, cnidarians, and ctenophores separated from the remaining animal lineages early in evolutionary history. Biologists have divided the remaining animals into two major lineages: the protostomes and the deuterostomes. Figure 32.1 shows the postulated order of divergence of the major animal groups that we will use in these three chapters.

Several differences in patterns of embryonic development provide clues to animal phylogeny. During the development of an animal from a single-celled zygote to a multicellular adult, distinct layers of cells form. The embryos of **diploblastic** animals have only two of these cell layers: an outer *ectoderm* and an inner *endoderm*. The embryos of **triploblastic** animals have, in addition to ectoderm and endoderm, a third layer, the *mesoderm*, which lies between the ectoderm and the endoderm. The existence of three cell layers distinguishes the protostomes and deuterostomes from those groups of simple animals that diverged from them earlier.

During early development in many animals, a cavity forms in a spherical embryo. The opening of this cavity is called the *blastopore*. Among the **protostomes** (from the Greek, "mouth first"), the mouth arises from the blastopore; the anus forms later. Among the **deuterostomes** ("mouth second"), the blastopore becomes the anus; the mouth forms later.

In the common ancestor of the protostomes and deuterostomes, the pattern of early cleavage of the fertilized egg was

32.1 A Current Phylogeny of the Animals The phylogenetic tree used in this and the following two chapters assumes that the animals are monophyletic. The characters highlighted by red circles on the tree will be explained as we discuss the different phyla.

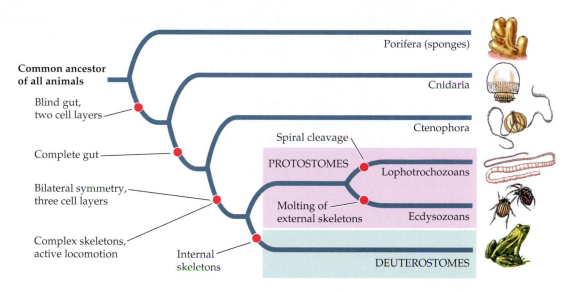

Common ancestor of all animals

Blind gut, two cell layers

Complete gut

Bilateral symmetry, three cell layers

Complex skeletons, active locomotion

Internal skeletons

Spiral cleavage

PROTOSTOMES

Molting of external skeletons

Porifera (sponges)

Cnidaria

Ctenophora

Lophotrochozoans

Ecdysozoans

DEUTEROSTOMES

radial (see Figure 20.4). This cleavage pattern persisted during the evolution of the deuterostomes and in many protostomate lineages, but spiral cleavage evolved in one major protostomate lineage, as we will see.

Body Plans: Basic Structural Designs

The general structure of an animal, its organ systems, and the integrated functioning of its parts are known as its **body plan**. A fundamental aspect of an animal's body plan is its overall shape, described in part by its **symmetry**. A symmetrical animal can be divided along at least one plane into similar halves. Animals that have no plane of symmetry are said to be **asymmetrical**. Many sponges are asymmetrical, but most animals have some kind of symmetry.

The simplest form of symmetry is **spherical symmetry**, in which body parts radiate out from a central point. An infinite number of planes passing through the central point can divide a spherically symmetrical organism into similar halves. Spherical symmetry is widespread among the protists, but most animals possess other forms of symmetry.

An organism with **radial symmetry** has one main axis around which its body parts are arranged. A perfectly radially symmetrical animal can be divided into similar halves by any plane that contains the main axis. Some simple sponges and a few other animals, such as sea anemones (Figure 32.2a), have radial symmetry. Most radially symmetrical animals are slightly modified so that fewer planes can divide them into identical halves. Two animal phyla—Cnidaria and Ctenophora—are composed primarily of radially symmetrical animals. These animals move slowly or not at all.

Bilateral symmetry is a common characteristic of animals that move rapidly through their environments. A bilaterally symmetrical animal can be divided into mirror images (left

and right sides) by a single plane that passes through the dorsoventral midline of its body from the front (*anterior*) to the back (*posterior*) end (Figure 32.2b). A plane at right angles to the first one divides the body into two dissimilar sides; the back side of a bilaterally symmetrical animal is its *dorsal* surface; the belly side is its *ventral* surface.

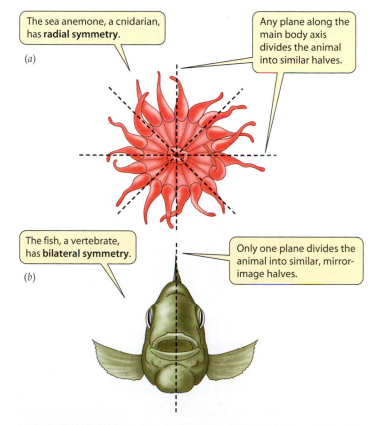

The sea anemone, a cnidarian, has **radial symmetry**.

(a)

Any plane along the main body axis divides the animal into similar halves.

The fish, a vertebrate, has **bilateral symmetry**.

(b)

Only one plane divides the animal into similar, mirror-image halves.

32.2 Body Symmetry Most animals are either radially or bilaterally symmetrical.

Bilateral symmetry is strongly correlated with **cephalization**: the concentration of sensory organs and nervous tissues in a head at the anterior end of the animal. Cephalization is favored because the anterior end of a freely moving animal typically encounters new environments first.

Fluid-filled spaces, called **body cavities**, lie between the ectoderm and endoderm of most protostomes and deuterostomes. The type of body cavity an animal has strongly influences the way it moves.

▶ Animals that lack an enclosed body cavity are called **acoelomates**. In these animals, the space between the gut and the body wall is filled with masses of cells called *mesenchyme* (Figure 32.3*a*).

▶ **Pseudocoelomate** animals have a body cavity called a *pseudocoel*, a liquid-filled space in which many of the internal organs are suspended. Their control over body shape is crude because the pseudocoel has muscles only on its outside; there is no inner layer of muscle surrounding the organs (Figure 32.3*b*).

▶ **Coelomate** animals have a *coelom*, a body cavity that develops within the mesoderm. It is lined with a special structure called the *peritoneum* and is enclosed on both the inside and the outside by muscles (Figure 32.3*c*).

The fluid-filled body cavities of simple animals function as **hydrostatic skeletons**. Because fluids are relatively incompressible, they move to another part of the cavity when the muscles surrounding them contract. If the body tissues around the cavity are flexible, fluids squeezed out of one region can cause some other region to expand. The moving fluids can thus move specific body parts. If a temporary attachment can be made to the substratum, the whole animals can move from one place to another.

In animals that have both circular muscles (encircling the body) and longitudinal muscles (running along the length of the body), the action of these antagonistic muscles on the fluid-filled body cavity gives the animal even greater control over its movement. A coelomate animal has better control over the movement of the fluids in its body cavity than does a pseudocoelomate animal, but its control is further improved if the coelom is separated into compartments or segments. Then muscles in each individual segment can change its shape independently of the other segments. Segmentation of the coelom evolved several different times among both protostomes and deuterostomes.

Other forms of skeletons developed in many animal lineages, either as substitutes for, or in combination with, hydrostatic skeletons. Some skeletons are internal (such as vertebrate bones); others are external (such as lobster shells). Some external skeletons consist of a single element (snail shells), others have two elements (clam shells), and still others have many elements (centipedes).

Sponges: Loosely Organized Animals

The lineage leading to modern sponges separated from the lineage leading to all other animals very early during animal evolution. The difference between protist colonies and simple multicellular animals is that the animal cells are differentiated and their activities are coordinated. However, sponge cells do not form true organs.

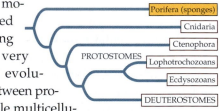

32.3 Animal Body Cavities There are three major types of body cavities among the animals. (*a*) Acoelomates do not have enclosed body cavities. (*b*) Pseudocoelomates have only one layer of muscle, and it lies outside the body cavity. (*c*) Coelomates have a peritoneum surrounding the internal organs. The body cavities of some coelomates, such as this earthworm, are segmented.

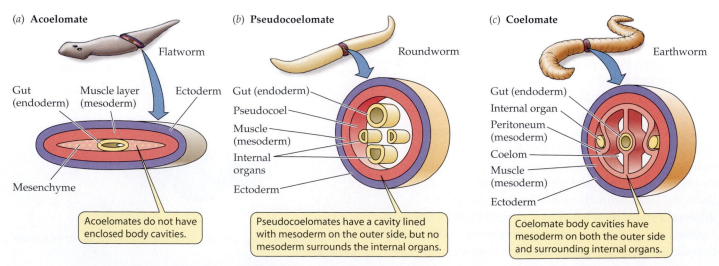

(*a*) **Acoelomate**
Flatworm
Gut (endoderm)   Muscle layer (mesoderm)   Ectoderm
Mesenchyme

Acoelomates do not have enclosed body cavities.

(*b*) **Pseudocoelomate**
Roundworm
Gut (endoderm)
Pseudocoel
Muscle (mesoderm)
Internal organs
Ectoderm

Pseudocoelomates have a cavity lined with mesoderm on the outer side, but no mesoderm surrounds the internal organs.

(*c*) **Coelomate**
Earthworm
Gut (endoderm)
Internal organ
Peritoneum (mesoderm)
Coelom
Muscle (mesoderm)
Ectoderm

Coelomate body cavities have mesoderm on both the outer side and surrounding internal organs.

Water, carrying food particles, enters through many small pores.

Water flows through the sponge's body and exits through a large opening, the osculum.

Pore

Osculum

Spicules

Choanocyte   Spicules

The inner wall is studded with specialized feeding cells called choanocytes.

32.4 The Body Plan of a Simple Sponge The flow of water through the sponge is shown by blue arrows. The inner body wall is studded with choanocytes, a type of specialized feeding cell that may be a link between animals and colonial protists (see Figure 28.28).

semblance to choanoflagellates (see Figure 28.28). By beating their flagella, choanocytes cause the surrounding water to flow through the animal. The water, along with any food particles it contains, enters by way of small pores and passes into the water canals, where food particles are captured by the choanocytes. Water then exits through one or more larger openings called *oscula* (Figure 32.4).

Between the thin epidermis and the choanocytes is another layer of cells, some of which are similar to amoebas and move about within the body. A supporting skeleton is also present, in the form of simple or branching spines, called *spicules*, and often an elastic, complex, network of fibers. A few species of sponges are carnivores that trap prey on hook-shaped spicules that protrude from the body surface. Sponges also have an extracellular matrix, composed of collagens, adhesive glycoproteins, and other molecules, that holds the cells together. This molecular adhesion system may also be involved in cell–cell signaling.

Thus, sponges are functionally more complex than a superficial look at their morphology might suggest. Nonetheless, sponges are loosely organized. Even if a sponge is completely disassociated by being strained through a filter, its cells can reassemble into a new sponge.

Most of the 5,500 species of sponges are marine animals; only about 50 species live in fresh water. Sponges come in a wide variety of sizes and shapes that are adapted to different movement patterns of water (Figure 32.5). Sponges living in intertidal or shallow subtidal environments, where they are subjected to strong wave action, hug the substratum. Many sponges that live in calm waters are simple, with a single large osculum on top of the body. Most sponges that live in slowly flowing water are flattened and are oriented at right

Sponges (phylum **Porifera**, from the Latin, "pore bearers"), the simplest of animals, are **sessile:** They live attached to the substratum and do not move about. The body plan of all sponges—even large ones, which may reach more than a meter in length—is an aggregation of cells built around a water canal system. Feeding cells called *choanocytes* line the inside of the internal chambers. These cells, with a collar of microscopic villi and a single flagellum, bear a striking re-

(a) *Euplectella aspergillum*

(b) *Aplysina lacunosa*

(c) *Asbestopluma* sp.

32.5 Sponges Differ in Size and Shape (a) Glass sponges are named after their glasslike spicules, which are formed of silicon. (b) The purple tube sponge is typical of many simple marine sponges. (c) This predatory sponge uses its hook-shaped spicules to capture small prey animals.

angles to the direction of current flow; they intercept water and the prey it contains as it flows past them.

Sponges reproduce both sexually and asexually. In most species, a single individual produces both eggs and sperm, but individuals do not self-fertilize. Water currents carry sperm from one individual to another. Asexual reproduction is by budding and fragmentation.

Cnidarians: Two Cell Layers and Blind Guts

Animals in all phyla other than Porifera have distinct cell layers and symmetrical bodies. The next lineage to diverge from the main line of animal evolution after the sponges led to a phylum of

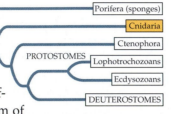

animals called the **cnidarians** (phylum **Cnidaria**). These animals are *diploblastic* (have two cell layers) and have a blind gut with only one entrance (the mouth/anus). Despite their relative structural simplicity, cnidarians have structural molecules (such as collagen, actin, and myosin) and homeobox genes.

Cnidarians are simple but specialized carnivores

Cnidarians appeared early in evolutionary history and radiated in the late Precambrian. About 11,000 cnidarian species—jellyfish, sea anemones, corals, and hydrozoans—live today (Figure 32.6), all but a few in the oceans. The smallest cnidarians can hardly be seen without a microscope; the largest known jellyfish is 2.5 meters in diameter. All cnidarians are carnivores; some gain additional nutrition from photosynthetic endosymbionts. The cnidarian body plan combines a low metabolic rate with the ability to capture large prey. These traits allow cnidarians to survive in environments where encounters with prey are infrequent.

All cnidarians possess tentacles covered with *cnidocytes*, specialized cells that contain stinging organelles called *nematocysts*, which can inject toxins into their prey (Figure 32.7). Cnidocytes allow cnidarians to capture large and complex prey, which are carried into the mouth by retracting the tentacles. Nematocysts are responsible for the stings that some jellyfish inflict on human swimmers.

The cnidarian body is based on a "sac plan," in which the mouth is connected to a blind sac called the *gastrovascular cavity*. The sac functions in digestion, circulation, and gas exchange and acts as a hydrostatic skeleton. The single opening serves as both mouth and anus. Cnidarians also have epithelial cells with muscle fibers whose contractions enable the animals to move, as well as simple *nerve nets* that integrate their body activities.

(a) Anthopleura elegantissima

(b) Ptilosarcus gurneyi

(c) Pelagia panopyra

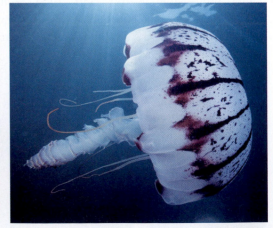

(d) Polyorchis penicillatus

32.6 Diversity among Cnidarians (*a*) The nematocyst-studded tentacles of this sea anemone from British Columbia are poised to capture large prey carried to the animal by water movement. (*b*) The orange sea pen is a colonial cnidarian that lives in soft bottom sediments and projects polyps above the substratum. (*c*) This purple jellyfish illustrates the complexity of a scyphozoan medusa. (*d*) The internal structure of the medusa of a North Atlantic colonial hydrozoan is visible here.

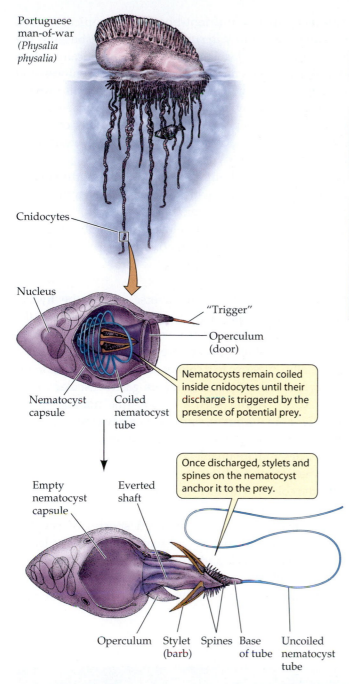

Portuguese
man-of-war
(*Physalia
physalia*)

Cnidocytes

Nucleus

"Trigger"

Operculum
(door)

Nematocysts remain coiled inside cnidocytes until their discharge is triggered by the presence of potential prey.

Nematocyst
capsule

Coiled
nematocyst
tube

Empty
nematocyst
capsule

Everted
shaft

Once discharged, stylets and spines on the nematocyst anchor it to the prey.

Operculum   Stylet   Spines   Base   Uncoiled
(barb)   of tube   nematocyst
tube

32.7 Nematocysts Are Potent Weapons Cnidarians such as the Portuguese man-of-war, which possesses a large number of nematocysts, can subdue and consume very large prey.

Cnidarian life cycles have two stages

The generalized cnidarian life cycle has two distinct stages (Figure 32.8), although many species lack one of these stages:

▶ The sessile **polyp** stage has a cylindrical stalk attached to the substratum. Tentacles surround a mouth/anus located at the end opposite from the stalk. Individual polyps may reproduce by budding, thereby forming a colony.

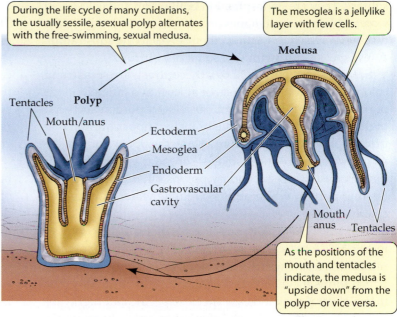

During the life cycle of many cnidarians, the usually sessile, asexual polyp alternates with the free-swimming, sexual medusa.

The mesoglea is a jellylike layer with few cells.

Medusa

Tentacles   **Polyp**

Mouth/anus

Ectoderm

Mesoglea

Endoderm

Gastrovascular
cavity

Mouth/
anus   Tentacles

As the positions of the mouth and tentacles indicate, the medusa is "upside down" from the polyp—or vice versa.

32.8 A Generalized Cnidarian Life Cycle Cnidarians typically have two body forms, one asexual (the polyp) and the other sexual (the medusa).

▶ The **medusa** (plural, medusae) is a free-swimming stage shaped like a bell or an umbrella. It typically floats with its mouth and tentacles facing downward. Medusae of many species produce eggs and sperm and release them into the water. When an egg is fertilized, it develops into a free-swimming, ciliated larva called a **planula**, which eventually settles to the bottom and develops into a polyp.

Although the polyp and medusa stages appear very different, they share a similar body plan. A medusa is essentially a polyp without a stalk. Most of the outward differences between polyps and medusae are due to the *mesoglea*, an internal mass of jellylike material that lies between the two cell layers. The mesoglea contains few cells and has a low metabolic rate. In polyps, the mesoglea is usually thin; in medusae it is very thick, constituting the bulk of the animal.

ANTHOZOANS. All 6,000 species of sea anemones and corals that constitute the **anthozoans** (class **Anthozoa**) are marine animals. Evidence from morphology, rRNA, and mitochondrial genes suggests that the anthozoans, which lack the medusa stage, are sister to the other classes of cnidarians, and that the medusa stage evolved after the anthozoa diverged from those other lineages. In the anthozoans, the polyp produces eggs and sperm, and the fertilized egg develops into a planula that develops directly into another polyp. Many species can also reproduce asexually by budding or fission. Sea anemones (see Figure 32.6*a*) are solitary. They are widespread in both warm and cold ocean waters. Many sea anemones are able to crawl slowly on the discs with which they attach themselves to the substratum. A few species can swim and some can burrow.

Sea pens (see Figure 32.6b), by contrast, are sessile and colonial. Each colony consists of at least two different kinds of polyps. The primary polyp has a lower portion anchored in the bottom sediment and a branched upper portion that projects above the substratum. Along the upper portion, the primary polyp produces smaller secondary polyps by budding. Some of these secondary polyps can differentiate into feeding polyps, while others circulate water through the colony.

Corals also are usually sessile and colonial. The polyps of most corals form a skeleton by secreting a matrix of organic molecules upon which they deposit calcium carbonate, which forms the eventual skeleton of the coral colony. The forms of coral skeletons are species-specific and highly diverse. The common names of coral groups—horn corals, brain corals, staghorn corals, and organ pipe corals, among others—describe their appearance (Figure 32.9a).

As a coral colony grows, old polyps die, but their calcareous skeletons remain. The living members form a layer on top of a growing bank of skeletal remains, eventually forming chains of islands and reefs (Figure 32.9b). The Great Barrier Reef along the northeastern coast of Australia is a system of coral formations more than 2,000 km long and as wide as 150 km. A reef hundreds of kilometers long in the Red Sea has been calculated to contain more material than all the buildings in the major cities of North America combined.

Corals flourish in nutrient-poor, clear, tropical waters. They can grow rapidly in such environments because the photosynthetic dinoflagellates that live symbiotically within their cells provide them with products of photosynthesis and contribute to calcium deposition. In turn, the corals provide the dinoflagellates with a place to live and nutrients. This symbiotic relationship explains why reef-forming corals are restricted to clear surface waters, where light levels are high enough to allow photosynthesis.

Coral reefs throughout the world are being threatened both by global warming, which is raising the temperatures of shallow tropical ocean waters, and by polluted runoff from development on adjacent shorelines. An overabundance of nitrogen in the runoff gives an advantage to algae, which overgrow and eventually smother the corals.

HYDROZOANS. Life cycles are diverse among the **hydrozoans** (class **Hydrozoa**). The polyp typically dominates the life cycle, but some species have only medusae and others only polyps. Most hydrozoans are colonial. A single planula eventually gives rise to a colony of many polyps, all interconnected and sharing a continuous gastrovascular cavity (Figure 32.10). Within such a colony (the man-of-war in Figure 32.7 is an example), some polyps have tentacles with many nematocysts; they capture prey for the colony. Others lack tentacles and are unable to feed, but are specialized for the production of medusae. Still others are fingerlike and defend the colony with their nematocysts.

SCYPHOZOANS. The several hundred species of **scyphozoans** (class **Scyphozoa**) are all marine. The mesoglea of their medusae is thick and firm, giving rise to their common names, jellyfish or sea jellies. The medusa, rather than the polyp, dominates the life cycle of scyphozoans. An indi-

32.9 Corals The South Pacific is home to many spectacular corals. (a) This unusually large formation of chalice coral was photographed off the coast of Fiji. (b) Many different species of corals and sponges grow together on this reef in Palau.

(a) *Montipora sp.*

(b)

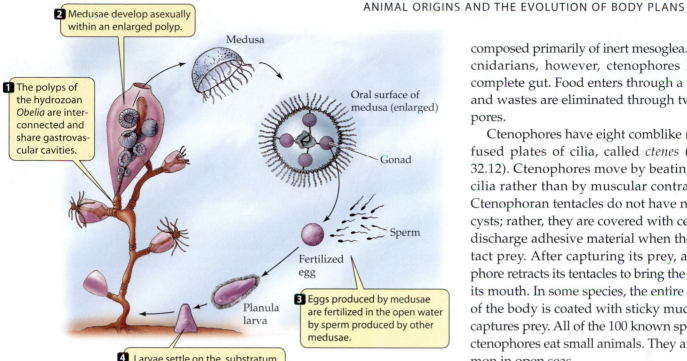

2 Medusae develop asexually within an enlarged polyp.

1 The polyps of the hydrozoan *Obelia* are interconnected and share gastrovascular cavities.

Medusa

Oral surface of medusa (enlarged)

Gonad

Sperm

Fertilized egg

3 Eggs produced by medusae are fertilized in the open water by sperm produced by other medusae.

Planula larva

4 Larvae settle on the substratum.

32.10 Hydrozoans Often Have Colonial Polyps The polyps within a hydrozoan colony may differentiate to perform specialized tasks. In the species whose life cycle is diagrammed here, the medusa is the sexual reproductive stage, producing eggs and sperm in organs called gonads.

vidual medusa is male or female, releasing eggs or sperm into the open sea. The fertilized egg develops into a small planula that quickly settles on a substratum and develops into a small polyp. This polyp feeds and grows and may produce additional polyps by budding. After a period of growth, the polyp begins to bud off small medusae (Figure 32.11). These medusae feed, grow, and transform themselves into adult medusae, which are commonly seen during summer in harbors and bays.

Ctenophores: Complete Guts and Tentacles

Ctenophores (phylum **Ctenophora**) were the next lineage to diverge from the lineage leading to all other animals. Ctenophores, also known as comb jellies, have body plans that are superficially similar to those of cnidarians. Both have two cell layers separated by a thick, gelatinous mesoglea, and both have radial symmetry and feeding tentacles. Like cnidarians, ctenophores have low metabolic rates because they are

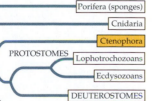

Porifera (sponges)
Cnidaria
Ctenophora
PROTOSTOMES — Lophotrochozoans
Ecdysozoans
DEUTEROSTOMES

composed primarily of inert mesoglea. Unlike cnidarians, however, ctenophores have a complete gut. Food enters through a mouth, and wastes are eliminated through two anal pores.

Ctenophores have eight comblike rows of fused plates of cilia, called *ctenes* (Figure 32.12). Ctenophores move by beating these cilia rather than by muscular contractions. Ctenophoran tentacles do not have nematocysts; rather, they are covered with cells that discharge adhesive material when they contact prey. After capturing its prey, a ctenophore retracts its tentacles to bring the food to its mouth. In some species, the entire surface of the body is coated with sticky mucus that captures prey. All of the 100 known species of ctenophores eat small animals. They are common in open seas.

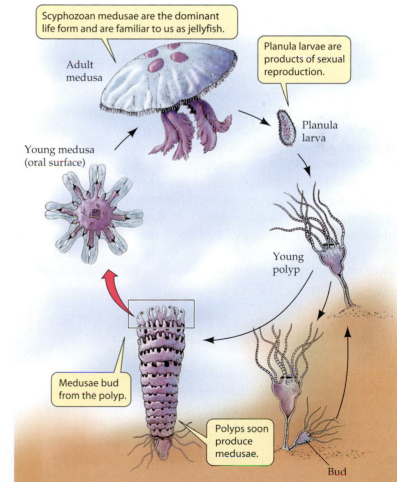

Scyphozoan medusae are the dominant life form and are familiar to us as jellyfish.

Planula larvae are products of sexual reproduction.

Adult medusa

Planula larva

Young medusa (oral surface)

Young polyp

Medusae bud from the polyp.

Polyps soon produce medusae.

Bud

32.11 Medusae Dominate Scyphozoan Life Cycles Scyphozoan medusae are the familiar jellyfish of coastal waters. The small, sessile polyps quickly produce medusae (see Figure 32.6c).

32.12 Comb Jellies Feed with Tentacles (a) The body plan of a typical ctenophore. The long, sticky tentacles sweep through the water, efficiently harvesting small prey. (b) A comb jelly photographed in Sydney Harbour, Australia, has short tentacles.

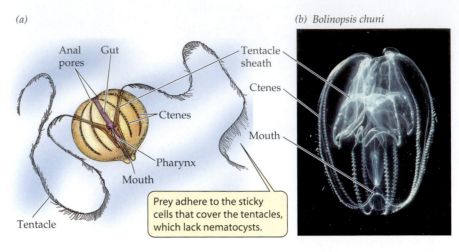

(a)

Anal pores
Gut
Tentacle sheath
Ctenes
Ctenes
Mouth
Pharynx
Mouth
Tentacle

(b) *Bolinopsis chuni*

Prey adhere to the sticky cells that cover the tentacles, which lack nematocysts.

Ctenophore life cycles are simple. Gametes are produced in structures called *gonads*, located on the walls of the gastrovascular cavity. The gametes are released into the cavity and then discharged through the mouth or the anal pores. Fertilization takes place in open seawater. In nearly all species, the fertilized egg develops directly into a miniature ctenophore that gradually grows into an adult.

The Evolution of Bilaterally Symmetrical Animals

The phylogenetic tree pictured in Figure 32.1 assumes that all bilaterally symmetrical animals share a common ancestor, but it does not tell us what that ancestor looked like. To infer the form of the earliest bilaterians, zoologists use evidence from the genes, development, and structure of existing animals. An important clue is provided by the fact that the development of all bilaterally symmetrical animals is controlled by homologous *Hox* and homeobox genes. Regulatory genes with similar functions are unlikely to have evolved independently in several different animal lineages.

Fossilized tracks from late Precambrian times provide additional clues to the nature of early bilaterians (Figure 32.13). The complexity of the movements recorded by these tracks suggests that early bilaterians had circulatory systems, systems of antagonistic muscles, and a tissue- or fluid-filled body cavity, structures that are also suggested by genetic data.

An early lineage split separated protostomes and deuterostomes

The next major split in the animal lineage after the divergence of the ctenophores occurred during the Cambrian period and separated two groups that have been evolving separately ever since. These two major lineages—the protostomes and the deuterostomes—dominate today's fauna. Members of both lineages are triploblastic (have three cell layers), bilaterally symmetrical, and cephalized. Because their skeletons and body cavities are more complex than those of the animals we have discussed so far, they are capable of more complex movements.

The most important shared, derived traits that unite the protostomes are

▶ An anterior brain that surrounds the entrance to the digestive tract
▶ A ventral nervous system consisting of paired or fused longitudinal nerve cords
▶ A free-floating larva with a food-collecting system consisting of compound cilia on multiciliate cells
▶ A blastopore that becomes the mouth
▶ Spiral cleavage (in some species)

32.13 The Trail of an Early Bilaterian These fossilized tracks indicate that their maker was able to crawl.

The major shared, derived traits that unite the deuterostomes are

▶ A dorsal nervous system
▶ A larva, if present, that has a food-collecting system consisting of cells with a single cilium
▶ A blastopore that becomes the anus
▶ Radial cleavage

The protostomes split into two lineages

Developmental, structural, and molecular data all suggest that the protostomes soon split into two major lineages that have been evolving independently since ancient times: lophotrochozoans and

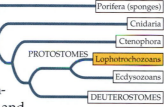

ecdysozoans. **Lophotrochozoans**, the animals we will discuss in the remainder of this chapter, grow by adding to the size of their skeletal elements. Some of them use cilia for locomotion, and many lineages have a type of free-living larva known as a **trochophore** (see Figure 32.23) The phylogeny of lophotrochozoans we will use in this chapter is shown in Figure 32.14. In contrast, **ecdysozoans**, the animals we will discuss in the next chapter, increase in size by molting their external skeletons. They move by mechanisms other than ciliary action, and they all have a common set of homeobox genes.

32.14 A Current Phylogeny of Lophotrochozoans   Three major lineages, including the lophophorate and spiralian phyla, dominate the tree. Some small phyla are not included here.

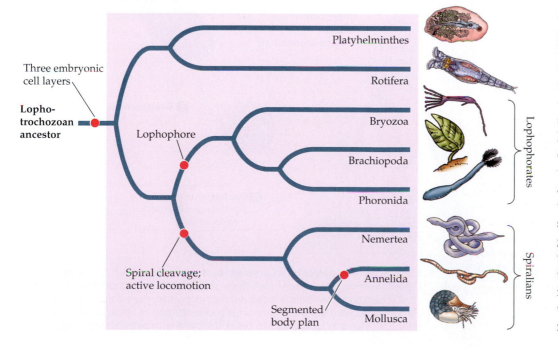

Simple Lophotrochozoans

The simplest lophotrochozoans—flatworms and rotifers—are small aquatic or parasitic animals. They move by rapidly beating cilia, and most have only simple organs.

Flatworms move by beating cilia

Members of the phylum **Platyhelminthes**, or **flatworms**, the simplest lophotrochozoans (Figure 32.15), are bilaterally symmetrical, unsegmented, acoelomate animals. They lack organs for transporting oxygen to internal tissues, and they have only simple cells for excreting metabolic wastes. Their lack of transport systems dictates that each cell must be near a body surface, a requirement met by their dorsoventrally flattened body form.

The digestive tract of a flatworm consists of a mouth opening into a blind sac. However, the sac is often highly branched, forming intricate patterns that increase the surface area available for the absorption of nutrients. Flatworms either feed on animal tissues (living or dead), or absorb nutrients from a host's gut. Free-living flatworms glide over surfaces, powered by broad bands of cilia. This form of movement is very slow, but it is sufficient for small, scavenging animals.

The flatworms that are probably most similar to the ancestral bilaterians are the turbellarians (class **Turbellaria**), which are small, free-living marine and freshwater animals (a few live in moist terrestrial habitats). At one end they have a head with chemoreceptor organs, two simple eyes, and a tiny brain composed of anterior thickenings of the longitudinal nerve cords.

Although the earliest flatworms were free-living (Figure 32.15a), many species evolved a parasitic existence. A likely evolutionary transition was from feeding on dead organisms to feeding on the body surfaces of dying hosts to invading and consuming parts of living, healthy hosts. Most of the 25,000 species of living flatworms—including the tapeworms (class **Cestoda**) and flukes (class **Trematoda**; Figure 32.15b)—are internal parasites. These flatworms absorb digested food from the digestive tracts of their hosts, so many of them lack digestive tracts. They inhabit the bodies of many vertebrates; some cause serious human diseases, such as schistoso-

32.15 Flatworms Live Freely and Parasitically (*a*) Some flatworm species are free-living, like this marine flatworm photographed in the oceans off Sulawesi, Indonesia. (*b*) The flatworm diagrammed here, which lives parasitically in the gut of sea urchins, is representative of parasitic flukes. Because their hosts provide all the nutrition they need, these intestinal parasites do not require elaborate feeding or digestive organs and can devote most of their bodies to reproduction.

The flatworm's body is filled primarily with sex organs.

(*b*) Anterior

Pharyngeal opening
Intestine
Egg capsule
Testis
Yolk gland
Seminal receptacle
Ovary
Vagina

Posterior

The flatworm gut has a single exterior opening. The pharyngeal opening serves as both "mouth" and "anus."

(*a*) *Pseudoceros bifurcus*

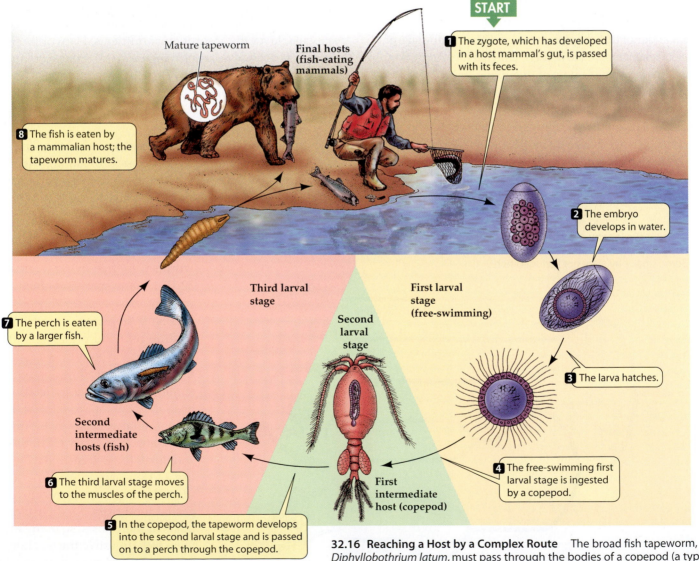

START

1 The zygote, which has developed in a host mammal's gut, is passed with its feces.

2 The embryo develops in water.

3 The larva hatches.

4 The free-swimming first larval stage is ingested by a copepod.

5 In the copepod, the tapeworm develops into the second larval stage and is passed on to a perch through the copepod.

6 The third larval stage moves to the muscles of the perch.

7 The perch is eaten by a larger fish.

8 The fish is eaten by a mammalian host; the tapeworm matures.

Mature tapeworm
Final hosts (fish-eating mammals)
Third larval stage
First larval stage (free-swimming)
Second larval stage
Second intermediate hosts (fish)
First intermediate host (copepod)

32.16 Reaching a Host by a Complex Route The broad fish tapeworm, *Diphyllobothrium latum*, must pass through the bodies of a copepod (a type of crustacean) and a fish before it can reinfect its primary host, a mammal. Such complex life cycles assist the flatworm's recolonization of hosts, but they also offer opportunities for humans to break the cycle with hygienic measures.

miasis. Monogeneans (class **Monogenea**) are external parasites of fishes and other aquatic vertebrates.

Parasites live in nutrient-rich environments where food is delivered to them, but they face other challenges. To complete their life cycle, parasites must overcome the defenses of their host. And because they die when their host dies, they must disperse their offspring to new hosts while their host is still living. The fertilized eggs of some parasitic flatworms are voided with the host's feces and later ingested directly by other host individuals. However, most parasitic species have complex life cycles involving one or more intermediate hosts and several larval stages (Figure 32.16). Such life cycles facilitate the transfer of individual parasites among hosts.

Rotifers are small but structurally complex

Rotifers (phylum **Rotifera**) are bilaterally symmetrical, unsegmented, pseudocoelomate lophotrochozoans. Most rotifers are tiny (50–500 μm long)—smaller than some ciliate protists—but they have highly developed internal organs (Figure 32.17). A complete gut passes from an anterior mouth to a posterior anus; the pseudocoel functions as a hydrostatic skeleton. Most rotifers propel themselves through the water by means of rapidly beating cilia rather than by muscular contraction. This type of movement is effective because rotifers are so small.

The most distinctive organs of rotifers are those they use to collect and process food. A conspicuous ciliated organ called the *corona* surmounts the head of many species. Coordinated beating of the cilia sweeps particles of organic matter from the water into the animal's mouth and down to a complicated structure called the *mastax*, in which food is ground into small pieces. By contracting the muscles around the pseudocoel, a few rotifer species that prey on protists and small animals can protrude the mastax through the mouth and seize small objects with it. Males and females are found in most species, but some species have only females that produce diploid eggs without being fertilized by a male.

Some rotifers are marine, but most of the 1,800 known species live in fresh water. Members of a few species rest on the surface of mosses or lichens in a desiccated, inactive state until it rains. When rain falls, they absorb water and become mobile, feeding in the films of water that temporarily cover the plants. Most rotifers live no longer than 1 or 2 weeks.

Lophophorates: An Ancient Body Plan

After the platyhelminthes and rotifers diverged from it, the lophotrochozoan lineage divided into two branches. The descendants of those branches became the modern **lophophorates**—the subject of this section—and the **spiralians**, which we will discuss in the following section.

About 4,850 living species of lophophorates are known, but many times that number of species existed during the Paleozoic and Mesozoic eras. Three lophophorate phyla survive today: Phoronida, Brachiopoda, and Ectoprocta. Nearly all members of these phyla are marine; only a few species of ectoprocts live in fresh water.

Lophophorate animals obtain food by filtering it from the surrounding water, a trait they share with many other protostomes. The most conspicuous feature of these animals is the **lophophore**, a circular or U-shaped ridge around the mouth that bears one or two rows of ciliated, hollow tentacles (Figure 32.18). This large and complex structure is an organ for both food collection and gas exchange. Nearly all adult lophophorate animals are sessile, and they use the tentacles and cilia of their lophophore to capture *plankton* (small floating organisms) from the water. Lophophorates also have a U-shaped gut; the anus is located close to the mouth, but outside the tentacles.

(a) Philadeina roseola

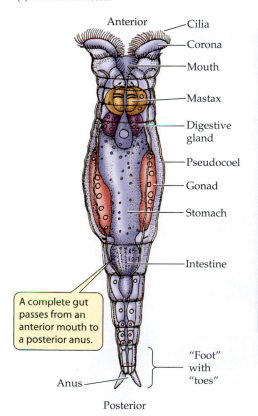

Anterior — Cilia
— Corona
— Mouth
— Mastax
— Digestive gland
— Pseudocoel
— Gonad
— Stomach
— Intestine

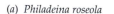
A complete gut passes from an anterior mouth to a posterior anus.

"Foot" with "toes"

Anus —

Posterior

(b) Stephanoceros fimbriatus

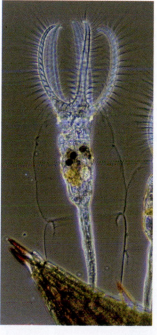

32.17 Rotifers (*a*) The rotifer diagrammed here reflects the general structure of many free-living species in the phylum. (*b*) A micrograph reveals the internal complexity of these living rotifers.

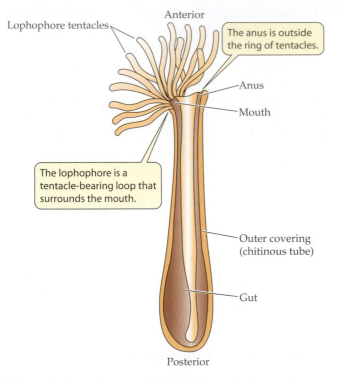

Anterior

Lophophore tentacles

The anus is outside the ring of tentacles.

Anus

Mouth

The lophophore is a tentacle-bearing loop that surrounds the mouth.

Outer covering (chitinous tube)

Gut

Posterior

32.18 Lophophore Artistry The lophophore dominates the anatomy of this phoronid. The phoronid gut is U-shaped.

Phoronids are sedentary lophophorates

The 20 known species of **phoronids** (phylum **Phoronida**) are sedentary worms that live in muddy or sandy sediments or attached to a rocky substratum. Phoronids are found in marine waters ranging from intertidal zones to about 400 meters deep. They range in size from 5 to 25 cm in length. They secrete chitinous tubes, in which they live (Figure 32.18). The lophophore is the most conspicuous external feature of the phoronids. Cilia drive water into the top of the lophophore, and water exits through the narrow spaces between the tentacles. Suspended food particles are caught and transported to the mouth by ciliary action. In most species, eggs are released into the water, where they are fertilized, but some species produce large eggs that are fertilized internally, where they are brooded until they hatch.

Ectoprocts are colonial lophophorates

Ectoprocts (phylum **Ectoprocta**) are colonial lophophorates that live in a "house" made of material secreted by the external body wall. A colony consists of many small (1–2 mm) individuals connected by strands of tissue along which materials can be moved (Figure 32.19a). Most of the 4,500 species of ectoprocts are marine, but a few live in fresh water. They are able to oscillate and rotate the lophophore to increase contact with prey (Figure 32.19b) and can retract it into the tube.

A colony of ectoprocts is created by the asexual reproduction of its founding members. A single colony may contain as many as 2 million individuals. In some species, individual colony members are specialized for feeding, reproduction, defense, or support. Ectoprocts reproduce sexually by releasing sperm into the water, where they are collected by other individuals. Eggs are fertilized internally, and developing embryos are brooded before they exit as larvae to seek suitable sites for attachment to the substratum.

Brachiopods superficially resemble bivalve mollusks

Brachiopods (phylum **Brachiopoda**) are solitary marine lophophorate animals. Their shells are divided into two parts

(a) *Lophopus crystallinus*

32.19 Ectoprocts (a) Branching colonies of ectoprocts may appear plantlike. (b) Ectoprocts have greater control over the movement of their lophophores than members of other lophophorate phyla.

(b)

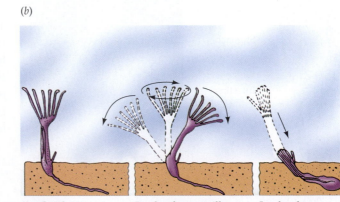

Lophophore spreads

Lophophore oscillates and rotates

Lophophore retracts

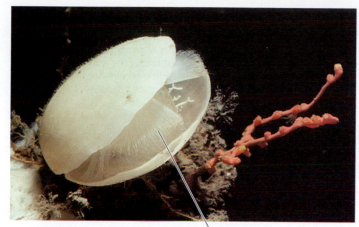

Laqueus sp.   Lophophore

32.20 Brachiopods   The lophophore of this North Pacific brachiopod can be seen between the valves of its shell.

that are connected by a ligament (Figure 32.20). The two halves can be pulled shut to protect the soft body. Brachiopods superficially resemble bivalve mollusks, but the brachiopod shell differs from that of mollusks in that the two halves are dorsal and ventral rather than lateral. The two-armed lophophore of a brachiopod is located within the shell. The beating of cilia on the lophophore draws water into the slightly opened shell. Food is trapped in the lophophore and directed to a ridge, along which it is transferred to the mouth. Most brachiopods are between 4 and 6 cm long, but some are as long as 9 cm.

Brachiopods live attached to a solid substratum or embedded in soft sediments. Most species are attached by means of a short, flexible stalk that holds the animal above the substratum. Gases are exchanged across body surfaces, especially the tentacles of the lophophore. Most brachiopods release their gametes into the water, where they are fertilized. The larvae remain among the plankton for only a few days before they settle and develop into adults.

Brachiopods reached their peak abundance and diversity in Paleozoic and Mesozoic times. More than 26,000 fossil species have been described. Only about 335 species survive, but they are common in some marine environments.

Spiralians: Spiral Cleavage and Wormlike Body Plans

The spiralian lineage, containing animals that typically have spiral cleavage patterns, gave rise to many phyla. Members of more than a dozen of these phyla are *wormlike*; that is, they are bilaterally symmetrical, legless, soft-bodied, and at least several times longer than they are wide. This body form enables animals to move efficiently through muddy and sandy marine sediments. Most of these phyla have no more than several hundred species. The most species-rich spiralian

phylum, the mollusks, shows significant modifications of the wormlike body plan.

Ribbon worms are unsegmented

The carnivorous **ribbon worms** (phylum **Nemertea**) are dorsoventrally flattened. They have nervous and excretory systems similar to those of flatworms, but unlike flatworms, they have a complete digestive tract with a mouth at one end and an anus at the other. Food moves in one direction through the digestive tract and is acted on by a series of digestive enzymes. Small ribbon worms move by beating their cilia. Larger ones employ waves of muscle contraction to move over the surface of sediments or to burrow. Movement by both of these methods is slow.

Within the body of nearly all of the 900 species of ribbon worms is a fluid-filled cavity called the *rhynchocoel*, within which lies a hollow, muscular *proboscis*. The proboscis, which is the feeding organ, may extend much of the length of the worm. Contraction of the muscles surrounding the rhynchocoel causes the proboscis to be everted explosively through an anterior opening (Figure 32.21) without moving the rest of the animal. The proboscis of most ribbon worms

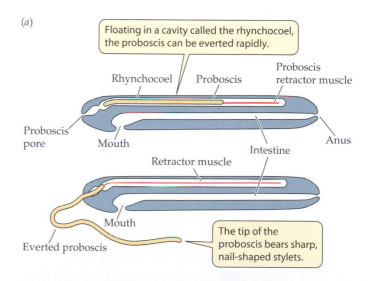

(a)

Floating in a cavity called the rhynchocoel, the proboscis can be everted rapidly.

Rhynchocoel   Proboscis   Proboscis retractor muscle

Proboscis pore   Mouth   Intestine   Anus

Retractor muscle

Mouth

Everted proboscis

The tip of the proboscis bears sharp, nail-shaped stylets.

(b) *Pelagonemertes* sp.

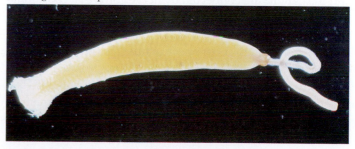

32.21 Ribbon Worms   (a) The proboscis is the ribbon worm's feeding organ. (b) This deep-water nemertean displays an everted proboscis.

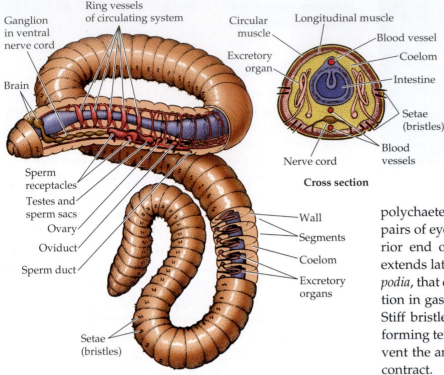

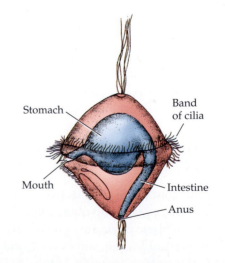

32.22 Annelids Have Many Body Segments The segmented structure of the annelids is apparent both externally and internally. Most organs of this earthworm are repeated serially.

Cross section

is armed with a sharp stylet that pierces the prey. Paralysis-causing toxins produced by the proboscis are discharged into the wound. Reproduction and development in ribbon worms is highly varied.

Segmentation improved locomotion in the annelids

Segmentation allows an animal to alter the shape of its body in complex ways and to control its movements more precisely. Fossils of segmented worms are known from the middle Cambrian; the earliest forms are thought to have been burrowing marine animals. Segmentation evolved several times among spiralians; we will discuss only one of the phyla with segmented members: the annelids.

The **annelids** (phylum **Annelida**) are a diverse group of segmented spiralian worms (Figure 32.22). The coelom in each segment is isolated from those in other segments. A separate nerve center called a *ganglion* controls each segment, and the ganglia are connected by nerve cords that coordinate their functioning. Most annelids lack a rigid, external protective covering. The body wall serves as a general surface for gas exchange in most species, but this thin, permeable body surface restricts annelids to moist environments; they lose body water rapidly in dry air. The approximately 16,500 described species live in marine, freshwater, and terrestrial environments.

POLYCHAETES. More than half of all annelid species are members of the class **Polychaeta** ("many hairs"). Nearly all

polychaetes are marine animals. Most have one or more pairs of eyes and one or more pairs of tentacles at the anterior end of the body. The body wall in most segments extends laterally as a series of thin outgrowths, called *parapodia*, that contain many blood vessels. The parapodia function in gas exchange, and some species use them to move. Stiff bristles called *setae* protrude from each parapodium, forming temporary attachments to the substratum that prevent the animal from slipping backward when its muscles contract.

Typically, males and female polychaetes release gametes into the water, where the eggs are fertilized and develop into trochophore larvae (Figure 32.23). The trochophore is a distinctive larval type found among polychaetes, mollusks, and several other marine lineages with spiral cleavage. The second half of the name "lophotrochozoans" is derived from this larva, which is believed by many researchers to represent an evolutionary link between the annelids and the mollusks.

As the polychaete trochophore develops, it forms body segments at its posterior end; eventually it becomes a small

32.23 The Trochophore Larva The trochophore ("wheel-bearer") is a distinctive larval form found in several animal lineages with spiral cleavage, most notably the marine polychaete worms and the mollusks.

(a) *Spirobranchus* sp.

(b) *Lumbricus* sp.

32.24 Diversity among the Annelids
(a) The feather duster worm is a marine poly-chaete with striking feeding tentacles. (b) Earthworms are hermaphroditic (each individual is simultaneously both male and female). When they copulate, each individual donates and receives sperm. (c) This Australian tiger leech is attached to a leaf by its posterior sucker as it waits for a mammalian host. (d) Vestimentiferans live around hydrothermal vents deep in the ocean. Their skin secretes chitin and other substances, forming tubes.

(c) *Microbdella* sp.

(d) *Riftia* sp.

adult worm. Many polychaete species live in burrows in soft sediments and filter prey from the surrounding water with elaborate, feathery tentacles (Figure 32.24a).

OLIGOCHAETES.   More than 90 percent of the approximately 3,000 described species of **oligochaetes** (class **Oligochaeta**) live in freshwater or terrestrial habitats. Oligochaetes ("few hairs") have no parapodia, eyes, or anterior tentacles, and they have relatively few setae. Earthworms—the most familiar oligochaetes (see Figure 32.22)—are scavengers and ingesters of soil, from which they extract food particles.

Unlike polychaetes, all oligochaetes are *hermaphroditic:* that is, each individual is both male and female. Sperm are exchanged simultaneously between two copulating individuals (Figure 32.24b). Eggs are laid in a cocoon outside the adult's body. The cocoon is shed, and when development is complete, miniature worms emerge and begin independent life.

LEECHES.   Leeches (class **Hirudinea**) probably evolved from oligochaete ancestors. Most species live in freshwater

or terrestrial habitats and, like oligochaetes, lack parapodia and tentacles. Like oligochaetes, leeches are hermaphroditic. The coelom of leeches is not divided into compartments; the coelomic space is largely filled with undifferentiated tissue. Groups of segments at each end of the body are modified to form suckers, which serve as temporary anchors that aid the leech in movement (Figure 32.24c). With its posterior sucker attached to a substratum, the leech extends its body by contracting its circular muscles. The anterior sucker is then attached, the posterior one detached, and the leech shortens itself by contracting its longitudinal muscles.

Many leeches are external parasites of other animals, but some species also eat snails and other invertebrates. A leech makes an incision in its host, from which blood flows. It can ingest so much blood in a single feeding that its body may enlarge several times. An anticoagulant secreted by the leech into the wound keeps the host's blood flowing. For hundreds of years leeches were widely employed in medicine. Even today they are used to reduce fluid pressure and prevent blood clotting in damaged tissues and to eliminate pools of coagulated blood.

VESTIMENTIFERANS. Members of one lineage of annelids, the **vestimentiferans** (class **Pogonophora**), evolved burrowing forms with a crown of tentacles through which gases are exchanged; they entirely lost their digestive systems (Figure 32.24*d*). Vestimentiferans secrete chitin and other substances to form the tubes in which they live.

A vestimentiferan's coelom consists of an anterior compartment into which the tentacles can be withdrawn, and a long, subdivided cavity that extends much of the length of its body. The posterior end of the body is segmented. Experiments using radioactively labeled molecules have shown that vestimentiferans take up dissolved organic matter at high rates from either the sediments in which they live or the surrounding water.

Vestimentiferans were not discovered until the twentieth century, when deep-sea exploration revealed them living many thousands of meters below the ocean surface. In these deep oceanic sediments, they are abundant, reaching densities of many thousands per square meter. About 145 species have been described. The largest and most remarkable vestimentiferans, which grow to 2 meters in length, live near deep-sea hydrothermal vents—volcanic openings in the sea floor through which hot, sulfide-rich water pours. The tissues of these species harbor endosymbiotic bacteria that fix carbon using energy obtained from oxidation of hydrogen sulfide (H_2S).

Mollusks evolved shells

Mollusks (phylum **Mollusca**) range in size from snails only a millimeter high to giant squids more than 18 meters long—the largest known invertebrates. Mollusks underwent one of the most dramatic of animal evolutionary radiations, based on a unique body plan with three major structural components: a foot, a mantle, and a visceral mass. Animals that appear very different, such as snails, clams, and squids, are all built from these components (Figure 32.25).

The molluscan *foot* is a large, muscular structure that originally was both an organ of locomotion and a support for the internal organs. In the lineage leading to squids and octopuses, the foot was modified to form arms and tentacles borne on a head with complex sense organs. In other groups, such as clams, the foot was transformed into a burrowing organ. In some lineages the foot is greatly reduced.

The *mantle* is a fold of tissue that covers the *visceral mass* of internal organs. In many mollusks, the mantle extends beyond the visceral mass to form a *mantle cavity*. The mantle secretes the hard, calcarous skeleton typical of most mollusks. The *gills*, which are used for gas exchange and, in some

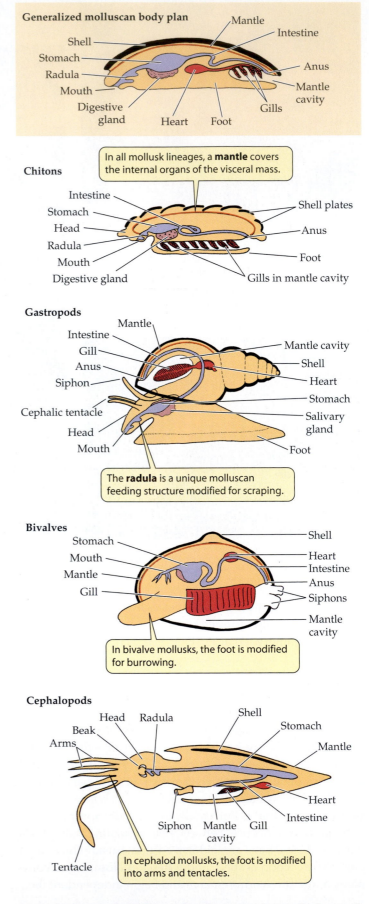

32.25 Molluscan Body Plans The diverse modern mollusks are all variations on a general body plan that includes a foot, a mantle, and a visceral mass of internal organs.

species, for feeding, lie in this cavity. When the cilia on the gills beat, they create a flow of water over the gills. The tissue of the gills, which is highly *vascularized* (contains many blood vessels), takes up O_2 from the water and releases CO_2.

Mollusks have an *open circulatory system* that empties into large fluid-filled cavities, through which fluids move around the animal and deliver O_2 to internal organs. Mollusks also developed a rasping feeding structure known as the *radula*. The radula was originally an organ for scraping algae from rocks, a function it retains in many living mollusks. However, in some mollusks, it has been modified into a drill or poison dart. In others, such as clams, it is absent.

Although individual components have been lost in some lineages, these three unique, shared derived characteristics—the foot, the mantle, and the visceral mass—lead zoologists to believe that all 95,000 species of mollusks have a common ancestor. A small sample of these species is shown in Figure 32.26.

MONOPLACOPHORANS. Monoplacophorans (class **Monoplacophora**) were the most abundant mollusks during the Cambrian period, 550 million years ago, but today there are only a few surviving species. Unlike all other living mollusks, the surviving monoplacophorans have respiratory organs, muscles, and excretory pores that are repeated over the length of the body. The respiratory organs are located in a large cavity under the shell, through which oxygen-bearing water circulates.

CHITONS. Chitons (class **Polyplacophora**) have multiple gills and shell plates, but the body is not truly segmented (Figure 32.26a). The chiton body is bilaterally symmetrical, and its internal organs, particularly the digestive and nervous systems, are relatively simple. The larvae of chitons are almost indistinguishable from those of annelids. Most chitons are marine herbivores that scrape algae from rocks with their sharp radulae. An adult chiton spends most of its life clinging tightly to rock surfaces with its large, muscular, mucus-covered foot. It moves slowly by means of rippling waves of muscular contraction in the foot. Fertilization in most chitons takes place in the water, but in a few species fertilization is internal and embryos are brooded within the body.

BIVALVES. One lineage of early mollusks developed a hinged, two-part shell that extended over the sides of the body as well as the top, giving rise to the **bivalves** (class **Bivalvia**), which include the familiar clams, oysters, scallops, and mussels (Figure 32.26b). Bivalves are largely sedentary and have greatly reduced heads. The foot is compressed, and in many clams, it is used for burrowing into mud and sand. Bivalves feed by taking in water through an opening called an *incurrent siphon* and extracting food from the water with their large gills, which are also the main sites of gas exchange. Water and gametes exit through the *excurrent siphon*. Fertilization takes place in open water in most species.

GASTROPODS. Another lineage of early mollusks gave rise to the **gastropods** (class **Gastropoda**), which include snails, whelks, limpets, slugs, abalones, and the often brilliantly ornamented nudibranchs. Gastropods, unlike bivalves, have one-piece shells. Most gastropods are motile, using the large foot to move slowly across the substratum or to burrow through it. Gastropods are the most species-rich and widely distributed of the molluscan classes (Figure 32.26c,d). Most species move by gliding on the muscular foot, but in a few species—the sea butterflies and heteropods—the foot is modified into a swimming organ with which the animal moves through open ocean waters. The only mollusks that live in terrestrial environments—land snails and slugs—are gastropods. In these terrestrial species, the mantle tissue is modified into a highly vascularized lung. Fertilization is internal in most species.

CEPHALOPODS. In one lineage of mollusks, the cephalopods (class **Cephalopoda**), the excurrent siphon became modified to allow the animal to control the water content of the mantle cavity. Ultimately, the modification of the mantle into a device for forcibly ejecting water from the cavity enabled these animals to move rapidly through the water. Furthermore, many early cephalopods had chambered shells into which gas could be secreted to adjust buoyancy. Together, these adaptations allow cephalopods to live in open water.

The cephalopods include the squids, octopuses, and nautiluses (Figure 32.26e, f). They first appeared about 600 million years ago, near the beginning of the Cambrian period. By the Ordovician period a wide variety of types were present. With their greatly enhanced mobility, some cephalopods, such as squids, became the major predators in the open waters of the Devonian oceans. They remain important marine predators today. Cephalopods capture and subdue their prey with their tentacles; octopuses also use their tentacles to move over the substratum. As is typical of active predators, cephalopods have a head with complex sensory organs, most notably eyes that are comparable to those of vertebrates in their ability to resolve images. The head is closely associated with a large, branched foot that bears the tentacles and a siphon. The large muscular mantle provides a solid external supporting structure. The gills hang in the mantle cavity. As is typical of behaviorally complex animals, many cephalopods have elaborate courtship behavior, which may involve striking color changes.

(*a*) *Mopalia* sp.

(*c*) *Phidiana hiltoni*

(*e*) *Octopus bimaculoides*

32.26 Diversity among the Mollusks (*a*) Chitons are common in the intertidal zones of the temperate zone coasts. (*b*) The giant clam of Indonesia is among the largest of the bivalve mollusks. (*c*) Slugs are gastropods that have lost their shells; this shell-less sea slug is very conspicuously colored. (*d*) Land snails are shelled, terrestrial gastropods. (*e*) Cephalopods such as the octopus are active predators. (*f*) The boundaries of its chambers are clearly visible on the outer surface of this shelled *Nautilus*, another cephalopod.

(*b*) *Tridacna gigas*

(*d*) *Helminthoglypta walkeriana*

(*f*) *Nautilus pompilius*

The earliest cephalopod shells were divided by partitions penetrated by tubes through which liquids could be moved. Nautiloids (genus *Nautilus*) are the only cephalopods with external chambered shells that survive today (Figure 32.26*f*).

Mollusks and brachiopods are among the lophotrochozoans that evolved hard shells that help to protect them from predators and the physical environment. A sturdy outer covering is the main feature of the second protostomate lineage, the ecdysozoans—the subject of the next chapter.

Chapter Summary

Animals: Descendants of a Common Ancestor

▶ All members of the kingdom Animalia are believed to have a common ancestor, which was a colonial flagellated protist.

▶ The specialization of cells by function made possible the complex, multicellular body plan of animals.

▶ Animals are multicellular heterotrophs. They take in complex organic molecules, expending energy to do so.

▶ Morphological, developmental, and molecular data all support similar animal phylogenies.

▶ The two major animal lineages—protostomes and deuterostomes—are believed to have diverged early in animal evolution; they differ in several components of their early development. **Review Figure 32.1**

Body Plans Are Basic Structural Designs

▶ Most animals have either radial or bilateral symmetry. Radially symmetrical animals move slowly or not at all. Bilateral symmetry is strongly correlated with more rapid movement and the concentration of sense organs at the anterior end of the animal. **Review Figure 32.2**

▶ The body cavity of an animal is strongly correlated with its ability to move. On the basis of their body cavities, animals are classified as acoelomates, pseudocoelomates, or coelomates. **Review Figure 32.3**

Sponges: Loosely Organized Animals

▶ Sponges (phylum Porifera) are simple animals that lack cell layers and true organs, but have several different cell types.

▶ Sponges feed by means of choanocytes, feeding cells that draw water through the sponge body and filter out food particles. **Review Figure 32.4**

▶ Sponges come in a variety of sizes and shapes that are adapted to different movement patterns of water.

Cnidarians: Two Cell Layers and Blind Guts

▶ Cnidarians (phylum Cnidaria) are radially symmetrical and diploblastic, but with their nematocyst-studded tentacles, they can capture prey larger and more complex than themselves. **Review Figure 32.7**

▶ Most cnidarian life cycles have a sessile polyp stage and a free-swimming, sexual, medusa stage, but some species lack one of the stages. **Review Figures 32.8, 32.10, 32.11**

See Web/CD Tutorial 32.1

Ctenophores: Complete Guts and Tentacles

▶ Ctenophores (phylum Ctenophora) are diploblastic marine carnivores with a complete gut and simple life cycles. **Review Figure 32.12**

The Evolution of Bilaterally Symmetrical Animals

▶ All bilaterally symmetrical animals probably share a common ancestor.

▶ Protostomes and deuterostomes are each monophyletic lineages that have been evolving separately since the Cambrian period. Their members are structurally more complex than cnidarians and ctenophores.

▶ Protostomes have a ventral nervous system, paired nerve cords, and larvae with compound cilia.

▶ Deuterostomes have a dorsal nervous system and larvae with a single cilium per cell.

▶ The protostomes split into two major groups: lophotrochozoans and ecdysozoans. **Review Figure 32.14**

Simple Lophotrochozoans

▶ Flatworms (phylum Platyhelminthes) are acoelomate, lack organs for oxygen transport, have only one entrance to the gut, and move by beating their cilia. Many species are parasitic. **Review Figures 32.15, 32.16**

▶ Although they are no larger than many ciliated protists, rotifers (phylum Rotifera) have highly developed internal organs. **Review Figure 32.17**

Lophophorates: An Ancient Body Plan

▶ The lophotrochozoan lineage split into two branches, whose descendants became the modern lophophorates and the spiralians.

▶ The lophophore dominates the anatomy of many lophophorate animals. **Review Figure 32.18**

▶ Ectoprocts are colonial lophophorates that can move their lophophores. **Review Figure 32.19**

▶ Brachiopods, which superficially resemble bivalve mollusks, were much more abundant in the past than they are today.

Spiralians: Spiral Cleavage and Wormlike Body Plans

▶ The spiralian lineage gave rise to many phyla, most of whose members are wormlike.

▶ Ribbon worms (phylum Nemertea) have a complete digestive tract and capture prey with an eversible proboscis. **Review Figure 32.21**

▶ Annelids (phylum Annelida) are a diverse group of segmented worms that live in marine, freshwater, and terrestrial environments. **Review Figures 32.22**

▶ Mollusks (phylum Mollusca) have a body plan with three basic components: foot, mantle, and visceral mass. **Review Figure 32.25**

▶ The molluscan body plan has been modified to yield a diverse array of animals that superficially appear very different from one another.

See Web/CD Activities 32.1 and 32.2 for a concept review of this chapter

Self-Quiz

1. The body plan of an animal is
   a. its general structure.
   b. the integrated functioning of its parts.
   c. its general structure and the integrated functioning of its parts.
   d. its general structure and its evolutionary history.
   e. the integrated functioning of its parts and its evolutionary history.

2. A bilaterally symmetrical animal can be divided into mirror images by
   a. any plane through the midline of its body.
   b. any plane from its anterior to its posterior end.
   c. any plane from its dorsal to its ventral surface.
   d. any plane through the midline of its body from its anterior to its posterior end.
   e. a single plane through the midline of its body from its dorsal to its ventral surface.

3. Among protostomes, cleavage of the fertilized egg is
   a. delayed while the egg continues to mature.
   b. always radial.
   c. spiral in some species and radial in others.
   d. triploblastic.
   e. diploblastic.

4. The sponge body plan is characterized by
   a. a mouth and digestive cavity but no muscles or nerves.
   b. muscles and nerves but no mouth or digestive cavity.
   c. a mouth, digestive cavity, and spicules.
   d. muscles and spicules but no digestive cavity or nerves.
   e. no mouth, digestive cavity, muscles, or nerves.

5. Which are phyla of diploblastic animals?
   a. Porifera and Cnidaria
   b. Cnidaria and Ctenophora
   c. Cnidaria and Platyhelminthes
   d. Ctenophora and Platyhelminthes
   e. Porifera and Ctenophora

6. Cnidarians have the ability to
   a. live in both salt and fresh water.
   b. move rapidly in the water column.
   c. capture and consume large numbers of small prey.
   d. survive where food is scarce, because of their low metabolic rate.
   e. capture large prey and to move rapidly.

7. Many parasites evolved complex life cycles because
   a. they are too simple to disperse readily.
   b. they are poor at recognizing new hosts.
   c. they were driven to it by host defenses
   d. complex life cycles increase the probability of a parasite's transfer to a new host.
   e. their ancestors had complex life cycles and they simply retained them.

8. Members of which phyla have lophophores?
   a. Phoronida, Brachiopoda, and Nemertea
   b. Phoronida, Brachiopoda, and Ectoprocta
   c. Brachiopoda, Ectoprocta, and Platyhelminthes
   d. Phoronida, Rotifera, and Ectoprocta
   e. Rotifera, Ectoprocta, and Brachiopoda

9. Which of the following is not part of the molluscan body plan?
   a. Mantle
   b. Foot
   c. Radula
   d. Visceral mass
   e. Jointed skeleton

10. Cephalopods control their buoyancy by
    a. adjusting salt concentrations in their blood.
    b. forcibly expelling water from the mantle.
    c. pumping water in and out of internal chambers.
    d. using the complex sense organs in their heads.
    e. swimming rapidly.

For Discussion

1. Differentiate among the members of each of the following sets of related terms:
   a. radial symmetry/bilateral symmetry
   b. protostome/deuterostome
   c. diploblastic/triploblastic
   d. coelomate/pseudocoelomate/acoelomate

2. In this chapter we listed some of the traits shared by all animals that convince most biologists that all animals are descendants of a single common ancestral lineage. In your opinion, which of these traits provides the most compelling evidence that animals are monophyletic?

3. Describe some features that allow animals to capture prey that are larger and more complex than they themselves are.

4. Why is bilateral symmetry strongly associated with cephalization, the concentration of sense organs in an anterior head?

5. Why might mollusks not have evolved segmentation, given that a segmented body enables improved control over locomotion?

33 *Ecdysozoans: The Molting Animals*

Early in animal evolution, the protostomate lineage split into two branches—the lophotrochozoans and the ecdysozoans—as we saw in the previous chapter. The distinguishing feature of the ecdysozoans is an **exoskeleton**, a nonliving covering that provides an animal with both protection and support. Once formed, however, an exoskeleton cannot grow. How, then, can ecdysozoans increase in size? Their solution is to shed, or **molt**, the exoskeleton and replace it with a new, larger one.

Before the animal molts, a new exoskeleton is already forming underneath the old one. When the old exoskeleton is shed, the new one expands and hardens. But until it has hardened, the animal is very vulnerable to its enemies both because its outer surface is easy to penetrate and because it can move only slowly.

The exoskeleton presented new challenges in other areas besides growth. Ecdysozoans cannot use cilia for locomotion, and most exdysozoans have hard exoskeletons that impede the passage of oxygen into the animal. To cope with these challenges, ecdysozoans evolved new mechanisms of locomotion and respiration.

Despite these constraints, the ecdysozoans—the molting animals—have more species than all other animal lineages combined. An increasingly rich array of molecular and genetic evidence, including a set of homeobox genes shared by all ecdysozoans, suggests that molting may have evolved only once during animal evolution.

In this chapter, we will review the diversity of the ecdysozoans. We will look at the characteristics of animals in the various ecdysozoan phyla and see how having an exoskeleton has influenced their evolution. The phylogeny we will follow is presented in Figure 33.1. In the first part of the chapter, we will look at several small phyla of wormlike ecdysozoans. Then we will detail the characteristics of the arthropods, an incredibly species-rich group of ecdysozoan phyla with hardened exoskeletons. We will close the chapter with an overview of evolutionary themes found in the evolution of the protostomate phyla, including both the lophotrochozoan and ecdysozoan lineages.

Cuticles: Flexible, Unsegmented Exoskeletons

Some ecdysozoans have wormlike bodies covered by exoskeletons that are relatively thin and flexible. Such an exoskeleton, called a **cuticle**, offers the animal some protection, but does not provide body support. The action of circular and longitudinal muscles on fluids in

Shedding the Exoskeleton This dragonfly has just gone through a molt, a shedding of the outer exoskeleton. Such molts are necessary in order for the insect to grow larger or to change its form.

33.1 A Current Phylogeny of the Ecdysozoans Those ecdysozoan phyla with jointed appendages are often placed in a single phylum, Arthropoda. Arthropods are the most numerous animals on the planet.

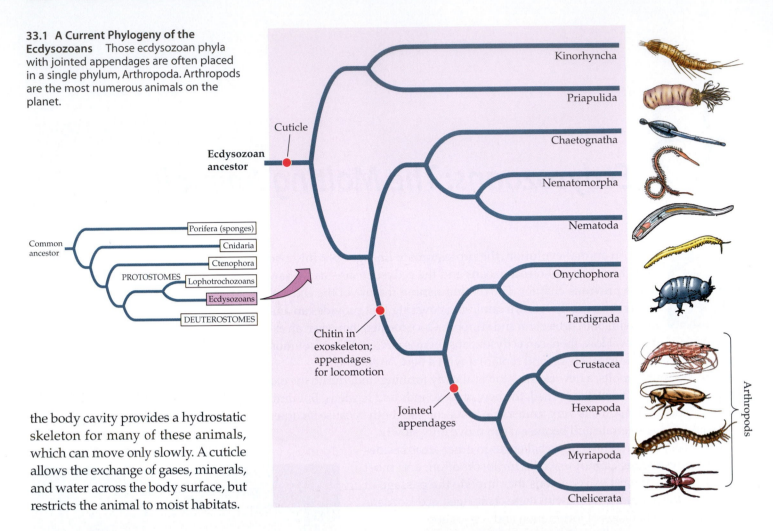

the body cavity provides a hydrostatic skeleton for many of these animals, which can move only slowly. A cuticle allows the exchange of gases, minerals, and water across the body surface, but restricts the animal to moist habitats.

Some marine ecdysozoan phyla have few species

Several phyla of marine wormlike animals branched off early within the ecdysozoan lineage. Each of these phyla contains only a few species. These animals have relatively thin cuticles that are molted periodically as the animals grow to full size.

PRIAPULIDS AND KINORHYNCHS. The 16 species of **priapulids** (phylum **Priapulida**) are cylindrical, unsegmented, worm-like animals that range in size from half a millimeter to 20 centimeters in length (Figure 33.2). They burrow in fine marine sediments and prey on soft-bodied invertebrates, such as polychaete worms. They capture prey with a toothed pharynx, a muscular organ that is everted through the mouth and then withdrawn into the body together with the grasped prey. Fertilization is external, and most species have a larval form that lives in the mud.

Priapulus caudatus

33.2 A Priapulid Priapulids are marine worms that live, usually as burrowers, on the ocean floor. They capture prey with a toothed pharynx that everts through the proboscis. They take their name from Priapus, the Greek god of procreation, who was typically portrayed with an oversize penis.

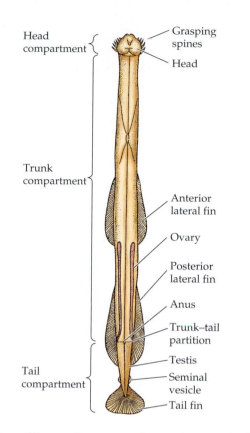

33.3 An Arrow Worm Arrow worms have a three-part body plan. Their fins and grasping spines are adaptations for a predatory lifestyle.

About 150 species of **kinorhynchs** (phylum **Kinorhyncha**) have been described. They are all less than 1 millimeter in length and live in marine sands and muds. Their bodies are divided into 13 segments, each with a separate cuticular plate. These plates are periodically molted during growth. Kinorhynchs feed by ingesting sediments and digesting the organic material found within them, which may include living algae as well as dead matter. Kinorhynchs have no distinct larval stage; fertilized eggs develop directly into juveniles, which emerge from their egg cases with 11 of the 13 body segments already formed.

ARROW WORMS. The phylogeny of the **arrow worms** (phylum **Chaetognatha**) is uncertain. Recent evidence indicates that these animals may in fact belong among the deuterostomes; however, this placement is still in question, and we continue to include them among the ecdysozoans.

The arrow worms body plan is based on a coelom divided into head, trunk, and tail compartments (Figure 33.3). Most arrow worms swim in the open sea, but a few live on the sea floor. Their abundance as fossils indicates that they were common more than 500 million years ago. The 100 or so living species of arrow worms are small enough—less than 12 cm long—that their gas exchange and excretion requirements are met by diffusion through the body surface, and they lack a circulatory system. Wastes and nutrients are moved around the body in the coelomic fluid, which is propelled by cilia

that line the coelom. There is no distinct larval stage. Miniature adults hatch directly from eggs that are fertilized internally following elaborate courtship.

Arrow worms are stabilized in the water by means of one or two pairs of lateral fins and a tail fin. They are major predators of small organisms in the open oceans, ranging in size from small protists to young fish as large as the arrow worms themselves. An arrow worm typically lies motionless in the water until water movement signals the approach of prey. The arrow worm then darts forward and grasps the prey with the stiff spines adjacent to its mouth.

Tough cuticles evolved in some unsegmented worms

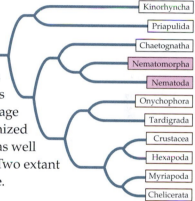

Tough external cuticles evolved in some members of another ecdysozoan lineage whose descendants colonized freshwater and terrestrial as well as marine environments. Two extant phyla represent this lineage.

HORSEHAIR WORMS. About 320 species of horsehair worms (phylum **Nematomorpha**) have been described. As their name implies, horsehair worms are extremely thin, and they range from a few millimeters up to a meter in length (Figure 33.4). Most adult horsehair worms live in fresh water among leaf litter and algal mats near the edges of streams and ponds. The larvae of horsehair worms are internal par-

Paragordius sp.

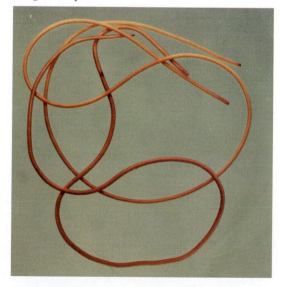

33.4 Horsehair Worms These worms get their name from their hair- or threadlike shape. They can grow to be up to a meter long.

asites of terrestrial and aquatic insects and freshwater crayfish. The horsehair worm's gut is greatly reduced, has no mouth opening, and is probably nonfunctional. These worms may feed only as larvae, absorbing nutrients from their hosts across the body wall, but many continue to grow after they have left their hosts, suggesting that adult worms may also absorb nutrients from their environment.

ROUNDWORMS. Roundworms (phylum **Nematoda**) have a thick, multilayered cuticle secreted by the underlying epidermis that gives their body its shape (Figure 33.5*a*). As a roundworm grows, it sheds its cuticle four times.

Roundworms exchange oxygen and nutrients with their environment through both the cuticle and the intestine, which is only one cell layer thick. Materials are moved through the gut by rhythmic contraction of a highly muscular organ, the *pharynx*, at the worm's anterior end. Roundworms move by contracting their longitudinal muscles.

Roundworms are one of the most abundant and universally distributed of all animal groups. About 25,000 species have been described, but the actual number of living species may be more than a million. Countless roundworms live as scavengers in the upper layers of the soil, on the bottoms of lakes and streams, in marine sediments (Figure 33.5*c*), and as parasites in the bodies of most kinds of plants and animals. The topsoil of rich farmland contains up to 3 billion nematodes per acre.

Many roundworms are predators, preying on protists and other small animals (including other roundworms). Many roundworms live parasitically within their hosts. The largest known roundworm, which reaches a length of 9 meters, is a parasite in the placentas of female sperm whales. The roundworms that are parasites of humans (causing serious tropical diseases such as trichinosis, filariasis, and elephantiasis), domestic animals, and economically important plants have been studied intensively in an effort to find ways of controlling them. One soil-inhabiting nematode, *Caenorhabitis elegans*, is a "model organism" in the laboratories of geneticists and developmental biologists.

The structure of parasitic roundworms is similar to that of free-living species, but the life cycles of many parasitic species have special stages that facilitate the transfer of individuals among hosts. *Trichinella spiralis*, the species that causes the human disease trichinosis, has a relatively simple life cycle. A person may become infected by eating the flesh of an animal (usually a pig) containing larvae of *Trichinella* encysted in its muscles. The larvae are activated in the digestive tract, emerge from their cysts, and attach to the person's intestinal wall, where they feed. Later, they bore through the intestinal wall and are carried in the bloodstream to muscles, where they form new cysts (Figure 33.5*b*). If present in great numbers, these cysts cause severe pain or death.

Arthropods and Their Relatives: Segmented External Skeletons

In Precambrian times, the cuticle of some wormlike ecdysozoan lineages became thickened by the incorporation of layers of protein and a strong, flexible, waterproof polysaccharide called **chitin**. This rigid body covering may originally have had a protective function, but eventually it acquired both support and locomotory functions as well.

A rigid body covering precludes wormlike movement. To move, the animal requires extensions of the body that can be

33.5 Roundworms (*a*) The body plan of *Trichinella spiralis*, a roundworm that causes trichinosis. (*b*) A cyst of *Trichinella spiralis* in the muscle tissue of a host. (*c*) This free-living roundworm moves through marine sediments.

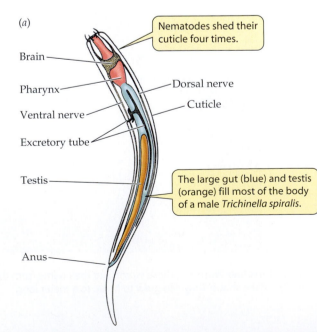

(*a*)

Nematodes shed their cuticle four times.

Brain

Pharynx

Dorsal nerve

Ventral nerve

Cuticle

Excretory tube

Testis

The large gut (blue) and testis (orange) fill most of the body of a male *Trichinella spiralis*.

Anus

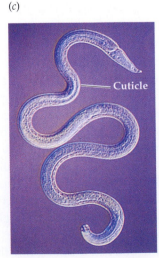

(*b*)

(*c*)

Cuticle

manipulated by muscles. Such **appendages** evolved several times in the late Precambrian, leading to the lineages collectively called the **arthropods** ("jointed foot"). Divisions among the arthropod lineages are ancient and have been the subject of much research in the past decade. These phylogenetic relationships are being examined daily in the light of a wealth of new information, much of it concerning gene expression. There is currently no consensus on an exact phylogeny, but most researchers agree that these important animal groups are monophyletic, and some taxonomists consider them as members of a single phylum: **Arthropoda**.

Before presenting one current view of arthropod phylogeny, let's look at some arthropod relatives that have segmented bodies but unjointed legs, and at an early arthropod lineage that disappeared but left an important fossil record.

Some relatives of the arthropods have unjointed legs

Although they were once thought to be closely related to annelid worms, recent molecular evidence links the 110 species of **onychophorans** (phylum **Onychophora**) to the arthropod lineages. Onychophorans have soft bodies that are covered by a thin, flexible cuticle that contains chitin. Onychophorans use their fluid-filled body cavities as hydrostatic skeletons. Their soft, fleshy, unjointed, claw-bearing legs are formed by outgrowths of the body (Figure 33.6*a*). These animals are probably similar in appearance to ancestral arthropods. Fertilization is internal, and the large, yolky eggs are brooded with the body of the female.

Kinorhyncha
Priapulida
Chaetognatha
Nematomorpha
Nematoda
Onychophora
Tardigrada
Crustacea
Hexapoda
Myriapoda
Chelicerata

Like the onychophorans, **water bears** (phylum **Tardigrada**) have fleshy, unjointed legs and use their fluid-filled body cavities as hydrostatic skeletons (Figure 33.6*b*). Water bears are extremely small (0.1–0.5 mm in length), and they lack circulatory systems and gas exchange organs. The 600 extant species of water bears live in marine sands and on temporary water films on plants. When these films dry out, the water bears also lose water and shrink to small, barrel-shaped objects that can survive for at least a decade in a dormant state. They have been found at densities as high as 2 million per square meter of moss.

33.6 Unjointed Legs (*a*) Onychophorans, also called "velvet worms," have unjointed legs and use the body cavity as a hydrostatic skeleton. (*b*) The appendages and general anatomy of water bears superficially resemble those of onychophorans.

Jointed legs appeared in the trilobites

The **trilobites** (phylum **Trilobita**) were among the earliest arthropods. They flourished in Cambrian and Ordovician seas, but disappeared in the great Permian extinction at the close of the Paleozoic era (245 mya). Because their heavy exoskeletons provided ideal material for fossilization, they left behind an abundant record of their existence (Figure 33.7).

Trilobites were heavily armored, and their body segmentation and appendages followed a relatively simple, repetitive plan. But their appendages were jointed, and some of them were modified for different functions. This specialization of appendage function became a theme as the evolution of the arthropod lineage continued.

Modern arthropods dominate Earth's fauna

Arthropod appendages have evolved an amazing variety of forms, and they serve many functions, including walking and swimming, gas exchange, food capture and manipulation, copulation, and sensory perception. The pattern of segmentation is similar among most arthropods because their development is governed by a common cascade of regula-

(*a*) *Peripatus* sp.

(*b*) *Echiniscus springer*

50 μm

Odontochile rugosa

33.7 A Trilobite The relatively simple, repetitive segments of the now-extinct trilobites are illustrated by a fossil trilobite from the shallow seas of the Devonian period, some 400 million years ago.

tory genes (see Figure 19.15), including homeotic genes that determine the kinds of appendages that are borne on each segment.

The bodies of arthropods are divided into segments. Their muscles are attached to the inside of the exoskeleton. Each segment has muscles that operate that segment and the appendages attached to it (Figure 33.8). The arthropod exoskeleton has had a profound influence on the evolution of these animals. Encasement within a rigid body covering provides support for walking on dry land, and the waterproofing provided by chitin keeps the animal from dehydrating in dry air. Aquatic arthropods were, in short, excellent candidates to invade terrestrial environments. As we will see, they did so several times.

| Kinorhyncha |
| Priapulida |
| Chaetognatha |
| Nematomorpha |
| Nematoda |
| Onychophora |
| Tardigrada |
| Crustacea |
| Hexapoda |
| Myriapoda |
| Chelicerata |

There are four major arthropod phyla living today: the crustaceans, hexapods (insects), myriapods, and chelicerates. Collectively, the arthropods (including both terrestrial and marine species) are the dominant animals on Earth, both in numbers of species (about 1.5 million described) and number of individuals (estimated at some 10^{18} individuals, or a billion billion).

Crustaceans: Diverse and Abundant

Crustaceans (phylum **Crustacea**) are the dominant marine arthropods today. The most familiar crustaceans belong to the class Malacostraca, which includes shrimp, lobsters, crayfish, and crabs (decapods; Figure 33.9a); and sow bugs (isopods; Figure 33.9b). Also included among the crustaceans are a vari-

ety of small species, many of which superficially resemble shrimp. The individuals of one group alone, the copepods (class Copepoda; Figure 33.9c), are so numerous that they may be the most abundant of all animals.

Barnacles (class Cirripedia) are unusual crustaceans that are sessile as adults (Figure 33.9d). With their calcareous shells, they superficially resemble mollusks but, as the zoologist Louis Agassiz remarked more than a century ago, a barnacle is "nothing more than a little shrimp-like animal, standing on its head in a limestone house and kicking food into its mouth."

Most of the 40,000 described species of crustaceans have a body that is divided into three regions: *head, thorax,* and *abdomen.* The segments of the head are fused together, and the head bears five pairs of appendages. Each of the multiple thoracic and abdominal segments usually bears one pair of appendages. In some cases, the appendages are branched, with different branches serving different functions. In many species, a fold of the exoskeleton, the *carapace,* extends dorsally and laterally back from the head to cover and protect some of the other segments (Figure 33.10a).

The fertilized eggs of most crustacean species are attached to the outside of the female's body, where they remain during their early development. At hatching, the young of some species are released as larvae; those of other species are released as juveniles that are similar in form to the adults. Still other species release eggs into the water or attach them to an object in the environment. The typical crustacean larva, called a **nauplius,** has three pairs of appendages and one central eye (Figure 33.10b). In many crustaceans, the nauplius larva develops within the egg before it hatches.

There is a growing recognition among researchers that a crustacean lineage may have been ancestral to all present-day arthropods. Therefore, the phylum Crustacea, as we recognize it here, may be paraphyletic (see Chapter 25). Molecular evidence points especially to a link between the crustaceans and another important lineage, the hexapods.

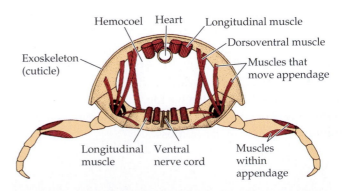

33.8 Arthropod Exoskeletons Are Rigid and Jointed This cross section through a thoracic segment of a generalized arthropod illustrates the arthropod body plan, which is characterized by a rigid exoskeleton with jointed appendages.

(a) *Opisthopus transversus*

(b) *Ligia occidentalis*

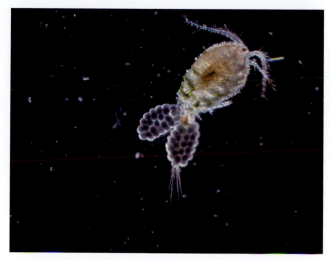

(c) *Cyclops* sp.

(d) *Lepas pectinata*

33.9 Crustacean Diversity (a) This mottled pea crab is a decapod crustacean. Its pigmentation depends on the food it ingests. (b) This isopod is found on the beaches of the California coast. (c) This microscopic freshwater copepod is only about 30 μm long. (d) Gooseneck barnacles attach to a substratum and feed by protruding and retracting feeding appendages from their shells.

Insects: Terrestrial Descendants of Marine Crustaceans

During the Devonian, more than 400 million years ago, arthropods made the leap from the marine environment onto land. Of the several groups who successfully colonized the terrestrial habitat, none is more prominent today than the six-legged individuals of the phylum **Hexapoda**—the insects.

Insects are found in most terrestrial and freshwater habitats, and they utilize nearly all species of plants and many species of animals as food. Some are internal parasites of plants and animals; others suck their host's blood or consume surface body tissues. The 1.4 million species of insects that have been described are believed to be only a small fraction of the total number of species living today.

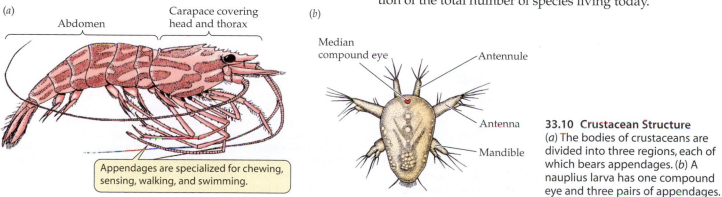

(a)

Abdomen

Carapace covering head and thorax

Appendages are specialized for chewing, sensing, walking, and swimming.

(b)

Median compound eye

Antennule

Antenna

Mandible

33.10 Crustacean Structure (a) The bodies of crustaceans are divided into three regions, each of which bears appendages. (b) A nauplius larva has one compound eye and three pairs of appendages.

Very few insect species live in the ocean. In freshwater environments, on the other hand, they are sometimes the dominant animals, burrowing through the substratum, extracting suspended prey from the water, and actively pursuing other animals. Insects were the first animals to achieve the ability to fly, and they are important pollinators of flowering plants.

Insects, like crustaceans, have three basic body regions: head, thorax, and abdomen. They have a single pair of antennae on the head and three pairs of legs attached to the thorax (Figure 33.11). Unlike the other arthropods, insects have no appendages growing from their abdominal segments (see Figure 21.5).

An insect exchanges gases by means of air sacs and tubular channels called *tracheae* (singular, trachea) that extend from external openings inward to tissues throughout the body. The adults of most flying insects have two pairs of stiff, membranous wings attached to the thorax. However, flies have only one pair of wings, and in beetles the forewings form heavy, hardened wing covers.

Wingless insects include springtails and silverfish (Figure 33.12). Of the modern insects, they are probably the most similar in form to insect ancestors. Apterygote insects have a simple life cycle, hatching from eggs as miniature adults.

Development in the winged insects (Figure 33.13) is complex. The hatchlings do not look like adults, and they undergo substantial changes at each molt. The immature stages of insects between molts are called **instars**. A substantial change that occurs between one developmental stage and another is called **metamorphosis**. If the changes between its instars are gradual, an insect is said to have **incomplete metamorphosis**.

Hydropodura aquatica

33.12 Wingless Insects
The wingless insects have a simple life cycle. They hatch looking like miniature adults, then grow by successive moltings of the cuticles as these springtails are doing.

In some insect groups, the larval and adult forms appear to be completely different animals. The most familiar example of such **complete metamorphosis** occurs in members of the order Lepidoptera, in which the larval caterpillar transforms itself into the adult butterfly (see Figure 1.1). During complete metamorphosis, the wormlike larva transforms itself during a specialized phase, called the **pupa**, in which many larval tissues are broken down and the adult form develops. In many of these groups, the different life stages are specialized for living in different environments and using different food sources. In many species, the larvae are adapted for feeding and growing, and the adults are specialized for reproduction and dispersal.

Entomologists divide the winged insects into about 29 different orders. We can make sense of this bewildering variety by recognizing three major lineages:

▶ Winged insects that cannot fold their wings against the body
▶ Winged insects that can fold their wings and that undergo incomplete metamorphosis
▶ Winged insects that can fold their wings and that undergo complete metamorphosis

Because they can fold their wings over their backs, flying insects belonging to the second and third lineages can tuck their wings out of the way upon landing and crawl into crevices and other tight places.

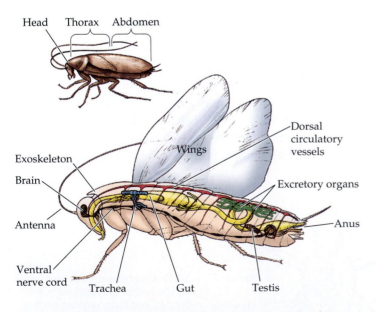

Head   Thorax   Abdomen

Wings

Exoskeleton

Brain

Antenna

Ventral nerve cord

Trachea   Gut   Testis

Dorsal circulatory vessels

Excretory organs

Anus

33.11 Structure of an Insect This diagram of a generalized insect illustrates its three-part body plan. The middle region, the thorax, bears three pairs of legs and, in most groups, two pairs of wings.

33.13 The Diversity of Insects (*a*) Unlike most flying insects, ▶ this dragonfly cannot fold its wings over its back. (*b*) The Mexican bush katydid represents the order Orthoptera. (*c*) Harlequin bugs are "true" bugs (order Hemiptera); (*d*) These mating mantophasmatodeans represent a recently discovered Hemipteran lineage found only in the Cape region of South Africa. (*e*) A predatory diving beetle (order Coleoptera). (*f*) The California dogface butterfly is a member of the Lepidoptera. (*g*) The flies, including this Mediterranean fruit fly, comprise the order Diptera. (*h*) Many genera in the order Hymenoptera, such as honeybees, are social insects.

(a) *Anax imperator*

(e) *Dytiscus marginalis*

(b) *Scudderia mexicana*

(f) *Colias eurydice*

(c) *Murgantia histrionica*

(g) *Ceratitis capitata*

(d) *Timema* sp.

(h) *Apis mellifera*

The only surviving members of the lineage whose members cannot fold their wings against the body are the orders Odonata (dragonflies and damselflies, Figure 33.13a) and Ephemeroptera (mayflies). All members of these two orders have aquatic larvae that transform themselves into flying adults after they crawl out of the water. Although many of these insects are excellent flyers, they require a great deal of open space in which to maneuver. Dragonflies and damselflies are active predators as adults, but adult mayflies lack functional digestive tracts and live only long enough to mate and lay eggs.

The second lineage, whose members can fold their wings and have incomplete metamorphosis, includes the orders Orthoptera (grasshoppers, crickets, roaches, mantids, and walking sticks; Figure 33.13b), Isoptera (termites), Plecoptera (stone flies), Dermaptera (earwigs), Thysanoptera (thrips), Hemiptera (true bugs; Figure 33.13c), and Homoptera (aphids, cicadas, and leafhoppers). In these groups, hatchlings are sufficiently similar in form to adults to be recognizable. They acquire adult organ systems, such as wings and compound eyes, gradually through several juvenile instars. Remarkably, a new insect order in this lineage, the Mantophasmatodea, was first described in 2002 (Figure 33.13d). These small insects are common in the Cape Region of southern Africa, an area of exceptional species richness and endemism for many animal and plant groups.

Insects belonging to the third lineage undergo complete metamorphosis. About 85 percent of all species of winged insects belong to this lineage. Familiar examples are the orders Neuroptera (lacewings and their relatives), Coleoptera (beetles; Figure 33.13e), Trichoptera (caddisflies), Lepidoptera (butterflies and moths; Figure 33.13f), Diptera (flies; Figure 33.13g), and Hymenoptera (sawflies, bees, wasps, and ants; Figure 33.13h).

Members of several orders of winged insects, including the Phthiraptera (lice) and Siphonaptera (fleas), are parasitic. Although descended from flying ancestors, these insects have lost the ability to fly.

Molecular data suggest that the lineage leading to the insects separated from the lineage leading to modern crustaceans about 450 million years ago, about the time of the appearance of the first land plants. These ancestral forms penetrated a terrestrial environment that was ecologically empty, which in part accounts for their remarkable success. But this success of the insects is also due to their wings, which arose only once early during insect evolution. Homologous genes control the development of insect wings and crustacean appendages, suggesting that that the insect wing evolved from a dorsal branch of a crustacean limb (Figure 33.14). The dorsal limb branch of crustaceans is used for respiration and osmoregulation. This finding suggests that the insect wing evolved from a gill-like structure that had a respiratory function.

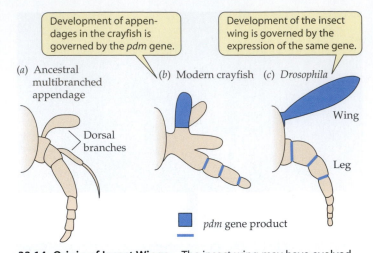

33.14 Origin of Insect Wings The insect wing may have evolved from an ancestral appendage similar to that of modern crustaceans. (a) A diagram of the ancestral, multibranched arthropod limb. (b, c) The *pdm* gene, a Hox gene, is expressed throughout the dorsal limb branch and walking leg of the thoracic limb of a crayfish (a) and in the wings and legs of *Drosophila* (b).

Arthropods with Two Body Regions

Insects and most crustaceans have tripartite body plans, with a head, thorax, and abdomen. In two other arthropod lineages, evolution resulted in a body plan with two regions—a head and a trunk.

Myriapods have many legs

Centipedes, millipedes, and the two other groups of animals comprise the phylum **Myriapoda**. Centipedes and millipedes have a well-formed head and a long, flexible, segmented trunk that bears many pairs of legs (Figure 33.15). Centipedes, which have one pair of legs per segment, prey on insects and other small animals. Millipedes, which have two pairs of legs per segment, scavenge and eat plants. More than 3,000 species of centipedes and 10,000 species of millipedes have been described; many more species probably remain unknown. Although most myriapods are less than a few centimeters long, some tropical species are ten times that size.

Most chelicerates have four pairs of walking legs

In the body plan of **chelicerates** (phylum **Chelicerata**), the anterior region (head) bears two pairs of appendages modified to form mouthparts. In addition, many chelicerates have four pairs of walking legs. The 63,000 described chelicerate species are usually placed in three classes: Pycnogonida, Merostomata, and Arachnida; most of them belong to Arachnida.

The **pycnogonids** (class **Pycnogonida**), or sea spiders, are a poorly known group of about 1,000 marine species (Figure 33.16a). Most are small, with leg spans less than 1 cm, but some deep-sea species have leg spans up to 60 cm. A few py-

(a) *Scolopendra heros*

(b) *Harpaphe haydeniana*

33.15 Myriapods
(a) Centipedes have powerful jaws for capturing active prey. (b) Millipedes, which are scavengers and plant eaters, have smaller jaws and legs. They have two pairs of legs per segment, in contrast to the one pair on each segment of centipedes.

cnogonids feed on algae, but most are carnivorous, feeding on a variety of small invertebrates.

The class **Merostomata** contains the horseshoe crabs (order Xiphosura), with five living species, and the extinct giant water scorpions (order Eurypterida). Horseshoe crabs, which have changed very little during their long fossil history, have a large horseshoe-shaped covering over most of the body. They are common in shallow waters along the eastern coasts of North America and Southeast Asia, where they scavenge and prey on bottom-dwelling invertebrates. Periodically they crawl into the intertidal zone in large numbers to mate and lay eggs (Figure 33.16b).

Arachnids (class **Arachnida**) are abundant in terrestrial environments. Most arachnids have a simple life cycle in which miniature adults hatch from internally fertilized eggs and begin independent lives almost immediately. Some arachnids retain their eggs during development and give birth to live young.

The most species-rich and abundant arachnids are the spiders, scorpions, harvestmen, mites, and ticks (Figure 33.17). The 30,000 described species of mites and ticks live in soil, leaf litter, mosses, and lichens, under bark, and as parasites of plants, invertebrates, and vertebrates. They are vectors for wheat and rye mosaic viruses, and they cause mange in domestic animals and skin irritation in humans.

Spiders are important terrestrial predators. Some have excellent vision that enables them to chase and seize their prey. Others spin elaborate webs made of protein threads in which they snare prey. The threads are produced by modified abdominal appendages connected to internal glands that secrete the proteins, which dry on contact with air. The webs of different groups of spiders are strikingly varied, and this variation enables the spiders to position their snares in many different environments. Spiders also use protein threads to construct safety lines during climbing and as homes, mating

(a) *Decalopoda sp.*

33.16 Minor Chelicerate Phyla (a) Although they are not spiders, it is easy to see why sea spiders were given their common name. (b) This spawning aggregation of horseshoe crabs was photographed on a sandy beach in Delaware.

(b) *Limulus polyphemus*

(a) *Phidippus formosus*

(b) *Pseudouroctonus minimus*

(c) *Hadrobunus maculosus*

(d) *Brevipalpus phoenicis*

33.17 Arachnid Diversity (*a*) The black jumping spider's bite produces an inflammatory reaction on mammalian skin. (*b*) Scorpions are nocturnal predators. (*c*) Harvestmen, also called daddy longlegs, are scavengers. (*d*) Mites are blood-sucking, external parasites on vertebrates.

structures, protection for developing young, and means of dispersal.

Themes in the Evolution of Protostomes

We end this chapter by reviewing some of the evolutionary trends we have seen in the animal groups we have discussed so far. Most of protostomate evolution took place in the oceans. Early protostomes used their fluid-filled body cavities as hydrostatic skeletons. Segmentation permitted different parts of the body to be moved independently of one another. Thus species in some protostomate lineages gradually evolved the ability to change their shape in complex ways and to move rapidly over and through the substratum or through the water.

During much of animal evolution, the only food in the water consisted of dissolved organic matter and very small organisms. Consequently, many different lineages of animals, including lophophorates, mollusks, tunicates, and some crustaceans, evolved feeding structures designed to filter small prey from water, as well as structures for moving water

through or over their prey-collecting devices. Animals that feed in this manner are abundant and widespread in marine waters today.

Because water flows readily, bringing food with it, sessile lifestyles also evolved repeatedly during lophotrochozoan and ecdysozoan evolution. Most phyla today have at least some sessile members. Being sessile presents certain challenges. For example, sessile animals cannot come together to mate. Some species eject both eggs and sperm into the water; others retain their eggs within their bodies and extrude only their sperm, which are carried by the water to other individuals. Species whose adults are sessile often have motile larvae, many of which have complicated mechanisms for locating suitable sites on which to settle.

A sessile animal gains access to local resources, but forfeits access to more distant resources. Many colonial sessile pro-

33.1 Anatomical Characteristics of the Major Protostomate Phyla

PHYLUM	BODY CAVITY	DIGESTIVE TRACT	CIRCULATORY SYSTEM
Lophotrochozoans			
Platyhelminthes	None	Dead-end sac	None
Rotifera	Pseudocoelom	Complete	None
Bryozoa	Coelom	Complete	None
Brachiopoda	Coelom	Complete in most	Open
Phoronida	Coelom	Complete	Closed
Nemertea	Coelom	Complete	Closed
Annelida	Coelom	Complete	Closed or open
Mollusca	Reduced coelom	Complete	Open except in cephalopods
Ecdysozoans			
Chaetognatha	Coelom	Complete	None
Nematomorpha	Pseudocoelom	Greatly reduced	None
Nematoda	Pseudoceolom	Complete	None
Crustacea	Hemocoel	Complete	Open
Hexapoda	Hemocoel	Complete	Open
Myriapoda	Hemocoel	Complete	Open
Chelicerata	Hemocoel	Complete	Open

Note: All protostomes have bilateral symmetry.

tostomes, however, are able to grow in the direction of better resources or into sites offering better protection. Individual members of colonies, if they are directly connected, can share resources. The ability to share resources enables some individuals to specialize for particular functions, such as reproduction, defense, or feeding. The nonfeeding individuals derive their nutrition from their feeding associates.

Predation may have been the major selective pressure for the development of hard, external body coverings. Such coverings evolved independently in many lophotrochozoan and ecdysozoan lineages. In addition to providing protection, they became key elements in the development of new systems of locomotion. Locomotory abilities permitted prey to escape more readily from predators, but also allowed predators to pursue their prey more effectively. Thus, the evolution of animals has been, and continues to be, a complex "arms race" among predators and prey.

Although we have concentrated on the evolution of greater complexity in animal lineages, many lineages whose members have remained simple have been very successful. Cnidarians are common in the oceans; roundworms are abundant in most aquatic and terrestrial environments. Parasites have lost complex body plans but have evolved complex life cycles.

The characteristics of the major existing phyla of protostomate animals are summarized in Table 33.1. Many major evolutionary trends were shared by protostomes and deuterostomes, the lineage that includes the chordates, the

group to which humans belong. We will consider the evolution of diversity among the deuterostomes in the next chapter.

Chapter Summary

▶ The ecdysozoan lineage is characterized by a nonliving external covering—an exoskeleton, or cuticle. **Review Figure 33.1**

▶ An animal with an exoskeleton grows by periodically shedding its exoskeleton and replacing it with a larger one, a process called molting.

Cuticles: Flexible, Unsegmented Exoskeletons

▶ Members of several phyla of marine worms with thin cuticles are descendants of an early split in the ecdysozoan lineage. **Review Figure 33.3**

▶ Tough cuticles are found in members of two phyla, the horsehair worms and the roundworms.

▶ Roundworms (phylum Nematoda) are one of the most abundant and universally distributed of all animal groups. Many are parasites. **Review Figure 33.5**

Arthropods and Their Relatives: Segmented External Skeletons

▶ Animals with rigid exoskeletons lack cilia for locomotion. To move, they have appendages that can be manipulated by muscles. **Review Figure 33.8**

▶ Although there is currently no consensus on an exact phylogeny, most researchers agree that the arthropod groups are monophyletic.

▶ Onychophorans and tardigrades have soft, unjointed legs. They are probably similar to ancestral arthropods.

▶ Trilobites flourished in Cambrian and Ordovician seas, but they became extinct at the close of the Paleozoic era.

Crustaceans: Species-Rich and Abundant

▶ The segments of the crustacean body are divided among three regions: head, thorax, and abdomen. **Review Figure 33.10**

▶ The most familiar crustaceans are shrimp, lobsters, crayfish, crabs, sow bugs, and sand fleas. Copepod crustaceans may be the most abundant animals on the planet.

▶ Recent molecular evidence indicates that the crustacean lineage may be ancestral to all the arthropods.

Insects: Terrestrial Descendants of Marine Crustaceans

▶ About 1.4 million species of insects (phylum Hexapoda) have been described, but that number is a small fraction of the total number of existing species. Although few species are found in marine environments, they are among the dominant animals in virtually all terrestrial and many freshwater habitats.

▶ Like crustaceans, insects have three body regions (head, thorax, abdomen). They bear a single pair of antennae on the head and three pairs of legs attached to the thorax. No appendages grow from their abdominal segments. **Review Figure 33.11**

▶ Wingless insects look like miniature adults when they hatch. Hatchlings of some winged insects resemble adults, but others undergo substantial changes at each molt.

▶ The winged insects can be divided into three major subgroups. Members of one subgroup cannot fold their wings back against the body. Members of the other two subgroups can.

▶ The wings of insects probably evolved from the dorsal branches of multibranched ancestral appendages. **Review Figure 33.14**

Arthropods with Two Body Regions

▶ Individuals of the remaining arthropod phyla generally have segmented bodies with two distinct regions, head and trunk.

▶ Myriapods (centipedes and millipedes) have many segments and many pairs of legs.

▶ Most chelicerates (phylum Chelicerata) have four pairs of legs.

▶ Arachnids—scorpions, harvestmen, spiders, mites, and ticks—are abundant in terrestrial environments.

Themes in the Evolution of Protostomes

▶ Most evolution of protostomes took place in the oceans.

▶ Early animals used fluid-filled body cavities as hydrostatic skeletons. Subdivision of the body cavity into segments allowed better control of movement.

▶ During much of animal evolution, the only food in the water consisted of dissolved organic matter and very small organisms.

▶ Flowing water brings food with it, allowing many aquatic animals to obtain food while being sessile.

▶ Predation may have been the major selective pressure for the development of hard, external body coverings.

See Web/CD Activities 33.1 and 33.2 for a concept review of this chapter.

Self-Quiz

1. The outer covering of ecdysozoans
   a. is always hard and rigid.
   b. is always thin and flexible.
   c. is present at some stage in the life cycle but not always among adults.
   d. ranges from very thin to hard and rigid.
   e. prevents the animals from changing their shapes.

2. The primary support for members of several small phyla of marine worms is
   a. their exoskeletons.
   b. their internal skeletons.
   c. their hydrostatic skeletons.
   d. the surrounding sediments.
   e. the bodes of other animals within which they live.

3. Roundworms are abundant and diverse because
   a. they are both parasitic and free-living and eat a wide variety of foods.
   b. they are able to molt their exoskeletons.
   c. their thick cuticle enables them to move in complex ways.
   d. their body cavity is a pseudocoelom.
   e. their segmented bodies enable them to live in many different places.

4. The arthropod exoskeleton is composed of a
   a. mixture of several kinds of polysaccharides.
   b. mixture of several kinds of proteins.
   c. single complex polysaccharide called chitin.
   d. single complex protein called arthropodin.
   e. mixture of layers of proteins and a polysaccharide called chitin.

5. Which phyla are arthropod relatives with unjointed legs?
   a. Trilobita and Onychophora

b. Onychophora and Tardigrada
c. Trilobita and Tardigrada
d. Onychophora and Chelicerata
e. Tardigrada and Chelicerata

6. The members of which crustacean group are probably the most abundant of all animals?
   a. Decapoda
   b. Amphipoda
   c. Copepoda
   d. Cirripedia
   e. Isopoda

7. The body plan of insects is composed of which of the three following regions?
   a. Head, abdomen, and trachea
   b. Head, abdomen, and cephalothorax
   c. Cephalothorax, abdomen, and trachea
   d. Head, thorax, and abdomen
   e. Abdomen, trachea, and mantle

8. Insects that hatch from eggs into juveniles that resemble miniature adults are said to have
   a. instars.
   b. neopterous development.
   c. accelerated development.
   d. incomplete metamorphosis.
   e. complete metamorphosis.

9. Which of the following groups of insects cannot fold their wings back against the body?
   a. Beetles
   b. True bugs
   c. Earwigs
   d. Stone flies
   e. Mayflies

10. Factors that may have contributed to the remarkable evolutionary diversification of insects include
    a. the terrestrial environments penetrated by insects lacked any other similar organisms.
    b. insects evolved the ability to fly.
    c. some lineages of insects evolved complete metamorphosis.
    d. insects evolved effective means of delivering oxygen to their internal tissues.
    e. All of the above

For Discussion

1. Segmentation has arisen several times during animal evolution. What advantages does segmentation provide? Given these advantages, why do so many unsegmented animals survive?

2. The British biologist J. B. S. Haldane is reputed to have quipped that "God was unusually fond of beetles." Beetles are, indeed, the most species-rich lineage of organisms. What features of beetles have contributed to the evolution and survival of so many species?

3. In Part Four of this book, we pointed out that major structural novelties have arisen infrequently during the course of evolution. Which of the features of protostomes do you think are major evolutionary novelties? What criteria do you use to judge whether a feature is a major as opposed to a minor novelty?

4. There are more described and named species of insects than of all other animal lineages combined. However, only a very few species of insects live in marine environments, and those species are restricted to the intertidal zone or the ocean surface. What factors may have contributed to the inability of insects to be successful in the oceans?

34 Deuterostomate Animals

Complex social systems, in which individuals associate with one another to breed and care for their offspring, characterize many species of fish, birds, and mammals—the most conspicuous and familiar deuterostomate animals. We tend to think of these social systems as having evolved relatively recently, but some amphibians, members of an ancient deuterostomate group, also have elaborate courtship and parental care behavior. For example, the male of the European midwife toad gathers eggs around his hind legs as the female lays them. He then carries the eggs until they are ready to hatch.

In the Surinam toad, mating and parental care are exquisitely coordinated, as an elaborate mating "dance" results in the female depositing eggs on the male's belly. The male fertilizes the eggs and, as the ritual ends, he presses them against the female's back, where they are carried until they hatch. The female poison dart frog lays clutches of eggs on a leaf or on the ground, which both parents then work to keep moist and protected. When the tadpoles hatch, they wiggle onto the back of one of their parents, who then carries the tadpoles to water.

There are fewer major lineages and many fewer species of deuterostomes than of protostomes (Table 34.1 on page 658), but we have a special interest in the deuterostomes because we are members of that lineage. In this chapter, we will describe and discuss the deuterostomate phyla: Echinodermata, Hemichordata, and Chordata. We close with a brief overview of some major themes in the evolution of animals.

Deuterostome Ancestors

A group of extinct animals known as the yunnanozoans are the likely ancestors of all deuterostomes. Many fossils of these animals have been discovered in China's Yunnan province. These well-preserved fossils show that the animals had a large mouth, six pairs of external gills, and a lightly cuticularized, segmented posterior body section (Figure 34.1). Later in deuterostome evolution, gills became internal and were connected to the exterior via slits in the body wall. These gill slits subsequently were lost in the lineage leading to the modern echinoderms.

Some Amphibian Parents Nurture Their Young Poison dart frogs (*Dendrobates reticulatus*) of the Amazon basin lay their eggs on land. Both parents protect and nurture the eggs until they hatch, at which time a parent carries the tadpoles to water on its back.

Yunnanozoan lividum

Mouth   External gills

34.1 The Ancestral Deuterostomes Had External Gills The extinct Yunnanozoan lineage is probably ancestral to all deuterostomes. This fossil, which dates from the Cambrian, shows the six pairs of external gills and segmented posterior body that characterized these animals.

Modern deuterostomes fall into two major clades (Figure 34.2). One clade, composed of echinoderms and hemichordates, is characterized by a three-part coelom and a bilaterally symmetrical, ciliated larva. The ancestors of the other clade, containing the chordates, had a distinctly different, nonfeeding, tadpole-like larva and a unique dorsal supporting structure.

Echinoderms: Pentaradial Symmetry

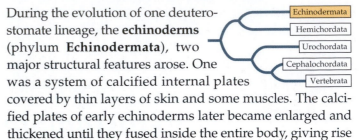

During the evolution of one deuterostomate lineage, the **echinoderms** (phylum **Echinodermata**), two major structural features arose. One was a system of calcified internal plates covered by thin layers of skin and some muscles. The calcified plates of early echinoderms later became enlarged and thickened until they fused inside the entire body, giving rise to an internal skeleton.

The other feature was a *water vascular system*, a network of water-filled canals leading to extensions called *tube feet*. This system functions in gas exchange, locomotion, and feeding (Figure 34.3*a*). Seawater enters the system through a perforated *madreporite*. A calcified canal leads from the madreporite to another canal that rings the *esophagus* (the tube leading from the mouth to the stomach). Other canals radiate from this *ring canal*, extending through the arms (in species that have arms) and connecting with the tube feet.

The development of these two structural innovations resulted in a striking evolutionary radiation. About 23 classes of echinoderms, of which only 6 survive today, have been described from fossils. The 13,000 species described from their fossil remains are probably only a small fraction of those that actually lived. Nearly all 7,000 species that survive today live only in marine environments. Some have bilaterally symmetrical, ciliated larvae (Figure 34.3*b*) that feed for some time as planktonic organisms before settling and transforming into adults with *pentaradial symmetry* (symmetry in five or multiples of five).

Living echinoderms are members of two lineages: subphylum **Pelmatozoa** and subphylum **Eleutherozoa**. These two groups differ in the form of their water vascular systems.

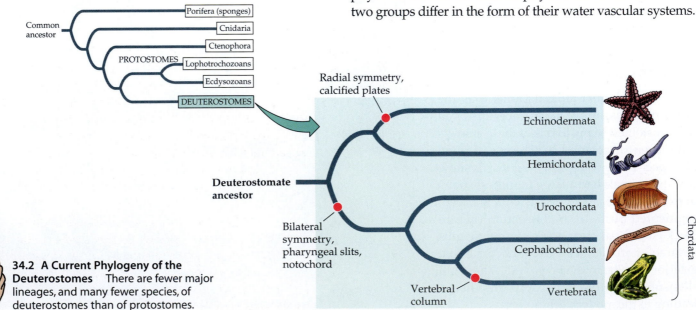

34.2 A Current Phylogeny of the Deuterostomes There are fewer major lineages, and many fewer species, of deuterostomes than of protostomes.

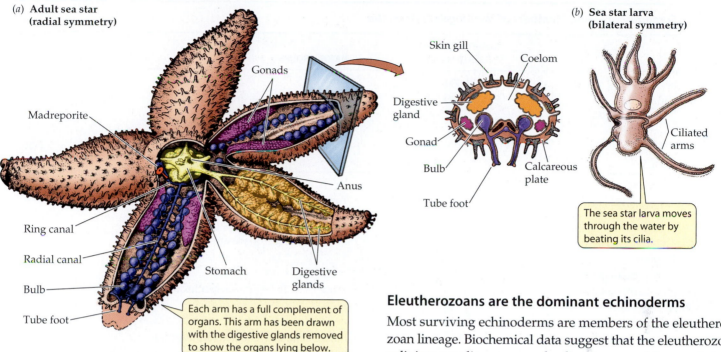

(*a*) **Adult sea star (radial symmetry)**

Gonads

Madreporite

Anus

Ring canal

Radial canal

Bulb

Stomach

Digestive glands

Tube foot

Each arm has a full complement of organs. This arm has been drawn with the digestive glands removed to show the organs lying below.

(*b*) **Sea star larva (bilateral symmetry)**

Skin gill

Coelom

Digestive gland

Gonad

Bulb

Calcareous plate

Tube foot

Ciliated arms

The sea star larva moves through the water by beating its cilia.

34.3 Echinoderms Display Two Evolutionary Innovations
(*a*) A dorsal view of a sea star displays the canals and tube feet of the echinoderm water vascular system, as well as a calcified internal skeleton. (*b*) The ciliated sea star larva has bilateral symmetry.

Pelmatozoans have jointed arms

Sea lilies and feather stars (class **Crinoidea**) are the only surviving pelmatozoans. Sea lilies were abundant 300–500 million years ago, but only about 80 species survive today. Most sea lilies attach to a substratum by means of a flexible stalk consisting of a stack of calcareous discs. The main body of the animal is a cup-shaped structure that contains a tubular digestive system. Five to several hundred arms, usually in multiples of five, extend outward from the cup. The jointed calcareous plates of the arms enable them to bend.

A sea lily feeds by orienting its arms in passing water currents. Food particles strike and stick to the tube feet, which are covered with mucus-secreting glands. The tube feet transfer these particles to grooves in the arms, where ciliary action carries the food to the mouth. The tube feet of sea lilies are also used for gas exchange and elimination of nitrogenous wastes.

Feather stars are similar to sea lilies, but they have flexible appendages with which they grasp the substratum (Figure 34.4*a*). Feather stars feed in much the same manner as sea lilies. They can walk on the tips of their arms or swim by rhythmically beating their arms. About 600 living species of feather stars have been described.

Eleutherozoans are the dominant echinoderms

Most surviving echinoderms are members of the eleutherozoan lineage. Biochemical data suggest that the eleutherozoa split into two lineages, one leading to sea urchins, sand dollars, and sea cucumbers, and the second leading to sea stars and brittle stars.

Sea urchins and sand dollars (class **Echinoidea**) lack arms, but they share a five-part body plan with all other echinoderms. Sea urchins are hemispherical animals that are covered with spines attached to the underlying skeleton via ball-and-socket joints (Figure 34.4*b*). The spines of sea urchins come in varied sizes and shapes; a few produce toxic substances. Many sea urchins consume algae, which they scrape from rocks with a complex rasping structure. Others feed on small organic debris that they collect with their tube feet or spines. Sand dollars, which are flattened and disc-shaped, feed on algae and fragments of organic matter found on the seafloor or suspended organic material.

The sea cucumbers (class **Holothuroidea**) lack arms, and their bodies are oriented in an atypical manner for an echinoderm. The mouth is anterior and the anus is posterior, not oral and aboral as in other echinoderms. Sea cucumbers use their tube feet primarily for attaching to the substratum rather than for moving. The anterior tube feet are modified into large, feathery, sticky tentacles that can be protruded from the mouth (Figure 34.4*c*). Periodically, the sea cucumber withdraws the tentacles, wipes off the material that has adhered to them, and digests it.

Sea stars (class **Asteroidea**; Figure 34.4*d*) are the most familiar echinoderms. Their digestive organs and gonads are located in the arms. Their tube feet serve as organs of locomotion, gas exchange, and attachment. Each tube foot of a sea star is also an adhesive organ consisting of an internal ampulla connected by a muscular tube to an external suction

34.1 Summary of Living Members of the Kingdom Animalia

PHYLUM	NUMBER OF LIVING SPECIES DESCRIBED	MAJOR GROUPS
Porifera: Sponges	10,000	
Cnidaria: Cnidarians	10,000	Hydrozoa: Hydras and hydroids Scyphozoa: Jellyfishes Anthozoa: Corals, sea anemones
Ctenophora: Comb jellies	100	
PROTOSTOMES		
Lophotrochozoans		
Platyhelminthes: Flatworms	20,000	Turbellaria: Free-living flatworms Trematoda: Flukes (all parasitic) Cestoda: Tapeworms (all parasitic) Monogenea (ectoparasites of fishes)
Rotifera: Rotifers	1,800	
Ectoprocta: Bryozoans	4,500	
Brachiopoda: Lamp shells	340	More than 26,000 fossil species described
Phoronida: Phoronids	20	
Nemertea: Ribbon worms	900	
Annelida: Segmented worms	15,000	Polychaeta: Polychaetes (all marine) Oligochaeta: Earthworms, freshwater worms Hirudinea: Leeches
Mollusca: Mollusks	50,000	Monoplacophora: Monoplacophorans Polyplacophora: Chitons Bivalvia: Clams, oysters, mussels Gastropoda: Snails, slugs, limpets Cephalopoda: Squids, octopuses, nautiloids
Ecdysozoans		
Kinorhyncha: Kinorhynchs	150	
Chaetognatha: Arrow worms*	100	
Nematoda: Roundworms	20,000	
Nematomorpha: Horsehair worms	230	
Onychophora: Onychophorans	80	
Tardigrada: Water bears	600	
Chelicerata: Chelicerates	70,000	Merostomata: Horseshoe crabs Arachnida: Scorpions, harvestmen, spiders, mites, ticks
Crustacea	50,000	Crabs, shrimps, lobsters, barnacles, copepods
Hexapoda	1,500,000	Insects
Myriapoda	13,000	Millipedes, centipedes
DEUTEROSTOMES		
Echinodermata: Echinoderms	7,000	Crinoidea: Sea lilies, feather stars Ophiuroidea: Brittle stars Asteroidea: Sea stars Concentricycloidea: Sea daisies Echinoidea: Sea urchins Holothuroidea: Sea cucumbers
Hemichordata: Hemichordates	95	Acorn worms and pterobranchs
Chordata: Chordates	50,000	Urochordata: Sea squirts Cephalochordata: Lancelets Agnatha: Lampreys, hagfishes Chondrichthyes: Cartilaginous fishes Osteichthyes: Bony fishes Amphibia: Amphibians Reptilia: Reptiles Aves: Birds Mammalia: Mammals

* The position of this phylum is uncertain. Many researchers place them in the deuterostomes.

(a) *Oxycomanthus bennetti*

(b) *Strongylocentrotus purpuratus*

(c) *Bohadschia argus*

(d) *Henricia leviuscula*

34.4 Diversity among the Echinoderms (a) The flexible arms of this golden feather star are clearly visible. (b) Purple sea urchins are important grazers on algae in the intertidal zone of the Pacific Coast of North America. (c) This sea cucumber lives on rocky substrata in the seas around Papua New Guinea. (d) The blood sea star is typical of many sea stars; some species, however, have more than five arms. (e) The arms of the brittle star are composed of hard but flexible plates.

(e) *Ophiothrix spiculata*

cup. The tube foot is moved by expansion and contraction of the circular and longitudinal muscles of the tube.

Many sea stars prey on polychaetes, gastropods, bivalves, and fish. They are important predators in many marine environments, such as coral reefs and rocky intertidal zones. With hundreds of tube feet acting simultaneously, a sea star can exert an enormous and continuous force. It can grasp a clam in its arms, anchor the arms with its tube feet, and, by steady contraction of the muscles in the arms, gradually exhaust the muscles the clam uses to keep its shell closed. Sea stars that feed on bivalves are able to push the stomach out through the mouth and then through the narrow space between the two halves of the bivalve's shell. The stomach secretes enzymes that digest the prey.

Brittle stars (class **Ophiuroidea**) are similar in structure to sea stars, but their flexible arms are composed of jointed hard plates (Figure 34.4e). Brittle stars generally have five arms, but each arm may branch a number of times. Most of the 2,000 species of brittle stars ingest particles from the upper regions of sediments and assimilate the organic material from them, but some species remove suspended food particles from the water; others capture small animals. Brittle stars eject the indigestible particles through their mouths because, unlike most other echinoderms, they have only one opening to the digestive tract.

An additional group, the sea daisies (class **Concentricycloidea**) were discovered only in 1986, and little is known about them. They have tiny disc-shaped bodies with a ring of marginal spines, and two ring canals, but no arms. Sea daisies are found on rotting wood in ocean waters. They apparently eat prokaryotes, which they digest outside their bodies and absorb either through a membrane that covers the oral surface or via a shallow, saclike stomach. Recent molecular data suggest that they are greatly modified sea stars.

Hemichordates: Conservative Evolution

Acorn worms and **pterobranchs** (phylum **Hemichordata**) are probably similar in form to the ancestor they share with the echinoderms. They have a three-part body plan, consisting of a proboscis, a collar, and a trunk.

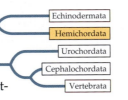

The 70 species of acorn worms range up to 2 meters in length. They live in burrows in muddy and sandy marine sediments. The large proboscis of an acorn worm is a digging organ (Figure 34.5a). It is coated with a sticky mucus that traps small organisms in the sediment. The mucus and its attached prey are conveyed by cilia to the mouth. In the esophagus, the food-laden mucus is compacted into a ropelike mass that is moved through the digestive tract by ciliary action. Behind the mouth is a muscular *pharynx*, a tube that connects the mouth to the intestine. The pharynx opens to the outside through a number of *pharyngeal slits* through which water can exit. Highly vascularized tissue surrounding the pharyngeal slits serves as a gas exchange apparatus. An acorn worm breathes by pumping water into its mouth and out through its pharyngeal slits.

The 10 living species of pterobranchs are sedentary animals up to 12 mm in length that live in a tube secreted by the proboscis. Some species are solitary; others form colonies of individuals joined together. Behind the proboscis is a collar

(a) *Saccoglossus kowalevskii*

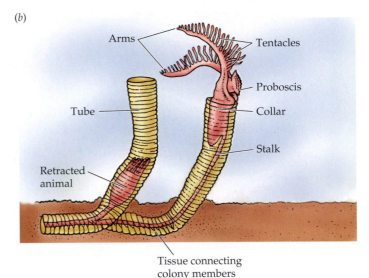

34.5 Hemichordates   (a) The proboscis of this acorn worm is modified for digging. This individual has been extracted from its burrow. (b) Pterobranchs may be colonial or solitary.

with 1–9 pairs of arms, bearing long tentacles that capture prey and function in gas exchange (Figure 34.5b).

Chordates: New Ways of Feeding

Members of the second major lineage of deuterostomes evolved several modifications of the coelom that provided new ways of capturing and handling food. They evolved a strikingly different body plan, characterized by an internal dorsal supporting structure. The pharyngeal slits, which originally functioned as sites for the uptake of O_2 and elimination of CO_2, and for eliminating water, were further enlarged. The result was a phylum (Chordata) of bilaterally

(a) *Rhopalaea crassa*

(b) *Pegea socia*

34.6 Urochordates (a) The tunic is clearly visible in this transparent sea squirt. (b) A chainlike colony of salps floats in tropical waters.

symmetrical animals with body plans characterized by the following shared features at some stage in their development:

▶ *Pharyngeal slits*
▶ A dorsal, hollow *nerve cord*
▶ A ventral *heart*
▶ A *tail* that extends beyond the anus
▶ A dorsal supporting rod, the *notochord*

The **notochord** is the distinctive derived trait of the lineage. It is composed of a core of large cells with turgid fluid-filled vacuoles that make it rigid but flexible. In some urochordates, the notochord is lost during metamorphosis to the adult stage. In vertebrates, it is replaced by other skeletal structures that provide support for the body.

The **tunicates** (subphylum **Urochordata**) may be similar to the ancestors of the chordates. All 2,500 species of tunicates are marine animals, most of which are sessile as adults. Their swimming, tadpole-like larvae reveal the close evolutionary relationship between tunicates and chordates (as Darwin realized; see Figure 25.4).

In addition to its pharyngeal slits, a tunicate larva has a dorsal, hollow nerve cord and a notochord that is restricted to the tail region. Bands of muscle surround the notochord, providing support for the body. After a short time swimming in the water, the larvae of most species settle to the seafloor and transform into sessile adults. The tunicate pharynx is enlarged into a *pharyngeal basket*, with which the animal feeds

by extracting plankton from the water. Some urochordates are solitary, but others produce colonies by asexual budding from a single founder.

There are three major urochordate groups: ascidians, thaliaceans, and larvaceans. More than 90 percent of the known species of tunicates are *ascidians* (sea squirts). Individual sea squirts range in size from less than 1 mm to 60 cm in length, but colonies may measure several meters across. The baglike body of an adult ascidian is enclosed in a tough tunic that is secreted by epidermal cells. The tunic is composed of proteins and a complex polysaccharide. Much of the body is occupied by a large pharyngeal basket lined with cilia, whose beating moves water through the animal (Figure 34.6a).

Thaliaceans (salps and others) float in tropical and subtropical oceans at all depths down to 1,500 meters (Figure 34.6b). They live singly or in chainlike colonies up to several meters long. *Larvaceans* are solitary planktonic animals usually less than 5 mm long. They retain their notochords and nerve cords throughout their lives.

The 25 species of **lancelets** (subphylum **Cephalochordata**) are small, fishlike animals that rarely exceed 5 cm in length. Their notochord extends the entire length of the body throughout their lives. Lancelets live partly buried in soft marine sediments. They extract small prey from the water with their pharyngeal baskets (Figure 34.7).

A jointed vertebral column replaced the notochord in vertebrates

In another chordate lineage, the enlarged pharyngeal basket came to be used to extract prey from mud. This lineage gave rise to the **vertebrates** (subphylum **Vertebrata**) (Figure 34.8). Vertebrates take their name from the jointed, dorsal **vertebral column** that replaced the notochord as their primary sup-

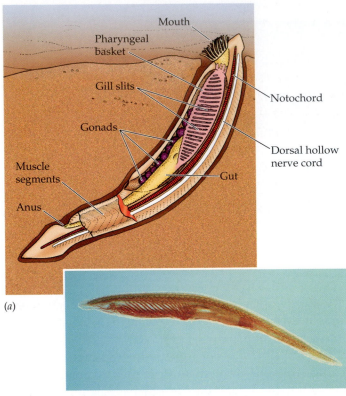

(a)

(b) *Branchiostoma* sp.

34.7 Lancelets (a) The internal structure of a lancelet. Note the large pharyngeal basket with gill slits. (b) This lancelet, which is about 6 cm long, has been excavated from the sediment to show its entire body.

port. The vertebrate body plan (Figure 34.9) can be characterized as follows:

▶ A rigid *internal skeleton*, with the vertebral column as its anchor, that provides support and mobility
▶ Two pairs of *appendages* attached to the vertebral column
▶ An anterior *skull* with a large *brain*
▶ Internal organs suspended in a large *coelom*
▶ A well-developed *circulatory system*, driven by contractions of a ventral *heart*

The ancestral vertebrates lacked jaws. They probably swam over the bottom, sucking up mud and straining it through the pharyngeal basket to extract microscopic food particles. The vascularized tissues of the basket also served a gas-exchange function. These animals gave rise to the jawless fishes.

One group of jawless fishes, called **ostracoderms** ("shell-skinned"), evolved a bony external armor that protected them from predators. With their heavy armor, these small fish could safely swim slowly above the substratum, which was easier than having to burrow through it, as all previous sediment feeders had done.

Jawless fishes could attach to dead organisms and use suction created by the pharynx to pull fluids and partly decomposed tissues into the mouth. Hagfishes and lampreys, the only jawless fishes to survive beyond the Devonian, feed on both dead and living organisms in this way (Figure 34.10). These fishes, often placed in the class **Agnatha**, have tough skins instead of external armor. They lack paired appendages

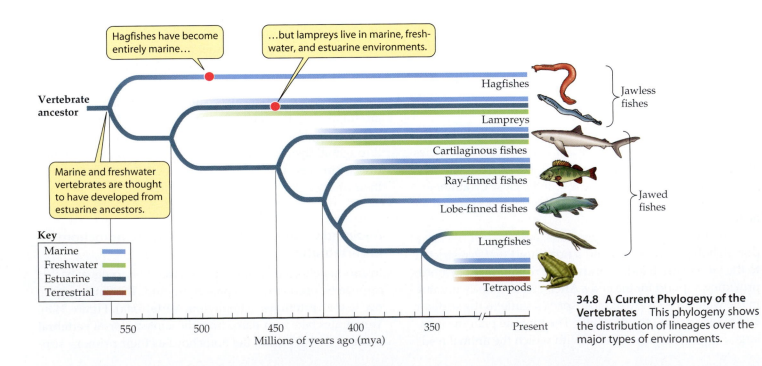

34.8 A Current Phylogeny of the Vertebrates This phylogeny shows the distribution of lineages over the major types of environments.

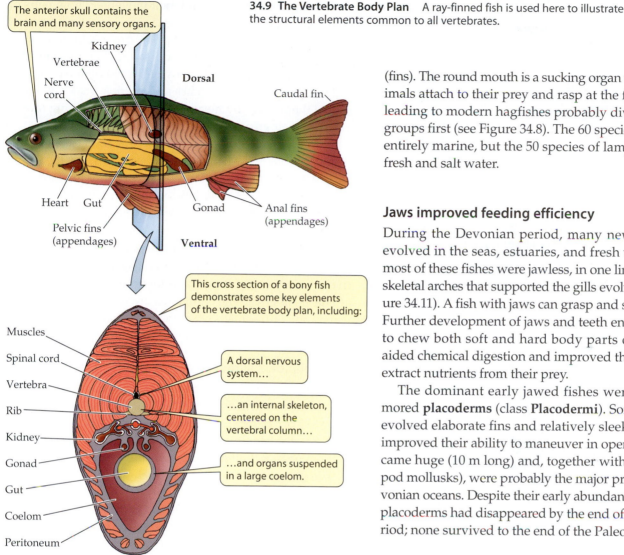

34.9 The Vertebrate Body Plan A ray-finned fish is used here to illustrate the structural elements common to all vertebrates.

The anterior skull contains the brain and many sensory organs.

Kidney

Vertebrae

Nerve cord

Dorsal

Caudal fin

Heart   Gut

Gonad

Anal fins (appendages)

Pelvic fins (appendages)

Ventral

This cross section of a bony fish demonstrates some key elements of the vertebrate body plan, including:

Muscles

Spinal cord

A dorsal nervous system…

Vertebra

Rib

…an internal skeleton, centered on the vertebral column…

Kidney

Gonad

Gut

…and organs suspended in a large coelom.

Coelom

Peritoneum

(fins). The round mouth is a sucking organ with which the animals attach to their prey and rasp at the flesh. The lineages leading to modern hagfishes probably diverged from other groups first (see Figure 34.8). The 60 species of hagfishes are entirely marine, but the 50 species of lampreys live in both fresh and salt water.

Jaws improved feeding efficiency

During the Devonian period, many new kinds of fishes evolved in the seas, estuaries, and fresh waters. Although most of these fishes were jawless, in one lineage, some of the skeletal arches that supported the gills evolved into jaws (Figure 34.11). A fish with jaws can grasp and subdue large prey. Further development of jaws and teeth enabled some fishes to chew both soft and hard body parts of prey. Chewing aided chemical digestion and improved the fishes' ability to extract nutrients from their prey.

The dominant early jawed fishes were the heavily armored **placoderms** (class **Placodermi**). Some of these fishes evolved elaborate fins and relatively sleek body forms that improved their ability to maneuver in open water. A few became huge (10 m long) and, together with squids (cephalopod mollusks), were probably the major predators in the Devonian oceans. Despite their early abundance, however, most placoderms had disappeared by the end of the Devonian period; none survived to the end of the Paleozoic era.

34.10 Modern Jawless Fishes (*a*) The Pacific hagfish. (*b*) Two sea lampreys using their large, jawless mouths to suck blood and flesh from a trout. The sea lamprey can live in either fresh or saltwater.

(*a*) *Eptatretus stouti*

(*b*) *Petromyzon marinus*

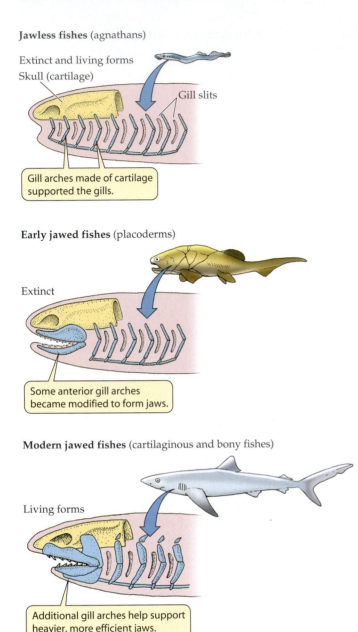

Jawless fishes (agnathans)

Extinct and living forms

Skull (cartilage)

Gill slits

Gill arches made of cartilage supported the gills.

Early jawed fishes (placoderms)

Extinct

Some anterior gill arches became modified to form jaws.

Modern jawed fishes (cartilaginous and bony fishes)

Living forms

Additional gill arches help support heavier, more efficient jaws.

34.11 Jaws from Gill Arches This series of diagrams illustrates one probable scenario for the evolution of jaws from the anterior gill arches of fishes.

Fins improved mobility

Several other groups of fishes became abundant during the Devonian period. **Cartilaginous fishes** (class **Chondrichthyes**)—the sharks, skates and rays, and chimaeras (Figure 34.12)—have a skeleton composed entirely of a firm but pliable material called *cartilage*. Their skin is flexible and leathery, sometimes bearing scales that give it the consistency of sandpaper.

Hagfishes
Lampreys
Cartilaginous fishes
Ray-finned fishes
Lobe-finned fishes
Lungfishes
Tetrapods

Cartilaginous fishes control their movement with pairs of unjointed appendages called *fins*: a pair of pectoral fins just behind the gill slits and a pair of pelvic fins just in front of the anal region (see Figure 34.9). A dorsal median fin stabilizes the fish as it moves. Sharks move forward by means of lateral undulations of their bodies and tail fins. Skates and rays propel themselves by means of vertical undulating movements of their greatly enlarged pectoral fins.

Most sharks are predators, but some feed by straining plankton from the water. The world's largest fish, the whale shark (*Rhincodon typhus*), is a filter feeder. It may grow to more than 12 meters in length and weigh more than 12,000 kilograms. Most skates and rays live on the ocean floor, where they feed on mollusks and other invertebrates buried in the sediments. Nearly all cartilaginous fishes live in the oceans, but a few are estuarine or migrate into lakes and rivers. One group of stingrays is found only in river systems of South America. The chimaeras are found in deep ocean waters and are seen less often than the sharks and rays.

Swim bladders allowed control of buoyancy

Ray-finned fishes (class **Actinopterygii**) have internal skeletons of calcified, rigid bone rather than flexible cartilage. The outer surface of most species of ray-finned fishes is covered with flat, thin, lightweight scales that provide some protection or enhance their movement through the water.

The gills of ray-finned fishes open into a single chamber covered by a hard flap. Movement of the flap improves the flow of water over the gills, where gas exchange takes place. Early ray-finned fishes also evolved gas-filled sacs that supplemented the action of the gills in respiration. These features enabled early ray-finned fishes to live where oxygen was periodically in short supply, as it often is in freshwater environments. The lunglike sacs evolved into *swim bladders*, which function as organs of buoyancy in most ray-finned fishes today. By adjusting the amount of gas in its swim bladder, a fish can control the depth at which it is suspended in the water without expending energy.

Ray-finned fishes radiated during the Tertiary into about 24,000 species, encompassing a remarkable variety of sizes, shapes, and lifestyles (Figure 34.13). The smallest are less than 1 cm long as adults; the largest weigh up to 900 kilograms. Ray-finned fishes exploit nearly all types of aquatic food sources. In the oceans they filter plankton from the water, rasp algae from rocks, eat corals and other colonial invertebrates, dig invertebrates from soft sediments, and prey upon virtually all other fishes. In fresh water they eat plankton, devour insects of all aquatic orders, eat fruits that fall into the water in flooded forests, and prey on other aquatic vertebrates and, occasionally, terrestrial vertebrates.

(a) *Triaenodon obesus*

(b) *Trygon pastinaca*

34.12 Cartilaginous Fishes (a) Most sharks, such as this whitetip reef shark, are active marine predators. (b) Skates and rays, represented here by a stingray, feed on the ocean bottom. Their modified pectoral fins are used for propulsion. (c) A chimaera, or ratfish. These deep-ocean fish often possess poisonous dorsal fins.

(c) *Chimaera* sp.

34.13 Diversity among Ray-Finned Fishes (a) The barracuda has the large teeth and powerful jaws of a predator. (b) The coral grouper lives on tropical coral reefs. (c) Commerson's frogfish can change its color over a range from pale yellow to orange-brown to deep red, thus enhancing its camouflage abilities. (d) This weedy sea dragon is difficult to see when it hides in vegetation. It is a larger relative of the more familiar seahorse.

(a) *Sphyraena barracuda*

(b) *Plectorhinchus chaetodonoides*

(c) *Antennarius commersonii*

(d) *Phyllopteryx taeniolatus*

Some fishes live buried in soft sediments, capturing passing prey or emerging at night to feed. Many fishes are solitary, but in open water others form large aggregations called *schools*. Many fishes perform complicated behaviors by means of which they maintain schools, build nests, court and choose mates, and care for their young.

Although ray-finned fishes can readily control their positions in open water, their eggs tend to sink. A few species produce small eggs that are buoyant enough to complete their development in the open water. However, most marine fishes move to food-rich shallow waters to lay their eggs, which is why coastal waters and estuaries are so important in the life cycles of many species. Some, such as salmon, actually abandon salt water when they breed, ascending rivers to spawn in freshwater streams and lakes.

Colonizing the Land: Obtaining Oxygen from the Air

The evolution of lunglike sacs in fishes appears to have been a response to the inadequacy of gills for respiration in oxygen-poor waters, but it also set the stage for the invasion of the land. Some early ray-finned fishes probably used their lungs to supplement their gills when oxygen levels in the water were low, as lungfishes do today. This ability would also have allowed them to leave the water temporarily and breathe air when pursued by predators unable to do so. But with their unjointed fins, these fishes could only flop around on land, as most fish out of water do today. Changes in the structure of the fins allowed these fishes to move on land.

The **lobe-finned fishes** (class **Actinistia**) were the first lineage to evolve jointed fins. Lobe-fins flourished from the Devonian period until about 65 million years ago, when they were thought to have become extinct. However, in 1938, a living lobe-fin was caught by commercial fishermen off South Africa. Since that time, several dozen specimens of this extraordinary fish, *Latimeria chalumnae*, have been collected. *Latimeria*, a predator on other fish, reaches a length of about 1.8 meters and weighs up to 82 kilograms (Figure 34.14*a*). Its skeleton is mostly composed of cartilage, not bone. A second species, *L. menadoensis*, was discovered in 1998 off the Indonesian island of Sulawesi.

Lungfishes (class **Dipnoi**) were important predators in shallow-water habitats in the Devonian, but most lineages died out. The three surviving species live in stagnant swamps and muddy waters in the Southern Hemisphere, one each in South America, Africa, and Australia (Figure 34.14*b*). Lungfishes have both gills and lungs. When ponds dry up, they can burrow deep into the mud and survive for many months in an inactive state.

It is believed that descendants of some lungfishes began to use terrestrial food sources, became more fully adapted to life on land, and eventually evolved to become the **tetrapods**—the four-legged amphibians, reptiles, birds, and mammals.

Amphibians invaded the land

During the Devonian period, **amphibians** (class **Amphibia**) arose from an ancestor they shared with lungfishes. In this lineage, stubby, jointed fins evolved into walking legs. The basic design of these legs has remained largely unchanged throughout the evolution of terrestrial vertebrates.

The Devonian predecessors of amphibians were probably able to crawl from one pond or stream to another by slowly pulling themselves along on their finlike legs, as do some modern species of catfishes. They gradually evolved the ability to live in swamps and, eventually, on dry land. Modern

34.14 Fishes with Jointed Fins (*a*) This lobe-fin fish, found in deep waters of the Indian Ocean, represents one of two surviving species of a lineage that was once thought to be extinct. (*b*) All surviving lungfish lineages live in the Southern Hemisphere.

(*a*) *Latimeria chalumnae*

(*b*) *Neoceratodus forsteri*

(a) *Dermophis mexicanus*

(b) *Gyrinophilus porphyriticus*

34.15 Diversity among the Amphibians (a) Burrowing caecilians superficially look more like worms than amphibians. (b) A Kentucky spring salamander. (c) This rare frog species was discovered in a national park on the island of Madagascar.

(c) *Scaphiophryne gottlebei*

amphibians have small lungs, and most species exchange gases through their skins as well. Most terrestrial species are confined to moist environments because they lose water rapidly through their skins when exposed to dry air, and because they require water for reproduction.

About 4,500 species of amphibians live on Earth today, many fewer than the number known only from fossils. Living amphibians belong to three orders (Figure 34.15): the wormlike, limbless, tropical, burrowing caecilians (order Gymnophiona), the frogs and toads (order Anura, which means "tailless"), and the salamanders (order Urodela, which means "tailed"). Most species of frogs and toads live in tropical and warm temperate regions, although a few are found at very high latitudes and altitudes. Some toads have tough skins that enable them to live for long periods of time in dry places. Salamanders are most diverse in temperate regions, but many species are found in cool, moist environments in Central American mountains. Many salamanders that live in rotting logs or moist soil lack lungs. They exchange gases entirely through the skin and mouth lining. Amphibians are the focus of much attention today because populations of many species are declining rapidly (see Chapter 1).

Most species of amphibians live in water at some time in their lives. In the typical amphibian life cycle, part or all of the adult stage is spent on land, but adults return to fresh wa-

ter to lay their eggs (Figure 34.16). Amphibian eggs can survive only in moist environments because they are enclosed within delicate envelopes that cannot prevent water loss in dry conditions. The fertilized eggs of most species give rise to larvae that live in water until they undergo metamorphosis to become terrestrial adults. Some amphibians, however, are entirely aquatic, never leaving the water at any stage of their lives. Others are entirely terrestrial, laying their eggs in moist places on land and skipping the aquatic larval stage.

Amniotes colonized dry environments

Two morphological changes contributed to the ability of one lineage of tetrapods to control water loss and, therefore, to exploit a wide range of terrestrial habitats:

▶ Evolution of an egg with a shell that is relatively impermeable to water
▶ A combination of traits that included a tough skin impermeable to water and kidneys that could excrete concentrated urine

The vertebrates that evolved both of these traits are called **amniotes**. They were the first vertebrates to become widely distributed over the terrestrial surface of Earth.

The amniote egg has a leathery or a brittle, calcium-impregnated shell that retards evaporation of the fluids inside but permits O_2 and CO_2 to pass through. Such an egg does not require a moist environment and can be laid anywhere. Within the shell and surrounding the embryo are membranes that protect the embryo from desiccation and assist its respiration and excretion of waste nitrogen. The egg also stores large quantities of food as *yolk*, permitting the embryo to attain a relatively advanced state of development before it hatches and must feed itself (Figure 34.17).

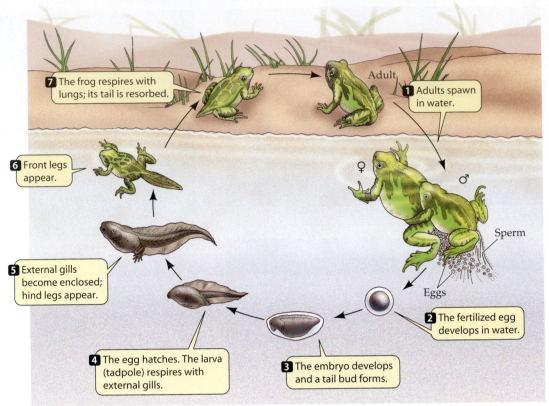

34.16 In and Out of the Water Most stages in the life cycle of temperate-zone frogs take place in water. The aquatic tadpole is transformed into a terrestrial adult through metamorphosis.

7 The frog respires with lungs; its tail is resorbed.

Adult

1 Adults spawn in water.

♀ ♂

Sperm

Eggs

2 The fertilized egg develops in water.

6 Front legs appear.

5 External gills become enclosed; hind legs appear.

4 The egg hatches. The larva (tadpole) respires with external gills.

3 The embryo develops and a tail bud forms.

An early amniote lineage, the **reptiles**, arose from a tetrapod ancestor in the Carboniferous period (Figure 34.18). The class "Reptilia," as we use the term here, is a paraphyletic group because some reptiles (crocodilians) are in fact more closely related to the birds than they are to lizards, snakes, and turtles (see Figure 25.8). However, because all members of "Reptilia" are structurally similar, it serves as a convenient group for discussing the characteristics of amniotes. Therefore, we use the traditional classification of "Reptilia" as a basis for our discussion while recognizing that, technically, the birds should be included within it.

Amphibia
"Reptilia"
Aves
Mammalia

About 6,000 species of reptiles live today. Most reptiles do not care for their eggs after laying them. In some species, the eggs do not develop shells, but are retained inside the female's body until they hatch. Some of these species evolved a structure called the *placenta* that nourishes the developing embryos.

The skin of a reptile is covered with horny scales that greatly reduce loss of water from the body surface. These scales, however, make the skin unavailable as an organ of gas exchange. In reptiles, gases are exchanged almost entirely by the lungs, which are proportionally much larger in surface area than those of amphibians. A reptile forces air into and out of its lungs by bellows-like movements of its ribs. The reptilian heart is divided into three and one-half or four chambers that partially separate oxygenated from unoxygenated blood. With this type of heart, reptiles can generate higher blood pressures than amphibians, which have three-chambered hearts, and can sustain higher levels of muscular activity.

Reptilian lineages diverged

The lineages leading to modern reptiles began to diverge about 250 mya. One lineage that has changed very little over the intervening millenia is the turtles (subclass **Testudines**). Turtles have a combination of ancestral traits and highly specialized characteristics that they do not share with any other vertebrate group. For this reason, their phylogenetic relationships are uncertain.

The dorsal and ventral bony plates of modern turtles and tortoises form a shell into which the head and limbs can be

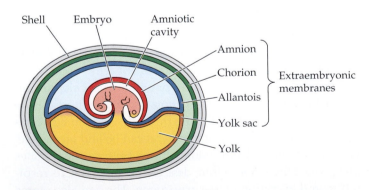

Shell Embryo Amniotic cavity

Amnion
Chorion — Extraembryonic membranes
Allantois
Yolk sac
Yolk

34.17 An Egg for Dry Places The evolution of the amniote egg, with its shell, four extraembryonic membranes, and embryo-nourishing yolk, was a major step in the colonization of the terrestrial environment.

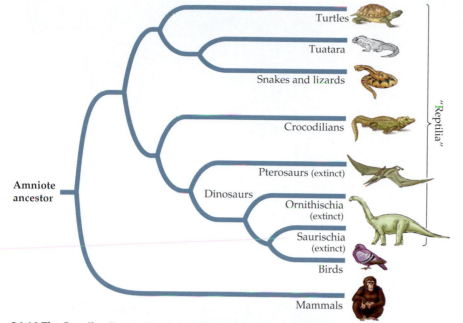

34.18 The Reptiles Form a Paraphyletic Group   The traditional classification of the amniotes creates the paraphyletic group "Reptilia." As used here, "Reptilia" does not include the birds (Aves), even though this major lineage split off from a dinosaur lineage relatively recently (in evolutionary terms).

withdrawn (Figure 34.19*a*). Most turtles live in lakes and ponds, but tortoises are terrestrial; some live in deserts. Sea turtles spend their entire lives at sea except when they come ashore to lay eggs. All seven species of sea turtles are endangered. A few species of turtles and tortoises are carnivores, but most species are omnivores that eat a variety of aquatic and terrestrial plants and animals.

The subclass **Squamata** includes lizards and snakes as well as the amphisbaenians (a group of legless, wormlike, burrowing animals with greatly reduced eyes). The tuataras (subclass **Sphenodontida**) are a sister group to the lizards and snakes. Sphenodontids were diverse dur-

(c) *Chamaeleo* sp.

(a) *Chelonia mydas*

(b) *Sphenodon punctatus*

(d) *Trimeresurus sumatranus*

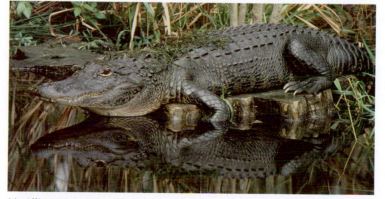

(e) *Alligator mississippiensis*

34.19 Reptilian Diversity   (*a*) The green sea turtle is widely distributed in tropical oceans. (*b*) This tuatara represents one of only two surviving species in a lineage that separated from lizards long ago. (*c*) The African chameleon, a lizard, has large eyes that move independ-ently in their sockets. (*d*) This venomous Sumatran pit viper is coiled to strike. (*e*) Alligators live in warm temperate environments in China and, like this one, in the southeastern United States.

ing the Mesozoic era, but today they are represented only by two species restricted to a few islands off the coast of New Zealand (Figure 34.19*b*). Tuataras superficially resemble lizards, but differ from them in tooth attachment and several internal anatomical features.

Most lizards are insectivores, but some are herbivores; a few prey on other vertebrates. The largest lizards, growing as long as 3 meters, are certain monitor lizards (such as the Komodo dragon) that live in the East Indies. Most lizards walk on four limbs (Figure 34.19*c*), but some are limbless, as are all snakes, which are descendants of burrowing lizards.

All snakes are carnivores; many can swallow objects much larger than themselves. This is the mode of feeding of the largest snakes, the pythons, which can grow to more than 10 meters long. Several snake lineages evolved a combination of venom glands and the ability to inject venom rapidly into their prey (Figure 34.19*d*).

A separate diverging lineage led to the crocodilians (subclass **Crocodylia**) and to the dinosaurs. The crocodilians—crocodiles, caimans, gharials, and alligators—are confined to tropical and warm temperate environments (Figure 34.19*e*). Crocodilians spend much of their time in water, but they build nests on land or on floating piles of vegetation. The eggs are warmed by heat generated by decaying organic matter that the parents place in the nest. Typically the female guards the eggs until they hatch. All crocodilians are carnivorous; they eat vertebrates of all classes, including large mammals.

The **dinosaurs** rose to prominence about 215 mya and dominated terrestrial environments for about 150 million years. During that time, virtually all terrestrial animals more than a meter in length were dinosaurs. Some of the largest dinosaurs weighed up to 100 tons. Many were agile and could run rapidly. The ability to breathe and run simultaneously, which we take for granted, was a major innovation in the evolution of terrestrial vertebrates. Not until the evolution of the lineages leading to the mammals, dinosaurs, and birds did the legs assume vertical positions directly under the body, which reduced the lateral forces on the body during locomotion. Special muscles that enabled the lungs to be filled and emptied while the limbs moved also evolved. We can infer the existence of such muscles in dinosaurs from the structure of the vertebral column in fossils and the capacity of many dinosaurs for bounding, bipedal (two-legged) locomotion.

Several fossil dinosaurs discovered recently in early Cretaceous deposits in Liaoning Province, in northeastern China, clearly show that in some small predatory dinosaurs, the scales had been highly modified to form feathers. One of these dinosaurs, *Microraptor gui*, had feathers on all four limbs, and those feathers were structurally similar to those of modern birds (Figure 34.20*a*).

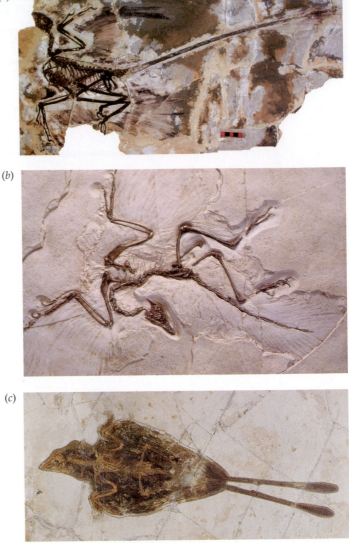

34.20 Mesozoic Birds and Their Ancestors Fossil remains demonstrate the probable evolution of birds from reptilian ancestors. (*a*) *Microraptor gui*, a feathered dinosaur from the early Cretaceous (about 140 mya). (*b*) *Archaeopteryx*, the oldest known bird. (*c*) The elongated tail feathers of a male *Confuciusornis sanctus* ("sacred bird of Confucius") fossil suggest that the males used them in courtship displays.

Birds: More Feathers and Better Flight

During the Mesozoic era, about 175 mya, a dinosaur lineage gave rise to the **birds** (subclass **Aves**). The oldest known avian fossil, *Archaeopteryx*, which lived about 150 mya, had teeth, unlike modern birds, but was covered with feathers that are virtually identical to those of modern birds. It also had well-developed wings, a long tail (Figure 34.20*b*), and a furcula, or "wishbone," to which some of the flight muscles were probably attached. *Archaeopteryx* had clawed fingers on its forelimbs, but it also had typical perching bird claws, suggesting that it lived in trees and shrubs and used the fingers to assist it in clambering over branches. Because the avian lineage separated from other reptiles long before *Archaeopteryx* lived, existing data are insufficient to identify the ancestors of birds with certainty. Most paleontologists believe that birds evolved from feathered terrestrial bipedal dinosaurs that used their forelimbs for capturing prey.

Many remains of other early birds have been discovered in 120–125-million-year-old fossil beds in northeastern China. One of these birds, *Confuciusornis sanctus*, is known from hundreds of complete specimens. The males had greatly elongated tail feathers (Figure 34.20c), which they probably used in communal courtship displays. Large numbers of individuals have been found together, as would be expected if many males assembled on communal display grounds, as some birds do today.

Birds range in size from the 2-gram bee hummingbird of the West Indies to the 150-kilogram ostrich (Figure 34.21). Some flightless birds of Madagascar and New Zealand known from fossils were even larger. These birds were exterminated by humans soon after they colonized those is-

lands. There are about 9,600 species of living birds, more than in any other major vertebrate group except ray-finned fishes.

As a group, birds eat almost all types of animal and plant material. A few aquatic species have bills modified for filtering small food particles from water. Insects are the most important dietary items for terrestrial species. Birds are major predators of flying insects during the day, and some species exploit that food source at night. In addition, birds eat fruits and seeds, nectar and pollen, leaves and buds, carrion, and other vertebrates. By eating the fruits and seeds of plants, birds serve as major agents of seed dispersal.

The feathers developed by some dinosaurs may originally have had thermoregulatory or display functions. Birds also use them for flying. Large quills that arise from the skin of the fore-

34.21 Diversity among the Birds (a) Penguins such as these gentoos are widespread in the cold waters of the Southern Hemisphere. They are expert swimmers, although they have lost the ability to fly. (b) Perching birds, represented here by a male northern cardinal, are the most species-rich of all the bird lineages. (c) Parrots are a diverse group of birds, especially in the Tropics of Asia, South America, and Australia. This king parrot is one member of Australia's rich parrot fauna. (d) The flightless ostrich is the largest bird species in existence today.

(a) *Pygoscelis papua*

(b) *Cardinalis cardinalis*

(c) *Alisterus scapularis*

(d) *Struthio camelus*

limbs create the flying surfaces of wings. Other strong feathers sprout like a fan from the shortened tail and serve as stabilizers during flight. The feathers that cover the body, along with an underlying layer of down feathers, provide insulation.

The bones of birds are modified for flight. They are hollow and have internal struts for strength. The *sternum* (breastbone) forms a large, vertical keel to which the flight muscles are attached. These muscles pull the wings downward during the main propulsive movement in flight. Flight is metabolically expensive. A flying bird consumes energy at a rate about 15–20 times faster than a running lizard of the same weight! Because birds have such high metabolic rates, they generate large amounts of heat. They control the rate of heat loss using their feathers, which may be held close to the body or elevated to alter the amount of insulation they provide.

The brain of a bird is larger in proportion to its body than a lizard or crocodile brain, primarily because the cerebellum, the center of sight and muscular coordination, is enlarged.

Most birds lay their eggs in a nest, where they are warmed by heat from an adult that sits on them. Because birds have such high body temperatures, the eggs of most species hatch within a few weeks. The offspring of many species are *altricial* (hatch at a relatively helpless stage) and are fed for some time by their parents. The young of other bird species, such as chickens, sandpipers, and ducks, are *precocial* (can feed themselves shortly after hatching). Adults of nearly all species attend their offspring for some time, warning them of and protecting them from predators, protecting them from bad weather, leading them to good foraging places, and feeding them.

The Origin and Diversity of Mammals

Mammals (class **Mammalia**) appeared in the early part of the Mesozoic era, about 225 million years ago, branching from a lineage of mammal-like reptiles. Small mammals coexisted with reptiles and dinosaurs for at least 150 million years. After the large reptiles and dinosaurs disappeared during the mass extinction at the close of the Mesozoic era, mammals increased dramatically in numbers, diversity, and size. Today, mammals range in size from tiny shrews and bats weighing only about 2 grams to the endangered blue whale, which measures up to 33 meters long and weighs up to 160,000 kilograms—the largest animal ever to live on Earth.

Skeletal simplification accompanied the evolution of early mammals from their larger reptilian ancestors. During mammalian evolution, some bones from the lower jaw were incorporated into the middle ear, leaving a single bone in the lower jaw. The number of bones in the skull also decreased. The bulk of both the limbs and the bony girdles from which they are suspended was reduced. Mammals have far fewer, but more highly differentiated, teeth than reptiles do. Differences in the number, type, and arrangement of teeth in mammals reflect their varied diets.

Skeletal features are readily preserved as fossils, but the soft parts of animals are seldom fossilized. Therefore, we do not know when mammalian features such as mammary glands, sweat glands, hair, and a four-chambered heart evolved. Mammals are unique among animals in supplying their young with a nutritive fluid (milk) secreted by mammary glands. Mammalian eggs are fertilized within the female's body, and the embryos undergo a period of development, called *gestation*, within a specialized organ, the *uterus*, prior to being born. In many species, the embryos are connected to the uterus and nourished by a placenta. In addition, mammals have a protective and insulating covering of hair, which is luxuriant in some species but has been almost entirely lost in whales, dolphins, and humans. In whales and dolphins, thick layers of insulating fat (blubber) replace hair as a heat-retention mechanism. Clothing assumes the same role for humans. The approximately 4,000 species of living mammals are divided into two major subclasses: Prototheria and Theria. The subclass **Prototheria** contains a single order, the Monotremata, with a total of three species, which are found only in Australia and New Guinea. These mammals, the duck-billed platypus and the spiny anteaters, or echidnas, differ from other mammals in lacking a placenta, laying eggs, and having legs that poke out to the side (Figure 34.22). Monotremes supply milk for their young, but they have no nipples on their mammary glands; rather, the milk simply oozes out and is lapped off the fur by the offspring.

Members of the other subclass, **Theria**, are further divided into two groups. In most species of the first group, the **Marsupialia**, females have a ventral pouch in which they carry and feed their offspring (Figure 34.23a). Gestation in marsupials is short; the young are born tiny but with well-developed forelimbs, with which they climb to the pouch. They attach to a nipple, but cannot suck. The mother ejects milk into the tiny offspring until they grow large enough to suckle. Once her offspring have left the uterus, a female marsupial may become sexually receptive again. She can then carry fertilized eggs capable of initiating development and replacing the offspring in her pouch should something happen to them.

There are about 240 living species of marsupials. At one time marsupials were found on all continents, but today the majority of species are restricted to the Australian region, with a modest representation in South America (Figure 34.23b). One species, the Virginia opossum, is widely distributed in the United States. Marsupials radiated to become terrestrial herbivores, insectivores, and carnivores, but no marsupial species live in the oceans or can fly, although some are gliders. The largest living marsupial is the red kangaroo of Australia (Figure 34.23a), which weighs up to 90 kilo-

(a) *Tachyglossus aculeata*

34.22 Monotremes (a) The short-beaked echidna is one of the two surviving species of echidnas. (b) The duck-billed platypus is the other surviving monotreme species.

(b) *Ornithorhynchus anatinus*

birth than are marsupials, and no external pouch houses them after birth. The nearly 4,000 species of eutherians are placed into 16 major groups (Figure 34.24), the largest of which is the rodents (order Rodentia) with about 1,700 species. The next largest group, the bats (order Chiroptera), has about 1,000 species, followed by the moles and shrews (order Insectivora) with slightly more than 400 species.

Eutherians are extremely varied in their form and ecology. Several lineages of terrestrial eutherians subsequently colonized marine environments to become whales, dolphins, seals, and sea lions. Eutherian mammals are—or

grams. Much larger marsupials existed in Australia until they were exterminated by humans soon after they reached the continent (about 50,000 years ago).

Most living mammals belong to the second therian group, the **eutherians**. (Eutherians are sometimes called *placental mammals*, but this name is not accurate because some marsupials also have placentas.) Eutherians are more developed at

were, until they were greatly reduced in numbers by humans—the most important grazers and browsers in most terrestrial ecosystems. Grazing and browsing have been an evolutionary force intense enough to select for the spines, tough leaves, and difficult-to-eat growth forms found in many plants—a striking example of coevolution.

Primates and the Origin of Humans

A eutherian lineage that has had dramatic effects on ecosystems worldwide is the **primate** lineage, which has undergone extensive recent evolutionary radiation. Primates probably

(a) *Macropus rufus*

(b) *Caluromys philander* (c) *Sarcophilus harrisii*

34.23 Marsupials (a) Australia's red kangaroos are the largest living marsupials. The marsupial radiation also produced (b) arboreal species, such as this South American opossum, and (c) carnivores, such as the Tasmanian devil.

(b) *Carollia perspicillata*

(a) *Citellus parryi*

(c) *Stenella longirostris*

(d) *Rangifer tarandus*

34.24 Diversity among the Eutherians (*a*) The Arctic ground squirrel is one of the many species of small, diurnal rodents found in North America. (*b*) Temperate-zone bats are all insectivores, but many tropical bats, such as this leaf-nosed bat, eat fruit. (*c*) These Hawaiian spinner dolphins represent a eutherian lineage that colonized the marine environment. (*d*) Large hoofed mammals are important herbivores in terrestrial environments. This caribou bull is grazing by himself, although caribou are often seen in huge herds.

Primates and the Origin of Humans

A eutherian lineage that has had dramatic effects on ecosystems worldwide is the **primate** lineage, which has undergone extensive recent evolutionary radiation. Primates probably descended from small *arboreal* (tree-living) insectivorous mammals early in the Cretaceous period. A nearly complete fossil of an early primate species, *Carpolestes*, from Wyoming, dated at 56 mya, had grasping feet with an opposable big toe that had a nail rather than a claw. Such grasping limbs are one of the major adaptations to arboreal life that distinguish primates from other mammals. However, *Carpolestes* did not have eyes positioned on the front of the face to provide good depth perception, as all modern primates do.

Early in its evolutionary history, the primate lineage split into two main branches, the prosimians and the anthropoids (Figure 34.25). **Prosimians**—lemurs, pottos, and lorises—once lived on all continents, but today they are restricted to Africa, Madagascar, and tropical Asia (Figure 34.26). All of the mainland prosimian species are arboreal and nocturnal. However, on the island of Madagascar, the site of a remarkable prosimian radiation, there are also diurnal and terrestrial species.

The **anthropoids**—tarsiers, monkeys, apes, and humans—evolved from an early primate lineage about 55 million years ago in Africa or Asia. New World monkeys diverged from Old World monkeys early enough that they could have reached South America from Africa when those two continents were still close to each other. All New World monkeys are arboreal (Figure 34.27*a*). Many of them have long, *prehensile* (grasping) tails with which they can hold onto branches. Many Old World primates are arboreal as well, but a number of species are terrestrial. Some of these species, such as baboons and macaques, live and travel in large groups (Figure 34.27*b*). No Old World primates have prehensile tails.

About 22 million years ago, the lineage that led to modern **apes** separated from the other Old World primates. Between 22 and 5.5 mya, as many as 100 species of apes ranged over Europe, Asia, and Africa. About 9 mya, members of one

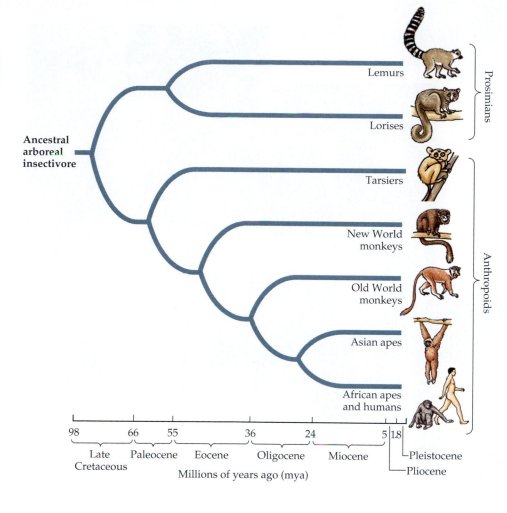

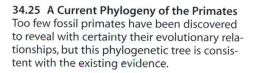

34.25 A Current Phylogeny of the Primates
Too few fossil primates have been discovered to reveal with certainty their evolutionary relationships, but this phylogenetic tree is consistent with the existing evidence.

Lemurs

Lorises

Prosimians

Tarsiers

New World monkeys

Old World monkeys

Asian apes

African apes and humans

Anthropoids

Ancestral arboreal insectivore

| 98 | 66 | 55 | 36 | 24 | 5 | 1.8 |

Late Cretaceous | Paleocene | Eocene | Oligocene | Miocene | Pleistocene
Pliocene

Millions of years ago (mya)

(a) *Leontopithecus rosalia*

(a) *Eulemur fulvus*

(b) *Loris tardigradus*

(b) *Macaca sylvanus*

34.26 Prosimians (a) The brown lemur is one of the many lemur species found in Madagascar, where they are part of a unique assemblage of plants and animals. (b) The slender loris is found in India. Its large eyes tell us that it is nocturnal.

34.27 Monkeys (a) Golden lion tamarins are endangered New World monkeys living in coastal Brazilian rainforests. (b) Many Old World species, such as these Barbary macaques, live in social groups. Here two members of a group groom each other.

(a) *Gorilla gorilla*

(b) *Pan troglodytes*

(c) *Hylobates lar*

(d) *Pongo pygmaeus*

34.28 Apes (a) Gorillas, the largest apes, are restricted to humid African forests. This male is a lowland gorilla. (b) Chimpanzees, our closest relatives, are found in forested regions of Africa. (c) Gibbons are the smallest of the apes. The common gibbon is found in Asia, from India to Borneo. (d) Orangutans live in the forests of Indonesia.

Human ancestors evolved bipedal locomotion

The **hominids**—the lineage that led to humans—separated from other ape lineages about 6 mya in Africa. The earliest protohominids, known as **ardipithecines**, had distinct morphological adaptations for **bipedalism**—locomotion in which the body is held erect and moved exclusively by movements of the hind legs. Bipedal locomotion frees the forelimbs to manipulate objects and to carry them while walking. It also elevates the eyes, enabling the animal to see over tall vegetation to spot predators and prey. At walking rates, bipedal movement is also energetically much more economical than quadrupedal (four-legged) locomotion. All three advantages were probably important for the ardipithecines and their descendants, the **australopithecines**.

The first australopithecine skull was found in South Africa in 1924. Since then, australopithecine fossils have been found in many sites in Africa. The most complete fossil skeleton of an australopithecine, approximately 3.5 million years old, was discovered in Ethiopia in 1974. That individual, a young female known to the world as Lucy, was assigned to the species *Australopithecus afarensis*. Fossil remains of more than 100 *A. afarensis* have now been discovered. During the past 5

years, fossils of other australopithecines that lived in Africa 4–5 million years ago have been unearthed.

Experts disagree over how many species are represented by the australopithecine fossils, but it is clear that several million years ago, at least two distinct types lived together over much of eastern Africa. The larger type (about 40 kilograms) is represented by at least two species (*Paranthropus robustus* and *P. boisei*), both of which died out suddenly about 1.5 million years ago.

Humans arose from australopithecine ancestors

Early members of the genus *Homo* lived contemporaneously with australopithecines for perhaps half a million years (Figure 34.29). The oldest fossils of the genus, an extinct species called *H. habilis*, were discovered in the Olduvai Gorge, Tanzania. These fossils are estimated to be 2 million years old. Other fossils of *H. habilis* have been found in Kenya and Ethiopia. Associated with the fossils are tools that these early hominids used to obtain food.

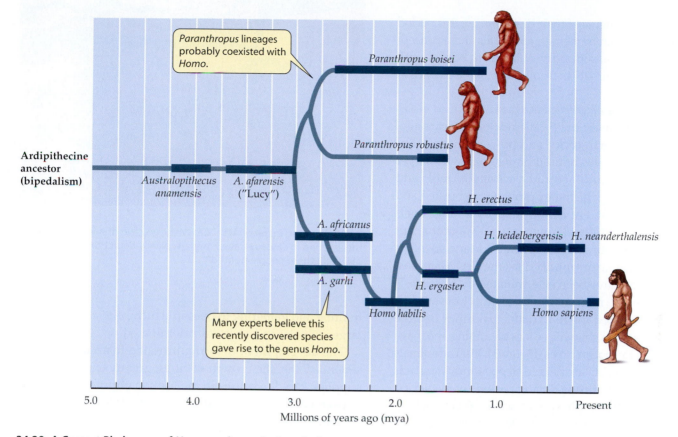

34.29 A Current Phylogeny of *Homo sapiens* At times in the past, more than one species of hominid lived on Earth. The heavy dark blue lines indicate the time frame over which each species lived.

Another extinct species of our genus, *Homo erectus*, evolved in Africa about 1.6 mya. Soon thereafter it had spread as far as eastern Asia. As it expanded its range and increased in abundance, *H. erectus* may have exterminated *H. habilis*. Members of *H. erectus* were as large as modern people, but their bones were considerably heavier. *Homo erectus* used fire for cooking and for hunting large animals, and made characteristic stone tools that have been found in many parts of the Old World. Although *H. erectus* survived in Eurasia until about 250,000 years ago, it was replaced in tropical regions by our species, *Homo sapiens*, about 200,000 years ago.

Human brains became larger

The earliest members of *Homo sapiens* had larger brains than members of the earlier species of *Homo*. Brain size in the lineage increased rapidly, reaching modern size by about 160,000 years ago. This striking change was probably favored by an increasingly complex social life. The ability of group members to communicate with one another would have been valuable in cooperative hunting and gathering and for improving one's status in the complex social interactions that must have characterized early human societies, just as they do ours today. But why did brains become larger only in the human lineage?

A clue to the answer is provided by brain chemistry. The human brain is a fat-rich organ. About 60 percent of its structural material is made up of lipids, most of them long-chain polyunsaturated omega-3 and omega-6 fatty acids. Humans must consume omega fatty acids in their diet because the body cannot synthesize these molecules fast enough from the other fatty acids found in vegetables, nuts, and seeds to supply their brains. Animal brains and livers contain omega fatty acids, but fish and shellfish are by far the best sources.

Therefore, because savannas and open woodlands provide few sources of omega fatty acids, the traditional view that early human evolution took place in those environments is being questioned. In contrast, the shores of Africa's many lakes would have been rich sources of fish and mollusks. Thus, access to fat-rich foods from aquatic environments may have been the key factor that supported the dramatic expansion of the human brain. The archeological record of the past 100,000 years includes hundreds of piles of mollusk shells and fish bones, as well as carved points used for fishing. Chimpanzees remained in the forest and ate fruits and nuts. They may have lacked food sources to support much larger brains.

Several *Homo* species existed during the mid-Pleistocene epoch, from about 1.5 million to about 300,000 years ago. All were skilled hunters of large mammals, but plants continued to be important components of their diets. During this period another distinctly human trait emerged: rituals and a concept of life after death. Deceased individuals were buried with tools and clothing, presumably supplies for their existence in the next world.

One species, *Homo neanderthalensis*, was widespread in Europe and Asia between about 75,000 and 30,000 years ago. Neanderthals were short, stocky, and powerfully built humans whose massive skulls housed brains somewhat larger than our own. They manufactured a variety of tools and hunted large mammals, which they probably ambushed and subdued in close combat. For a short time, their range overlapped that of the *H. sapiens* known as Cro-Magnons, but then the Neanderthals abruptly disappeared. Many scientists believe that they were exterminated by the Cro-Magnons, just as *H. habilis* may have been exterminated by *H. erectus*.

Cro-Magnon people made and used a variety of sophisticated tools. They created the remarkable paintings of large mammals, many of them showing scenes of hunting, that have been discovered in European caves (Figure 34.30). The animals depicted were characteristic of the cold steppes and grasslands that occupied much of Europe during periods of glacial expansion. Cro-Magnon people spread across Asia, reaching North America perhaps as early as 20,000 years ago, although the date of their arrival in the New World is still uncertain. Within a few thousand years, they had spread southward through North America to the southern tip of South America.

Humans evolved language and culture

As our ancestors evolved larger brains, their behavioral capabilities increased, especially the capacity for language. Most animal communication consists of a limited number of signals, which refer mostly to immediate circumstances and are associated with changed emotional states induced by those circumstances. Human language is far richer in its symbolic character than any other animal vocalizations. Our words can refer to past and future times and to distant places. We are capable of learning thousands of words, many of them referring to abstract concepts. We can rearrange words to form sentences with complex meanings.

The expanded mental abilities of humans are largely responsible for the development of **culture**, the process by which knowledge and traditions are passed along from one generation to another by teaching and observation. Culture can change rapidly because genetic changes are not necessary for a cultural trait to spread through a population. A potential disadvantage of culture is that its norms must be taught to each generation.

Cultural learning greatly facilitated the spread of domestic plants and animals and the resultant conversion of most human societies from ones in which food was obtained by

34.30 Hunting Inspires Art Cro-Magnon cave drawings such as those found in Lascaux Cave, France, typically depict the large mammals that these people hunted.

hunting and gathering to ones in which *pastoralism* (herding large animals) and *agriculture* dominated.

The development of agriculture led to an increasingly sedentary life, the growth of cities, greatly expanded food supplies, rapid increases in the human population, and the appearance of occupational specializations, such as artisans, shamans, and teachers.

Deuterostomes and Protostomes: Shared Evolutionary Themes

The evolution of deuterostomes paralleled the evolution of protostomes in several important ways. Both lineages exploited the abundant food supplies buried in soft marine substrata, attached to rocks, or suspended in water. Many groups of both lineages developed elaborate structures for moving water and extracting prey from it.

In some lineages of both groups, the body cavity became divided into compartments that allowed better control of shape and movement. Some members of both groups evolved mechanisms for controlling their buoyancy in water using gas-filled internal spaces. Planktonic larval stages evolved in marine members of many protostomate and deuterostomate phyla.

Both protostomes and deuterostomes colonized the land, but the consequences were very different. The jointed external skeletons of arthropods, although they provide excellent support and protection in air, cannot support large animals, as the internal skeletons developed by deuterostomes can.

Terrestrial deuterostomes recolonized aquatic environments a number of times. Suspension feeding evolved once again in several of these lineages. The largest living animals, baleen (toothless) whales, feed upon small prey only a few centimeters long, which they extract from the water with large straining structures in their mouths.

Chapter Summary

Origins of the Deuterostomes

▶ The deuterostomate lineage separated from the protostomate lineage early in animal evolution. The ancestral deuterostome had external gills. **Review Figure 34.1**

▶ There are only two major deuterostomate lineages, and there are fewer species of deuterostomes than protostomes, but as members of the lineage, we have a special interest in its members. **Review Figure 34.2. See Web/CD Activity 34.1**

Echinoderms: Pentaradial Symmetry

▶ Echinoderms have a pentaradially symmetrical body plan, a unique water vascular system, and a calcified internal skeleton. **Review Figure 34.3***a*

▶ Nearly all living species of echinoderms have a bilaterally symmetrical, ciliated larva that feeds as a planktonic organism. **Review Figure 34.3***b*

▶ Six major groups of echinoderms survive today, but 23 other lineages existed in the past. Some groups of echinoderms have arms, but others do not.

Hemichordates: Conservative Evolution

▶ Acorn worms and pterobranchs are similar to ancestral deuterostomes. **Review Figure 34.5**

Chordates: New Ways of Feeding

▶ Members of another deuterostomate lineage evolved enlarged pharyngeal slits used as feeding devices and a dorsal supporting rod, the notochord.

▶ Most urochordates are sessile as adults and filter prey from seawater with large pharyngeal baskets. But some species retain their notochords and nerve cords as planktonic adults.

Evolution of the Chordates

▶ Cephalochordates probably resemble the ancestors of all other chordates. **Review Figure 34.7**

▶ Vertebrates evolved jointed internal skeletons that enabled them to swim rapidly. Early vertebrates used the pharyngeal basket to filter small animals from mud. **Review Figures 34.8, 34.9**

▶ Jaws, which evolved from anterior gill arches, enabled their possessors to grasp and chew their prey. Jawed fishes rapidly became dominant animals in both marine and fresh waters. **Review Figure 34.11**

▶ Fishes evolved two pairs of unjointed fins, with which they control their swimming movements and stabilize themselves in the water, and swim bladders, which help keep them suspended in open water.

▶ Ray-finned fishes come in a wide variety of sizes and shapes. Many species have complex social systems.

Colonizing the Land: Obtaining Oxygen from the Air

▶ Two lineages of fishes—lobe-finned fishes and lungfishes—evolved jointed fins.

▶ Amphibians, the first terrestrial vertebrates, arose from lungfish ancestors.

▶ The 4,500 species of amphibians living today belong to three groups: caecilians, frogs and toads, and salamanders.

▶ Most amphibians live in water at some time in their lives, and their eggs must remain moist. **Review Figure 34.16. See Web/CD Tutorial 34.1**

▶ Amniotes evolved eggs with shells impermeable to water and thus became the first vertebrates to be independent of water for reproduction. **Review Figure 34.17. See Web/CD Activity 34.2**

▶ Modern reptiles are members of four lineages: snakes and lizards, tuataras, turtles and tortoises, and crocodilians. **Review Figure 34.18**

▶ Dinosaurs rose to dominance about 215 mya and dominated terrestrial environments for about 150 million years until they became extinct about 65 mya.

▶ Some dinosaurs evolved feathers and were capable of flight.

Birds: More Feathers and Better Flight

▶ Birds arose about 175 mya from feathered dinosaur ancestors.

▶ The 9,600 species of birds are characterized by feathers, high metabolic rates, and parental care.

The Origin and Diversity of Mammals

▶ Mammals evolved during the Mesozoic era, about 225 mya.

▶ The eggs of mammals are fertilized within the body of the female, and embryos develop for some time within a uterus before being born. Mammals are unique in suckling their young with milk secreted by mammary glands.

▶ The three species of mammals in subclass Prototheria lay eggs, but all other mammals give birth to live young.

▶ Therian mammals are divided into two major groups: the marsupials, which give birth to tiny young that are, in most species, raised in a pouch on the female's belly, and the eutherians, which give birth to relatively well-developed offspring.

Primates and the Origin of Humans

▶ The primates split into two major lineages, one leading to the prosimians (lemurs and lorises) and the other leading to the tarsiers, monkeys, apes, and humans. **Review Figure 34.25**

▶ Hominids evolved in Africa from terrestrial, bipedal ancestors. **Review Figure 34.29**

▶ Early humans evolved large brains, language, and culture. They manufactured and used tools, developed rituals, and domesticated plants and animals. In combination, these traits enabled humans to increase greatly in number and to transform the face of Earth.

Deuterostomes and Protostomes: Shared Evolutionary Themes

▶ Both protostomes and deuterostomes evolved structures to filter prey from the water, mechanisms to control their buoyancy in water, and planktonic larval stages.

See Web/CD Activity 34.3 for a concept review of this chapter.

Self-Quiz

1. Which of the following deuterostomate groups have a three-part body plan?
   a. Acorn worms and tunicates
   b. Acorn worms and pterobranchs
   c. Pterobranchs and tunicates
   d. Pterobranchs and lancelets
   e. Tunicates and lancelets

2. The structure used by adult ascidians to capture food is a
   a. pharyngeal basket.
   b. proboscis.
   c. lophophore.
   d. mucus net.
   e. radula.

3. The pharyngeal gill slits of chordates originally functioned as sites for
   a. uptake of oxygen only.

b. release of carbon dioxide only.

c. both uptake of oxygen and release of carbon dioxide.

d. removal of small prey from the water.

e. forcible expulsion of water to move the animal.

4. The key to the vertebrate body plan is a
   a. pharyngeal basket.
   b. vertebral column to which internal organs are attached.
   c. vertebral column to which two pairs of appendages are attached.
   d. vertebral column to which a pharyngeal basket is attached.
   e. pharyngeal basket and two pairs of appendages.

5. Which of the following fishes do *not* have a cartilaginous skeleton?
   a. Chimaeras
   b. Lungfishes
   c. Sharks
   d. Skates
   e. Rays

6. In most fishes, lunglike sacs evolved into
   a. pharyngeal gill slits.
   b. true lungs.
   c. coelomic cavities.
   d. swim bladders.
   e. none of the above

7. Most amphibians return to water to lay their eggs because
   a. water is isotonic to egg fluids.
   b. adults must be in water while they guard their eggs.
   c. there are fewer predators in water than on land.
   d. amphibians need water to produce their eggs.
   e. amphibian eggs quickly lose water and desiccate if their surroundings are dry.

8. The horny scales that cover the skin of reptiles prevent them from
   a. using their skin as an organ of gas exchange.
   b. sustaining high levels of metabolic activity.
   c. laying their eggs in water.
   d. flying.
   e. crawling into small spaces.

9. Which statement about bird feathers is *not* true?
   a. They are highly modified reptilian scales.
   b. They provide insulation for the body.
   c. They exist in two layers.
   d. They help birds fly.
   e. They are important sites of gas exchange.

10. Monotremes differ from other mammals in that they
    a. do not produce milk.
    b. lack body hair.
    c. lay eggs.
    d. live in Australia.
    e. have a pouch in which the young are raised.

11. Bipedalism is believed to have evolved in the human lineage because bipedal locomotion is
    a. more efficient than quadrupedal locomotion.
    b. more efficient than quadrupedal locomotion, and it frees the forelimbs to manipulate objects.
    c. less efficient than quadrupedal locomotion, but it frees the forelimbs to manipulate objects.
    d. less efficient than quadrupedal locomotion, but bipedal animals can run faster.
    e. less efficient than quadrupedal locomotion, but natural selection does not act to improve efficiency.

For Discussion

1. In what animal phyla has the ability to fly evolved? How do the structures used for flying differ among these animals?

2. Extracting suspended food from the water column is a common mode of foraging among animals. Which groups contain species that extract prey from the air? Why is this mode of obtaining food so much less common than extracting prey from the water?

3. Large size both confers benefits and poses certain risks. What are these risks and benefits?

4. Amphibians have survived and prospered for many millions of years, but today many species are disappearing and populations of others are declining seriously. What features of amphibian life histories might make them especially vulnerable to the kinds of environmental changes now happening on Earth?

5. The body plan of most vertebrates is based on four appendages. Describe the varied forms that these appendages take and how they are used. How do the vertebrates that have kept their four appendages move?

6. Compare the ways that different animal lineages colonized the land. How were those ways influenced by the body plans of animals in the different lineages?

What is our duty to nature?

- by Holmes Rolston, III -

Environmental ethics seeks appropriate respect for values in and duties regarding nature. This starts with human concerns for a healthy environment. If people have a right to life, they also have a right to a quality environment, needed for human welfare.

Environmental ethics then turns in nonhuman directions. What about the whooping cranes or the sequoia trees, the myriad species with which we co-inhabit Earth? Is there some intrinsic value in their lives we ought to protect? Surely we, *Homo sapiens*, the wise species, the only self-consciously moral species, are less wise than we ought to be if we act only in our collective self-interest.

Western ethics, philosophy, religion, politics, and economics have been dominantly humanistic, or anthropocentric. Contemporary ethics seeks to be inclusive: the poor as well as the rich, women as well as men, indigenous cultures as well as modern ones, future generations beyond the present. Environmental ethics is even more inclusive, concerned about whales slaughtered, whooping cranes and their habitats, ancient forests cut, Earth threatened by global warming.

Science alone does not teach us what we most need to know about nature: how to value it. Still, biology confronts every biologist (researcher and student alike) with an urgent moral concern—caring for life on Earth. Somewhat ironically, just when humans, with their increasing industry and technology, seemed further and further from nature, the natural world has emerged as a focus of ethical concern.

Ought not biologists (above all!) celebrate and cherish Earth's biodiversity?

This concern arises, ironically again, despite somewhat uncertain relations between science and ethics, how to move from what *is* (description of biological facts) to what *ought to be* (prescription of duty). It is not simply what a society does to its women, racial minorities, handicapped, children, or future generations, but what it does to its fauna, flora, species, ecosystems, and landscapes that reveals the character of that society.

Animals hunt and howl, care for their young, flee from threats, value their lives. There is "somebody there" behind the fur and feathers. "Man is the measure of things," said Protagoras, an ancient Greek philosopher. But wild animals do not make man the measure at all. Human values may override animal values, but we ought to justify such overriding—especially if we eat animals, exploit them, or experiment on them. Biology teaches that we and they are kin; ethically, their pains and pleasures count morally, too.

Most of the biological world, however, has yet to be taken into account: Plants, lower animals, insects, microbes, all are quite alive with vital interests. Every living organism has a *good-of-its-kind*; it defends its own kind as a *good kind*. Maybe "life" is a better measure of value than "man," or "vertebrate."

Life goes on at multiple levels. An inclusive ethic will be concerned for any ongoing species, for lifelines regenerating. Extinction is a sort of super-killing, a shutdown of life. In threatening Earth's biodiversity, humans are stopping the historical vitality of life.

We reach a "land ethic" (Aldo Leopold) with concern for ecosystems, for living communities, for life processes. Individual animals and plants are what they

Holmes Rolston, III, University Distinguished Professor and Professor of Philosophy at Colorado State University, is the author of *Environmental Ethics* (Temple University Press, 1988) and *Conserving Natural Value* (Columbia University Press, 1994). He is past president of the International Society for Environmental Ethics. In 2003 he was awarded the Templeton Prize in Religion, recognizing his work on respect for nature and reverence for life

are not as mere individuals (as though in a zoo or botanical garden), but they flourish in species lines and live in niches in habitats. An organism, a species, is what it is where it is, adapted for living in ongoing ecological and evolutionary systems. The most appropriate unit for moral concern is the whole system, the fundamental unit of development and survival.

Now we can put humans back in the picture. After all, ecology is about living at home (Greek *oikos*, "house"), the inclusive system again. Humans have entwined destinies with the natural world; their richest quality of life requires identifying with these communities.

Environmental ethics becomes Earth ethics. Humans are the only evaluators who can reflect at global scales. When humans do this, they must set up the scales. Animals, plants, insects, species, ecosystems, cannot take part in such inclusive and comprehensive concern for biodiversity on Earth, But they are what is to be measured. Earth (as seen from space) is quite a wonder. We Earthlings ought to care for this home planet.

Discussion Questions

1. Are good biologists always conservationists?

2. Can environmental ethics always be "win–win," people and nature?

3. Does an environmental ethic need to be science-based?

4. In wild nature is there anything bad? Anything ugly? Or that ought not to be respected or conserved?

Web Links

International Society for Environmental Ethics On-Line Bibliography
www.phil.unt.edu/bib/

Environmental Ethics, Systematic Works
www.cep.unt.edu/theo.html

Environmental Ethics, Anthologies
www.cep.unt.edu/anthol.html

Environmental Ethics, Introductory Articles
www.cep.unt.edu/intro.html

53 Behavioral Ecology

 Spices have played a major role in human history. The Gothic leader Alaric, who laid siege to Rome nearly 2,500 years ago, demanded (in addition to large quantities of precious metals) 1,364 kilograms of pepper as a ransom. The voyages of Marco Polo, Ferdinand Magellan, and Christopher Columbus were underwritten by kings and undertaken at great risk to sailors to find new and faster routes to spice-growing countries.

Why do humans crave spices? We know that spices enhance the flavors and colors of foods. However, that simple answer quickly suggests other questions. Why do we find foods more appealing when they contain pungent plant products? Why do people use dozens of different kinds of spices? Why are the foods of some cultures spicier than others? These questions are typical of the kinds of questions biologists ask about how and why animals make choices about what kinds of foods to eat. Such questions are the concern of behavioral ecology, a field that merges two areas of study within the life sciences.

Ecology is the science that deals with all kinds of biological interactions in the living world. Interactions among individuals of the same species may give rise to complex social behaviors and elaborate social systems. Biological interactions also include those between individuals of different species and between organisms and their physical environment. These interactions, in turn, influence the structure of communities (the organisms living together in the same area), ecosystems (all organisms in an area and their physical environment), and the biosphere (see Figure 1.6).

In this first chapter of Part Eight, we will look at the field of behavioral ecology. We will discuss how organisms respond to changes in the environment, decide where to carry out their activities, select the resources they need (food, water, shelter, nest sites), respond to predators and competitors, and associate with other members of their own species. Individual behavioral choices are the foundation of much of ecology because changes in the densities and distributions of populations are the cumulative results of the decisions of many individuals.

Our use of the words "decision" and "decide" here does not imply that the behavioral choices animals make are conscious. Rather, we mean that behavioral choices have been molded by natural selection such that individuals act "as if" they knew how their choices would influence their survival and reproductive success.

The term **environment**, as used by ecologists, includes both **abiotic** (physical and chemical) factors, such as water, nutrients,

The Quest for Spice Over the centuries humans have traveled far and endured great risks in order to provide themselves with pungent spices from tropical plants. What is it about spices that produces such a profound effect on our behavior?

light, temperature, and wind, and **biotic** factors, which includes all other organisms living in an area. Interactions between organisms and their environments are two-way processes: Organisms both influence and are influenced by their environments. Indeed, dealing with environmental changes caused by our own species is one of the major challenges facing organisms in the modern world. For this reason, ecologists are often asked to help analyze causes of environmental problems and to assist in finding solutions for them. However, it is important not to confuse the science of ecology with "environmentalism," or with the term "ecology," as it is often used in popular writing, to describe nature as some kind of superorganism.

Responding to Environmental Variation

Any organism that reaches old age has made decisions throughout its life history as it grew to maturity, reproduced, and suffered the effects of aging. Animals choose where to settle, how long to stay there, and when, if ever, to leave. They also select places for specific activities, such as resting and nesting, and they choose which things to eat from among the rich array of potential food sources in their immediate environments. Most animals also choose with whom to associate and for what purposes. And they make these choices in an environment that is continually changing. Plants, because they lack nervous systems and (except as seeds or spores) generally can move only by growing, make fewer choices than animals, but the same principles apply to them.

Environmental changes to which organisms must respond happen at many different time scales. Some changes, such as the approach of a fire, storm, or predator, require immediate responses; other changes allow time for a more gradual response. Some plants detach their leaves when storm winds reach a critical velocity and regrow new leaves afterward (Figure 53.1). Many plants reduce water loss and overheating by shifting the position of their leaves during the day so that they intercept sunlight early and late in the day, but do not overheat at midday. Lizards bask in sunshine in the morning to raise their body temperature but move into the shade when it gets too hot (see Figure 41.8).

All organisms have the ability to change their locations, either actively or passively, at some time during their lives; that is, few individuals die exactly where they were born. Individuals may leave the site of their birth to find a place where they can reproduce. Others may seek new locations when local conditions deteriorate. If repeated seasonal changes alter an environment, organisms may evolve life cycles that appear to anticipate those changes. Migration is one response to cyclical environmental changes. Most insectivorous birds, for example, leave high (temperate) latitudes in autumn for more favorable wintering grounds at low latitudes. Grazing mammals migrate away from seasonally dry areas, following the rains that produce lush grass (Figure 53.2). Other animals enter a resting state (hibernation; see Figure 41.19) before adverse conditions materialize and remain in that state until environmental signals indicate that conditions have improved.

Animals choose where to live

Selecting a place in which to live is one of the most important decisions an individual makes. The environment in which an organism lives is called its **habitat**. Once a habitat is chosen, an animal seeks its food, resting places, nest sites, and escape routes within that habitat. Choice of habitat may strongly influence survival and reproductive success, but some of the ways in which organisms make their choices are surprisingly simple. The cues most organisms use to select suitable habitats have a common feature: They are good predictors of general conditions suitable for future survival and reproduction.

A young red abalone, a kind of gastropod mollusk, begins its life as an egg that is fertilized in the open ocean. About

53.1 Plants Can Respond to Environmental Changes These palm trees in Quintana Roo on the Yucatan Peninsula dropped their leaves during a hurricane, which saved their trunks from being blown out of the ground. New leaves will now grow from the top of the trunk.

53.2 Migration Is a Response to Predictable Seasonal Changes East African wildebeest live in large herds that follow the rains to places with fresh grass.

Connochaetes taurinus

14 hours after fertilization, the egg hatches, but the swimming larva has enough yolk to continue developing for another 7 days without eating. Then the larva stops developing, swims to the seafloor, chooses a place in which to settle, and metamorphoses.

Red abalone larvae settle only on coralline algae, upon which they feed. They recognize coralline algae by a chemical these algae produce. In the laboratory, abalone larvae will settle on any surface on which this molecule has been placed, but in nature only coralline algae produce it. By using this simple cue, the larvae always settle on a surface that is suitable for their future development.

Many animals use the presence and success of already settled individuals as an indication that the habitat may be good. After collared flycatchers arrive on their breeding grounds in spring, they regularly peer into the nests of other individuals. Seeing this behavior, researchers hypothesized that the flycatchers were assessing the quality of the habitat by seeing how well their neighbors were doing. To test this hypothesis, the researchers created some areas with super-sized broods by taking young birds from some nests and adding them to nests in another area. The next year, flycatchers preferentially settled in those areas where broods had been artificially enlarged (Figure 53.3).

An animal may leave an area, either temporarily or permanently, if its population has grown too large to be supported by the local resources or if the environment has deteriorated. When a colony of the ant *Lepidothorax albipennis* has grown too large for its nest site (or the nest site has been damaged), recruiter ants—all of which are female workers (which we will describe later in this chapter)—look for potential new nest sites. When an ant finds a suitable site, she returns to the nest and releases a pheromone that attracts another recruiter. The two run together back to the site. If the other recruiter likes the site better than others she has visited, she returns to the nest and recruits another worker. The behavior of recruiters is governed by only two rules. First, if a site is not very attractive, a recruiter delays her return to it. As a result, ants gather at mediocre sites more slowly than at better sites. Second, once a threshold number of workers have been re-

cruited to a particular site, the recruiters change their behavior and begin carrying eggs and larvae from the colony to the new site. In this way the colony reaches agreement on the best site, which may not be the first one discovered.

Defending a territory may improve fitness

In many cases, an animal can improve its survival and reproductive success by establishing exclusive use of the resources of part of its habitat. The most common way of

53.3 Flycatchers Use Neighbors' Success to Assess Habitat Quality Collared flycatchers (*Ficedula albicollis*) settled at higher densities in areas where experimenters had artificially enlarged the broods of other flycatchers.

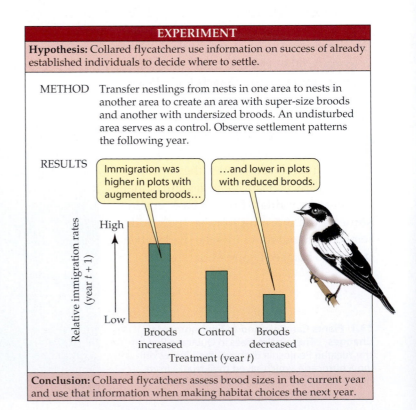

EXPERIMENT

Hypothesis: Collared flycatchers use information on success of already established individuals to decide where to settle.

METHOD   Transfer nestlings from nests in one area to nests in another area to create an area with super-size broods and another with undersized broods. An undisturbed area serves as a control. Observe settlement patterns the following year.

RESULTS

Immigration was higher in plots with augmented broods...

...and lower in plots with reduced broods.

Relative immigration rates (year *t* + 1)

High

Low

Broods increased | Control | Broods decreased

Treatment (year *t*)

Conclusion: Collared flycatchers assess brood sizes in the current year and use that information when making habitat choices the next year.

doing so is to establish a **territory** from which the resident excludes *conspecifics* (other individuals of the same species)—and sometimes individuals of other species as well—by advertising that it owns the area and, if necessary, chasing others away. But advertising and chasing take time and energy that could have been used for other beneficial purposes, such as finding food and watching out for predators.

To understand the evolution of these kinds of behavior, ecologists often use a *cost–benefit approach*. This approach assumes that an animal has only a limited amount of time and energy to devote to its activities. Animals seldom perform behaviors whose total costs are greater than the sum of their benefits—the improvements in survival and reproductive success that the animal achieves by performing the behavior. A cost–benefit approach provides a framework that behavioral ecologists can use to design experiments and make observations that enable them to understand why behavior patterns evolve as they do.

The total cost of any particular behavior has three components:

▶ **Energetic cost** is the difference between the energy the animal would have expended had it rested and the energy expended in performing the behavior.

▶ **Risk cost** is the increased chance of being injured or killed as a result of performing the behavior, compared with resting.

▶ **Opportunity cost** is the sum of the benefits the animal forfeits by not being able to perform other behaviors during the same time interval. An animal that devotes all of its time to foraging, for example, cannot achieve high reproductive success!

An experiment estimated the costs incurred by male lizards when defending a territory. Male Yarrow's spiny lizards defend territories, from which they exclude conspecific males. They normally do so most vigorously during September and October, when females are most receptive to mating (in this case, potential mates living in the territory are a resource for the males). To assess the costs of territorial behavior, experimenters inserted small capsules containing testosterone, a hormone that the lizards normally produce in the fall and which induces territorial behavior, beneath the skin of some males. They performed the experiment in June and July, a time of year when the lizards are normally only weakly territorial. Control males were also captured and released, but they received no testosterone implants.

Males with implanted testosterone capsules patrolled their territories more, performed more advertising displays, and expended about one-third more energy (energetic cost) than control males. As a result, they had less time to feed (opportunity cost), captured fewer insects, stored less energy, and died at a faster rate (risk cost) (Figure 53.4). This experiment

demonstrated that the costs of active territorial defense are high. In June and July, when females are less receptive, these costs probably outweigh the benefits. Probably that is why the lizards normally reduce their territorial behavior at that time of year.

Animals choose what foods to eat

After choosing a habitat, individuals use the resources of that habitat, including food. Because food is so important, we consider it here in some detail. When an animal *forages* (looks for food), how much time should it spend searching an area

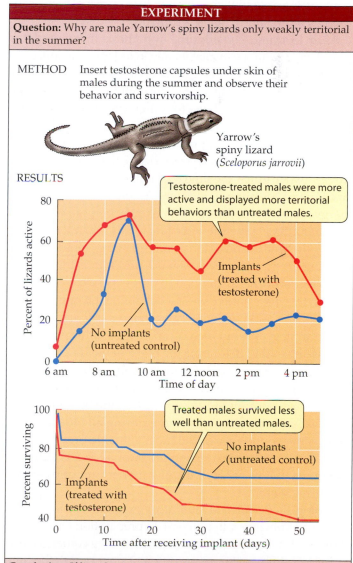

EXPERIMENT

Question: Why are male Yarrow's spiny lizards only weakly territorial in the summer?

METHOD Insert testosterone capsules under skin of males during the summer and observe their behavior and survivorship.

Yarrow's spiny lizard (*Sceloporus jarrovii*)

RESULTS

Testosterone-treated males were more active and displayed more territorial behaviors than untreated males.

Implants (treated with testosterone)

No implants (untreated control)

Treated males survived less well than untreated males.

No implants (untreated control)

Implants (treated with testosterone)

Conclusion: If lizards were territorial in the summer, they would die at a higher rate than nonterritorial lizards.

53.4 The Costs of Defending a Territory By using hormone implants to increase territorial behavior, experimenters measured the costs to male lizards of defending a territory during the summer months.

before moving to another site? When many different types of prey are available, which ones should a predator take, and which ones should it ignore? **Foraging theory** helps us answer these questions.

To predict how a foraging animal should behave, a scientist first specifies the objective of the behavior and then attempts to determine the behavioral choices that would best achieve that objective. This approach to foraging theory is known as **optimality modeling**. Its underlying assumption is that natural selection molds the behavior of animals so that, generally, they make the best choices available to them. A number of hypotheses can be proposed because a forager may have a number of objectives: It may attempt to maximize the rate at which it obtains energy (calories), vitamins, or minerals, to avoid toxins, or to reduce its risk of being captured by a predator while it is foraging.

As an example, consider the hypothesis that a predator should choose among available prey in order to maximize the rate at which it obtains energy. This is a plausible hypothesis because the more rapidly a predator captures food, the more time and energy it will have for other activities, such as reproduction or avoiding its own predators. To determine how a predator should choose prey if its objective is to maximize its energy intake rate, we characterize each type of available prey by two features: the time it takes the predator to pursue, capture, and consume an individual prey, and the amount of energy an individual prey contains. We then rank the prey types according to the amount of energy the predator gets relative to the time the predator spends pursuing, capturing, and handling the prey. The most valuable prey type is the one that yields the most energy per unit of time expended.

With this information, we can determine the rate at which a predator would obtain energy given a particular prey selection strategy. We can then compare alternative prey selection strategies and determine the one that yields the highest rate of energy intake. Such calculations show that, if the most valuable prey type is abundant enough, a predator gains the most energy per unit of time spent foraging by taking only the most valuable prey type and ignoring all others. However, as the abundance of the most valuable prey type decreases, an energy-maximizing predator adds less valuable prey to its diet in order of the energy per unit of time that those prey yield.

Ecologists performed laboratory experiments to test the energy maximization hypothesis. In preparation for their experiments, the scientists measured the energy content of water fleas of different sizes (the different prey types), the time bluegill sunfish (the predators) needed to capture and eat different prey types, the energy they spent pursuing and capturing prey, and rates at which they encountered prey under different prey densities. The scientists then stocked experimental environments with different densities and propor-

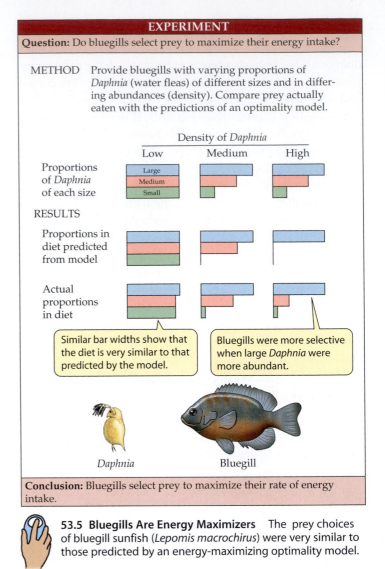

EXPERIMENT

Question: Do bluegills select prey to maximize their energy intake?

METHOD Provide bluegills with varying proportions of *Daphnia* (water fleas) of different sizes and in differing abundances (density). Compare prey actually eaten with the predictions of an optimality model.

Density of *Daphnia*
Low   Medium   High

Proportions of *Daphnia* of each size
Large
Medium
Small

RESULTS

Proportions in diet predicted from model

Actual proportions in diet

Similar bar widths show that the diet is very similar to that predicted by the model.

Bluegills were more selective when large *Daphnia* were more abundant.

Daphnia   Bluegill

Conclusion: Bluegills select prey to maximize their rate of energy intake.

53.5 Bluegills Are Energy Maximizers The prey choices of bluegill sunfish (*Lepomis macrochirus*) were very similar to those predicted by an energy-maximizing optimality model.

tions of large, medium, and small water fleas. They made two predictions from the hypothesis: (1) that in an environment stocked with low densities of all three sizes of prey, the fish would eat every water flea they encountered, and (2) that in an environment with abundant large water fleas, the fish would ignore smaller water fleas. The proportions of large, medium, and small water fleas taken by the fish under different conditions were close to those predicted by the hypothesis (Figure 53.5).

In the bluegill example, only the energy content of prey mattered, but minerals are also important to foragers. Many species of mammals and birds, all of which are primarily or exclusively herbivores or seed eaters, get mineral nutrients by eating soil at particular sites where mineral-rich soil is exposed (Figure 53.6). Humans, especially pregnant women in many traditional societies, consume soil, either by itself or mixed with otherwise toxic or bitter foods, such as acorns and wild potatoes. About 500 tons per year of mineral-rich clay are extracted from the ground or from termite mounds in Nigeria and exported for sale in markets throughout West Africa.

Ara chloroptera and *A. macao*

53.6 Mineral Seekers These macaws are obtaining mineral nutrients from the clay at this mineral lick in the Amazon jungle of Peru.

Like minerals, the spices humans add to our food contain little energy. They must provide some other benefit for us to value them so highly. One hypothesis proposed to explain why we find spicy foods tasty is the antimicrobial hypothesis. This hypothesis is based on the fact that spices are chemicals known to protect the plants that produce them against bacteria and fungi. It is reasonable to assume that they might also protect food from attacks by bacteria and fungi, and thus it might be adaptive for people to use spices in cooking to protect themselves from contaminated food.

The prediction that spices used in cooking should have antimicrobial activity has been tested experimentally in the laboratory by challenging food-borne bacteria and fungi with chemicals found in spices. In one such experiment, scientists prepared alcohol extracts of spices and added them to cultures of food-borne bacteria. Most of the commonly used spices were found to inhibit the growth of more than one kind of food-borne bacteria (Figure 53.7). These findings support the antimicrobial hypothesis.

Although these tests support the antimicrobial hypothesis, they do not exclude the possibility that an alternative hypothesis might also explain the human taste for spices. One such alternative hypothesis is that we enjoy spices because they disguise the smell and taste of spoiled foods. This hypothesis cannot be tested easily. However, some data suggest

53.7 Most Spices Have Antimicrobial Activity Laboratory tests show that most commonly used spices have moderate to strong antimicrobial activity.

that toxins from food-borne bacteria kill thousands of people every year and debilitate millions more. Therefore, eating spoiled food by covering up its bad flavors is a dangerous thing to do, even for a starving person. Natural selection is not likely to have favored people who ate rancid food because the flavors that signaled danger were disguised.

Choice of associates influences fitness

Most animals do not lead solitary lives. They associate with other individuals for a variety of reasons. Consider, for example, one important decision made by individuals of sexually reproducing species: the choice of mating partners. These

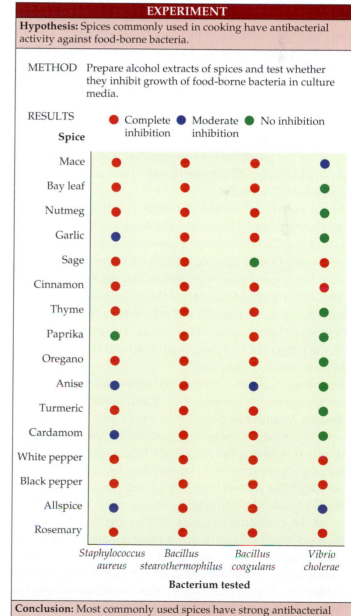

choices may be based on the inherent qualities of a potential mate, on the resources it controls (food, nest sites, escape places), or on a combination of the two. How do individuals choose mates?

The reproductive behaviors of males and females are often very different. Males usually initiate courtship, and they often fight for opportunities to mate with females. Females seldom fight over males, and they often reject courting males. Why are these sexual roles so different?

The answer lies in part in the costs of producing sperm and eggs. Because sperm are small and cheap to produce, one male produces enough to father a very large number of offspring—usually many more than the number of eggs a female can produce or the number of young she can nourish. Therefore, males of most species can increase their reproductive success by mating with many females.

Eggs, on the other hand, are typically much larger than sperm and are expensive to produce. Consequently, a female is unlikely to increase her reproductive output very much by increasing the number of males with which she mates. The reproductive success of a female depends primarily upon the quality of the genes she receives from her mate, the resources he controls, or the amount of assistance he provides in the care of her offspring. By their choices among males, females may cause the evolution of exaggerated traits that reliably signal male quality. This process, called *sexual selection*, was described in Chapter 23.

Males employ a variety of tactics to induce females to copulate with them. If a male controls no resources, he may use courtship behavior that signals in some way that he is in good health, that he is a good provider of parental care, or that he has a good genotype. For example, males of some species of hangingflies court females by offering them dead insects. A female hangingfly will mate with a male only if he provides her with food in this manner. The bigger the food item, the longer she copulates with him, and the more of her eggs he fertilizes (Figure 53.8).

Females can improve their reproductive success if they can correctly assess the genetic quality and health of potential mates, the quantity of parental care they may provide, and the quality of the resources they control. But how can females make such assessments accurately when all males would benefit by attempting to signal that they excel in all three of these traits? The answer is that by paying particular attention to those signals at which males cannot cheat, females have favored the evolution of "reliable" signals. Possession of a large dead insect reliably indicates that a male hangingfly is a good forager.

Biologists were slow to discover some of the signals used by animals in mate choice because they could not be detected with humans' unaided sense organs. For example, scientists did not discover until the 1970s that ultraviolet (UV) vision

53.8 A Male Wins His Mate   The male hangingfly on the left has just presented a moth to his mate, thus demonstrating his foraging skills. She feeds on the moth while they copulate. The bigger the moth, the longer they copulate, and the more eggs he fertilizes.

is widespread among birds, because humans cannot see UV light. Experiments with the bluethroat, a small bird that breeds in northern Europe and Asia, showed that females respond to UV light reflected by the bright blue throat patches of males. They prefer normal males rather than males whose throat patches have been dulled by applying a sunscreen chemical that absorbs UV wavelengths (Figure 53.9). Why should females use UV reflectance to assess potential mates? Laboratory research has shown that a bird's physical condition can influence the intensity with which its plumage reflects UV. Thus, UV reflectance is a reliable indicator of a male's health.

The throat feathers of a male bluethroat reflect ultraviolet light.

Luscinia svecica

53.9 Ultraviolet-Reflecting Plumage Affects Female Choice
Female bluethroats are attracted to males whose throat feathers have high UV reflectance, which signals a healthy, high-quality male.

The Evolution of Animal Societies

Social behavior evolves when cooperation among conspecifics produces, on average, higher rates of survival and reproduction than solitary individuals can achieve. Associations for reproduction may consist of little more than a coming together of eggs and sperm, but individuals of many species associate for longer times to provide care for their offspring. Associating with conspecifics may also improve survival in ways unrelated to reproduction, such as by reducing the risk of being captured by a predator.

The social systems of many animals are very simple: Males court females, the fertilized females disperse and lay eggs, and the eggs and larvae grow to maturity untended. Other systems—such as the elaborate colonies of ants, bees, and wasps or the social groups of lions and primates—are very complex. How did these complex animal societies evolve?

Although today's social systems are the result of long periods of evolution, behavior leaves few traces in the fossil record. Biologists must infer possible routes of the evolution of social systems by studying current patterns of social organization. Fortunately, many degrees of social system complexity exist among living species; the simpler systems suggest stages through which the more complex ones may have passed.

We will describe only a few animal social systems, but as we look at these examples, we will keep in mind three important concepts:

▶ Social systems are best understood not by asking how they benefit the species as a whole, but by asking how the individuals that join together benefit by the association.
▶ Social systems are dynamic; individuals constantly communicate with one another and adjust their relationships.
▶ The costs and benefits experienced by individuals in a social system differ according to their age, sex, physiological condition, and status.

Group living confers benefits but also imposes costs

Living in groups may confer many types of benefits. It may improve hunting success or expand the range of prey that can be captured. For example, by hunting in groups, our ancestors were able to kill large mammals they could not have subdued as individual hunters. These social humans could also defend their prey and themselves from other carnivores, and could tell one another about, for example, the locations of food and predators.

Many small birds forage in flocks. To test whether flocking provides protection against predators, an investigator released a trained goshawk near wood pigeons in England. The hawk was most successful when it attacked solitary pigeons. Its success in capturing a pigeon decreased as the number of pigeons in the flock increased (Figure 53.10). The larger the flock of pigeons, the sooner some individual in the flock spotted the hawk and flew away. This escape behavior stimulated other individuals in the flock to take flight as well. But foraging in a flock also imposes a cost: The pigeons in a flock interfere with one another's ability to find seeds.

Social behavior has many costs as well as benefits. In some social species, individuals inhibit one another's reproduction or injure one another's offspring. An almost universal cost associated with group living is higher exposure to diseases and parasites. Long before the causes of diseases were known, people knew that association with sick persons increased their chances of getting sick. Quarantine has been used to combat the spread of illness for as long as we have

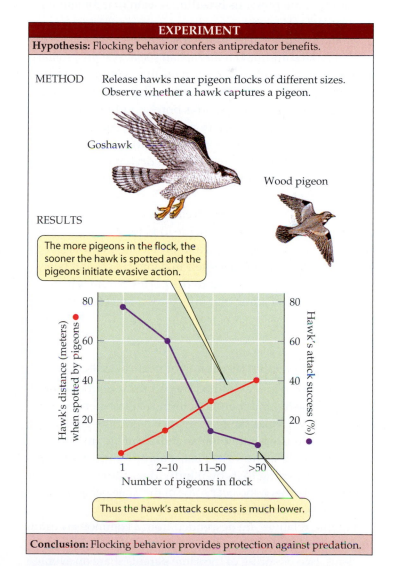

EXPERIMENT

Hypothesis: Flocking behavior confers antipredator benefits.

METHOD   Release hawks near pigeon flocks of different sizes. Observe whether a hawk captures a pigeon.

Goshawk

Wood pigeon

RESULTS

The more pigeons in the flock, the sooner the hawk is spotted and the pigeons initiate evasive action.

Thus the hawk's attack success is much lower.

Conclusion: Flocking behavior provides protection against predation.

53.10 Groups Provide Protection from Predators   The larger a flock of pigeons, the greater the distance at which they detect an approaching hawk, and the less likely the hawk is to succeed in capturing a pigeon.

written records. The diseases of wild animals are not as well known, but they too are spread mostly by close contact.

In some species, parents care for their offspring

The most widespread form of social system is the family, an association of one or more adults and their dependent offspring. If parental care lasts a long time, or if the breeding season is longer than the time it takes for offspring to mature, adults may still be caring for younger offspring when older offspring reach parenting age. These older offspring may help their parents care for their younger siblings. Among birds, many communal breeding systems probably evolved by this route. Florida scrub jays live all year on territories, each of which contains a breeding pair and up to six helpers that bring food to the nest. Nearly all helpers are offspring from the previous breeding season that remain with their parents.

Most mammals also evolved social systems via an extended family. In simple mammalian social systems, solitary females or male–female pairs care for their young. As the period of parental care increases, older offspring are still present when the next generation is born, and they often help rear their younger siblings. In most social mammal species, female offspring remain in the group in which they were born, but males tend to leave, or are driven out, and must seek other social groups. Therefore, among mammals, most helpers are females.

Raising a family involves tremendous costs for parents and helpers. Animals who provide food for their young may sacrifice food for themselves, and protecting the young may involve the animal putting itself in danger. Acts that benefit another individual at a cost to the performer are **altruistic acts**. How can behavior that inflicts a cost on the performer evolve?

Altruism can evolve by means of natural selection

Altruistic behaviors exhibited by parents toward their offspring are easily understood in terms of close genetic relatedness. Genetic relatedness extends beyond the parent–offspring relationship, allowing an individual to influence its fitness in two different ways. First, it may produce its own offspring, contributing to its own **individual fitness**. Second, it may help relatives (who bear some of the same genes) in ways that increase their fitness.

Because relatives are descended from a common ancestor, they are likely to bear some of the same alleles. In diploid organisms, two offspring of the same parents share on average 50 percent of the same alleles; an individual is likely to share 25 percent of its alleles with its sibling's offspring. Therefore, by helping its relatives, an individual can increase the representation of some of its own alleles in the population. This process is called **kin selection**. Together, individual fitness and fitness gained through helping non-descendent kin determine the **inclusive fitness** of an individual. Occasional altruistic acts may eventually evolve into altruistic behavior patterns if the benefits of increasing the reproductive success of relatives exceed the costs of decreasing the altruist's own reproductive success.

Many social groups consist of some individuals that are close relatives and others that are unrelated or distantly related. Individuals of some species recognize their relatives and adjust their behavior accordingly. White-fronted bee-eaters are African birds that nest colonially. Most breeding pairs are assisted by nonbreeding adults that help incubate eggs and feed nestlings. Nearly all of these helpers assist close relatives (Figure 53.11). When helpers have a choice of two nests at which to help, about 95 percent of the time they choose the nest with the young more closely related to them.

Several other pieces of evidence suggest that the helping behavior of white-fronted bee-eaters evolved through kin selection. First, both males and females help to care for nestlings, but males help more often than females. Males remain in the social group in which they were born, but females join other social groups when they mature. Therefore, females typically live in social groups composed primarily of nonrelatives.

Merops bullockoides

Closely related birds—brothers or sisters—are more likely to help than distantly related or unrelated birds.

53.11 White-Fronted Bee-Eaters are Altruists Bee-eaters that help to care for nestlings preferentially help close relatives.

Second, individual bee-eaters do not appear to gain anything in addition to inclusive fitness by helping—helpers do not gain experience that improves their performances when they become breeders. Finally, nests with helpers produce more fledglings than do nests without helpers, showing that helpers do increase the number of fledglings produced by their close relatives. Notice that all these patterns are consistent with the principle that bee-eaters behave in ways that improve their individual fitness, not in ways that benefit the species.

Eusociality is extreme social behavior

Species whose social groups include sterile individuals are said to be **eusocial**. This extreme form of social behavior has evolved in termites and many hymenopterans (ants, bees, and wasps). In these species, most females are *workers* that forage for the colony and/or defend it against predators, but do not reproduce. Workers may include soldiers with large, specialized defensive weapons (Figure 53.12), which may be killed while defending the colony. Only a few females, known as *queens*, are fertile, and they produce all the offspring of the colony.

Both genetic and environmental factors facilitate the evolution of eusociality. The British evolutionist W. D. Hamilton first suggested that eusociality evolved among the Hymenoptera because its members have an unusual sex determination system in which males are haploid but females are diploid. Among the Hymenoptera, a fertilized (diploid) egg hatches into a female; an unfertilized (haploid) egg hatches into a male.

If a female copulates with only one male, all the sperm she receives are identical because a haploid male has only one set of chromosomes, all of which are transmitted to every sperm cell. Therefore, a female's daughters share all of their father's genes. They also share, on average, half of the genes they receive from their mother. As a result, they share 75 percent of their alleles on average, rather than the 50 percent they would share if both parents were diploid. Since workers are more genetically similar to their sisters than they would be to their own offspring, they can increase their fitness more by caring for their sisters than by producing and caring for their own offspring.

Eusociality may also be favored if establishment of new colonies is difficult and dangerous. Nearly all eusocial animals construct elaborate nests or burrow systems within which their offspring are reared. Naked mole-rats—the most eusocial mammals—live in underground colonies containing 70 to 80 individuals. The colony's tunnel systems are maintained by sterile workers. Breeding is restricted to a single queen and several kings that live in a nest chamber in the center of the colony. Individuals attempting to found new colonies are at high risk of being captured by predators, and most founding events fail. Thus, high predation rates, which favor cooperation among founding individuals, may facilitate the evolution of eusociality.

Inbreeding—the mating of individuals who are genetically related—can generate increased genetic relatedness within a group. Even if two parents are unrelated, but each is the product of generations of intense inbreeding, all of their offspring may be genetically nearly identical. Such offspring would increase their fitness by helping to rear siblings. Genetic similarity generated by inbreeding could explain the evolution of eusociality among the many hymenopteran species in which queens mate with many males and among termites and naked mole-rats, in which both sexes are diploid.

Behavioral Ecology, Population Dynamics, and Community Structure

The ways in which organisms make decisions about habitats, food, and associates have many important implications for the structure and functioning of ecological systems. We will describe two of those implications here. First, animals with complex social organization often achieve remarkably high abundances. Second, the ways in which animals select habitats and food, combined with their interactions with individuals of other species, may influence the range of habitat and foods a species uses in nature.

Social animals may achieve great abundances

The abundances achieved by some social animals are impressive. For example, up to 94 percent of the individuals and 86 percent of the biomass of arthropods in the *canopies* (tree-

Eciton burchelli

53.12 Sterile Workers are Extreme Altruists Eusocial insect species contain classes of sterile worker individuals. These soldier army ants from Panama protect their colonies with their large, powerful jaws.

53.13 Termite Mounds Are Large and Complex These immense Australian termite mounds are constructed over many years by millions of worker termites. Elaborate nests or burrows, which are very costly to construct and maintain, characterize nearly all eusocial animals.

tops) of tropical rainforests are social ants. Termites (also social insects) are the primary consumers of plant tissues in the savannas of Africa. They live in and build large mounds, within which many other species of animals live. Termites may extract nutrients from the soil at depths as great as 80 meters. In parts of Australia, termite density may reach 1,000 colonies per hectare (Figure 53.13).

Ants and termites have achieved these remarkable abundances in part because their social organization allows them to exploit the services of other organisms in harvesting vital resources. The most abundant and highly productive ants and termites actively cultivate fungi that break down difficult-to-digest plant tissues, including wood. Some ants tend aphids and other insects that tap phloem fluids, protecting the phloem-suckers from predators. Because phloem is rich in carbohydrates but poor in proteins, phloem-suckers ingest more carbohydrates than they can use. They eject the excess in the form of sugar-rich anal drops (see Figure 36.13), which the ants eat. Because the ants can easily obtain enough carbohydrates in this manner, they need to get only proteins in

other ways, such as by eating other insects. Moreover, with their high, sugar-based metabolic rates, the ants can expend the energy needed to drive other predatory insects away from their food sources. In this way, ants strongly influence the community of insects in tropical forest canopies.

Social living also enables organisms to find and use temporally and spatially patchy foods. The wildebeest, which travels in large herds, is the most abundant large mammal in Africa (see Figure 53.2). More than a million individuals are found in the herd that migrates between the Masai Mara in Kenya and the Serengeti in Tanzania to feed on the rapidly growing grass that follows seasonal rains in each area.

Even more striking is the abundance achieved by our own species (Figure 53.14). Social living enabled members of human groups to specialize in different activities. Among the benefits of specialization were domestication of plants and animals and cultivation of land. These innovations enabled our ancestors to increase the resources at their disposal dramatically. Those increases, in turn, stimulated rapid population growth up to the limit determined by the agricultural productivity that was possible with human- and animal-powered tools. Agricultural machines and artificial fertilizers, made possible by the tapping of fossil fuels, greatly increased agricultural productivity and removed that earlier limit. In addition, the development of modern medicine reduced the mortality rate in human populations. Medicine and better hygiene have also allowed people to live in large numbers in areas where diseases formerly kept numbers very low. However, these successes have been accompanied by many problems, some of which we will discuss in subsequent chapters.

Interspecific interactions influence animal distributions

As we have seen, animals assess habitat quality and settle preferentially in better places. They also select the food items that give them the best return for the time and energy they expend in getting them. The optimality modeling approach used to develop and test hypotheses about how such choices are made has yielded an important general "rule of thumb" of behavioral ecology: As much as possible, organisms concentrate on doing what they do best and avoid doing what they do poorly.

However, interspecific interactions may prevent animals from living in those environments in which they would do best. Individuals of a behaviorally dominant species may be able to exclude individuals of a subordinate species from its preferred foraging areas. How such behavioral dominance influences use of foraging areas can be illustrated by observing hummingbird behavior.

Hummingbirds extract nectar from flowers and often defend patches of flowers from other hummingbirds. In an ex-

53.14 Social Organization Allows Humans to Live at High Densities Human cities such as Benidorm on Spain's Costa Blanca are examples of how social organization allows our species to achieve and sustain extreme population densities.

periment done in southeastern Arizona, investigators created artificial "flower patches" by setting up an array of feeders. Some feeders contained artificial nectar that was rich in sucrose; others contained a more dilute solution that was a poorer source of sucrose. Hummingbirds quickly learned which were the high-quality feeders because the rich ones had blue bee guards; the poor ones had yellow bee guards.

Males of three hummingbird species visited the feeders. Interactions were strongest between two of them: Male blue-throated hummingbirds, which weigh about 8.3 g on average, behaviorally dominated the smaller male black-chinned hummingbirds, which weighed only about 3.2 g. When no male blue-throats were present, black-chinned males fed almost exclusively at the rich feeders; but when male blue-throats were present, black-chinned males fed at poor feeders as often as they fed at rich feeders. Even though the nectar at the poor feeders was more dilute, the black-chins achieved about the same rate of energy from them as from the rich feeders because they were able to feed longer at the poor feeders without being chased away by the larger blue-throat males.

These kinds of observations show us that the ways in which animals choose what to eat, where to seek food, and with whom to associate influence the sizes and distributions of populations of many species and how they interact in nature. These aspects of populations will form the focus of the next chapter, on population ecology.

Chapter Summary

▶ Behavioral ecology is the study of how animals decide where to carry out different activities, select the resources they need, respond to predators and competitors, and interact with other members of their own species.

▶ An organism's environment includes both abiotic and biotic components.

Responding to Environmental Variation

▶ Organisms respond adaptively to environmental changes.

▶ The cues animals use to select habitats may be simple, but they must be good predictors of conditions suitable for future survival and reproduction.

▶ The success of already settled individuals may provide evidence of habitat quality. **Review Figure 53.3**

▶ Cost–benefit analyses of behavior are based on the principle that animals have only limited amounts of time and energy to devote to their activities.

▶ Behaviors such as defending a territory may have three kinds of costs: energetic cost, opportunity cost, and risk cost. **Review Figure 53.4. See Web/CD Tutorial 53.1**

▶ Foraging theory was developed to understand how animals select foods from those present in the environment. **Review Figure 53.5. See Web/CD Tutorial 53.2**

▶ The human taste for spices may have evolved because spices have antimicrobial activity. **Review Figures 53.7**

▶ Because males produce enough sperm to fertilize many more eggs than a single female can produce, males typically increase their reproductive success by mating with many females. The reproductive success of females, on the other hand, is typically limited by the cost of producing eggs. As a result, males usually initiate courtship and often fight for opportunities to mate with females. Females seldom fight over males and often reject courting males.

▶ Courting males perform behaviors that signal their desirability as mating partners By paying particular attention to those signals at which males cannot cheat, females have favored the evolution of "reliable" signals.

The Evolution of Animal Societies

▶ Social systems are best understood not by asking how they benefit the species as a whole, but by asking how the individuals that join together benefit by the association.

▶ Social systems are dynamic; individuals constantly communicate with one another and adjust their relationships.

▶ Living in a group may provide protection against predators. **Review Figure 53.10**

▶ The origin of most animal societies is the family, an association of one or more adults and their dependent offspring.

▶ Altruism among closely related individuals can evolve by means of kin selection because individuals who help close relatives can improve their inclusive fitness. **Review Figure 53.11**

▶ Eusocial systems with sterile individuals have evolved among hymenopterans (ants, bees, and wasps), termites, and one mammal, the naked mole-rat.

▶ The more closely related the individuals in a colony are to one another, and the greater the difficulty of establishing independent colonies, the more likely eusociality is to evolve.

Behavioral Ecology, Population Dynamics, and Community Structure

▶ Social animals may achieve great abundances.

▶ Interspecific interactions, as well as habitat and food choices, influence animal distributions.

See Web/CD Activity 53.1 for a concept review of this chapter.

Self-Quiz

1. Which of the following is *not* a component of the cost of performing a behavior?
   a. Its energetic cost
   b. The risk of being injured
   c. Its opportunity cost
   d. The risk of being attacked by a predator
   e. Its information cost

2. An almost universal cost associated with group living is
   a. increased risk of predation.
   b. interference with foraging.
   c. higher exposure to diseases and parasites.
   d. poorer access to mates.
   e. poorer access to sleeping sites.

3. Which is *not* an important assumption of foraging behavior theory?
   a. Efficient foragers spend less time fulfilling their energy needs than inefficient ones.
   b. Superior foragers will generally produce more surviving offspring.
   c. A successful predator will choose its prey in such a way as to maximize its energy intake.
   d. An efficient predator will always choose the most abundant prey.
   e. The ability of a predator to discriminate among prey items has a genetic basis.

4. The basic components of an optimality model of behavior are
   a. the type of behavior and its neural control mechanisms.
   b. the objective of the behavior and the choices that would best achieve it.
   c. the objective of the behavior and its neural control mechanisms.
   d. the goal of the behavior and the constraints imposed by the animal's structure.
   e. the objective to be maximized and the currency used to measure it.

5. The choice of a mating partner may be based on
   a. the inherent qualities of a potential mate.
   b. the resources held by a potential mate.
   c. both the inherent qualities of a potential mate and the resources it holds.
   d. the success of individuals of the opposite sex in courtship.
   e. All of the above

6. Altruistic behavior
   a. confers a benefit on the performer by inflicting some cost on some other individual.
   b. confers a benefit both on the performer and on some other individual.
   c. inflicts a cost both on the performer and on some other individual.
   d. confers a benefit on another individual at some cost to the performer.
   e. confers a cost on the performer without benefiting any other individual.

7. Kin selection is
   a. mating between relatives.
   b. the adoption of young by an unrelated adult.
   c. the ability to recognize one's relatives in a social group.
   d. a behavior that increases the survivorship of an individual's relatives.
   e. only found among social mammals.

8. Species whose social groups include sterile individuals are said to be
   a. eusocial.
   b. semisocial.
   c. oligosocial.
   d. sterisocial.
   e. supersocial.

For Discussion

1. Most hawks are solitary hunters. Swallows often hunt in groups. What are some plausible explanations for this difference? How could you test your ideas?

2. Among birds, males of species that mate with many females and perform communal courtship displays are usually much larger and more brightly colored than females, whereas among species that form monogamous pairs, males are usually similar in size to females, whether or not they are more brightly colored. What hypotheses can be advanced to explain this difference?

3. Many animals defend space, but the sizes of the territories they defend and the resources these areas provide vary enormously. Why don't all animals defend the same type and size of territory?

4. When frogs mate, a male clasps a female behind her front legs and stays with her until she lays her eggs, at which time he fertilizes them. In most species of frogs, the male remains clasped to the female for a short time, usually no longer than a few hours. However, in some species, pairs may remain together for up to several weeks. In view of the fact that a male cannot court or mate with any other female while clasping one, and that a female lays only a single clutch of eggs, why is it advantageous for males to behave this way? What can you guess about the breeding ecology of frogs that remain clasped for long periods? Why should females permit males to clasp them for so long? (Females do not struggle!)

5. Among vertebrates, helpers are individuals capable of reproducing, and most of them later breed on their own. Among eusocial insects, sterile castes have evolved repeatedly. What differences between vertebrates and insects might explain the failure of sterile castes to evolve in the former?

6. The use of DNA fingerprinting technology (see Chapter 16) has shown that in many species, social partners and genetic partners differ. Under what conditions do individuals benefit from copulating with individuals other than their social mates? Do males and females benefit equally from this behavior?

54 *Population Ecology*

The bay checkerspot butterfly lives in the hills south of San Francisco, California. If you were to travel through this region in search of butterflies, you would find checkerspots in some areas, but not in others. If you looked for butterflies in the region over a number of years, you would find more butterflies in some years than in others. In some places where you found checkerspots one year, you would go back the following year and find none, and in other places where there were previously no checkerspots, you would find them. What is the reason for these variations?

Ecologists who study populations attempt to answer this and a number of other questions: Why do the numbers of individuals of a species in a certain area fluctuate? Why do the geographic ranges of species vary so much? Why is a species abundant in some parts of its geographic range and rare in others, and why does its range change over time?

To understand how and why population sizes fluctuate, ecologists count individuals in different places and try to understand the relative importance of the processes that determine the number of individuals of a species in any particular location. These processes are influenced by individuals of the same species and by individuals of other species living in the same environment, as well as by abiotic environmental factors.

In this chapter we will discuss how and why the sizes of populations of species vary over space and time, and show how this knowledge is used to predict and manage the growth of populations of special interest to people. To set the stage for answering questions about populations, we first describe the kinds of information ecologists gather about the populations they study and how they use that information to answer the questions we just posed. Then we describe how populations of different species interact and how those interactions influence numbers of individuals and where they live. In the next chapter we will describe how interactions among populations and between populations and the physical environment influence the structure of ecological communities.

A Case Study in Subpopulations Patchy subpopulations of the rare bay checkerspot butterfly, *Euphydryas editha bayensis*, provide a well-studied example of population dynamics. This individual was photographed in Morgan Hill, a large patch of suitable habitat for this species in the San Francisco Bay area (see Figure 54.13).

Populations in Space and Time

The individuals of a species within a given area constitute a **population**. At any given moment, an individual organism occupies only one spot in space, and is of a particular age and size. The members of a population, however, are distributed over space, and they differ in age and size. The distribution of the ages of individuals in a population and the way those individuals are distributed over the environment describe **population structure**. Ecologists study population structure because the spatial distributions of individuals and their ages influence the stability of populations and affect interactions among species.

The number of individuals of a species per unit of area (or volume) is its **population density**. Ecologists are interested in population densities because dense populations often exert strong influences on their members and on populations of other species. Scientists working in agriculture, conservation, and medicine typically try to maintain high population densities of some species (crop plants, game animals, aesthetically attractive species, threatened or endangered species) and reduce the densities of others (agricultural pests, disease organisms). To manage populations, we need to know what factors cause populations to grow or decrease in size and how these factors work.

Because organisms and their environments differ, population densities may be measured in more than one way. Ecologists usually measure the densities of organisms in terrestrial environments as the number of individuals per unit of area. However, number per unit of volume is generally a more useful measure for organisms living in the water column. For species whose members differ markedly in size, as do most plants and some animals (such as mollusks, fishes, and reptiles), the percentage of ground covered or the total mass of individuals may be more useful measures of density than the number of individuals.

The structure of a population changes continually because **demographic events**—births, deaths, immigration (movement of individuals into the area), and emigration (movement of individuals out of the area)—are common occurrences. Knowledge of when individuals are born and when they die provides a surprising amount of information about a population. Let's examine how ecologists measure birth and death rates and use that information to understand **population dynamics**—the change in population density through time and space. The study of birth, death, and movement rates that give rise to population dynamics is known as **demography**.

Births, deaths, and movements drive population dynamics

Ecologists measure the *rates* (number per unit of time) at which births, deaths, and movements take place in a population, and they study how these rates are influenced by environmental factors, life histories, and population densities.

The number of individuals in a population at any given time is equal to the number present at some time in the past, plus the number born between then and now, minus the number that died, plus the number that immigrated, minus the number that emigrated. That is, the number of individuals at a given time, N_1, is given by the equation

$$N_1 = N_0 + B - D + I - E$$

where N_1 is the number of individuals at time 1; N_0 is the number of individuals at time 0; B is the number of individuals born, D the number that died, I the number that immigrated, and E the number that emigrated between time 0 and time 1. If we measure these rates over many time intervals, we can determine how a population's density changes over time.

Life tables summarize patterns of births and deaths

Life tables provide summaries of births and deaths in a population. Life tables were developed by the Romans nearly 2,000 years ago to determine how much money needed to be set aside to compensate families of soldiers that might be killed in battle. Today, life insurance companies use life tables to determine how much to charge people for insurance policies. Biologists use life tables to predict future trends in populations.

We can construct a life table by determining for a group of individuals born at the same time (called a *cohort*) the number that are still alive at later dates (*survivorship*). Some life tables also include the number of offspring produced by the cohort during each time interval. An example of a life table based on an intensive study of the cactus finch carried out on Isla Daphne in the Galápagos archipelago, is shown in Table 54.1.

The data in Table 54.1 come from 210 birds that hatched in 1978 and were followed until 1991, when only 3 individuals were still alive. The table shows that the mortality rate for these birds was high during the first year of life. It then dropped dramatically for several years, followed by a general increase in later years. Mortality rates fluctuated among years because the survival of these birds depends on seed production, which strongly correlates with rainfall. The Galápagos archipelago experiences both drought years and years of heavy rain. During drought years, plants produce few seeds, birds do not nest, and adult survival is poor. In years when rainfall is heavy, seed production is high, most birds breed several times, and adult survival is high. The survival rates in the table reflect these rainfall fluctuations. Variation in seed production resulting from the alternation of wet and dry years is a major reason why the cactus finch population fluctuates so greatly.

54.1 Life Table of the 1978 Cohort of the Cactus Finch (Geospiza scandens) on Isla Daphne

AGE IN YEARS (X)	NUMBER ALIVE	SURVIVORSHIP[a]	SURVIVAL RATE[b]	MORTALITY RATE[c]
0	210	1.000	0.434	0.566
1	91	0.434	0.857	0.143
2	78	0.371	0.898	0.102
3	70	0.333	0.928	0.072
4	65	0.309	0.955	0.045
5	62	0.295	0.678	0.322
6	42	0.200	0.548	0.452
7	23	0.109	0.652	0.348
8	15	0.071	0.933	0.067
9	14	0.067	0.786	0.214
10	11	0.052	0.909	0.091
11	10	0.048	0.400	0.600
12	4	0.019	0.750	0.250
13	3	0.014	0.996	

[a]Survivorship = the proportion of newborns who survive to age x.
[b]Survival rate = the proportion of individuals of age x who survive to age x + 1.
[c]Mortality rate = the proportion of individuals of age x who die before the age of x + 1.

Ecologists often use graphs to highlight the most important changes in populations. Graphs of survivorship in relation to age show when individuals survive well and when they do not. Survivorship curves in many populations fall into one of three patterns. In some populations, most individuals survive for most of their potential life span, then die at about the same age. For example, because of intensive parental care and the availability of medical services, the survivorship of humans in the United States is high for many decades, but then declines rapidly in older individuals (Figure 54.1a). In a second pattern, which is characteristic of many songbirds, the probability of surviving is about the same over most of the life span once individuals are a few months old (Figure 54.1b). A third widespread survivorship

pattern is found among organisms that produce a large number of offspring, each of which receives little energy or parental care. In these species, high death rates of young individuals are followed by high survival rate during the middle part of the life span. *Spergula vernalis*, an annual plant that grows on sand dunes in Poland, illustrates this pattern (Figure 54.1c).

The age distribution of individuals in a population reveals much about the recent history of births and deaths in the population. The timing of births and deaths can influence age distributions for many years in populations of long-lived species. The human population of the United States is a good example. Between 1947 and 1964, the United States experienced what is known as the post-World War II baby boom. During these years, average family size grew from 2.5 to 3.8 children; an unprecedented 4.3 million babies were born in 1957. Birth rates declined during the 1960s, but Americans born during the baby boom still constitute the dominant age class in the first part of the twenty-first century (Figure 54.2). "Baby boomers" became parents in the 1980s, producing another bulge in the age distribution—a "baby boom echo"—but they had, on average, fewer children than their parents did, so the bulge is not as large.

By summarizing information on when individuals are born and die, life tables help us understand why population densities change over time. Life table data can also be used to determine how heavily a population can be harvested and which age groups should be the focus of our efforts to save rare species. We will discuss the management of populations later in this chapter, but first we'll see how interactions among species influence the dynamics of particular populations.

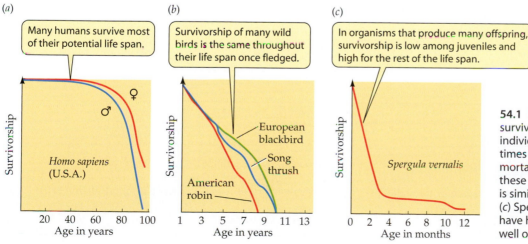

(a) **Many humans survive most of their potential life span.**

Homo sapiens (U.S.A.)

Survivorship / Age in years (20 40 60 80 100) ♀ ♂

(b) **Survivorship of many wild birds is the same throughout their life span once fledged.**

European blackbird
Song thrush
American robin

Survivorship / Age in years (1 3 5 7 9 11 13)

(c) **In organisms that produce many offspring, survivorship is low among juveniles and high for the rest of the life span.**

Spergula vernalis

Survivorship / Age in months (0 2 4 6 8 10 12)

54.1 Survivorship Curves Three common survivorship curves show the number of individuals in a cohort still alive at different times over the life span. (a) For this curve, mortality is highest at advanced ages. (b) For these species, the probability of survivorship is similar throughout much of the life span. (c) Species with this survivorship pattern have high mortality at early ages, but survive well once past a critical point.

54.2 Age Distributions Change over Time The graphs shows age distributions for the human population of the United States from 1960 to 2020. The high birth rates during the "baby boom" have influenced the structure of the population over many decades.

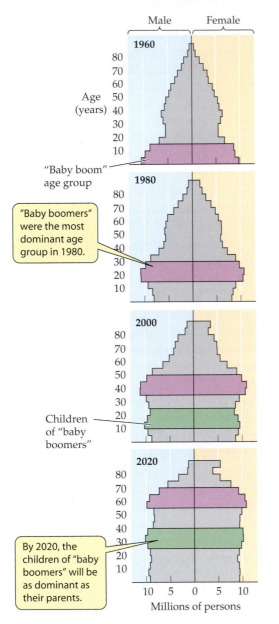

Male   Female

1960

Age (years) 80 70 60 50 40 30 20 10

"Baby boom" age group

1980

"Baby boomers" were the most dominant age group in 1980.

80 70 60 50 40 30 20 10

2000

80 70 60 50 40 30 20 10

Children of "baby boomers"

2020

80 70 60 50 40 30 20 10

By 2020, the children of "baby boomers" will be as dominant as their parents.

10 5   0   5   10
Millions of persons

Types of Ecological Interactions

So far, we have considered only survival and reproductive rates of single species. Before we can answer the questions we have posed about populations, we also need to look at the ways in which populations of different species interact with one another (Table 54.2) These interactions fall into five general categories:

► A **mutualism** is an interaction in which both participants benefit (+/+ interaction).
► A **commensalism** is an interaction in which one participant benefits but the other is unaffected (+/0 interaction).
► An **amensalism** is an interaction in which one participant is harmed but the other is unaffected (0/− interaction).
► A **predator–prey** or **parasite–host interaction** is one in which one participant is harmed, but the other benefits (+/− interaction).
► If two organisms use the same resources and those resources are insufficient to supply their combined needs, the organisms are called competitors, and their interactions constitute **competition** (−/− interactions).

Mutualistic interactions exist between plants and microorganisms, protists and fungi, plants and insects, among animals, and among plants. Most plants have beneficial associations with soil-inhabiting fungi, called mycorrhizae, which enhance the plant's ability to extract minerals from the soil (see Figure 31.16). Some plants have mutualistic relationships with nitrogen-fixing bacteria of the genus *Rhizobium* (see Figure 37.7).

Animals have important mutualistic interactions with protists, plants, and other animals. Corals and some tunicates gain most of their energy from photosynthetic protists living within their tissues. In exchange, they provide the protists with nutrients from the small animals they capture. Termites have nitrogen-fixing protists in their guts that help them digest the cellulose in the wood they eat. The termites provide the protists with a suitable environment in which to live and an abundant supply of cellulose.

54.2 *Types of Ecological Interactions*

		EFFECT ON ORGANISM 2		
		HARM	BENEFIT	NO EFFECT
EFFECT ON ORGANISM 1	HARM	Competition (−/−)	Predation or parasitism (−/+)	Amensalism (−/0)
	BENEFIT	Predation or parasitism (+/−)	Mutualism (+/+)	Commensalism (+/0)
	NO EFFECT	Amensalism (0/−)	Commensalism (0/+)	—

54.3 Commensalism Benefits One Partner Cattle egrets (*Bubulcus ibis*) capture more insects with less effort when they forage around large grazing mammals such as this Cape buffalo (*Syncerus caffer*). The buffaloes are neither harmed nor helped by the egrets.

An example of a commensalism is the relationship between cattle egrets and grazing mammals. Cattle egrets are found throughout the tropics and subtropics. They typically forage on the ground around cattle or other large mammals, concentrating their attention near the mammals' heads and feet, where they capture insects flushed by their hooves and mouths (Figure 54.3). Cattle egrets foraging close to grazing mammals capture more food for less effort than egrets foraging away from grazing mammals. The benefit to the egrets is clear; the mammals neither gain nor lose.

Amensalisms are widespread and important interactions. Mammals, for example, may congregate around water holes, trampling and killing many plants. The mammals benefit by drinking water, but not by trampling and killing the plants. Leaves and branches falling from trees often damage smaller plants beneath them. The trees drop their old structures regardless of whether or not they damage other plants.

Predation and competition have particularly important influences on population dynamics. For that reason, we will illustrate several examples of these interactions later in this chapter. All five types of interactions, combined with the effects of the physical environment, determine the range of environmental conditions under which a species can persist. If there were no competitors, predators, or pathogens in its environment, a species would be able to persist under a broader array of physical conditions than it can in the presence of other species that negatively affect it. On the other hand, the presence of beneficial species may increase the range of physical conditions in which a species can persist.

With this background information on population structure and dynamics and interactions among populations, we can now turn to the questions that we posed at the beginning of this chapter. We will begin with abundance and rarity.

Factors Influencing Population Densities

You have probably observed that in a particular area, some species are much more abundant than others. Some locally rare species may be abundant somewhere else; other species may exist at low population densities everywhere. Some species that are rare at a given time may be abundant at some later time, or vice versa. Four factors—resource abundance, the size of individuals, the length of time a species has lived in an area, and social organization—exert strong influences on population density.

▶ *Species that use abundant resources often reach higher population densities than species that use scarce resources.* Thus, on average, animals that eat plants are typically more common than animals that eat other animals. We will explore this pattern in greater detail in Chapter 55.

▶ *Species with small individuals generally reach higher population densities than species with large individuals.* In general, population density decreases as body size increases, because small individuals require less energy to survive than large individuals.

This relationship can be demonstrated by a logarithmic plot of population density against body size for a variety of mammals worldwide (Figure 54.4). Although there is a strong relationship between population density and body size, the great scatter of points on the graph shows that some small

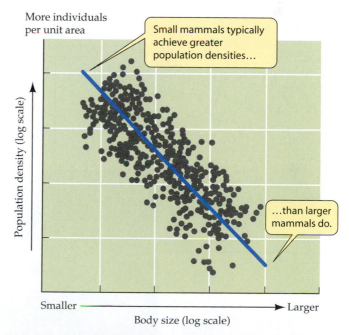

54.4 Population Density Decreases as Body Size Increases This trend is illustrated by a logarithmic plot (that is, each tick mark is 10 times greater than the one before it) of population density against body size for mammals of different sizes; each dot represents a different species, and the resulting slope (straight line) is determined algebraically.

species may use scarce resources and some large species may use abundant resources.

▶ *Some newly introduced species reach high population densities.* Species that have recently escaped control by the factors that normally prevent them from becoming more abundant may achieve temporarily high population densities. Species that are introduced into a new region, where their normal predators and diseases are absent, sometimes reach population densities much higher than ever found in their native ranges.

The zebra mussel, *Dreissena polymorpha*, whose larvae were carried from Europe in the ballast water of commercial cargo ships, became established in the Great Lakes in about 1985. Zebra mussels spread rapidly, and today they occupy much of the Great Lakes and the Mississippi River drainage (Figure 54.5). In some places these mussels have reached densities as high as 400,000 individuals per square meter; such

densities are never found in Europe. Densities of zebra mussels in North America are likely to decrease in the future as local predators and diseases begin to attack them.

▶ *Complex social organization may facilitate high densities.* As we saw in Chapter 53, some highly social species, including ants, termites, and humans, can achieve remarkably high population densities.

The factors that influence population density may strengthen or weaken over time. When this happens, population densities change. Let's look now at how and why population densities fluctuate.

Fluctuations in Population Densities

Although some populations fluctuate markedly in density, even the most dramatic fluctuations are much less than those that are theoretically possible. To visualize those possibilities, consider a single bacterium selected at random from the surface of this book. If all its descendants were able to grow and reproduce in an unlimited environment, explosive population growth would result. In a month, this bacterial colony would weigh more than the visible universe and would be expanding outward at the speed of light. Similarly, a single pair of Atlantic cod and their descendants, reproducing at the maximum rate of which they are capable, would fill the Atlantic Ocean basin in 6 years if none of them died. Obviously, such dramatic population growth does not occur in nature. What prevents it from happening?

All populations have the potential for exponential growth

Bacteria and cod illustrate the fact that all populations have the potential for explosive growth. As the number of individuals in a population increases, the number of new individuals added per unit of time accelerates, even if the rate of increase expressed on a per individual basis—called the *per capita growth rate*—remains constant. If births and deaths occur continuously and at constant rates, a graph of the population size over time forms a continuous, J-shaped curve (Figure 54.6). This form of explosive increase is called **exponential growth**. It can be expressed mathematically in the following way:

Rate of increase in number of individuals

$$= \left(\begin{array}{l} \text{Average per capita birth rate} \\ - \text{ Average per capita death rate} \end{array} \right)$$
$$\times \text{ Number of individuals}$$

or, more concisely,

$$= \frac{\Delta N}{\Delta t} = (b - d)N$$

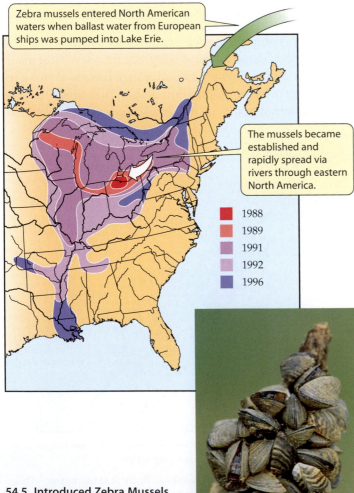

Zebra mussels entered North American waters when ballast water from European ships was pumped into Lake Erie.

The mussels became established and rapidly spread via rivers through eastern North America.

■ 1988
■ 1989
■ 1991
■ 1992
■ 1996

54.5 Introduced Zebra Mussels Have Spread Rapidly
Between 1989 and 1991, the range of zebra mussels in North America more than doubled.

Dreissena polymorpha

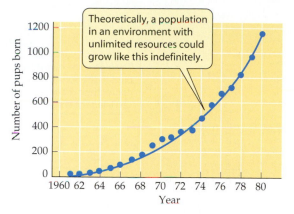

Theoretically, a population in an environment with unlimited resources could grow like this indefinitely.

54.6 Exponential Population Growth The growth of the elephant seal population on Año Nuevo Island, California, between 1960 and 1980 illustrates the exponential population growth curve. Theoretically, a population in a habitat with unlimited resources (including space) could continue to grow indefinitely. Since no resource on Earth is unlimited, this pattern cannot continue indefinitely for any species (including humans).

Mirounga angustirostris

where $\Delta N/\Delta t$ is the rate of change in the size of the population (ΔN = change in number of individuals; Δt = change in time).

The difference between the average per capita birth rate in a population (b) and its average per capita death rate (d) is the *net reproductive rate* (r). (In these equations, b includes both births and immigrations, and d includes both deaths and emigrations.) When conditions are optimal for the population, the net reproductive rate has its highest value, called r_{max}, or the *intrinsic rate of increase; r_{max}* has a characteristic value for each species. Therefore, the rate of growth of a population under optimal conditions is

$$\frac{\Delta N}{\Delta t} = r_{max} N$$

For very short time periods, some populations may grow at rates close to the intrinsic rate of increase. For example, northern elephant seals were hunted nearly to extinction in the late nineteenth century. In 1890, only about 20 animals remained, confined to Isla Guadalupe off the northwestern coast of Mexico. Once the hunting was stopped, the population was protected from its major predator, and ample elephant seal habitat remained available, so the population began to increase rapidly. Elephant seals recolonized Año Nuevo Island near Santa Cruz, California, in 1960. In the 20 years after colonization, the population breeding on the island expanded exponentially (see Figure 54.6).

Population growth is influenced by environmental limits

No real population can maintain exponential growth for very long. As a population increases in size, environmental limits cause birth rates to drop and death rates to rise. In fact, over long time periods, the densities of most populations fluctu-

ate around a relatively constant number. The simplest way to picture the limits imposed by the environment is to assume that an environment can support no more than a certain number of individuals of any particular species per unit of area. This number, called the **environmental carrying capacity** (K), is determined by the availability of resources—food, nest sites, shelter—as well as by disease, predators, and, in some cases, social interactions.

Because of environmental limits, the growth of a population typically slows down as its density approaches the environmental carrying capacity. A graph of the population size over time results in an S-shaped curve (Figure 54.7). This pattern is called **logistic growth**. The simplest way to generate an S-shaped growth curve is to add to the equation for exponential growth a term, $(K-N)/K$, that slows the population's growth as it approaches the carrying capacity. This

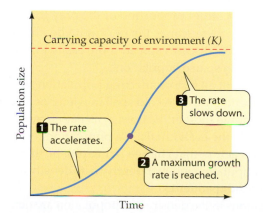

54.7 Logistic Population Growth Typically, a population in an environment with limited resources stops growing exponentially long before it reaches the environmental carrying capacity.

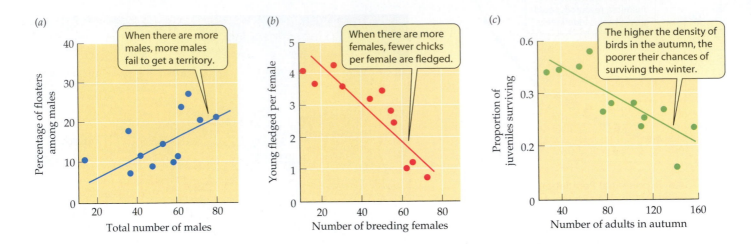

54.8 Regulation of an Island Population of Song Sparrows
The size of the population of song sparrows (*Melospiza melodia*) on Mandarte Island, British Columbia, is determined in part by the severity of winter weather. In addition, the population is regulated by (a) the territorial behavior of males, (b) the reproductive success of females, and (c) the survival of juveniles in relation to population density.

term implies that each individual added to the population depresses population growth by an equal amount:

$$\frac{\Delta N}{\Delta t} = r\left(\frac{K - N}{K}\right)N$$

Population growth stops when $N = K$ because then $(K - N) = 0$, so $(K - N)/K = 0$, and thus $\Delta N/\Delta t = 0$.

Population densities influence birth and death rates

Because each additional individual typically makes things worse for other members of the population in a limited environment, per capita birth and death rates usually change in response to population density; that is, they are **density-dependent**. Birth and death rates may be density-dependent for several reasons. First, as a species increases in abundance, it may deplete its food supply, reducing the amount of food available to each individual. Poorer nutrition may increase death rates and decrease birth rates. Second, predators may be attracted to areas with high densities of their prey. If predators capture a larger proportion of the prey than they did when the prey were scarce, the per capita death rate of the prey rises. Third, diseases spread more easily in dense populations than in sparse populations.

However, not all factors affecting population size act in a density-dependent way. A cold spell in winter or a hurricane that blows down most of the trees in its path may kill a large proportion of the individuals in a population regardless of its density. Factors that change per capita birth and death rates in a population independently of its density are said to be **density-independent**.

Fluctuations in the density of a population are determined by all the density-dependent and density-independent factors acting on it. The combined action of these factors is shown by the dynamics of a population of song sparrows on Mandarte Island, off the coast of British Columbia, Canada.

Over a period of 12 years, the number of song sparrows fluctuated between 4 and 72 breeding females and between 9 and 100 territorial males. Death rates are high during particularly cold, snowy winters, regardless of the density of the population. Several density-dependent factors also contribute to fluctuations in the density of the population. The number of breeding males, for example, is limited by territorial behavior: The larger the number of males, the larger the number that fail to gain territories and must live as "floaters" with little chance of reproducing (Figure 54.8a). Also, the larger the number of breeding females, the fewer offspring each female fledges (raises to the age when it can leave the nest) (Figure 54.8b). And the more birds alive in autumn, the poorer the chances of juveniles born in that year surviving the winter (Figure 54.8c). Thus, the number of males and females breeding each year is influenced by both density-independent and density-dependent factors.

Population Fluctuations

The cactus finch, which we met in our earlier discussion of life tables, is a small, short-lived seed-eating bird that lives only in the Galápagos archipelago. The south polar skua is a long-lived carnivorous seabird with a broad geographic range in the southern oceans. Over several decades, the number of cactus finches fluctuated widely, as we have already noted. The number of skuas fluctuated very little over an even longer time period (Figure 54.9). Why did the population of skuas fluctuate so much less than the population of cactus finches?

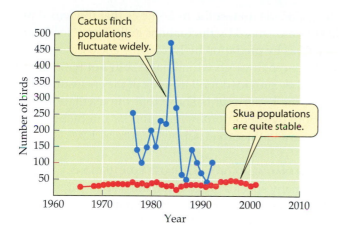

54.9 Population Sizes May Be Stable or Highly Variable
Populations of small, short-lived species, such as that of cactus finch-es (*Geospiza scandens*) on Isla Daphne in the Galápagos Archipelago, tend to fluctuate much more than do populations of larger, longer-lived species, such as that of south polar skuas (*Catharacta mac-cormicki*).

black cherry trees in a Wisconsin forest in 1971 became established between 1931 and 1941 (Figure 54.10*b*). Population densities increase following years of good reproductive success, but they decrease following years of poor reproduction.

RESOURCE FLUCTUATIONS GENERATE CONSUMER FLUCTUATIONS.
Densities of populations of species that depend on a single or just a few resources are likely to fluctuate more than those of species that use a greater variety of resources. As we have seen, cactus finch populations fluctuate with the annual production of the seeds they eat, which varies greatly. Similarly, several species of birds and mammals that live in northern coniferous forests depend on seeds in conifer cones. Most trees in northern coniferous forests reproduce synchronously and episodically; consequently, over large areas, there are years of massive seed production and years of little or no seed production. Some birds (such as cross-bills) wander over large areas, looking for places where

All populations fluctuate less than the theoretical maximum, but the sizes of some populations fluctuate remarkably little. The comparison between south polar skuas and cactus finches illustrates one cause of such differences: Species with long-lived individuals that have low reproductive rates, such as south polar skuas, typically have more stable populations than species with short-lived individuals that have high reproductive rates, such as cactus finches. Small, short-lived individuals are generally more vulnerable to environmental changes than long-lived individuals. That is why insect population densities tend to fluctuate much more than those of birds and mammals, and population densities of annual plants fluctuate much more than those of trees.

EPISODIC REPRODUCTION GENERATES FLUC-TUATIONS. For most species, some years are better for reproducing than other years. In Lake Erie, 1944 was such an excellent year for reproduction of whitefish that individuals born in that year dominated whitefish catches in the lake for several years (Figure 54.10*a*). Similarly, most of the individuals found in a population of

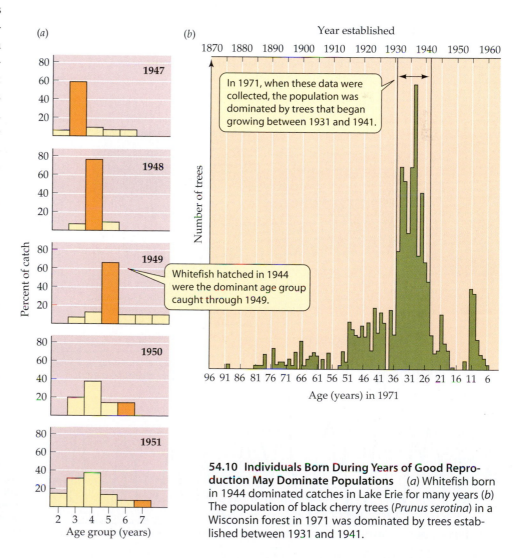

54.10 Individuals Born During Years of Good Repro-duction May Dominate Populations (*a*) Whitefish born in 1944 dominated catches in Lake Erie for many years (*b*) The population of black cherry trees (*Prunus serotina*) in a Wisconsin forest in 1971 was dominated by trees established between 1931 and 1941.

cones have been produced. Other birds (such as jays and nutcrackers) and some mammals (squirrels) store cones during years of high production, but they often suffer high mortality rates during years when the trees in their area produce few or no cones.

POPULATION INTERACTIONS GENERATE FLUCTUATIONS   Predators often cause fluctuations in the densities of their prey because predator population growth nearly always lags behind growth in populations of their prey. The predator population grows and eats most of its prey population, followed by a crash in the predator population, which no longer has enough food. Population oscillations among small mammals and their predators living at high latitudes, where many predators depend on only one or a few prey species, are the best-known examples of fluctuations in population densities driven by predator–prey interactions. Populations of Arctic lemmings and their chief predators—snowy owls, jaegers, and Arctic foxes—oscillate with a 3- to 4-year periodicity. Populations of Canadian lynx and their principal prey, snowshoe hares, oscillate on a 9- to 11-year cycle (Figure 54.11).

For many years, ecologists thought that hare–lynx oscillations were caused only by interactions between hares and lynx. Recently, ecologists performed experiments in Yukon Territory, Canada, to test the hypothesis that the lynx–hare oscillations are caused by fluctuations in the hares' food supply as well as by predation by lynx. They enclosed some areas with fences through which hares, but not lynx, could pass, and they provided food in some of the enclosures. The results of the experiments show that the oscillations are driven both by predation by lynx and by interactions between hares and their food supply (Figure 54.12).

POPULATION FRAGMENTATION GENERATES FLUCTUATIONS.   Populations of many species are divided into separated, discrete subpopulations living in distinct habitat patches, among which some exchange of individuals occurs. Each subpopulation has a probability of "birth" (colonization) and "death" (extinction). Within each subpopulation, growth occurs in the ways we have just described, but because the subpopulations are much smaller than the population as a whole, local disturbances and random fluctuations in numbers of individuals are more likely to cause the extinction of subpopulations than the extinction of an entire population. However, if individuals move frequently between subpopulations, immigrants may prevent declining subpopulations from becoming extinct. This process is known as the **rescue effect**.

EXAMPLES OF SUBPOPULATION DYNAMICS.   The bay checkerspot butterfly (*Euphydryas editha bayensis*) provides a good illustration of the dynamics of subpopulations. The caterpillars (larvae) of this butterfly feed on only a few species of annual plants, which are restricted to outcrops of serpentine rock on hills south of San Francisco, California. The bay checkerspot has been studied for many years by Stanford University biologists. During drought years, most host plants die early in spring, before the caterpillars have developed far enough to be able to enter their summer rest-

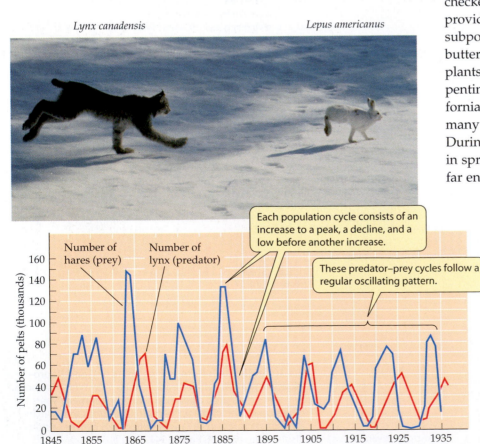

Lynx canadensis   *Lepus americanus*

Each population cycle consists of an increase to a peak, a decline, and a low before another increase.

These predator–prey cycles follow a regular oscillating pattern.

Number of hares (prey)

Number of lynx (predator)

54.11 Hare and Lynx Populations Cycle in Nature   The 9–11-year population cycle of the snowshoe hare and its major predator, the Canadian lynx, was revealed in records of the number of pelts that were sold to the Hudson's Bay Company by fur trappers.

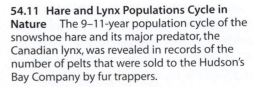

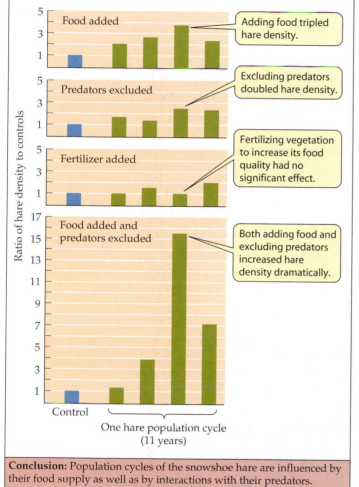

EXPERIMENT

Hypothesis: Population cycles of hares are influenced by both food supply and predators.

METHOD   Select 9 1-km² blocks of undisturbed coniferous forest. In two of the blocks give the hares supplemental food year-round. Erect an electric fence around two other blocks, with mesh large enough to allow hares, but not lynxes, to pass. Provide extra food in one of these blocks. In two other blocks add fertilizer to increase food quality. Use three other blocks as unmanipulated controls.

RESULTS

Ratio of hare density to controls

Food added — *Adding food tripled hare density.*

Predators excluded — *Excluding predators doubled hare density.*

Fertilizer added — *Fertilizing vegetation to increase its food quality had no significant effect.*

Food added and predators excluded — *Both adding food and excluding predators increased hare density dramatically.*

Control

One hare population cycle (11 years)

Conclusion: Population cycles of the snowshoe hare are influenced by their food supply as well as by interactions with their predators.

54.12 Prey Population Cycles May Have Multiple Causes Experiments showed that both food supply and predation (but not food quality) affect the population densities of snowshoe hares.

ing stage. At least three butterfly subpopulations became extinct during a severe drought in 1975–1977. The largest patch of suitable butterfly habitat, Morgan Hill, typically supports thousands of butterflies (Figure 54.13). It probably served as a source of individuals that dispersed to and recolonized small patches where the butterflies had become extinct.

In another study, ecologists manipulated the habitat of tiny arthropods (springtails—tiny insects without wings—and mites) to investigate the subpopulation dynamics of these animals. In one experiment, they created isolated patches of the animals' habitat—mosses growing on rocks—by clearing moss from parts of the rock surface (Figure 54.14, Experiment 1). The number of species present in these patches declined about 40 percent within a year, with more rare species than common species disappearing from the patches. The experiment illustrated that small, isolated populations are more likely to become extinct than large populations are.

In a second experiment, the investigators created similar patches, but these patches were connected by narrow corridors of moss that were either intact or disrupted by a barrier only 10 mm wide (Figure 54.14, Experiment 2). Moss patches connected by unbroken corridors contained more species of arthropods a year later than patches whose corridors were discontinuous. Thus, a gap of only 10 mm was sufficient to reduce the rescue effect.

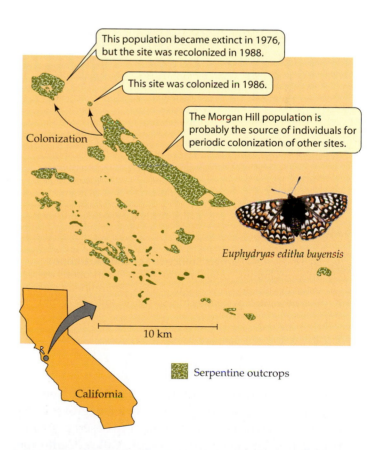

This population became extinct in 1976, but the site was recolonized in 1988.

This site was colonized in 1986.

The Morgan Hill population is probably the source of individuals for periodic colonization of other sites.

Colonization

Euphydryas editha bayensis

10 km

California

Serpentine outcrops

54.13 Subpopulation Dynamics The bay checkerspot butterfly population is divided into a number of subpopulations confined to patches of habitat (serpentine rock) that contain the food plants of its larvae. Extinction of these subpopulations is common.

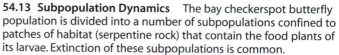

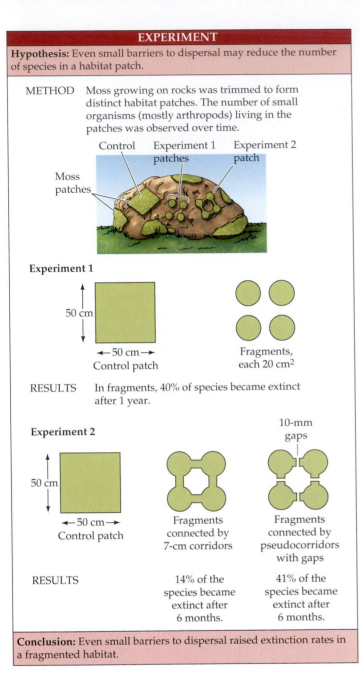

EXPERIMENT

Hypothesis: Even small barriers to dispersal may reduce the number of species in a habitat patch.

METHOD   Moss growing on rocks was trimmed to form distinct habitat patches. The number of small organisms (mostly arthropods) living in the patches was observed over time.

Control   Experiment 1 patches   Experiment 2 patch

Moss patches

Experiment 1

50 cm

←50 cm→
Control patch

Fragments, each 20 cm²

RESULTS   In fragments, 40% of species became extinct after 1 year.

Experiment 2

10-mm gaps

50 cm

←50 cm→
Control patch

Fragments connected by 7-cm corridors

Fragments connected by pseudocorridors with gaps

RESULTS   14% of the species became extinct after 6 months.   41% of the species became extinct after 6 months.

Conclusion: Even small barriers to dispersal raised extinction rates in a fragmented habitat.

54.14 Narrow Barriers Suffice to Separate Arthropod Subpopulations   Many species of small arthropods went extinct in isolated habitat patches. Recolonization of patches was prevented by barriers to dispersal as small as 10 mm.

Variations in Species' Ranges

Douglas firs are widespread in western North America, whereas giant sequoias are restricted to a few groves in the southern Sierra Nevada of California. The desert pupfish is restricted to a single spring in Death Valley, California, whereas smallmouth bass live in most of the rivers and lakes in eastern North America. Why do the geographic ranges of species vary so much? The factors contributing to this varia-

tion include speciation processes, dispersal abilities, and interactions with other species. As you might suspect, not all of the factors that influence geographic ranges are important for all species.

SPECIATION PROCESSES INFLUENCE RANGE SIZES.   As we saw in Chapter 24, there are several ways in which a new species can originate. A species that arises by polyploidy inevitably begins with a very small range. Because many polyploid plant species have formed only recently and have not spread much beyond the site of their origin, many of these plants have small ranges. Similarly, species that arise through founder events begin their history with small ranges. In contrast, most species that arise via allopatric speciation begin with large ranges. Finally, as a species declines toward extinction, as may be happening to giant sequoias (Figure 54.15), its range shrinks until it vanishes when the last individual dies.

DISPERSAL ABILITIES RESTRICT GEOGRAPHIC RANGES.   As we also saw in Chapter 24, the dispersal abilities of different species vary greatly. The experiments with small arthropods living in mosses on rocks show that even narrow barriers may prevent some species from reaching and colonizing an area. The solitary spring that is home to the desert pupfish is isolated from other bodies of fresh water, so the fish cannot disperse. Thus, the absence of many species from an area is simply due to a failure to get there. Zebra mussels, for

Sequoiadendron giganteum

54.15 The Last Refuge   The range of giant sequoias has progressively shrunk to a few remaining groves of trees scattered in the southern Sierra Nevada mountains of California.

example, were not found in North America before 1985 because they were unable to disperse across the Atlantic Ocean from Europe. Lack of suitable habitat was not the reason for their absence, as demonstrated by their dramatic population growth in North America once they were transported there by human activities. Once they reached North America, they were able to disperse rapidly because the larvae are free-swimming and the adults can attach to moving objects, such as boat hulls.

PREDATORS MAY RESTRICT SPECIES' RANGES. Predators may eliminate their prey in some places, but not in others. For example, chorus frogs (*Pseudacris triseriata*) are found in only some of the ponds on islands in Lake Superior. Three major predators—the larvae of a salamander, the nymphs of a large dragonfly, and dytiscid beetles—eat chorus frog tadpoles. An ecologist noticed that the tadpoles were common in ponds with beetles, but rare in ponds with salamander larvae and dragonfly nymphs. In laboratory experiments, he established that the salamander larvae could eat only small tadpoles, but that dragonfly nymphs could eat tadpoles of all sizes. Therefore, he hypothesized that dragonfly nymphs were responsible for eliminating chorus frogs from many ponds. He tested his hypothesis by manipulating densities of predators and prey in ponds. The results showed that dragonfly nymphs can eliminate chorus frogs from ponds that would otherwise be suitable for them (Figure 54.16).

COMPETITION MAY RESTRICT SPECIES' RANGES. How competitive interactions may restrict the ranges of species is illustrated by interactions between two species of barnacles, *Balanus balanoides* and *Chthamalus stellatus*, on rocky North Atlantic seashores. These barnacles have planktonic larvae, which settle between high and low tide levels on the shoreline and become sessile adults. Adult *Chthamalus* generally live higher in the intertidal zone than do adult *Balanus*, and there is little overlap between the two species. What keeps their ranges so distinct?

By experimentally removing one or the other species, researchers have shown that the vertical ranges of adults of both species are greater in the absence of the other species. *Chthamalus* larvae normally settle in large numbers in the *Balanus* zone. If *Balanus* are absent, young *Chthamalus* survive and grow well in the *Balanus* zone, but if *Balanus* are present, they smother, crush, or undercut the *Chthamalus*. *Balanus* larvae also settle in the *Chthamalus* zone, but the young *Balanus* grow slowly there because they lose water rapidly when exposed to air, so *Chthamalus* outcompete *Balanus* in that zone. The result of the competitive interaction between the two species is intertidal zonation, with *Chthamalus* growing above *Balanus* (Figure 54.17).

54.16 Predators Exclude Prey from Some Habitats The speed with which dragonfly nymphs can eliminate tadpoles of the chorus frog from a pond is illustrated by the results of experiments in which populations of predators and prey were manipulated.

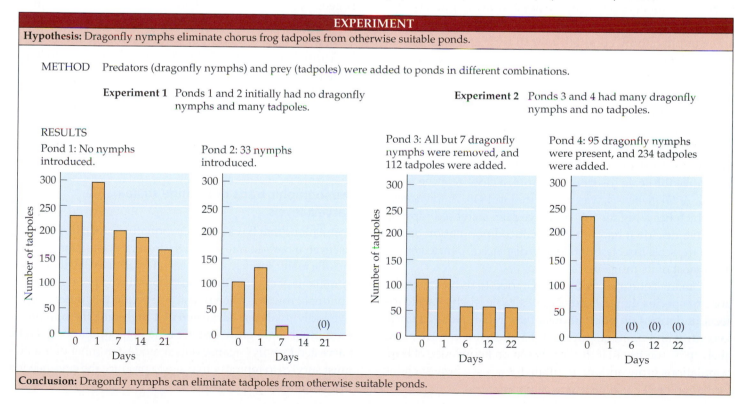

EXPERIMENT

Hypothesis: Dragonfly nymphs eliminate chorus frog tadpoles from otherwise suitable ponds.

METHOD   Predators (dragonfly nymphs) and prey (tadpoles) were added to ponds in different combinations.

Experiment 1 Ponds 1 and 2 initially had no dragonfly nymphs and many tadpoles.

Experiment 2 Ponds 3 and 4 had many dragonfly nymphs and no tadpoles.

RESULTS

Pond 1: No nymphs introduced.

Pond 2: 33 nymphs introduced.

Pond 3: All but 7 dragonfly nymphs were removed, and 112 tadpoles were added.

Pond 4: 95 dragonfly nymphs were present, and 234 tadpoles were added.

Conclusion: Dragonfly nymphs can eliminate tadpoles from otherwise suitable ponds.

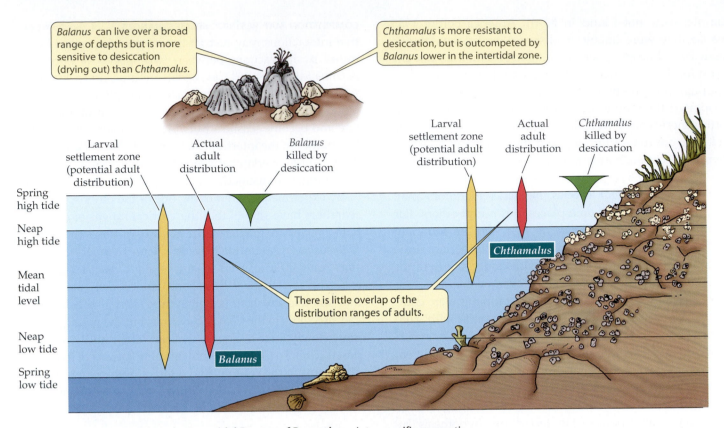

Balanus can live over a broad range of depths but is more sensitive to desiccation (drying out) than *Chthamalus*.

Chthamalus is more resistant to desiccation, but is outcompeted by *Balanus* lower in the intertidal zone.

Larval settlement zone (potential adult distribution)

Actual adult distribution

Balanus killed by desiccation

Larval settlement zone (potential adult distribution)

Actual adult distribution

Chthamalus killed by desiccation

There is little overlap of the distribution ranges of adults.

Chthamalus

Balanus

Spring high tide

Neap high tide

Mean tidal level

Neap low tide

Spring low tide

54.17 Competition Restricts the Intertidal Ranges of Barnacles Interspecific competition between *Balanus* and *Chthamalus* makes the zone each species occupies smaller than the zone it could occupy in the absence of the other species. The width of the red and gold bars is proportional to the density of the populations.

Sessile animals such as barnacles and many plants compete for space, but most mobile animals compete for food. As an example of how competition can restrict the ranges of such species, consider the distribution of two species of wasps in California. These wasps lay their eggs on scale insects, and the larvae that hatch from those eggs burrow into, eat, and kill the scale insects. Both wasps were introduced to control outbreaks of scale insects that were damaging citrus orchards. The Mediterranean wasp *Aphytis chrysomphali* was introduced to southern California around 1900, but it failed to control the scale insects. Therefore, a close relative from China, *A. lingnanensis*, was introduced in 1948. *A. lingnanensis*, which has a higher reproductive rate, increased rapidly. Within a decade it had not only reduced population densities of the scale insects, but had also displaced *A. chrysomphali* from most of its range in California.

For many centuries, people have tried to reduce populations of species they consider undesirable, such as scale insects, and maintain populations of desirable species. Efforts to control and manage populations of organisms are more likely to be successful if they are based on knowledge of how populations grow and are regulated. Let's see how such information can be used to manage populations.

Managing Populations

A general principle of population dynamics is that both the total number of births and the growth rates of individuals tend to be highest when a population is well below its carrying capacity (see Figure 54.8). Therefore, if we wish to maximize the number of individuals that can be harvested from a population, we should manage the population so that it is far enough below carrying capacity to have high birth and growth rates. Hunting seasons for game birds and mammals are established with this objective in mind.

Demographic traits determine sustainable harvest levels

Populations that have high reproductive capacities can persist even if harvest rates are high. In such populations (which include many species of fish), each female may lay thousands or millions of eggs. In these fast-reproducing populations, individual growth is often density-dependent. If prereproductive individuals are harvested at a high rate, the remaining individuals may grow faster. Some fish populations can be harvested heavily because only a modest number of females must survive to reproductive age to produce the eggs needed to maintain the population.

Fish can, of course, be overharvested. Many fish populations have been greatly reduced because so many individuals were harvested that too few reproductive adults survived to maintain the population. The Georges Bank off the coast of New England—a source of cod, halibut, and other prime food fishes—was exploited so heavily during the twentieth century that many fish stocks were reduced to levels insufficient to support a commercial fishery. The fishery has remained closed into the twenty-first century.

The whaling industry has also engaged in excessive harvests. The blue whale, Earth's largest animal, was the first whale species to be hunted nearly to extinction. The industry then turned to smaller species of whales that were still numerous enough to support commercially viable whaling operations (Figure 54.18).

Management of whale populations is difficult for two reasons. First, unlike fish, whales reproduce at very low rates. They have long prereproductive periods before they mature, produce only one offspring at a time, and have long intervals between births. Thus, many adult whales are needed to produce even a small number of offspring. Second, because whales are distributed widely throughout Earth's oceans, they are an international resource whose conservation and wise management depends upon cooperative action by all whaling nations. This goal continues to be difficult to achieve.

Demographic information is used to control populations

The same management principles apply if we wish to reduce the size of populations of undesirable species and keep them at low densities. At densities well below carrying capacity, populations typically have high birth rates, and can therefore withstand higher death rates than they can when they are closer to carrying capacity.

When population dynamics are influenced primarily by factors that operate in a density-dependent manner, killing part of a population typically reduces it to a density at which it reproduces at a higher rate. A more effective approach to reducing such a population is to remove its resources, thereby lowering the carrying capacity of its environment. We can rid our dumps and cities of rats more easily by making garbage unavailable (reducing the carrying capacity of the rats' environment) than by poisoning rats (which only increases their reproductive rate). However, this option may not exist in agriculture, in which a high density of the crop is the management objective.

Similarly, if we wish to preserve a rare species, the most important step usually is to provide it with suitable habitat. If habitat is available, the species will usually reproduce at rates sufficient to maintain its population. If the habitat is insufficient, preserving the species usually requires expensive and continuing intervention, such as providing extra food.

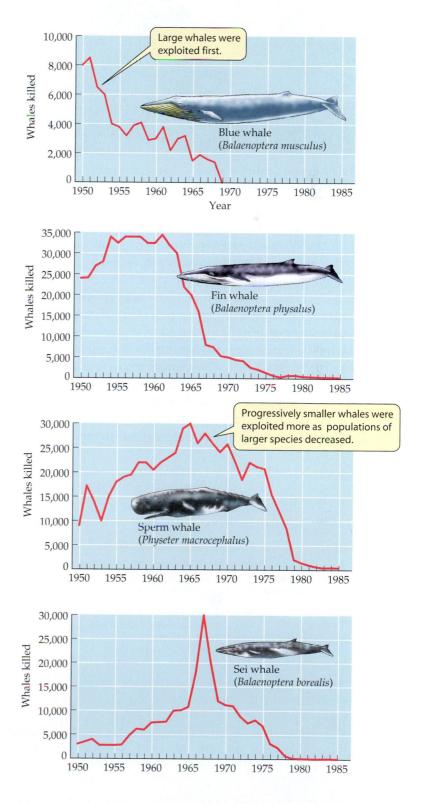

54.18 Overexploitation of Whales These graphs show the numbers of whales of four species killed each year from 1950 to 1985. All four species were driven to very low population levels by sustained hunting.

54.19 Biological Control of a Pest   These *Cactoblastis* caterpillars are consuming an *Opuntia* cactus in Australia.

Humans often attempt to reduce the populations of introduced species that have dramatically increased in density by introducing their predators and parasites. For example, the cactus *Opuntia*, introduced into Australia from South America, spread rapidly and became a pest over vast expanses of valuable sheep-grazing land. *Opuntia* was controlled in Australia by the introduction of a moth (*Cactoblastis cactorum*) whose larvae eat *Opuntia*. Once female moths find a patch of cactus and lay their eggs on it, the larvae that hatch from those eggs completely destroy the patch (Figure 54.19). However, new patches of cactus arise in other places from seeds dispersed by birds. These new patches flourish until they are found and destroyed by *Cactoblastis*. Over a large region, the numbers of both *Opuntia* and *Cactoblastis* are now fairly constant and low, but in local areas, there are extreme oscillations caused by the extermination of first the cactus and then the moth. Today, both *Opuntia* and *Cactoblastis* in Australia are distributed as scattered subpopulations among which individuals occasionally disperse.

Can we manage our own population?

Managing our own population has become a matter of great concern because the size of the human population is responsible for most of the environmental problems we are facing today, from pollution to extinctions of other species. For thousands of years, Earth's carrying capacity for human populations was set at a low level by food and water supplies and disease. We saw in Chapter 53 how human social behavior and specialization has allowed us to develop technologies for increasing our resources and combating diseases. The domestication of plants and animals, improved crops and farm yields, mining and use of fossil fuels, and the development of modern medicine have all contributed to the staggering increase in Earth's human population (Figure 54.20).

What is Earth's present carrying capacity for people? Today's carrying capacity is set in part by Earth's ability to absorb the by-products, especially carbon dioxide, of our enormous consumption of fossil fuel energy; by water availability (in many areas); and by whether we are willing to cause the extinction of millions of other species to accommodate our increasing use of Earth's resources. We will explore some of the consequences of high human population densities and high per capita use of resources for the survival of other species in Chapter 57.

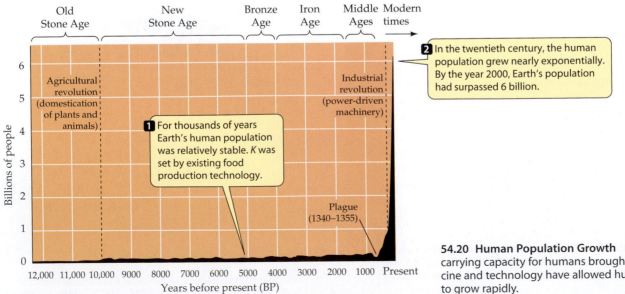

54.20 Human Population Growth   Increases in Earth's carrying capacity for humans brought about by medicine and technology have allowed human populations to grow rapidly.

Regional and Global Processes Influence Local Population Dynamics

Between 1950 and 1980, annual counts of breeding birds were conducted in Eastern Wood, in southeastern England. During that time, populations of some species increased while others decreased (Figure 54.21). For example, the population of wood pigeons more than doubled, but the population of garden warblers decreased to zero in 1971; no more than two pairs have bred in the wood since then. The population of blue tits increased from just a few pairs to an average of more than 15 pairs. Why did these populations change so differently?

No matter how intensively ecologists might have studied the birds of Eastern Wood, they could not have answered that question, because populations of two of these three species were strongly influenced by events remote from Eastern Wood. Wood pigeons increased greatly over most of southern England during this 30-year period because of the widespread adoption of oilseed rape as an agricultural crop. Rape fields provide wood pigeons with abundant winter food. Garden warblers decreased because their overwinter survival was poor due to a severe drought on their wintering grounds in West Africa.

The population of blue tits was influenced primarily by changes within Eastern Wood itself. Until the early 1950s, trees in Eastern Wood were periodically felled and sold for timber. After the cutting stopped, more holes, in which blue tits nest, became available in mature and dead trees.

Local population dynamics are often influenced both by local interactions and by more distant or remote events, per-haps even on different continents. How processes occurring at varying spatial and temporal scales influence the structure of ecological communities is the focus of the next chapter.

Chapter Summary

Populations in Space and Time

▶ A population consists of all the individuals of a species within a given area.

▶ The number of individuals of a species per unit of area (or volume) is its population density. Dense populations often exert strong influences on populations of other species.

▶ Life tables summarize information about births and deaths in populations. **Review Table 54.1**

▶ Graphs of survivorship in relation to age show when individuals survive well and when they do not. **Review Figure 54.1**

▶ The age distribution of individuals in a population reveals much about the recent history of births and deaths in the population. The timing of births and deaths may influence age distributions for many years. **Review Figure 54.2**

Types of Ecological Interactions

▶ Individuals of two populations may interact in ways that may benefit or harm either or both participants. **Review Table 54.2. See Web/CD Activity 54.1**

Factors Influencing Population Densities

▶ Species with small individuals typically achieve higher population densities than species with large individuals. **Review Figure 54.4**

▶ Introduced species sometimes achieve great population densities. **Review Figure 54.5**

Fluctuations in Population Densities

▶ All populations have the potential to grow exponentially under optimal conditions. **Review Figure 54.6. See Web/CD Tutorial 54.1**

▶ No population can maintain exponential growth for very long because environmental limits cause birth rates to drop and death rates to rise.

▶ The number of individuals of a particular species that an environment can support—called the carrying capacity (K)—is determined by the availability of resources and by disease and predators.

▶ A population in a limited environment shows a logistic growth pattern, in which growth rates decrease as the carrying capacity is approached. **Review Figure 54.7. See Web/CD Activity 54.2 and Tutorial 54.2**

▶ The density of a population is influenced by the combined effects of all density-dependent and density-independent factors affecting it. **Review Figure 54.8**

Population Fluctuations

▶ Populations do not fluctuate as much as theoretically possible, but some fluctuate much more than others. The amount of fluctuation is influenced by body size, reproductive rate, and range size. **Review Figure 54.9**

▶ Population fluctuations may be strongly influenced by years of good reproduction. **Review Figure 54.10**

▶ Predator–prey interactions may generate population cycles. **Review Figures 54.11, 54.12**

▶ Populations of many species exist as small, fragmented subpopulations. Extinction of subpopulations is common, but indi-

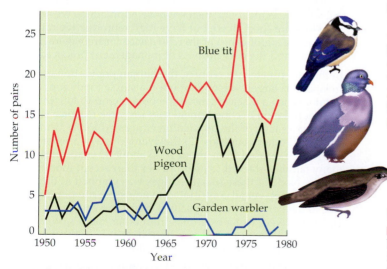

54.21 Populations May Be Influenced by Remote Events
Populations of some birds increased in Eastern Wood, England, while others decreased. The wood pigeon (*Columba palumbus*) and garden warbler (*Sylvia borin*) population shifts were strongly influenced by different events that took place far from Eastern Wood. Only the blue tit (*Parus caeruleus*) population was affected most strongly by events within Eastern Wood itself.

viduals from other fragments may recolonize them. **Review Figures 54.13, 54.14. See Web/CD Tutorial 54.4**
See Web/CD Tutorial 54.3

Variation in Species' Ranges

▶ Some species are restricted to very small areas, whereas others are widely distributed on Earth.

▶ Species' ranges are influenced by speciation processes, dispersal abilities, predators, and competition. **Review Figures 54.16, 54.17**

Managing Populations

▶ Humans can use the principles of population dynamics to control and manage populations of desirable and undesirable species. Nevertheless, humans have overexploited many populations. **Review Figure 54.18**

Regional and Global Processes Influence Local Population Dynamics

▶ Population densities may be influenced by both local conditions and remote events. **Review Figure 54.20**

Self-Quiz

1. The number of individuals of a species per unit of area is known as its
   a. population size.
   b. population density.
   c. population structure.
   d. subpopulation.
   e. biomass.

2. The age distribution of a population is determined by
   a. the timing of births.
   b. the timing of deaths.
   c. the timing of both births and deaths.
   d. the rate at which the population is growing.
   e. All of the above

3. Which of the following is *not* a demographic event?
   a. Growth
   b. Birth
   c. Death
   d. Immigration
   e. Emigration

4. A group of individuals born at the same time is known as a
   a. deme.
   b. subpopulation.
   c. Mendelian population.
   d. cohort.
   e. taxon.

5. Two organisms that use the same resources when those resources are in short supply are said to be
   a. predators.
   b. competitors.
   c. mutualists.
   d. commensalists.
   e. amensalists.

6. Damage caused to shrubs by branches falling from overhead trees is an example of
   a. interference competition.
   b. partial predation.
   c. amensalism.
   d. commensalism.
   e. diffuse coevolution.

7. A population grows at a rate closest to its intrinsic rate of increase when
   a. its birth rates are the highest.
   b. its death rates are the lowest.
   c. environmental conditions are optimal.
   d. it is close to the environmental carrying capacity.
   e. it is well below the environmental carrying capacity.

8. Immigrants that prevent a subpopulation from becoming extinct result in a
   a. colonization effect.
   b. rescue effect.
   c. metapopulation effect.
   d. genetic drift effect.
   e. salvage effect.

9. Density-dependent population regulation is strongest when
   a. only birth rates change in response to density.
   b. only death rates change in response to density.
   c. diseases spread in populations at all densities.
   d. both birth and death rates change in response to density.
   e. population densities fluctuate very little.

10. The best way to reduce the population of an undesirable species in the long term is to
    a. reduce the carrying capacity of the environment for the species.
    b. selectively kill reproducing adults.
    c. selectively kill pre-reproductive individuals.
    d. attempt to kill individuals of all ages.
    e. sterilize individuals.

For Discussion

1. Why are big, fierce animals rare?

2. Why do predator–prey interactions often generate cycles or great fluctuations in population densities? Would you expect lynx populations to fluctuate as much as they do if lynx had a variety of abundant prey species available to them?

3. Most organisms whose populations we wish to manage for higher densities are long-lived and have low reproductive rates, whereas most organisms whose populations we attempt to reduce are short-lived, but have high reproductive rates. What is the significance of this difference for management strategies and the effectiveness of management practices?

4. In the mid-nineteenth century, the human population of Ireland was largely dependent upon a single food crop, the potato. When a disease caused the potato crop to fail, the Irish population declined drastically for three reasons: (1) a large percentage of the population emigrated to the United States and other countries; (2) the average age of a woman at marriage increased from about 20 to about 30 years; and (3) many families starved to death rather than accept food from Britain. None of these social changes was planned at the national level, yet all contributed to adjusting the population size to the new carrying capacity. Discuss the ecological principles involved, using examples from other species. What would you have done had you been in charge of the national population policy for Ireland at that time?

5. Because some species introduced to control a pest have become pests themselves, some scientists argue that species introductions should not be used under any circumstances to control pests. Others argue that, provided they are properly researched and controlled, we should continue to use introductions as part of our set of tools for managing pest populations. Which view do you support? Why?

55 Communities and Ecosystems

It is interesting to contemplate an entangled bank, clothed with many plants of many kinds, with birds singing on the bushes, with various insects flitting about, and with worms crawling through the damp earth, and to reflect that these elaborately constructed forms, so different from each other, and dependent on each other in so complex a manner, have all been produced by the laws acting around us.

—CHARLES DARWIN, 1859

 Charles Darwin is remembered mostly for his contributions to evolutionary theory, but this quote from *The Origin of Species* shows that he was also a pioneering ecologist who understood the nature and complexity of the interactions among the species of organisms that live in a particular place.

The species that live and interact in an area constitute an **ecological community**. Darwin's "entangled bank" was an ecological community that had obvious boundaries defined by adjacent crops, pastures, and gardens. But the organisms living in the bank were not confined within those boundaries. Some of the seeds that landed in the bank and grew into trees and shrubs came from parent plants living far away. The insects and birds Darwin observed must have flown into and out of the bank from a large area. To understand which species live in such a bank and how they interact, he would have needed to know about such movements, just as we need to know about drought in Africa in order to understand fluctuations in the population of garden warblers in Eastern Wood, as we saw at the end of Chapter 54.

Darwin's entangled bank had also changed over time. Glaciers had covered the area 10,000 years earlier. The plant species Darwin observed colonized Britain at different times over the several thousand years since the glaciers melted. Ecological communities are not assemblages of organisms that move together as units when environmental conditions change. Rather,

"Clothed with Many Plants of Many Kinds" Charles Darwin's eloquent description of an "entangled bank" reveals his understanding of the nature and complexity of interactions among species living in the same community.

each species has unique interactions with its biotic and abiotic environments.

The species that form an ecological community, together with the physical environment, constitute an **ecosystem**. To understand the processes that influence ecosystems and the patterns they produce, we must study both the interactions of organisms with one another and their interactions with the physical environment; that is, we must study both communities and ecosystems.

Communities: Loose Assemblages of Species

Ecological communities contain many species that interact with one another via the processes we discussed in Chapter 54: competition, predation, mutualism, commensalism, and amensalism. The importance of these interactions changes as a result of changes in the physical environment, gains and losses of species, and changes in the population densities of species. These interactions also change over longer times as the interacting species evolve.

Early in the twentieth century, there was a major debate between two leading North American plant ecologists over the nature of communities. In 1926, Henry Gleason argued that plant communities were loose assemblages of species, each of which was individualistically distributed according to its unique interactions with the physical environment. In a paper published in 1936, Frederick Clements argued that plant communities were tightly integrated "superorganisms." He believed that there were places where one group of species dropped out and was replaced by a very different group.

The debate was resolved by detailed studies of the distributions of plants. Especially influential were analyses of the vegetation of the Siskiyou Mountains of Oregon. Those studies showed that different combinations of plant species were found at different locations. Species entered and dropped out of communities independently over environmental gradients (Figure 55.1). These and other results generally supported Gleason's view of the nature of communities. However, where environmental conditions change abruptly, as they do at the edges of lakes and streams, the ranges of many species may terminate at the same place.

Ecologists try to understand ecological communities and ecosystems by asking several general questions: What patterns exist in ecological communities and ecosystems? How does the physical environment influence those patterns? What are the relative roles of historical accident and current interactions in determining those patterns? How does evolution, acting on members of the community, influence the assemblage of species that live together?

Given that each community contains a multitude of species, each interacting with many other species, this goal might appear to be unattainable. Fortunately, we do not need

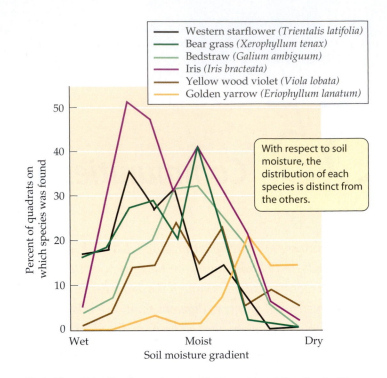

55.1 Plant Distributions along an Environmental Gradient The abundances of different plant species change gradually and individually along a soil moisture gradient in Oregon's Siskiyou Mountains. Each species performs best at a different optimum soil moisture. (*Quadrats* are sample plots of ground designated and marked for an ecological study.)

to know all of the details to make considerable progress, because a few interactions often determine most of the features of an ecosystem. Let's begin by looking at such interactions in one community.

A few interactions may determine the features of a community

Although hundreds of species live in oak forest communities in the eastern United States, only a few species dominate the ecological interactions in those communities: oak trees (and their leaves and acorns) white-footed mice (*Peromyscus maniculatus*), gypsy moths (*Lymantria dispar*), and white-tailed deer (*Odocoileus virginianus*). Oak trees produce large crops of acorns every 2–5 years, but produce few acorns during intervening years. Acorns are a critical food for white-footed mice and deer. During years of heavy acorn production, white-footed mice survive well during winter and have great reproductive success. During years of poor acorn crops, the mice survive poorly.

Gypsy moth larvae eat oak leaves. When they have completed their development, the larvae crawl to the tree trunks, where they pupate. While the larvae inside the pupae are transforming themselves into moths, they may be eaten by white-footed mice, which search for them on the tree trunks. Every 6 or 7 years, gypsy moths become extremely abundant.

The population cycles of gypsy moths and white-footed mice are of interest to us because these species greatly affect humans and their resources. In years when gypsy moth caterpillars are extremely abundant, they defoliate millions of hectares of oak forest, damaging and killing many trees. Also, white-footed mice, along with white-tailed deer, are the primary hosts of the black-legged tick (*Ixodes scapularis*), the vector of the spirochete bacterium (*Borrelia burgdorferi*) that causes Lyme disease in humans.

Ecologists wanted to test the hypothesis that white-footed mice generate fluctuations in gypsy moth populations by eating their pupae. They performed an experiment during a year of low gypsy moth but high white-footed mouse population densities. They removed mice from three experimental areas by trapping them, but did nothing to the mice in three other areas, which served as controls. On the control plots, which had high mouse densities, they searched tree trunks, but found no moth pupae. To determine that predation was responsible for the absence of pupae, they attached previously collected freeze-dried pupae to tree trunks. Within 2–4 days, all of the pupae in the control plots had been eaten, most of them by white-footed mice. In the experimental plots, from which white-footed mice had been removed, 22 percent of the freeze-dried pupae remained uneaten for at least 13 days. That would have been long enough for the pupae to complete their metamorphosis into adult moths.

If white-footed mice control gypsy moth populations, what controls mouse populations? In another experiment, the ecologists added more than 811,000 acorns to experimental plots during a year when oak trees were producing very few acorns. Mouse populations became much more dense on the plots with added acorns than on the control plots.

These experiments demonstrated that when white-footed mice are abundant, they can prevent gypsy moths from completing their life cycle. The experiments also demonstrated that white-footed mouse population densities are determined primarily by acorn density (Figure 55.2). Without performing those experiments, investigators could only speculate about the interactions of the species in oak forest communities and their importance.

With this background, we will now discuss the major processes that influence communities and ecosystems and the patterns they generate.

Process and Pattern in Communities and Ecosystems

We begin our discussion of processes with solar energy input and precipitation because nearly all ecological processes depend, either directly or indirectly, on the amount and seasonal pattern of solar energy input and supply of water. Next we'll discuss what ecologists have learned by studying interactions among species and how the influences of various factors on community patterns change over space and time. As you read, be aware that the structure of human language forces us to discuss these factors one at a time, but several of them typically operate simultaneously.

Solar energy and precipitation drive ecosystem processes

All organisms depend on inputs of energy (in the form of sunlight or high-energy molecules), water, and nutrients for their metabolism and growth. With the exception of a few ecosystems (some caves, deep-sea hydrothermal vent systems) in which solar energy is not the main energy source, all energy utilized by organisms comes (or once came) from the sun. Even the fossil fuels—coal, oil, and natural gas—upon which the economy of modern human civilization is based are reserves of captured solar energy locked up in the remains of organisms that lived millions of years ago.

Solar energy enters ecosystems by way of plants and other photosynthetic organisms. Only about 5 percent of the solar energy that arrives on Earth is captured by photosynthesis. The remaining energy is either radiated back into the atmosphere as heat or consumed by the evaporation of water from plants and other surfaces. **Gross primary productivity** is the rate at which energy is incorporated into the bodies of pho-

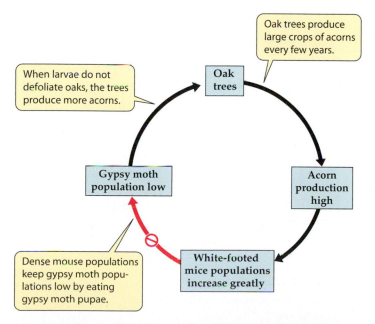

55.2 Interactions within Communities Control Populations Experiments in oak forests in eastern North America have shown that acorn abundance affects the abundance of the white-footed mouse, and that the mice in turn are a major control on the gypsy moth population.

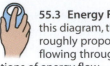

55.3 Energy Flow through an Ecosystem In this diagram, the width of each channel is roughly proportional to the amount of energy flowing through it. The arrows indicate directions of energy flow.

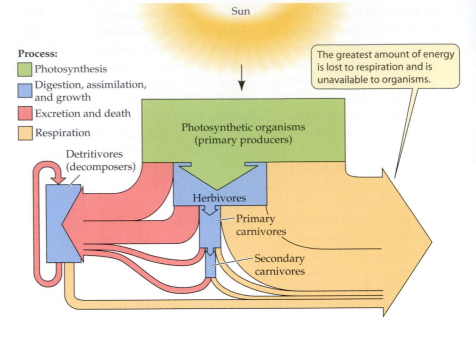

tosynthetic organisms. The accumulated energy is called **primary production**; that is, productivity is a rate, and production is a product. Plants use some of this energy for their own metabolism; the rest is stored in their bodies or used for their growth and reproduction. The energy available to organisms that eat plants, called **net primary production**, is gross primary production minus the energy expended by the plants on their respiration. Only the energy content of an organism's net production—its growth plus reproduction—is available to other organisms that consume it (Figure 55.3).

The distribution of the total amount of energy that plants assimilate by means of photosynthesis reflects the distribution of land masses, temperature, and moisture on Earth (Figure 55.4). Close to the equator at sea level, temperatures are high throughout the year, and the water supply typically is adequate for plant growth much of the time. In these climates, productive forests thrive. In lower-latitude and mid-

55.4 Primary Production in Different Ecosystems The primary production of Earth's ecosystems can be measured (a) by the geographic extent of the different ecosystems; (b) by net annual primary production; and (c) by the percentage of Earth's total primary production contributed by each ecosystem.

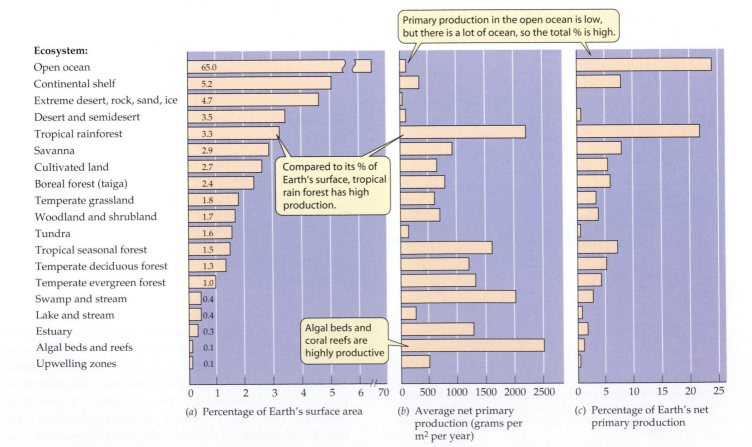

(a) Percentage of Earth's surface area
(b) Average net primary production (grams per m² per year)
(c) Percentage of Earth's net primary production

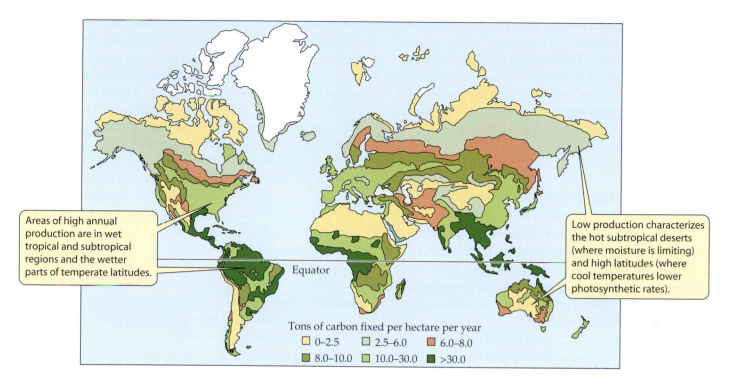

Areas of high annual production are in wet tropical and subtropical regions and the wetter parts of temperate latitudes.

Low production characterizes the hot subtropical deserts (where moisture is limiting) and high latitudes (where cool temperatures lower photosynthetic rates).

Equator

Tons of carbon fixed per hectare per year

| ☐ 0–2.5 | ☐ 2.5–6.0 | ■ 6.0–8.0 |
| ■ 8.0–10.0 | ■ 10.0–30.0 | ■ >30.0 |

55.5 Net Primary Production of Terrestrial Ecosystems Variations in temperature and water availability over Earth's land surface influence the annual production of ecosystems.

latitude deserts, where plant growth is limited by lack of moisture, annual primary production is low. At still higher latitudes, even though moisture is generally available, annual primary production is also low because it is cold much of the year (Figure 55.5). Production in aquatic systems is limited by light, which decreases rapidly with depth; by nutrients, which sink and must be replaced by upwelling of water; and by temperature (see Chapter 58). Primary productivity strongly influences two other important features of ecological communities: species richness and food web structure.

Species richness is influenced by primary productivity

The number of species living in a community (its **species richness**) is correlated with gross primary productivity, but the relationship between these two factors is complex. Ecologists first observed that species richness often increases with productivity up to a point, but then decreases (Figure 55.6). The increase occurs because the number of individuals an area can support increases with productivity, and with larger population sizes, species extinction rates are lower. But why should species richness decrease when productivity is still higher?

One hypothesis proposed to explain this decrease postulates that interspecific competition becomes more intense when productivity is higher, resulting in **competitive exclusion** of some species. This hypothesis is supported by the re-

sults of a long-term experiment at the Rothamstead Experiment Station in England in which fertilizer has been added to some plots of land to increase productivity. Fertilized and unfertilized plots have been monitored continuously at Rothamstead since 1856. Over this time period, the number of plant species in unfertilized plots has remained roughly constant, whereas species richness has declined in the fertilized plots.

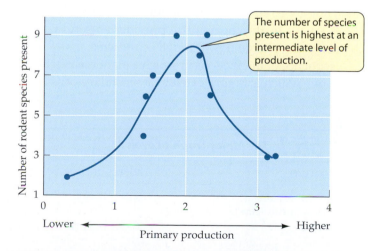

The number of species present is highest at an intermediate level of production.

55.6 Local Species Richness Peaks at Intermediate Productivity The number of species of rodents that live on gravel and rocky plains in the Gobi desert peaks at intermediate levels of primary production. Ecologists believe that when productivity is highest, interspecific competition becomes stronger, resulting in the exclusion of some rodent species by others.

All photosynthetic plants depend on sunlight and the same set of mineral nutrients, so we would expect them to compete with one another. However, all ecological communities are composed of species of different taxonomic groups that depend on many different resources. Such species may be less likely to compete with one another than plants are. How is the pattern of their relationships influenced by productivity?

Food web structure is influenced by productivity

Energy flows through ecosystems when organisms eat one another. The organisms in a community can be divided into **trophic levels** based on the way in which they obtain their energy (Table 55.1). A trophic level is defined by the number of steps through which energy passes to reach the organisms in it. Photosynthetic plants (*autotrophs*) get their energy directly from sunlight. Collectively, they constitute a trophic level called *photosynthesizers*, or *primary producers*. They produce the energy-rich organic molecules that nearly all other organisms consume.

In most ecological communities, all nonphotosynthetic organisms (*heterotrophs*) consume, either directly or indirectly, the energy-rich organic molecules produced by primary producers. Organisms that eat plants constitute a trophic level called *herbivores* or *primary consumers*. Organisms that eat herbivores are called *secondary consumers*. Those that eat secondary consumers are called *tertiary consumers*, and so on. Organisms that eat the dead bodies of organisms or their waste products are called *detritivores* or *decomposers*. Organisms that obtain their food from both primary producers and another trophic level are called *omnivores*. Because many species are omnivores, trophic levels are often not clearly distinct, but if we remember that boundaries between trophic levels are fuzzy, the concept still provides a useful way of characterizing energy flow through ecosystems.

A sequence of interactions in which a plant is eaten by an herbivore, which is in turn eaten by a secondary consumer,

and so on, can be diagrammed as a **food chain**. Food chains are usually interconnected to make a **food web** because most species in a community eat and are eaten by more than one other species. Ecological communities contain so many species that it is impossible to show all of them in a food web. Therefore, all such diagrams are simplified, as shown by the food web of Isle Royale National Park (Figure 55.7).

Despite their considerable differences, most communities have only three to five trophic levels. Why are there so few levels? Loss of energy between trophic levels is partly responsible (see Figure 55.3). To show how energy decreases as it flows from lower to higher trophic levels, ecologists construct diagrams called *energy pyramids* (Figure 55.8). Another factor affecting community structure is the amount of living matter, or *biomass*, at each trophic level. To show the biomass of organisms existing at different trophic levels, ecologists construct *biomass pyramids*. A biomass pyramid illustrates the amount of biomass that is available at a given time for organisms at the next trophic level.

Pyramids of energy and biomass for a particular ecosystem usually have similar shapes, but sometimes they do not. Their shapes depend on the dominant organisms and how they allocate their energy. In most terrestrial ecosystems, the dominant photosynthetic plants are large. They store energy for long periods, some of it in difficult-to-digest forms (such as cellulose and lignin). Therefore, the primary producer level in these systems contains a large biomass. However, grasslands and forests have strikingly different patterns of energy flow. Trees store a great deal of their energy as wood, which is composed of difficult-to-digest material. Wood is rarely eaten unless the tree is diseased or otherwise weakened. In contrast, grassland plants produce few hard-to-digest woody tissues. Mammals may consume 30–40 percent of the annual aboveground net primary production of grasslands, and insects may consume an additional 5–15 percent. Soil organisms, primarily nematodes, may consume 6–40 percent of the belowground production (Figure 55.8a,b). Thus, the herbivore level has a relatively larger biomass in grasslands than in forests.

55.1 The Major Trophic Levels

TROPHIC LEVEL	SOURCE OF ENERGY	EXAMPLES
Photosynthesizers (primary producers)	Solar energy	Green plants, photosynthetic bacteria and protists
Herbivores	Tissues of primary producers	Termites, grasshoppers, gypsy moth larvae, anchovies, deer, geese, white-footed mice
Primary carnivores	Herbivores	Spiders, warblers, wolves, copepods
Secondary carnivores	Primary carnivores	Tuna, falcons, killer whales
Omnivores	Several trophic levels	Humans, opossums, crabs, robins
Detritivores (decomposers)	Dead bodies and waste products of other organisms	Fungi, many bacteria, vultures, earthworms

Trophic level

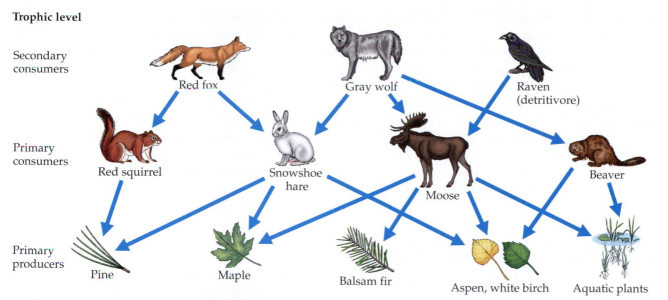

Secondary consumers

Red fox   Gray wolf   Raven (detritivore)

Primary consumers

Red squirrel   Snowshoe hare   Moose   Beaver

Primary producers

Pine   Maple   Balsam fir   Aspen, white birch   Aquatic plants

55.7 Food Web of Isle Royale National Park   This food web includes only large vertebrates and the plants on which they depend. Even with these restrictions, the web is complex. The arrows show who eats whom.

In most aquatic ecosystems, the dominant photosynthesizers are bacteria and protists. Those unicellular organisms have such high rates of cell division that a small biomass of photosynthesizers can feed a much larger biomass of herbivores, which grow and reproduce much more slowly. This pattern can produce an inverted biomass pyramid, even though the energy pyramid for the same ecosystem has the typical shape (Figure 55.8c).

Much of the energy ingested by organisms is converted to biomass that is eventually consumed by detritivores, such as bacteria, fungi, worms, mites, and insects. These organisms transform the dead remains and waste products of organisms into free mineral nutrients that can again be taken up by plants. If there were no detritivores, most nutrients would eventually be tied up in dead bodies, where they would be unavailable to plants. Continued ecosystem productivity depends on the rapid decomposition of detritus.

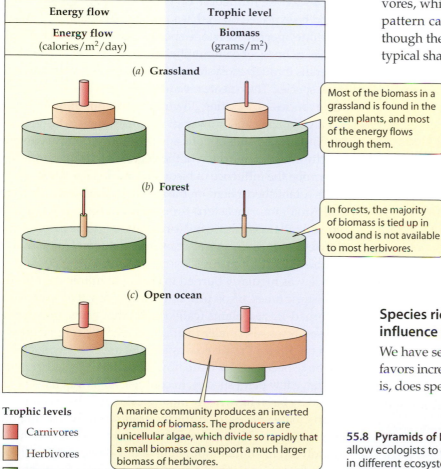

Energy flow	Trophic level
Energy flow (calories/m^2/day)	**Biomass** (grams/m^2)

(a) **Grassland**

Most of the biomass in a grassland is found in the green plants, and most of the energy flows through them.

(b) **Forest**

In forests, the majority of biomass is tied up in wood and is not available to most herbivores.

(c) **Open ocean**

A marine community produces an inverted pyramid of biomass. The producers are unicellular algae, which divide so rapidly that a small biomass can support a much larger biomass of herbivores.

Trophic levels

■ Carnivores
■ Herbivores
■ Producers (photosynthesizers)

Species richness and productivity influence ecosystem stability

We have seen that, up to a point, high primary productivity favors increased species richness, but is the reverse true? That is, does species richness also influence ecosystem productiv-

55.8 Pyramids of Biomass and Energy   Energy pyramids (left column) allow ecologists to compare patterns of energy flow through trophic levels in different ecosystems. Biomass pyramids (right column) allow them to compare the amount of material present in living organisms at different trophic levels.

ity? Ecologists hypothesized that species richness might enhance ecosystem productivity because no two species in a community have the same relationship with the environment. Therefore, a richer mixture of species should result in a more complete use of the available resources. In addition, if the environment changes, a species-rich ecosystem is more likely to contain species that are already adapted to the new conditions than is a species-poor ecosystem. Therefore, ecologists hypothesized that a species-rich ecosystem should also be more stable—that is, over time it should change less in both productivity and species composition than a species-poor ecosystem.

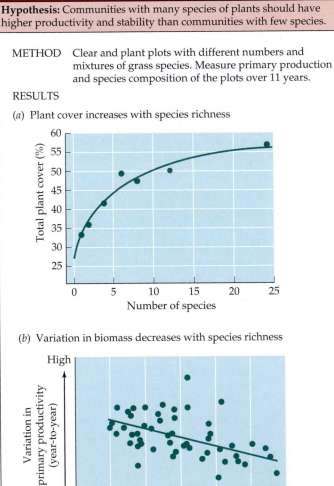

EXPERIMENT

Hypothesis: Communities with many species of plants should have higher productivity and stability than communities with few species.

METHOD   Clear and plant plots with different numbers and mixtures of grass species. Measure primary production and species composition of the plots over 11 years.

RESULTS

(a) Plant cover increases with species richness

(b) Variation in biomass decreases with species richness

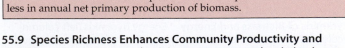

Conclusion: Plots with more species were more productive and varied less in annual net primary production of biomass.

55.9 Species Richness Enhances Community Productivity and Stability   Net primary productivity was greater, and variation in net primary productivity from one year to the next was less, in species-rich than in species-poor plots of grasses.

To test this hypothesis, ecologists cleared several outdoor plots, in which they planted grasses in a variety of mixtures of different species richness, from a few to 25 species. At the end of each growing season, they measured grass biomass (a measure of net primary production) and the population densities of all the grasses in the plots. These measurements were made over a period of 11 years, which included a serious drought. The plots with more species were more productive and their productivity varied less from one year to another, supporting the hypothesis (Figure 55.9). However, the population densities of individual species in the plots were not stable over the years (regardless of a plot's species richness), because different species performed better during drought years and during wet years.

Although species richness and productivity are often positively correlated, as they were in this experiment, such a correlation could result if only one or a few species exerted very strong influences on the flow of matter and energy through an ecosystem. Ecologists have made an effort to identify and study such species.

Individual species may influence community processes

Species whose influences on ecosystems are greater than would be expected on the basis of their abundance are called **keystone species**. Keystone species may influence both the species richness of communities and the flow of energy and materials through ecosystems. Beavers, for example, are keystone species. They create meadows and ponds—habitats for other species—by cutting down trees and building dams.

Large grazing and browsing mammals, such as bison (Figure 55.10a), also are keystone species. They often change the structure and composition of the vegetation dramatically. To determine the influence of bison on prairie vegetation, ecologists established a herd on the Konza Prairie Research Natural Area in northeastern Kansas, the largest tract of unplowed tallgrass prairie in North America. The bison were allowed to graze in some areas, but excluded from others. The bison herd numbered about 200 animals in 2003. The prairie was regularly burned in spring to mimic the fires of prehistoric times. Bison primarily eat grasses; they eat few of the broad-leaved plants (called *forbs*) that grow among the grasses. Bison also prefer to graze on recently burned areas.

In the Konza Prairie ecosystem, the areas from which bison have been excluded are now dominated by tall grasses and have few plant species. In contrast, areas that have been grazed by bison have many more species of forbs because the bison create spaces for forbs by preferentially grazing on grasses (Figure 55.10b). Furthermore, urea in bison urine is hydrolyzed to ammonium within a few days, and the nitrogen in the ammonium is immediately available for plants. In contrast, decomposing leaf litter releases nitrogen much more

Bison bison

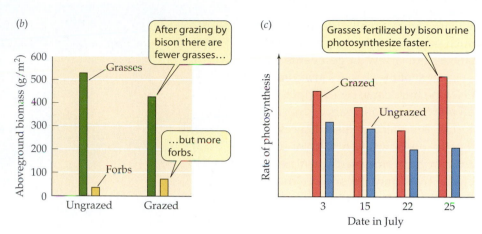

55.10 Grazing Increases Plant Species Richness and Productivity
(a) Bison are keystone species in the tallgrass prairie ecosystem. By grazing preferentially on grasses, they increase both the density of forbs (b) and overall plant productivity (c).

slowly. Therefore, plants in patches grazed by bison have higher leaf nitrogen levels and photosynthesize faster than plants growing in ungrazed patches (Figure 55.10c). From these observations, the investigators concluded that bison are keystone herbivores that influence species composition, nutrient cycling, and energy flow by their grazing.

Another keystone species is the sea star *Pisaster ochraceous*, which lives in rocky intertidal ecosystems on the Pacific coast of North America. Its preferred prey is the mussel *Mytilus californianus*. In the absence of sea stars, these mussels crowd out other competitors in a broad belt of the intertidal zone. By consuming mussels, *Pisaster* creates bare spaces that are taken over by a variety of other species (Figure 55.11).

The influence of *Pisaster* on species richness was demonstrated by experimentally removing sea stars from selected parts of the intertidal zone repeatedly over a 5-year period. Two major changes occurred in the areas from which sea stars were removed. First, the lower edge of the mussel bed extended lower down into the intertidal zone, showing that sea stars are able to eliminate mussels completely where they are covered with water most of the time. Second, and more dramatically, 28 species of animals and algae disappeared from the sea star removal zone. Eventually only *Mytilus*, the dominant competitor, occupied the entire substratum. By altering competitive relationships, predation by *Pisaster* largely determines which species live in these rocky intertidal ecosystems.

Disturbance and Community Structure

By their activities, keystone species generate disturbances. A *disturbance* is an event that changes the survival rate of one or more species in an ecological community. Logs carried by waves may crush algae and animals attached to rocks in an intertidal community or a windstom may blow down a tree, crushing shrubs and herbs. The effects of such disturbances are typically limited to a small area. Other kinds of disturbances, such as hurricanes and volcanic eruptions, may af-

55.11 Sea Stars are Keystone Predators Ochre sea stars (*Pisaster ochraceus*) have harvested all the mussels from the lower parts of these rocks on the Olympic Peninsula of Washington. By consuming mussels, *Pisaster* creates bare spaces that are taken over by a variety of other species, thus exerting a keystone influence on the intertidal ecosystems of the Pacific Northwest.

fect large areas. Small disturbances are much more common than large disturbances, but a few large events may cause most of the changes in a community. One hurricane, for example, may fell more trees than years of "normal" storms. The effects of disturbances also depend on how often they occur. If strong windstorms are frequent, for example, trees may never have the opportunity to grow tall.

A particular type of disturbance can have a variety of effects. For example, in 1988, massive fires burned one-third of Yellowstone National Park, but they created a mosaic that included unburned patches, areas where only herbs and shrubs were burned, and areas where all trees were consumed (Figure 55.12).

Although the consequences of various kinds of disturbances are highly variable, their results conform to a general pattern: Communities with very high levels of disturbance and those with very low levels of disturbance have fewer species than communities subjected to intermediate levels of disturbance. The discovery of this general pattern generated the **intermediate disturbance hypothesis**. This hypothesis explains the low species richness in areas with high disturbance levels by suggesting that only species with great dispersal abilities and rapid reproductive rates can persist in such areas. Conversely, the hypothesis explains the decline in species richness where disturbance levels are low by suggesting that competitively dominant species displace other species, as mussels did when sea stars were removed.

The intermediate disturbance hypothesis was tested using boulders on intertidal beaches in California. An ecologist observed that boulders of intermediate size had more species of algae and barnacles on them than either larger or smaller boulders did. He hypothesized that this pattern existed because waves move small boulders more easily, more often, and farther than large boulders. When a wave moves a boulder, its motion destroys organisms living on the boulder's surface. To test his hypothesis, the ecologist altered disturbance levels by gluing small boulders to the substratum. These secured small boulders accumulated species more rapidly than unsecured small boulders, supporting the hypothesis (Figure 55.13). The experiment also showed that species not normally found on small boulders can survive there if the boulders are not moved often.

This experiment also demonstrated that the number of species in a community changes over time following a disturbance. A change in community composition following a disturbance is called **ecological succession**. The changes on

55.12 Fires Create Mosaics of Burned and Unburned Patches This view of Mount Washburn in Yellowstone National Park was taken 10 years after the massive forest fires of 1988.

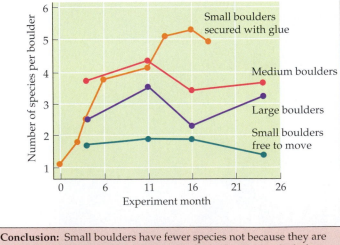

EXPERIMENT

Hypothesis: Medium-sized boulders have more species growing on them than either smaller or larger boulders because they are subjected to intermediate levels of disturbance.

METHOD   Sterilize a number of small boulders. Secure some of them to the natural substratum with glue. Leave other boulders unsecured to serve as controls. Observe accumulation of species on the boulders over time.

RESULTS   Secured small boulders accumulate many more species than unsecured small boulders.

Conclusion: Small boulders have fewer species not because they are unsuitable for local species, but because the high rates at which they are moved by waves prevent many species from surviving on them.

55.13 Species Richness Is Greatest at Intermediate Levels of Disturbance By gluing boulders to the substratum, an ecologist showed that small boulders can support more species if they are not disturbed at high rates.

the intertidal boulders were rapid: The secured small boulders had more species than the unsecured boulders within 16 months. Most ecological successions progress much more slowly.

Ecologists divide successions into two major types: primary succession and secondary succession. A **primary succession** begins on sites that lack living organisms. A **secondary succession** begins on sites where some organisms have survived the most recent disturbance. The patterns and causes of ecological succession are varied, but the species that colonize a site soon after the disturbance often alter environmental conditions so that they become favorable for other species.

A good example of a primary succession is the changes in the plant community that followed the retreat of a glacier in Glacier Bay, Alaska, over the last 200 years. The melting and retreating glacier left a series of *moraines*—gravel deposits formed where the glacial front was stationary for a number of years. No human observer was present to measure changes over the 200-year period, but ecologists have inferred the temporal pattern of succession by measuring plant communities on moraines of different ages. The youngest moraines, close to the current glacial front, are populated with bacteria, fungi, and photosynthetic microorganisms. Slightly older moraines farther from the glacial front have lichens, mosses, and a few species of shallow-rooted herbs. Still farther from the glacial front, successively older moraines have shrubby willows, alders, and spruces.

By comparing moraines of different ages, ecologists deduced the pattern of plant succession and changes in soil nitrogen content at Glacier Bay (Figure 55.14). Succession is caused in part by changes in the soil brought about by the plants themselves. Nitrogen is virtually absent from glacial moraines, so the only plants that can grow on recently exposed moraines at Glacier Bay are a herbaceous plant (*Dryas*) and alder trees (*Alnus*), both of which have nitrogen-fixing bacteria in nodules on their roots. Nitrogen fixation by *Dryas* and alders improves the soil so that spruces can grow. Spruces then outcompete and displace the alders and *Dryas*. If the local climate does not change dramatically, a forest community dominated by spruces is likely to persist for many centuries on old moraines at Glacier Bay.

A secondary succession may begin with the dead parts of organisms. The succession of fungal species that decompose pine needles in litter beneath Scots pines (*Pinus sylvestris*) is shown in Figure 55.15. New needles continuously fall from the pines, so the surface layer of litter is young and deeper layers are progressively older. Decomposition begins when the first group of fungi starts consuming the needles soon after they fall. Each group of fungi derives its energy by decomposing certain compounds, converting them to other compounds that are used by the next group of species. This process continues over about 7 years, by which time the last group of fungi—basidiomycetes—has decomposed the last remaining compounds.

Dispersal, Extinction, and Community Structure

When we discussed population ecology in Chapter 54, we noted that local subpopulations often become extinct, but are reestablished by immigrants from other subpopulations. We know that immigration and emigration influence the structure of communities because species deliberately or inadvertently introduced by humans, such as gypsy moths and zebra mussels, may come to dominate the communities they invade. The rate of introduction of new species and the extinction of existing species has been greatly increased by human activities over the past few centuries, but throughout the history of life on Earth, species have colonized new areas and gone extinct. The composition of ecological communities is influenced not only by the relatively

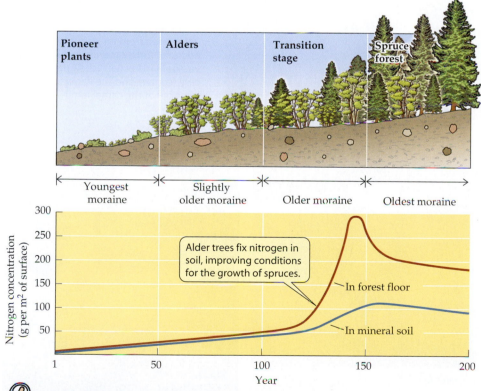

55.14 Primary Succession on a Glacial Moraine   As the plant community occupying an Alaskan glacial moraine changes from pioneer plants to a spruce forest, nitrogen accumulates in the mineral soil.

55.15 Secondary Succession on Pine Needles As indicated by the widths of the bars, the abundances of ten types of fungi in pine needle litter change with time, which increases with depth within the layer.

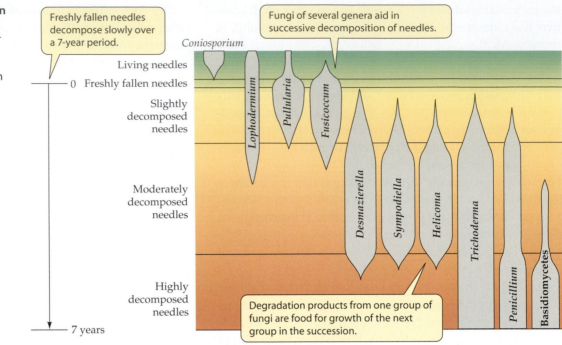

Freshly fallen needles decompose slowly over a 7-year period.

Fungi of several genera aid in successive decomposition of needles.

Coniosporium

Living needles
0 Freshly fallen needles
Slightly decomposed needles
Moderately decomposed needles
Highly decomposed needles
7 years

Lophodermium
Pullularia
Fusicoccum
Desmazierella
Sympodiella
Helicoma
Trichoderma
Penicillium
Basidiomycetes

Degradation products from one group of fungi are food for growth of the next group in the succession.

(a)

Dasypus novemcinctus

Erethizon dorsatum

55.16 North and South America Exchanged Mammals (a) The nine-banded armadillo and the porcupine are among only a handful of species that colonized North American from South America. (b) Some species that exist today only in South America are descended from ancestors who migrated from North America, including the Patagonian fox (left), the Chacoan peccary (center), and the llama (right), a member of the camel family.

(b)

Dusicyon griseus

Catagonus wagneri

Lama glama

short-term interactions we have already discussed, but also by these long-term, large-scale ecological and evolutionary processes.

A massive colonization of new areas by mammals occurred when the Central American land bridge formed about 4 million years ago. This land bridge connected North and South America for the first time in about 65 million years. While the two continents were separated, their mammals evolved independently of one another because terrestrial mammals (with the exception of bats) are poor dispersers across water barriers. South America evolved a distinctive mammalian fauna dominated by marsupials, primates, edentates (armadillos, sloths, ground sloths, and anteaters), and caviomorph rodents (porcupines, capybaras, pacas, agoutis, guinea pigs, chinchillas, and others) (Figure 55.16a).

Many species of mammals dispersed across the newly established land bridge. Only a few South American species—the porcupine, nine-banded armadillo, and Virginia opossum—became established north of the tropical forests of Central America, but many North America mammals—rabbits, mice, foxes, bears, raccoons, weasels, cats, horses, tapirs, peccaries, camels, and deer—successfully colonized South America. The North American invasion apparently caused the extinction of several kinds of large marsupial carnivores and the large herbivores that were their prey. Subsequently, the northern invaders formed new species that today exist only in South America (Figure 55.16b).

The exchange of mammals between North and South America is but one example of the profound influence of immigration, extinction, and subsequent evolution on the patterns of distribution of life on Earth. We will explore these patterns and their causes in detail in Chapter 56.

Chapter Summary

▶ Ecological communities are assemblages of species, each of which interacts in unique ways with its environment. In most cases, species drop out of and are added to communities gradually across environmental gradients. **Review Figure 55.1**

Communities: Loose Assemblages of Species

▶ Experiments can tell us which interactions among species exert the strongest effects on community structure. **Review Figure 55.2**

Process and Pattern in Communities and Ecosystems

▶ Most of the energy incorporated by an organism is used in its respiration. Only a small proportion can be captured by other organisms that consume it. **Review Figure 55.3. See Web/CD Activity 55.1**

▶ Primary production is determined by temperature and precipitation. Therefore, it varies over Earth's surface. **Review Figures 55.4, 55.5**

▶ Species richness increases with primary production up to a point, after which it declines. **Review Figure 55.6**

▶ A trophic level consists of those organisms whose major food source has passed through the same number of steps. **See Web/CD Activity 55.2**

▶ Food webs are diagrams of who eats whom in ecological communities. Most food webs have only three to five trophic levels. **Review Figure 55.7**

▶ Energy pyramids show the flow of energy through trophic levels. Biomass pyramids show the amount of living matter at each trophic level. **Review Figure 55.8**

▶ Communities with more species are generally more productive and more stable than communities with fewer species. **Review Figure 55.9**

▶ Keystone species influence community structure and dynamics out of proportion to their abundances. **Review Figure 55.10**

Disturbance and Community Structure

▶ All ecological communities are subjected to a variety of disturbances. Typically, small disturbances are much more common than large ones.

▶ Communities subjected to moderate levels of disturbance typically have more species than communities subjected to lower or higher levels of disturbance. **Review Figure 55.13**

▶ After a disturbance, the structure and composition of an ecological community changes as organisms modify the physical environment and interact with one another. **Review Figures 55.14, 55.15. See Web/CD Tutorial 55.1**

Dispersal, Extinction, and Community Structure

▶ The composition of ecological communities is influenced by ecological and evolutionary events taking place over long time periods and large spatial scales.

Self-Quiz

1. An ecological community is
   a. all the species of organisms that live and interact with one another in an area.
   b. all the species that live and interact in an area together with the abiotic environment.
   c. all the species in an area that belong to a particular trophic level.
   d. all the species that are members of a local food web.
   e. all of the above

2. What is the difference between primary productivity and primary production?
   a. Primary productivity is always greater than primary production.
   b. Primary productivity is always less that primary production.
   c. Primary productivity is a rate, whereas primary production is a product.
   d. Primary productivity is a product, but primary production is a rate.
   e. There is no real difference between primary productivity and primary production.

3. The total amount of energy that plants assimilate by photosynthesis is called
   a. gross primary production.
   b. net primary production.

c. biomass.

d. a pyramid of energy.

e. succession.

4. The amount of energy reaching a higher trophic level is determined by
   a. net primary production.
   b. net primary production and the efficiencies with which food energy is converted to biomass.
   c. gross primary production.
   d. gross primary production and the efficiencies with which food energy is converted to biomass.
   e. gross primary production and net primary production.

5. The pyramids of energy and biomass of forests and grasslands differ because
   a. forests are more productive than grasslands.
   b. forests are less productive than grasslands.
   c. large mammals avoid living in forests.
   d. trees store much energy in difficult-to-digest wood, whereas grassland plants produce few difficult-to-digest tissues.
   e. grasses grow faster than trees.

6. Keystone species
   a. influence the structure of the communities in which they live more than expected on the basis of their abundance.
   b. strongly influence the species composition of communities.
   c. may speed up the rate of nutrient cycling.
   d. may be herbivores or carnivores.
   e. all of the above

7. What is the general relationship between species richness and disturbance?
   a. Species richness peaks at low levels of disturbance.
   b. Species richness peaks at high levels of disturbance.
   c. Species richness peaks at intermediate levels of disturbance.
   d. Species richness is less at intermediate levels of disturbance.
   e. There is no general relationship between species richness and level of disturbance.

8. Ecological succession is
   a. the changes in species over time.
   b. the gradual process by which the species composition of a community changes.
   c. the changes in a forest as the trees grow larger.
   d. the process by which a species becomes abundant.
   e. the buildup of soil nutrients.

9. Primary succession begins
   a. soon after a disturbance ends.
   b. at varying times after a disturbance ends.
   c. at sites where some species survived the disturbance.
   d. at sites were no species survived the disturbance.
   e. at sites where only primary producers survived the disturbance.

10. The South American mammals that became established in North America after crossing the Central American land bridge include
    a. porcupines, armadillos, and caviomorph rodents.
    b. porcupines, caviomorph rodents, and Virginia opossums.
    c. Virginia opossums, porcupines, and anteaters.
    d. Virginia opossums, porcupines, and armadillos.
    e. armadillos, anteaters, and caviomorph rodents.

For Discussion

1. Some evidence suggests that interspecific competition may be responsible for the decrease in species richness at high levels of productivity. What other hypotheses might explain this puzzling relationship? How would you test them?

2. The increased productivity and stability of species-rich communities could be explained by ecological differences among the species or by the fact that the more species in a community, the greater the chance that it will contain an unusually productive species. How could you distinguish between these competing hypotheses?

3. If species-rich communities are more productive than species-poor communities, how can modern agriculture, which is based almost entirely on cultivating a single species on a plot, be so productive?

4. We illustrated succession with two examples from forests. How might ecological succession differ in grasslands? In deserts? In the rocky intertidal zone?

5. Many conservationists believe that our greatest efforts should be expended to save undisturbed environments. Many users of natural resources, on the other hand, argue that disturbing environments to extract resources will actually improve species preservation. Is the latter view an appropriate invocation of the intermediate disturbance hypothesis?

56 Biogeography

[I]t is...those [species] which range widely over the world, are the most diffused within their own country, and are the most numerous in individuals, which oftenest produce well-marked varieties, or as I consider them, incipient species.
—Charles Darwin, 1859

In this passage from the second chapter of *The Origin of Species*, Charles Darwin was reporting the results of his tabulations of several well-studied regional floras. Darwin never published those data, but his suggestion that species that are widespread tend to be both more abundant and more variable than species with narrower ranges has been supported by recent evidence. Thus, in addition to his contributions to evolutionary theory and ecology, Darwin anticipated modern advances in the field of biogeography.

Widespread species are often abundant locally, but no species is found everywhere. The study of the distribution of organisms over Earth's surface began when eighteenth-century European explorers, settlers, and travelers started to take note of the vast differences among the biota on the different continents and attempted to understand them.

When the first Europeans arrived in Australia, they saw plants and animals that differed in perplexing ways from the ones they knew at home. Among these oddities were flowers pollinated by brush-tongued parrots and mammals that hopped around on their hind legs, carrying their offspring in pouches. In contrast, the first Europeans to visit North America felt at home because the plants (such as oaks, elms, and pines) and animals (such as deer, rabbits, foxes, thrushes, and crows) of North America were similar to those of Europe. Why was North America's biota so similar to Europe's while Australia's was so bizarrely different?

Biogeography is the science that documents and attempts to explain the patterns of distribution of populations, species, and ecological communities across Earth. In this chapter, we will show how biogeographers identify the processes that influence the distributions of species, both those that operated in the remote past and those that are operating today. We will also review Earth's major biogeographic regions. Finally, we will look at the factors that influence the number of species that live together.

An Australian Endemic The numbat (*Myrmecobius fasciatus*) is an Australian marsupial mammal that uses its highly specialized tongue to feed on ants and termites. Unrelated and different-appearing animals on other continents also feed on these insects, but the numbat, like many other mammals, is unique to Australia.

Earth's Biogeographic Regions

Explaining species' distributions might seem to be a simple matter, because the question of why a species is or is not found in a certain location has only a few possible answers:

▶ If a species occupies a particular area, either it evolved there, or it evolved elsewhere and dispersed to the area.

▶ If a species is not found in a particular area, either it evolved elsewhere and never dispersed to the area, or it was once present in the area but no longer lives there.

Determining which of these possible answers is correct requires information about the evolutionary histories of species, which comes from fossils and from knowledge of their phylogenetic relationships. It also requires information about changes in Earth's geography (such as continental drift, glacial advances and retreats, sea level changes, and mountain building) that occurred as the organisms were evolving. Such geological information can tell us whether organisms evolved where they are currently found or dispersed and colonized new areas from a distant area of origin.

The biotas of the continents differ enough to allow the division of Earth into several major **biogeographic regions**.

Biogeographic regions are based on the taxonomic similarities of the organisms living in them. The boundaries of biogeographic regions are set where species compositions change dramatically over short distances (Figure 56.1). The biotas of the biogeographic regions differ because oceans, mountains, deserts, and other barriers restrict the dispersal of organisms. Although there has been dispersal of organisms between adjacent biogeographic regions (as happened between North and South America, as we described at the end of Chapter 55), such interchanges have not been frequent enough to eliminate the striking differences that have resulted from speciation and extinction within each region.

A species found only within a certain region is said to be **endemic** to that region. Remote islands typically have distinctive endemic biotas because water barriers greatly restrict immigration. If the islands are large enough or form part of an archipelago, allopatric speciation often produces unique species and communities, as we saw in Chapter 24. For example, nearly all the tracheophytes and vertebrates of Madagascar, a large island off the eastern coast of Africa, are endemic to that island (Figure 56.2). Madagascar by itself could be called a biogeographic region, but because dozens of is-

56.1 Major Biogeographic Regions The biotas of Earth's major biogeographic regions differ strikingly from one another.

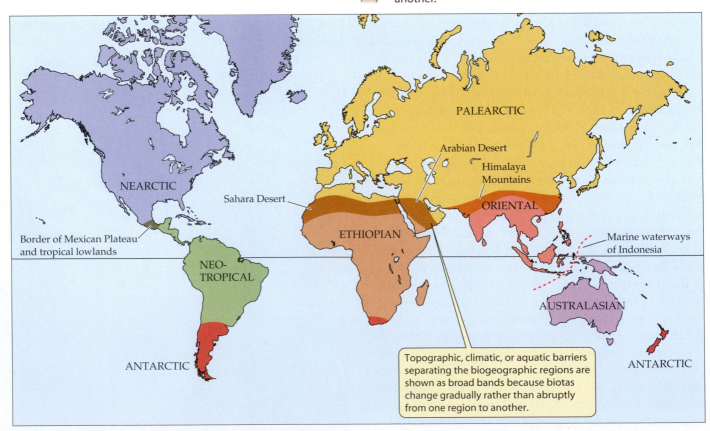

Chamaeleo pantheri (panther chameleon)

Hemicentetes semispinosus (yellow-streaked tenrec)

Cryptoprocta ferox (fossa)

Lemur catta (ring-tailed lemur)
climbing *Alluaudia procera* (Madagascar ocotillo)

Adansonia grandidieri (giant baobob tree)

56.2 Madagascar Abounds with Endemic Species The majority of tracheophyte and vertebrate species found on the island of Madagascar are found nowhere else on Earth.

Pachypodium rosulatum (elephant's foot)

lands would qualify as biogeographic regions on the basis of the distinctness of their biotas, islands are treated differently from continents.

Most species are confined to a single biogeographic region. The most widespread species today is probably *Homo sapiens*, but a few other species—for example, the great egret, the osprey, the peregrine falcon, and the barn owl—are found on all continents except Antarctica.

Next we will discuss the influence of speciation, extinction, and other historical processes on biogeographic patterns, and then consider the influence of processes operating today.

History and Biogeography

Before 1850, as we saw in Chapter 22, most people, including biogeographers, believed in a relatively unchanging Earth that was too young for long-term processes to account for the diversity and distribution of life. Linnaeus (1758), for example, believed that all organisms had been created in one place, which he called Paradise, from which they later dispersed. Indeed, because most people believed that the continents were fixed in their positions, the only way to account for the current distributions of organisms was to invoke massive dispersal.

The notion that the continents might have moved was not seriously considered until 1912, when Alfred Wegener, a German meteorologist, argued that the continents had drifted over time. Wegener based his theory on several observations:

▶ The shapes of continents (the outlines of Africa and South America seem to fit together like pieces of a puzzle)
▶ The alignment of mountain chains, rock strata, coal beds, and glacial deposits on different continents
▶ The distributions of closely related species that were shared between Africa and South America, which were difficult to explain if the continents had always been where they are now

When Wegener proposed that the continents had moved, few scientists took him seriously. There were no known mechanisms to move continents, and no convincing geological evidence of such movements existed. As we learned in Chapter 22, geological evidence and plausible mechanisms were eventually discovered. The broad pattern of continental movement, which continues today, is now clear.

About 280 million years ago, the continents were united to form a single land mass, called Pangaea (see Figure 22.13). The continents then began to separate from one another, but when the continents were still very close to one another (about 245 mya), many groups of terrestrial and freshwater organisms, such as insects, freshwater fishes, frogs, and tracheophytes, had already evolved. The ancestors of some organisms that live on widely separated continents today were probably present on those land masses when they were part of Pangaea.

By 100 mya, continental drift had separated Pangaea into northern (Laurasia) and southern (Gondwana) land masses, and the southern continents were moving away from each other (see Figure 22.15). Over time, India separated from Africa and slammed into southern Asia, Australia moved closer to Southeast Asia, and South America, which had drifted as an island for 60 million years, came into contact with North America. Throughout the history of life, continental drift has both separated and combined biotas, thus greatly influencing the distribution and evolution of species.

Biogeographers convert phylogenies to "area phylogenies"

As the age of Earth, the geological processes that shaped it, and the mechanics of evolution became better understood, biogeographers were able to ask questions such as, Where and when did evolutionary lineages originate? How did they spread? What do the present-day distributions of organisms tell us about their past histories?

A technique that was developed to help answer these questions was the creation of *area phylogenies*. To generate an area phylogeny, biogeographers alter a taxonomic phylogeny by replacing the names of the taxa with the names of the places where those taxa live or lived. For example, an area phylogeny suggests that horses speciated as they moved from Asia to Africa, whereas the speciation of zebras took place entirely in Africa (Figure 56.3).

Biogeographers use several approaches to infer the approximate times of separation of taxa within a lineage. First, if a molecular clock has been ticking at a relatively constant rate, the amount of difference in the molecules of species should be strongly correlated with the length of time their lineages have been evolving independently, as we saw in Chapter 26. Second, fossils can help to show how long a taxon has been present in an area and whether its members formerly lived in areas where they are no longer found. A third valuable source of information is the distribution of living species. Much more information can be gathered on current distributions than will ever be available from fossils. Similarities in the distributions of many lineages of organisms provide clues about past events that affected them.

Vicariant events and dispersal both influence distributions

The appearance of a barrier that splits the range of a species is called a **vicariant event**. A vicariant event divides the population of a species even though no individuals have dis-

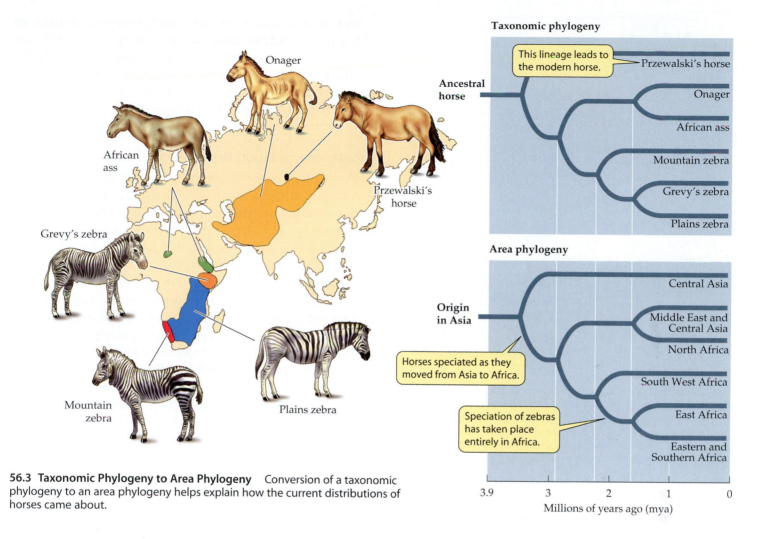

56.3 Taxonomic Phylogeny to Area Phylogeny   Conversion of a taxonomic phylogeny to an area phylogeny helps explain how the current distributions of horses came about.

persed to new areas. If, however, members of a species cross an already existing barrier and establish a new population, the species' disjunct range is a result of **dispersal**.

By studying a single evolutionary lineage, a biogeographer may discover evidence suggesting that the distribution of an ancestral species was influenced by a vicariant event such as a change in sea level or mountain building (see Figure 24.4). If that inference is correct, then species in other lineages are likely to have been influenced by the same event; that is, a number of lineages may have similar distribution patterns. Differences in distribution patterns among lineages may indicate that the lineages responded differently to the same vicariant events, that the lineages separated at different times, or that the lineages have had very different dispersal histories. By analyzing such similarities and differences, biogeographers can discover the relative roles of vicariant events and dispersal in determining today's distribution patterns.

The longer an area has been isolated from other areas by a vicariant event, such as continental drift, the more endemic taxa it is likely to have, because there has been more time for

evolutionary divergence to take place. Australia, which has been separated the longest from the other continents (about 65 million years), has the most distinctive biota of any continent. South America has the next most distinctive biota, having been isolated from other continents for nearly 60 million years. North America and Eurasia, which were joined together for much of Earth's history, have very similar biotas. That is why the early European travelers felt more at home in North America than in Australia.

When several hypotheses can explain a pattern, scientists typically prefer the most *parsimonious* one—the one that requires the smallest number of unobserved events to account for it. We saw how the parsimony principle is used in the reconstruction of phylogenies in Chapter 25. To see how it is applied to biogeography, consider the distribution of the New Zealand flightless weevil *Lyperobius huttoni*, a species that is found in the mountains of South Island and on sea cliffs at the extreme southwestern corner of North Island (Figure 56.4). If you knew only its current distribution and the current positions of the two islands, you might surmise that, even though this weevil cannot fly, it had somehow managed

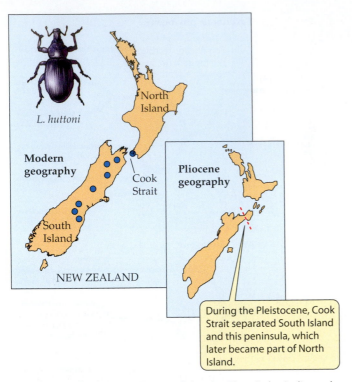

During the Pleistocene, Cook Strait separated South Island and this peninsula, which later became part of North Island.

56.4 A Vicariant Distribution Explained Blue circles indicate the current distribution of the weevil *Lyperobius huttoni*. A comparison of the present New Zealand geography with that of the Pliocene, when the southern part of today's North island was part of South Island, suggests that a vicariant event—a physical split separating populations—is the most parsimonious explanation for this distribution.

to cross Cook Strait, the 25-kilometer body of water that separates the two islands.

However, more than 60 other animal and plant species, including other species of flightless insects, live on both sides of Cook Strait. Although organisms do cross marine and terrestrial barriers, it is unlikely that all of these species made the same ocean crossing. In fact, we do not need to make that assumption. Geological evidence indicates that the present-day southwestern tip of North Island was formerly united with South Island. Therefore, none of the 60 species need have made a water crossing. A single vicariant event, the separation of the northern tip of South Island from the remainder of the island by the newly formed Cook Strait, could have split all of the distributions.

As we have just seen, the distributions of species today have been determined, in part, by history. However, because all organisms must be able to survive in today's environmental conditions, Earth's climates also exert a powerful influence on the distributions, abundances, and evolution of species.

Ecology and Biogeography

The **climate** of a region is the average of the atmospheric conditions (temperature, precipitation, and wind velocity) found there over time. Climates vary greatly from place to place on Earth, primarily because different places receive different amounts of solar energy. We will first examine how these differences in solar energy input determine atmospheric and oceanic circulation. Then we will show how climates influence the geographic distributions of organisms.

Solar energy inputs drive global climates

Every place on Earth receives the same total number of hours of sunlight each year—an average of 12 hours per day—but not the same amount of *energy*. The rate at which solar energy arrives per unit of Earth's surface depends primarily on the angle of sunlight. If the sun is low in the sky, a given amount of solar energy is spread over a larger area (and is thus less intense) than if the sun is directly overhead. In addition, when the sun is low in the sky, sunlight must pass through more of Earth's atmosphere, so more of its energy is absorbed and reflected before it reaches the ground. Thus, at higher latitudes (closer to the poles), there is greater variation in both day length and the angle of arriving solar energy over the course of a year than at latitudes closer to the equator. On average, mean annual air temperature decreases about 0.4°C for every degree of latitude (about 110 kilometers) at sea level.

Air temperature also decreases with elevation. As a parcel of air rises, it expands, its molecules move farther apart, its pressure and temperature drop, and it releases moisture. When a parcel of air descends, it is compressed, its pressure rises, its temperature increases, and it takes up moisture.

Earth's climates are strongly influenced by global air circulation patterns, which result from the global variation in solar energy input that we have just described and from the spinning of Earth on its axis. Air rises when heated by the sun. Warm air rises in the Tropics, which receive the greatest solar energy input. This rising air is replaced by air that flows toward the equator from the north and south. The coming together of these air masses produces the *intertropical convergence zone*. Typically, heavy rains fall in a region when it is close to the intertropical convergence zone as the rising air releases its moisture. The intertropical convergence zone shifts latitudinally with the seasons, following the shift in the zone of greatest solar energy input. This shift results in a characteristic latitudinal pattern of distribution of rainy and dry seasons in tropical and subtropical regions (Figure 56.5).

The air that moves into the intertropical convergence zone to replace the rising air is replaced, in turn, by air from aloft that descends at roughly 30° north and south latitudes after having traveled away from the equator at great heights. This air, which cooled and lost its moisture while rising at the equator, now descends, warms, and takes up rather than releasing moisture. Many of Earth's deserts, such as the Sahara

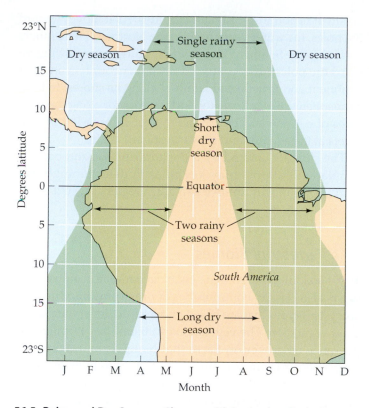

56.5 Rainy and Dry Seasons Change with Latitude In the Tropics and Subtropics, rainy and dry months are highly predictable based on the region's latitude.

ern Hemisphere and to the left in the Southern Hemisphere. Winds blowing toward the equator from the north and south veer to become the northeast and southeast trade winds, respectively. Winds blowing away from the equator also veer and become the westerly winds that prevail at mid-latitudes. The blue arrows in Figure 56.6 show the average directions of these surface winds.

When wind patterns bring air into contact with a mountain range, the air rises to pass over the mountains, cooling as it does so. Because cool air cannot hold as much moisture as warm air, clouds frequently form and release moisture as rain or snow. On the leeward side of the mountain range, the now-dry air descends, warms, and once again picks up moisture. This pattern often results in a dry area called a *rain shadow* on the leeward sides of moutain ranges (Figure 56.7).

Global oceanic circulation is driven by wind patterns

The global pattern of wind circulation drives the circulation of ocean water. Ocean water generally moves in the direction of the prevailing winds (Figure 56.8). Winds blowing toward the equator from the northeast and southeast cause water to converge at the equator and move westward until it encounters a continental land mass. At that point the water splits, some of it moving north and some of it moving south along continental shores. This poleward movement of ocean water that has been warmed in the Tropics is a major mech-

and the Australian deserts, are located at these latitudes where dry air descends.

At about 60° north and south latitudes, air rises again and moves either toward or away from the equator. At the poles, where there is little input of solar energy, air descends. The black arrows around the edge of Figure 56.6 show these vertical air circulation patterns. These movements of air masses are responsible, in part, for global wind patterns.

The spinning of Earth on its axis also influences surface winds because Earth's velocity is rapid at the equator, where its diameter is greatest, but relatively slow close to the poles. An air mass at a particular latitude has the same velocity as Earth has at that latitude. As an air mass moves toward the equator, it confronts an increasingly faster spin, and its rotational movement is slower than that of Earth beneath it. Conversely, as an air mass moves poleward, it confronts an increasingly slower spin, and it speeds up relative to Earth beneath it. Therefore, air masses moving latitudinally are deflected to the right in the North-

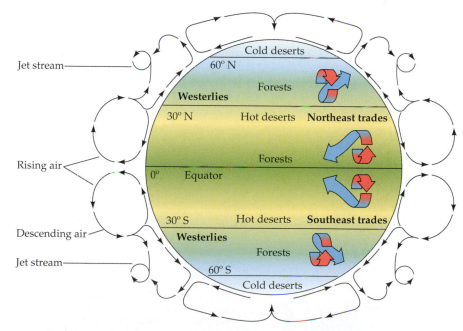

56.6 The Circulation of Earth's Atmosphere If we could stand outside Earth and observe its air movements, we would see vertical air circulation patterns similar to those indicated by the black arrows and surface winds similar to those shown by the blue arrows. Both the vertical and horizontal circulation patterns shift to the north during the northern summer and to the south during the northern winter.

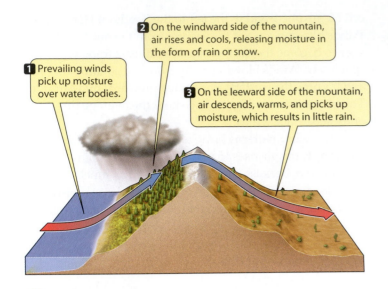

1 Prevailing winds pick up moisture over water bodies.

2 On the windward side of the mountain, air rises and cools, releasing moisture in the form of rain or snow.

3 On the leeward side of the mountain, air descends, warms, and picks up moisture, which results in little rain.

56.7 A Rain Shadow Average annual rainfall tends to be lower on the leeward side of a mountain range than on the windward side.

anism of heat transfer to high latitudes. As it moves toward the poles, the water veers right in the Northern Hemisphere and left in the Southern Hemisphere. Thus, water flowing toward the poles turns eastward until it encounters another continent and is deflected laterally along its shores. In both hemispheres, water flows toward the equator along the west sides of continents, continuing to veer right or left until it meets at the equator and flows westward again.

The climates created by these atmospheric and oceanic circulation patterns play key roles in determining what kinds of organisms can live in a given region, as we'll see in the next section.

Terrestrial Biomes

In addition to recognizing biogeographic regions, ecologists also classify communities of organisms into **biomes**, ecosystem types that are based on the structure of their dominant vegetation. The vegetation of a biome has a similar appearance wherever on Earth that biome is found, but the plant species in these communities, despite their physical similarities, may not be evolutionarily closely related. Biomes are named for and identified by their characteristic vegetation, sometimes supplemented by a description of their location or climate, but each biome contains many species in all other taxonomic groups.

56.8 Global Oceanic Circulation To see that ocean currents are driven primarily by winds, compare the surface currents shown here with the prevailing surface winds shown in Figure 56.6. Deep ocean currents differ strikingly from the surface ones shown here.

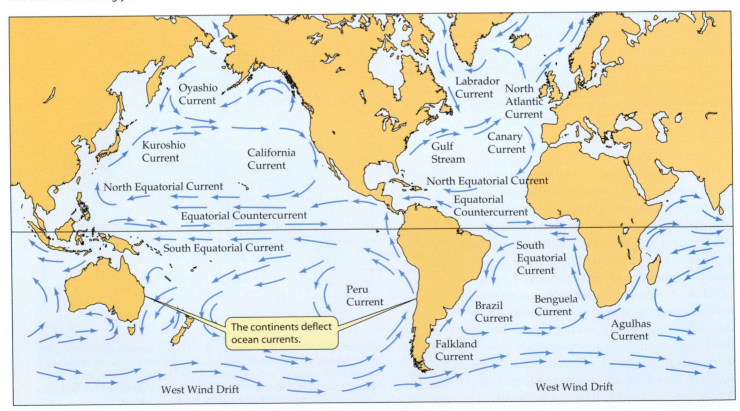

Oyashio Current

Labrador Current
North Atlantic Current

Kuroshio Current

California Current

Gulf Stream

Canary Current

North Equatorial Current

North Equatorial Current

Equatorial Countercurrent

Equatorial Countercurrent

South Equatorial Current

Peru Current

The continents deflect ocean currents.

South Equatorial Current

Brazil Current

Benguela Current

Agulhas Current

Falkland Current

West Wind Drift

West Wind Drift

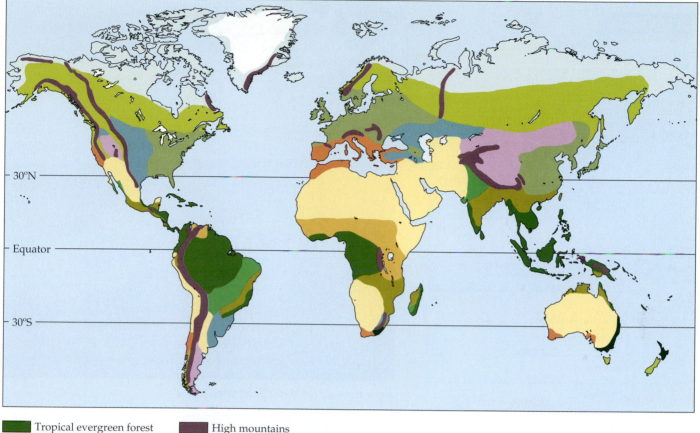

Tropical evergreen forest

Tropical deciduous forest

Thorn forest

Savanna

Hot desert

Chaparral

Cold desert

High mountains
(boreal forest and tundra)

Temperate evergreen forest

Temperate deciduous forest

Boreal forest

Arctic tundra

Temperate grassland

Polar ice cap

56.9 Biomes Have Distinct Geographic Distributions The distribution of biomes is strongly influenced by patterns of temperature and rainfall.

The distribution of biomes on Earth is strongly influenced by annual patterns of temperature and rainfall (Figure 56.9). In some biomes, such as temperate deciduous forest, precipitation is relatively constant throughout the year, but temperature varies strikingly between summer and winter. In other biomes, both temperature and precipitation change seasonally. In still other biomes, such as tropical rainforest, temperatures are nearly constant, but rainfall varies seasonally. In the Tropics, where seasonal temperature fluctuations are small, annual cycles are dominated by wet and dry seasons (see Figure 56.5).

It is easiest to grasp the similarities and differences among terrestrial biomes by means of a combination of photographs and graphs of temperature, precipitation, and biological ac-

tivity, supplemented by a few words that describe the species richness and other attributes of those biomes. We use this method in the following pages to describe the major terrestrial biomes of the world.

▶ Each biome is represented by a map showing its locations and two photographs that illustrate either the biome at different times of year or representatives of the biome in different places on Earth.

▶ One set of graphs plots seasonal patterns of temperature and precipitation at a site in the biome.

▶ Other graphs show how active different kinds of organisms are during the year. (For high-latitude biomes, patterns in the Southern Hemisphere are six months out of phase with those shown, which represent the Northern Hemisphere.) Levels of biological activity, shown by the width of the horizontal bars, change either because resident organisms become more or less active (produce leaves, come out of hibernation, hatch, or reproduce) or because organisms migrate into and out of the biome at different times of the year.

▶ A small box describes the growth forms of the plants that dominate the vegetation in the biome and its patterns of species richness.

These descriptions are very general and cannot describe the variation that exists within each biome.

TUNDRA

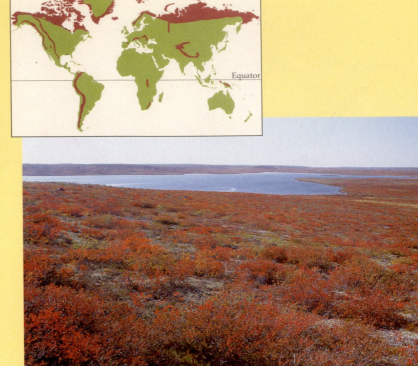

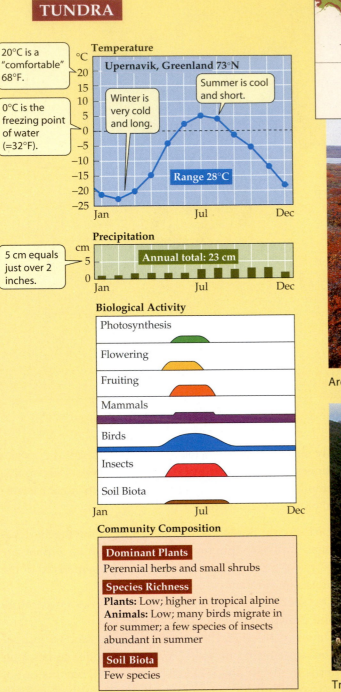

Temperature

20°C is a "comfortable" 68°F.

0°C is the freezing point of water (=32°F).

Upernavik, Greenland 73°N

Winter is very cold and long.

Summer is cool and short.

Range 28°C

Precipitation

5 cm equals just over 2 inches.

Annual total: 23 cm

Biological Activity

Photosynthesis
Flowering
Fruiting
Mammals
Birds
Insects
Soil Biota

Community Composition

Dominant Plants
Perennial herbs and small shrubs

Species Richness
Plants: Low; higher in tropical alpine
Animals: Low; many birds migrate in for summer; a few species of insects abundant in summer

Soil Biota
Few species

Arctic tundra, Northwest Territories, Canada

Tropical alpine tundra, Teleki Valley, Mt. Kenya

Tundra is found at high latitudes and in high mountains

The **tundra** biome is found in the Arctic and high in mountains at all latitudes. Arctic tundra vegetation, which consists of short perennial plants, is underlain by *permafrost*—soil whose water is permanently frozen. The top few centimeters of soil thaw during the short summers, when the sun shines 24 hours a day. Even though there is little precipitation, lowland Arctic tundra is very wet because water cannot drain down through the permafrost. Plants grow for only a few months each year. Most Arctic tundra animals either migrate into the area only for the summer or are dormant for most of the year.

Tropical alpine tundra is not underlain by permafrost, so photosynthesis and most other biological activities continue (albeit slowly) throughout the entire year. As the photo of alpine vegetation on Mt. Kenya shows, more plant growth forms are present in tropical alpine than in arctic tundra vegetation.

BOREAL FOREST

Temperature

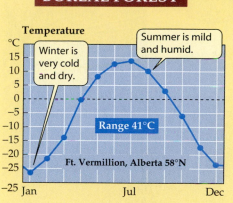

Winter is very cold and dry.

Summer is mild and humid.

Range 41°C

Ft. Vermillion, Alberta 58°N

Precipitation

Annual total: 31 cm

Biological Activity

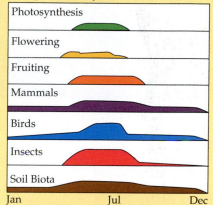

Photosynthesis
Flowering
Fruiting
Mammals
Birds
Insects
Soil Biota

Community Composition

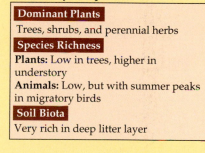

Dominant Plants
Trees, shrubs, and perennial herbs
Species Richness
Plants: Low in trees, higher in understory
Animals: Low, but with summer peaks in migratory birds
Soil Biota
Very rich in deep litter layer

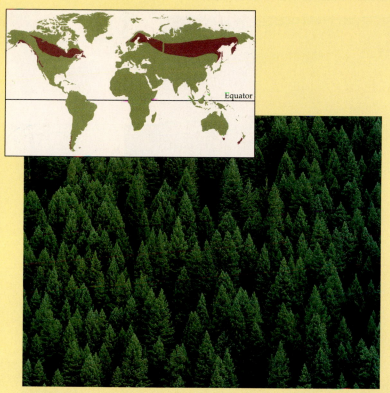

Equator

Northern conifer forest, Gunnison National Forest, Colorado

Southern conifer forest, Fiordland National Park, New Zealand

Most boreal forests are dominated by evergreen trees

Boreal forest is found equatorward from tundra and at lower elevations on temperate-zone mountains. Boreal forest winters are long and very cold; summers are short (although often warm). The shortness of the summers favors trees with evergreen leaves because these trees are ready to photosynthesize as soon as temperatures warm in spring.

The boreal forests of the Northern Hemisphere are dominated by evergreen coniferous gymnosperms. In the South-

ern Hemisphere the dominant trees are southern beeches (*Nothofagus*), some of which are evergreen. Evergreen forests also grow along the west coasts of continents at middle to high latitudes in both hemispheres, where winters are mild but very wet and summers are cool and dry. These forests are home to Earth's tallest trees.

Boreal forests have only a few tree species. The dominant animals (e.g., moose, hares) eat leaves. The seeds in the cones of conifers support a fauna of rodents and birds.

TEMPERATE DECIDUOUS FOREST

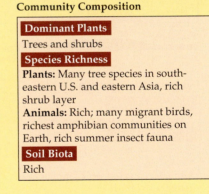

Equator

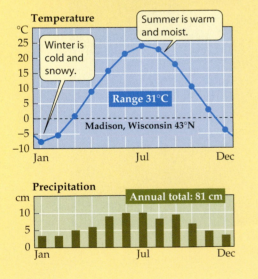

A Rhode Island forest in summer and…

…in winter

Temperature

°C
25
20
15
10
5
0
-5
-10

Winter is cold and snowy.

Summer is warm and moist.

Range 31°C

Madison, Wisconsin 43°N

Jan   Jul   Dec

Precipitation

cm
10
5
0

Annual total: 81 cm

Jan   Jul   Dec

Biological Activity

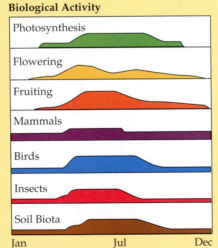

Photosynthesis

Flowering

Fruiting

Mammals

Birds

Insects

Soil Biota

Jan   Jul   Dec

Community Composition

Dominant Plants
Trees and shrubs

Species Richness
Plants: Many tree species in southeastern U.S. and eastern Asia, rich shrub layer
Animals: Rich; many migrant birds, richest amphibian communities on Earth, rich summer insect fauna

Soil Biota
Rich

Temperate deciduous forests change with the seasons

The **temperate deciduous forest** biome is found in eastern North America, eastern Asia, and Europe. Temperatures in these regions fluctuate dramatically between summer and winter. Precipitation is relatively evenly distributed throughout the year.

Deciduous trees, which dominate these forests, lose their leaves during the cold winters and produce leaves that pho-

tosynthesize rapidly during the warm, moist summers. Many more tree species live here than in boreal forests. The temperate forests richest in species are in the southern Appalachian Mountains of the United States and in eastern China and Japan—areas that were not covered by glaciers during the Pleistocene. Many genera of plants and animals are shared among the three geographically separate deciduous forest biomes.

TEMPERATE GRASSLANDS

Temperature

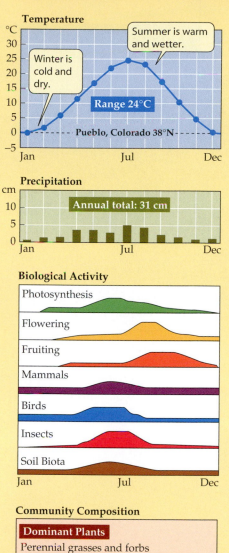

Winter is cold and dry.

Summer is warm and wetter.

Range 24°C

Pueblo, Colorado 38°N

Precipitation

Annual total: 31 cm

Biological Activity

Photosynthesis
Flowering
Fruiting
Mammals
Birds
Insects
Soil Biota

Jan   Jul   Dec

Community Composition

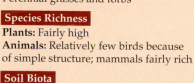

Dominant Plants
Perennial grasses and forbs

Species Richness
Plants: Fairly high
Animals: Relatively few birds because of simple structure; mammals fairly rich

Soil Biota
Rich

Equator

Nebraska prairie in spring

The Veldt, Natal, South Africa

Temperate grasslands are widespread

The **temperate grassland** biome is found in many parts of the world, all of which are relatively dry for much of the year. Most grasslands, such as the pampas of Argentina, the veldt of South Africa, and the Great Plains of North America, have hot summers and relatively cold winters. Most of this biome has been converted to agriculture. In some grasslands, most of the precipitation falls in winter (California grasslands); in others, the majority falls in summer (Great Plains, Russian steppe).

Grassland vegetation is structurally simple, but it is rich in species of perennial grasses, sedges, and forbs. Grasslands are often riots of color when forbs are in bloom. Grassland plants are adapted to grazing and fire. They store much of their energy underground and quickly resprout after they are burned or grazed.

COLD DESERT

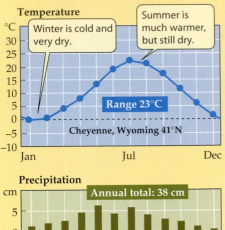

Temperature

Winter is cold and very dry.

Summer is much warmer, but still dry.

Range 23°C

Cheyenne, Wyoming 41°N

Precipitation

Annual total: 38 cm

Biological Activity

Photosynthesis

Flowering

Fruiting

Mammals

Birds

Insects

Soil Biota

Jan   Jul   Dec

Community Composition

Dominant Plants
Low stature shrubs and herbaceous plants

Species Richness
Plants: Few species
Animals: Rich in seed-eating birds, ants, and rodents; low in all other taxa

Soil Biota
Poor in species

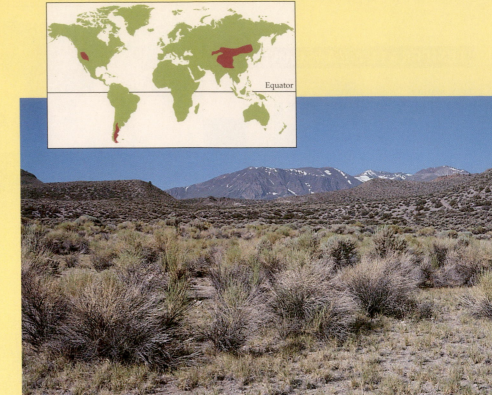

Sagebrush steppe near Mono Lake, California

Los Glaciares National Park, Argentina

Cold deserts are high and dry

The **cold desert** biome is found in dry regions at middle to high latitudes, especially in the interiors of large continents in the rain shadows of mountain ranges. Seasonal changes in temperature are great.

Cold deserts are dominated by a few species of low-growing shrubs. The surface layers of the soil are recharged with moisture in winter, and plant growth is concentrated in spring. Because soils dry rapidly in spring, annual primary production is low.

Cold deserts are relatively poor in species of most taxonomic groups, but the plants of this biome tend to produce large numbers of seeds, supporting a rich fauna of seed-eating birds, ants, and rodents.

HOT DESERT

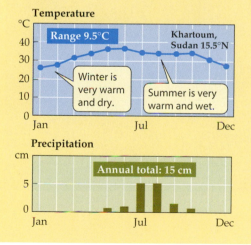

Anzo Borrego Desert, California

Simpson Desert, following rain, Australia

Temperature

°C
Range 9.5°C   Khartoum, Sudan 15.5°N
40
30
20   Winter is very warm and dry.
10   Summer is very warm and wet.
0
Jan   Jul   Dec

Precipitation

cm
Annual total: 15 cm
5
0
Jan   Jul   Dec

Biological Activity

Photosynthesis
Flowering
Fruiting
Mammals
Birds
Insects
Soil Biota
Jan   Jul   Dec

Community Composition

Dominant Plants
Many different growth forms
Species Richness
Plants: Fairly high; many annuals
Animals: Very rich in rodents; richest bee communities on Earth; very rich in reptiles and butterflies
Soil Biota
Poor in species

Hot deserts form around 30° latitude

The **hot desert** biome is found in two belts, centered around 30° north and 30° south latitudes, where air descends, warms, and picks up moisture. Hot deserts receive most of their rainfall in summer, but they also receive winter rains from storms that form over the mid-latitude oceans. The driest large regions, where summer and winter rains rarely penetrate, are in the center of Australia and the middle of the Sahara Desert of Africa.

Except in these driest regions, hot deserts have richer and structurally more diverse vegetation than cold deserts. Succulent plants that store large quantities of water in their expandable stems are conspicuous in some hot deserts. Annual plants germinate in abundance and grow when rain falls. Pollination and dispersal of fruits by animals are common. Rodents, termites, and ants are often remarkably abundant, and lizards and snakes typically are rich in species and abundant.

CHAPARRAL

Temperature

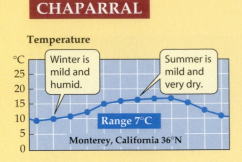

Winter is mild and humid.

Summer is mild and very dry.

Range 7°C

Monterey, California 36°N

Precipitation

Annual total: 42 cm

Biological Activity

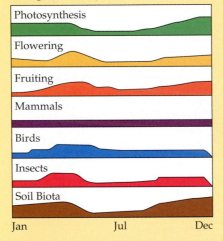

Photosynthesis

Flowering

Fruiting

Mammals

Birds

Insects

Soil Biota

Jan   Jul   Dec

Community Composition

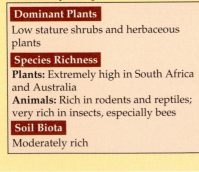

Dominant Plants
Low stature shrubs and herbaceous plants

Species Richness
Plants: Extremely high in South Africa and Australia
Animals: Rich in rodents and reptiles; very rich in insects, especially bees

Soil Biota
Moderately rich

Fynbos vegetation, Cape of Good Hope, South Africa

Mendocino, California

The chaparral climate is dry and pleasant

The **chaparral** biome is found on the west sides of continents at moderate latitudes (around 30°), where cool ocean waters flow offshore. Winters in this biome are cool and wet; summers are warm and dry. Such climates are found in the Mediterranean region of Europe, coastal California, central Chile, extreme southern Africa, and southwestern Australia.

The dominant plants of chaparral vegetation are low-growing shrubs and trees with tough, evergreen leaves. The shrubs carry out most of their growth and photosynthesis in early spring, when insects are active and birds breed. Annual plants are abundant and produce copious seeds that fall onto the soil. This biome thus supports large populations of small rodents, most of which store seeds in underground burrows. Chaparral vegetation is naturally adapted to survive periodic fires. Many shrubs of Northern Hemisphere chaparral produce bird-dispersed fruits that ripen in the late fall, when large numbers of migrant birds arrive from the north.

THORN FOREST and TROPICAL SAVANNA

Temperature

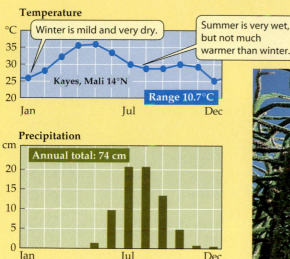

°C

Winter is mild and very dry.

Summer is very wet, but not much warmer than winter.

35
30
25
20

Kayes, Mali 14°N

Range 10.7°C

Jan   Jul   Dec

Precipitation

cm

Annual total: 74 cm

20
15
10
5
0

Jan   Jul   Dec

Biological Activity

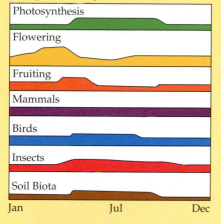

Photosynthesis

Flowering

Fruiting

Mammals

Birds

Insects

Soil Biota

Jan   Jul   Dec

Community Composition

Dominant Plants
Shrubs and small trees; grasses

Species Richness
Plants: Moderate in thorn forest; low in savanna
Animals: Rich mammal faunas; moderately rich in birds, reptiles, and insects

Soil Biota
Rich

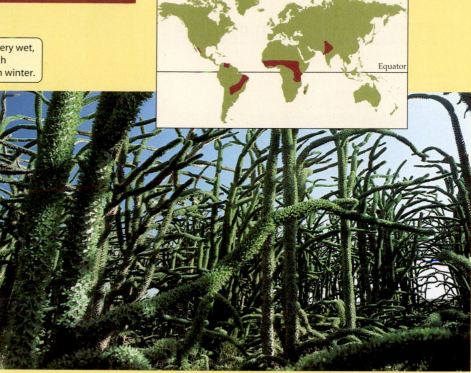

Equator

Thorn forest in Madagascar

Savanna in Tanzania

Thorn forests and savannas have similar climates

Thorn forests are found on the equatorial sides of hot deserts. The climate is semiarid; little or no rain falls during winter, but rainfall may be heavy during summer. Thorn forests contain many plants similar to those found in hot deserts. The dominant plants are spiny shrubs and small trees, many of which drop their leaves during the long dry winter. Members of the genus *Acacia* are common in thorn forests worldwide.

The dry tropical and subtropical regions of Africa, South America, and Australia have extensive areas of **savannas**—expanses of grasses and grasslike plants with scattered trees. The largest savannas are found in central and eastern Africa, where the biome supports huge numbers of grazing and browsing mammals and many large carnivores that prey on them. The grazers and browsers maintain the savannas. If savanna vegetation is not grazed, browsed, or burned, it typically reverts to dense thorn forest.

TROPICAL DECIDUOUS FOREST

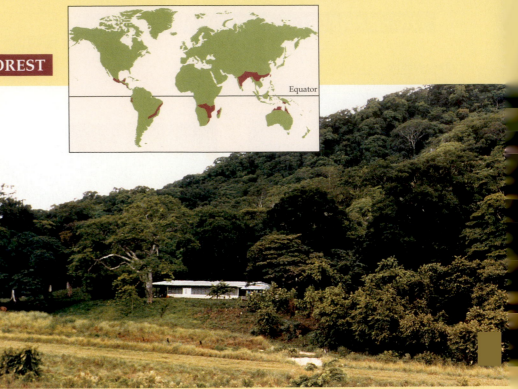

Temperature

Winter is very warm and dry.

Summer is warm and wet.

°C
30
25
20

Range 5.4°C

Timbo, Guinea 10°N

Jan Jul Dec

Precipitation

cm
35
30
25
20
15
10
5
0

Annual total: 163 cm

Jan Jul Dec

Biological Activity

Photosynthesis

Flowering

Fruiting

Mammals

Birds

Insects

Soil Biota

Jan Jul Dec

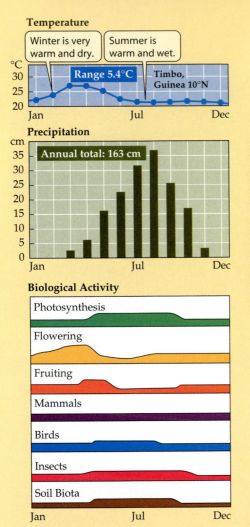

Community Composition

Dominant Plants
Deciduous trees

Species Richness
Plants: Moderately rich in tree species
Animals: Rich mammal, bird, reptile, and amphibian communities; rich in insects

Soil Biota
Rich, but poorly known

Palo Verde National Park, Costa Rica, in the rainy season...

...and in the dry season

Tropical deciduous forests occur in hot lowlands

As the length of the rainy season increases toward the equator, **tropical deciduous forests** replace thorn forests. These forests have taller trees and fewer succulent plants than thorn forests, and they are much richer in plant and animal species. Most of the trees, except for those growing along rivers, lose their leaves during the long, hot dry season. Many of them flower while they are leafless, and most species are pollinated by animals. During the hot rainy season, biological activity is intense.

The soils of the tropical deciduous biome are some of the best soils in the tropics for agriculture, because they are less leached of nutrients than the soils of wetter areas. As a result, most tropical deciduous forests have been cleared for agriculture and cattle grazing. Restoration efforts are underway on several continents.

TROPICAL EVERGREEN FOREST

Equator

The exterior of lowland wet forest…

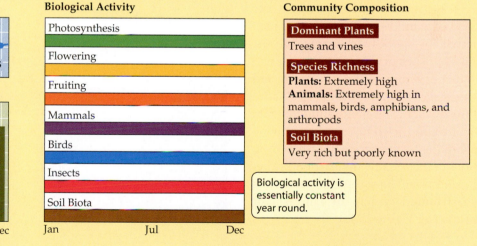

…and its interior, Cocha Cashu, Peru

Temperature

°C

Warm and rainy all year.

25
20
15
10

Range 2.2°C   Equitos, Peru 3°S

Precipitation

cm

Annual total: 262 cm

30
25
20
15
10
5
0

Jan   Jul   Dec

Biological Activity

Photosynthesis

Flowering

Fruiting

Mammals

Birds

Insects

Soil Biota

Jan   Jul   Dec

Biological activity is essentially constant year round.

Community Composition

Dominant Plants

Trees and vines

Species Richness

Plants: Extremely high
Animals: Extremely high in mammals, birds, amphibians, and arthropods

Soil Biota

Very rich but poorly known

Tropical evergreen forests are species-rich

Tropical evergreen forests are found in equatorial regions where total rainfall exceeds 250 cm annually and the dry season lasts no longer than 2 or 3 months. They are the richest of all biomes in number of species of both plants and animals, with up to 500 species of trees per km². Along with their immense species richness, tropical evergreen forests have the highest overall productivity of all ecological communities. However, most mineral nutrients are tied up in the vegetation. The soils usually cannot support agriculture without massive applications of fertilizers.

On the slopes of tropical mountains, trees are shorter than lowland tropical trees. Their leaves are smaller, and there are more *epiphytes* (plants that grow on other plants, deriving their nutrients and moisture from air and water rather than soil).

Aquatic Biogeography

Three-fourths of Earth's surface is covered by water, most of it in the oceans. Earth's oceans form one large, interconnected water mass with only partial barriers to dispersal. Fresh waters, in contrast, are divided into river basins and thousands of relatively isolated lakes. For organisms that cannot survive out of water, terrestrial habitats between bodies of water are barriers to dispersal. However, some aquatic species have flying adults that can disperse widely among water bodies. Others have windborne, desiccation-resistant spores and seeds. Still others are small enough to be transported by mud on the feet of birds.

Freshwater environments have little water but many species

Although only about 2.5 percent of Earth's water is found in ponds, lakes, and streams, about 10 percent of all aquatic species live in freshwater habitats. Many freshwater taxa that are capable of dispersing across terrestrial barriers are found over several continents. Prominent among freshwater taxa are the more than 25,000 species of insects that have at least one aquatic stage in their life cycle. Typically, eggs and larvae are aquatic; the adults have wings. Some of these insects, such as dragonflies, are powerful flyers, but mayflies and some other species are weak flyers, desiccate rapidly in air, and live no longer than a few days. As you would expect, oceanic islands have few, if any, species of these weak flyers.

Similarly, fishes unable to live in salt water can disperse only within the connected rivers and lakes of a river basin. Most families of freshwater fishes that cannot tolerate salt water are restricted to a single continent. Those families with species distributed on both sides of major saltwater barriers are believed to be ancient lineages whose ancestors were distributed widely in Laurasia or Gondwana (see Figure 25.11).

Water temperature defines marine biogeographic regions

As we saw in Figure 56.8, ocean water moves in great circular patterns—clockwise in the Northern Hemisphere and counterclockwise in the Southern Hemisphere. Even organisms with limited swimming abilities can move long distances simply by floating with ocean currents. Nevertheless, most marine organisms have restricted ranges. Why is this true?

The oceans may be connected, but water temperatures, salinities, and food supplies all may change spatially. Living successfully in different regions of the ocean requires different physiological tolerances and morphological attributes. Ocean temperatures, for example, can be barriers to dispersal because many marine organisms function well in only a relatively narrow range of temperatures. The main biogeographic divisions of the ocean coincide with regions where the surface water temperatures and salinities change relatively abruptly as a result of horizontal and vertical ocean currents (Figure 56.10). These temperature changes, in combination with seasonal changes in the amount of daylight, determine the seasons of maximum primary production. Species of marine algae photosynthesize either in summer or in winter, but not during both seasons.

Deep ocean waters prevent the dispersal of marine organisms that live only in shallow water. The distance that eggs and larvae of many marine organisms can be carried by ocean currents is determined in large part by the time it takes for larvae to metamorphose into sedentary adults. Relatively few species have eggs and larvae that survive long enough to dis-

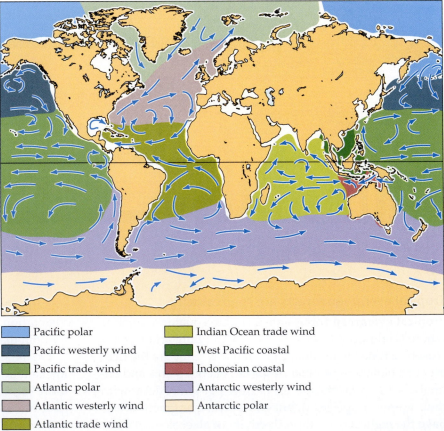

56.10 Oceanic Biogeographic Regions are Determined by Ocean Currents The arrows represent ocean currents. Different biogeographic regions, in which photosynthesis is maximized at different seasons, are indicated by different colors.

Pacific polar	Indian Ocean trade wind
Pacific westerly wind	West Pacific coastal
Pacific trade wind	Indonesian coastal
Atlantic polar	Antarctic westerly wind
Atlantic westerly wind	Antarctic polar
Atlantic trade wind	

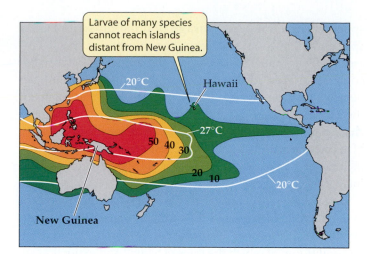

56.11 Generic Richness of Reef-Building Corals Declines with Distance from Indonesia The colored zones represent areas with equal numbers of coral genera. The 20° and 27° mean annual temperature isotherms are also shown.

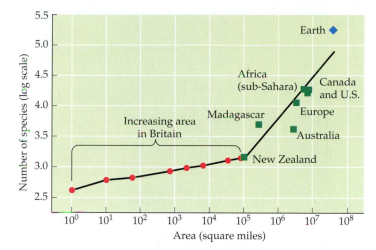

56.12 Species Richness Increases with Area Sampled Plotted here are the number of species of vascular plants in eight increasingly large samples in Britain; in several larger areas, and, finally, on the entire Earth. Recall that in a logarithmic scale, each increment of measurement is 10 times larger than the preceding one.

perse across wide barriers of deep water. As a result, the richness of shallow-water species in the intertidal and subtidal zones of isolated islands in the Pacific Ocean decreases with distance from the larger islands of Indonesia (Figure 56.11).

Regional Patterns of Species Richness

As we saw in Chapter 55, local species richness is often positively correlated with both productivity and disturbance level. Other patterns of species richness appear at larger spatial scales. As we increase the area we are sampling, the number of species we record increases slowly (Figure 56.12). However, if our sampling area crosses a biogeographic boundary, the rate at which we add new species suddenly increases. At that point, we have added to our sample another biogeographic region with a different evolutionary history and a different biota.

One of the first geographic patterns of species richness observed by biologists was that more species are found in low-latitude than in high-latitude regions. Figure 56.13 shows this latitudinal gradient in species richness for mammals in North and Central America. Similar pat-

terns exist for birds, frogs, and trees and for many marine taxa. The figure also shows that more species are found in mountainous regions than in relatively flat areas because more vegetation types and climates exist within these topographically complex areas.

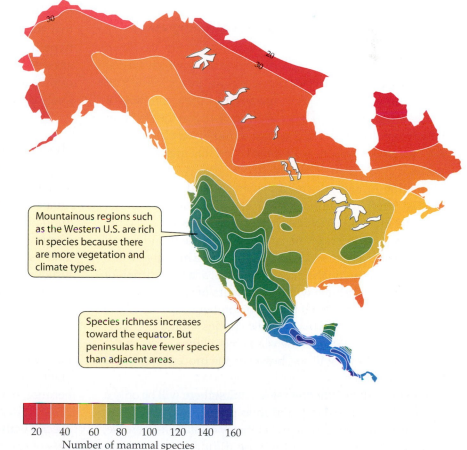

Mountainous regions such as the Western U.S. are rich in species because there are more vegetation and climate types.

Species richness increases toward the equator. But peninsulas have fewer species than adjacent areas.

56.13 The Latitudinal Gradient of Species Richness of North American Mammals The colored zones represent regions with equal numbers of species. An increase in species richness toward the equator typifies many other taxa, such as birds, amphibians, and trees.

Number of mammal species

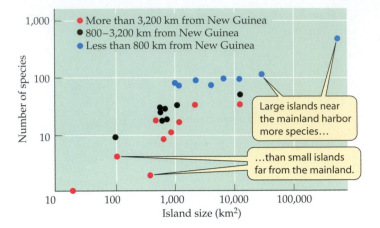

56.14 Small, Distant Islands Have Fewer Bird Species The dots show the numbers of land and freshwater bird species on islands of different sizes in the Moluccas, Melanesia, Micronesia, and Polynesia. These islands have been divided into three groups according to their distance from the "mainland," which in this case is New Guinea.

Species richness on islands and peninsulas is always less than that in an equivalent area on the nearest mainland. On islands, species richness is positively correlated with island size, but inversely correlated with distance from the mainland (Figure 56.14). An influential model relates this pattern to the island's history of immigrations and extinctions.

Species richness is related to rates of immigration and extinction

Over periods of a few hundred years (during which speciation is unlikely), the species richness of an area is influenced by the immigration of new species and the extinction of species already present. It is easiest to visualize the effects of these two processes if we consider, as did Robert MacArthur and Edward O. Wilson, an oceanic island that initially has no species.

Imagine a newly formed oceanic island that receives colonists from a mainland area. The list of species on the mainland that might possibly colonize the island is called the *species pool*. The first colonists to arrive on the island are all "new" species because no species live there initially. As the number of species on the island increases, a larger fraction of colonizing individuals will be members of species already present. Therefore, even if the same number of species arrive as before, the rate of arrival of new species should decrease, until it reaches zero when the island has all the species in the species pool. As we will see, however, the process is unlikely to proceed that far.

Now consider extinction rates. At first there will be only a few species on the island, and their populations may grow large. As more species arrive and their populations increase, the resources of the island will be divided among more

species. Therefore, the average population size of each species will become smaller as the number of species increases. The smaller a population, the more likely it is to become extinct. In addition, the number of species that can possibly become extinct increases as species accumulate on the island. Furthermore, new arrivals on the island may include pathogens and predators that increase the probability of extinction for other species. For all these reasons, the rate of extinction increases as the number of species on the island increases.

Because the rate of arrival of new species decreases and the extinction rate increases as the number of species increases, eventually the number of species on the island should reach an equilibrium at which the rates of arrival and extinction are equal (Figure 56.15a). If there are more species than the equilibrium number, extinctions should exceed arrivals, and species richness should decline. If there are fewer species than the equilibrium number, arrivals should exceed extinctions, and species richness should increase. The equilibrium is dynamic because if either rate fluctuates, as they generally do, the equilibrium number of species shifts up and down.

MacArthur and Wilson's model can also be used to predict how species richness should differ among islands of different sizes and different distances from the mainland. We expect extinction rates to be higher on small islands than on large islands because species' populations are, on average, smaller there. Similarly, we expect fewer immigrants to reach islands that are more distant from the mainland. Figure 56.15b gives hypothetical relative species richnesses for islands of different sizes and distances from the mainland. As you can see, the number of species should be highest for islands that are relatively large and relatively close to the mainland.

The MacArthur-Wilson model has been tested

Major disturbances, which serve as "natural experiments," sometimes permit colonization and extinction rates to be estimated directly. In August 1883, Krakatau, an island in the Sunda Strait between Sumatra and Java, was devastated by a series of volcanic eruptions that destroyed all life on the island's surface. After the lava cooled, Krakatau was colonized rapidly by plants and animals from Sumatra to the east and Java to the west. By 1933, the island was again covered with a tropical evergreen forest, and 271 species of plants and 27 species of resident land birds were found there.

During the 1920s, when a forest canopy was developing, there were high rates of colonization by both birds and plants (Table 56.1). Birds probably brought the seeds of many plants because, between 1908 and 1934, both the percentage (from 20% to 25%) and the absolute number (from 21 to 54) of plant species with bird-dispersed seeds increased. Today the numbers of species of plants and birds are not increasing as fast

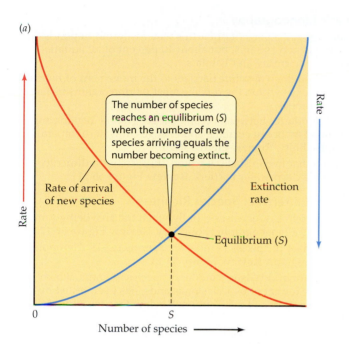

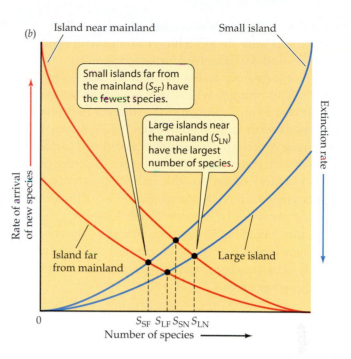

56.15 MacArthur and Wilson's Model of Species Richness on Islands (*a*) The rate of arrival of new species and the rate of extinction of species already present determine the equilibrium number of species on an island. (*b*) These rates are affected by the size of the island and its distance from the mainland.

as they did during the 1920s, but colonizations and extinctions continue, as predicted by the model.

Biogeography and Human History

The distributions of land masses and species on Earth have had a strong influence on human history. Humans first evolved in Africa, but eventually dispersed throughout the world. In recent times, human cultures from Eurasia came to dominate other cultures. Biogeography gave these cultures a number of advantages.

56.1 Number of Species of Resident Land Birds on Krakatau

PERIOD	NUMBER OF SPECIES	EXTINCTIONS	COLONIZATIONS
1908	13		
1908–1919		2	17
1919–1921	28		
1921–1933		3	4
1933–1934	29		
1934–1951		3	7
1951	33		
1952–1984		4	7
1984–1996	36		

Eurasia happened to have a large number of species of plants and animals that were suitable for domestication. Eurasia was home to 39 species of large-seeded grasses, many more than were found in Africa or the Americas. It also had 72 species of large mammals, compared with 51 in sub-Saharan Africa and 24 in the Americas. Thirteen of these species, including pigs, horses, cattle, sheep, goats, and camels, were domesticated in Eurasia. None were domesticated in Africa, and only one, the llama, in the Americas.

To be amenable to domestication, large mammals had to have three important social characteristics: They had to live in herds, have well-developed male dominance hierarchies, and not defend territories. These traits enabled humans to tame the animals, exert behavioral dominance over them, and keep them in herds. All of the large mammals of Africa lacked one or more of these traits.

Besides providing people with food, the domestication of large mammals had other important influences on human history. Many human diseases, such as smallpox and measles, were acquired from domesticated mammals. Eurasian people acquired immunity to these diseases, but people on other continents did not. Thus, when Europeans colonized the New World, they brought with them diseases that devastated the indigenous people. Those unfortunate people transmitted no fatal diseases to the Europeans. In addition, the Europeans had horses, the only domesticated mammals capable of carrying a person at high speeds. Throughout human history, cultures with horses have defeated and dominated cultures without them.

In Eurasia, most mountain ranges are oriented in an east-west direction. Therefore, dispersal of people and their domesticated plants and animals was relatively easy, and dispersing individuals were always within climates with similar temperatures and day lengths. Humans dispersed only recently into North America across the high-latitude Bering Land Bridge. They brought with them no domesticated plants or animals, except dogs. North America had few species of grasses with large seeds. Maize, the grass that came to dominate American agriculture, was difficult to domesticate. Its eventual spread northward from its center of domestication in Mexico was possible only after extensive genetic changes that adapted the plants to the very different day lengths and climates of temperate North America.

Human history would have been very different if the continents and their biotas had been distributed differently. Thus, the study of biogeography can help us understand ourselves as well as other species.

Chapter Summary

▶ Biogeography is the science that attempts to describe and explain patterns in the distribution of life on Earth.

Earth's Biogeographic Regions

▶ If a species occupies a particular area, either it evolved there, or it evolved elsewhere and dispersed to that area.

▶ If a species is not found in a particular area, either it evolved elsewhere and never dispersed to that area, or it was once present in that area but no longer lives there.

▶ Biogeographers divide Earth into several major biogeographic regions. **Review Figure 56.1. See Web/CD Activity 56.1**

▶ Remote islands contain many endemic species.

History and Biogeography

▶ Continental drift has influenced the distributions of organisms throughout Earth's history.

▶ Biogeographers often analyze species distributions by converting phylogenies into area phylogenies. **Review Figure 56.3**

▶ Distribution patterns can result from vicariant events or from dispersal.

▶ The principle of parsimony is used to explain distribution patterns. **Review Figure 56.4**

Ecology and Biogeography

▶ Global atmospheric circulation is driven by solar energy input. **Review Figure 56.6**

▶ Wind circulation patterns influence the amount and seasonal nature of rainfall. **Review Figures 56.5, 56.7. See Web/CD Tutorial 56.1**

▶ Global oceanic circulation is driven by winds. **Review Figure 56.8**

Terrestrial Biomes

▶ Terrestrial biomes are major ecosystem types that differ from one another in the structure of their dominant vegetation.

▶ The distribution of biomes on Earth is strongly influenced by annual patterns of temperature and rainfall. **Review Figure 56.9**

▶ The major terrestrial biomes are tundra, boreal forest, temperate deciduous forest, temperate grassland, cold desert, hot desert, chaparral, thorn forest and savanna, tropical deciduous forest, and tropical evergreen forest. **Review Pages 1076–1085.**

See Web/CD Tutorial 56.2

Aquatic Biogeography

▶ Only about 2.5 percent of Earth's water is found in ponds, lakes, rivers and streams, but about 10 percent of all aquatic species live in freshwater habitats.

▶ Even though no absolute barriers to the movement of marine organisms exist within the oceans, most marine organisms have restricted ranges.

▶ The boundaries between oceanic biogeographic regions are determined by ocean currents and changes in water temperature and salinity. **Review Figure 56.10**

▶ Species that live in shallow waters disperse with difficulty across wide deep-water barriers. **Review Figure 56.11**

Regional Patterns of Species Richness

▶ Species richness increases rapidly when the area being sampled crosses a biogeographic boundary. **Review Figure 56.12**

▶ The number of species in most lineages increases from polar to tropical regions. **Review Figure 56.13**

▶ MacArthur and Wilson's model of species richness, which predicts the number of species on islands, has been tested by examining patterns of distribution. **Review Figures 56.14, 56.15, Table 56.1**

See Web/CD Tutorial 56.3

Biogeography and Human History

▶ The distributions of plants, animals, and continents have exerted powerful influences on human history.

Self-Quiz

1. Biogeography as a science began when
    a. eighteenth-century travelers first noted intercontinental differences in the distributions of organisms.
    b. Europeans went to the Middle East during the Crusades.
    c. phylogenetic methods were developed.
    d. the fact of continental drift was accepted.
    e. Charles Darwin proposed the theory of natural selection.

2. Vicariant events
    a. are infrequent in nature.
    b. were common in the past but are rare today.
    c. separate species ranges in the absence of dispersal.
    d. were rare in the past but are common today.
    e. caused most of today's disjunct distributions.

3. Marine biogeographic regions exist even though the oceans are all connected because
    a. primary production is low in the oceans.
    b. ocean currents keeps organisms close to where they were born.
    c. most families and higher taxa of marine organisms evolved before the oceans were separated by continental drift.
    d. water temperatures and salinities often change abruptly where ocean currents meet.
    e. oceanic circulation is too slow to carry marine organisms from one ocean to another.

4. A parsimonious interpretation of a distribution pattern is one that
    a. requires the smallest number of undocumented vicariant events.
    b. requires the smallest number of undocumented dispersal events.
    c. requires the smallest total number of undocumented vicariant plus dispersal events.
    d. accords with the phylogeny of a lineage.
    e. accounts for centers of endemism.

5. The only major biogeographic region that today is completely isolated by water from other regions is
   a. Greenland.
   b. Africa.
   c. South America.
   d. Australasia.
   e. North America.

6. In MacArthur and Wilson's model, equilibrium species richness is reached when
   a. immigration rates of new species and extinction rates of species are equal.
   b. immigration rates of all species and extinction rates of species are equal.
   c. the rate of vicariant events equals the rate of dispersal.
   d. the rate of island formation equals the rate of island loss.
   e. No equilibrium number of species exists in that model.

7. Chaparral vegetation is dominated by
   a. deciduous trees.
   b. evergreen trees.
   c. deciduous shrubs.
   d. evergreen shrubs.
   e. grasses.

8. Which of the following is *not* true of tropical evergreen forests?
   a. They have large numbers of species of trees.
   b. Most plant species are animal-pollinated.
   c. Most plant species have animal-dispersed fruits.
   d. Biological energy flow is very high.
   e. High productivity depends on a rich supply of soil nutrients.

9. Cold deserts
   a. are dominated by a few species of low-growing shrubs.
   b. are dominated by a rich flora of low-growing shrubs.
   c. have few species of woody plants, but of varied growth forms.
   d. have many species of woody plants of varied growth forms.
   e. are dominated by a few species of tall shrubs.

10. Biogeography exerted a strong influence on human history because
    a. humans first evolved in Africa.
    b. Eurasia had more species of plants and animals that were easily domesticated.
    c. Old World mountain ranges are oriented in an east-west direction.
    d. horses were found only in Eurasia.
    e. All of the above

For Discussion

1. Horses evolved in North America, but subsequently became extinct there. They survived to modern times only in Africa and Asia. In the absence of a fossil record, we would probably infer that horses originated in the Old World. Today, the Hawaiian Islands have by far the greatest number of species of fruit flies (*Drosophila*). Would you conclude that the genus *Drosophila* originally evolved in Hawaii and spread to other regions? Under what circumstances do you think it is safe to conclude that a group of organisms evolved close to where the greatest number of species live today?

2. Processes in nature do not always conform to the parsimony principle. Why, then, do biogeographers often use the parsimony principle to infer geographic histories of species and lineages?

3. A well-known legend states that Saint Patrick drove the snakes out of Ireland. Give some alternative explanations, based on sound biogeographic principles, for the absence of indigenous snakes in that country.

4. Most of the world's flightless birds are either nocturnal and secretive (such as the kiwi of New Zealand) or large, swift, and powerful (such as the ostrich of Africa). The exceptions are found primarily on islands. Many of these island species have become extinct with the arrival of humans and their domestic animals. What special conditions on islands might permit the survival of flightless birds? Why has human colonization so often resulted in the extinction of such birds? The power of flight has been lost secondarily in representatives of many groups of birds and insects; what are some possible evolutionary advantages of flightlessness that might offset its obvious disadvantages?

5. MacArthur and Wilson's model of species richness on islands incorporates almost nothing about the biology of the species. What traits of species should be incorporated into more realistic models of rates of colonization and extinction of species on islands?

6. A legislator introduces a controversial bill into the U.S. Congress that would ban all introductions of exotic species to the Hawaiian Islands. Would you vote in favor of this bill if you were in Congress? Why or why not?

57 *Conservation Biology*

 In 1998, botanists of the Massachusetts Natural Heritage and Endangered Species Program published a booklet called *A Guide to Invasive Plants in Massachusetts*. The purpose of the booklet was to warn citizens about certain non-native plants that were becoming established and interfering with the state's natural ecosystems. But many nursery owners were upset; some of these invasive exotic plants were big money-makers for the horticultural industry. Horticulturalists lobbied the state government and succeeded in getting the booklet withdrawn from publication.

Fortunately, many people in the horticultural world now recognize that even though most introduced plants do not become invasive, some plant species have become serious pests on several continents. Indeed, colonization of new areas by introduced plants, animals, and microorganisms that become abundant in their new ranges is second only to habitat loss as a threat to Earth's biodiversity.

Humans have caused extinctions for thousands of years. When people first crossed the Bering Land Bridge and arrived in North America about 20,000 years ago, they encountered a rich fauna of large mammals. Most of those species were exterminated—probably by overhunting—within a few thousand years. A similar extermination of large animals followed the human colonization of Australia, about 40,000 years ago. At that time, Australia had 13 genera of marsupials larger than 50 kg, a genus of gigantic lizards, and a genus of heavy, flightless birds. All the species in 13 of those 15 genera had become extinct by 18,000 years ago. When Polynesian people settled in Hawaii about 2,000 years ago, they exterminated, probably by overhunting, at least 39 species of endemic land birds. Among them were 7 species of geese, 2 species of flightless ibises, a sea eagle, a small hawk, 7 flightless rails, 3 species of owls, 2 large crows, a honeyeater, and at least 15 species of finches.

The pace of human-caused extinction of species is accelerating rapidly. Most of the human activities that are currently causing extinctions are not new—but today there are many more humans living on Earth, doing more things that endanger species. Current extinction rates have raised serious concerns about the future of biological diversity on Earth. These concerns led to the rapid development during the 1980s of the applied discipline of **conservation biology**: the scientific study of how to preserve the diversity of life. Conservation biologists study the factors that threaten species with extinction, and they develop methods to help preserve genes, species, communities, and ecosystems. The

A Successful Invasion Introduced into the northeastern United States from Europe during the 1800s, *Lythrum salicaria*—purple loosestrife—was sold as an ornamental plant and for medicinal uses. Loosestrife establishes itself readily in natural wetlands, such as this riverbank in Massachusetts, where it outcompetes native species and changes the habitat of waterfowl and other animals.

science of conservation biology draws heavily on concepts and knowledge from population genetics, evolution, ecology, biogeography, wildlife management, economics, and sociology. In turn, the needs of conservation are stimulating new research in those fields.

In this chapter, we will see how conservation biologists estimate rates of species extinction and determine the causes of extinctions. We will learn how science is used to reduce extinction rates and help populations recover. But why should we care about species extinctions?

Why Care about Species Extinctions?

Extinction is forever. If we purposely or inadvertently exterminate a species, we have irreversibly destroyed a resource of unknown value. But people value biodiversity for many reasons:

▶ Humans depend on other species for food, fiber, and medicine. More than half the medical prescriptions written in the United States contain a natural plant or animal product.

▶ Humans derive enormous aesthetic pleasure from interacting with other organisms. Many people would consider a world with far fewer species to be a less desirable place in which to live.

▶ Living in ways that cause the extinction of other species raises serious ethical issues. These issues are receiving increased attention from philosophers, ethicists, and religious leaders.

▶ Extinctions deprive us of opportunities to study and understand ecological relationships among organisms. The more species are lost, the more difficult it will be to understand the structure and functioning of ecological communities and ecosystems.

▶ Species are necessary for the functioning of the ecosystems of which they are a part and the many benefits those ecosystems provide to humanity.

Among the benefits provided by ecosystems are generation and maintenance of fertile soils, prevention of soil erosion, detoxification and recycling of waste products, regulation of the hydrological cycle and the composition of the atmosphere, control of agricultural pests, and pollination of plants.

The benefits provided to humans by functioning ecosystems are very hard to calculate, but their value can be estimated. The benefits provided by the native vegetation of the Western Cape Province, South Africa, were estimated by a group of economists, ecologists, and land managers. The native vegetation of the highlands of this area is a species-rich community of shrubs, known as *fynbos* (pronounced "fainbos"). These shrubs can survive regular summer droughts, nutrient-poor soils, and the fires that periodically sweep

through the highlands (Figure 57.1*a*). The fynbos-clad highlands provide about two-thirds of the Western Cape's water requirements. In addition, some species of the endemic flora

(*a*)

(*b*) **Stream flow from fynbos watersheds**

As biomass has increased—largely from the increase in invasive species—stream flow has decreased proportionately.

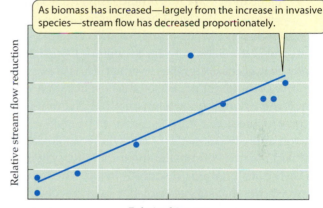

Relative stream flow reduction

Relative biomass

(*c*) **Computer simulation**

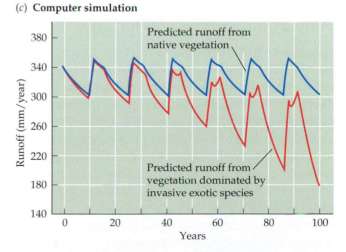

Runoff (mm/year)

Predicted runoff from native vegetation

Predicted runoff from vegetation dominated by invasive exotic species

Years

57.1 Invasive Species Disrupt Ecosystem Function (*a*) The unique fynbos ecosystems of South Africa provide much of the area's water. (*b*) Stream flow from fynbos watersheds is inversely proportional to plant biomass. (*c*) A computer simulation of stream flows from watersheds that have and have not been invaded by exotic trees.

are harvested for cut and dried flowers and thatching grass. The combined value of these harvests in 1993 was about $19 million. Some of the income from tourism in the region comes from people who want to see the fynbos. About 400,000 people visit the Cape of Good Hope Nature Reserve each year, primarily to see the many endemic plants.

During recent decades, a number of plants introduced into South Africa from other continents have invaded the fynbos. Because they are taller and grow faster than the native plants, these exotics increase the intensity and severity of fires. By transpiring larger quantities of water, they decrease stream flows to less than half the amount flowing from mountains covered with native plants, reducing the water supply (Figure 57.1b). Removing the exotic plants by felling and digging out invasive trees and shrubs and managing fire is estimated to cost between $140 and $830 per hectare, depending on the densities of invasive plants. Annual follow-up operations cost about $8 per hectare.

When natural ecosystems are lost, the services they provided must be replaced, often at a much higher cost. A sewage purification plant that would deliver the same volume of water to the Western Cape Province as a well-managed watershed of 10,000 hectares would cost $135 million to build and $2.6 million per year to operate. Desalination of seawater would cost four times as much. Thus, the available alternatives would deliver water at a cost between 1.8 and 6.7 times more than the cost of maintaining natural vegetation in the watershed.

Modern industrial societies often favor technologically sophisticated methods of substituting for lost ecosystem services. The study of water resources in the Western Cape Province shows that simple but labor-intensive methods—cutting and burning—may, in some cases, be cheaper.

Estimating Current Rates of Extinction

We do not know how many species will become extinct during the next 100 years because we do not know how many species live on Earth, and because the number of extinctions will depend both on what we do and on unexpected events.

Nevertheless, several methods exist for estimating probable rates of extinction resulting from human actions. For example, conservation biologists often use the well-established relationship between the size of an area and the number of species present to estimate the number of species extinctions likely to result from habitat destruction. We saw in Chapter 56 that the number of species on an island increases with the size of the island. This **species–area relationship** can be applied to habitat patches on the mainland as well. Biologists have measured the rate at which species richness tends to decrease with decreasing patch size. Their findings suggest that, on average, a 90 percent loss of habitat will result in the loss of half of the

species living in that habitat. The current rate of loss of tropical evergreen forests—Earth's richest biome—is about 2 percent of the remaining forest each year. If this rate of loss continues, about 1 million species that live in tropical evergreen forests will become extinct during the coming century.

To estimate the risk that a population will become extinct, conservation biologists develop models that incorporate information about a population's size, its genetic variation, and the morphology, physiology, and behavior of its members. Species in imminent danger of extinction over all or a significant part of their range are labeled *endangered species*. *Threatened species* are those that are likely to become endangered in the near future. Although rarity in and of itself is not always a cause for concern, species whose populations are shrinking rapidly usually are at risk. Species with only a few individuals confined to a small range are likely to be eliminated by local disturbances such as fires, unusual weather, disease, and predators.

In an example of such a population study, an ecologist constructed a quantitative model of the dynamics of the grizzly bear population in Yellowstone National Park, using detailed data collected over a 12-year period. The model kept track of individual bears and incorporated the effects of chance events, such as fires. The output of the model suggested that for the grizzly bear population to have a 95 percent chance of persisting for a century, there must be enough habitat to support 70–90 bears. To achieve a higher probability of survival, or the same probability of survival for 200 years, more bear habitat would be needed.

Preserving Biodiversity

The human activities that threaten species include habitat destruction, the introduction of invasive species, overexploitation, disease, alteration of disturbance patterns, and climate change. Conservation biologists determine how these activities are affecting species and use that information to devise actions to preserve species that are endangered or threatened.

Habitat loss is studied by observation and experimentation

Habitat loss is the most important cause of endangerment of species in the United States, especially species that live in fresh waters (Figure 57.2). As habitats are progressively destroyed by human activities, the remaining habitat patches become smaller and more isolated. In other words, the habitat becomes **fragmented**. Small habitat patches are qualitatively different from larger patches of the same habitat in ways that affect the survival of species. Small patches cannot maintain populations of species that require large areas, and they can support only small populations of many of the species that can survive in them.

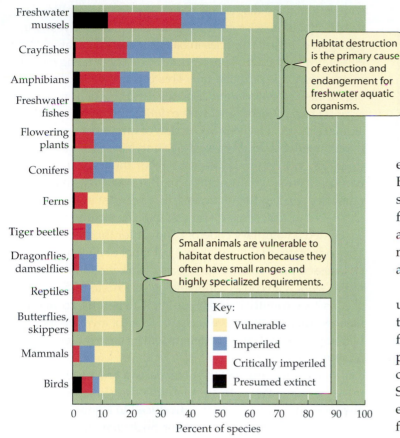

57.2 Proportions of U.S. Species Extinct or In Peril The groups of species that are most endangered—mussels, crayfishes, amphibians, and fishes—live in freshwater habitats, which have been extensively destroyed and polluted.

higher than they are farther inside the forest. Species from surrounding habitats often colonize the edges of patches to compete with or prey upon the species living there.

Usually we do not know which organisms lived in an area before its habitats became fragmented. To address this problem, a major research project in a tropical evergreen forest near Manaus, Brazil, was launched before logging took place. Landowners agreed to preserve forest patches of certain sizes and configurations (Figure 57.4). Biologists counted species in those patches while they were still part of the continuous forest. Soon after the surrounding forest was cut and converted to pasture, species began to disappear from isolated patches. The first species to be eliminated were monkeys that travel over large areas. Army ants and the birds that follow army ant swarms also disappeared.

Species that become extinct in small habitat fragments are unlikely to become reestablished because the more isolated the patches are, the less likely dispersing individuals are to find them. However, as we saw in Chapter 54, a species may persist in a small patch if it is connected to other patches by corridors of habitat through which individuals can disperse. Some of the pastures that surrounded the experimental forest fragments in Brazil have been abandoned, and a young forest is growing on them. Within 7–9 years of abandonment, some ant-following birds recolonized forest fragments connected to larger forest patches by young forests. Other species of birds that forage in the forest canopy also reestablished themselves. The young forest is not a suitable permanent habitat for most of these species, but it is an environment through which individuals can disperse to find new places where they can live.

Introduced predators, competitors, and pathogens have eliminated many species

Some species that have been introduced to regions outside their original range have become *invasive*—that is, they have spread widely and become unduly abundant, at a cost to the native species of the region. Invasive species are a major component of human-caused environmental change. Deliberately or accidentally, people move many species of organisms from one continent to another. Hundreds of species of plants have been introduced to new areas as ornamentals, as we saw at the beginning of this chap-

In addition, the fraction of a patch that is influenced by effects originating outside the habitat—**edge effects**—increases rapidly as patch size decreases (Figure 57.3). Close to the edges of forest patches, for example, winds are stronger, temperatures are higher, humidity is lower, and light levels are

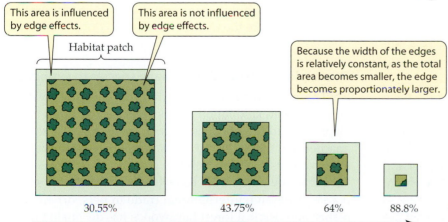

30.55%   43.75%   64%   88.8%

Percentage of patch influenced by edge effects

57.3 Edge Effects The smaller a patch of habitat, the greater the proportion of that patch that is influenced by conditions in the surrounding environment.

1 Isolated patches lose species much more quickly...

2 ...than patches connected to the main forest.

3 Even larger patches are strongly influenced by edge effects.

57.4 Brazilian Forest Fragments Studied for Species Loss
Isolated patches lost species much more quickly than patches connected to the main forest. Even the larger patches, such as the one in the foreground, were too small to maintain populations of some species.

ter. Weed seeds have been carried around the world accidentally in sacks of crop seeds. Europeans deliberately introduced rabbits and foxes to Australia for sport hunting. Nearly half of the small to medium-sized marsupials and rodents of Australia have been exterminated during the last 100 years by a combination of competition with introduced rabbits and predation by introduced domestic cats and foxes.

Some pathogens have proliferated quickly following their introduction to new continents. Exotic disease-causing organisms have decimated populations of several eastern North American forest trees. The chestnut blight, caused by a European fungus, virtually eliminated the American chestnut, formerly an abundant tree in Appalachian Mountain forests. Nearly all American elms over large areas of the East and Midwest have been killed by Dutch elm disease, caused by the fungus *Ceratocystis ulmi*, which reached North America in 1930. Ecologists suspect that intercontinental movement of disease organisms caused extinctions in the past, but evidence of such disease outbreaks is not usually preserved in the fossil record.

The best way to reduce the damage caused by invasive species is to prevent their establishment in the first place. For example, the shipping industry often spreads invasive species (bacteria, dinoflagellates, invertebrates, and fish) in ballast water, which is pumped into a ship at one port and discharged at another. (That is how zebra mussels were introduced into North America from Europe, as we saw in Chapter 54.) San Francisco Bay is now home to at least 234 exotic species, most of which arrived in ballast water, and

some of them are displacing native species. Controlling invasive aquatic species costs millions of dollars per year, but transport of invasive species in ballast water could largely be eliminated by the simple procedure of deoxygenating ballast water before it is pumped out. This practice both kills most organisms in the water and extends the life of ballast tanks.

Strict rules already govern the deliberate introduction of animal species, but the introduction of ornamental plants is poorly regulated. In 1998, Australia and New Zealand began to require a weed risk assessment for the importation of plants not already in the country or not on a "clean list" of permitted species. Regulations do not yet exist in the United States, but in 2002 some members of the horticultural industry crafted a voluntary code of conduct for their profession. The code states that the invasive potential of a plant should be assessed prior to introducing and marketing it. Horticulturists work with conservation biologists to determine which species are currently invasive, or likely to become so, and to identify suitable alternative species. Stocks of invasive species will be phased out, and gardeners will be encouraged to use noninvasive plants.

But how can we assess the potential of a species to become invasive? One way is to compare the traits of species that have become invasive when introduced to a new area with those of other species that have not. Such comparisons show that a plant species is more likely to become invasive if it has a short generation time, small seeds, is dispersed by vertebrates, has a large range in its native continent, depends on nonspecific mutualists (root symbionts, pollinators, and seed dispersers), and is not evolutionarily closely related to plants in the area to which it is introduced. The best predictor, however, is whether the species is already known to be invasive elsewhere.

Using the traits that characterize most invasive species, conservation biologists have developed a decision tree to be used to determine whether an exotic species should be introduced into North America (Figure 57.5). Using such a decision tree cannot eliminate the introduction of all potentially invasive species, but if used conscientiously, its application can greatly reduce the risk.

Overexploitation has driven many species to extinction

Until recently, humans caused extinctions primarily by overhunting. Overexploitation of other species continues today. Elephants and rhinoceroses are threatened in Africa because poachers kill them for their tusks and horns, which are used for ornaments and knife handles, and because some men believe that powdered rhinoceros horn enhances their sexual potency. Massive international trade in pets, ornamental plants, and tropical forest hardwoods has decimated many species of orchids, tropical fishes, corals, parrots, and reptiles.

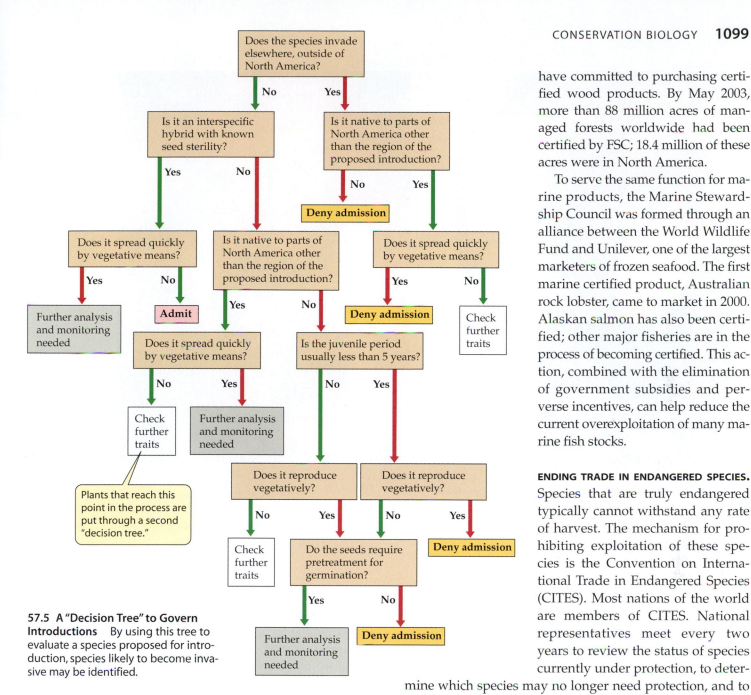

57.5 A "Decision Tree" to Govern Introductions By using this tree to evaluate a species proposed for introduction, species likely to become invasive may be identified.

Several programs have been initiated to help us continue to use species in ways that do not threaten their survival.

CERTIFICATION PROGRAMS. Many purchasers of wood products would like to buy only products that have been harvested in ways that protect biodiversity and ecosystem productivity. To enable them to exercise that choice, the Forest Stewardship Council (FSC) was established in 1993 by a consortium of environmental organizations and members of the forest product industry. FSC establishes criteria that a forest product company must meet for its products to be certified. Certification companies determine whether a forestry operation meets the criteria and ensure that there is a chain of custody that tracks certified products on their way to market. More than 400 companies in 18 countries have committed to purchasing certified wood products. By May 2003, more than 88 million acres of managed forests worldwide had been certified by FSC; 18.4 million of these acres were in North America.

To serve the same function for marine products, the Marine Stewardship Council was formed through an alliance between the World Wildlife Fund and Unilever, one of the largest marketers of frozen seafood. The first marine certified product, Australian rock lobster, came to market in 2000. Alaskan salmon has also been certified; other major fisheries are in the process of becoming certified. This action, combined with the elimination of government subsidies and perverse incentives, can help reduce the current overexploitation of many marine fish stocks.

ENDING TRADE IN ENDANGERED SPECIES. Species that are truly endangered typically cannot withstand any rate of harvest. The mechanism for prohibiting exploitation of these species is the Convention on International Trade in Endangered Species (CITES). Most nations of the world are members of CITES. National representatives meet every two years to review the status of species currently under protection, to determine which species may no longer need protection, and to add new species to its lists. CITES rules currently prohibit international trade in items such as whale meat, rhinoceros horn, and many species of parrots and orchids.

Some species depend on particular disturbance patterns

In Chapter 56, we saw that local species richness is sometime greatest at intermediate levels of disturbance. Many species depend on particular patterns of disturbance to persist. Some plant species, for example, germinate only after a fire; others depend on flooding to open sites where they can become established. Humans often try to reduce the frequency and intensity of such disturbances for their own purposes. Conservation biologists work to assess whether reestablishment of historic disturbance patterns may help preserve biodiversity.

57.6 The Frequency and Intensity of Fires Affect Ecosystems
(*a*) As revealed by scars (arrows) in tree growth rings, low-intensity ground fires were frequent in the pine forests of the southwestern United States prior to fire suppression. (*b*) Fire suppression results in the buildup of large quantities of fuel, so that subsequent fires are likely to spread to the canopy and kill most trees.

Many species require periodic fires for successful establishment and survival, but for many years the official policy in the United States, symbolized by Smokey Bear, was to suppress all forest fires. It is now generally regarded as appropriate to use controlled burning as a management tool, particularly in Western North America. But to determine how to do so, we need to know the historical pattern of fires in an area.

Scars in the annual growth rings of trees preserve evidence of past fires that did not kill them. Therefore, tree-ring researchers can determine when fires occurred, how severe they were, and when fire patterns changed. Annual growth rings on ponderosa pines show that low-intensity ground fires were common near Los Alamos, New Mexico, until about 1900 (Figure 57.6*a*). After that date, cattle and sheep grazing in pine forests and fire suppression greatly reduced the frequency of low-intensity fires. Without these fires, dead branches and needles accumulated in the forest. The buildup of these fuels meant that when fires inevitably did occur, they were much more likely to become intense, tree-consuming canopy fires (Figure 57.6*b*). Today, ground fires are deliberately started in many areas to keep fuel loads low and to mimic historic fire patterns, to which many native species are adapted.

Rapid climate change may cause species extinctions

Scientists from many fields believe that Earth's climate is rapidly becoming warmer as a result of human-caused changes in Earth's atmosphere. We will examine the causes of this global warming in Chapter 58. Conservation biologists cannot alter rates of global warming, but their research can help us to predict how the resulting climate changes will affect organisms and find ways of mitigating those effects. Such research activities include analyses of past climatic events and studies of sites currently undergoing rapid climate change.

Atmospheric scientists predict that temperatures in North America will increase 2° to 5°C by the end of the twenty-first century. If the climate warms by only 1°C, the average temperature currently found at any particular location in North America will be found 150 km to the north. If the climate warmed 2°–5°C, species would need to shift their ranges as much as 500 to 800 km in a single century. Some habitats, such as alpine tundra, could be eliminated as forests expand up mountains.

Knowledge of how organisms responded to past climate changes can help us predict the effects of the current warming trend. Biologists are studying how rapidly species ranges

(a)

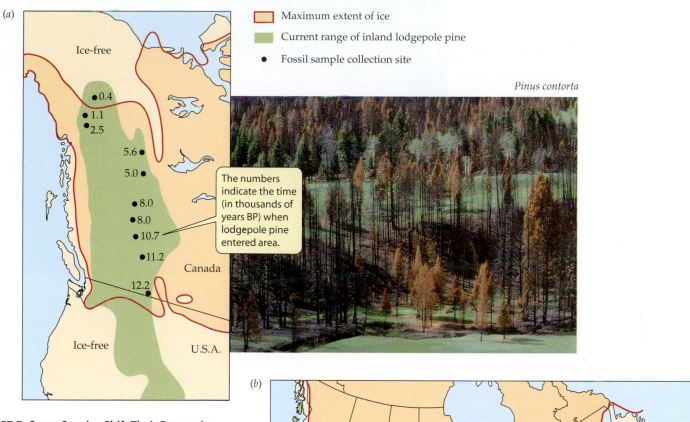

- ☐ Maximum extent of ice
- ▩ Current range of inland lodgepole pine
- ● Fossil sample collection site

Pinus contorta

Ice-free

● 0.4
● 1.1
● 2.5

5.6 ●

5.0 ●

The numbers indicate the time (in thousands of years BP) when lodgepole pine entered area.

● 8.0
● 8.0
● 10.7

● 11.2

Canada

12.2 ●

Ice-free   U.S.A.

57.7 Some Species Shift Their Ranges in Response to Climate Change (a) The range of lodgepole pines in North America expanded north nearly as fast as glaciers retreated. (b) Some native earthworm species disperse so slowly they have hardly moved into glaciated regions.

(b)

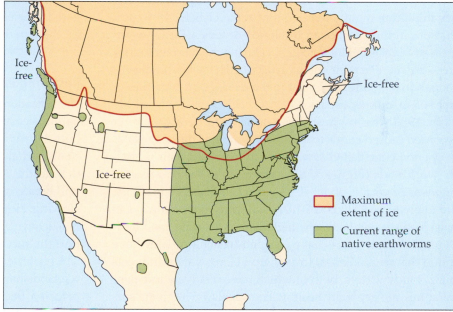

Ice-free

Ice-free

Ice-free

- ☐ Maximum extent of ice
- ▩ Current range of native earthworms

shifted during the last 10,000 years of postglacial warming, which species were and were not able to keep pace with climate change by shifting their ranges, and how past ecological communities differed from those of today. Some organisms with good dispersal abilities, such as birds, can shift their ranges as rapidly as the climate changes, provided that appropriate habitat exists in new areas. However, the ranges of many species with sedentary adults are likely to shift slowly. As the glaciers retreated in North America, the ranges of some coniferous trees expanded northward, so that today they grow as far north as the current climate permits (Figure 57.7a). Some species of earthworms, on the other hand, spread very slowly into the areas that had been covered by ice (Figure 5.7b). Introduced European earthworms survive well in parts of Canada north of the ranges of native earthworms, indicating that slow dispersal, not lack of suitable habitat, is responsible for the range limitations of this group.

If Earth's surface warms as predicted, climatic zones will not simply shift northward. In addition to such shifts, entirely new climates will develop, and some existing climates will disappear. New climates are certain to develop at low elevations in the Tropics because a warming of even 2°C would result in climates near sea level that are warmer than those found anywhere in the humid Tropics today. Adaptation to those climates may prove difficult for many tropical organisms. Although there has been little recent climate warming in tropical regions, nights are now slightly warmer than they

(a)

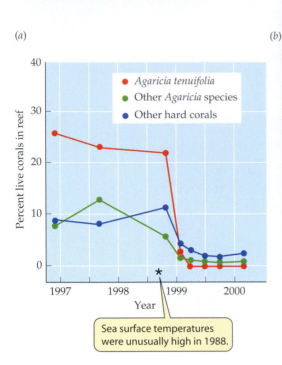

Sea surface temperatures were unusually high in 1988.

(b)

57.8 Global Warming Affects Corals (a) Unusually high sea surface temperatures in 1988 caused massive bleaching and death of corals on a reef in Belize. (b) Large areas of coral reefs in Florida Bay have been bleached.

were only a few decades ago. Since the mid-1980s, the average minimum nightly temperature at the La Selva Biological Station, in the Caribbean lowlands of Costa Rica, has increased from about 20°C to 22°C. During the warmer nights, trees use more of their energy reserves. The result has been a reduction of about 20 percent in the average growth rates of trees of six different species.

In 1988, the highest sea surface temperatures ever recorded caused corals to lose their symbiotic dinoflagellates (a phenomenon called *bleaching*) and increased their mortality worldwide (Figure 57.8). If warming of the oceans continues as predicted, about 40 percent of coral reefs worldwide are likely to be killed by 2010. To identify possible ways to help preserve coral reefs, biologists are measuring conditions in places where corals have escaped bleaching. They have found that reefs adjacent to cool, upwelling waters and reefs with cloudy waters, both of which have relatively low temperatures, are generally healthy. These reefs are receiving special protection because corals are likely to continue to survive well there. Corals from those reefs could be sources of colonists for reestablishing reefs where the corals have died if cooler ocean temperatures return in the future.

Habitat Restoration and Species Recovery

If the cause of a species' endangerment is the loss or modification of its habitat , conservation biologists can attempt to find ways of restoring that habitat. A field called **restoration ecology** has developed to study methods of restoring natural habitats. Such methods are needed because many ecosystems will not recover, or will do so only very slowly, without assistance. Biologists can also attempt to maintain endangered species in captivity until suitable habitat is available for them in the wild.

Restoring ecosystem processes is difficult

Conservation biologists have only a limited ability to restore natural ecosystems. In the United States, the false belief that humans can create functioning ecosystems has resulted in policies that make it easy to get permits for developments that destroy habitats. Developers need only state that they will create habitats to substitute for the ones they are destroying. However, even the most experienced wetland ecologists have great difficulty creating new wetlands that support the species that live in those being destroyed.

In southern California, where 90 percent of the coastal wetlands have been destroyed, wetland restoration is a high priority. Because species have been lost from degraded coastal wetlands, restoration requires species introductions, but which species should be introduced? In early attempts at restoration, only one or two common, easily grown wetland species were planted. Many wetland-associated species failed to recolonize these "rehabilitated" wetlands. To understand why, biologists established a large field experiment at the Tijuana Estuary to examine the effects of plant species richness on several factors that might affect the success of wetland restoration. They found that experimental plots planted with species-rich mixtures developed a complex vegetation structure, which is important to insects and birds. The species-rich plots also accumulated nitrogen faster than species-poor experimental communities (Figure 57.9).

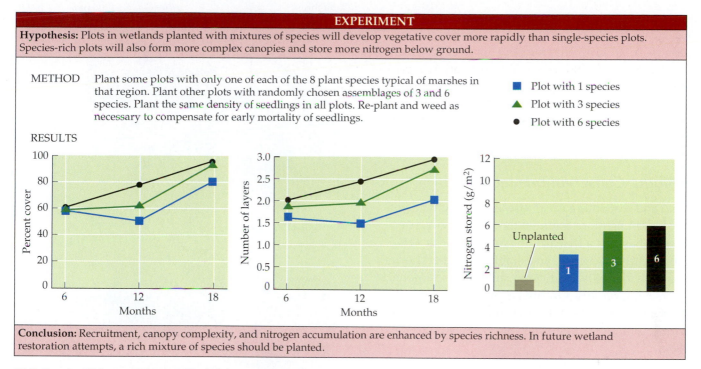

EXPERIMENT

Hypothesis: Plots in wetlands planted with mixtures of species will develop vegetative cover more rapidly than single-species plots. Species-rich plots will also form more complex canopies and store more nitrogen below ground.

METHOD   Plant some plots with only one of each of the 8 plant species typical of marshes in that region. Plant other plots with randomly chosen assemblages of 3 and 6 species. Plant the same density of seedlings in all plots. Re-plant and weed as necessary to compensate for early mortality of seedlings.

■ Plot with 1 species
▲ Plot with 3 species
● Plot with 6 species

RESULTS

Conclusion: Recruitment, canopy complexity, and nitrogen accumulation are enhanced by species richness. In future wetland restoration attempts, a rich mixture of species should be planted.

57.9 Species Richness Enhances Wetland Restoration   Both vegetation complexity and nitrogen accumulation are greater in species-rich than in species-poor experimental plots.

Captive propagation can prevent some species from becoming extinct

Sometimes an endangered species can be maintained in captivity while the external threats to its existence are reduced or removed. However, captive propagation is only a temporary measure that buys conservation biologists time to deal with those threats. Existing zoos, aquariums, and botanical gardens do not have enough space to maintain adequate populations of more than a small fraction of Earth's rare and endangered species. Nonetheless, captive propagation can play an important role by maintaining species during critical periods and by providing a source of individuals for reintroduction into the wild. Captive propagation projects in zoos also have raised public awareness of threatened and endangered species.

Captive propagation is helping to save the California condor, North America's largest bird (Figure 57.10). Two hundred years ago, condors ranged from southern British Columbia to northern Mexico, but by 1978, the wild population was plunging toward extinction—only 25 to 30 birds remained in southern California. Many birds were poisoned by ingesting carcasses containing lead shot.

To save the condor from certain extinction, biologists initiated a captive propagation program in 1983. The first chick conceived in captivity hatched in 1988. By 1993, nine captive pairs were producing chicks, and the captive population had increased to more than 60 birds. The captive population was large enough that six captive-bred birds could be released in the mountains north of Los Angeles in 1992. These birds are provided with lead-free food in remote areas, and they are using the same roosting sites, bathing pools, and mountain

Gymnogyps californianus

57.10 Soaring High Once More   Captive propagation has enabled California condor populations to be reestablished. Captive-reared birds have successfully survived after being released into the wild in California and Arizona.

ridges as did their predecessors. Captive-reared birds also were released late in 1996 in northern Arizona. It is still too early to pronounce the program a success, but as of February 2003, there were 81 wild condors in California and Arizona. Lead poisoning is still a problem, but an effort to encourage hunters to use non-lead ammunition is under way. Without captive propagation, the California condor would probably be extinct today.

Healing Biotas: Conservation Medicine

On both land and sea, outbreaks of diseases among wild organisms are becoming more common threats to biodiversity. The Caribbean Basin is a disease hot spot. *Diadema antillarum*, a dominant sea urchin, and staghorn and elkhorn corals have been virtually eradicated, and disease among corals is increasing rapidly. Outbreaks of several diseases have affected large areas of corals in the Indo-Pacific region. Mortality rates of marine mammals are increasing in the North Atlantic.

The impressive endemic bird fauna of the Hawaiian Islands has been decimated by habitat destruction, overhunting, and introduced predators and diseases (Figure 57.11). For example, wild pigs introduced by the native Polynesians damage the ground cover and soils of Hawaiian forests. A side effect of this habitat destruction is that indentations left by the pigs' foraging fill with water and are breeding grounds for mosquitoes that carry avian malaria. Below 1,000 meters elevation, nearly all endemic Hawaiian bird species have been eliminated by this disease, which was introduced to the islands with exotic birds. The native birds, never having been exposed to malaria, were highly susceptible. Species that inhabit altitudes above the current range of mosquitoes have fared better, but the insects' range may be expanding upward as the climate warms.

Another disease of birds, the mosquito-borne West Nile virus, has exploded across the United States, where it has killed more than 250,000 birds. The virus primarily infects birds, but can be transmitted to humans. First detected in New York in autumn 1999, within 4 years the virus was found in 43 of the contiguous 48 states and 6 of Canada's 10 provinces. By November 2003, there had been a reported 11,516 human cases in the U.S., with 439 deaths. How West Nile virus spread so rapidly is not understood. To find out, biologists are studying where its mosquito vectors feed, how long they survive, and where they hibernate.

A new field of **conservation medicine** is developing to help identify the causes of such increases in wildlife diseases and to devise effective solutions. Molecular techniques are being used to identify species, strains, and life cycle stages of microbial pathogens. Life histories of disease vectors are be-

Myadestes oahuensis
(Oahu thrush)

Loxops rufa
('Akepa)

Psittirostra psittacea
(O'u)

Hemignathus lucidus
(Nukupu'u)

Hemignathus obscurus
('Akialoa)

Moho apicalis
(O'o)

57.11 Extinct Hawaiian Honeycreepers Shown here are just six of the many Hawaiian bird species that have disappeared over the past 150 years. The O'o was among the birds native Polynesians hunted for their feathers, hundreds of thousands of which were used in ceremonial capes for the chiefs. Since 1900, many honeycreeper species have become extinct largely due to avian malaria, an introduced disease to which most endemic birds have no resistance.

ing studied to discover the vulnerable stages where interventions are most likely to prevent transmission of the pathogen and limit its effects.

Setting Limits: The Legacy of Samuel Plimsoll

During the nineteenth century, many British merchant ships sailed Earth's oceans. At that time, there were no undersea telegraph cables or shipboard radios. Once a ship left a harbor, it was out of contact with the rest of the world; in the case of a shipwreck, rescue was impossible. Owners could maximize their profits by overloading their ships, even though this caused some of them to be unseaworthy and sink. Samuel Plimsoll, a member of England's Parliament, became concerned about the rate of loss of British ocean-going vessels and sailors. He convinced Parliament to require that a "load line" be painted on the hull of every large vessel. The position of the line was calculated using factors such as the structural strength of the vessel and the shape of its hull. If the load line was under water, the ship was not permitted to leave the harbor. The "Plimsoll line," as it has come to be known, dramatically reduced the rate of loss of British ships and sailors at sea.

The increasing loss of Earth's species suggests that the load of human activities has pushed the hull of Noah's Ark below the Plimsoll line. But where and how should society draw that line? The decision should be based on scientific information, but just as in Samuel Plimsoll's time, science cannot determine an "acceptable rate of loss." Moreover, we must be concerned not only with species extinctions and ecosystem functioning, but with the overall functioning of the biosphere as well. To help you think more about how society should decide where to draw its Plimsoll line, we turn in the next and final chapter of this book to the functioning of the entire Earth system and how human activities are changing its processes at a global scale.

Chapter Summary

▶ Humans have caused extinctions of species for thousands of years, but the rate of human-caused extinctions is rising rapidly today.

Why Care about Species Extinctions?

▶ Species provide the food, fiber, medicines, and aesthetic opportunities upon which the quality of human life depends.
▶ The extinction of species as a result of human activities raises serious ethical issues.
▶ Extinctions deprive us of opportunities to understand ecological relationships among organisms.
▶ Ecosystems provide valuable services that can be replaced only by expensive and continuing human effort. **Review Figure 57.1**

Estimating Current Rates of Extinction

▶ Estimates of current rates of extinction are based primarily on species–area relationships and population models.

Preserving Biodiversity

▶ Habitat destruction is the most important cause of species extinction today. **Review Figure 57.2**
▶ A greater proportion of small than large habitat patches is affected by external influences. **Review Figures 57.3, 57.4**
▶ Invasive species are major causes of extinction. Biologists use information on species that have become invasive to identify species likely to become invasive if introduced. **Review Figure 57.5**
▶ Certification programs enable consumers to purchase materials produced in ways that do not harm biodiversity.
▶ Overexploitation, which historically resulted in most human-caused extinctions, is still an important cause of extinctions today.
▶ Information on how species are affected by disturbances helps conservation biologists decide where to reestablish historic disturbance patterns.

▶ Species have responded at different rates to past climate changes. **Review Figure 57.7**

Habitat Restoration and Species Recovery

▶ Restoration of habitats is often necessary to preserve species. Restoration of some ecosystem types, especially wetlands, is difficult. **Review Figure 57.9**
▶ Captive propagation plays a useful but limited role in conservation.

Healing Biotas: Conservation Medicine

▶ Disease outbreaks among wild species are increasing. Some of these diseases can be transmitted to humans. The new field of conservation medicine is helping to identify the causes of increases in diseases and to devise effective solutions.

Setting Limits: The Legacy of Samuel Plimsoll

▶ Like an overloaded merchant ship, the "Noah's Ark" of Earth's biodiversity may be in danger of sinking from an overload of stresses and extinctions attributable to human activities.

See Web/CD Activity 57.1 for a concept review of this chapter.

Self-Quiz

1. Which of the following is *not* currently a major cause of species extinctions?
   a. Habitat destruction
   b. Rising sea levels
   c. Overexploitation
   d. Introduction of predators
   e. Introduction of diseases

2. The most important cause of endangerment of species in the United States currently is
   a. pollution.
   b. exotic species.
   c. overexploitation.
   d. habitat loss.
   e. loss of mutualists.

3. People care about species extinctions because
   a. more than half of the medical prescriptions written in the United States contain a natural plant or animal product.
   b. people derive aesthetic pleasure from interacting with other organisms.
   c. causing species extinctions raises serious ethical issues.
   d. biodiversity helps maintain ecosystem services.
   e. All of the above

4. As a habitat patch gets smaller, it
   a. cannot support populations of species that require large areas.
   b. supports only small populations of many species.
   c. is influenced to an increasing degree by edge effects.
   d. is invaded by species from surrounding habitats.
   e. All of the above

5. A plant species is most likely to become invasive when introduced to a new area if it
   a. grows tall.
   b. has become invasive in other places where it has been introduced.
   c. is closely related to species living in the area into which it is introduced.
   d. has specialized dispersers of its seeds.
   e. has a long life span.

6. Conservation biologists are concerned about global warming because
   a. the rate of change in climate is projected to be faster than the rate at which many species can shift their ranges.
   b. it is already too hot in the Tropics.
   c. climates have been so stable for thousands of years that many species lack the ability to tolerate variable temperatures.
   d. climate change will be especially harmful to rare species.
   e. None of the above

7. Scientists can determine the historical frequency of fires in an area by
   a. examining charcoal in sites of ancient villages.
   b. measuring carbon in soils.
   c. radioactively dating fallen tree trunks.
   d. examining fire scars in growth rings of living trees.
   e. determining the age structure of forests.

8. Captive propagation is a useful conservation tool, provided that
   a. there is space in zoos, aquariums, and botanical gardens for breeding a few individuals.
   b. the genetic pedigree of all individuals is known.
   c. the threats that endangered the species are being alleviated so that captive-reared individuals can later be released back into the wild.
   d. there are sufficient caretakers.
   e. Captive propagation should not be used because it directs attention away from the need to protect the species in their natural habitats.

9. Restoration ecology is an important field because
   a. many areas have been highly degraded.
   b. many areas are vulnerable to global climate change.
   c. many species suffer from demographic stochasticity.
   d. many species are genetically impoverished.
   e. fire is a threat to many areas.

10. The new discipline of conservation medicine has developed because
   a. the frequency of diseases has increased among marine organisms.
   b. the frequency of diseases has increased among terrestrial organisms.
   c. the frequency of diseases has increased among both marine and terrestrial organisms.
   d. scientists can better control diseases today than they previously could.
   e. diseases can be readily diagnosed today.

For Discussion

1. Most species driven to extinction by humans in the past were large vertebrates. Do you expect this pattern to persist into the future? If not, why not?

2. Conservation biologists have debated extensively which is better: many small nature reserves or a few large ones. What ecological processes should be evaluated in making judgments about the size and location of reserves? To what extent should we be concerned with preserving the largest number of species rather than those species judged to be of unusual importance for scientific, aesthetic, or commercial reasons?

3. During World War I, French doctors adopted a "triage" system for dealing with wounded soldiers. The wounded were divided into three categories: those almost certain to die no matter what was done to help them, those likely to recover even if not assisted, and those whose probability of survival was greatly increased if they were given immediate medical attention. Limited medical resources were directed primarily at the third category. What are some implications of adopting a similar attitude toward species preservation?

4. Utilitarian arguments dominate discussions about the importance of preserving the biological richness of the planet. In your opinion, what role should ethical and moral arguments play?

5. The desert bighorn sheep of the southwestern United States is endangered. Its major predator, the puma, is also threatened in the region. Under what conditions, if any, would it be appropriate to suppress the population of one rare species to assist another rare species?

58 Earth System Science

An atom of phosphorus "X" had marked time in the limestone ledge since the Paleozoic seas covered the land. Time, to an atom locked in a rock, does not pass. The break came when a bur-oak root nosed down a crack and began prying and sucking. In the flash of a century the rock decayed, and X was pulled out and up into the world of living things. He helped build a flower, which became an acorn, which fattened a deer, which fed an Indian all in a single year.

—Aldo Leopold, A Sand County Almanac

 In this account, Leopold, a pioneer wildlife manager and promoter of environmental stewardship, vividly portrays the flow of atoms between the physical and biological environment and among different organisms. A few of Earth's atoms have changed by radioactive decay, a few have escaped to space, and meteors and meteorites have delivered a few new atoms to Earth. Nevertheless, virtually all of the atoms in our own bodies, and in all other living organisms, have been present on Earth since its formation about 4.5 billion years ago, cycling among its various components.

We began Part One of this book with an introduction to the non-living atoms and molecules that are the building blocks of Earth and of the life found here. It is perhaps fitting that the final chapter should return to these atoms, including them now as part of a larger story. For scientists trained to read them, the atmosphere, rocks, soil, and living organisms harbor clues that tell us about their histories, what was happening on Earth when they were formed, and how they were subsequently transformed. Knowledge of Earth's history helps us understand the changes taking place on Earth today.

Earth system science has emerged as a new field of inquiry that focuses on Earth as a whole. A *system* is a group of entities that interact to yield some product. For example, an individual animal is made up of systems of organs that work together to perform some function, such as digestion. Earth's system is composed of cycles of materials, inputs of solar energy, and interactions between living organisms and the physical environment. Interactions among these components determine how Earth as a planet functions.

"Time, to an Atom Locked in a Rock, Does Not Pass." The "pillar peaks" of Zhangjiajie National Forest Park in China's Hunan province stand like sentinels in the mist. The atoms in these rocks move through biogeochemical cycles as inorganic atoms from the rock-based soil are taken up by plants.

In this chapter we will describe the major cycles of materials among the compartments of Earth's system. We will show how life has modified Earth's features throughout evolutionary history. We will also show how human alterations of the great biogeochemical cycles continue to modify Earth's system today.

Earth's System Has Four Compartments

Earth is an essentially closed system with respect to atomic matter, but it is an open system with respect to energy. The sun delivers a nearly constant amount of energy to Earth every day, and has done so for billions of years. Energy from the sun, combined with the energy from radioactive decay that melts the magma in Earth's interior, drives the processes that move materials around the planet. Many of these processes are cyclic (Figure 58.1). Almost all the rocks that compose the continents have been processed at least once through a chemical and physical cycle involving weathering, formation of sediments, and movement into Earth's interior through continental drift. Deep within Earth, these rocks are subjected to great heat and pressure to form new rocks. The water in the oceans has been evaporated, condensed, precipitated, and returned to the oceans via rivers and groundwater flow many thousands of times. The carbon (C), nitrogen (N), oxygen (O), and sulfur (S) in atmospheric gases have been cycled repeatedly through living organisms.

We take for granted many features of Earth, but Earth is actually a very unusual planet. Its unusual properties include the presence of life, an ocean, a moderate surface temperature, continental drift, and a large moon. The moon has had a profound effect on Earth's history. First, it stabilized the tilt of Earth on its axis. The degree of tilt strongly influences climate; for example, if the tilt were 50° or more, equatorial regions would receive less solar energy over the year than would the polar regions! The moon also plays a major role in producing ocean tides, as well as slowing Earth's rotation.

Planets without life are in perpetual thermodynamic equilibrium, but Earth is far from thermodynamic equilibrium. Oxygen gas (O_2), nitrogen gas (N_2), and water vapor (H_2O) are the main molecules of Earth's atmosphere, but without living organisms, these gases would

produce nitric acid (HNO_3), which would subsequently dissolve in the ocean and remain there. It is because of the activities of living organisms that this does not happen.

As we saw in Chapter 55, energy flows through ecosystems from producers to consumers. At each transformation, much of this energy is used to power metabolism and is dissipated as heat, a form that cannot be used by other organisms to power their metabolism. Chemical elements, on the other hand, are not altered when they are transferred among organisms. The availability of the chemical elements of which living organisms are composed—carbon, nitrogen, phosphorus, calcium, sodium, sulfur, hydrogen, oxygen, and a few others—is strongly influenced by how organisms get them, how long they retain them, and what they do with them while they have them.

To understand the cycling of elements, it is convenient to divide the physical environment into four interacting com-

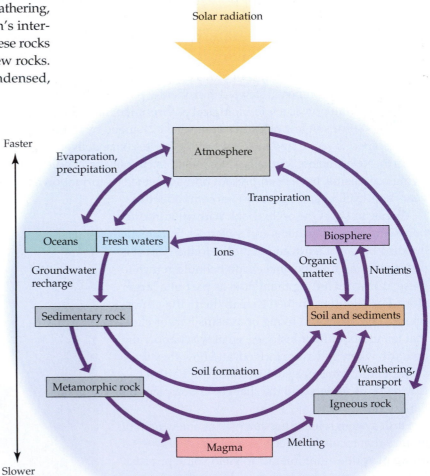

58.1 A Generalized Biogeochemical Cycle The arrows indicate fluxes of material between various compartments. Fluxes between rocks and soils are slower than fluxes between organisms, oceans, fresh waters, and the atmosphere.

partments: oceans, fresh waters, atmosphere, and land. These four compartments, and the types of organisms living in them, are very different. Therefore, the quantities of different elements in each compartment, what happens to them, and the rates at which they enter and leave the compartments differ strikingly. After we have described the four compartments, we will discuss how elements cycle among them.

Oceans receive materials from the land and atmosphere

Oceans receive materials from land primarily in runoff from rivers. Over time scales of hundreds to thousands of years, most materials that cycle through the four compartments end up in the oceans. The oceans are enormous, and they exchange materials with the atmosphere only at their surface, so they respond very slowly to inputs from other compartments.

Except near land on continental shelves, ocean waters mix slowly. Most elements that enter the oceans from other compartments gradually sink to the seafloor. They may remain there for millions of years until the seafloor sediments are elevated above sea level by uplifts of Earth's crust. Therefore, concentrations of mineral nutrients are very low in most ocean waters. Some elements are brought back to the surface near the coasts of continents where offshore winds push surface waters away from shore, causing cold bottom water to rise to the surface (Figure 58.2). Waters in these *zones of upwelling* are rich in mineral nutrients, and most of the world's great fisheries are concentrated there.

Earth's water moves rapidly through lakes and rivers

Only a small fraction of Earth's water resides at any one time in lakes and rivers. However, water moves rapidly through them. Some mineral nutrients enter the freshwater compartment in rainfall, but most are released by the weathering of rocks and are carried to lakes and rivers by surface flow or the by movement of *groundwater* (water in soil and rocks).

After entering rivers, mineral nutrients are usually carried rapidly to lakes or to the oceans. In lakes, they are taken up by organisms and incorporated into their cells. These organisms eventually die and sink to the bottom, taking the nutrients with them. Decomposition of their tissues consumes the O_2 in the bottom water. The surface waters of lakes thus quickly become depleted of nutrients, while deeper waters become depleted of O_2. This process is countered, however, by vertical movements of water called *turnover*. Turnover brings nutrients and dissolved CO_2 to the surface and O_2 to deeper water. Wind is an important agent of turnover in shallow lakes, but in deeper lakes it usually mixes only surface waters. Deep lakes in temperate climates have an annual turnover cycle that is driven by temperature (Figure 58.3).

Lakes in temperate regions turn over because water is most dense at 4°C; above and below that temperature, it expands. The spring sun warms the surface layer of a lake, with the depth of the warm layer gradually increasing as spring and summer progress. However, there is still a well-defined thermocline where the temperature drops abruptly to about 4°C. Only if the lake is shallow enough to warm right to the bottom does the temperature of the deepest water rise above 4°C.

In autumn, as the surface of the lake cools, the cooler surface water—which is denser than the warmer water below it—sinks and is replaced by warmer water from below. This process continues until the entire lake has reached 4°C. At this point, the density of the water is uniform throughout the lake, and even modest winds readily mix the entire water column. As colder weather then cools the surface water below 4°C, that water becomes less dense than the 4°C water below it. Therefore, it floats at the top. Another turnover occurs in spring, when the surface layers above the thermocline warm to 4°C and the water column, again being of uniform density throughout, is easily mixed by wind.

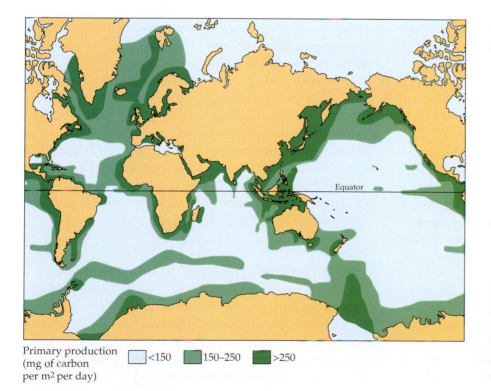

Primary production (mg of carbon per m² per day) □ <150 ▨ 150–250 ▨ >250

58.2 Primary Production Is High in Zones of Upwelling Primary production in the oceans is highest near continents where surface waters, driven by prevailing winds, move offshore and are replaced by cool, nutrient-rich water upwelling from below.

58.3 Annual Temperature and Oxygen Cycles in a Temperate Lake These vertical temperature and oxygen profiles are typical of temperate-zone lakes that freeze in the winter.

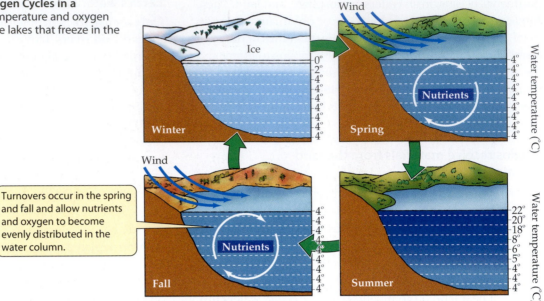

Turnovers occur in the spring and fall and allow nutrients and oxygen to become evenly distributed in the water column.

Arctic lakes turn over only once each year. Deep tropical and subtropical lakes may be permanently stratified because they never become cool enough to have uniformly dense water. Their bottom waters lack oxygen because decomposition quickly depletes any oxygen that reaches them. However, many tropical lakes are overturned at least periodically by strong winds so that their deeper waters are occasionally oxygenated. When this does not happen, lake productivity can be drastically affected. Fishermen of Lake Tanganyika, an extremely large and deep tropical lake in East Africa, have seen their catches decline by as much as 50 percent over the past 30 years. This decline has been attributed at least in part to warmer water temperatures and a decade of slack wind conditions.

The atmosphere regulates temperatures close to Earth's surface

The third major compartment of Earth's system, the **atmosphere**, is a thin layer of gases surrounding Earth. About 80 percent of the mass of the atmosphere lies in its lowest layer, the *troposphere*, which extends upward from Earth's surface about 17 km in the Tropics and Subtropics, but only about 10 km at higher latitudes. Most global air circulation takes place within the troposphere, and virtually all atmospheric water vapor is found there.

The *stratosphere*, which extends from the top of the troposphere up to about 50 kilometers above Earth's surface, contains very little water vapor. Most materials enter the stratosphere from the troposphere near the equator, where air rises to high altitudes, as we saw in Chapter 56. These materials tend to remain in the stratosphere for a relatively long time.

Ozone (O_3) in the stratosphere absorbs most of the biologically damaging short-wavelength ultraviolet radiation that enters the atmosphere. (Ozone released into the troposphere by human activities, however, may contribute to global warming, as we will see below.)

The atmosphere is 78.08 percent N_2, 20.95 percent O_2, 0.93 percent argon, and 0.03 percent carbon dioxide (CO_2). It also contains traces of hydrogen gas, neon, helium, krypton, xenon, ozone, and methane. The atmosphere contains Earth's biggest pool of nitrogen as well as large supplies of O_2. Although CO_2 constitutes a very small fraction of the atmosphere, it is the source of the carbon used by terrestrial photosynthetic organisms and of the dissolved carbonate in water used by marine producers.

The atmosphere plays a decisive role in regulating temperatures at and close to Earth's surface. Without an atmosphere, the average surface temperature of Earth would be about −18°C, rather than its actual +17°C. Earth has this warm temperature because the atmosphere is relatively transparent to visible light. However, it traps a large part of the heat that Earth radiates back to space. Water vapor, CO_2, O_3, and certain other gases, known as **greenhouse gases**, are especially important trappers of heat.

Land covers about one-fourth of Earth's surface

About one-fourth of Earth's surface, most of it in the Northern Hemisphere, is currently above sea level. Even though the global supply of chemical elements is constant, regional and local deficiencies of particular elements strongly affect ecosystem processes on land, where elements move slowly and usually over only short distances.

The land compartment is connected to the atmospheric compartment by organisms that take chemical elements from and release them to the air. Chemical elements in soils are carried in solution into groundwater and eventually into the oceans, where they are unavailable to organisms until an episode of uplifting raises marine sediments and a new cycle of erosion and weathering begins. The type of soil that exists in an area depends on the underlying rock, as well as on climate, topography, the organisms living there, and the length of time that soil-forming processes have been acting. Very old soils are much less fertile than most young soils because nutrients leach out of them over time.

Although land covers such a small proportion of Earth's surface, human life depends intimately on soil fertility and the productivity of terrestrial ecosystems. And the land is the Earth system compartment that has been most strongly affected by human activities.

Biogeochemical Cycles, Water, and Fire

The chemical elements organisms need in large quantities—carbon, hydrogen, oxygen, nitrogen, phosphorus, and sulfur—cycle through organisms to the physical environment and back again. The pattern of movement of a chemical element through organisms and the four compartments of the physical environment is called its **biogeochemical cycle**. Each chemical element used by organisms has a distinctive biogeochemical cycle whose properties depend on the physical and chemical nature of the element and how organisms use it. All chemical elements cycle quickly through organisms because no individual, even of the longest-lived species, lives very long in geological terms.

Before we describe the cycles of these elements, however, we must discuss water and fire. As we have just seen, the movement of water transfers many elements between the atmosphere, land, fresh waters, and oceans. Fire is a powerful agent that speeds the cycling of chemical elements.

Water transfers materials from one compartment to another

The cycling of water through the oceans, atmosphere, fresh waters, and land is known as the **hydrological cycle**. The hydrological cycle operates because more water is evaporated from the surface of the oceans than is returned to them as precipitation (rain or snow). The excess evaporated water is carried by winds over the land, where it falls as precipitation.

Water also evaporates from soils, from lakes and rivers, and from the leaves of plants (transpiration), but the total amount evaporated from those surfaces is less than the amount that falls on them as precipitation. The excess terrestrial precipitation eventually returns to the oceans via rivers, coastal runoff, and groundwater flows (Figure 58.4).

Earth's 16 largest rivers account for more than one-third of total water discharge; more than half of the discharge comes from the three largest rivers (Amazon, Congo, and Yangtze). Despite their relatively small volume, rivers play a disproportionate role in the hydrological cycle because the average residence time of water in lakes and rivers is only 4.3 years, compared with 2,640 years in the oceans. The average turnover time of water in living things is much shorter—about 5.6 days.

By building dams, canals, and reservoirs and by diverting huge quantities of water to irrigated fields, humans have had major effects on the temporal and spatial distribution of fresh water on Earth. The most important consequence of human activities is that more water now evaporates from land and less flows to the oceans than before the Industrial Revolution. In addition, freshwater flow patterns are being seriously altered. For example, dams on the Columbia River in Washington State are

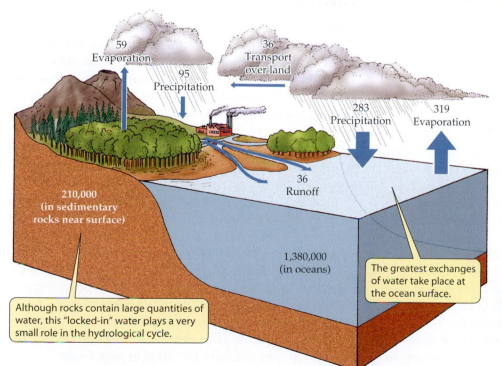

59 Evaporation

36 Transport over land

95 Precipitation

283 Precipitation

319 Evaporation

36 Runoff

210,000 (in sedimentary rocks near surface)

1,380,000 (in oceans)

The greatest exchanges of water take place at the ocean surface.

Although rocks contain large quantities of water, this "locked-in" water plays a very small role in the hydrological cycle.

58.4 The Global Hydrological Cycle The numbers show the relative amounts of water (expressed as units of 10^{18} g) held in or exchanged annually by ecosystem compartments. The widths of the arrows are proportional to the sizes of the fluxes.

(a)

(b)

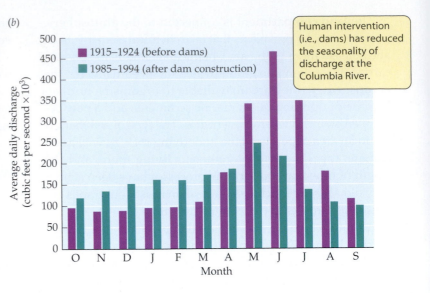

Human intervention (i.e., dams) has reduced the seasonality of discharge at the Columbia River.

58.5 Columbia River Flows Have Been Massively Altered
(a) Dozens of dams built along the Columbia River and its tributaries regulate the river's flow and generate hydroelectric power. This is the Bonneville Dam near Portland, Oregon. (b) Before the construction of dams, rates of freshwater flow from the Columbia River to the Pacific Ocean varied greatly with the seasons. Human intervention has substantially reduced this variation.

managed to reduce the variation in water flow rates during the year (Figure 58.5). An unintended result of this practice is reduced autumn and winter surface salinity from the mouth of the river north along the coast to the Aleutian Islands, with negative consequences for the growth and survival of salmon in that part of the ocean.

In some places, groundwater is being seriously depleted because humans are using it more rapidly than it can be replaced, primarily by pumping it for irrigation. In some areas of the High Plains of the central United States, more than half of the groundwater has been removed. On the North China Plain, depletion of shallow aquifers is forcing people to sink wells more than 1,000 meters deep to reach water. Much of the groundwater we are using today was deposited during the last Ice Age, when regional precipitation was much greater than it is today. Flows of water to the oceans have been increased by the pumping of groundwater, thereby contributing to sea level rises during the past century.

The World Resources Institute's *Pilot Analysis of Global Ecosystems*, published in 2000, predicts that if current water consumption patterns continue, by 2025, at least 3.5 billion people (48% of the world population) will live in areas with inadequate water supplies, primarily in Asia. Fortunately, current water consumption patterns need not persist. Per capita consumption in the United States and Europe is dropping because of increasing use of water-efficient home appliances, increasing prices charged for water, and development of regulations that restrict water use. If programs to improve water use efficiency are implemented vigorously, global water use could be even less in 2025, despite continued population growth.

Fire is a major mover of elements

Fire is an important disturber of ecosystems worldwide. Every year, 200–400 hectares of savannas, 5–15 million hectares of boreal forests, and lesser amounts of other biomes burn. Lightning ignites some fires, but most fires are started by humans as a way of managing vegetation. Fires consume the energy of, and release chemical elements from, the vegetation they burn. Some nutrients, such as nitrogen, sulfur, and selenium, are easily vaporized by fire. They are discharged to the atmosphere in smoke or carried to groundwater by rain falling on the burned ground.

Fires also release large amounts of carbon into the atmosphere. The global annual flux of carbon to the atmosphere from savanna and forest fires is estimated to be in the range of 1.7 to 4.1 petagrams (one petagram = 10^{15} grams). Biomass burning is responsible for about 40 percent of Earth's annual production of CO_2. It is also a significant contributor to the production of other greenhouse gases, such as carbon monoxide (CO) (32%), methane (CH_4) (10%), and tropospheric O_3 (38%).

Because humans deliberately start most fires, we can take steps to reduce their frequency and intensity. As we saw in Chapter 57, periodic burning of fire-prone vegetation can prevent the buildup of large fuel loads and reduce the frequency of high-intensity canopy fires, which discharge great quantities of materials to the atmosphere. Improvements in the productivity of tropical agriculture can reduce the rate at which forests are cleared and burned to create more agricultural land. Expanded use of solar cookers can reduce the need to cut trees and burn the wood to cook meals.

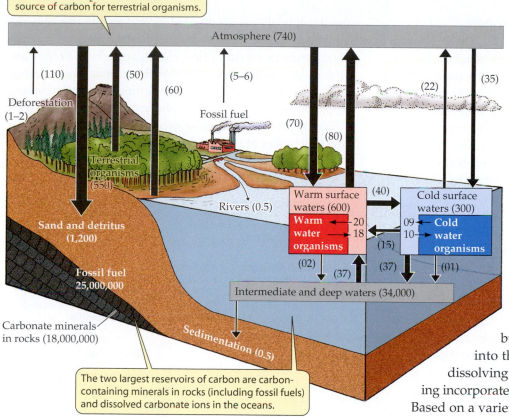

Atmospheric CO_2 is the immediate source of carbon for terrestrial organisms.

Atmosphere (740)

(110)   (50)   (5–6)   (22)   (35)

(60)

Deforestation (1–2)

Fossil fuel (70)

(80)

Terrestrial organisms (550)

Rivers (0.5)

Warm surface waters (600)   (40)   Cold surface waters (300)

Warm water organisms   20   18   (15)   09 10   Cold water organisms

Sand and detritus (1,200)

Fossil fuel 25,000,000

(02)   (37)   (37)   (01)

Intermediate and deep waters (34,000)

Carbonate minerals in rocks (18,000,000)

Sedimentation (0.5)

The two largest reservoirs of carbon are carbon-containing minerals in rocks (including fossil fuels) and dissolved carbonate ions in the oceans.

58.6 The Global Carbon Cycle
The numbers show the quantities of carbon (expressed as units of 10^{15} g) held in or exchanged annually by ecosystem compartments. The widths of the arrows are proportional to the sizes of the fluxes.

Our discussion of water and fire shows that the biogeochemical cycles of different chemical elements are connected to one another. However, it is easier to understand the interactions of these cycles if we discuss each one separately.

The Carbon Cycle

Organisms are triumphs of organic chemistry; to survive, they must have access to carbon atoms. Nearly all the carbon in organisms comes from CO_2 in the atmosphere or dissolved carbonate ions (HCO_3^-) in water. Carbon is incorporated into organic molecules by photosynthesis in the cells of autotrophs. All heterotrophic organisms get their carbon by consuming autotrophs or other heterotrophs, their remains, or their waste products. Biological processes move carbon between the atmospheric and terrestrial compartments, removing it from the atmosphere during photosynthesis and returning it to the atmosphere during respiration (Figure 58.6). Most of Earth's carbon is stored in the oceans; on land, most carbon that is available to organisms is stored in soils.

At times in the remote past, great quantities of carbon were removed from the global carbon cycle when organisms died in large numbers and were buried in sediments lacking O_2. In such anaerobic environments, detritivores do not reduce organic carbon to CO_2. Instead, organic molecules accumulate and eventually are transformed into de-

posits of oil, natural gas, coal, or peat. Humans have discovered and used these deposits, known as **fossil fuels**, at ever-increasing rates during the past 150 years. As a result, CO_2, the final product of the burning of these fuels, is being released into the atmosphere faster today than it is dissolving in surface waters of the oceans or being incorporated into terrestrial biomass.

Based on a variety of calculations, atmospheric scientists believe that 150 years ago, before the Industrial Revolution, the concentration of atmospheric CO_2 was probably about 265 parts per million. Today it is 350 parts per million (Figure 58.7). This difference represents a rate of increase more than 10 times faster than at any other time for millions of years. How will global climate and Earth's ecosystems change in response to this rapid CO_2 enrichment? What are the main factors driving these changes in the carbon cycle, and what processes may tend to stabilize them?

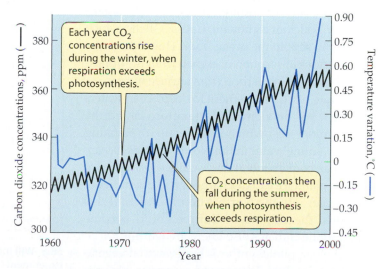

Each year CO_2 concentrations rise during the winter, when respiration exceeds photosynthesis.

CO_2 concentrations then fall during the summer, when photosynthesis exceeds respiration.

Carbon dioxide concentrations, ppm (——)

Temperature variation, °C (——)

Year

58.7 Atmospheric Carbon Dioxide Concentrations Are Increasing
These carbon dioxide concentrations, expressed as parts per million by volume of dry air, were recorded on top of Mauna Loa, Hawaii.

Carbon concentrations in the atmosphere influence Earth's climate

The buildup of atmospheric CO_2 that has resulted from the burning of fossil fuels is warming Earth. The concentration of CO_2 in air trapped in the Antarctic and Greenland ice caps during the last Ice Age—between 15,000 and 30,000 years ago—was as low as 200 parts per million. During a warm interval 5,000 years ago, atmospheric CO_2 may have been slightly higher than it is today. This long-term record shows that Earth was warmer when CO_2 levels were higher and cooler when they were lower.

Complex computer models of Earth's system indicate that a doubling of the atmospheric CO_2 concentration would shift climates toward the poles, would probably cause droughts in the central regions of continents, and would increase precipitation in coastal areas. Global warming could result in the melting of the Greenland and Antarctic ice caps, and would warm the oceans. If so, the oceans would expand, raising sea levels and flooding coastal cities and agricultural lands.

Both physical and biological processes control the carbon cycle

Over decades to centuries, the oceans, which have 50 times the amount of dissolved inorganic carbon as the atmosphere, determine atmospheric CO_2 concentrations. The rate at which CO_2 moves from the atmosphere to the oceans depends, in part, on photosynthesis by plankton in the surface waters. These organisms remove carbon from the water, thereby increasing its absorption of carbon from the atmosphere. In addition, many marine organisms form calcium carbonate ($CaCO_3$) shells, which eventually sink to the ocean floor. This sedimentation increases absorption of carbon from the atmosphere by removing carbon from surface waters.

The rate of CO_2 movement to deep ocean waters also depends on a circulation pattern called the *ocean conveyor belt*, which is driven by the sinking of dense, saline water in the North Atlantic Ocean. The ocean conveyor belt may weaken if melting of the Greenland ice cap discharges great quantities of fresh water into the North Atlantic Ocean. If this happens, the climate of Europe could become colder while climates elsewhere are warming.

Each year, photosynthesis by terrestrial vegetation, principally in forests and savannas, absorbs about 60 billion metric tons of carbon. About the same amount of carbon is released by respiration,

about half by the plants themselves and half by microbes decomposing organic matter produced by the plants. Currently photosynthetic consumption of CO_2 appears to exceed respiratory production of CO_2 by 2 billion metric tons of carbon per year. Thus, Earth's forests are storing carbon that would otherwise be increasing atmospheric CO_2 concentrations.

Ecologists are conducting experiments to determine the effects of higher atmospheric concentrations of CO_2 on rates

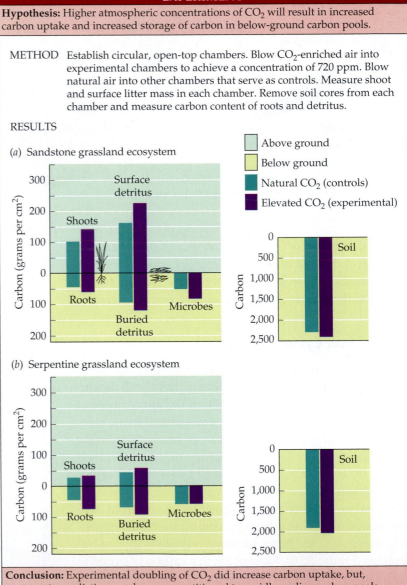

EXPERIMENT

Hypothesis: Higher atmospheric concentrations of CO_2 will result in increased carbon uptake and increased storage of carbon in below-ground carbon pools.

METHOD Establish circular, open-top chambers. Blow CO_2-enriched air into experimental chambers to achieve a concentration of 720 ppm. Blow natural air into other chambers that serve as controls. Measure shoot and surface litter mass in each chamber. Remove soil cores from each chamber and measure carbon content of roots and detritus.

RESULTS

Conclusion: Experimental doubling of CO_2 did increase carbon uptake, but, contrary to predictions, carbon was partitioned to rapidly cycling carbon pools below ground. As a result, carbon storage did not increase. Thus, increased productivity may not lead to increased long-term storage of carbon.

58.8 Will Increased CO_2 Levels Increase Carbon Storage? The results of these experiments, which took place over 3 years on two different grassland ecosystems, suggest that in these ecosystems elevated levels of atmospheric CO_2 result in higher rates of photosynthesis (carbon uptake), but the amount of carbon stored in above-ground plant parts does not increase significantly.

of photosynthesis and carbon storage in ecosystems (Figure 58.8). In the experiment shown, photosynthetic rates increased but long-term carbon storage did not. However, this lack of increased carbon storage in grassland ecosystems does not mean that carbon storage in forests—where carbon is stored in wood—will not increase.

Humans must try to influence the carbon cycle

The adverse consequences of increased atmospheric concentrations of CO_2 are likely to be severe. Therefore, scientists are exploring various methods to reduce the rate of release of CO_2 to the atmosphere and to increase the rate of its removal from the atmosphere and its transfer to long-term storage. No single, simple action can solve the problem, but many steps can be taken that would help to reduce atmospheric concentrations of CO_2.

For example, we can increase fuel efficiency of motor vehicles and airplanes, the efficiency of appliances, and improve mass public transportation. We could also impose carbon taxes that would encourage more efficient use of fossil fuels.

The Nitrogen Cycle

Nitrogen gas (N_2) makes up 78 percent of the atmosphere, but most organisms cannot use nitrogen in its gaseous form. Only a few species of microorganisms can convert atmospheric N_2 into forms that are usable by plants, a process called *nitrogen fixation*. *Denitrification*, the principal process that removes nitrogen from the biosphere and returns it to the atmosphere, is also carried out by microorganisms. These movements of nitrogen among organisms and between organisms and the atmosphere were detailed in Chapter 37 (see Figure 37.8). They account for about 95 percent of all nitrogen fluxes on Earth (Figure 58.9).

Biologically usable nitrogen is often in short supply in ecosystems. That is why nearly all fertilizers contain compounds of nitrogen. Populations of nitrogen-fixing organisms do not increase to such an extent that nitrogen is no

longer limiting because nitrogen tends to be lost rapidly from ecosystems by leaching, vaporization of ammonia, and denitrification. Also, fixing nitrogen requires large amounts of energy. As a result, nitrogen-fixing organisms often lose out in competition with non-fixers when nitrogen becomes more readily available.

As a result of extensive use of fertilizers on agricultural crops and the burning of fossil fuels (which generates nitric oxide), the total nitrogen fixation by humans today is about equal to global natural nitrogen fixation. The human-caused flux is expected to continue to increase during the coming decades. A variety of adverse effects are associated with these large perturbations of the nitrogen cycle. They include the contamination of groundwater by nitrate (NO_3^-) from agricultural runoff, increases in atmospheric nitrous oxide (N_2O) and tropospheric O_3 (both greenhouse gases), and smog production.

When more nitrogen is applied to croplands than is taken up by plants, the excess nitrogen moves downward into groundwater and, eventually, into rivers, lakes, and oceans. The addition of nutrients to these bodies of water, known as **eutrophication**, can have a number of negative effects on aquatic ecosystems, as we'll see in our discussion of the phosphorus cycle. The "dead zone" in the Gulf of Mexico that has formed around the mouth of the Mississippi River is a result of flows of nitrogen-enriched water from agricultural fields in the Upper Midwest (Figure 58.10). Outbreaks of the toxic dinoflagellate *Pfiesteria* in estuaries on the Atlantic coast of North America are another example of the adverse consequences of nitrogen enrichment of ocean waters.

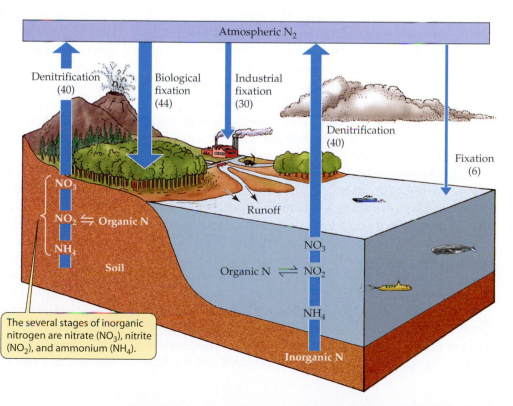

58.9 The Global Nitrogen Cycle The numbers show the quantities of nitrogen (expressed as units of 10^9 kg) in organisms and in various reservoirs, and the amounts that move annually between ecosystem compartments. The widths of the arrows are proportional to the sizes of the fluxes.

The several stages of inorganic nitrogen are nitrate (NO_3), nitrite (NO_2), and ammonium (NH_4).

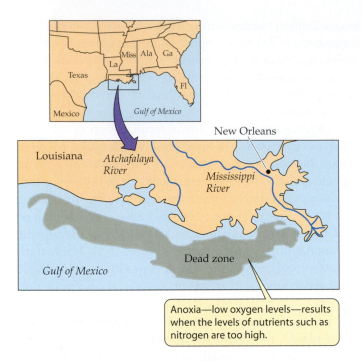

58.10 A "Dead Zone" at the Mouth of the Mississippi River Rising amounts of nitrogen and phosphorus in the runoff from agricultural lands in the midwest United States reach the Gulf of Mexico. Resulting algal growth from this nutrient enrichment depletes the oxygen in the water, creating a deoxygenated "dead zone" in which most aquatic organisms cannot survive.

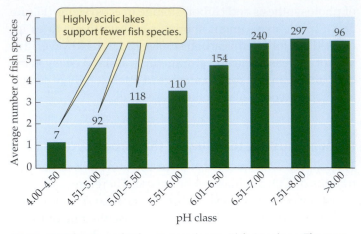

58.11 Acidification of Lakes Exterminates Fish Species The number of fish species found in lakes in the Adirondack region of New York State is inversely correlated with pH. (Recall from Chapter 2 that lower numbers indicate greater acidity, and a pH of 7 is neutral. The numbers above the bars indicate the number of lakes in each pH category.)

The Sulfur Cycle

Emissions of the gases sulfur dioxide (SO_2) and hydrogen sulfide (H_2S) from volcanoes and fumaroles (vents for hot gases) are the only significant natural nonbiological fluxes of sulfur. These emissions release, on average, between 10 and 20 percent of the total natural flux of sulfur to the atmosphere, but they vary greatly in time and space. Large volcanic eruptions spread great quantities of sulfur over broad areas, but they are rare events. Terrestrial and marine organisms also emit compounds of sulfur. Certain marine algae produce large amounts of dimethyl sulfide (CH_3SCH_3), which accounts for about half the biotic component of the sulfur cycle. Sulfur is apparently always abundant enough to meet the needs of living organisms.

Sulfur plays an important role in global climate. Even if air is moist, clouds do not form readily unless there are small particles around which water can condense. Dimethyl sulfide is the major component of such particles. Therefore, increases or decreases in atmospheric sulfur levels can change cloud cover and hence climate.

Humans have altered the sulfur cycle, as well as the nitrogen cycle, by the burning of fossil fuels. An important regional effect of these alterations is **acid precipitation**—rain or snow whose pH is lowered by the presence of sulfuric acid (H_2SO_4) and nitric acid (HNO_3), derived in large part from the burning of fossil fuels. These acids enter the atmosphere and may travel hundreds of kilometers before they settle to Earth in precipitation or as dry particles.

Acid precipitation now characterizes all major industrial countries and is particularly widespread in eastern North America and Europe. The normal pH of precipitation in New England is about 5.6, but precipitation there now averages about pH 4.4, and there are occasional storms with a precipitation pH as low as 3.0. Precipitation with a pH of about 3.5 or lower causes direct damage to the leaves of plants and reduces photosynthetic rates. Acidification of lakes in the Adirondack region of New York has reduced fish species richness by causing the extinction of acid-sensitive species (Figure 58.11). Fortunately, as a result of the establishment of a flexible regulatory system under the 1990 Clean Air Act Amendments, precipitation in much of the eastern United States is less acid today than it was 18 years ago, primarily because of reductions in sulfur emissions (Figure 58.12).

Ecologists in Canada studied the effects of acid precipitation on small lakes by adding enough H2SO4 to two lakes to reduce their pH from about 6.6 to 5.2. In both lakes, nitrifying bacteria failed to adapt to these moderately acidic conditions. As a result, the nitrogen cycle was blocked, and ammonium accumulated in the water. When the ecologists stopped adding acid to one of the lakes, its pH increased to 5.4, and nitrification resumed. After about 1 year, the pH of the lake returned to its original value. These experiments show that lakes are very sensitive to acidification, but pH can return rapidly to normal values because water in lakes is exchanged at a rapid rate.

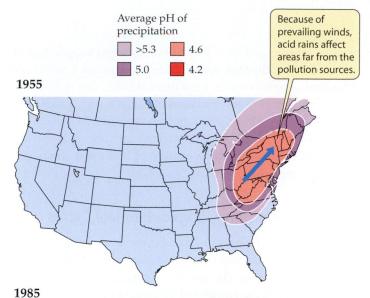

Average pH of precipitation

▨	>5.3	▨	4.6
▨	5.0	▧	4.2

1955

Because of prevailing winds, acid rains affect areas far from the pollution sources.

1985

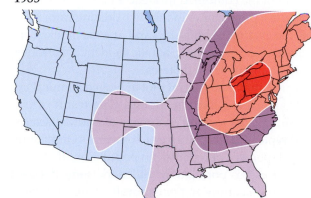

1998

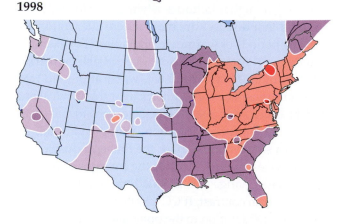

58.12 Acid Precipitation Is Decreasing in the Eastern United States   Thanks to emission controls, precipitation in many parts of the eastern United States is less acid than it was three decades ago.

The Phosphorus Cycle

Phosphorus accounts for only about 0.1 percent of Earth's crust, but it is an essential nutrient for all life forms. It is a key component of DNA and ATP. Unlike the other elements discussed in this section, phosphorus lacks a gaseous phase. Some phosphorus is transported on dust particles, but in general the atmospheric compartment plays a very minor

role in the phosphorus cycle. The global phosphorus cycle takes millions of years to complete because the processes of rock formation on the ocean bottom, subsequent uplifting, and weathering of rock into soil all act slowly (Figure 58.13). However, an atom of phosphorus may cycle rapidly among organisms, as did X in Aldo Leopold's account at the beginning of this chapter.

Human activity has radically accelerated some parts of the phosphorus cycle. About 90 percent of the phosphorus that is mined is used to produce fertilizers and animal feeds. One consequence of our massive use of phosphorus for fertilizer is that between 10.5 and 15.5 million metric tons of phosphorus are accumulating in soils each year, primarily on agricultural lands (Figure 58.14). Increasing concentrations of phosphorus in agricultural soils is certain to lead to increased flows of phosphorus to streams and lakes.

Phosphorus is often a limiting nutrient in soils and lakes, which is why adding phosphorus to farmland and lakes increases their biological productivity. Most of this extra phosphorus enters lakes in the form of phosphates derived from fertilizers and household detergents. The resulting eutrophication allows algae and bacteria to multiply, forming blooms that turn the water green. The decomposition of dead cells produced by this increased biological activity consumes all the O_2 in the lake, and anaerobic organisms come to dominate the bottom sediments. These anaerobic organisms do not break down carbon compounds all the way to CO_2. Their metabolic end products build up; many of these products have unpleasant odors.

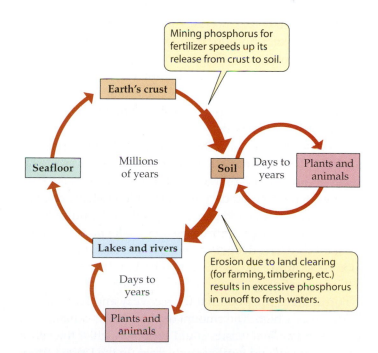

Mining phosphorus for fertilizer speeds up its release from crust to soil.

Earth's crust

Seafloor

Millions of years

Soil

Days to years

Plants and animals

Lakes and rivers

Days to years

Plants and animals

Erosion due to land clearing (for farming, timbering, etc.) results in excessive phosphorus in runoff to fresh waters.

58.13 The Phosphorus Cycle   The width of the arrows is proportional to flux rates. The two great increases are due to human activities.

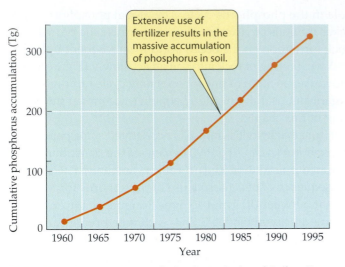

58.14 Phosphorus Is Accumulating in Agricultural Soils Between 10.5 and 15.5 million metric tons of phosphorus are being added to agricultural soils each year. This graph shows the cumulative effects of this excess over a period of 40 years. (A teragram, Tg, equals 10^{12} grams.)

Lake Erie, one of the Great Lakes on the border between the United States and Canada, is a eutrophic lake today. Two hundred years ago, it had only moderate rates of photosynthesis and clear, oxygenated water. More than 15 million people live in the Lake Erie drainage basin. Nearby cities pour more than 250 billion liters of domestic and industrial wastes into the lake annually. The entire basin is intensely farmed and heavily fertilized.

In the early part of the twentieth century, nutrient concentrations in the lake increased greatly, and algae proliferated. At the water filtration plant in Cleveland, Ohio, algae increased from 81 per milliliter in 1929 to 2,423 per milliliter in 1962. Algal blooms and populations of bacteria also increased. The numbers of the colon bacterium *Escherichia coli* increased enough to cause the closing of many of the lake's beaches as health hazards.

Since 1972, the United States and Canada have invested more than 9 billion dollars to improve municipal waste facilities and reduce discharges of phosphorus into Lake Erie. As a result, the amount of phosphate added to Lake Erie has decreased more than 80 percent from the maximum level, and phosphorus concentrations in the lake have declined substantially. The deeper waters of Lake Erie still become poor in O_2 during the summer months, but the rate of O_2 depletion is declining.

Fortunately, the potential for recovery and recycling of phosphorus is high. The amount of phosphorus contained in sewage and animal wastes could supply much of the needs of the detergent and fertilizer industries. In the United Kingdom, for example, if only 50 percent of the phosphorus in 25 percent of the sewage and 15 percent of the animal wastes

were recovered and recycled, recycling could supply half of the country's industrial demand for phosphorus. Careful application of fertilizers on agricultural lands can reduce the rate of phosphorus accumulation in soils without reducing crop yields. However, reducing phosphorus in soils takes many decades after remedial actions are initiated. During that time, increased eutrophication of lakes and streams is certain to happen.

Interactions among Biogeochemical Cycles

Our discussion of biogeochemical cycles shows that they interact strongly, and that they are being substantially altered by human activities. The acidity of precipitation, for example, is a result of the combined influences of SO_2, NO_3^-, and NH_3. Nitrate is a powerful oxidant. By oxidizing iron, NO_3^- influences the cycling of both iron and arsenic in lakes, increasing the mobility of arsenic, a toxic element of considerable importance in the United States.

Human alterations of several global biogeochemical cycles are also warming Earth's climate. One result of this global warming is increasing outbreaks of many diseases. Winter cold typically kills pathogens, sometimes eliminating as much as 99 percent of a pathogen population. If climate warming reduces this population bottleneck, some diseases will become more common. For example, dengue fever is now spreading to higher latitudes where it was formerly absent. Several plant diseases are more severe after mild winters or during periods of warmer temperatures. For example, during a 39-year study in Maine, beech bark cankering caused by a fungus (*Nectria* spp.) was worse after mild winters or dry autumns. These conditions favor the survival and spread of the beech scale insect, which weakens the trees and predisposes them to fungal infection. The spread of this infection poses a serious problem for the timber industry.

The 1990 Assessment Report of the Intergovernmental Panel on Climate Change (IPCC) paid little attention to risks to human health. In contrast, IPCC's 1996 Assessment Report gave detailed consideration to the potential effects of climate change on human health. Our increasingly interconnected global society enables disease organisms to travel rapidly around the world, as in the case of the SARS virus, which was carried by infected people from China to Canada within a few days. The combination of human mobility and climate warming poses serious challenges to human health throughout the world.

Visions of the Future

As we saw in Chapter 54, the size of Earth's human population is a factor in many environmental problems, and it is increasing rapidly (see Figure 54.20). Anyone 40 years old or

58.15 Earth from Space   The view from space reminds us that Earth is a small planet. The future of humanity, as well as that of other species, will depend on how we function as stewards of Earth and its resources.

▶ Primary production in oceans is highest in zones of upwelling adjacent to continents, where nutrient-rich waters rise to the surface. **Review Figure 58.2**

▶ Gases in the atmosphere are important regulators of temperatures on Earth.

▶ Deep lakes in the temperate zone have an annual turnover cycle that is driven by temperature. **Review Figure 58.3**

Biogeochemical Cycles, Water, and Fire

▶ The hydrological cycle is driven by evaporation of water, most of it from ocean surfaces. **Review Figure 58.4. See Web/CD Tutorial 58.1**

▶ Human activities have altered the flux of water from the land to the oceans. **Review Figure 58.5**

▶ Biomass burning contributes great quantities of carbon to the atmosphere.

The Carbon Cycle

▶ Atmospheric carbon dioxide is the immediate source of carbon for terrestrial organisms, but only a small part of Earth's carbon is in the atmosphere. **Review Figure 58.6. See Web/CD Tutorial 58.2**

▶ Increasing concentrations of CO_2 in the atmosphere are changing climates and influencing ecological processes. **Review Figure 58.7, 58.8**

▶ The ocean conveyor belt carries great quantities of carbon to the deep ocean. **See Web/CD Tutorial 58.3**

The Nitrogen Cycle

▶ Nitrogen makes up 78 percent of Earth's atmosphere, but nitrogen can be converted into biologically useful forms only by a few species of microorganisms. **Review Figure 58.9. See Web/CD Tutorial 58.4**

▶ Runoff of nitrogen from agricultural lands causes eutrophication in aquatic ecosystems. **Review Figure 58.10**

The Sulfur Cycle

▶ Acid precipitation results from the combined effects of human alterations in the nitrogen and sulfur cycles. **Review Figures 58.11, 58.12**

The Phosphorus Cycle

▶ The phosphorus cycle differs from the cycles of carbon, nitrogen, and sulfur in that it lacks a gaseous phase. **Review Figure 58.13**

▶ As a result of high fertilization rates, large amounts of phosphorus are accumulating in agricultural soils. **Review Figure 58.14**

▶ Eutrophication resulting from human inputs of phosphorus has damaged ecosystems in many lakes.

Interactions among Biogeochemical Cycles

▶ Human alterations of global biogeochemical cycles are changing Earth's climate. Outbreaks of diseases are one effect of this global warming.

older has lived through a doubling of the human population. But around 1965, the population growth rate peaked and began to decline. The most recent decline before that time, in the fourteenth century, was caused by increased death rates from plagues, wars, and famines. In contrast, the current decline is being caused by voluntary reductions in fertility. Fertility rates have been declining in rich countries for more than a century, but they began to fall in poor countries as well in 1965. Nobody had predicted this change, but the fact that it was voluntary gives us hope that humanity may be able, at least in part, to choose the kind of world in which its children and grandchildren will live.

If humans are to live on Earth in a sustainable manner that provides opportunities for the development of human creativity and rewarding lifestyles, major changes will need to be made in how we use environmental resources. Because we have limited abilities to regenerate and restore degraded natural ecosystems, we will need to develop an increased appreciation of the roles of living organisms in Earth's system. We must not load our ecosystems so heavily that their Plimsoll lines sink below the water line.

Chapter Summary

Earth's System Has Four Compartments

▶ The elements on which life depends cycle among the four compartments of Earth's physical environment: oceans, fresh waters, atmosphere, and land. **Review Figure 58.1**

Visions of the Future

▶ Actions people might undertake today could greatly influence the future quality of human life and the welfare of other species that share our small planet with us.

Self-Quiz

1. Earth is in chemical disequilibrium because
   a. Earth has a moon.
   b. organisms dissipate energy as heat.
   c. most continents are in the Northern Hemisphere.
   d. Earth has living organisms that maintain the O_2, N_2, and H_2O of the atmosphere.
   e. Earth is tilted on its axis.

2. Zones of marine upwelling are important because
   a. they help scientists measure the chemistry of deep ocean water.
   b. they bring to the surface organisms that are difficult to observe elsewhere.
   c. ships can sail faster in these zones.
   d. they increase marine productivity by bringing nutrients back to surface ocean waters.
   e. they bring oxygenated water to the surface.

3. Which of the following is *not* true of the troposphere?
   a. It contains nearly all of the atmospheric water vapor.
   b. Materials enter it primarily at the intertropical convergence zone.
   c. It is about 17 km deep in the Tropics.
   d. Most global air circulation takes place there.
   e. It contains about 80 percent of the mass of the atmosphere.

4. The hydrological cycle is driven by the
   a. flow of water into the oceans via rivers.
   b. evaporation (transpiration) of water from the leaves of plants.
   c. evaporation of water from the surface of the oceans.
   d. precipitation falling on the land.
   e. fact that less water falls on the ocean as precipitation than evaporates from its surface.

5. Carbon dioxide is called a greenhouse gas because
   a. it is used in greenhouses to increase plant growth.
   b. it is transparent to heat, but traps sunlight.
   c. it is transparent to sunlight, but traps heat.
   d. it is transparent to both sunlight and heat.
   e. it traps both sunlight and heat.

6. The ocean conveyor belt carries water to the deep ocean in the
   a. North Pacific Ocean
   b. South Pacific Ocean
   c. North Atlantic Ocean
   d. South Atlantic Ocean
   e. Indian Ocean

7. The cycle of phosphorus differs from the cycles of carbon and nitrogen in that
   a. it lacks a gaseous phase.
   b. it lacks a liquid phase.
   c. only phosphorus is cycled through marine organisms.
   d. living organisms do not need phosphorus.
   e. The phosphorus cycle does not differ importantly from the carbon and nitrogen cycles.

8. The sulfur cycle influences the global climate because
   a. sulfur compounds are important greenhouse gases.
   b. sulfur compounds help transfer carbon from the atmosphere to the oceans.
   c. sulfur compounds in the atmosphere are particles around which water condenses to form clouds.
   d. sulfur compounds contribute to acid precipitation.
   e. The sulfur cycle does not influence the global climate.

9. Acid precipitation results from human modifications of
   a. the carbon and nitrogen cycles.
   b. the carbon and sulfur cycles.
   c. the carbon and phosphorus cycles.
   d. the nitrogen and sulfur cycles.
   e. the nitrogen and phosphorus cycles.

10. Acid precipitation may change lakes by
    a. making phosphorus more available to plants, thereby increasing productivity.
    b. making nitrogen more available to plants, thereby increasing productivity.
    c. increasing populations of nitrogen-fixing microorganisms.
    d. accelerating the carbon cycle within lakes.
    e. causing the local extinction of species that cannot tolerate the lower pH.

For Discussion

1. A powerful hurricane strikes the East Coast of the United States. Some people claim that this disaster was due to warming of the oceans caused by greenhouse gases in the atmosphere. Others assert that global warming is not responsible because hurricanes have occurred for many centuries. How would you evaluate these conflicting claims?

2. The waters of Lake Washington, adjacent to the city of Seattle, rapidly returned to their pre-industrial condition when sewage was diverted from the lake to Puget Sound, an arm of the Pacific Ocean. Would all lakes being polluted with sewage clean themselves up as quickly as Lake Washington if pollutant inputs were stopped? What characteristics of a lake are most important to its rate of recovery following reduction of pollutant inputs?

3. Tropical forests currently are being cut at a very rapid rate. Does this necessarily mean that deforestation is a major source of carbon dioxide to the atmosphere? If not, why not?

4. What types of experiments would you conduct to assess the likely consequences of fertilization of the oceans with iron to increase rates of photosynthesis? At what spatial and temporal scales should they be conducted?

5. A government official authorizes the construction of a large coal-burning power plant in a former wilderness area. Its smokestacks discharge great quantities of combustion wastes. List and describe all likely effects on Earth's system at local, regional, and global levels. Now suppose the wastes were thoroughly scrubbed from the stack gases. Which of the effects you have just outlined would still happen?

6. Many nations recently signed the Kyoto Accord, which obligates them to reduce their emissions of carbon dioxide to the atmosphere. The United States refused to sign the treaty, claiming it was against the country's interests. The United States has been severely criticized for abandoning the Kyoto process. Is such criticism warranted? Justify your answer.

Toward economic principles for sustainable ecosystems management

– by William E. Rees –

Most economic analysis deals only with those things of direct use to humans that can be readily measured in monetary terms. All other values, including most of those associated with nature, are deemed *externalities* (outside the compass of the market) by economists.

Environmental economists have responded to this limitation by trying to "price" certain features of ecosystems that have not historically been perceived to have monetary value. This effort at least recognizes that nature makes substantial contributions to the economy. The question is, Do these evaluations enable us to incorporate all the values associated with ecosystems into the prices people pay for commodities derived from "natural capital"? Regrettably, the answer is no.

Many of the values associated with ecosystems are difficult to quantify, let alone price. In particular, consider those ecosystem components and processes whose precise contribution to life support and the economy is not known, and may not be even be suspected until they are lost or destroyed. An economics that does not reflect this reality would be dangerously misleading in the decision process.

The emerging discipline of ecological economics attempts to address this problem, recognizing that the human economy is an open, growing, *dependent* subsystem of the materially closed, nongrowing ecosphere. Ecological economics sees economic activity as a tranformation process subject to biophysical laws, particularly the second law of thermodynamics. In this light, both ecosystems and the economy are revealed as structures that grow and develop by extracting available energy and material from their host systems and exporting their wastes (entropy) back into their hosts. In short, they maintain their internal order at the expense of increasing disorder elsewhere in the hierarchy.

The problem is that, while the ecosphere *evolves* by dissipating solar energy, the economy *grows* by dissipating the ecosphere. In other words, thermodynamically, the economy is positioned to consume and pollute the ecosphere from the inside out—a second law explanation for the degradation of the ecosphere.

Moreover, recognizing that a certain minimal amount of nature must be conserved to support the human enterprise, ecological economists have advanced the "constant capital stocks criterion" for sustainability. In its strongest form, this criterion states that *each generation must inherit from the previous generation at least an equivalent adequate per capita stock of both human-made and natural capital.*

But how can we determine what constitutes an "adequate" supply of nature? The behavior of ecosystems under stress is often chaotic and characterized by multiple possible equilibrium states. An overexploited ecosystem may approach some critical threshold with little warning that the system is about to flip into a new equilibrium—and it is possible that the new state may not be compatible with human needs.

This is not mere conjecture. Studies show that industrial fishing has reduced the predatory fish biomass of the oceans

William E. Rees received his Ph.D. in bioecology from the University of Toronto. He has taught at the University of British Columbia since 1969 and is currently a professor in the University's School of Community and Regional Planning. His recent work is in ecological economics, urban ecology, and human carrying capacity. He is co-author of *Our Ecological Footprint: Reducing Human Impact on Earth*, published in 1996.

to about 10 percent of preindustrial levels in just 50 years. Many depleted fish stocks, such as Canada's northern cod (which once supported one of the richest fisheries in the world) show no sign of recovery after a decade or more of suspended fishing effort. Taming humanity's expansionist tendencies must be the primary goal of sustainability economics. Humans cannot harvest more than ecosystems can produce sustainably without being pushed beyond some unforeseen, dangerous threshold.

In summary, sustainability economics must be a risk-averse economics designed to keep human society at safe distance from treacherous boundaries. But mere economics is not enough. Sustainable ecosystems management must factor nonmarket values into the decision-making equation. People living in today's techno-industrial world would benefit greatly from a renewed sense of *biophilia*, a feeling of oneness with nature. The simple fact is that humans do not conserve what they do not love.

Discussion Questions

1. Environmental economists argue that a total economic valuation of nature may show that an ecosystem or species has sufficient dollar value in its natural state to protect it against any "development" that would produce less value. Therefore, economic valuation achieves the conservationist's purpose without having to resort to moral, ethical, spiritual, or other arguments. A counter argument is that a proper consideration of moral and spiritual values and ecological uncertainty would enable us to come to the right decision without having to resort to fundamentally flawed economic analyses. Discuss the pros and cons of these arguments.

2. It can be argued that, like other species, *Homo sapiens* has an innate tendency to expand into all the ecological space available to it regardless of resultant habitat destruction and displacement of other species. Economic growth as promoted by conventional economics is therefore merely a reflection of biological reality, and since the dynamics are "natural" there's not much we can do about it. Eventually nature will have her revenge. Discuss.

Web Links

International Society for Ecological Economics www.ecologicaleconomics.org/

United States Society for Ecological Economics www.ussee.org/

Post-Autistic Economics Network www.paecon.net/

The following site is a rich source of classic papers and data on ecological decay and ecological economics: www.dieoff.com

Notes

Notes

Notes

Appendix:
Some Measurements Used in Biology

QUANTITY	NAME OF UNIT	SYMBOL	DEFINITION
Length	meter (*also* metre)	m	A base unit. 1 m = 100 cm (39.37 inches)
	kilometer	km	1 km = 1000 m = 10^3 m (0.62 miles)
	centimeter	cm	1 cm = $\frac{1}{100}$ m = 10^{-2} m
	millimeter	mm	1 mm = $\frac{1}{1000}$ m = 10^{-3} m
	micrometer	μm	1 μm = $\frac{1}{1000}$ mm = 10^{-6} m
	nanometer	nm	1 nm = $\frac{1}{1000}$ μm = 10^{-9} m
Area	square meter	m^2	Area encompassed by a square, each side of which is 1 m in length
	hectare	ha	1 ha = 10,000 m^2 = 10^4 m^2 (2.47 acres)
	square centimeter	cm^2	1 cm^2 = $\frac{1}{10,000}$ m^2 = 10^{-4} m^2
Volume	liter (*also* litre)	l	1 l = $\frac{1}{1000}$ m^3 = 10^{-3} m^3 (1.057 qts)
	milliliter	ml	1 ml = $\frac{1}{1000}$ l = 10^{-3} l = 1 cm^3 = 1 cc
	microliter	μl	1 μl = $\frac{1}{1000}$ ml = 10^{-3} ml = 10^{-6} l
Mass	kilogram	kg	A base unit. 1 kg = 1000 g = 2.20 lbs
	gram	g	1 g = $\frac{1}{1000}$ kg = 10^{-3} kg
	milligram	mg	1 mg = $\frac{1}{1000}$ g = 10^{-3} g = 10^{-6} kg
Time	second	s	A base unit. 1 s = $\frac{1}{60}$ min
	minute	min	1 min = 60 s
	hour	h	1 h = 60 min = 3,600 s
	day	d	1 d = 24 h = 86,400 s
Temperature	kelvin	K	A base unit. 0 K = −273.15°C = absolute zero
	degree Celsius	°C	0°C = 273.15 K = melting point of ice
Heat, work	calorie	cal	1 cal = heat necessary to raise 1 gram of pure water from 14.5°C to 15.5°C = 4.184 J
	kilocalorie	kcal	1 kcal = 1000 cal = 10^3 cal = (in nutrition) 1 Calorie
	joule	J	1 J = 0.2389 cal (The joule is now the accepted unit of heat in most sciences.)
Electric potential	volt	V	A unit of potential difference or electromotive force
	millivolt	mV	1 mV = $\frac{1}{1000}$ V = 10^{-3} V

Glossary

Abdomen (ab' duh mun) [L. *abdomin*: belly] In arthropods, the posterior segments of the body; in mammals, the part of the body containing the intestines and most other internal organs, posterior to the thorax.

Abiotic (a' bye ah tick) [Gk. *a*: not + *bios*: life] Nonliving.

Abscisic acid (ab sighs' ik) A plant growth substance having growth-inhibiting action. Causes stomata to close.

Abscission (ab sizh' un) [L. *abscissio*: break off] The process by which leaves, petals, and fruits separate from a plant.

Absolute temperature scale Also known as the Kelvin scale. A temperature scale in which zero is the state of no molecular motion, or "absolute zero" (–273° on the Celsius scale).

Absorption (1) Of light: complete retention, without reflection or transmission. (2) Of liquids: soaking up (taking in through pores or cracks).

Absorption spectrum A graph of light absorption versus wavelength of light; shows how much light is absorbed at each wavelength.

Abyssal zone (uh biss' ul) [Gr. *abyssos*: bottomless] The deep ocean, below the point that light can penetrate.

Accessory pigments Pigments that absorb light and transfer energy to chlorophylls for photosynthesis.

Acetylcholine A neurotransmitter substance that carries information across vertebrate neuromuscular junctions and some other synapses.

Acetylcholinesterase An enzyme that breaks down acetylcholine.

Acetyl coenzyme A (acetyl CoA) Compound that reacts with oxaloacetate to produce citrate at the beginning of the citric acid cycle; a key metabolic intermediate in the formation of many compounds.

Acid [L. *acidus*: sharp, sour] A substance that can release a proton in solution. (Contrast with base.)

Acid precipitation Precipitation that has a lower pH than normal as a result of acid-forming precursors introduced into the atmosphere by human activities.

Acidic Having a pH of less than 7.0 (a hydrogen ion concentration greater than 10^{-7} molar).

Acoelomate Lacking a coelom.

Acquired Immune Deficiency Syndrome See AIDS.

Acrosome (a' krow soam) [Gr. *akros*: highest + *soma*: body] The structure at the forward tip of an animal sperm which is the first to fuse with the egg membrane and enter the egg cell.

ACTH (adrenocorticotropin) A pituitary hormone that stimulates the adrenal cortex.

Actin [Gr. *aktis*: ray] One of the two major proteins of muscle; it makes up the thin filaments.

Forms the microfilaments found in most eukaryotic cells.

Action potential An impulse in a neuron taking the form of a wave of depolarization or hyperpolarization imposed on a polarized cell surface.

Action spectrum A graph of a biological process versus light wavelength; shows which wavelengths are involved in the process.

Activating enzymes Also called aminoacyl-tRNA synthetases; these enzymes catalyze the addition of amino acids to their appropriate tRNAs.

Activation energy (E_a) The energy barrier that blocks the tendency for a set of chemical substances to react.

Active site The region on the surface of an enzyme where the substrate binds, and where catalysis occurs.

Active transport The energy-dependent transport of a substance across a biological membrane against a concentration gradient—that is, from a region of low concentration (of that substance) to a region of high concentration. (See primary active transport, secondary active transport; contrast with facilitated diffusion.)

Adaptation (a dap tay' shun) In evolutionary biology, a particular structure, physiological process, or behavior that makes an organism better able to survive and reproduce. Also, the evolutionary process that leads to the development or persistence of such a trait.

Adenine (A) (a' den een) A nitrogen-containing base found in nucleic acids, ATP, NAD, and other compounds.

Adenosine triphosphate See ATP.

Adenylate cyclase Enzyme catalyzing the formation of cyclic AMP (cAMP) from ATP.

Adrenal (a dree' nal) [L. *ad*: toward + *renes*: kidneys] An endocrine gland located near the kidneys of vertebrates, consisting of two glandular parts, the cortex and medulla.

Adrenaline See epinephrine.

Adrenocorticotropin See ACTH.

Adsorption Binding of a gas or a solute to the surface of a solid.

Aerobic (air oh' bic) [Gr. *aer*: air + *bios*: life] In the presence of oxygen; requiring oxygen.

Afferent (af' ur unt) [L. *ad*: toward + *ferre*: to carry] Carrying to, as in a neuron that carries impulses to the central nervous system, or a blood vessel that carries blood to a structure. (Contrast with efferent.)

AIDS (acquired immune deficiency syndrome) Condition caused by a virus (HIV) in which the body's helper T lymphocytes are reduced, leaving the victim subject to opportunistic diseases.

Aldehyde (al' duh hide) A compound with a —CHO functional group. Many sugars are aldehydes. (Contrast with ketone.)

Aldosterone (al dohs' ter own) A steroid hormone produced in the adrenal cortex of mammals. Promotes secretion of potassium and reabsorption of sodium in the kidney.

Alga (al' gah) (plural: algae) [L.: seaweed] Any one of a wide diversity of mostly photosynthetic protists.

Allantois (al lan' to is) A sac-like extraembryonic membrane that contains nitrogen waste from embryo.

Allele (a leel') [Gr. *allos*: other] The alternate forms of a genetic character found at a given locus on a chromosome.

Allele frequency The relative proportion of a particular allele in a specific population.

Allergy [Ger. *allergie*: altered reaction] An overreaction to amounts of an antigen that do not affect most people; often involves IgE antibodies.

Allometric growth A pattern of growth in which some parts of the body of an organism grow faster than others, resulting in a change in body proportions as the organism grows.

Allopatric speciation (al' lo pat' rick) [Gr. *allos*: other + *patria*: homeland] Also called geographical speciation, this is the formation of two species from one when reproductive isolation occurs because of the interposition of (or crossing of) a physical geographic barrier such as a river. (Contrast with parapatric speciation, sympatric speciation.)

Allostery (al' lo steer y) [Gr. *allos*: other + *stereos*: structure] Regulation of the activity of a protein by the binding of an effector molecule at a site other than the active site.

Alpha (α) helix Type of protein secondary structure; a right-handed spiral.

Alternation of generations The succession of multicellular haploid and diploid phases in some sexually reproducing organisms, notably plants.

Altruism Behavior that harms the individual who performs it but benefits other individuals.

Alveolus (al ve' o lus) (plural: alveoli) [L. *alveus*: cavity] A small, baglike cavity, especially the blind sacs of the lung.

Ambient That which surrounds; the immediate environment.

Amensalism (a men' sul ism) Interaction in which one animal is harmed and the other is unaffected. (Contrast with commensalism, mutualism.)

Amine An organic compound with an amino group (see Amino acid).

Amino acid An organic compound containing both NH_2 and COOH groups. Proteins are polymers of amino acids.

Ammonotelic (am moan' o teel' ic) [Gr. *telos*: end] Describes an organism in which the final product of breakdown of nitrogen-containing compounds (primarily proteins) is ammonia. (Contrast with ureotelic, uricotelic.)

Amnion (am' nee on) The fluid-filled sac in which the embryos of reptiles, birds, and mammals develop.

Amniote Any of the vertebrate animals whose embryos are enclosed in an amnion: reptiles, birds, and mammals.

Amniote egg A shelled egg surrounding four extraembryonic membranes and embryo-nourishing yolk. This adaptation allowed animals to colonize the terrestrial environment.

Amphipathic (am' fi path' ic) [Gr. *amphi*: both + *pathos*: emotion] Of a molecule, having both hydrophilic and hydrophobic regions.

Amylase (am' ill ase) Any of a group of enzymes that digest starch.

Anabolism (an ab' uh liz' em) [Gr. *ana*: upward + *ballein*: to throw] Synthetic reactions of metabolism, in which complex molecules are formed from simpler ones. (Contrast with catabolism.)

Anaerobic (an ur row' bic) [Gr. *an*: not + *aer*: air + *bios*: life] Occurring without the use of molecular oxygen, O_2.

Anagenesis Evolutionary change in a single lineage over time.

Analogy (a nal' o jee) [Gr. *analogia*: resembling] A resemblance in function, and often appearance as well, between two structures that is due to convergent evolution rather than to common ancestry. (Contrast with homology.)

Anaphase (an' a phase) [Gr. *ana*: upward progress] The stage in nuclear division at which the first separation of sister chromatids (or, in the first meiotic division, of paired homologues) occurs.

Anaphylactic shock A precipitous drop in blood pressure caused by loss of fluid from capillaries because of an increase in their permeability stimulated by an allergic reaction.

Ancestral trait Trait shared by a group of organisms as a result of descent from a common ancestor.

Androgens (an' dro jens) The male sex steroids.

Aneuploidy (an' you ploy dee) A condition in which one or more chromosomes or pieces of chromosomes are either lacking or present in excess.

Angiosperm (an' jee oh spurm) [Gr. *angion*: vessel + *sperma*: seed] One of the flowering plants; literally, one whose seed is carried in a "vessel" (i.e., the fruit).

Angiotensin (an' jee oh ten' sin) A peptide hormone that raises blood pressure by causing peripheral vessels to constrict. Also maintains glomerular filtration by constricting efferent vessels and stimulates thirst and the release of aldosterone.

Animal [L. *animus*: breath, soul] A member of the kingdom Animalia. In general, a multicellular eukaryote that obtains its food by ingestion.

Animal hemisphere The metabolically active upper portion of some animal eggs, zygotes, and embryos; does not contain the dense nutrient yolk. (Contrast with vegetal hemisphere.)

Anion (an' eye on) [Gk. *ana*: upward progress] A negatively charged ion. (Contrast with cation.)

Anisogamy (an eye sog' a mee) [Gr. *aniso*: unequal + *gamos*: marriage] The existence of two dissimilar gametes (egg and sperm).

Annual Referring to a plant whose life cycle is completed in one growing season. (Contrast with biennial, perennial.)

Antenna system In photosynthesis, a group of different molecules that cooperate to absorb light energy and transfer it to a reaction center.

Anterior pituitary The portion of the vertebrate pituitary gland that derives from gut epithelium and produces tropic hormones.

Anther (an' thur) [Gr. *anthos*: flower] A pollen-bearing portion of the stamen of a flower.

Antheridium (an' thur id' ee um) (plural: antheridia) [Gr. *antheros*: blooming] The multicellular structure that produces the sperm in bryophytes and ferns.

Antibody One of the millions of proteins produced by the immune system that specifically binds to a foreign substance and initiates its removal from the body.

Anticodon The three nucleotides in transfer RNA that pair with a complementary triplet (a codon) in messenger RNA.

Antidiuretic hormone A hormone that controls water reabsorption in the mammalian kidney. Also called vasopressin.

Antigen (an' ti jun) Any substance that stimulates the production of an antibody or antibodies in the body of a vertebrate.

Antigenic determinant A specific region of an antigen, which is recognized by and binds to a specific antibody.

Antiparallel Pertaining to molecular orientation in which a molecule or parts of a molecule have opposing directions.

Antipodal cell At one end of the megagametophyte, one of the three cells which eventually degenerate.

Antiport A membrane transport process that carries one substance in one direction and another in the opposite direction. (Contrast with symport.)

Antisense nucleic acid A single-stranded RNA or DNA complementary to and thus targeted against the mRNA transcribed from a harmful gene such as an oncogene.

Anus (a' nus) Opening through which digestive wastes are expelled, located at the posterior end of the gut.

Aorta (a or' tah) [Gr. *aorte*: aorta] The main trunk of the arteries leading to the systemic (as opposed to the pulmonary) circulation.

Apex (a' pecks) The tip or highest point of a structure, as the apex of a growing stem or root.

Apical (a' pi kul) Pertaining to the apex, or tip, usually in reference to plants.

Apical dominance Inhibition by the apical bud of the growth of axillary buds.

Apical meristem The meristem at the tip of a shoot or root; responsible for the plant's primary growth.

Apomixis (ap oh mix' is) [Gr. *apo*: away from + *mixis*: sexual intercourse] The asexual production of seeds.

Apoplast (ap' oh plast) in plants, the continuous meshwork of cell walls and extracellular spaces through which material can pass without crossing a plasma membrane. (Contrast with symplast.)

Apoptosis (ay pu toh' sis) A series of genetically programmed events leading to cell death.

Aquaporin A transport protein in plant and animal cells through which water passes in osmosis.

Aquatic (a kwa' tic) [L. *aqua*: water] Living in water. (Compare with marine, terrestrial.)

Aqueous (a' kwee us) [L. *aqua*: water] Pertaining to water or a watery solution.

Archegonium (ar' ke go' nee um) [Gr. *archegonos*: first, foremost] The multicellular structure that produces eggs in bryophytes, ferns, and gymnosperms.

Archenteron (ark en' ter on) [Gr. *archos*: first + *enteron*: bowel] The earliest primordial animal digestive tract.

Arteriosclerosis See atherosclerosis.

Artery A muscular blood vessel carrying oxygenated blood away from the heart to other parts of the body. (Contrast with vein.)

Ascus (ass' cuss) [Gr. *askos*: bladder] In ascomycete fungi (sac fungi), the club-shaped sporangium within which spores (ascospores) are produced by meiosis.

Asexual Without sex.

Assortative mating A breeding system in which mates are selected on the basis of a particular trait or group of traits.

Atherosclerosis (ath' er oh sklair oh' sis) [Gk. *athero*: gruel, porridge + *skleros*: hard] A disease of the lining of the arteries characterized by fatty, cholesterol-rich deposits in the walls of the arteries. When fibroblasts infiltrate these deposits and calcium precipitates in them, the disease become arteriosclerosis, or "hardening of the arteries."

Atmosphere The gaseous mass surrounding our planet. Also a unit of pressure, equal to the normal pressure of air at sea level.

Atom [Gr. *atomos*: indivisible] The smallest unit of a chemical element. Consists of a nucleus and one or more electrons.

Atomic mass The average mass of an atom of an element; the average depends on the relative amounts of different isotopes of the element on Earth. Also called atomic weight.

Atomic number The number of protons in the nucleus of an atom; also equals the number of electrons around the neutral atom. Determines the chemical properties of the atom.

ATP (adenosine triphosphate) An energy-storage compound containing adenine, ribose, and three phosphate groups. When it is formed from ADP, useful energy is stored; when it is broken down (to ADP or AMP), energy is released to drive endergonic reactions.

ATP synthase An integral membrane protein that couples the transport of proteins with the formation of ATP.

Atrium (a' tree um) [L. *atrium*: central hall] An internal chamber. In the hearts of vertebrates, the thin-walled chamber(s) entered by blood on its way to the ventricle(s). Also, the outer ear.

Autoimmune disease A disorder in which the immune system attacks the animal's own antigens.

Autonomic nervous system The system that controls such involuntary functions as those of guts and glands.

Autosome Any chromosome (in a eukaryote) other than a sex chromosome.

Autotroph (au' tow trow' fik) [Gr. *autos*: self + *trophe*: food] An organism that is capable of living exclusively on inorganic materials, water, and some energy source such as sunlight or chemically reduced matter. (Contrast with heterotroph.)

Auxin (awk' sin) [Gr. *auxein*: to grow] In plants, a substance (the most common being indoleacetic acid) that regulates growth and various aspects of development.

Auxotroph (awks' o trofe) [Gr. *auxein*: to grow + *trophe*: food] A mutant form of an organism that requires a nutrient or nutrients not usually required by the wild type. (Contrast with prototroph.)

Axon [Gr. *axon*: axle] The part of a neuron that conducts action potentials away from the cell body.

Axon hillock The junction between an axon and its cell body, where action potentials are generated.

Axon terminals The endings of an axon; they form synapses and release neurotransmitter.

Axoneme (ax' oh neem) The complex of microtubules and their crossbridges that forms the motile apparatus of a cilium.

Bacillus (bah sil' us) [L: little rod] Any of various rod-shaped bacteria.

Bacteria (bak teer' ee ah) (singular: bacterium) [Gr. *bakterion*: little rod] Prokaryote in the Domain Bacteria. The chromosomes of bacteria are not contained in nuclear envelopes.

Bacteriophage (bak teer' ee o fayj) [Gr. *bakterion*: little rod + *phagein*: to eat] One of a group of viruses that infect bacteria and ultimately cause their disintegration.

Bacteroids Nitrogen-fixing organelles that develop from endosymbiotic bacteria.

Balanced polymorphism [Gr. *polymorphos*: many forms] The maintenance of more than one form, or the maintenance at a given locus of more than one allele, at frequencies of greater than 1 percent in a population. Often results when heterozygotes are more fit than either homozygote.

Bark All tissues outside the vascular cambium of a plant.

Baroreceptor [Gr. *baros*: weight] A pressure-sensing cell or organ.

Barr body In mammals, an inactivated X chromosome.

Basal Pertaining to one end—the base—of an axis.

Basal body Centriole found at the base of a eukaryotic flagellum or cilium.

Basal metabolic rate (BMR) The minimum rate of energy turnover in an awake (but resting) bird or mammal that is not expending energy for thermoregulation.

Base (1) A substance which can accept a hydrogen ion in solution. (Contrast with acid.) (2) In nucleic acids, the purine or pyrimidine that is attached to each sugar in the backbone.

Base pairing See complementary base pairing.

Basic Having a pH greater than 7.0 (i.e., having a hydrogen ion concentration lower than 10^{-7} molar).

Basidium (bass id' ee yum) In basidiomycete fungi, the characteristic sporangium in which four spores are formed by meiosis and then borne externally before being shed.

Basophils One type of phagocytic white blood cell that releases histamine and may promote T cell development.

B cell A type of lymphocyte involved in the humoral immune response of vertebrates. Upon recognizing an antigenic determinant, a B cell develops into a plasma cell, which secretes an antibody. (Contrast with T cell.)

Benefit An improvement in survival and reproductive success resulting from performing a behavior or having a trait. (Contrast with cost.)

Benign (be nine') A tumor that grows to a certain size and then stops, uaually with a fibrous capsule surrounding the mass of cells. Benign tumors do not spread (metastasize) to other organs.

Benthic zone [Gr. *benthos*: bottom] The bottom of the ocean.

Beta (β) pleated sheet Type of protein secondary structure; results from hydrogen bonding between polypeptide regions running antiparallel to each other.

Biennial Referring to a plant whose life cycle includes vegetative growth in the first year and flowering and senescence in the second year. (Contrast with annual, perennial.)

Bilateral symmetry The condition in which only the right and left sides of an organism, divided exactly down the back, are mirror images of each other. (Contrast with biradial symmetry.)

Bile A secretion of the liver delivered to the small intestine via the common bile duct. In the intestine, bile emulsifies fats.

Binocular cells Neurons in the visual cortex that respond to input from both retinas; involved in depth perception.

Binomial (bye nome' ee al) Consisting of two elements; for example, the binomial nomenclature of biology in which each species has two names (the genus name followed by the species name).

Biodiversity crisis The current high rate of loss of species, caused primarily by human activities.

Biogeochemical cycles Movement of elements through living organisms and the physical environment.

Biogeographic region A continental-scale part of Earth that has a biota distinct from that of other such regions.

Bioluminescence The production of light by biochemical processes in an organism.

Biomass The total weight of all the living organisms, or some designated group of living organisms, in a given area.

Biome (bye' ome) A major division of the ecological communities of Earth; characterized by distinctive vegetation.

Biota (bye oh' tah) All of the organisms, including animals, plants, fungi, and microorganisms, found in a given area.

Biradial symmetry Radial symmetry modified so that only two planes can divide the animal into similar halves.

Blastocoel (blass' toe seal) [Gr. *blastos*: sprout + *koilos*: hollow] The central, hollow cavity of a blastula.

Blastocyst (blass' toe cist) An early embryo formed by the first divisions of the fertilized egg (zygote). In mammals, a hollow ball of cells.

Blastodisc (blass' toe disk) A disk of cells forming on the surface of a large yolk mass, comparable to a blastula, but occurring in animals such as birds and reptiles, in which the massive yolk restricts cleavage to one side of the egg only.

Blastomere A cell produced by the division of a fertilized egg.

Blastopore The opening from the archenteron to the exterior of a gastrula.

Blastula (blass' chu luh) An early stage in animal embryology; in many species, a hollow sphere of cells surrounding a central cavity, the blastocoel. (Contrast with blastodisc.)

Blood–brain barrier A property of the blood vessels of the brain that prevents most chemicals from diffusing from the blood into the brain.

Blue light receptor Molecule in plants that absorbs blue light (400–500 nm). Mediates many plant responses including phototropism, stomatal movements, and expression of some genes.

Body cavity Membrane-lined, fluid-filled compartment that lies between the cell layers of many animals.

Body plan A basic structural design that includes an entire animal, its organ systems, and the integrated functioning of its parts.

Bottleneck Refers to stressful periods that only a few individuals of a once-large population survive, resulting in substantially reduced genetic variation in the population.

Bowman's capsule An elaboration of kidney tubule cells that surrounds a know of capillaries (the glomerulus). Blood is filtered across the walls of these capillaries and the filtrate is collected into Bowman's capsule.

Brain stem The portion of the vertebrate brain between the spinal cord and the forebrain.

Brassinosteroids Plant steroid hormones that mediate light effects promoting the elongation of stems and pollen tubes.

Bronchus (plural: bronchi) The major airway(s) branching off the trachea into the vertebrate lung.

Brown fat Fat tissue in mammals that is specialized to produce heat. It has many mitochondria and capillaries, and a protein that uncouples oxidative phosphorylation.

Browser An animal that feeds on the tissues of woody plants.

Bryophyte (bry' uh fite) [Gk. *bryon*: moss; *phyton*: plant] A moss. This term was once frequently used to refer to all nontracheophyte plants.

Budding Asexual reproduction in which a more or less complete new organism simply grows from the body of the parent organism and eventually detaches itself.

Buffer A substance that can transiently accept or release hydrogen ions and thereby resist changes in pH.

C_3 photosynthesis The form of photosynthesis in which 3-phosphoglycerate is the first stable product, and ribulose bisphosphate is the CO_2 receptor.

C_4 photosynthesis The form of photosynthesis in which oxaloacetate is the first stable product, and phosphoenolpyruvate is the CO_2 acceptor. C_4 plants also perform the reactions of C_3 photosynthesis.

Calcitonin A hormone produced by the thyroid gland; it lowers blood calcium and promotes bone formation. (Compare with parathyroid hormone.)

Calmodulin (cal mod' joo lin) A calcium-binding protein found in all animal and plant cells; mediates many calcium-regulated processes.

calorie [L. *calor*: heat] The amount of heat required to raise the temperature of one gram of water by one degree Celsius (1°C) from 14.5°C to 15.5°C. Calorie spelled with a capital C refers to the kilocalorie (1 kcal = 1,000 cal).

Calvin–Benson cycle The stage of photosynthesis in which CO_2 reacts with RuBP to form 3PG, 3PG is reduced to a sugar, and RuBP is regenerated, while other products are released to the rest of the plant.

Calyx (kay' licks) [Gr. *kalyx*: cup] All of the sepals of a flower, collectively.

CAM See Crassulacean acid metabolism.

Cambium (kam' bee um) [L. *cambiare*: to exchange] A meristem that gives rise to radial rows of cells in stem and root, increasing them in girth; commonly applied to the vascular cambium which produces wood and phloem, and the cork cambium, which produces bark.

cAMP (cyclic AMP) A compound formed from ATP that mediates the effects of numerous animal hormones.

Canopy The leaf-bearing part of a tree. Collectively, the aggregate of the leaves and branches of the larger woody plants of an ecological community.

Capillaries [L. *capillaris*: hair] Very small tubes, especially the smallest blood-carrying vessels of animals between the termination of the arteries and the beginnings of the veins.

Carbohydrates Organic compounds containing carbon, hydrogen, and oxygen in the ratio 1:2:1 (i.e., with the general formula $C_nH_{2n}O_n$). Common examples are sugars, starch, and cellulose.

Carboxylic acid (kar box sill' ik) An organic acid containing the carboxyl group, —COOH, which dissociates to the carboxylate ion, —COO⁻.

Carcinogen (car sin' oh jen) A substance that causes cancer.

Cardiac (kar' dee ak) [Gr. *kardia*: heart] Pertaining to the heart and its functions.

Carnivore [L. *carn*: flesh + *vovare*: to devour] An organism that eats animal tissues. (Contrast with detritivore, herbivore, omnivore.)

Carotenoid (ka rah' tuh noid) A yellow, orange, or red lipid pigment commonly found as an accessory pigment in photosynthesis; also found in fungi.

Carpel (kar' pel) [Gr. *karpos*: fruit] The organ of the flower that contains one or more ovules.

Carrier (1) In facilitated diffusion, a membrane protein that binds a specific molecule and transports it through the membrane. (2) In respiratory and photosynthetic electron transport, a participating substance such as NAD that exists in both oxidized and reduced forms. (3) In genetics, a person heterozygous for a recessive trait.

Carrying capacity. In ecology, the maximum number of individuals of a particular species that can be supported on a sustained basis in a suitable habitat.

Cartilage In vertebrates, a tough connective tissue found in joints, the outer ear, and elsewhere. Forms the entire skeleton in some animal groups.

Casparian strip A band of cell wall containing suberin and lignin, found in the endodermis. Restricts the movement of water across the endodermis.

Catabolism [Gr. *kata*: to break down + *ballein*: to throw] Degradational reactions of metabolism, in which complex molecules are broken down. (Contrast with anabolism.)

Catabolite repression In the presence of abundant glucose, the diminished synthesis of catabolic enzymes for other energy sources.

Catalyst (cat' a list) [Gr. *kata*: to break down] A chemical substance that accelerates a reaction without itself being consumed in the overall course of the reaction. Catalysts lower the activation energy of a reaction. Enzymes are biological catalysts.

Cation (cat' eye on) An ion with one or more positive charges. (Contrast with anion.)

Caudal [L. *cauda*: tail] Pertaining to the tail, or to the posterior part of the body.

cDNA See complementary DNA.

Cecum (see' cum) [L. *caecus*: blind] A blind branch off the large intestine. In many nonruminant mammals, the cecum contains a colony of microorganisms that contribute to the digestion of food.

Cell adhesion molecules Molecules on animal cell surfaces that affect the selective association of cells during development of the embryo.

Cell cycle The stages through which a cell passes between one division and the next. Includes all stages of interphase and mitosis.

Cell division The reproduction of a cell to produce two new cells. In eukaryotes, this process involves nuclear division (mitosis) and cytoplasmic division (cytokinesis).

Cell junctions Specialized structures associated with the plasma membranes of epithelial cells. Some contribute to cell adhesion, others to intercellular communication.

Cell plate Following mitosis in plant cells, the initial wall-like structure that forms separating the nuclei from the surrounding cytoplasm. Later becomes the cell wall.

Cell theory The theory, well established, that all organisms consist of cells, and that all cells come from preexisting cells.

Cell wall A relatively rigid structure that encloses cells of plants, fungi, many protists, and most prokaryotes. Gives these cells their shape and limits their expansion in hypotonic media.

Cellular immune response Action of the immune system based on the activities of T cells. Directed against parasites, fungi, intracellular viruses, and foreign tissues (grafts). (Contrast with humoral immune system.)

Cellular respiration See respiration.

Cellulose (sell' you lowss) A straight-chain polymer of glucose molecules, used by plants as a structural supporting material.

Central dogma The statement that information flows from DNA to RNA to polypeptide (in retroviruses, there is also information flow from RNA to cDNA).

Central nervous system That part of the nervous system which is condensed and centrally located, e.g., the brain and spinal cord of vertebrates; the chain of cerebral, thoracic and abdominal ganglia of arthropods.

Centrifuge [L. *centrum*: center + *fugere*: to flee] A laboratory device in which a sample is spun around a central axis at high speed. Used to separate suspended materials of different densities.

Centriole (sen' tree ole) A paired organelle that helps organize the microtubules in animal and protist cells during nuclear division.

Centromere (sen' tro meer) [Gr. *centron*: center + *meros*: part] The region where sister chromatids join.

Centrosome (sen' tro soam) The major microtubule organizing center of an animal cell.

Cephalization (sef ah luh zay' shun) [Gr. *kephale*: head] The evolutionary trend toward increasing concentration of brain and sensory organs at the anterior end of the animal.

Cerebellum (sair uh bell' um) [L.: diminutive of *cerebrum*, brain] The brain region that controls muscular coordination; located at the anterior end of the hindbrain.

Cerebral cortex The thin layer of gray matter (neuronal cell bodies) that overlays the cerebrum.

Cerebrum (su ree' brum) [L. *cerebrum*: brain] The dorsal anterior portion of the forebrain, making up the largest part of the brain of mammals. In mammals, the chief coordination center of the nervous system; consists of two cerebral hemispheres.

Cervix (sir' vix) [L. *cervix*: neck] The opening of the uterus into the vagina.

cGMP (cyclic guanosine monophosphate) An intracellular messenger that is part of signal transmission pathways involving G proteins. (See G protein.)

Channel protein A membrane protein that forms an aqueous passageway though which specific solutes may pass.

Chaperonins A group of proteins that limit inappropriate interactions between cellular proteins under denaturing conditions such as high temperature.

Chaperone protein A protein that assists a newly forming protein in adopting its appropriate tertiary structure.

Chemical bond An attractive force stably linking two atoms.

Chemical reaction The change in the composition or distribution of atoms of a substance with consequent alterations in properties.

Chemiosmosis The formation of ATP in mitochondria and chloroplasts, resulting from a pumping of protons across a membrane (against a gradient of electrical charge and of pH), followed by the return of the protons through a protein channel with ATPase activity.

Chemoautotroph See Chemolithotroph.

Chemoheterotroph An organism that must obtain both carbon and energy from organic substances. (Contrast with chemolithotroph, photoautotroph, photoheterotroph.)

Chemolithotroph [Gk. *lithos*: stone, rock] An organism that uses carbon dioxide as a carbon source and obtains energy by oxidizing inorganic substances from its environment. (Contrast with chemoheterotroph, photoautotroph, photoheterotroph.)

Chemoreceptor A cell or tissue that senses specific substances in its environment.

Chemosynthesis Synthesis of food substances, using the oxidation of reduced materials from the environment as a source of energy.

Chiasma (kie az′ muh) (plural: chiasmata) [Gr. *chiasmata*: cross] An X-shaped connection between paired homologous chromosomes in prophase I of meiosis. A chiasma is the visible manifestation of crossing over between homologous chromosomes.

Chitin (kye′ tin) [Gr. *kiton*: tunic] The characteristic tough but flexible organic component of the exoskeleton of arthropods, consisting of a complex, nitrogen-containing polysaccharide. Also found in cell walls of fungi.

Chlorophyll (klor′ o fill) [Gr. *kloros*: green + *phyllon*: leaf] Any of a few green pigments associated with chloroplasts or with certain bacterial membranes; responsible for trapping light energy for photosynthesis.

Chloroplast [Gr. *kloros*: green + *plast*: a particle] An organelle bounded by a double membrane containing the enzymes and pigments that perform photosynthesis. Chloroplasts occur only in eukaryotes.

Choanocyte (ko′ an uh site) The collared, flagellated feeding cells of sponges.

Cholecystokinin (ko′ luh sis tuh kai′ nin) A hormone produced and released by the lining of the duodenum when it is stimulated by undigested fats and proteins. It stimulates the gallbladder to release bile and slows stomach activity.

Chorion (kor′ ee on) [Gr. *khorion*: afterbirth] The outermost of the membranes protecting mammal, bird, and reptile embryos; in mammals it forms part of the placenta.

Chromatid (kro′ ma tid) Each of a pair of new sister chromosomes from the time at which the molecular duplication occurs until the time at which the centromeres separate at the anaphase of nuclear division.

Chromatin The nucleic acid–protein complex found in eukaryotic chromosomes.

Chromatophore (krow mat′ o for) [Gr. *kroma*: color + *phoreus*: carrier] A pigment-bearing cell that expands or contracts to change the color of the organism.

Chromosomal mutation Loss of or changes in position/direction of a DNA segment on a chromosome.

Chromosome (krome′ o sowm) [Gr. *kroma*: color + *soma*: body] In bacteria and viruses, the DNA molecule that contains most or all of the genetic information of the cell or virus. In eukaryotes, a structure composed of DNA and proteins that bears part of the genetic information of the cell.

Chylomicron (ky low my′ cron) Particles of lipid coated with protein, produced in the gut from dietary fats and secreted into the extracellular fluids.

Chyme (kime) [Gr. *kymus*: juice] Created in the stomach; a mixture of ingested food with the digestive juices secreted by the salivary glands and the stomach lining.

Cilium (sil′ ee um) (plural: cilia) [L.: eyelash] Hairlike organelle used for locomotion by many unicellular organisms and for moving water and mucus by many multicellular organisms. Generally shorter than a flagellum.

Circadian rhythm (sir kade′ ee an) [L. *circa*: approximately + *dies*: day] A rhythm in behavior, growth, or some other activity that recurs about every 24 hours under constant conditions.

Circannual rhythm [L. *circa*: approximately + *annus*: year) A rhythm of behavior, growth, or some other activity that recurs on a yearly basis.

Citric acid cycle A set of chemical reactions in cellular respiration, in which acetyl CoA is oxidized to carbon dioxide, and hydrogen atoms are stored as NADH and FADH$_2$. Also called the Krebs cycle.

Clade (Gk. *klados*: branch] In taxonomy, a monophyletic group made up of an ancestor and all of its descendants.

Class I MHC molecules These cell surface proteins participate in the cellular immune response directed against virus-infected cells.

Class II MHC molecules These cell surface proteins participate in the cell-cell interactions (of helper T cells, macrophages, and B cells) of the humoral immune response.

Class switching The process whereby a plasma cell changes the class of immunoglobulin that it synthesizes by changing the DNA region coding for the C segment.

Clathrin A fibrous protein on the inner surfaces of animal cell membranes that strengthens coated vesicles and thus participates in receptor-mediated endocytosis.

Cleavage First divisions of the fertilized egg of an animal.

Cline A gradual change in the traits of a species over a geographical gradient.

Cloaca (klo ay′ kuh) [L. *cloaca*: sewer] In some invertebrates, the posterior part of the gut; in many vertebrates, a cavity receiving material from the digestive, reproductive, and excretory systems.

Clonal anergy Prevention of the synthesis of antibodies against the body's own antigens. When a T cell binds to a self-antigen, it does not receive signals from an antigen-presenting cell; thus the T cell dies (becomes anergic) rather than yielding a clone of active cells.

Clonal deletion The inactivation or destruction of lymphocyte clones that would produce immune reactions against the animal's own body.

Clonal selection The mechanism by which exposure to antigen results in the activation of selected T- or B-cell clones, resulting in an immune response.

Clone [Gr. *klon*: twig, shoot] Genetically identical cells or organisms produced from a common ancestor by asexual means.

Cnidocytes (nye′ duh sites) The feeding cells of cnidarians, within which nematocysts are housed.

Coacervate (ko as′ er vate) [L. *coacervare*: to heap up] An aggregate of colloidal particles in suspension.

Coated vesicle Cytoplasmic vesicle containing distinctive proteins, including clathrin.

Coccus (kock′ us) [Gr. *kokkos*: berry, pit] Any of various spherical or spheroidal bacteria.

Cochlea (kock′ lee uh) [Gr. *kokhlos* snail] A spiral tube in the inner ear of vertebrates; it contains the sensory cells involved in hearing.

Codominance A condition in which two alleles at a locus produce different phenotypic effects and both effects appear in heterozygotes.

Codon Three nucleotides in messenger RNA that direct the placement of a particular amino acid into a polypeptide chain. (Contrast with anticodon.)

Coelom (see′ lum) [Gr. *koiloma*: cavity] The body cavity of certain animals; the coelom is lined with cells of mesodermal origin.

Coelomate Having a coelom.

Coenocyte (seen′ a sight) [Gr. *koinos*: common + *kytos*: container] A "cell" enclosed by a single plasma membrane but containing many nuclei.

Coenzyme A nonprotein organic molecule that plays a role in catalysis by an enzyme.

Cofactor An inorganic ion that is weakly bound to an enzyme and required for its activity.

Cohesin Proteins involved in binding chromatids together.

Coevolution Concurrent evolution of two or more species that are mutually affecting each other's evolution.

Cohort (co′ hort) [L. *cohors*: company of soldiers] A group of similar-age organisms, considered as it passes through time.

Cold hardening Increased capacity of some plant species to withstand cold spells by their repeated exposure to cool but not damaging temperatures.

Coleoptile A sheath that surrounds and protects the apical meristem and young primary leaves of a seedling as they move through the soil.

Collagen [Gr. *kolla*: glue] A fibrous protein found extensively in bone and connective tissue.

Collecting duct In vertebrates, a tubule that receives urine produced in the nephrons of the kidney and delivers that fluid to the ureter for excretion.

Collenchyma (cull eng′ kyma) [Gr. *kolla*: glue + *enchyma*: infusion] A type of plant cell, living at functional maturity, which lends flexible support by virtue of primary cell walls thickened at the corners. (Contrast with parenchyma, sclerenchyma.)

Colon [Gr. *kolon*: large intestine] The large intestine.

Common bile duct A single duct that delivers bile from the gallbladder and secretions from the pancreas into the small intestine.

Communication A signal from one organism (or cell) that alters the functioning or behavior of another organism (or cell).

Community Any ecologically integrated group of species of microorganisms, plants, and animals inhabiting a given area.

Companion cell Specialized cell found adjacent to a sieve tube member in flowering plants.

Comparative genomics Computer-aided comparison of DNA sequences between different organisms to reveal genes with related functions.

Comparative method An approach to studying evolution and ecology in which hypotheses are tested by measuring the distribution of states among a large number of species.

Compensation point The light intensity at which the rates of photosynthesis and of cellular respiration are equal.

Competition In ecology, use of the same resource by two or more species, when the resource is present in insufficient supply for the combined needs of the species.

Competitive exclusion A result of competition between species for a limiting resource in which one species completely eliminates the other.

Competitive inhibitor A nonsubstrate that binds to the active site of an enzyme and thereby inhibits binding of substrate and reaction from part of the environment.

Complement system A group of eleven proteins that play a role in some reactions of the immune system. The complement proteins are not immunoglobulins.

Complementary base pairing The AT (or AU), TA (or UA), CG, and GC pairing of bases in double-stranded DNA, in transcription, and between tRNA and mRNA.

Complementary DNA (cDNA) DNA formed by reverse transcriptase acting with an RNA template; essential intermediate in the reproduction of retroviruses; used as a tool in recombinant DNA technology; lacks introns.

Complete metamorphosis A change of state during the life cycle of an organism in which the body is almost completely rebuilt to produce an individual with a very different body form. Characteristic of insects such as butterflies, moths, beetles, ants, wasps, and flies.

Compound (1) A substance made up of atoms of more than one element. (2) A structure made up of many units, as the compound eyes of arthropods.

Condensation reaction A reaction in which two molecules become connected by a covalent bond and a molecule of water is released. (AH + BOH → AB + H_2O.)

Conditional mutations Mutations that show characteristic phenotype only under certain environmental conditions such as temperature.

Conformation. The three-dimensional shape of a protein or other macromolecule.

Cones (1) In the vertebrate retina: photoreceptors responsible for color vision. (2) In gymnosperms: reproductive structures consisting of many sporophylls packed relatively tightly.

Conidium (ko nid' ee um) [Gr. *konis*: dust] An asexual fungus spore borne singly or in chains either apically or laterally on a hypha.

Conifer (kahn' e fer) [Gr. *konos*: cone + *phero*: carry] One of the cone-bearing gymnosperms, mostly trees, such as pines and firs.

Conjugation (kon ju gay' shun) [L. *conjugare*: yoke together] The close approximation of two cells during which they exchange genetic material, as in *Paramecium* and other ciliates, or during which DNA passes from one to the other through a tube, as in bacteria.

Connective tissue An animal tissue that connects or surrounds other tissues; its cells are embedded in a collagen-containing matrix.

Connexon In a gap junction, a protein channel linking adjacent animal cells.

Consensus sequences Short stretches of DNA that appear, with little variation, in many different genes.

Constant region For a particular class of immunoglobulin molecules, the region with identical amino acid composition.

Constitutive enzyme An enzyme that is present in approximately constant amounts in a system, whether its substrates are present or absent. (Contrast with inducible enzyme.)

Consumer An organism that eats the tissues of some other organism.

Continental drift The gradual movements of the world's continents that has occurred over billions of years.

Convergent evolution The evolution of similar features independently in unrelated taxa from different ancestral structures.

Copulation Reproductive behavior that results in a male depositing sperm in the reproductive tract of a female.

Corepressor A low-molecular-weight compound that unites with a protein (the repressor) to prevent transcription in a repressible operon.

Cork A waterproofing tissue in plants, with suberin-containing cell walls. Produced by a cork cambium.

Corolla (ko role' lah) [L. *corolla*: a small crown] All of the petals of a flower, collectively.

Coronary (kor' oh nair ee) (L. *corona*: crown) Referring to the blood vessels of the heart.

Corpus luteum (kor' pus loo' tee um) [L.: yellow body] A structure formed from a follicle after ovulation; it produces hormones important to the maintenance of pregnancy.

Cortex [L. *cortex*: bark, rind] (1) In plants, the tissue between the epidermis and the vascular tissue of a stem or root. (2) In animals, the outer tissue of certain organs, such as the adrenal cortex and cerebral cortex.

Corticosteroids Steroid hormones produced and released by the cortex of the adrenal gland.

Cotyledon (kot' ul lee' dun) [Gr. *kotyledon*: hollow space] A "seed leaf." An embryonic organ that stores and digests reserve materials; may expand when seed germinates.

Countercurrent exchange An adaptation that promotes maximum exchange of heat or any diffusible substance between two fluids by the fluids flow in opposite directions through parallel tubes close together

Covalent bond A chemical bond that arises from the sharing of electrons between two atoms. Usually a strong bond.

Crassulacean acid metabolism (CAM) A metabolic pathway enabling the plants that possess it to store carbon dioxide at night and then perform photosynthesis during the day with stomata closed.

Crista (plural: cristae) A small, shelflike projection of the inner membrane of a mitochondrion; the site of oxidative phosphorylation.

Critical night length In the photoperiodic flowering response of short-day plants, the length of night above which flowering occurs and below which the plant remains vegetative. (The reverse applies in the case of long-day plants.)

Critical period The age during which some particular type of learning must take place or during which it occurs much more easily than at other times. Typical of song learning among birds.

Cross section A section taken perpendicular to the longest axis of a structure. Also called a transverse section.

Crossing over The mechanism by which linked markers undergo recombination. In general, the term refers to the reciprocal exchange of corresponding segments between two homologous chromatids.

Cryptic [Gr. *kryptos*: hidden] The resemblance of an animal to some part of its environment, which helps it to escape detection by predators.

Cryptochromes [Gr. *kryptos*: hidden + *kroma*: color] Photoreceptors mediating some blue-light effects in plants and animals.

Culture (1) A laboratory association of organisms under controlled conditions. (2) The collection of knowledge, tools, values, and rules that characterize a human society.

Cuticle A waxy layer on the outer surface of a plant or an insect, tending to retard water loss.

Cyanobacteria (sigh an' o bacteria) [Gr. *kuanos*: blue] A lineage of photosynthetic bacteria, formerly referred to as blue-green algae; they use chlorophyll *a* in photosynthesis.

Cyclic AMP See cAMP.

Cyclic electron transport In photosynthetic light reactions, the flow of electrons that produces ATP but no NADPH or O_2.

Cyclins Proteins that activate cyclin-dependent kinases, bringing about transitions in the cell cycle.

Cyclin-dependent kinase (cdk) A *kinase* catalyzes the addition of phosphate groups from ATP to target molecules. Cdk's target proteins are involved in transitions in the cell cycle and are active only when complexed to additional protein subunits, cyclins.

Cytochromes (sy' toe chromes) [Gr. *kytos*: container + *chroma*: color] Iron-containing red proteins, components of the electron-transfer chains in photophosphorylation and respiration.

Cytokinesis (sy' toe kine ee' sis) [Gr. *kytos*: container + *kinein*: to move] The division of the cytoplasm of a dividing cell. (Compare with mitosis.)

Cytokinin (sy' toe kine' in) A member of a class of plant growth substances playing roles in senescence, cell division, and other phenomena.

Cytoplasm The contents of the cell, excluding the nucleus.

Cytoplasmic determinants In animal development, gene products whose spatial distribution may determine such things as embryonic axes.

Cytosine (C) (site' oh seen) A nitrogen-containing base found in DNA and RNA.

Cytoskeleton The network of microtubules and microfilaments that gives a eukaryotic cell its shape and its capacity to arrange its organelles and to move.

Cytosol The fluid portion of the cytoplasm, excluding organelles and other solids.

Cytotoxic T cells (T$_C$) Cells of the cellular immune system that recognize and directly eliminate virus-infected cells. (Compare with helper T cells.)

DAG See Diacylglycerol.

Deciduous [L. *deciduus*: falling off] Refers to a woody plant that sheds it leaves but does not die.

Decomposer See detritivore.

Degeneracy The situation in which a single amino acid may be represented by any of two or more different codons in messenger RNA. Most of the amino acids can be represented by more than one codon.

Deletion A mutation resulting from the loss of a continuous segment of a gene or chromosome. Such mutations never revert to wild type. (Contrast with duplication, point mutation.)

Deme (deem) [Gr. *demos*: the people] Any local population of individuals belonging to the same species that interbreed with one another.

Demographic processes The events—such as births, deaths, immigration, and emigration—that determine the number of individuals in a population.

Demographic stochasticity Random variations in the factors influencing the size, density, and distribution of a population.

Denaturation Loss of activity of an enzyme or nucleic acid molecule as a result of structural changes induced by heat or other means.

Dendrite [Gr. *dendron*: tree] A fiber of a neuron which often cannot carry action potentials. Usually much branched and relatively short compared with the axon, and commonly carries information to the cell body of the neuron.

Denitrification Metabolic activity by which inorganic nitrogen-containing ions are reduced to form nitrogen gas and other products; carried on by certain soil bacteria.

Density dependence Change in the severity of action of agents affecting birth and death rates within populations that are directly or inversely related to population density.

Density independence The state where the severity of action of agents affecting birth and death rates within a population does not change with the density of the population.

Deoxyribonucleic acid See DNA.

Depolarization A change in the electric potential across a membrane from a condition in which the inside of the cell is more negative than the outside to a condition in which the inside is less negative, or even positive, with reference to the outside of the cell. (Contrast with hyperpolarization.)

Derived trait A trait found among members of a lineage that was not present in the ancestors of that lineage.

Dermal tissue system The outer covering of a plant, consisting of epidermis in the young plant and periderm in a plant with extensive secondary growth. (Contrast with ground tissue system and vascular tissue system.)

Desmosome (dez' mo sowm) [Gr. *desmos*: bond + *soma*: body] An adhering junction between animal cells.

Determination Process whereby an embryonic cell or group of cells becomes fixed into a predictable developmental pathway.

Detritivore (di try' ti vore) [L. *detritus*: worn away + *vorare*: to devour] An organism that obtains its energy from the dead bodies and/or waste products of other organisms.

Deuterostome A major evolutionary lineage in animals, characterized by radial cleavage, enterocoelous development, and other traits. (Compare with protostome.)

Development Progressive change, as in structure or metabolism; in most kinds of organisms, development continues throughout the life of the organism.

Diacylglycerol (DAG) In hormone action, the second messenger produced by hydrolytic removal of the head group of certain phospholipids.

Diaphragm (dye' uh fram) [Gr. *diaphrassein*: barricade] A sheet of muscle that separates the thoracic and abdominal cavities in mammals; responsible for breathing. (2) A method of birth control in which a sheet of rubber is fitted over the woman's cervix, blocking the entry of sperm.

Diastole (dye ass' toll ee) [Gr. : dilation] The portion of the cardiac cycle when the heart muscle relaxes. (Contrast with systole.)

Dicot (short for dicotyledon) [Gr. *di*: two + *kotyledon*: a hollow space] This term, not used in this book, formerly referred to all angiosperms other than the monocots. (See eudicot, monocot.)

Differentiation Process whereby originally similar cells follow different developmental pathways. The actual expression of determination.

Diffusion Random movement of molecules or other particles, resulting in even distribution of the particles when no barriers are present.

Digestion Enzyme-catalyzed process by which large, usually insoluble, molecules (foods) are hydrolyzed to form smaller molecules of soluble substances.

Dihybrid cross A mating in which the parents differ with respect to the alleles of two loci of interest.

Dikaryon (di care' ee ahn) [Gr. *di*: two + *karyon*: kernel] A cell or organism carrying two genetically distinguishable nuclei. Common in fungi.

Dioecious (die eesh' us) [Gr.: *di*: two + *oikos*: house] Refers to organisms in which the two sexes are "housed" in two different individuals, so that eggs and sperm are not produced in the same individuals. Examples: humans, fruit flies, date palms. (Contrast with monoecious.)

Diploblastic Having two cell layers. (Contrast with triploblastic.)

Diploid (dip' loid) [Gr. *diplos*: double] Having a chromosome complement consisting of two copies (homologues) of each chromosome. Designated 2*n*.

Directional selection Selection in which phenotypes at one extreme of the population distribution are favored. (Contrast with disruptive selection, stabilizing selection.)

Disaccharide A carbohydrate made up of two monosaccharides (simple sugars).

Displacement activity Apparently irrelevant behavior performed by an animal under conflict situations, especially when tendencies to attack and escape are closely balanced.

Display A behavior that has evolved to influence the actions of other individuals.

Disruptive selection Selection in which phenotypes at both extremes of the population distribution are favored. (Contrast with directional selection; stabilizing selection.)

Distal Away from the point of attachment or other reference point. (Contrast with proximal.)

Disturbance A short-term event that disrupts populations, communities, or ecosystems by changing the environment.

Disulfide bridge The covalent bond between twosulfur atoms (–S—S–) linking to molecules or remote parts of the same molecule.

Diverticulum (di ver tik' u lum) [L. *divertere*: turn away] A small cavity or tube that connects to a major cavity or tube.

Division A term used by some microbiologists and formerly by botanists, corresponding to the term phylum.

DNA (deoxyribonucleic acid) The fundamental hereditary material of all living organisms. In eukaryotes, stored primarily in the cell nucleus. A nucleic acid using deoxyribose rather than ribose.

DNA chip A small glass or plastic square onto which thousands of single-stranded DNA sequences are fixed. Hybridization of cell-derived RNA or DNA to the target sequences can be performed. (See DNA hybridization.)

DNA fingerprint An individual's unique DNA fragments produced by action of restriction endonucleases and separated by electrophoresis.

DNA helicase. An enzyme that functions during DNA replication to unwind the double helix.

DNA hybridization A process by which DNAs from two species are mixed and heated so that interspecific double helixes are formed.

DNA ligase Enzyme that unites Okazaki fragments of the lagging strand during DNA replication; also mends breaks in DNA strands. It connects pieces of a DNA strand and is used in recombinant DNA technology.

DNA methylation Addition of methyl groups to DNA; plays role in regulation of gene expression; protects a bacterium's DNA against its restriction endonucleases.

DNA polymerase Any of a group of enzymes that catalyze the formation of DNA strands from a DNA template.

DNA sequencing Determining the precise sequence of nucleotides in DNA.

DNA topoisomerase Enzymes that introduce positive or negative supercoils into the double-stranded DNA of continuous (circular) chromosomes.

Domain The largest unit in the current taxonomic nomenclature. Members of the three domains (Bacteria, Archaea, and Eukarya) are believed to have been evolving independently of each other for at least a billion years.

Dominance In genetics, the ability of one allelic form of a gene to determine the phenotype of a heterozygous individual, in which the homologous chromosomes carries both it and a different (recessive) allele. (Contrast with recessive.)

Dormancy A condition in which normal activity is suspended, as in some seeds and buds.

Dorsal [L. *dorsum*: back] Pertaining to the back or upper surface. (Contrast with ventral.)

Dorsal lip In amphibian embryo, the dorsal part of the blastpore which directs the development of nearby regions.

Double fertilization Process virtually unique to angiosperms in which one sperm nucleus combines with the egg to produce a zygote, and the other sperm nucleus combines with the two polar nuclei to produce the first cell of the triploid endosperm.

Double helix In DNA, the natural, right-handed coil configuration of two complementary, antiparallel strands.

Duodenum (do' uh dee' num) The beginning portion of the vertebrate small intestine. (Compare with ileum, jejunum.)

Dynein [Gr. *dynamis*: power] A protein that plays a part in the movement of eukaryotic flagella and cilia by means of conformational changes.

Ecdysone (eck die' sone) [Gr. *ek*: out of + *dyo*: to clothe] In insects, a hormone that induces molting.

Ecological community The species living together at a particular site.

Ecological niche (nitch) [L. *nidus*: nest] The functioning of a species in relation to other species and its physical environment.

Ecological succession The sequential replacement of one assemblage of populations by another in a habitat following some disturbance.

Ecology [Gr. *oikos*: house + *logos*: study] The scientific study of the interaction of organisms with their living and nonliving (abiotic) environment.

Ecosystem (eek' oh sis tum) The organisms of a particular habitat, such as a pond or forest, together with the physical environment in which they live.

Ectoderm [Gr. *ektos*: outside + *derma*: skin] The outermost of the three embryonic tissue layers first delineated during gastrulation. Gives rise to the skin, sense organs, nervous system, etc.

Ectotherm [Gr. *ektos*: outside + *thermos*: heat] An animal unable to control its body temperature. (Contrast with endotherm.)

Edema (i dee' mah) [Gr. *oidema*: swelling] Tissue swelling caused by the accumulation of fluid.

Edge effect The changes in ecological processes in a community caused by physical and biological factors originating in an adjacent community.

Effector Any organ, cell, or organelle that moves the organism through the environment or else alters the environment; for example, muscle, exocrine glands, chromatophores.

Effector phase Stage of the immune response, when cytotoxic T cells attack virus-infected cells, and helper T cells assist B cells to differentiate into plasma cells.

Efferent [L. *ex*: out + *ferre*: to bear] In physiology, conducting outward or away from an organ or structure. (Contrast with afferent.)

Egg In all sexually reproducing organisms, the female gamete; in birds, reptiles, and some other vertebrates, a structure witin which early embryonic development occurs. (Compare with amniote egg.)

Elasticity The property of returning quickly to a former state after a disturbance.

Electrocardiogram (EKG) A graphic recording of electrical potentials from the heart.

Electroencephalogram (EEG) A graphic recording of electrical potentials from the brain.

Electromyogram (EMG) A graphic recording of electrical potentials from muscle.

Electron transport. The passage of electrons through a series of proteins with a release of energy which may be captured in a concentration gradient or chemical form such as NADH or ATP.

Electronegativity The tendency of an atom to attract electrons when it occurs as part of a compound.

Electrostatic Pertaining to the attraction and repulsion of negative and positive charges on atoms due to the number and distribution of electrons.

Electrophoresis (e lek' tro fo ree' sis) [L. *electrum*: amber + Gr. *phorein*: to bear] A separation technique in which substances are separated from one another on the basis of their electric charges and molecular weights.

Elongation Growth of a plant axis or cell primarily in the longitudinal direction.

Embolus (em' buh lus) [Gr. *embolos*: inserted object; stopper] A circulating blood clot. Blockage of a blood vessel by an embolus or by a bubble of gas is referred to as an embolism. (Contrast with thrombus.)

Embryo [Gr. *en*: within + *bryein*: to grow] A young animal, or young plant sporophyte, while it is still contained within a protective structure such as a seed, egg, or uterus.

Embryo sac In angiosperms, the female gametophyte. Found within the ovule, it consists of eight

or fewer cells, membrane bounded, but without cellulose walls between them.

Emergent property A property of a complex system that is not exhibited by its individual component parts.

Emigration The deliberate and usually oriented departure of an organism from the habitat in which it has been living.

3′ End (3 prime) The end of a DNA or RNA strand that has a free hydroxyl group at the 3′ carbon of the sugar (deoxyribose or ribose).

5′ End (5 prime) The end of a DNA or RNA strand that has a free phosphate group at the 5′ carbon of the sugar (deoxyribose or ribose).

Endemic (en dem' ik) [Gr. *endemos*: native, dwelling in] Confined to a particular region, thus often having a comparatively restricted distribution.

Endergonic reaction A chemical reaction that requires the input of energy in order to proceed. (Contrast with exergonic reaction.)

Endocrine gland (en' doh krin) [Gr. *endo*: within + *krinein*: to separate] Any gland, such as the adrenal or pituitary gland of vertebrates, that secretes certain substances, especially hormones, into the body through the blood.

Endocytosis A process by which liquids or solid particles are taken up by a cell through invagination of the plasma membrane. (Contrast with exocytosis.)

Endoderm [Gr. *endo*: within + *derma*: skin] The innermost of the three embryonic tissue layers delineated during gastrulation. Gives rise to the digestive and respiratory tracts and structures associated with them.

Endodermis In plants, a specialized cell layer marking the inside of the cortex in roots and some stems. Frequently a barrier to free diffusion of solutes.

Endomembrane system Endoplasmic reticulum plus Golgi apparatus; also lysosomes, when present. A system of membranes that exchange material with one another.

Endoplasmic reticulum (ER) [Gr. *endo*: within + L. *plasma*: form + L. *reticulum*: net] A system of membranous tubes and flattened sacs found in the cytoplasm of eukaryotes. Exists in two forms: rough ER, studded with ribosomes; and smooth ER, lacking ribosomes.

Endorphins Naturally occurring, opiate-like substances in the mammalian brain.

Endoskeleton [Gr. *endo*: within + *skleros*: hard] An internal skeleton covered by other, soft body tissues. (Contrast with exoskeleton.)

Endosperm [Gr. *endo*: within + *sperma*: seed] A specialized triploid seed tissue found only in angiosperms; contains stored nutrients for the developing embryo.

Endosymbiosis [Gr. *endo*: within + *sym*: together + *bios*: life] Two species living together, with one living inside the body (or even the cells) of the other.

Endosymbiotic theory The theory that the eukaryotic cell evolved from a prokaryote that contained other endosymbiotic prokaryotes.

Endotherm [Gr. *endo*: within + *thermos*: heat] An animal that can control its body temperature by the expenditure of its own metabolic energy. (Contrast with ectotherm.)

End product inhibition A control capacity of some metabolic pathways in which the final product produced inhibits an early enzyme in the pathway.

Energetic cost The difference between the energy an animal expends in performing a behavior and the energy it would have expended had it rested.

Energy The capacity to do work or move matter against an opposing force. The capacity to accomplish change.

Enhancer In eukaryotes, a DNA sequence, lying on either side of the gene it regulates, that stimulates a specific promoter.

Enterocoelous development A pattern of development in which the coelum is formed by an outpocketing of the embryonic gut (enteron).

Enterokinase (ent uh row kine' ase) An enzyme secreted by the mucosa of the duodenum. It activates the zymogen trypsinogen to create the active digestive enzyme trypsin.

Entrainment With respect to circadian rhythms, the process whereby the period is adjusted to match the 24-hour environmental cycle.

Entropy (en' tro pee) [Gr. *tropein*: to change] A measure of the degree of disorder in any system. Spontaneous reactions in a closed system are always accompanied by an increase in disorder and entropy.

Environment Whatever surrounds and interacts with a population, an organism or cell. May be external or internal.

Enzyme (en' zime) [Gr. *en*: in + *zyme*: yeast] A protein, on the surface of which are chemical groups so arranged as to make the enzyme a catalyst for a chemical reaction.

Eosinophils Phagocytic white blood cells that attack multicellular parasites once they have been coated with antibodies.

Epi- [Gr.: upon, over] A prefix used to designate a structure located on top of another; for example: epidermis, epiphyte.

Epiblast [Gr. *epi*: upon, over] The upper or overlying portion of the avian blastula which is joined to the hypoblast at the margins of the blastodisc.

Epicotyl (epp' i kot' il) [Gr. *epi*: over + *kotyle*: something hollow] That part of a plant embryo or seedling that is above the cotyledons.

Epidermis [Gr. *epi*: over + *derma*: skin] In plants and animals, the outermost cell layers. (Only one cell layer thick in plants.)

Epididymis (epuh did' uh mus) [Gr. *epi*: over + *didymos*: testicle] Coiled tubules in the testes that store sperm and conduct sperm from the seiminiferous tubules to the vas deferens.

Epinephrine (ep i nef' rin) [Gr. *epi*: over + *nephros*: kidney] The "fight or flight" hormone produced by the medulla of the adrenal gland; it also functions as a neurotransmitter. (Also known as adrenaline.)

Epiphyte (ep' e fyte) [Gr. *epi*: over + *phyton*: plant] A specialized plant that grows on the surface of other plants but does not parasitize them.

Epistasis Interaction between genes in which the presence of a particular allele of one gene determines whether another gene will be expressed.

Epithelium In animals, a layer of cells covering or lining an external surface or a cavity.

Equatorial plate In a cell undergoing mitosis, the region in the middle of a cell where the centromeres will align during metaphase.

Equilibrium Any state of balanced opposing forces and no net change.

ER See Endoplasmic reticulum.

Erythrocyte (ur rith' row site) [Gr. *erythros*: red + *kytos*: container] A red blood cell.

Esophagus (i soff' i gus) [Gr. *oisophagos*: gullet] That part of the gut between the pharynx and the stomach.

Ester linkage A condensation (water-releasing) reaction in which the carboxyl group of a fatty acid reacts with the hydroxyl group of an alcohol. Lipids are formed in this way.

Estivation (ess tuh vay' shun) [L. *aestivalis*: summer] A state of dormancy and hypometabolism that occurs during the summer; usually a means of surviving drought and/or intense heat. Contrast with hibernation.

Estrogen Any of several steroid sex hormones; produced chiefly by the ovaries in mammals.

Estrus (es' truss) [L. *oestrus*: frenzy] The period of heat, or maximum sexual receptivity, in some female mammals. Ordinarily, the estrus is also the time of release of eggs in the female.

Ethylene One of the plant hormones, the gas $H_2C=CH_2$.

Euchromatin Chromatin that is diffuse and non-staining during interphase; may be transcribed. (Contrast with heterochromatin.)

Eudicots (yew die' kots) [Gr. *eu*: true + *di*: two + *kotyledon*: a cup-shaped hollow] The most diverse and abundant lineage of angiosperms. Eudicot embryos have two cotyledons, and eudicot flowers usually have parts (sepals, petals, etc.) in fours and fives.

Eukaryotes (yew car' ree oats) [Gr. *eu*: true + *kary-on*: kernel or nucleus] Organisms whose cells contain their genetic material inside a nucleus. Includes all life other than the viruses, archaea, and bacteria.

Eusocial Term applied to insects, such as termites, ants, and many bees and wasps, in which individuals cooperate in the care of offspring, there are sterile castes, and generations overlap.

Eutrophication (yoo trofe' ik ay' shun) [Gr. *eu*: truly + *trephein*: to flourish] The addition of nutrient materials to a body of water, resulting in changes in ecological processes and species composition therein.

Evolution Any gradual change. Organic evolution, often referred to as evolution, is any genetic and resulting phenotypic change in organisms from generation to generation.

Evolutionary agent Any factor that influences the direction and rate of evolutionary changes.

Evolutionarily conserved Refers to traits that have evolved very slowly and are similar or even identical in individuals of many different phyla.

Evolutionary radiation The proliferation of species within a single evolutionary lineage.

Evolutionary reversal The reappearance of the ancestral state of a trait in a lineage in which that trait had acquired a derived state.

Excision repair The removal and damaged DNA and its replacement by the appropriate nucleotides.

Excitatory postsynaptic potential (EPSP) A change in the resting potential of a postsynaptic membrane in a positive (depolarizing) direction. (Contrast with inhibitory postsynaptic potential.)

Excretion Release of metabolic wastes by an organism.

Exergonic reaction A reaction in which free energy is released. (Contrast with endergonic reaction.)

Exocrine gland (eks' oh krin) [Gr. *exo*: outside + *krinein*: to separate] Any gland, such as a salivary gland, that secretes to the outside of the body or into the gut. (Contrast with endocrine gland.)

Exocytosis A process by which a vesicle within a cell fuses with the plasma membrane and releases its contents to the outside. (Contrast with endocytosis.)

Exon A portion of a DNA molecule, in eukaryotes, that codes for part of a polypeptide. (Contrast with intron.)

Exoskeleton (eks' oh skel' e ton) [Gr. *exos*: outside + *skleros*: hard] A hard covering on the outside of the body to which muscles are attached. (Contrast with endoskeleton.)

Exotoxins Highly toxic proteins released by living, multiplying bacteria.

Experiment The basis of the scientific method, in which particular factors are manipulated while other factors are held constant so that the potential influences of the manipulated factors can be determined.

Exponential growth Growth, especially in the number of organisms in a population, which is a geometric function of the size of the growing entity: the larger the entity, the faster it grows. (Contrast with logistic growth.)

Expression vector A DNA vector, such as a plasmid, that carries a DNA sequence that includes the adjacent sequences for its expression into mRNA and protein in a host cell.

Expressivity The degree to which a genotype is expressed in the phenotype; may be affected by the environment.

Extensor A muscle the extends an appendage.

Extinction The termination of a lineage of organisms.

Extrinsic protein A membrane protein found only on the surface of the membrane. (Contrast with intrinsic protein.)

Extracellular matrix. In animal tissues, a material of heterogeneous composition surrounding cells and performing many functions including adhesion of cells.

Extraembryonic membranes. The four membranes that support the developing embryo of reptiles, birds, and mammals but are not part of the embryo (amnion, allantois, chorion, and yolk sac)

F₁ (first filial generation) The immediate progeny of a parental (P) mating.

F₂ (second filial generation) The immediate progeny of a mating between members of the F₁ generation.

Facilitated diffusion Passive movement through a membrane involving a specific carrier protein; does not proceed against a concentration gradient. (Contrast with active transport, diffusion.)

Facultative anaerobes Prokaryotes that can shift their metabolism between anaerobic and aerobic operations depending on the presence or absence of O_2.

FAD See Flavin adenine dinucleotide.

Fat A triglyceride that is solid at room temperature. (Contrast with oil.)

Fate map. A map of the blastula showing which blastomers will contribute to specific tissues and organs in the mature body.

Fatty acid A molecule with a long hydrocarbon tail and a carboxyl group at the other end. Found in many lipids.

Fauna (faw' nah) All of the animals found in a given area. (Contrast with flora.)

Feces [L. *faeces*: dregs] Waste excreted from the digestive system.

Feedback control Control of a particular process induced, directly or indirectly, by the presence or absence of a product of that process.

Fermentation (fur men tay' shun) [L. *fermentum*: yeast] The anaerobic degradation of a substance such as glucose to smaller molecules with the extraction of energy.

Fertilization Union of gametes. Also known as syngamy.

Fertilization membrane A membrane surrounding an animal egg which becomes rapidly raised above the egg surface within seconds after fertilization, serving to prevent entry of a second sperm.

Fetus The latter stages of an embryo that is still contained in an egg or uterus; in humans, the unborn young from the eighth week of pregnancy to the moment of birth.

Fiber An elongated, tapering cell of flowering plants, usually with a thick cell wall. Serves a support function.

Fibrin A protein that polymerizes to form long threads that provide structure to a blood clot.

Filter feeder An organism that feeds upon much smaller organisms, that are suspended in water or air, by means of a straining device.

Filtration In the excretory physiology of some animals, the process by which the initial urine is formed; water and most solutes are transferred into the excretory tract, while proteins are retained in the blood or hemolymph.

First law of thermodynamics Energy can be neither created nor destroyed.

Fission Reproduction of a prokaryote by division of a cell into two comparable progeny cells.

Fitness The contribution of a genotype or phenotype to the genetic composition of subsequent generations, relative to the contribution of other genotypes or phenotypes. (See inclusive fitness.)

Fixed action pattern A behavior that is genetically programmed.

Flagellum (fla jell' um) (plural: flagella) [L. *flagellum*: whip] Long, whiplike appendage that propels cells. Prokaryotic flagella differ sharply from those found in eukaryotes.

Flavin adenine dinucleotide (FAD) A coenzyme involved in redox reactions and containing the vitamin riboflavin (B₂).

Flexor A muscle that flexes an appendage.

Flora (flore' ah) All of the plants found in a given area. (Contrast with fauna.)

Floral meristem Meristem that forms the sexual parts of flowering plants (sepals, petals, stamens, and carpels).

Florigen A plant hormone (not yet isolated) involved in the conversion of a vegetative shoot apex to a flower.

Flower The total reproductive structure of an angiosperm; its basic parts include the calyx, corolla, stamens, and carpels.

Fluid mosaic model A molecular model for the structure of biological membranes consisting of a fluid phospholipid bilayer in which suspended proteins are free to move in the plane of the bilayer.

Fluorescence The emission of a photon of visible light by an excited atom or molecule.

Follicle [L. *folliculus*: little bag] In female mammals, an immature egg surrounded by nutritive cells.

Follicle-stimulating hormone A gonadotropic hormone produced by the anterior pituitary.

Food chain A portion of a food web, most commonly a simple sequence of prey species and the predators that consume them.

Food vacuole Membrane enclosed structure formed by phagocytosis in which engulfed food particles are digested by the action of lysosomal enzymes.

Food web The complete set of food links between species in a community; a diagram indicating which ones are the eaters and which are eaten.

Forb Any broad-leaved herbaceous plant. Especially applied to such plants growing in grasslands.

Fossil Any recognizable structure originating from an organism, or any impression from such a structure, that has been preserved over geological time.

Fossil fuel A fuel (particularly petroleum products) composed of the remains of organisms that lived in the remote past.

Founder effect Random changes in allele frequencies resulting from establishment of a population by a very small number of individuals.

Fovea [L. *fovea*; a small pit] The area, in the vertebrate retina, of most distinct vision.

Frame-shift mutation A mutation resulting from the addition or deletion of one or two consecutive base pairs in the DNA sequence of a gene, resulting in misreading mRNA during translation and production of a nonfunctional protein. (Contrast with missense mutation, nonsense mutation, synonymous mutation.)

Free energy That energy which is available for doing useful work, after allowance has been made for the increase or decrease of disorder.

Frequency-dependent selection Selection that changes in intensity with the proportion of individuals in a population having the trait.

Fruit In angiosperms, a ripened and mature ovary (or group of ovaries) containing the seeds. Sometimes applied to reproductive structures of other groups of plants.

Fruiting body A structure that bears spores.

Functional genomics The assignment of functional roles to genes first identified by sequencing entire genomes.

Functional group A characteristic combination of atoms that contribute specific properties when attached to larger molecules.

Functional mRNA Eukaryotic mRNA that has been modified after transcription by the removal of introns and the addition of a 5' cap and a 3' poly(A) tail.

Fungus A member of the kingdom Fungi, a (usually) multicellular eukaryote with absorptive nutrition. (Yeasts are unicellular fungi.)

G cap A chemically modified GTP added to the 5' end of mRNA; facilitates binding of mRNA to ribosome and prevents mRNA breakdown.

G₁ phase In the cell cycle, the gap between the end of mitosis and the onset of the S phase.

G₂ phase In the cell cycle, the gap between the S (synthesis) phase and the onset of mitosis.

G protein A membrane protein involved in signal transduction; characterized by binding GDP or GTP.

Gametangium (gam uh tan' gee um) [Gr. *gamos*: marriage + *angeion*: vessel] Any plant or fungal structure within which a gamete is formed.

Gamete (gam' eet) [Gr. *gamete/gametes*: wife, husband] The mature sexual reproductive cell: the egg or the sperm.

Gametogenesis (ga meet' oh jen' e sis) [Gr. *gamete/gametes*: wife, husband + *genesis*: source] The specialized series of cellular divisions that leads to the production of sex cells (gametes). (Contrast with oogenesis and spermatogenesis.)

Gametophyte (ga meet' oh fyte) In plants and photosynthetic protists with alternation of generations, the multicellular haploid phase that produces the gametes. (Contrast with sporophyte.)

Ganglion (gang' glee un) [Gr.: tumor] A group or concentration of neuron cell bodies.

Gap genes During insect development, the first step of segmentation genes to act organizing the anterior-posterior axis.

Gap junction A 2.7-nanometer gap between plasma membranes of two animal cells, spanned by protein channels. Gap junctions allow chemical substances or electrical signals to pass from cell to cell.

Gas exchange In animals, the process of taking up oxygen from the environment and releasing carbon dioxide to the environment.

Gastrovascular cavity Serving for both digestion (gastro) and circulation (vascular); in particular, the central cavity of the body of jellyfish and other cnidarians.

Gastrula (gas' true luh) [Gr. *gaster*: stomach] An embryo forming the characteristic three cell layers (ectoderm, endoderm, and mesoderm) which will give rise to all of the major tissue systems of the adult animal.

Gastrulation Development of a blastula into a gastrula.

Gated channel A membrane protein that opens and closes in response to binding of specific molecules or to changes in membrane potential. When open, it allows specific ions to move across the membrane.

Gene [Gr. *genes*: to produce] A unit of heredity. Used here as the unit of genetic function which carries the information for a single polypeptide or RNA.

Gene amplification Creation of multiple copies of a particular gene, allowing the production of large amounts of the RNA transcript (as in rRNA synthesis in oocytes).

Gene cloning Formation of a clone of bacteria or yeast cells containing a particular foreign gene.

Gene family A set of identical, or once-identical, genes, derived from a single parent gene; need not be on the same chromosomes; classic example is the globin family in vertebrates.

Gene flow The exchange of genes between different species (an extreme case referred to as hybridization) or between different populations of the same species caused by migration following breeding.

Gene frequency See Allele frequency.

Gene library All of the cloned DNA fragments generated by action of a restriction endonuclease on a genome or chromosome.

Gene pool All of the alleles of all of the genes in a population.

Gene therapy Treatment of a genetic disease by providing patients with cells containing functioning alleles of the genes that are nonfunctional in their bodies.

Generative cell In a pollen tube, a haploid nucleus that undergoes mitosis to produce the two sperm nuclei that participate in double fertilization. (Contrast with tube cell.)

Genetics The study of the structure, functioning, and inheritance of genes, the units of hereditary information.

Genetic drift Changes in gene frequencies from generation to generation in a small population as a result of random (chance) processes.

Genetic screening The application of medical tests to determine whether an individual carries a specific allele.

Genetic stochasticity Random variation in the frequencies of alleles and genotypes in a population over time. (Compare with demographic stochasticity.)

Genome (jee' nome) All the genes in a complete haploid set of chromosomes. (Compare with proteome.)

Genomics The study of entire sets of genes and their interactions.

Genomic imprinting When a given gene's phenotype is determined by whether that gene is inherited from the male or the female parent.

Genotype (jean' oh type) [Gr. *gen*: to produce + *typos*: impression] An exact description of the genetic constitution of an individual, either with respect to a single trait or with respect to a larger set of traits. (Contrast with phenotype.)

Genus (jean' us) (plural: genera) [Gr. *genos*: stock, kind] A group of related, similar species.

Geotropism See gravitropism.

Germ cell [L. *germen*: to beget] A reproductive cell or gamete of a multicellular organism. Contrast with somatic cell.

Germ layers The three embryonic tissue layers formed during gastrulation (ectoderm, mesoderm, endoderm).

Germination The sprouting of a seed or spore.

Gestation (jes tay' shun) [L. *gestare*: to bear] The period during which the embryo of a mammal develops within the uterus. Also known as pregnancy.

Gibberellin (jib er el' lin) A class of plant growth substances playing roles in stem elongation, seed germination, flowering of certain plants, etc. Named for the fungus *Gibberella*.

Gill An organ for gas exchange in aquatic organisms.

Gill arch A skeletal structure that supports gill filaments and the blood vessels that supply them.

Gizzard (giz' erd) [L. *gigeria*: cooked chicken parts] A muscular port of the stomach of birds that grinds up food, sometimes with the aid of fragments of stone.

Gland An organ or group of cells that produces and secretes one or more substances.

Glans penis Sexually sensitive tissue at the tip of the penis.

Glia (glee' uh) [Gr. *glia*: glue] Cells, found only in the nervous system, that do not conduct action potentials.

Glomerulus (glo mare' yew lus) [L. *glomus*: ball] Sites in the kidney where blood filtration takes place. Each glomerulus consists of a knot of capillaries served by afferent and efferent arterioles.

Glucocorticoids Steroid hormones produced by the adrenal cortex. Secreted in response to ACTH, they inhibit glucose uptake by many tissues in addition to mediating other stress responses.

Glucagon A hormone produced and released by cells in the islets of Langerhans of the pancreas. It stimulates the breakdown of glycogen in liver cells.

Gluconeogenesis The biochemical synthesis of glucose from other substances, such as amino acids, lactate, and glycerol.

Glycerol (gliss' er ole) A three-carbon alcohol with three hydroxyl groups; a component of phospholipids and triglycerides.

Glycogen (gly' ko jen) An energy storage polysaccharide found in animals and fungi; a branched-chain polymer of glucose, similar to starch.

Glycolysis (gly kol' li sis) [Gr. *gleukos*: sugar + *lysis*: break apart] The enzymatic breakdown of glucose to pyruvic acid. One of the evolutionarily oldest of the cellular energy-yielding mechanisms.

Glycosidic linkage The bond between sugar molecules through an intervening oxygen atom (–O–).

Glyoxysome (gly ox' ee soam) An organelle found in plants, in which stored lipids are converted to carbohydrates.

Golgi apparatus (goal' jee) A system of concentrically folded membranes found in the cytoplasm of eukaryotic cells; functions in secretion from cell by exocytosis.

Gonad (go' nad) [Gr. *gone*: seed] An organ that produces sex cells in animals: either an ovary (female gonad) or testis (male gonad).

Gonadotropin A hormone that stimulates the gonads.

Gondwana The large southern land mass that existed from the Cambrian (540 mya) to the Jurassic (138 mya). Present-day remnants are South America, Africa, India, Australia, and Antarctica.

Graft A bud or stem segment from one plant artificially and viably attached to another plant. A form of asexual reproduction.

Gram stain A differential purple stain useful in characterizing bacteria.

Granum (plural: grana) Within a chloroplast, a stack of thylakoids.

Gravitropism A directed plant growth response to gravity.

Grazer An animal that eats the vegetative tissues of herbaceous plants.

Green gland An excretory organ of crustaceans.

Greenhouse effect The heating of Earth's atmosphere by gases that are transparent to sunlight but opaque to heat.

Greenhouse gases Gases that contribute climate warming (water vapor, carbon dioxide, and methane) because they are transparent to light but opaque to heat.

Gross primary production The total energy captured by plants growing in a particular area.

Ground meristem That part of an apical meristem that gives rise to the ground tissue system of the primary plant body.

Ground tissue system Those parts of the plant body not included in the dermal or vascular tissue systems. Ground tissues function in storage, photosynthesis, and support.

Group transfer The exchange of atoms between molecules.

Growth Irreversible increase in volume (probably the most accurate definition, but at best a dangerous oversimplification).

Growth factors A group of proteins that circulate in the blood and trigger the normal growth of cells. Each growth factor acts only on certain target cells.

Guanine (G) (gwan' een) A nitrogen-containing base found in DNA, RNA, and GTP.

Guard cells In plants, specialized, paired epidermal cells that surround and control the opening of a stoma (pore). See stoma.

Gut An animal's digestive tract.

Guttation The extrusion of liquid water through openings in leaves, caused by root pressure.

Gymnosperm (jim' no sperm) [Gr. *gymnos*: naked + *sperma*: seed] A plant, such as a pine or other conifer, whose seeds do not develop within an ovary (hence, the seeds are "naked").

Gyrus The raised or ridged portion of the convoluted surface of the brain. (Contrast to sulcus.)

Habit The form or pattern of growth characteristic of an organism.

Habitat The environment in which an organism lives.

Habituation (ha bich' oo ay shun) The simplest form of learning, in which an animal presented with a stimulus without reward or punishment eventually ceases to respond.

Hair cell A type of mechanoreceptor in animals. Detects sound waves and other forms of motion in air or water.

Half-life The time required for half of a sample of a radioactive isotope to decay to its stable, nonradioactive form.

Halophyte (hal' oh fyte) [Gr. *halos*: salt + *phyton*: plant] A plant that grows in a saline (salty) environment.

Haploid (hap' loid) [Gr. *haploeides*: single] Having a chromosome complement consisting of just one copy of each chromosome; designated $1n$ or n. (Contrast with diploid.)

Hardy–Weinberg equilibrium The allele frequency at a given locus in a sexually reproducing population that is not being acted on by agents of evolution.

Haustorium (haw stor' ee um) [L. *haustus*: draw up] A specialized hypha or other structure by which fungi and some parasitic plants draw food from a host plant.

Haversian systems Units of organization in compact bone that reflect the action of intercommunicating osteoblasts.

Heat-shock proteins Chaperone proteins expressed in cells exposed to high temperatures or other forms of environmental stress.

Helper T cells (T_H) T cells that participate in the activation of B cells and of other T cells; targets of the HIV-I virus, the agent of AIDS. (Contrast with cytotoxic T cells.)

Hematocrit (heme at' o krit) [Gr. *heaema*: blood + *krites*: judge] The proportion of 100 cc of blood that consists of red blood cells.

Hemizygous (hem' ee zie' gus) [Gr. *hemi*: half + *zygotos*: joined] In a diploid organism, having only one allele for a given trait, typically the case for X-linked genes in male mammals and Z-linked genes in female birds. (Contrast with homozygous, heterozygous.)

Hemoglobin (hee' mo glow bin) [Gr. *heaema*: blood + L. *globus*: globe] Oxygen-transporting protein found in the red blood cells of vertebrates (and found in some invertebrates).

Hensen's node In avian embryos, a structure at the anterior end of the primitive groove; determines the fates of cells passing over it during gastrulation.

Hepatic (heh pat' ik) [Gr. *hepar*: liver] Pertaining to the liver.

Hepatic duct Duct that conveys bile from the liver to the gallbladder.

Herbivore (ur' bi vore) [L. *herba*: plant + *vorare*: to devour] An animal that eats plant tissues. (Contrast with carnivore, detritivore, omnivore.)

Heritable Able to be inherited; in biology refers to genetically influenced traits.

Hermaphroditism (her maf' row dite ism) [Gr. Hermes (messenger god) + Aphrodite (goddess of love)] The coexistence of both female and male sex organs in the same organism.

Hertz (abbreviated Hz) Cycles per second.

Hetero- [Gr.: *heteros*: other, different] A prefix specifying that two or more different conditions are involved; for example, heterotroph, heterozygous.

Heterochromatin Chromatin that retains its coiling during interphase; generally not transcribed. (Contrast with euchromatin.)

Heterocyst A large, thick-walled cell in the filaments of certain cyanobacteria; performs nitrogen fixation.

Heterogeneous nuclear RNA (hnRNA) The product of transcription of a eukaryotic gene, including transcripts of introns.

Heterokaryon In fungi, hypha containing two genetically different nuclei.

Heteromorphic (het' er oh more' fik) [Gr. *heteros*: different + *morphe*: form] having a different form or appearance, as two heteromorphic life stages of a plant. (Contrast with isomorphic.)

Heterosporous (het' er os' por us) Producing two types of spores, one of which gives rise to a female megaspore and the other to a male microspore. (Contrast with homosporous.)

Heterosis Situation in which heterozygous genotypes are superior to homozygous genotypes with respect to growth, survival, or fertility. Also called hybrid vigor.

Heterotherm An animal that regulates its body temperature at a constant level at some times but not others, such as a hibernator.

Heterotroph (het' er oh trof) [Gr. *heteros*: different + *trophe*: food] An organism that requires preformed organic molecules as food. (Contrast with autotroph.)

Heterozygous (het' er oh zie' gus) [Gr. *heteros*: different + *zygotos*: joined] Of a diploid organism having different alleles of a given gene on the pair of homologues carrying that gene. (Contrast with homozygous.)

Hibernation [L. *hibernum*: winter] The state of inactivity of some animals during winter; marked by a drop in body temperature and metabolic rate.

Hierarchical sequencing An approach to DNA sequencing in which markers are mapped and DNA sequences are aligned by matching overlapping sites of known sequence.

Highly repetitive DNA Short DNA sequences present in millions of copies in the genome, next to each other (in tandem). In reassociation experiments, denatured highly repetitive DNA reanneals very quickly.

Hippocampus A part of the forebrain that takes part in long-term memory formation.

Histamine (hiss' tah meen) A substance released by damaged tissue, or by mast cells in response to allergens. Histamine increases vascular permeability, leading to edema (swelling).

Histology The study of tissues.

Histone Any one of a group of basic proteins forming the core of a nucleosome, the structural unit of a eukaryotic chromosome. (Compare with nucleosome.)

Hierarchical sequencing An approach to DNA sequencing in which markers are mapped and DNA sequences are aligned by matching overlapping sites of known sequence.

hnRNA See heterogeneous nuclear RNA.

Homeobox A 180-base-pair segment of DNA found in certain genes (called Hox genes), perhaps regulating the expression of other genes and thus controlling large-scale developmental processes.

Homeostasis (home' ee o sta' sis) [Gr. *homos*: same + *stasis*: position] The maintenance of a steady state, such as a constant temperature or a stable social structure, by means of physiological or behavioral feedback responses.

Homeotherm (home' ee o therm) [Gr. *homos*: same + *thermos*: heat] An animal that maintains a constant body temperature by its own internal heating and cooling mechanisms. (Contrast with heterotherm, poikilotherm.)

Homeotic genes (home ee ot' ic) Genes that determine the developmental fate of entire segments of an animal.

Homeotic mutations Mutations in homeotic genes that drastically alter the characteristics of a particular body segment, giving it the characteristics of other segments (as when wings grow from a *Drosophila* thoracic segment that should have produced legs).

Homo- [Gr. *homos*: same] Prefix indicating two or more similar conditions, structures, or processes. Contrast to hetero-.

Homolog (home' o log') [Gr. *homos*: same + *logos*: word] One of a pair (or larger set, of chromosomes having the same overall genetic composition and sequence. In diploid organisms, each chromosome inherited from one parent is matched by an identical (except for mutational changes) chromosome—its homolog—from the other parent.

Homology (ho mol' o jee) [Gr. *homologia*: of one mind; agreement] A similarity between two structures that is due to inheritance from a common ancestor. The structures are said to be homologous. (Contrast with analogy.)

Homoplasy (home' uh play zee) [Gr. *homos*: same + *plastikos*: shape, mold] The presence in several species of a trait not present in their most common ancestor. Can result from convergent evolution, reverse evolution, or parallel evolution.

Homosporous Producing a single type of spore that gives rise to a single type of gametophyte, bearing both female and male reproductive organs. (Contrast with heterosporous.)

Homozygous (home' oh zie' gus) [Gr. *homos*: same + *zygotos*: joined] In a diploid organism, having identical alleles of a given gene on both homologous chromosomes. An individual may be a homozygote with respect to one gene and a heterozygote with respect to another. (Contrast with heterozygous.)

Hormone (hore' mone) [Gr. *hormon*: to excite, stimulate] A substance produced in minute amount at one site in a multicellular organism and transported to another site where it acts on target cells.

Host An organism that harbors a parasite or symbiont and provides it with nourishment.

Hox genes See homeobox.

Humoral immune response The part of the immune system mediated by B cells that produce circulating antibodies active against extracellular bacterial and viral infections.

Humus (hew' muss) The partly decomposed remains of plants and animals on the surface of a soil.

Hyaluronidase ((high' uh loo ron' uh dase) An enzyme that digests proteoglycans. In sperm cells, it digests the coatings surrounding an egg so the sperm can enter.

Hybrid (high' brid) [L. *hybrida*: mongrel] The offspring of genetically dissimilar parents. In molecular biology, a double helix formed of nucleic acids from different sources.

Hybrid vigor See heterosis.

Hybridoma A cell produced by the fusion of an antibody-producing cell with a myeloma cell; it produces monoclonal antibodies.

Hybrid zone A narrow zone where two populations interbreed, producing hybrid individuals.

Hydrocarbon A compound containing only carbon and hydrogen atoms.

Hydrogen bond A weak electrostatic bond which arises from the attraction between the slight positive charge on a hydrogen atom and a slight negative charge on a nearby oxygen or nitrogen atom.

Hydrological cycle The movement of water from the oceans to the atmosphere, to the soil, and back to the oceans.

Hydrolysis (high drol' uh sis) [Gr. *hydro*: water + *lysis*: break apart] A chemical reaction that breaks a bond by inserting the components of water: AB + $H_2O \rightarrow AH + BOH$.

Hydrophilic (high dro fill' ik) [Gr. *hydro*: water + *philia*: love] Having an affinity for water. (Contrast with hydrophobic.)

Hydrophobic (high dro foe' bik) [Gr. *hydro*: water + *phobia*: fear] Having no affinity for water. Uncharged and nonpolar groups of atoms are hydrophobic, for example fats and side chain of the amino acid phenylalanine. (Contrast with hydrophilic.)

Hydrostatic pressure Pressure generated by compression of liquid in a confined space. Generated in plants, fungi, and some protists with cell walls by the osmotic uptake of water. Generated in animals with closed circulatory systems by the beating of a heart.

Hydrostatic skeleton The incompressible internal liquids of some animals that transfer forces from one part of the body to another when acted upon by the surrounding muscles.

Hydroxyl group The —OH group found on alcohols and sugars.

Hyper- [Gk. *hyper*: above, over] Prefix indicating above, higher, more.

Hyperpolarization A change in the resting potential of a membrane so the inside of a cell becomes more electronegative. (Contrast with depolarization.)

Hypersensitive response A defensive response of plants to microbial infection; it results in a "dead spot."

Hypertension High blood pressure.

Hypertonic Having a greater solute concentration. Said of one solution compared to another. (Contrast with hypotonic, isotonic.)

Hypha (high' fuh) (plural: hyphae) [Gr. *hyphe*: web] In the fungi and oomycetes, any single filament.

Hypo- [Gk. *hypo*: beneath, under] Prefix indicating underneath, below, less.

Hypoblast The lower tissue portion of the avian blastula which is joined to the epiblast at the margins of the blastodisc.

Hypocotyl [Gk. *hypo*: beneath + *kotyledon*: hollow space] That part of the embryonic or seedling plant shoot that is below the cotyledons.

Hypothalamus The part of the brain lying below the thalamus; it coordinates water balance, reproduction, temperature regulation, and metabolism.

Hypothesis A tentative answer to a question, from which testable predictions can be generated. (Contrast with theory.)

Hypothesis-prediction method A method of science in which hypotheses are generated, predictions are made from them, and experiments and observations are performed to test the predictions.

Hypotonic Having a lesser solute concentration. Said of one solution in comparing it to another. (Contrast with hypotonic, isotonic.)

Imaginal disc [L. *imagos*: image, form] In insect larvae, groups of cells that develop into specific adult organs.

Imbibition Water uptake by a seed; first step in germination.

Immune system [L. *immunis*: exempt from] A system in vertebrates that recognizes and attempts to eliminate or neutralize foreign substances (e.g., bacteria, viruses, pollutants).

Immunization The deliberate introduction of antigen to bring about an immune response.

Immunoglobulins A class of proteins, with a characteristic structure, active as receptors and effectors in the immune system.

Immunological memory The capacity to more rapidly and massively respond to a second exposure to an antigen than occurred on first exposure.

Immunological tolerance A mechanism by which an animal does not mount an immune response to the antigenic determinants of its own macromolecules.

Imprinting (1) In genetics, the differential modification of a gene depending on whether it is present in a male or a female. (2) In animal behavior, a rapid form of learning in which an animal comes to make a particular response, which is maintained for life, to some object or other organism.

Inclusive fitness The sum of an individual's genetic contribution to subsequent generations both via production of its own offspring and via its influence on the survival of relatives who are not direct descendants.

Incomplete dominance Condition in which the heterozygous phenotype is intermediate between the two homozygous phenotypes.

Incomplete metamorphosis Insect development in which changes between instars are gradual.

Incus (in' kus) [L. *incus*: anvil] The middle of the three bones that conduct movements of the eardrum to the oval window of the inner ear. (See malleus, stapes.)

Independent assortment During meiosis, the random separation of genes carried on nonhomologous chromosomes. Articulated by Mendel as his second law.

Individual fitness That component of inclusive fitness resulting from an organism producing its own offspring. (Contrast with kin selection component.)

Indoleacetic acid See auxin.

Induced fit A change in enzyme conformation upon binding to substrate with an increase in the rate of catalysis.

Induced mutation A mutation resulting from treatment with a chemical or other agent.

Inducer (1) In enzyme systems, a small molecule which, when added to a growth medium, causes a large increase in the level of some enzyme. (2) In embryology, a substance that causes a group of target cells to differentiate in a particular way.

Inducible enzyme An enzyme that is present in much larger amounts when a particular compound (the inducer) has been added to the system. (Contrast with constitutive enzyme.)

Inflammation A nonspecific defense against pathogens; characterized by redness, swelling, pain, and increased temperature.

Inflorescence A structure composed of several flowers.

Inflorescence meristem A meristem that produces floral meristems as well as other small leafy structures (bracts).

Inhibitor A substance that binds to the surface of an enzyme and interferes with its action on its substrates.

Inhibitory postsynaptic potential A change in the resting potential of a postsynaptic membrane in the hyperpolarizing (negative) direction.

Initial cells In plant meristems, undifferentiated cells that retain the capacity to divide producing both undifferentiated cells (initials) and cells committed to differentiation. (Compare with stem cells.)

Initiation complex Combination of a ribosomal light subunit, an mRNA molecule, and the tRNA charged with the first amino acid coded for by the mRNA; formed at the onset of translation.

Initiation factors Proteins that assist in forming the translation initiation complex at the ribosome.

Inner cell mass Derived from the mammalian blastula (bastocyst), the inner cell mass will give rise to the yolk sac (via hypoblast) and embryo (via epiblast).

Inositol triphosphate (IP_3) An intracellular second messenger derived from membrane phospholipids.

Instar (in' star) An immature stage of an insect between molts.

Insulin (in' su lin) [L. *insula*: island] A hormone synthesized in islet cells of the pancreas that promotes the conversion of glucose into the storage material, glycogen.

Integral membrane protein A membrane protein embedded in the bilayer of the membrane. (Contrast with peripheral membrane protein.)

Integrase An enzyme that integrates retroviral cDNA into the genome of the host cell.

Integrated pest management Control of pests by the use of natural predators and parasites in conjunction with sparing use of chemicals; an attempt to limit environmental damage.

Integument [L. *integumentum*: covering] A protective surface structure. In gymnosperms and angiosperms, a layer of tissue around the ovule which will become the seed coat.

Intercalary meristem A meristematic region in plants which occurs not apically, but between two regions of mature tissue. Intercalary meristems occur in the nodes of grass stems, for example.

Intercostal muscles Muscles between the ribs that can augment breathing movements by elevating and suppressing the rib cage.

Interferon A glycoprotein produced by virus-infected animal cells; increases the resistance of neighboring cells to the virus.

Interleukins Regulatory proteins, produced by macrophages and lymphocytes, that act upon other lymphocytes and direct their development.

Intermediate filaments Cytoskeletal component with diameters between the larger microtubules and smaller microfilaments.

Internode The region between two nodes of a plant stem.

Interphase The period between successive nuclear divisions during which the chromosomes are diffuse and the nuclear envelope is intact. It is during this period that the cell is most active in transcribing and translating genetic information.

Interspecific competition Competition between members of two or more species. (Contrast with intraspecific competition.)

Intertropical convergence zone The tropical region where the air rises most strongly; moves north and south with the passage of the sun overhead.

Intraspecific competition Competition among members of the same species. (Contrast with interspecific competition.)

Intrinsic protein A membrane protein that is embedded in the phospholipid bilayer of the membrane. (Contrast with extrinsic protein.)

Intrinsic rate of increase The rate at which a population can grow when its density is low and environmental conditions are highly favorable.

Intron A portion of a DNA molecule that, because of RNA splicing, is not involved in coding for part of a polypeptide molecule. (Contrast with exon.)

Invagination An infolding of cells during animal embryonic development.

Inversion A rare 180° reversal of the order of genes within a segment of a chromosome.

Invertebrate Any animal that is not a vertebrate, that is, whose nerve cord is not enclosed in a backbone of bony segments.

In vitro [L.: in glass] In a test tube, rather than in a living organism. (Contrast with in vivo.)

In vivo [L.: in the living state] In a living organism. Many processes that occur in vivo can be reproduced in vitro with the right selection of cellular components. (Contrast with in vitro.)

Ion (eye' on) [Gr. *ion*: wanderer] An atom or group of atoms with electrons added or removed, giving it a negative or positive electrical charge.

Ion channel A membrane protein that can let ions diffuse across the membrane. The channel can be ion-selective, and it can be voltage-gated or ligand-gated.

Ionic bond An electrostatic attraction between positively and negatively charged ions. Usually a strong bond.

Iris (eye' ris) [Gr. *iris*: rainbow] The round, pigmented membrane that surrounds the pupil of the eye and adjusts its aperture to regulate the amount of light entering the eye.

Irruption A rapid increase in the density of a population. Often followed by massive emigration.

Islets of Langerhans Clusters of hormone-producing cells in the pancreas.

Iso- [Gr. *isos*: equal] Prefix used two separate entities that share some element of identity.

Isogamous Describes male and female gametes that are morphologically identical.

Isolating mechanism Geographical, physiological, ecological, or behavioral mechanisms that lead to a reduction in the frequency of successful matings between individuals in separate populations of a species. Can lead to the eventual evolution of separate species.

Isomers Molecules consisting of the same numbers and kinds of atoms, but differing in the bonding patterns by which the atoms are held together.

Isomorphic (eye so more' fik) [Gr. *isos*: equal + *morphe*: form] Having the same form or appearance, as when the haploid and diploid life stages of an organism appear identical. (Contrast with heteromorphic.)

Isotonic Having the same solute concentration; said of two solutions. (Contrast with hypertonic, hypotonic.)

Isotope (eye' so tope) [Gr. *isos*: equal + *topos*: place] Isotopes of a given chemical element have the same number of protons in their nuclei (and thus are in the same position on the periodic table), but differ in the number of neutrons

Isozymes Forms of an enzyme that have somewhat different amino acid sequences but catalyze the same reaction.

Jasmonates Plant hormones that trigger defenses against pathogens and herbivores.

Jejunum (jih jew' num) The middle division of the small intestine, where most absorption of nutrients occurs. (See duodenum, ileum.)

Joule (jool, or jowl) A unit of energy, equal to 0.24 calories.

Juvenile hormone In insects, a hormone maintaining larval growth and preventing maturation or pupation.

Karyotype The number, forms, and types of chromosomes in a cell.

Kelvin temperature scale See absolute temperature scale.

Keratin (ker' a tin) [Gr. keras: horn] A protein which contains sulfur and is part of such hard tissues as horn, nail, and the outermost cells of the skin.

Ketone (key' tone) A compound with a C=O group attached to two other groups, neither of which is an H atom. Many sugars are ketones. (Contrast with aldehyde.)

Keystone species A species that exerts a major influence on the composition and dynamics of the community in which it lives.

Kidneys A pair of excretory organs in vertebrates.

Kin selection The component of inclusive fitness resulting from helping the survival of relatives containing the same alleles by descent from a common ancestor.

Kinase (kye' nase) An enzyme that transfers a phosphate group from ATP to another molecule. Protein kinases transfer phosphate from ATP to specific proteins, playing important roles in cell regulation.

Kinesin Motor protein having the capacity to attach to organelles or vesicles and move them along microtubules of the cytoskeleton.

Kinetic energy The energy associated with movement.

Kinetochore (kin net' oh core) [Gr. *kinetos*: moving] Specialized structure on a centromere to which microtubules attach.

Koch's posulates Four rules for establishing that a particular microorganism causes a particular disease.

Krebs cycle See citric acid cycle.

Lactic acid fermentation Fermentation whose end product is lactic acid (lactate).

Lagging strand In DNA replication, the daughter strand that is synthesized in discontinuous stretches. (See Okazaki fragments.)

Lamella (la mell' ah) (L. *lamina*: thin sheet] Layer.

Larva (plural: larvae) [L. *lares*: guiding spirits] An immature stage of any invertebrate animal that differs dramatically in appearance from the adult.

Larynx (lar' inks) [Gk. *larynx*: voice box] A structure between the pharynx and the trachea that includes the vocal cords.

Lateral Pertaining to the side.

Lateral bud Located above the point of attachment of leaf to stem, an axillary meristem, short stem, immature leaves, and covering scales.

Lateral gene transfer The transfer of genes from one prokaryotic species to another.

Lateral meristems The vascular cambium and cork cambium, which give rise to secondary tissue in plants.

Laticifers (luh tiss' uh furs) In some plants, elongated cells containing secondary plant products such as latex.

Leader sequence A sequence of amino acids at the amino-terminal end of a newly synthesized protein; determines where the protein will be placed in the cell.

Leading strand In DNA replication, the daughter strand that is synthesized continuously. (Contrast with lagging strand.)

Lenticel (len' ti sill) Spongy region in a plant's periderm, allowing gas exchange.

Leukocyte (loo' ko sight) [Gr. *leukos*: clear + *kytos*: container] A white blood cell.

Lichen (lie' kun) An organism resulting from the symbiotic association of a true fungus and either a cyanobacterium or a unicellular alga.

Life cycle The entire span of the life of an organism from the moment of fertilization (or asexual generation) to the time it reproduces in turn.

Life history The stages an individual goes through during its life.

Life table A table showing, for a group of equal-aged individuals, the proportion still alive at different times in the future and the number of offspring they produce during each time interval.

Ligament A band of connective tissue linking two bones in a joint.

Ligand (lig' and) Any molecule that binds to a receptor site of another (usually larger) molecule.

Lignin The principal noncarbohydrate component of wood, a polymer that binds together cellulose fibrils in some plant cell walls.

Limbic system A group of primitive vertebrate forebrain nuclei that form a network and are involved in emotions, drives, instinctive behaviors, learning, and memory.

Limiting resource The required resource whose supply most strongly influences the size of a population.

Linkage Association between genetic markers on the same chromosome such that they do not show random assortment and seldom recombine; the closer the markers, the lower the frequency of recombination.

Lipase (lip' ase; lye' pase) An enzyme that digests fats.

Lipids (lip' ids) [Gr. *lipos*: fat] Substances in a cell which are easily extracted by organic solvents; fats, oils, waxes, steroids, and other large organic molecules, including those which, with proteins, make up the cell membranes. (Compare with phospholipids.)

Littoral zone The coastal zone from the upper limits of tidal action down to the depths where the water is thoroughly stirred by wave action.

Liver A large digestive gland. In vertebrates, it secretes bile and is involved in the formation of blood.

Lobes Regions of the human cerebral hemispheres; includes the temporal, frontal, parietal, and occipital lobes.

Locus In genetics, a specific location on a chromosome. May be considered to be synonymous with *gene*.

Logistic growth Growth, especially in the size of an organism or in the number of organisms that constitute a population, which slows steadily as the entity approaches its maximum size. (Contrast with exponential growth.)

Long-day plants. A plant that requires long day to flower.

Loop of Henle (hen' lee) Long, hairpin loop of the mammalian renal tubule that runs from the cortex down into the medulla, and back to the cortex. Creates a concentration gradient in the interstitial fluids in the medulla.

Lophophore A U-shaped fold of the body wall with hollow, ciliated tentacles that encircles the mouth of animals in several different phyla. Used for filtering prey from the surrounding water.

Lordosis (lor doe' sis) [Gk. *lordosis*: curving forward] A posture assumed by females of some mammalian species (especially rodents) to signal sexual receptivity.

Lumen (loo' men) [L. *lumen*: light] The cavity inside any tubular organ or structure, such as the gut or a kidney tubule.

Luteinizing hormone A gonadotropin produced by the anterior pituitary. It stimulates the gonads to produce sex hormones.

Lymph [L. *lympha*: liquid] A clear, watery fluid that is formed as a filtrate of blood; it contains white blood cells; it collects in a series of special vessels and is returned to the bloodstream.

Lymph nodes Specialized tissue regions that act as filters for cells, bacteria and foreign matter.

Lymphocyte A major class of white blood cells. Includes T cells, B cells, and other cell types important in the immune response.

Lymphoid tissue Tissues of the immune defense system dispersed throughout the body and consisting of: thymus, spleen, bone marrow, lymph nodes, blood, and lymph.

Lysis (lie' sis) [Gr. *lysis*: break apart] Bursting of a cell.

Lysogenic cycle A form of viral replication in which the virus becomes incorporated into the bacterial chromosome and the host cell is not killed. (Contrast with lytic cycle.)

Lysosome (lie' so soam) [Gr. *lysis*: break away + *soma*: body] A membrane-enclosed organelle found in eukaryotic cells (other than plants). Lysosomes contain a mixture of enzymes that can digest most of the macromolecules found in the rest of the cell.

Lysozyme (lie' so zyme) An enzyme in saliva, tears, and nasal secretions that attacks bacterial cell walls, as one of the body's nonspecific defense mechanisms.

Lytic cycle A form of viral reproduction that lyses the host bacterium releasing the new viruses. (Contrast with lysogenic cycle.)

M phase The portion of the cell cycle in which mitosis takes place.

Macroevolution [Gr. *makros*: large, long] Evolutionary changes occurring over long time spans and usually involving changes in many traits. (Contrast with microevolution.)

Macromolecule A giant polymeric molecule. The macromolecules are proteins, polysaccharides, and nucleic acids.

Macronutrient A mineral element required by plant tissues in concentrations of at least 1 milligram per gram of their dry matter.

Macrophage (mac' roh faj) A type of white blood cell that endocytoses bacteria and other cells.

Major histocompatibility complex (MHC) A complex of linked genes, with multiple alleles, that control a number of cell surface antigens that identify self and can lead to graft rejection.

Malleus (mal' ee us) [L. *malleus*: hammer] The first of the three bones that conduct movements of the eardrum to the oval window of the inner ear. (See incus, stapes.)

Malpighian tubule (mal pee' gy un) A type of protonephridium found in insects.

Mammal [L. *mamma*: breast, teat] Any animal of the class Mammalia. Mammals are characterized by the production of milk by the female mammary glands and the possession of hair for body covering.

Mantle A sheet of specialized tissues that covers most of the viscera of mollusks; provides protection to internal organs and secretes the shell.

Mapping In genetics, determining the order of genes on a chromosome and the distances between them.

Marine [L. *mare*: sea, ocean] Pertaining to or living in the ocean. (Contrast with aquatic, terrestrial.)

Marker A gene of identifiable phenotype that indicates the presence on another gene, DNA segment, or chromosome fragment.

Marsupial (mar soo' pee al) A mammal belonging to the subclass Metatheria, such as opossums and kangaroos. Most have a pouch (marsupium) that contains the milk glands and serves as a receptacle for the young.

Mass extinctions Geological periods during which rates of extinction were much higher than during intervening times.

Mass number The sum of the number of protons and neutrons in an atom's nucleus.

Mast cells Typically found in connective tissue, mast cells can be provoked by antigens or inflammation to release histamine.

Maternal effect genes These genes code for morphogens that determine the polarity of the egg and larva in the fruit fly *Drosophila melanogaster*.

Maternal inheritance Inheritance in which the mother's phenotype is exclusively expressed. Mitochondria and chloroplasts are maternally inherited via egg cytoplasm. Also known as cytoplasmic inheritance.

Mating types A mating system in which the sexes are morphologically identical but carry different alleles and will mate.

Mechanoreceptor A cell that is sensitive to physical movement and generates action potentials in response.

Medulla (meh dull' luh) (1) The inner, core region of an organ, as in the adrenal medulla (adrenal gland) or the renal medulla (kidneys). (2) The portion of the brain stem that connects to the spinal cord.

Megagametophyte A female gametophyte that produces eggs only.

Megaspore [Gr. *megas*: large + *spora*: to sow] In plants, a haploid spore that produces a female gametophyte.

Meiosis (my oh' sis) [Gr. *meiosis*: diminution] Division of a diploid nucleus to produce four haploid daughter cells. The process consists of two successive nuclear divisions with only one cycle of chromosome replication.

Membrane potential The difference in electrical charge between the inside and the outside of a cell, caused by a difference in the distribution of ions.

Memory cells Long-lived lymphocytes produced by exposure to antigen. They persist in the body and are able to mount a rapid response to subsequent exposures to the antigen.

Mendelian population A local population of individuals belonging to the same species and exchanging genes with one another.

Mendel's first law See Segregation.

Mendel's second law See Independent assortment.

Menstrual cycle The monthly sloughing off of the uterine lining if fertilization does not occur in the female. Occurs between puberty and menopause.

Meristem [Gr. *meristos*: divided] Plant tissue made up of undifferentiated actively dividing cells.

Mesenchyme (mez' en kyme) [Gr. *mesos*: middle + *enchyma*: infusion] Embryonic or unspecialized cells derived from the mesoderm.

Mesoderm [Gr. *mesos*: middle + *derma*: skin] The middle of the three embryonic tissue layers first delineated during gastrulation. Gives rise to skeleton, circulatory system, muscles, excretory system, and most of the reproductive system.

Mesophyll (mez' uh fill) [Gr. *mesos*: middle + *phyllon*: leaf] Chloroplast-containing, photosynthetic cells in the interior of leaves.

Mesosome (mez' uh soam') [Gr. *mesos*: middle + *soma*: body] A localized infolding of the plasma membrane of a bacterium.

Messenger RNA (mRNA) A transcript of one of the strands of DNA; carries information (as a sequence of codons) for the synthesis of one or more proteins.

Meta- [Gr.: between, along with, beyond] A prefix used in biology to denote a change or a shift to a new form or level; for example, as used in metamorphosis.

Metabolism (meh tab' a lizm) [Gr. *metabole*: to change] The sum total of the chemical reactions that occur in an organism, or some subset of that total (as in respiratory metabolism).

Metabolic compensation Changes in metabolic properties of an organism that render it less sensitive to temperature changes.

Metabolic pathway A series of enzyme-catalyzed reactions so arranged that the product of one reaction is the substrate of the next.

Metamorphosis (met' a mor' fo sis) [Gr. *meta*: between + *morphe*: form, shape] A change occur-

ring between one developmental stage and another, as for example from a tadpole to a frog. (See complete metamorphosis, incomplete metamorphosis.)

Metaphase (met' a phase) The stage in nuclear division at which the centromeres of the highly supercoiled chromosomes are all lying on a plane (the metaphase plane or plate) perpendicular to a line connecting the division poles.

Metapopulation A population divided into subpopulations, among which there are occasional exchanges of individuals.

Metastasis (meh tass' tuh sis) The spread of cancer cells from their original site to other parts of the body.

Methanogen Any member of a group of archaea that release methane as a metabolic product. This group is considered to be an extremely ancient one.

Methylation The addition of a methyl group (—CH₃) to a molecule. Extensive methylation of cytosine in DNA is correlated with reduced transcription.

MHC See Major histocompatibility complex.

Microbiology [Gr. *mikros*: small + *bios*: life + *logos*: discourse] The scientific study of microscopic organisms, particularly bacteria, protists, and viruses.

Microevolution The small evolutionary changes typically occurring over short time spans; generally involving a small number of traits and minor genetic changes. (Contrast with macroevolution.)

Microfilament Minute fibrous structure generally composed of actin found in the cytoplasm of eukaryotic cells. They play a role in the motion of cells.

Microgametophyte A male gametophyte that produces sperm only.

Micronutrient A mineral element required by plant tissues in concentrations of less than 100 micrograms per gram of their dry matter.

Micropyle (mike' roh pile) [Gr. *mikros*: small + *pylon*: gate] Opening in the integument(s) of a seed plant ovule through which pollen grows to reach the female gametophyte within.

Microspore [Gr. *mikros*: small + *spora*: to sow] In plants, a haploid spore that produces a male gametophyte.

Microtubules Minute tubular structures found in centrioles, spindle apparatus, cilia, flagella, and cytoskeleton of eukaryotic cells. These tubules play roles in the motion and maintenance of shape of eukaryotic cells.

Microvilli (singular: microvillus) The projections of epithelial cells, such as the cells lining the small intestine, that increase their surface area.

Middle lamella A layer of polysaccharides that separates plant cells; a shared middle lamella lies outside the primary walls of the two cells.

Migration The regular, seasonal movements of animals.

Mineral An inorganic substance other than water.

Mineral nutrients Inorganic ions required by organisms for normal growth and reproduction.

Mismatch repair When a single base in DNA is changed into a different base, or the wrong base

inserted during DNA replication, there is a mismatch in base pairing with the base on the opposite strand. A repair system removes the incorrect base and inserts the proper one for pairing with the opposite strand.

Missense mutation A nonsynonymous mutation, or one that changes a codon for one amino acid to a codon for a different amino acid. (Contrast with frame-shift mutation, nonsense mutation, synonymous mutation.)

Mitochondrial matrix The fluid interior of the mitochondrion, enclosed by the inner mitochondrial membrane.

Mitochondrion (my' toe kon' dree un) [Gr. *mitos*: thread + *chondros*: grain] An organelle in eukaryotic cells that contains the enzymes of the citric acid cycle, the respiratory chain, and oxidative phosphorylation.

Mitosis (my toe' sis) [Gr. *mitos*: thread] Nuclear division in eukaryotes leading to the formation of two daughter nuclei each with a chromosome complement identical to that of the original nucleus.

Mitotic center Cellular region that organizes the microtubules for mitosis. In animals a centrosome serves as the mitotic center.

Moderately repetitive DNA DNA sequences that appear hundreds to thousands of times in the genome. They include the DNA sequences coding for rRNAs and tRNAs, as well as the DNA at telomeres.

Modular organism An organism which grows by producing additional units of body construction (modules) that are very similar to the units of which it is already composed.

Mole A quantity of a compound whose weight in grams is numerically equal to its molecular weight expressed in atomic mass units. Avogadro's number of molecules: 6.023×10^{23} molecules.

Molecular clock The theory that macromolecules diverge from one another over evolutionary time at a constant rate; this rate may provide insight into the phylogenetic relationships among organisms.

Molecular weight The sum of the atomic weights of the atoms in a molecule.

Molecule A particle made up of two or more atoms joined by covalent bonds or ionic attractions.

Molting The process of shedding part or all of an outer covering, as the shedding of feathers by birds or of the entire exoskeleton by arthropods.

Monoclonal antibody Antibody produced in the laboratory from a clone of hybridoma cells, each of which produces the same specific antibody.

Monocot [Gr. *mono*: one + *kotyledon*: a cup-shaped hollow] Any member of the angiosperm lineage in which the embryo produces a single cotyledon (seed leaf). Leaves of most monocots have their major veins arranged parallel to each other.

Monocytes White blood cells that produce macrophages.

Monoecious (mo nee' shus) [Gr. *mono*: one + *oikos*: house] Describes organisms in which both sexes are "housed" in a single individual that produces both eggs and sperm. (In some plants, these are found in different flowers within the same plant.)

Examples: corn, peas, earthworms, hydras. (Contrast with dioecious, perfect flower.)

Monohybrid cross A mating in which the parents differ with respect to the alleles of only one locus of interest.

Monomer [Gr. *mono*: one + *meros*: unit] A small molecule, two or more of which can be combined to form oligomers (consisting of a few monomers) or polymers (consisting of many monomers).

Monophyletic (mon' oh fih leht' ik) [Gk. *mono*: one + *phylon*: tribe] Descended from a single ancestral stock.

Monosaccharide A simple sugar. Oligosaccharides and polysaccharides are made up of monosaccharides.

Monosynaptic reflex A neural reflex that begins in a sensory neuron and makes a single synapse before activating a motor neuron.

Morphogen A diffusible substances whose concentration gradients determine patterns of development in animals and plants.

Morphogenesis (more' fo jen' e sis) [Gr. *morphe*: form + *genesis*: origin] The development of form; the overall consequence of determination, differentiation, and growth.

Morphology (more fol' o jee) [Gr. *morphe*: form + *logos*: study, discourse] The scientific study of organic form, including both its development and function.

Mosaic development Pattern of animal embryonic development in which each blastomere contributes a specific part of the adult body. (Contrast with regulative development.)

Motor end plate The modified area on a muscle cell membrane where a synapse is formed with a motor neuron.

Motor neuron A neuron carrying information from the central nervous system to an effector such as a muscle fiber.

Motor proteins Specialized proteins that use energy to change shape and move cells or structures within cells. See dynein, kinesin.

Motor unit A motor neuron and the set of muscle fibers it controls.

mRNA See messenger RNA.

Mucosa (mew koh' sah) An epithelial membrane containing cells that secrete mucus. The inner cell layers of the digestive and respiratory tracts.

Muscle Contractile tissue containing actin and myosin organized into polymeric chains called microfilaments. Muscle fiber A single muscle cell. In the case of striated muscle, a syncitial, multinucleate cell.

Muscle spindle Modified muscle fibers encased in a connective sheat and functioning as stretch receptors.

Mutagen (mute' ah jen) [L. *mutare*: change + Gr. *genesis*: source] Any agent (e.g., chemicals, radiation) that increases the mutation rate.

Mutation A detectable, heritable change in the genetic material not caused by recombination.

Mutation pressure Evolution (change in gene proportions) by different mutation rates alone (i.e., without the influence of natural selection).

Mutualism The type of symbiosis, such as that exhibited by fungi and algae or cyanobacteria in forming lichens, in which both species profit from the association.

Mycelium (my seel' ee yum) [Gr. *mykes*: fungus] In the fungi, a mass of hyphae.

Mycorrhiza (my' ko rye' za) [Gr. *mykes*: fungus + *rhiza*: root] An association of the root of a plant with the mycelium of a fungus.

Myelin (my' a lin) A material forming a sheath around some axons. Formed by Schwann cells that wrap themselves about the axon, myelin insulates the axon electrically and increases the rate of transmission of a nervous impulse.

Myofibril (my' oh fy' bril) [Gr. *mys*: muscle + L. *fibrilla*: small fiber] A polymeric unit of actin or myosin in a muscle.

Myogenic (my oh jen' ik) [Gr. *mys*: muscle + *genesis*: source] Originating in muscle.

Myoglobin (my' oh globe' in) [Gr. *mys*: muscle + L. *globus*: sphere] An oxygen-binding molecule found in muscle. Consists of a heme unit and a single globiin chain, and carrys less oxygen than hemoglobin.

Myosin One of the two major proteins of muscle, it makes up the thick filaments. (See actin.)

NAD (nicotinamide adenine dinucleotide) A compound found in all living cells, existing in two interconvertible forms: the oxidizing agent NAD^+ and the reducing agent $NADH + H^+$.

NADP (nicotinamide adenine dinucleotide phosphate) A compound similar to NAD, but possessing another phosphate group; plays similar roles but is used by different enzymes.

Natural killer cells A nonspecific defensive cell (lymphocyte) that attacks tumor cells and virus infected cells.

Natural selection The differential contribution of offspring to the next generation by various genetic types belonging to the same population. The mechanism of evolution proposed by Charles Darwin.

Necrosis (nec roh' sis) [Gk. *nekros*: death] Tissue damage resulting from cell death.

Negative control The situation in which a regulatory macromolecule (generally a repressor) functions to turn off transcription. In the absence of a regulatory macromolecule, the structural genes are turned on.

Nematocyst (ne mat' o sist) [Gr. *nema*: thread + *kystis*: cell] An elaborate, threadlike structure produced by cells of jellyfish and other cnidarians, used chiefly to paralyze and capture prey.

Nephridium (nef rid' ee um) [Gr. *nephros*: kidney] An organ which is involved in excretion, and often in water balance, involving a tube that opens to the exterior at one end.

Nephron (nef' ron) [Gr. *nephros*: kidney] The functional unit of the kidney, consisting of a structure for receiving a filtrate of blood, and a tubule that absorbs selected parts of the filtrate back into the bloodstream.

Nephrostome (nef' ro stome) [Gr. *nephros*: kidney + *stoma*: opening] An opening in a nephridium through which body fluids can enter.

Nerve A structure consisting of many neuronal axons and connective tissue.

Net primary production Total photosynthesis minus respiration by plants.

Neural plate A thickened strip of ectoderm along the dorsal side of the early vertebrate embryo; gives rise to the central nervous system.

Neural tube An early stage in the development of the vertebrate nervous system consisting of a hollow tube created by two opposing folds of the dorsal ectoderm along the anterior–posterior body axis.

Neuromuscular junction The region where a motor neuron contacts a muscle fiber, creating a synapse.

Neuron (noor' on) [Gr. *neuron*: nerve] A nervous system cell that can generate and conduct action potentials along an axon to a synapse with another cell.

Neurotransmitter A substance produced in and released by one a neuron (the presynaptic cell) that diffuses across a synapse and excites or inhibits another cell (the postsynaptic cell).

Neurula (nure' you la) Embryonic stage during the dorsal nerve cord forms from two ectodermal ridges.

Neutral allele An allele that does not alter the functioning of the proteins for which it codes.

Neutral theory A view of molecular evolution that postulates that most mutations do not affect the amino acid being coded for, and that such mutations accumulate in a population at rates driven by genetic drift and mutation rates.

Neutron (new' tron) One of the three most fundamental particles of matter, with mass approximately 1 amu and no electrical charge.

Neutrophils Abundant, short-lived phagocytic leukocytes that attack antibody-coated antigens.

Niche See ecological niche.

Nitrate reduction The process by which nitrate (NO_3^-) is reduced to ammonia (NH_3).

Nitric oxide (NO) An unstable molecule (a gas) that serves as a second messenger causing smooth muscle to relax. In the nervous system it operates as a neurotransmitter.

Nitrification The oxidation of ammonia to nitrite and nitrate ions, performed by certain soil bacteria.

Nitrogenase In nitrogen-fixing organisms, an enzyme complex that mediates the stepwise reduction of atmospheric N_2 to ammonia.

Nitrogen fixation Conversion of nitrogen gas to ammonia, which makes nitrogen available to living things. Carried out by certain prokaryotes, some of them free-living and others living within plant roots.

Node [L. *nodus*: knob, knot] In plants, a (sometimes enlarged) point on a stem where a leaf is or was attached.

Node of Ranvier A gap in the myelin sheath covering an axon; the point where the axonal membrane can fire action potentials.

Noncompetitive inhibitor An inhibitor that binds the enzyme at a site other than the active site. (Contrast with competitive inhibitor.)

Nondisjunction Failure of sister chromatids to separate in meiosis II or mitosis, or failure of homologous chromosomes to separate in meiosis I. Results in aneuploidy.

Nonpolar molecule A molecule whose electric charge is evenly balanced from one end of the molecule to the other.

Nonsense mutation Mutations that prematurely terminate a polypeptide by changing a codon for an amino acid to one of the codons (UAG, UAA, or UGA) that signal termination of translation. (Contrast with frame-shift mutation, missense mutation, synonymous mutation.)

Nonspecific defenses Immunologic responses directed against any invading agent without reacting to apecific antigens.

Nonsynonymous mutation A nucleotide substitution that that changes the amino acid specified (i.e., AGC → AGA, or serine → arginine). (Contrast with synonymous mutation.)

Nonsynonymous substitution The situation when a nonsynonymous mutation becomes dominant in a population. (Contrast with synonymous substitution.)

Nontracheophytes Those plants lacking well-developed vascular tissue; the liverworts, hornworts, and mosses. (Contrast with tracheophytes.)

Norepinephrine A neurotransmitter found in the central nervous system and also at the postganglionic nerve endings of the sympathetic nervous system. Also called noradrenaline.

Notochord (no′ tow kord) [Gr. *notos*: back + *chorde*: string] A flexible rod of gelatinous material serving as a support in the embryos of all chordates and in the adults of tunicates and lancelets.

Nuclear envelope The surface, consisting of two layers of membrane, that encloses the nucleus of eukaryotic cells.

Nuclear pore complex Protein structure situated in nuclear pores through which RNA and proteins enter and leave the nucleus.

Nucleic acid (new klay′ ik) A long-chain alternating polymer of deoxyribose or ribose and phosphate groups, with nitrogenous bases—adenine, thymine, uracil, guanine, or cytosine (A, T, U, G, or C)—as side chains. DNA and RNA are nucleic acids.

Nucleoid (new′ klee oid) The region that harbors the chromosomes of a prokaryotic cell. Unlike the eukaryotic nucleus, it is not bounded by a membrane.

Nucleolar organizer (new klee′ o lar) A region on a chromosome that is associated with the formation of a new nucleolus following nuclear division. The site of the genes that code for ribosomal RNA.

Nucleolus (new klee′ oh lus) A small, generally spherical body found within the nucleus of eukaryotic cells. The site of synthesis of ribosomal RNA.

Nucleoplasm (new′ klee o plazm) The fluid material within the nuclear envelope of a cell, as opposed to the chromosomes, nucleoli, and other particulate constituents.

Nucleosome A portion of a eukaryotic chromosome, consisting of part of the DNA molecule wrapped around a group of histone molecules, and held together by another type of histone molecule. The chromosome is made up of many nucleosomes.

Nucleotide The basic chemical unit in a nucleic acid. A nucleotide in RNA consists of one of four nitrogenous bases linked to ribose, which in turn is linked to phosphate. In DNA, deoxyribose is present instead of ribose.

Nucleoside A nucleotide without the phosphate group.

Nucleus (new′ klee us) [L. *nux*: kernel or nut] In cells, the centrally located compartment of eukaryotic cells that is bounded by a double membrane and contains the chromosomes.

Null hypothesis The assertion that an effect proposed by its companion hypothesis does not in fact exist.

Nutrient A food substance; or, in the case of mineral nutrients, an inorganic element required for completion of the life cycle of an organism.

Obligate anaerobe An anaerobic prokaryote that cannot survive exposure to O_2.

Oil A triglyceride that is liquid at room temperature. (Contrast with fat.)

Okazaki fragments Newly formed DNA making up the lagging strand in DNA replication. DNA ligase links Okazaki fragments together to give a continuous strand.

Olfactory [L. *olfacere*: to smell] Having to do with the sense of smell.

Oligomer [Gr.: *oligo*: a few + *meros*: units] A compound molecule of intermediate size, made up of two to a few monomers. (Contrast with monomer, polymer.)

Oligosaccharins Plant hormones, derived from the plant cell wall, that trigger defenses against pathogens.

Ommatidium [Gr. *omma*: eye] One of the units which, collected into groups of up to 20,000, make up the compound eye of arthropods.

Omnivore [L. *omnis*: everything + *vorare*: to devour] An organism that eats both animal and plant material. (Contrast with carnivore, detritivore, herbivore.)

Oncogene [Gr. *onkos*: mass, tumor + *genes*: born] Genes that greatly stimulate cell division, giving rise to tumors.

Oocyte (oh′ eh site) [Gr. *oon*: egg + *kytos*: container] The cell that gives rise to eggs in animals.

Oogenesis (oh′ eh jen e sis) [Gr. *oon*: egg + *genesis*: source] Female gametogenesis, leading to production of the egg.

Oogonium (oh′ eh go′ nee um) In some algae and fungi, a cell in which an egg is produced.

Operator The region of an operon that acts as the binding site for the repressor.

Operon A genetic unit of transcription, typically consisting of several structural genes that are transcribed together; the operon contains at least two control regions: the promoter and the operator.

Opportunity cost The sum of the benefits an animal forfeits by not being able to perform some other behavior during the time when it is performing a given behavior.

Opsin (op′ sin) [Gr. *opsis*: sight] The protein portion of the visual pigment rhodopsin. (See rhodopsin.)

Optic chiasm [Gr. *chiasma*: cross] Structure on the lower surface of the vertebrate brain where the two optic nerves come together.

Optical isomers Two isomers that are mirror images of one another.

Organ [Gk. *organon*: tool] A body part, such as the heart, liver, brain, root, or leaf. Organs are composed of different tissues integrated to perform a distinct function. Organs are in turn often integrated into systems, such as the digestive or reproductive system.

Organ identity genes Plant genes that specify the various parts of the flower. See homeotic genes.

Organ of Corti Structure in the inner ear that transforms mechanical forces produced from pressure waves ("sound waves") into action potentials that are sensed as sound.

Organelles (or gan els′) Organized structures that are found in or on cells. Examples: ribosomes, nuclei, mitochrondria, chloroplasts, cilia, and contractile vacuoles.

Organic Pertaining to any aspect of living matter, e.g., to its evolution, structure, or chemistry. The term is also applied to any chemical compound that contains carbon.

Organism Any living being.

Organizer Region of an early embryo that directs the development of nearby regions. In amphibian early gastrulas, the dorsal lip of the blastopore is the organizer.

Organogenesis The formation of organs and organ systems during development.

Origin of replication A DNA sequence at which helicase unwinds the DNA double helix and DNA polymerase binds to initiate DNA replication.

Osmolarity The concentration of osmotically active particles in a solution.

Osmoregulation Regulation of the chemical composition of the body fluids of an organism.

Osmoreceptor A neuron that converts changes in the osmotic potential of interstial fluids into action potentials.

Osmosis (oz mo′ sis) [Gr. *osmos*: to push] The movement of water across a differentially permeable membrane, from one region to another region where the water potential is more negative.

Ossicle (oss′ ick ul) [L. *os*: bone] The calcified construction unit of echinoderm skeletons.

Osteoblasts (oss′ tee oh blast) [Gk. *osteon*: bone + *blastos*: sprout] Cells that lay down the protein matrix of bone.

Osteoclasts (oss′ tee oh clast) [Gk. *osteon*: bone + *klastos*: broken] Cells that dissolve bone.

Otolith (oh′ tuh lith) [Gk. *otikos*: ear + *lithos*: stone[Structures in the vertebrate vestibular apparatus that mechanically stimulate hair cells when the head moves or changes position.

Oval window The flexible membrane that, when moved by the bones of the middle ear, produces pressure waves in the inner ear

Ovary (oh′ var ee) [L. *ovum*: egg] Any female organ, in plants or animals, that produces an egg.

Oviduct [L. *ovum*: egg + *ducere*: to lead] In mammals, the tube serving to transport eggs to the uterus or to outside of the body.

Oviparous (oh vip' uh rus) Reproduction in which eggs are released by the female and development is external to the mother's body. (Contrast with viviparous.)

Ovulation The release of an egg from an ovary.

Ovule (oh' vule) In plants, a structure that contains a gametophyte and, within the gametophyte, an egg; when it matures, an ovule becomes a seed.

Ovum (oh' vum) [L. *ovum*: egg] The egg, the female sex cell.

Oxidation (ox i day' shun) Relative loss of electrons in a chemical reaction; either outright removal to form an ion, or the sharing of electrons with substances having a greater affinity for them, such as oxygen. Most oxidation, including biological ones, are associated with the liberation of energy. (Contrast with reduction.)

Oxidative phosphorylation ATP formation in the mitochondrion, associated with flow of electrons through the respiratory chain.

Oxidizing agent A substance that can accept electrons from another. The oxidizing agent becomes reduced; its partner becomes oxidized.

P generation Parental generation. The individuals that mate in a genetic cross. Their immediate offspring are the F_1 generation.

Pacemaker That part of the heart which undergoes most rapid spontaneous contraction, thus setting the pace for the beat of the entire heart. In mammals, the sinoatrial (SA) node. Also, an artificial device, implanted in the heart, that initiates rhythmic contraction of the organ.

Pacinian corpuscle A modified nerve ending that senses touch and vibration.

Pair rule genes Segmentation genes that divide the *Drosophila* larva into two segments each.

Paleontology (pale' ee on tol' oh jee) [Gr. *palaios*: ancient + *logos*: discourse] The scientific study of fossils and all aspects of extinct life.

Pancreas (pan' cree us) A gland located near the stomach of vertebrates that secretes digestive enzymes into the small intestine and releases insulin into the bloodstream.

Pangaea (pan jee' uh) [Gk. *pan*: all, every] The single land mass formed when all the continents came together in the Permian period.

Para- [Gk. *para*: akin to, beside] Prefix indicating association in being along side or accessory to.

Parabronchi Passages in the lungs of birds through which air flows.

Paracrine A hormone that acts locally, near the site of its secretion. (Compare with endocrine gland.)

Parallel evolution Evolutionary patterns that exist in more than one lineage. Often the result of underlying developmental processes.

Parapatric speciation [Gr. *para*: along side + *patria*: homeland] Reproductive isolation between subpopulations arising from some non-geographic but physical condition, such as soil nutrient content. (Contrast with allopatric speciation, sympatric speciation.)

Paraphyletic taxon A taxon that includes some, but not all, of the descendants of a single ancestor.

Parasite An organism that attacks and consumes parts of an organism much larger than itself. Parasites sometimes, but not always, kill the host.

Parasympathetic nervous system A portion of the autonomic (involuntary) nervous system. (Contrast with sympathetic nervous system.)

Parathyroids Four glands on the posterior surface of the thyroid that produce and release parathormone.

Parathyroid hormone Hormone secreted by the parathyroid glands. Stimulates osteoclast activity and raises blood calcium levels.

Parenchyma (pair eng' kyma) A plant tissue composed of relatively unspecialized cells without secondary walls.

Parsimony The principle of preferring the simplest among a set of plausible explanations of any phenomenon.

Parthenocarpy Formation of fruit from a flower without fertilization.

Parthenogenesis (par' then oh jen' e sis) [Gr. parthenos: virgin + genesis: source] The production of an organism from an unfertilized egg.

Partial pressure The portion of the barometric pressure of a mixture of gases that is due to one component of that mixture. For example, the partial pressure of oxygen at sea level is 20.9% of barometric pressure.

Passive transport Diffusion across a membrane; may or may not require a channel or carrier protein. Contrast to active transport.

Patch clamping A technique for isolating a tiny patch of membrane to allow the study of ion movement through a particular channel.

Pathogen (path' o jen) [Gr. *pathos*: suffering + *genesis*: source] An organism that causes disease.

Pathway Any sequence of enzyme-catalyzed, chemical reactions in which the product(s) of a reaction is/are the substrate(s) for another reaction.

Pattern formation In animal embryonic development, the organization of differentiated tissues into specific structures such as wings.

Pedigree The pattern of transmission of a genetic trait within a family.

Penetrance Of a genotype, the proportion of individuals with that genotype who show the expected phenotype.

PEP carboxylase The enzyme that combines carbon dioxide with PEP to form a 4-carbon dicarboxylic acid at the start of C4 photosynthesis or of crassulacean acid metabolism (CAM).

Pepsin [Gr. *pepsis*: digestion] An enzyme, in gastric juice, that digests protein.

Peptide linkage The bond between amino acid residues in a protein. Formed between a carboxyl group and amino group ($CO—NH^-$) with the loss of water molecules.

Peptidoglycan The cell wall material of many prokaryotes, consisting of a single enormous molecule that surrounds the entire cell.

Perennial (per ren' ee al) [L. *per*: throughout + *annus*: year] Refers to a plant that survives from year to year. (Contrast with annual, biennial.)

Perfect flower A flower with both stamens and carpels, therefore hermaphroditic.

Pericycle [Gr. *peri*: around + *kyklos*: ring or circle] In plant roots, tissue just within the endodermis, but outside of the root vascular tissue. Meristematic activity of pericycle cells produces lateral root primordia.

Periderm The outer tissue of the secondary plant body, consisting primarily of cork.

Period (1) A minor category in the geological time scale. (2) The duration of a single cycle in a cyclical event, such as a circadian rhythm.

Peripheral membrane protein Membrane protein not embedded in the bilayer. Contrast to integral membrane protein.

Peripheral nervous system Neurons that transmit information to and from the central nervous system and whose cell bodies reside outside the brain or spinal cord.

Peristalsis (pair' i stall' sis) [Gr. *peri*: around + *stellein*: place] Wavelike muscular contractions proceeding along a tubular organ, propelling the contents along the tube.

Peritoneum The mesodermal lining of the body cavity among coelomate animals.

Permease A membrane protein that specifically transports a compound or family of compounds across the membrane.

Peroxisome An organelle that houses reactions in which toxic peroxides are formed. The peroxisome isolates these peroxides from the rest of the cell.

Petal [Gk. *petalon*: spread out] In an angiosperm flower, a sterile modified leaf, nonphotosynthetic, frequently brightly colored, and often serving to attract pollinating insects.

Petiole (pet' ee ole) [L. *petiolus*: small foot] The stalk of a leaf.

pH The negative logarithm of the hydrogen ion concentration; a measure of the acidity of a solution. A solution with pH = 7 is said to be neutral; pH values higher than 7 characterize basic solutions, while acidic solutions have pH values less than 7.

Phage (fayj) Short for bacteriophage. A virus that infects bacteria.

Phagocyte [Gk. *phagein*: to eat + *kystos*: sac] A white blood cell that ingests microorganisms by endocytosis.

Pharynx [Gr. *pharynx*: throat] The part of the gut between the mouth and the esophagus.

Phenotype (fee' no type) [Gr. *phanein*: to show] The observable properties of an individual resulting from both genetic and environmental factors. (Contrast with genotype.)

Phenotypic plasticity Refers to the fact that the phenotype of a developing organism is determined by a complex series of processes that are affected by both its genotype and its environment.

Pheromone (feer' o mone) [Gr. *pheros*: carry + *hormon*: excite, arouse] A chemical substance used in communication between organisms of the same species.

Phloem (flo' um) [Gr. *phloos*: bark] In vascular plants, the tissue that transports sugars and other solutes from sources to sinks. It consists of sieve cells or sieve tubes, fibers, and other specialized cells.

Phosphate group The functional group —OPO$_3$H$_2$. The transfer of energy from one compound to another is often accomplished by the transfer of a phosphate group.

Phosphodiester linkage The connection in a nucleic acid strand, formed by linking two nucleotides.

Phospholipids Lipids containing a phosphate group; important constituents of cellular membranes. (See lipids.)

Phosphorylation The addition of a phosphate group.

Photoautotroph An organism that obtains energy from light and carbon from carbon dioxide. (Contrast with chemolithotroph, chemoheterotroph, photoheterotroph.)

Photoheterotroph An organism that obtains energy from light but must obtain its carbon from organic compounds. (Contrast with chemolithotroph, chemoheterotroph, photoautotroph.)

Photon (foe' ton) [Gr. *photos*: light] A quantum of visible radiation; a "packet" of light energy.

Photoperiod (foe' tow peer' ee ud) The duration of a period of light, such as the length of time in a 24-hour cycle in which daylight is present.

Photoreceptor (1) A pigment that triggers a physiological response when it absorbs a photon. (2) A cell that senses and responds to light energy.

Photorespiration Light-driven uptake of oxygen and release of carbon dioxide, the carbon being derived from the early reactions of photosynthesis.

Photosynthesis (foe tow sin' the sis) [literally, "synthesis from light"] Metabolic processes, carried out by green plants, by which visible light is trapped and the energy used to synthesize compounds such as ATP and glucose.

Phototropism [Gr. *photos*: light + *trope*: turning] A directed plant growth response to light.

Phylogenetic tree Graphic representation of lines of descent among organisms.

Phylogeny (fy loj' e nee) [Gr. *phylon*: tribe, race + *genesis*: source] The evolutionary history of a particular group of organisms; also, the diagram of the "family tree" that shows genetic linkages between ancestors and descendants.

Phylum (plural: phyla) In taxonomy, a high-level category just beneath kingdom and above the class; a group of related, similar classes.

Physiology (fiz' ee ol' o jee) [Gr. *physis*: natural form + *logos*: discourse, study] The scientific study of the functions of living organisms and the individual organs, tissues, and cells of which they are composed.

Phytoalexins Substances toxic to pathogens, produced by plants in response to fungal or bacterial infection.

Phytochrome (fy' tow krome) [Gr. *phyton*: plant + *chroma*: color] A plant pigment regulating a large number of developmental and other phenomena in plants.

Pigment A substance that absorbs visible light.

Pilus (pill' us) [L. *pilus*: hair] A surface appendage by which some bacteria adhere to one another during conjugation.

Pistil [L. *pistillum*: pestle] The structure of an angiosperm flower within which the ovules are borne. May consist of a single carpel, or of several carpels fused into a single structure. Usually differentiated into ovary, style, and stigma.

Pith In plants, relatively unspecialized tissue found within a cylinder of vascular tissue.

Pituitary A small gland attached to the base of the brain in vertebrates. Its hormones control the activities of other glands. Also known as the hypophysis.

Pits Recessed cavities in the cell walls of a plant vascular element where only the primary wall is present. facilitating the movement of sap between cells.

Placenta (pla sen' ta) [Gr. *plax*: flat surface] The organ found in most mammals that provides for the nourishment of the fetus and elimination of the fetal waste products.

Placental (pla sen' tal) Pertaining to mammals of the subclass Eutheria, a group characterized by the presence of a placenta; contains the majority of living species of mammals.

Plant A member of the kingdom Plantae. Multicellular, gaining its nutrition by photosynthesis.

Planula (plan' yew la) [L. *planum*: flat] The free-swimming, ciliated larva of the cnidarians.

Plaque (plack) [Fr.: a metal plate or coin] (1) A circular clearing in a turbid layer (lawn) of bacteria growing on the surface of a nutrient agar gel. (2) An accumulation of prokaryotic organisms on tooth enamel. Acids produced by these microorganisms can cause tooth decay. (3) A region of arterial wall invaded by fibroblasts and fatty deposits (see atherosclerosis).

Plasma (plaz' muh) [Gr. *plassein*: to mold] The liquid portion of blood, in which blood cells and other particulates are suspended.

Plasma cell An antibody-secreting cell that developed from a B cell. The effector cell of the humoral immune system.

Plasma membrane The membrane that surrounds the cell, regulating the entry and exit of molecules and ions. Every cell has a plasma membrane.

Plasmid A DNA molecule distinct from the chromosome(s); that is, an extrachromosomal element. May replicate independently of the chromosome.

Plasmodesma (plural: plasmodesmata) [Gr. *plassein*: to mold + *desmos*: band] A cytoplasmic strand connecting two adjacent plant cells.

Plasmolysis (plaz mol' i sis) Shrinking of the cytoplasm and plasma membrane away from the cell wall, resulting from the osmotic outflow of water. Occurs only in cells with rigid cell walls.

Plastid Organelle in plants that serves for food manufacture (by photosynthesis) or food storage; bounded by a double membrane.

Plastoquinone A mobile electron carrier within the thylakoid membrane of the chloroplast linking photosystems I and II of photosynthesis.

Platelet A membrane-bounded body without a nucleus, arising as a fragment of a cell in the bone marrow of mammals. Important to blood-clotting action.

Pleiotropy (plee' a tro pee) [Gr. *pleion*: more] The determination of more than one character by a single gene.

Pleural membrane [Gk. *pleuras*: rib, side] The membrane lining the outside of the lungs and the walls of the thoracic cavity. Inflammation of these membranes is a condition known as pleurisy.

Podocytes Cells of Bowman's capsule of the nephron that cover the capillaries of the glomerulus, forming filtration slits.

Poikilotherm (poy' kill o therm) [Gr. *poikilos*: varied + *thermos*: heat] An animal whose body temperature tends to vary with the surrounding environment. (Contrast with homeotherm, heterotherm.)

Point mutation A mutation that results from a small, localized alteration in the chemical structure of a gene; can revert to wild type. (Contrast with deletion.)

Polar body A nonfunctional nucleus produced by meiosis, accompanied by very little cytoplasm. The meiosis which produces the mammalian egg produces in addition three polar bodies.

Polar molecule A molecule in which the electric charge is not distributed evenly in the covalent bonds.

Polar nuclei In flowering plants, the two nuclei in the central cell of the megagametophyte; following fertilization they give rise to the endosperm.

Polarity In development, the difference between one end and the other. In chemistry, the property that makes a polar molecule.

Pollen [L. *pollin*: fine, powdery flour] In seed plants, the microscopic grains containing the male gametophyte (microgametophyte) and gamete (microspore).

Pollination The process of transferring pollen from the anther to the receptive surface (stigma) of the pistil in plants.

Poly- [Gr. *poly*: many] A prefix denoting multiple entities.

Poly(A) tail A long sequence of adenine nucleotides (50–250) added after transcription to the 3' end of most eukaryotic mRNAs.

Polygenes Multiple loci whose alleles increase or decrease a continuously variable phenotypic trait.

Polymer [Gr. *poly*: many + *meros*: unit] A large molecule made up of similar or identical subunits called monomers. (Contrast with monomer, oligomer.)

Polymerase chain reaction (PCR) An enzymatic technique for the rapid production of millions of copies of a particular stretch of DNA.

Polymerization reactions Chemical reactions that generate polymers by linking monomers.

Polymorphism (pol' lee mor' fiz um) [Gr. *poly*: many + *morphe*: form, shape] In genetics, the coexistence in the same population of two distinct hereditary types based on different alleles.

Polyp The sessile asexual stage in the life cycle of most cnidarians.

Polypeptide A large molecule made up of many amino acids joined by peptide linkages. Large polypeptides are called proteins.

Polyphyletic group A group containing taxa, not all of which share the most recent common ancestor.

Polyploid (pol' lee ploid) A cell or an organism in which the number of complete sets of chromosomes is greater than two.

Polysaccharide A macromolecule composed of many monosaccharides (simple sugars). Common examples are cellulose and starch.

Polysome (polyribosome) A complex consisting of a threadlike molecule of messenger RNA and several (or many) ribosomes. The ribosomes move along the mRNA, synthesizing polypeptide chains as they proceed.

Polytene (pol' lee teen) [Gr. *poly*: many + *taenia*: ribbon] An adjective describing giant interphase chromosomes, such as those found in the salivary glands of fly larvae. The characteristic pattern of bands and bulges seen on these chromosomes provided a method for preparing detailed chromosome maps of several organisms.

Pons [L. *pons*: bridge] Region of the brain stem anterior to the medulla.

Population Any group of organisms coexisting at the same time and in the same place and capable of interbreeding with one another.

Population bottleneck See bottleneck.

Population density The number of individuals (or modules) of a population in a unit of area or volume.

Population genetics The study of genetic variation and its causes within populations.

Population structure The proportions of individuals in a population belonging to different age classes (age structure). Also, the distribution of the population in space.

Portal vein [L. *portal*: gate] A vein connecting two capillary beds, as in the hepatic portal system.

Positional cloning A technique for isolating a gene associated with a disease on the basis of its approximate chromosomal location.

Positional information Signals by which genes regulate cell functions to locate cells in a tissue during development.

Positive control The situation in which a regulatory macromolecule is needed to turn transcription of structural genes on. In its absence, transcription will not occur.

Positive cooperativity Occurs when a molecule can bind several ligands and each one that binds alters the conformation of the molecule so that it can bind the next ligand more easily. The binding of four molecules of O_2 by hemoglobin is an example of positive cooperativity.

Post [L. *postere*: behind, following after] Prefix denoting something that comes after.

Postabsorptive period When there is no food in the gut and no nutrients are being absorbed.

Postsynaptic cell The cell whose membranes receive neurotransmitter after its release by another cell (the presynaptic cell) at a synapse.

Postzygotic reproductive barrier Any mechanism that prevents the hybrid gametes of two different species from developing into viable reproductive adults. (Contrast with prezygotic reproductive barrier.)

Pre-mRNA (precursor mRNA) Initial gene transcript before it is modified to produce functional mRNA. Also known as the primary transcript.

Predator An organism that kills and eats other organisms.

Pressure flow model An effective model for phloem transport in angiosperms. It holds that sieve element transport is driven by an osmotically driven pressure gradient between source and sink.

Pressure potential The hydrostatic pressure of an enclosed solution in excess of the surrounding atmospheric pressure.

Presynaptic excitation/inhibition Occurs when a neuron modifies activity at a synapse by releasing a neurotransmitter onto the presynaptic nerve terminal.

Prezygotic reproductive barrier Any barrier to gene exchange that operates before mating. (Contrast with postzygotic reproductive barrier.)

Prey [L. *praeda*: booty] An organism consumed as an energy source.

Primary active transport Form of active transport in which ATP is hydrolyzed, yielding the energy required to transport ions against their concentration gradients. (Contrast with secondary active transport.)

Primary embryonic organizer. See Organizer.

Primary growth In plants, growth produced by the apical meristems. (Contrast with secondary growth.)

Primary producer A photosynthetic or chemosynthetic organism that synthesizes complex organic molecules from simple inorganic ones.

Primary succession Succession that begins in an area initially devoid of life, such as on recently exposed glacial till or lava flows.

Primary structure The specific sequence of amino acids in a protein.

Primary wall Cellulose-rich cell wall layers laid down by a growing plant cell.

Primate (pry' mate) A member of the order Primates: a prosimian, monkey, ape, or human.

Primer A short, single-stranded segment of DNA that is the necessary starting material for the synthesis of a new DNA strand, which is synthesized from the 3' end of the primer.

Primitive streak A line running axially along the blastodisc, the site of inward cell migration during formation of the three-layered embryo. Formed in the embryos of birds and fish.

Primordium [L. *primordium*: origin] The most rudimentary stage of an organ or other part.

Prion An infectious protein that can proliferate by converting other proteins.

Pro- [L.: first, before, favoring] A prefix often used in biology to denote a developmental stage that comes first or an evolutionary form that appeared earlier than another. For example, prokaryote, prophase.

Probe A segment of single stranded nucleic acid used to identify DNA molecules containing the complementary sequence.

Procambium Primary meristem that produces the vascular tissue.

Progesterone [L. *pro*: favoring + *gestare*: to bear] A vertebrate female sex hormone that maintains pregnancy.

Prokaryotes (pro kar' ry otes) [L. *pro*: before + Gk. *karyon*: kernel, nucleus] Organisms whose genetic material is not contained within a nucleus: the bacteria and archaea. Considered an earlier stage in the evolution of life than the eukaryotes.

Prometaphase The phase of nuclear division that begins with the disintegration of the nuclear envelope.

Promoter The region of an operon that acts as the initial binding site for RNA polymerase.

Proofreading The correction of an error in DNA replication just after an incorrectly paired base is added to the growing polynucleotide chain.

Prophage (pro' fayj) The noninfectious units that are linked with the chromosomes of the host bacteria and multiply with them but do not cause dissolution of the cell. Prophage can later enter into the lytic phase to complete the virus life cycle.

Prophase (pro' phase) The first stage of nuclear division, during which chromosomes condense from diffuse, threadlike material to discrete, compact bodies.

Prostaglandin Any one of a group of specialized lipids with hormone-like functions. It is not clear that they act at any considerable distance from the site of their production.

Prosthetic group Any nonprotein portion of an enzyme.

Proteasome In the eukaryotic cytoplasm, a huge protein structure that binds to and digests cellular proteins that have been tagged by ubiquitin.

Protein (pro' teen) [Gr. *protos*: first] One of the most fundamental building substances of living organisms. A long-chain polymer of amino acids with twenty different common side chains. Occurs with its polymer chain extended in fibrous proteins, or coiled into a compact macromolecule in enzymes and other globular proteins.

Proteolysis [protein + Gk. *lysis*: break apart] An enzymatic digestion of a protein or polypeptide.

Proteome The total of the different proteins that can be made by an organism. Because of alternate splicing of pre-mRNA, the number of proteins that can be made is usually much larger than the number of protein-coding genes present in the organism's genome.

Protobiont [Gr. *protos*: first, before + *bios*: life] Aggregates of abiotically produced molecules that cannot reproduce but do maintain internal chemical environments that differ from their surroundings.

Protoderm Primary meristem that gives rise to the plant epidermis.

Proton (pro' ton) [Gr. *protos*: first, before] (1) A subatomic particle with a single positive charge. The number of protons in the nucleus of an atom determine its element. (2) A hydrogen ion, H^+.

Proton pump An active transport system that uses ATP energy to move hydrogen ions across a membrane generating an electric potential (voltage).

Proton motive force A force generated across a membrane expressed in millivolts having two components: a chemical potential (difference in proton concentration) plus an electrical potential due to the electrostatic charge on the proton.

Proto-oncogenes The normal alleles of genes possessing oncogenes (cancer-causing genes) as mutant alleles. Proto-oncogenes encode growth factors and receptor proteins.

Protostome [Gr. *protos*: first + *stoma*: mouth] One of the major lineages of animal evolution. Characterized by spiral, determinate cleavage of the egg, and by schizocoelous development. (Compare with deuterostome.)

Prototroph (pro' tow trofe') [Gr. *protos*: first + *trophein*: to nourish] The nutritional wild type, or reference form, of an organism. Any deviant form that requires growth nutrients not required by the prototrophic form is said to be a nutritional mutant, or auxotroph.

Protozoa (pro to zoe' ah) [Gk. *protos*: first, before + *zoon*: animal] A term formerly used for a single polyphyletic phylum of single-celled eukaryotic organisms including the flagellates, amoebas, and ciliates. This book does not use the term.

Provirus Viral DNA inserted into a bacterial host genome. (See Lysogenic cycle.)

Proximal Near the point of attachment or other reference point. (Contrast with distal.)

Pseudocoelom [Gr. *pseudes*: false] A body cavity not surrounded by a peritoneum. Characteristic of nematodes and rotifers.

Pseudogene [Gr. *pseudes*: false] A DNA segment that is homologous to a functional gene but contains a nucleotide change that prevents its expression.

Pseudopod (soo' do pod) [Gr. *pseudes*: false + *podos*: foot] A temporary, soft extension of the cell body that is used in location, attachment to surfaces, or engulfing particles.

Pulmonary [L. *pulmo*: lung] Pertaining to the lungs.

Punctuated equilibrium An evolutionary pattern in which periods of rapid change are separated by longer periods of little or no change.

Pupa (pew' pa) [L. *pupa*: doll, puppet] In certain insects (the Holometabola), the encased developmental stage between the larva and the adult.

Pupil The opening in the vertebrate eye through which light passes.

Purine (pure' een) One of the types of nitrogenous bases. The purines adenine and guanine are found in nucleic acids. (Contrast with pyrimidine.)

Purkinje fibers Specialized heart muscle cells that conduct excitation throughout the ventricular muscle.

Pyramid of biomass Graphical representation of the total body masses at different trophic levels in an ecosystem.

Pyramid of energy Graphical representation of the total energy contents at different trophic levels in an ecosystem.

Pyrimidine (per im' a deen) A type of nitrogenous base. The pyrimidines cytosine, thymine, and uracil are found in nucleic acids.

Pyruvate A three-carbon acid; the end product of glycolysis and the raw material for the citric acid cycle.

Q_{10} A value that compares the rate of a biochemical process or reaction over a 10°C range of temperature. A process that is not temperature-sensitive has a Q_{10} of 1; values of 2 or 3 mean the reaction speeds up as temperature increases.

Quantum (kwon' tum) [L. *quantus*: how great] An indivisible unit of energy.

Quaternary structure The specific three dimensional arrangement of protein subunits.

Quiescent center In root meristem, central region where cells do not divide or divide very slowly.

R factor (resistance factor) A plasmid that contains one or more genes that encode resistance to antibiotics.

R gene Resistance gene that functions in plant defenses against bacteria, fungi, and nematodes.

R group The distinguishing group of atoms of a particular amino acid.

Radial symmetry The condition in which two halves of a body are mirror images of each other regardless of the angle of the cut, providing the cut is made along the center line. Thus, a cylinder cut lengthwise down its center displays this form of symmetry. (Contrast with biradial symmetry.)

Radioisotope A radioactive isotope of an element. Examples are carbon-14 (^{14}C) and hydrogen-3, or tritium (^{3}H).

Radiometry The use of the regular, known rates of decay of radioisotopes of elements to determine dates of events in the distant past.

Reactant A chemical substance that enters into a chemical reaction with another substance.

Reaction A chemical change in which changes take place in the kind, number, or position of atoms making up a substance.

Reaction center A group of electron transfer proteins that receive energy from light-absorbing pigments and convert it to chemical energy by redox reactions.

Receptacle The end of a plant stem to which the parts of the flower are attached.

Receptor A site or protein on the outer surface of the plasma membrane or in the cytoplasm to which a specific ligand from another cell binds.

Receptor-mediated endocytosis Endocytosis initiated by macromolecular binding to a specific membrane receptor.

Receptor potential The change in the resting potential of a sensory cell when it is stimulated.

Recessive In genetics, an allele that does not determine phenotype in the presence of a dominant allele. Contrast with dominance.

Reciprocal crosses A pair of crosses, in one of which a female of genotype A mates with a male of genotype B and in the other of which a female of genotype B mates with a male of genotype A.

Recognition site See restriction site.

Recombinant An individual, meiotic product, or single chromosome in which genetic materials originally present in two individuals end up in the same haploid complement of genes. The reshuffling of genes can be either by independent segregation, or by crossing over between homologous chromosomes.

Recombinant DNA DNA generated in vitro, from more than one source.

Recombinant DNA technology The application of restriction endonucleases, plasmids, and transformation to alter and assemble recombinant DNA, with the goal of producing specific proteins.

Rectum The terminal portion of the gut, ending at the anus.

Redox reaction A chemical reaction in which one reactant becomes oxidized and the other becomes reduced.

Reducing agent A substance that can donate electrons to another substance. The reducing agent becomes oxidized, and its partner becomes reduced.

Reduction Gain of electrons by a chemical reactant; any reduction is accompanied by an oxidation. (Contrast with oxidation.)

Reflex An automatic action, involving only a few neurons (in vertebrates, often in the spinal cord), in which a motor response swiftly follows a sensory stimulus.

Refractory period Of a neuron, the time interval after an action potential, during which another action potential cannot be elicited.

Regulative development A pattern of animal embryonic development in which the fates of the first blastomeres are not absolutely fixed. (Contrast with mosaic development.)

Regulatory gene A gene that codes for a protein that controls the transcription of another gene(s).

Releaser A sensory stimulus that triggers a fixed action pattern.

Releasing hormone One of several hypothalamic hormones that stimulates the secretion of anterior pituitary hormone.

REM sleep A sleep state characterized by dreaming, skeletal muscle relaxation, and rapid eye movements.

Renal [L. *renes*: kidneys] Relating to the kidneys.

Replication Pertaining to the duplication of genetic material.

Replication complex The close association of several proteins operating in the replication of DNA.

Replication fork A point at which a DNA molecule is replicating. The fork forms by the unwinding of the parent molecule.

Reporter gene Marker genes included in recombinant DNA to indicate the presence of the recombinant DNA in a host cell.

Repressible enzyme An enzyme whose synthesis can be decreased or prevented by the presence of a particular compound. A repressible operon often controls the synthesis of such an enzyme.

Repressor A protein coded by the regulatory gene. The repressor can bind to a specific operator and prevent transcription of the operon.

Reproductive isolating mechanism Any trait that prevents individuals from two different populations from producing fertile hybrids.

Reproductive isolation The condition in which a population is not exchanging genes with other populations of the same species.

Rescue effect The process by which a few individuals moving among declining subpopulations of a species and reproducing may prevent their extinction.

Resolving power Of an optical device such as a microscope, the smallest distance between two lines that allows the lines to be seen as separate from one another.

Resource Something in the environment required by an organism for its maintenance and growth that is consumed in the process of being used.

Respiration (res pi ra' shun) [L. *spirare*: to breathe] (1) Cellular respiration; the catabolic pathways by which electrons are removed from various molecules and passed through intermediate electron carriers to O_2, generating H_2O and releasing energy. (2) Breathing.

Respiratory chain The terminal reactions of cellular respiration, in which electrons are passed from NAD or FAD, through a series of intermediate carriers, to molecular oxygen, with the concomitant production of ATP.

Resting potential The membrane potential of a living cell at rest. In cells at rest, the interior is negative to the exterior. (Contrast with action potential, electrotonic potential.)

Restoration ecology The science and practice of restoring damaged or degraded ecosystems.

Restriction endonuclease Any one of several enzymes, produced by bacteria, that break foreign DNA molecules at very specific sites. Some produce "sticky ends." Extensively used in recombinant DNA technology.

Restriction fragment length polymorphism See RFLP.

Restriction site A specific DNA base sequence recognized and acted on by a restriction endonuclease cutting the DNA.

Reticular system A central region of the vertebrate brain stem that includes complex fiber tracts conveying neural signals between the forebrain and the spinal cord, with collateral fibers to a variety of nuclei that are involved in autonomic functions, including arousal from sleep.

Retina (rett' in uh) [L. *rete*: net] The light-sensitive layer of cells in the vertebrate or cephalopod eye.

Retinal The light-absorbing portion of visual pigment molecules. Derived from β-carotene.

Retrovirus An RNA virus that contains reverse transcriptase. Its RNA serves as a template for cDNA production, and the cDNA is integrated into a chromosome of the mammalian host cell.

Reverse transcriptase An enzyme that catalyzes the production of DNA (cDNA), using RNA as a template; essential to the reproduction of retroviruses.

RFLP (Restriction fragment length polymorphism) Coexistence of two or more patterns of restriction fragments (patterns produced by restriction enzymes), as revealed by a probe. The polymorphism reflects a difference in DNA sequence on homologous chromosomes.

Rhizoids (rye' zoids) [Gr. *rhiza*: root] Hairlike extensions of cells in mosses, liverworts, and a few vascular plants that serve the same function as roots and root hairs in vascular plants. The term is also applied to branched, rootlike extensions of some fungi and algae.

Rhizome (rye' zome) A special underground stem (as opposed to root) that runs horizontally beneath the ground.

Rhodopsin A photopigment used in the visual process of transducing photons of light into changes in the membrane potential of photoreceptor cells.

Ribonucleic acid See RNA.

Ribosomal RNA (rRNA) Several species of RNA that are incorporated into the ribosome. Involved in peptide bond formation.

Ribosome A small organelle that is the site of protein synthesis.

Ribozyme An RNA molecule with catalytic activity.

Risk cost The increased chance of being injured or killed as a result of performing a behavior, compared to resting.

RNA (ribonucleic acid) An often single stranded nucleic acid whose nucleotides use ribose rather than deoxyribose and in which the base uracil replaces thymine found in DNA. Serves as genome from some viruses. (See rRNA, tRNA, mRMA, and ribozyme.)

RNA editing The alteration of bases on mRNA prior to its translation.

RNA polymerase An enzyme that catalyzes the formation of RNA from a DNA template.

RNA primase A replication complex enzyme that makes the primer strand of DNA needed to initiate DNA replication.

RNA splicing The last stage of RNA processing in eukaryotes, in which the transcripts of introns are excised through the action of small nuclear ribonucleoprotein particles (snRNP).

Rods Light-sensitive cells (photoreceptors) in the retina. (Contrast with cones.)

Root The organ responsible for anchoring the plant in the soil, absorbing water and minerals, and producing certain hormones. Some roots are storage organs.

Root cap A thimble-shaped mass of cells, produced by the root apical meristem, that protects the meristem; the organ that perceives the gravitational stimulus in root gravitropism.

Root hair A long, thin process from a root epidermal cell that absorbs water and minerals from the soil solution.

Rough ER That portion of the endoplasmic reticulum whose outer surface has attached ribosomes. Compare with smooth ER.

rRNA See ribosomal RNA.

Rubisco (Ribulose bisphosphate carboxylase/oxygenase) Acronym for the enzyme that combines carbon dioxide or oxygen with ribulose bisphosphate to catalyze the first step of the Calvin-Benson cycle.

Rumen (rew' mun) The first division of the ruminant stomach. It stores and initiates bacterial fermentation of food. Food is regurgitated from the rumen for further chewing.

Ruminant An herbivorous, cud-chewing mammal such as a cow, sheep, or deer, having a stomach consisting of four compartments.

S phase In the cell cycle, the stage of interphase during which DNA is replicated. (Contrast with G_1 phase, G_2 phase, M phase.)

Saprobe [Gr. *sapros*: rotten] An organism (usually a bacterium or fungus) that obtains its carbon and energy directly from dead organic matter.

Sarcomere (sark' o meer) [Gr. *sark*: flesh + *meros*: unit] The contractile unit of a skeletal muscle.

Saturated fatty acid A fatty acid usually containing from 12 to 18 carbon atoms and no double bonds.

Schizocoelous development [Gk. *schizo*: split + *koiloma*: cavity] Formation of a coelom during embryological development by a splitting of mesodermal masses.

Schwann cell A glial cell that wraps around part of the axon of a peripheral neuron, creating a myelin sheath.

Sclereid [Gr. *skleros*: hard] A type of sclerenchyma cell, commonly found in nutshells, that is not elongated.

Sclerenchyma (skler eng' kyma) [Gr. *skleros*: hard + *kymus*: juice] A plant tissue composed of cells with heavily thickened cell walls, dead at functional maturity. The principal types of sclerenchyma cells are fibers and sclereids.

Second law of thermodynamics States that in any real (irreversible) process, there is a decrease in free energy and an increase in entropy.

Second messenger A compound, such as cAMP, that is released within a target cell after a hormone (first messenger) has bound to a surface receptor on a cell; the second messenger triggers further reactions within the cell.

Secondary active transport Form of active transport which does not use ATP as an energy source; rather, transport is coupled to ion diffusion down a concentration gradient established by primary active transport.

Secondary growth In plants, growth produced by vascular and cork cambia, contributing to an increase in girth. (Contrast with primary growth.)

Secondary metabolite A compound synthesized by a plant that is not needed for basic cellular metabolism. Typically has an antiherbivore or antiparasite function.

Secondary structure Of a protein, localized regularities of structure, such as the α helix and the β pleated sheet.

Secondary succession Ecological succession after a disturbance that does not eliminate all the organisms that originally lived on the site.

Secondary wall Wall layers laid down by a plant cell that has ceased growing; often impregnated with lignin or suberin.

Secretin (si kreet' in) A peptide hormone secreted by the upper region of the small intestine when acidic chyme is present. Stimulates the pancreatic duct to secrete bicarbonate ions.

Section A thin slice, usually for microscopy, as a tangential section or a transverse section.

Seed A fertilized, ripened ovule of a gymnosperm or angiosperm. Consists of the embryo, nutritive tissue, and a seed coat.

Seed plant Plants in which the embryo is protected and nourished within a seed; the gymnosperms and angiosperms.

Seedling A young plant that has grown from a seed (rather than by grafting or by other means.)

Segmentation genes In insect larvae, genes that determine the number and polarity of larval segments.

Segment polarity genes Genes that determine the boundaries and front-to-back organization of the segments in the *Drosophila* larva.

Segregation In genetics, the separation of alleles, or of homologous chromosomes, from one another during meiosis so that each of the haploid daughter nuclei produced by meiosis contains one or the other member of the pair found in the diploid mother cell, but never both. This principle was articulated by Mendel as his "first law."

Selective permeability Allowing certain substances to pass through while other substances are excluded; a characteristic of membranes.

Self incompatability In plants, the rejection of their own pollen; promotes genetic variation and limits inbreeding.

Selfish act A behavioral act that benefits its performer but harms the recipients.

Semen (see' men) [L. *semin*: seed] The thick, whitish liquid produced by the male reproductive organ in mammals, containing the sperm.

Semiconservative replication The common way in which DNA is synthesized. Each of the two partner strands in a double helix acts as a template for a new partner strand. Hence, after replication, each double helix consists of one old and one new strand.

Seminiferous tubules The tubules within the testes within which sperm production occurs.

Senescence [L. *senescere*: to grow old] Aging; deteriorative changes with aging; the increased probability of dying with increasing age.

Sensory neuron A neuron leading from a sensory cell to the central nervous system. (Contrast with motor neuron.)

Sepal (see' pul) [L. *sepalum*: covering] One of the outermost structures of the flower, usually protective in function and enclosing the rest of the flower in the bud stage.

Septum [L. *saeptum*: partition, fence] A membrane or wall between two cavities.

Sertoli cells Cells in the seminiferous tubules that nuture the developing sperm.

Sessile (sess' ul) [L. *sedere*: to sit] Permanently attached; not moving.

Set point In a regulatory system, the threshold sensitivity to the feedback stimulus.

Sex chromosome In organisms with a chromosomal mechanism of sex determination, one of the chromosomes involved in sex determination.

Sex linkage The pattern of inheritance characteristic of genes located on the sex chromosomes of organisms having a chromosomal mechanism for sex determination.

Sexual reproduction Reproduction involving union of gametes.

Sexual selection Selection by one sex of characteristics in individuals of the opposite sex. Also, the favoring of characteristics in one sex as a result of competition among individuals of that sex for mates.

Shared derived trait A trait that arose in the ancestor of a phylogenetic group and is present

(sometimes in modified form) in all of its members, thus helping define that group. Also called a synapomorphy.

Shoot system The aerial parts of a vascular plant, consisting of the leaves, stem(s), and flowers.

Short-day plant (SDP) A plant that requires short days (or long nights) in order to flower.

Sieve tube A column of specialized cells found in the phloem, specialized to conduct organic matter from sources (such as photosynthesizing leaves) to sinks (such as roots). Found principally in flowering plants.

Sieve tube element A single cell of a sieve tube, containing cytoplasm but relatively few organelles, with highly specialized perforated end walls leading to elements above and below.

Signal A chemical (neurotransmitter or hormone) or light message emitted from one cell/cells or organism(s) and received by others to cause some change in function or behavior.

Signal recognition particle (SRP) A complex of RNA and protein that recognizes both the signal sequence on a growing polypeptide and receptor protein on the surface of the ER.

Signal sequence The sequence of a protein that directs the protein through a particular cellular membrane.

Signal transduction pathway The series of biochemical steps whereby a stimulus to a cell (such as a hormone or neurotransmitter binding to a receptor) is translated into a response of the cell.

Silencer sequence A sequence of eukaryotic DNA that binds proteins that inhibit the transcription of an associated gene.

Silent mutation A change in gene sequence that, due to the redundancy of the genetic code, has no effect on the amino acid produced, and thus no effect on the protein phenotype. See synonymous mutation.

Similarity matrix A matrix to compare the structures of two molecules constructed by adding the number of their amino acids that are identical or different.

Sinoatrial node (sigh' no ay' tree al) [L. *sinus*: curve + *atrium*: hall, chamber] The pacemaker of the mammalian heart.

Sink In plants, any organ that imports the products of photosynthesis, such as roots, developing fruits, immature leaves. Contrast with source.

Sinus (sigh' nus) [L. *sinus*: curve, hollow] A cavity in a bone, a tissue space, or an enlargement in a blood vessel.

Sister chromatid In the eukaryotic cell, a chromatid resulting from chromosome replication during interphase.

Sister group Two phylogenetic groups that are each other's closes relative.

Skeletal muscle See striated muscle.

Sliding filament theory A proposed mechanism of muscle contraction based on formation and breaking of crossbridges between actin and myosin filaments, causing them to slide together.

Small intestine The portion of the gut between the stomach and the colon, consisting of the duodenum, the jejunum, and the ileum.

Small nuclear ribonucleoprotein particle (snRNP) A complex of an enzyme and a small nuclear RNA molecule, functioning in RNA splicing.

Smooth muscle One of three types of muscle tissue. Usually consists of sheets of mononucleated cells innervated by the autonomic nervous system.

Sodium–potassium pump The complex protein in plasma membranes that is responsible for primary active transport; it pumps sodium ions out of the cell and potassium ions into the cell, both against their concentration gradients.

Solute A substance that is dissolved in a liquid (solvent).

Solute potential A property of any solution, resulting from its solute contents; it may be zero or have a negative value.

Solution A liquid (solvent) and its dissolved solutes.

Solvent A liquid that has dissolved or can dissolve one or more solutes.

Somatic [Gr. *soma*: body] Pertaining to the body. Somatic cells are cells of the body (as opposed to germ cells).

Somite (so' might) One of the segments into which an embryo becomes divided longitudinally, leading to the eventual segmentation of the animal as illustrated by the spinal column, ribs, and associated muscles.

Source In plants, an organ exporting photosynthetic products in excess of its own needs. For example, a mature leaf or storage organ. Contrast with sink.

Spatial summation In the production or inhibition of action potentials in a postsynaptic neuron, the interaction of depolarizations and hyperpolarizations produced by several terminal boutons.

Spawning The direct release of sex cells into the water.

Speciation (spee' shee ay' shun) The process of splitting one population into two populations that are reproductively isolated from one another.

Species (spee' shees) [L. *specie*: kind] The basic lower unit of classification, consisting of a population or series of populations of closely related and similar organisms. The more narrowly defined "biological species" consists of individuals capable of interbreeding freely with each other but not with members of other species.

Species diversity A weighted representation of the species of organisms living in a region; large and common species are given greater weight than are small and rare ones. (Contrast with species richness.)

Species richness The total number of species living in a region. (Contrast with species diversity.)

Specific defenses Defensive reactions of the immune system that are based on antibody reaction with a specific antigen.

Specific heat The amount of energy that must be absorbed by a gram of a substance to raise its temperature by one degree centigrade. By convention, water is assigned a specific heat of one.

Sperm [Gr. *sperma*: seed] A male gamete (reproductive cell).

Spermatocyte (spur mat' oh site) [Gr. *sperma*: seed + *kytos*: container] The cell that gives rise to the sperm in animals.

Spermatogenesis (spur mat' oh jen' e sis) [Gr. *sperma*: seed + *genesis*: source] Male gametogenesis, leading to the production of sperm.

Spermatogonia Undifferentiated germ cells that give rise to primary spermatocytes and hence to sperm.

Sphincter (sfink' ter) [Gr. *sphinkter*: something that binds tightly] A ring of muscle that can close an orifice, for example at the anus.

Spindle apparatus An array of microtubules stretching from pole to pole of a dividing nucleus and playing a role in the movement of chromosomes at nuclear division. Named for its shape.

Spiracle (spy' rih kel) [L. *spirare*: to breathe] An opening of the treacheal respiratory system of terrestrial arthorpods.

Spliceosome An RNA–protein complex that splices out introns from eukaryotic pre-mRNAs.

Splicing The removal of introns and connecting of exons in eukaryotic pre-mRNAs.

Spontaneous generation The idea that life is generated continually from nonliving matter. **Spontaneous reaction** A chemical reaction which will proceed on its own, without any outside influence. A spontaneous reaction need not be rapid.

Sporangium (spor an' gee um) [Gr. *spora*: seed + *angeion*: vessel or reservoir] In plants and fungi, any specialized stucture within which one or more spores are formed.

Spore [Gr. *spora*: seed] Any asexual reproductive cell capable of developing into an adult organism without gametic fusion. In plants, haploid spores develop into gametophytes, diploid spores into sporophytes. In prokaryotes, a resistant cell capable of surviving unfavorable periods.

Sporocyte Specialized cells of the diploid sporophyte that will divide by meiosis to produce four haploid spores. Germination of these spores produces the haploid gametophyte.

Sporophyte (spor' o fyte) [Gr. *spora*: seed + *phyton*: plant] In plants and protists with alternation of generations, the diploid phase that produces the spores. (Contrast with gametophyte.)

Stabilizing selection Selection against the extreme phenotypes in a population, so that the intermediate types are favored. (Contrast with disruptive selection.)

Stamen (stay' men) [L. *stamen*: thread] A male (pollen-producing) unit of a flower, usually composed of an anther, which bears the pollen, and a filament, which is a stalk supporting the anther.

Starch [O.E. *stearc*: stiff] A polymer of glucose; used by plants to store energy.

Start codon The mRNA triplet (AUG) that acts as a signal for the beginning of translation at the ribosome. (Compare with stop codons.)

Stasis [Gk. *stasis*: to stop, stand still] Period during which little or no evolutionary change takes place within a lineage or groups of lineages.

Statocyst (stat' oh sist) [Gk. *statos*: stationary + *kystos*: cell] An organ of equilibrium in some invertebrates.

Statolith (stat' oh lith) [Gk. *statos*: stationary + *lithos*: stone] A solid object that responds to gravity or movement and stimulates the mechanoreceptors of a statocyst.

Stele (steel) [Gr. *stylos*: pillar] The central cylinder of vascular tissue in a plant stem.

Stem Plant structure that holds leaves and/or flowers; it is the site for transporting and distributing material throughout the plant.

Stem cells In animals, undifferentiated cells that are capable of extensive proliferation. A stem cell generates more stem cells and a large clone of differentiated progeny cells. Compare with initial cells.

Steroid Any of numerous lipids based on a 17-carbon atom ring system.

Sticky ends On a piece of two-stranded DNA, short, complementary, one-stranded regions produced by the action of a restriction endonuclease. Sticky ends allow the joining of segments of DNA from different sources.

Stigma [L. *stigma*: mark, brand] The part of the pistil at the apex of the style that is receptive to pollen, and on which pollen germinates.

Stimulus [L. *stimulare*: to goad] Something causing a response; something in the environment detected by a receptor.

Stolon [L. *stolon*: branch, sucker] A horizontal stem that forms roots at intervals.

Stoma (plural: stomata) [Gr. *stoma*: mouth, opening] Small opening in the plant epidermis that permits gas exchange; bounded by a pair of guard cells whose osmotic status regulates the size of the opening.

Stop codons The mRNA codons that signal the end of protein translation at the ribosome: UAG, UGA, UAA.

Stratosphere The upper part of Earth's atmosphere, above the troposphere; extends from approximately 18 kilometers upward to approximately 50 kilometers above the surface.

Stratum (plural strata) [L. *stratos*: layer] A layer or sedimentary rock laid down at a particular time in a past.

Striated muscle Contractile tissue characterized by multinucleated cells containing highly ordered arrangements of actin and myosin microfilaments. Also known as skeletal muscle.

Stroma The fluid contents of an organelle such as a chloroplast.

Stromatolites Composite, flat-to-domed structures composed of successive mineral layers produced by the action of cyanobacteria in water; ancient ones provide evidence for early life on the earth.

Structural gene A gene that encodes the primary structure of a protein.

Style [Gr. *stylos*: pillar or column] In flowering plants, a column of tissue extending from the tip of the ovary, and bearing the stigma or receptive surface for pollen at its apex.

Sub- [L. *sub*: under] A prefix often used to designate a structure that lies beneath another or is less than another. For example, subcutaneous (beneath the skin); subspecies.

Suberin A waxlike lipid that acts as a barrier to water and solute movement across the Casparian strip of the endodermis. Suberin is the waterproofing element in the cell walls of cork.

Submucosa (sub mew koe' sah) The tissue layer just under the epithelial lining of the lumen of the digestive tract.

Substrate (sub' strayte) The molecule or molecules on which an enzyme exerts catalytic action.

Substratum The base material on which a sessile organism lives.

Succession In ecology, the gradual, sequential series of changes in species composition of a community following a disturbance.

Sulcus [L. *sulcare*: to plow] The valleys or creases between the raised portions of the convoluted surface of the brain. (Contrast with gyrus.)

Summation The ability of a neuron to fire action potentials in response to numerous subthreshold postsynaptic potentials arriving simultaneously at differentiated places on the cell, or arriving at the same site in rapid succession.

Surface area-to-volume ratio For any cell, organism, or geometrical solid, the ratio of surface area to volume; this is an important factor in setting an upper limit on the size a cell or organism can attain.

Surfactant A substance that decreases the surface tension of a liquid. Lung surfactant, secreted by cells of the alveoli, is mostly phospholipid and decreases the amount of work necessary to inflate the lungs.

Suspensor In the embryos of seed plants, the stalk of cells that pushes the embryo into the endosperm and is a source of nutrient transport to the embryo.

Symbiosis (sim' bee oh' sis) [Gr. *sym*: together + *bios*: living] The living together of two or more species in a prolonged and intimate ecological relationship. (Compare with parasitism and mutualism.)

Symmetry Describes an attribute of an animal body in which at least one plane can divide the body into similar, mirror-image halves. (See bilateral symmetry, biradial symmetry, radial symmetry.)

Sympathetic nervous system A division of the autonomic (involuntary) nervous system. (Contrast with parasympathetic nervous system.)

Sympatric speciation (sim pat' rik) [Gr. *sym*: same + *patria*: homeland] Speciation due to reproductive isolation without any physical separation of the subpopulation. (Contrast with allopatric speciation, parapatric speciation.)

Symplast The continuous meshwork of the interiors of living cells in the plant body, resulting from the presence of plasmodesmata. (Contrast with apoplast.)

Symport A membrane transport process that carries two substances in the same direction across the membrane. (Contrast with antiport.)

Synapse (sin' aps) [Gr. *syn*: together + *haptein*: to fasten] The narrow gap between the terminal bouton of one neutron and the dendrite or cell body of another.

Synapsis (sin ap' sis) The highly specific parallel alignment (pairing) of homologous chromosomes during the first division of meiosis.

Synaptic vesicle A membrane-bounded vesicle, containing neurotransmitter, which is produced in and discharged by the presynaptic neuron.

Synapomorphy See shared derived trait.

Synergids [Gk. *syn*: together + *ergos*: performing work] In flowering plants, the two cells accompanying the egg cell at one end of the megmagametophyte.

Syngamy (sing' guh mee) [Gr. *syn*: together + *gamos*: marriage] Union of gametes. Also known as fertilization.

Synonymous mutation A mutation that substitutes one nucleotide for another but does not change the amino acid specified (i.e., UUA → UUG, both specifying leucine). (Compare with frame-shift mutation, missense mutation, nonsense mutation.)

Synonymous substitution The situation when a synonymous mutation becomes widespread in a population. Typically not influenced by natural selection, these substitutions can accumulate in a population. (Contrast with nonsynonymous substitution.)

Systematics The scientific study of the diversity of organisms, and of their relationships. Includes both taxonomy (classification) and phylogeny (evolutionary relationships).

Systemic circulation The part of the circulatory system serving those parts of the body other than the lungs or gills.

Systemic acquired resistance A general resistance to many plant pathogens following infection by a single agent.

Systemin The only polypeptide plant hormone; participates in response to tissue damage.

Systole (sis' tuh lee) [Gr. *systole*: contraction] Contraction of a chamber of the heart, driving blood forward in the circulatory system.

T cell A type of lymphocyte, involved in the cellular immune response. The final stages of its development occur in the thymus gland. (Contrast with B cell; see also cytotoxic T cell, helper T cell, suppressor T cell.)

T cell receptor A protein on the surface of a T cell that recognizes the antigenic determinant for which the cell is specific.

T tubules A system of tubules that runs throughout the cytoplasm of muscle fibers, through which action potentials spread.

Target cell A cell with the appropriate receptors to bind and respond to a particular hormone or other chemical mediator.

Taste bud A structure in the epithelium of the tongue that includes a cluster of chemoreceptors innervated by sensory neurons.

TATA box An eight-base-pair sequence, found about 25 base pairs before the starting point for transcription in many eukaryotic promoters, that binds a transcription factor and thus helps initiate transcription.

Taxis (tak' sis) [Gr. *taxis*: arrange, put in order] The movement of an organism or its part directly toward or away from the stimulus. For example, positive phototaxis is movement toward a light source, negative geotaxis is movement away from gravity).

Taxon A unit such as genus, family, class, or order in a taxonomic system.

Taxonomy (taks on' oh me) [Gr. *taxis*: arrange, put in order] The science of classifying organisms.

Telomeres (tee' lo merz) [Gr. *telos*: end + *meros*: units, segments] Repeated DNA sequences at the ends of eukaryotic chromosomes.

Telophase (tee' lo phase) [Gr. *telos*: end] The final phase of mitosis or meiosis during which chromosomes became diffuse, nuclear envelopes reform, and nucleoli begin to reappear in the daughter nuclei.

Template In biochemistry, a molecule or surface upon which another molecule is synthesized in complementary fashion, as in the replication of DNA. In the brain, a pattern that responds to a normal input but not to incorrect inputs.

Template strand In a stretch of double-stranded DNA, the strand that is transcribed.

Temporal summation [L. *tempus*: time; *summus*: highest amount] In the production or inhibition of action potentials in a postsynaptic neuron, the interaction of depolarizations or hyperpolarizations produced by rapidly repeated stimulation of a single point.

Tendon A collagen-containing band of tissue that connects a muscle with a bone.

Termination The end of protein synthesis triggered by a stop codon which binds release factor that causes the polypeptide to release from the ribosome.

Terrestrial (ter res' tree al) [L. *terra*: earth] Pertaining to the land. (Contrast with aquatic, marine.)

Territory A fixed area from which an animal or group of animals excludes other members of the same (and sometimes other) species by aggressive behavior or display.

Tertiary structure In reference to a protein, the relative locations in three-dimensional space of all the atoms in the molecule. The overall shape of a protein. (Contrast with primary, secondary, and quaternary structures.)

Test cross A cross of a dominant-phenotype individual (which may be either heterozygous or homozygous) with a homozygous-recessive individual.

Testis (tes' tis) (plural: testes) [L. *testis*: witness] The male gonad; the organ that produces the male sex cells.

Testosterone (tes toss' tuhr own) A male sex steroid hormone.

Tetanus [Gr. *tetanos*: stretched] (1) A state of sustained maximal muscular contraction caused by rapidly repeated stimulation. (2) In medicine, an often fatal disease ("lockjaw") caused by the bacterium *Clostridium tetani*.

Tetrad [Gr. *tettares*: four] During prophase I of meiosis, the association of a pair of homologous chromosomes or four chromatids.

Thalamus [Gk. *thalamos*: chamber] A region of the vertebrate forebrain; involved in integration of sensory input.

Thallus (thal' us) [Gr. *thallos*: sprout] Any algal body which is not differentiated into root, stem, and leaf.

Theory [Gk. *theoria*: analysis of facts] A far-reaching explanation of observed facts that is supported by such a wide body of evidence, with no significant contradictory evidence, that it is scientifically accepted as a factual framework. Examples are Newton's theory of gravity and Darwin's theory of evolution. (Contrast with hypothesis.)

Thermoneutral zone The range of temperatures over which an endotherm does not have to expend extra energy to thermoregulate.

Thermoreceptor A cell or structure that responds to changes in temperature.

Thoracic cavity [Gk. *thorax*: breastplate] The portion of the mammalian body cavity bounded by the ribs, shoulders, and diaphragm. Contains the heart and the lungs.

Thorax [Gk. *thorax*: breastplate] In an insect, the middle region of the body, between the head and abdomen. In mammals, the part of the body between the neck and the diaphragm.

Thrombin An enzyme that converts fibrinogen to fibrin, thus triggering the formation of blood clots.

Thrombus (throm' bus) [Gk. *thrombos*: clot] A blood clot that forms within a blood vessel and remains attached to the wall of the vessel. (Contrast with embolus.)

Thylakoid (thigh la koid) [Gk. *thylakos*: sack or pouch] A flattened sac within a chloroplast. Thylakoid membranes contain all of the chlorophyll in a plant, in addition to the electron carriers of photophosphorylation. Thylakoids stack to form grana.

Thymine (T) A nitrogen-containing base found in DNA.

Thymus [Gr. *thymos*: warty] A ductless, glandular portion of the lymphoid system, involved in development of the immune system of vertebrates.

Thyroid [Gr. *thyreos*: door-shaped] A two-lobed gland in vertebrates. Produces the hormone thyroxin.

Tight junction A junction between epithelial cells, in which there is no gap whatever between the adjacent cells. Materials may pass through a tight junction only by entering the epithelial cells themselves.

Tissue A group of similar cells organized into a functional unit; usually integrated with other tissues to form part of an organ.

Tonus (toe' nuss) [L. *tonus*: tension] A low level of muscular tension that is maintained even when the body is at rest.

Topsoil The uppermost soil layer; contains most of the organic matter of soil, but may be depleted of most mineral nutrients.

Totipotency [L. *toto*: whole, entire + *potens*: powerful] In a cell, the condition of possessing all the genetic information and other capacities necessary to form an entire individual.

Toxic [L. *toxicum*: poison] Injurious to the tissues of the host organism.

Trachea (tray' kee ah) [Gr. *trakhoia*: tube] A tube that carries air to the bronchi of the lungs of vertebrates, or to the cells of arthropods.

Tracheary element Refers to either or both types of xylem cells: tracheids and vessel elements.

Tracheid (tray' kee id) A distinctive conducting and supporting cell found in the xylem of nearly all vascular plants, characterized by tapering ends and walls that are pitted but not perforated.

Tracheophytes [Gr. *trakhoia*: tube + *phyton*: plant] Those plants with xylem and phloem, including psilophytes, club mosses, horsetails, ferns, gymnosperms, and angiosperms. (Contrast with non-trachoephytes.)

Trait One form of a character: Eye color is a character; brown eyes and blue eyes are traits.

Transcription The synthesis of RNA, using one strand of DNA as the template.

Transcription factors Proteins that assemble on a eukaryotic chromosome, allowing RNA polymerase II to perform transcription.

Transduction (1) Transfer of genes from one bacterium to another, with a bacterial virus acting as the carrier of the genes. (2) In sensory cells, the transformation of a stimulus (e.g., light energy, sound pressure waves, chemical or electrical stimulants) into action potentials.

Transfection Uptake, incorporation, and expression of recombinant DNA.

Transfer cell A modified parenchyma cell that transports solutes from its cytoplasm into its cell wall, thus moving the solutes from the symplast into the apoplast.

Transfer RNA (tRNA) A family of double stranded RNA molecules. Each kind of tRNA carries a specific amino acid and anticodon that will pair with the complementary codon in mRNA during translation.

Transformation Mechanism for transfer of genetic information in bacteria in which pure DNA extracted from bacteria of one genotype is taken in through the cell surface of bacteria of a different genotype and incorporated into the chromosome of the recipient cell.

Transgenic organism An organism containing recombinant DNA incorporated into its genetic material.

Translation The synthesis of a protein (polypeptide). This occurs on ribosomes, using the information encoded in messenger RNA.

Translocation (1) In genetics, a rare mutational event that moves a portion of a chromosome to a new location, generally on a nonhomologous chromosome. (2) In vascular plants, movement of solutes in the phloem.

Transpiration [L. *spirare*: to breathe] The evaporation of water from plant leaves and stem, driven by heat from the sun, and providing the motive force to raise water (plus ions) from the roots.

Transposable element A segment of DNA that can move to, or give rise to copies at, another locus on the same or a different chromosome.

Transposon A mobile DNA segment that can insert into a chromosome and cause genetic change.

Triglyceride A simple lipid in which three fatty acids are combined with one molecule of glycerol.

Triplet See codon.

Triplet repeat Occurrence of repeated triplet of bases in a gene, often leading to genetic disease, as does excessive repetition of CGG in the gene responsible for fragile-X syndrome.

Triploblastic Having three cell layers. (Contrast with diploblastic.)

Trisomic Containing three rather than two members of a chromosome pair.

tRNA See transfer RNA.

Trophoblast At the 32-cell stage of mammalian development, the outer group of cells that will become part of the placenta. See also Inner cell mass.

Trochophore (troke' o fore) [Gr. *trochos*: wheel + *phoreus*: bearer] The free-swimming larva of some annelids and mollusks. Distinguished by a wheel-like band of cilia around the middle, the trochophore suggests an evolutionary relationship between these two groups.

Trophic level A group of organisms united by obtaining their energy from the same part of the food web of a biological community.

Tropic hormones Hormones of the anterior pituitary that control the secretion of hormones by other endocrine glands.

Tropism [Gr. *tropos*: to turn] In plants, growth toward or away from a stimulus such as light (phototropism) or gravity (gravitropism).

Tropomyosin [troe poe my' oh sin] A protein that, along with actin, constitutes the thin filaments of myofibrils. It controls the interactions of actin and myosin necessary for muscle contraction.

roposphere The lowest atmospheric zone, reaching upward from the Earth's surface approximately 17 km in the tropics and subtropics but only to about 10 km at higher latitudes. The zone in which virtually all the water vapor in the atmosphere is located.

Trypsin A protein-digesting enzyme. Secreted by the pancreas in its inactive form (trypsinogen), it becomes active in the duodenum of the small intestine.

T-tubules A set of transverse tubes that penetrates skeletal muscle fibers and terminates in the sarcoplasmic reticulum. The T-system transmits impulses to the sacs, which then release Ca^{2+} to initiate muscle contraction.

Tube cell The larger of the two cells in a pollen grain; responsible for growth of the pollen tube. See Generative cell.

Tubulin A protein that polymerizes to form microtubules.

Tumor [L. *tumor*: a swollen mass] A disorganized mass of cells, often growing out of control. Malignant tumors spread to other parts of the body.

Tumor suppressor genes Genes which, when homozygous mutant, result in cancer. Such genes code for protein products that inhibit cell proliferation.

Turgor pressure [L. *turgidus*: swollen] See Hydrostatic pressure.

Twitch A single unit of muscle contraction.

Tympanic membrane [Gr. *tympanum*: drum] The eardrum.

Ubiquinone A mobile electron carrier of the mitochondrial respiratory chain. Similar to plastoquinone found in chloroplasts.

Ubiquitin A small protein that is covalently linked to other cellular proteins identified for breakdown by the proteosome.

Umbilical cord Tissue made up of embryonic membranes and blood vessels that connects the embryo to the placenta in eutherian mammals.

Understory The aggregate of smaller plants growing beneath the canopy of dominant plants in a forest.

Unicellular (yoon' e sell' yer ler) [L. *unus*: one + *cella*: chamber] Consisting of a single cell; as for example a unicellular organism. (Contrast with multicellular.)

Uniport A membrane transport process that carries a single substance. (Contrast with antiport, symport.)

Unsaturated hydrocarbon A compound containing only carbon and hydrogen atoms, with one or more pairs of carbon atoms that are connected by double bonds.

Upwelling The upward movement of nutrient-rich, cooler water from deeper layers of the ocean.

Uracil (U) A pyrimidine base found in nucleotides of RNA.

Urea A compound serving as the main excreted form of nitrogen by many animals, including mammals.

Ureotelic Describes an organism in which the final product of the breakdown of nitrogen-containing compounds (primarily proteins) is urea. (Contrast with ammonotelic, uricotelic.)

Ureter (your' uh tur) A long duct leading from the vertebrate kidney to the urinary bladder or the cloaca.

Urethra (you ree' thra) In most mammals, the canal through which urine is discharged from the bladder and which serves as the genital duct in males.

Uric acid A compound that serves as the main excreted form of nitrogen in some animals, particularly those which must conserve water, such as birds, insects, and reptiles.

Uricotelic Describes an organism in which the final product of the breakdown of nitrogen-containing compounds (primarily proteins) is uric acid. (Contrast with ammonotelic, ureotelic.)

Urine (you' rin) [Gk. *ouron*: urine] In vertebrates, the fluid waste product containing the toxic nitrogenous by-products of protein and amino acid metabolism.

Uterus (yoo' ter us) [L. *utero*: womb] The uterus or womb is a specialized portion of the female reproductive tract in certain mammals. It receives the fertilized egg and nurtures the embryo in its early development.

Vaccination Injection of virus or bacteria or their proteins into the body, to induce immunization. The injected material is usually attenuated (weakened) before injection.

Vacuole (vac' yew ole) [Fr.: small vacuum] A liquid-filled, membrane-enclosed compartment in cytoplasm; may function as digestive chambers, storage chambers, waste bins.

Vagina (vuh jine' uh) [L.: sheath] In female mammals, the passage leading from the external genital orifice to the uterus; receives the copulatory organ of the male in mating.

van der Waals forces Weak attractions between atoms resulting from the interaction of the electrons of one atom with the nucleus of another. This type of attraction is about one-fourth as strong as a hydrogen bond.

Variable regions The part of an immunoglobulin molecule or T-cell receptor that includes the antigen-binding site.

Vascular (vas' kew lar) [L. *vasculum*: a small vessel] Pertaining to organs and tissues that conduct fluid, such as blood vessels in animals and phloem and xylem in plants.

Vascular bundle In vascular plants, a strand of vascular tissue, including conducting cells of xylem and phloem as well as thick-walled fibers.

Vascular cambium A lateral meristem giving rise to secondary xylem and phloem.

Vascular rays In vascular plants, radially oriented sheets of cells produced by the vascular cambium, carrying materials laterally between the wood and the phloem.

Vascular system The conductive system of the plant, consisting primarily of xylem and phloem.

Vasopressin See antidiuretic hormone.

Vector (1) An agent, such as an insect, that carries a pathogen affecting another species. (2) A plasmid or virus that carries an inserted piece of DNA into a bacterium for cloning purposes in recombinant DNA technology.

Vegetal hemisphere The lower portion of some animal eggs, zygotes, and embryos, in which the dense nutrient yolk settles. The vegetal pole refers to the very bottom of the egg or embryo. (Contrast with animal hemisphere.)

Vegetative Nonreproductive, or nonflowering, or asexual.

Vegetative reproduction Asexual reproduction.

Vein [L. *vena*: channel] A blood vessel that returns blood to the heart. (Contrast with artery.)

Ventral [L. *venter*: belly, womb] Toward or pertaining to the belly or lower side. (Contrast with dorsal.)

Ventricle A muscular heart chamber that pumps blood through the body.

Vernalization [L. *vernalis*: spring] Events occurring during a required chilling period, leading eventually to flowering.

Vertebral column The jointed, dorsal column that is the primary support structure of vertebrates.

Vertebrate An animal whose nerve cord is enclosed in a backbone of bony segments, called vertebrae. The principal groups of vertebrate animals are the fishes, amphibians, reptiles, birds, and mammals.

Vesicle A membrane enclosed compartment within the cytoplasm.

Vessel elements In plants, nonliving water conducting cells with perforated end walls. Compare with tracheids.

Vestibular apparatus (ves tib' yew lar) [L. *vestibulum*: an enclosed passage] Structures associated with the vertebrate ear; these structures sense changes in position or momentum of the head, affecing balance and motor skills.

Vestigial (ves tij' ee al) [L. *vestigium*: footprint, track] The remains of body structures that are no longer of adaptive value to the organism and therefore are not maintained by selection.

Vicariant distribution A population distribution resulting from the disruption of a formerly continuous range by a vicariant event.

Vicariant event (vye care' ee unce) [L. *vicus*: change] The splitting of the range of a taxon by the imposition of some barrier to interchange among its members.

Villus (vil' lus) (plural: villi) [L. *villus*: shaggy hair or beard] A hairlike projection from a membrane; for example, from many gut walls.

Virion (veer' e on) The virus particle, the minimum unit capable of infecting a cell.

Viroid (vye' roid) An infectious agent consisting of a single-stranded RNA molecule with no protein coat; produces diseases in plants.

Virulent [L. *virus*: poison, slimy liquid] Causing or capable of causing disease and death.

Virus Any of a group of ultramicroscopic infectious particles constructed of nucleic acid and protein (and, sometimes, lipid) that can reproduce only in living cells.

Vitamins [L. *vita*: life] Organic compounds that an organism cannot synthesize, but nevertheless requires in small quantity for normal growth and metabolism.

Viviparous (vye vip' uh rus) [L. *vivus*: alive] Reproduction in which fertilization of the egg and development of the embryo occur inside the mother's body. (Contrast with oviparous.)

VNTRs (variable number of tandem repeats) In the human genome, short DNA sequences that are repeated a characteristic number of times in related individuals. Can be used to make a DNA fingerprint.

Waggle dance The running movement of a working honey bee on the hive, during which the worker traces out a repeated figure eight. The dance contains elements that transmit to other bees the location of the food.

Water potential In osmosis, the tendency for a system (a cell or solution) to take up water from pure water, through a differentially permeable membrane. Water flows toward the system with a more negative water potential. (Contrast with osmotic potential, turgor pressure.)

Water vascular system The array of canals and tubelike appendages that serves as the circulatory system, locomotory system, and food-capturing system of many echinoderms; is in direct connection with the surrounding sea water.

Wavelength The distance between successive peaks of a wave train, such as electromagnetic radiation.

Wild type Geneticists' term for standard or reference type. Deviants from this standard, even if the deviants are found in the wild, are said to be mutant.

Wood Secondary xylem tissue.

Xanthophyll (zan' tho fill) [Gr. *xanthos*: yellowish-brown + *phyllon*: leaf] A yellow or orange pigment commonly found as an accessory pigment in photosynthesis, but found elsewhere as well. An oxygen-containing carotenoid.

X-linked A character that is coded for by a gene on the X chromosome; a sex-linked trait.

Xerophyte (zee' row fyte) [Gr. *xerox*: dry + *phyton*: plant] A plant adapted to an environment with a limited water supply.

Xylem (zy' lum) [Gr. *xylon*: wood] In vascular plants, the tissue that conducts water and minerals; xylem consists, in various plants, of tracheids, vessel elements, fibers, and other highly specialized cells.

Yeast artificial chromosome (YAC) A laboratory-made DNA molecule containing sequences of yeast chromosomes (origin of replication, telomeres, centromere, and selectable markers) so that it can be used as a vector in yeast.

Yolk [M.E. *yolke*: yellow] The stored food material in animal eggs, usually rich in protein and lipid.

Yolk sac In embryonic development of reptiles, birds, and mammals, the extraembryonic membrane that forms from the endoderm of the hypoblast; it encloses and digests the yolk.

Z-DNA A form of DNA in which the molecule spirals to the left rather than to the right.

Zeaxanthin A carotenoid pigment that is a blue-light receptor.

Zoospore (zoe' o spore) [Gr. *zoon*: animal + *spora*: seed] In algae and fungi, any swimming spore. May be diploid or haploid.

Zygote (zye' gote) [Gr. *zygotos*: yoked] The cell created by the union of two gametes, in which the gamete nuclei are also fused. The earliest stage of the diploid generation.

Zymogen An inactive precursor of a digestive enzyme secreted into the lumen of the gut, where a protease cleaves it to form the active enzyme.

Answers to Self-Quizzes

Chapter 2
1.	b	6.	a
2.	d	7.	d
3.	c	8.	a
4.	c	9.	e
5.	d	10.	b

Chapter 3
1.	e	6.	a
2.	e	7.	c
3.	c	8.	e
4.	d	9.	a
5.	b	10.	d

Chapter 4
1.	b	6.	e
2.	d	7.	a
3.	c	8.	d
4.	e	9.	b
5.	a	10.	d

Chapter 5
1.	e	6.	c
2.	c	7.	c
3.	a	8.	b
4.	d	9.	e
5.	c	10.	c

Chapter 6
1.	c	6.	e
2.	e	7.	d
3.	b	8.	b
4.	c	9.	a
5.	c	10.	e

Chapter 7
1.	d	6.	d
2.	d	7.	a
3.	e	8.	b
4.	e	9.	a
5.	c	10.	e

Chapter 8
1.	c	6.	c
2.	b	7.	c
3.	d	8.	d
4.	b	9.	d
5.	e	10.	b

Chapter 9
1.	d	6.	d
2.	c	7.	e
3.	d	8.	d
4.	d	9.	c
5.	c	10.	c

Chapter 10*
1.	e	6.	d
2.	a	7.	b
3.	d	8.	b
4.	d	9.	b
5.	d	10.	e

Chapter 11
1.	c	6.	b
2.	a	7.	d
3.	c	8.	d
4.	b	9.	c
5.	e	10.	c

Chapter 12
1.	c	6.	d
2.	d	7.	b
3.	e	8.	d
4.	b	9.	d
5.	a	10.	a

Chapter 13
1.	b	6.	d
2.	e	7.	d
3.	a	8.	c
4.	c	9.	b
5.	c	10.	d

Chapter 14
1.	c	6.	c
2.	c	7.	c
3.	a	8.	b
4.	a	9.	e
5.	c	10.	d

Chapter 15
1.	d	6.	a
2.	d	7.	e
3.	c	8.	b
4.	c	9.	c
5.	d	10.	a

Chapter 16
1.	b	6.	b
2.	a	7.	c
3.	a	8.	a
4.	c	9.	e
5.	e	10.	e

*Answers to Chapter 10 Genetics Problems

1. Each of the eight boxes in the Punnett squares should contain the genotype Tt, regardless of which parent was tall and which dwarf.

2. Yellow parent = $s^Y s^b$; offspring 3 yellow (s^Y–): 1 black ($s^b s^b$). Black parent = $s^b s^b$; offspring all black ($s^b s^b$). Orange parent = $s^O s^b$; offspring 3 orange (s^O–): 1 black ($s^b s^b$). Both s^O and s^Y are dominant to s^b.

3. See Figure 10.4, page 192.

4. The trait is autosomal. Mother $dp\ dp$, father $Dp\ dp$. If the trait were sex-linked, all daughters would be wild-type and sons would be *dumpy*.

5. All females wild-type; all males spotted.

6. F_1 all wild-type, $PpSwsw$; F_2 9:3:3:1 in phenotypes. See Figure 10.7, page 194, for analogous genotypes.

7a. Ratio of phenotypes in F_2 is 3:1 (double dominant to double recessive).

7b. The F_1 are $Pby\ pB^Y$; they produce just two kinds of gametes (Pby and pBy). Combine them carefully and see the 1:2:1 phenotypic ratio fall out in the F_2.

7c. Pink-blistery.

7d. See Figures 9.14 and 9.16 (pages 178–180). Crossing over took place in the F_1 generation.

8. The genotypes are:
   $PpSwsw$
   $Ppswsw$
   $ppSwsw$
   $ppswsw$
   Ratio: 1:1:1:1

The phenotypes are:
wild eye, long wing	pink eye, long wing
wild eye, short wing	pink eye, short wing

Ratio: 1:1:1:1

9a. 1 black:2 blue:1 splashed white

9b. Always cross black with splashed white.

10a. $w^+ > w^e > w$

10b. Parents $w^e w$ and $w^+ Y$. Progeny $w+w^e$, $w+w$, $w^e Y$, and wY.

11. All will have normal vision because they inherit dad's wild-type X chromosome, but half of them will be carriers.

12. Agouti parent $AaBb$. Albino offspring $aaBb$ and $aabb$; black offspring $Aabb$; agouti offspring $AaBb$.

13. Because the gene is carried on mitochondrial DNA, it is passed through the mother only. Thus if the woman does not have the disease but her husband does, their child will not be affected. On the other hand, if the woman has the disease but her husband does not, their child *will* have the disease.

Chapter 17
1. a   6. b
2. c   7. e
3. b   8. d
4. b   9. c
5. d   10. b

Chapter 18
1. a   6. a
2. b   7. d
3. a   8. d
4. e   9. a
5. c   10. d

Chapter 19
1. c   6. c
2. a   7. d
3. a   8. b
4. b   9. a
5. b   10. b

Chapter 20
1. a   6. c
2. c   7. b
3. e   8. d
4. c   9. b
5. d   10. a

Chapter 21
1. e   6. b
2. d   7. a
3. a   8. c
4. b   9. e
5. c   10. b

Chapter 22
1. d   6. a
2. b   7. c
3. e   8. b
4. c   9. c
5. a   10. e

Chapter 23
1. b   7. d
2. d   8. b
3. d   9. e
4. c   10. b
5. d   11. c
6. e

Chapter 24
1. c   7. d
2. e   8. a
3. d   9. a
4. c   10. c
5. a   11. e
6. b

Chapter 25
1. c   7. a
2. a   8. d
3. d   9. b
4. e   10. d
5. b   11. e
6. c

Chapter 26
1. d   7. a
2. b   8. a
3. e   9. a
4. c   10. a
5. a   11. c
6. e   12. b

Chapter 27
1. e   6. b
2. e   7. d
3. b   8. a
4. c   9. c
5. e   10. b

Chapter 28
1. a   6. d
2. e   7. c
3. c   8. b
4. d   9. b
5. a   10. d

Chapter 29
1. d   6. e
2. c   7. c
3. e   8. b
4. b   9. b
5. b   10. d

Chapter 30
1. d   6. c
2. c   7. a
3. d   8. e
4. a   9. c
5. d   10. a

Chapter 31
1. b   6. a
2. d   7. e
3. e   8. a
4. c   9. c
5. d   10. c

Chapter 32
1. c   6. d
2. d   7. d
3. b   8. b
4. e   9. e
5. b   10. c

Chapter 33
1. d   6. c
2. c   7. d
3. a   8. d
4. e   9. e
5. b   10. e

Chapter 34
1. b   7. e
2. a   8. a
3. c   9. e
4. c   10. c
5. b   11. b
6. d

Chapter 35
1. d   6. b
2. b   7. b
3. e   8. c
4. e   9. a
5. a   10. d

Chapter 36
1. c   6. d
2. d   7. d
3. b   8. e
4. b   9. e
5. b   10. a

Chapter 37
1. d   6. c
2. d   7. e
3. c   8. a
4. a   9. d
5. a   10. e

Chapter 38
1. a   6. c
2. e   7. e
3. c   8. c
4. d   9. a
5. b   10. b

Chapter 39
1. d   6. e
2. b   7. a
3. e   8. b
4. b   9. c
5. d   10. d

Chapter 40
1. e   6. a
2. b   7. b
3. c   8. c
4. c   9. d
5. d   10. a

Chapter 41
1. c   6. e
2. a   7. a
3. d   8. e
4. b   9. e
5. b   10. c

Chapter 42
1. b   6. b
2. a   7. d
3. b   8. d
4. e   9. c
5. e   10. c

Chapter 43
1. e   6. d
2. e   7. d
3. a   8. c
4. d   9. d
5. d   10. a

Chapter 44
1. d   6. e
2. a   7. e
3. d   8. c
4. c   9. d
5. c   10. d

Chapter 45
1. d   6. e
2. d   7. b
3. a   8. c
4. b   9. c
5. e   10. d

Chapter 46
1. c   6. c
2. a   7. a
3. e   8. c
4. d   9. a
5. d   10. a

Chapter 47
1. e   6. d
2. a   7. e
3. b   8. a
4. c   9. a
5. b   10. a

Chapter 48
1. e   6. b
2. d   7. c
3. a   8. c
4. b   9. a
5. c   10. d

Chapter 49
1. d   6. d
2. a   7. b
3. c   8. d
4. d   9. c
5. c   10. e

Chapter 50
1. b   6. d
2. e   7. a
3. c   8. b
4. a   9. d
5. b   10. d

Chapter 51

1.	b	6.	b
2.	a	7.	e
3.	d	8.	a
4.	c	9.	c
5.	d	10.	e

Chapter 52

1.	c	6.	d
2.	a	7.	d
3.	c	8.	d
4.	e	9.	b
5.	d	10.	b

Chapter 53

1.	e	5.	c
2.	c	6.	d
3.	d	7.	d
4.	b	8.	a

Chapter 54

1.	b	6.	c
2.	c	7.	c
3.	a	8.	b
4.	d	9.	d
5.	b	10.	a

Chapter 55

1.	a	6.	e
2.	c	7.	c
3.	a	8.	a
4.	b	9.	d
5.	d	10.	d

Chapter 56

1.	a	6.	a
2.	c.	7.	d
3.	d	8.	e
4.	c	9.	a
5.	d	10.	e

Chapter 57

1.	b	6.	a
2.	d	7.	d
3.	e	8.	c
4.	e	9.	a
5.	b	10.	c

Chapter 58

1.	d	6.	c
2.	d	7.	a
3.	e	8.	c
4.	e	9.	d
5.	c	10.	e

Illustration Credits

Inc. 9.7b: Conly L. Rieder/Biological Photo Service. 9.8: Andrew S. Bajer, U. Oregon. 9.10a: T. E. Schroeder/Biological Photo Service. 9.10b: B. A. Palevitz & E. H. Newcomb/Biological Photo Service. 9.11: Garry T. Cole/Biological Photo Service. 9.12 left: Andrew Syred/SPL/Photo Researchers, Inc. 9.12 center: E. Webber/Visuals Unlimited. 9.12 right: Bill Kamin/Visuals Unlimited. 9.13: Courtesy of Dr. Thomas Ried & Dr. Evelin Schröck, NIH. 9.14: C. A. Hasenkampf/Biological Photo Service. 9.15: Klaus W. Wolf, U. West Indies. 9.19: Gopal Murti/Photo Researchers, Inc.

Chapter 10 *Opener*: David H. Wells/Corbis. 10.2: Wally Eberhart/Visuals Unlimited. 10.12: Courtesy the American Netherland Dwarf Rabbit Club. 10.15: NCI/Photo Researchers, Inc. 10.16: Courtesy of Pioneer Hi-Bred International, Inc. 10.24: Science VU/Visuals Unlimited. *Bay scallops*: Barbara J. Miller/Biological Photo Service.

Chapter 11 *Opener*: Universal City Studios/Shooting Star. 11.2: Biozentrum, Universtiy of Basel/SPL/Photo Researchers, Inc. 11.4: Courtesy of Prof. M. H. F. Wilkins, Dept. of Biophysics, King's College, U. London. 11.6a: A. Barrington Brown/Photo Researchers, Inc. 11.6b: Structure prepared by Jason Kahn with InsightII (Accelrys); data from S. Arnott & D. W. Hukins. 1972. *Biochem. Biophys. Res. Commun.* 47(6): 1504. 11.18: Dr. Peter Lansdorp/Visuals Unlimited.

Chapter 12 *Opener*: Alan L. Detrick/Photo Researchers, Inc. 12.7: Data from PDB 1EHZ, H. Shi & P. B. Moore. 2000. *RNA* 6: 1091. 12.13b: Courtesy of J. E. Edström and *EMBO J.* 12.17: Stanley Flegler/Visuals Unlimited.

Chapter 13 *Opener*: Dennis Kunkel Microscopy, Inc. 13.1a: Dennis Kunkel Microscopy, Inc. 13.1b: E.O.S./Gelderblom/Photo Researchers, Inc. 13.1c: Dennis Kunkel Microscopy, Inc. 13.8: Courtesy of L. Caro & R. Curtiss. 13.21: Based on an illustration by Anthony R. Kerlavage, Institute for Genomic Research. *Science* 269: 449 (1995).

Chapter 14 *Opener*: Inga Spence/Visuals Unlimited. 14.6: Tiemeier et al. 1978. *Cell* 14: 237. 14.17: Courtesy of Murray L. Barr, U. Western Ontario. 14.19: Courtesy of O. L. Miller, Jr.

Chapter 15 *Opener*: Ryan McVay, Photodisc Green/Getty Images. 15.2 *inset*: Biophoto Associates/Photo Researchers, Inc. 15.3: From de Vos et al., 1992. *Science* 255: 306. 15.12: Stephen A. Stricker, courtesy of Molecular Probes, Inc.

Chapter 16 *Opener*: Keith V. Wood/Science VU/Visuals Unlimited. 16.2: Philippe Plailly/Photo Researchers, Inc. 16.15 *left*: Courtesy of Ingo Potrykus, Swiss Federal Institute of Technology. 16.15 *right*: Joan Gemme & David McIntyre. 16.16: Courtesy of Eduardo Blumwald. 16.18: Bettmann/Corbis.

Chapter 17 *Opener*: Data from PDB 1IEP, B. Nagar et al. 2002. *Cancer Res.* 62: 4236. 17.5: C. Harrison et al., 1983. *J. Med. Genet.* 20: 280. 17.10: Courtesy of Harvey Levy & Cecelia Walraven, New England Newborn Screening Program. 17.13: Dennis Kunkel Microscopy, Inc. 17.18b: David M. Martin, M.D./SPL/Photo Researchers, Inc. 17.24: From P. H. O'Farrell. 1975. High resolution two-dimensional electrophoresis of proteins. *J. Biol. Chem.* 250: 4007–4021. Courtesy of Patrick H. O'Farrell.

Chapter 18 *Opener*: Francis G. Mayer/Corbis. 18.3: Dennis Kunkel Microscopy, Inc. 18.9: Dr. Gopal Murti/SPL/Photo Researchers, Inc. 18.14: Dr. Andrejs Liepins/SPL/Photo Researchers, Inc. 18.16: David Phillips/Science Source/Photo Researchers, Inc.

Chapter 19 *Opener*: Courtesy of Advanced Cell Technology, Worcester, Mass. USA. 19.4: Roddy Field, the Roslin Institute. 19.5: Courtesy of T. Wakayama & R. Yanagimachi. 19.10: J. E. Sulston & H. R. Horvitz, 1977. *Dev. Bio.* 56: 100. 19.12b: Courtesy of J. Bowman. 19.13 *left*: Courtesy of J. Bowman. 19.13 *right*: Courtesy of Detlef Weigel. 19.14: Courtesy of W. Driever & C. Nüsslein-Vollhard. 19.16: Courtesy of F. R. Turner, Indiana U.

Chapter 20 *Opener*: Dave B. Fleetham/Tom Stack & Assoc. 20.1: D. M. Phillips/Science Source/Photo Researchers, Inc. 20.2: Courtesy of Richard Elinson, U. Toronto. 20.5: Courtesy of J. G. Mulnard. 20.21a: Dr. G. Moscoso/SPL/Photo Researchers, Inc. 20.21b: Tissuepix/SPL/Photo Researchers, Inc.

Chapter 21 *Opener*: LogicStock/Painet Inc. 21.1 *upper left*: Colin Milkins/Oxford Scientific Films. 21.1 *upper right*: Jan Hinsch/SPL/Photo Researchers, Inc. 21.1 *lower left*: Oxford Scientific Films. 21.1 *lower right*: Robert Brons/Biological Photo Service. 21.2: Courtesy of W. J. Gehring & G. Halder. From Halder et al., 1995. *Science* 267: 1788. 21.4a: Courtesy of E. B. Lewis. 21.4b: Courtesy of H. Le Mouellic, Y. Lallemand, & P. Brûlet. From Le Mouellic et al. 1992. *Cell* 69: 251. 21.5 *cladogram*: After R. Galant and S. Carroll, 2002. *Nature* 415: 910. 21.5 *beetle*: Stockbyte/PictureQuest. 21.5 *centipede*: Burke/Triolo/Brand X Pictures/PictureQuest. 21.6: Courtesy of J. Hurle & E. Laufer. 21.7: Courtesy of J. Hurle. 21.9: Courtesy of S. Carroll & P. Brakefield. 21.10: Erick Greene. 21.11: Courtesy of A. A. Agrawal. 21.12: Nigel Cattlin, Holt Studios International/Photo Researchers, Inc.

Chapter 22 *Opener*: PhotoLink/Photodisc/PictureQuest. 22.1: Robert Fried/Tom Stack & Assoc. 22.3a: Tom & Therisa Stack/Tom Stack & Assoc. 22.3b: Stanley M. Awramik/Biological Photo Service. 22.6: François Gohier/Photo Researchers, Inc. 22.7: Jeff J. Daly/Visuals Unlimited. 22.8 *left*: Ken Lucas/Biological Photo Service. 22.8 *right*: Stanley M. Awramik/Biological Photo Service. 22.9: Chip Clark. 22.10: Hans Steur/Visuals Unlimited. 22.11: Tom McHugh/Field Museum, Chicago/Photo

Researchers, Inc. 22.12: Chase Studios, Cedarcreek, MO. 22.14: Chris Butler/SPL/Photo Researchers, Inc. 22.16: K. Simons & David Dilcher. 22.18a: Kjell Sandved/Visuals Unlimited. 22.18b: Hans Reinhard/Okapia/Photo Researchers, Inc. 22.21: Calvin Larsen/Photo Researchers, Inc.

Chapter 23 *Snake*: Joseph T. Collins/Photo Researchers, Inc. *Newt*: Robert Clay/Visuals Unlimited. 23.1a,b: SPL/Photo Researchers, Inc. 23.2: Levi, W. 1965. *Encyclopedia of Pigeon Breeds*. T. F. H. Publications, Jersey City, NJ. (a,b: photos by R. L. Kienlen, courtesy of Ralston Purina Company; c,d: photos by Stauber.). 23.9: S. Maslowski/Visuals Unlimited. 23.11: Judith Worley/Painet Inc. 23.17b: Tony Tilford/Oxford Scientific Films. 23.22a: Marilyn Kazmers/Dembinsky Photo Assoc. 23.22b: Paul Osmond/Painet Inc. 23.23: Anup Shah/Dembinsky Photo Assoc.

Chapter 24 *Opener*: Rob Simpson/Painet Inc. 24.2a *left*: Gary Meszaros/Dembinsky Photo Assoc. 24.2a *right*: Lior Rubin/Peter Arnold, Inc. 24.2b: Fi Rust/Painet Inc. 24.9a: Virginia P. Weinland/Photo Researchers, Inc. 24.9b: José Manuel Sánchez de Lorenzo Cáceres. 24.11a: D. Cavagnaro/DRK Photo. 24.11b: Charles Webber/California Academy of Sciences. 24.13a: Stephen Dalton/NHPA. 24.13b: Daniel Heuclin/NHPA. 24.14 *upper, lower*: Peter J. Bryant/Biological Photo Service. 24.14 *center*: Courtesy of Kenneth Y. Kaneshiro, U. Hawaii. 24.16 *Madia*: Peter K. Ziminsky/Visuals Unlimited. 24.16 *Argyroxiphium*: Elizabeth N. Orians. 24.16 *Dubautia*: Noble Proctor/Photo Researchers, Inc. 24.16 *Wilkesia*: Gerald D. Carr.

Chapter 25 *Opener*: Lorne Resnick/AGE Fotostock. 25.3 *left*: Adam Jones/Dembinsky Photo Assoc. 25.3 *center*: Brian Parker/Tom Stack & Assoc. 25.3 *right*: Joe McDonald/DRK Photo. 25.11a : Mark Smith/Photo Researchers, Inc. 25.11b: After E. Verheyen et al., 2003. *Science* 300: 328.

Chapter 26 *Opener*: Courtesy of James Gathany/CDC. 26.4: Courtesy of Richard Alexander, U. Pennsylvania. 26.6a: Belinda Wright/DRK Photo. 26.6b: M. Graybill/J. Hodder/Biological Photo Service.

Chapter 27 *Opener*: Thomas Dressler/DRK Photo. 27.1: Kari Lounatmaa/Photo Researchers, Inc. 27.3a: D. M. Phillips/Photo Researchers, Inc. 27.3b: R. Kessel-G. Shih/Visuals Unlimited. 27.3c: Courtesy of Janice Carr/NCID/CDC. 27.4a: J. A. Breznak & H. S. Pankratz/Biological Photo Service. 27.4b: J. Robert Waaland/Biological Photo Service. 27.5: USDA/Visuals Unlimited. 27.6a *left*: D. M. Phillips/Visuals Unlimited. 27.6a *right*: Courtesy of Peter Hirsch & Stuart Pankratz. 27.6b *left*: Courtesy of the CDC. 27.6b *right*: Courtesy of Peter Hirsch & Stuart Pankratz. 27.10: P. Gates/Biological Photo Service. 27.11a: Paul W. Johnson/Biological Photo Service. 27.11b: H. S. Pankratz/Biological Photo Service. 27.11c: Bill Kamin/Visuals Unlimited. 27.12:

Courtesy of David Cox/CDC. 27.13: Randall C. Cutlip. 27.14: T. J. Beveridge/Biological Photo Service. 27.15: Dr. Gary Gaugler/Visuals Unlimited. 27.16: D. M. Phillips/Visuals Unlimited. 27.17: M. Gabridge/Visuals Unlimited. 27.19: Krafft/Hoa-qui/Photo Researchers, Inc. 27.20: Martin G. Miller/Visuals Unlimited.

Chapter 28 *Opener*: London School of Hygiene/SPL/Photo Researchers, Inc. 28.1*a*: Dennis Kunkel Microscopy, Inc. 28.1*b*: J. Paulin/Visuals Unlimited. 28.1*c*: Randy Morse/Tom Stack & Assoc. 28.4: Mike Abbey/Visuals Unlimited. 28.7*a*: Christian Gautier/Jacana/Photo Researchers, Inc. 28.7*b*: David Patterson/SPL/Photo Researchers, Inc. 28.7*c*: James Solliday/Biological Photo Service. 28.8: David Patterson, Linda Amaral Zettler, Mike Peglar, & Tom Nerad/micro*scope. 28.12: Oliver Meckes/Photo Researchers, Inc. 28.13: Sanford Berry/Visuals Unlimited. 28.15*a*: Mike Abbey/Visuals Unlimited. 28.15*b*: Dennis Kunkel Microscopy, Inc. 28.15*c*: Paul W. Johnson/Biological Photo Service. 28.15*d*: M. Abbey/Photo Researchers, Inc. 28.18*a*: Manfred Kage/Peter Arnold, Inc. 28.18*b*: Biophoto Associates/Photo Researchers, Inc. 28.20*a*: Joyce Photographics/Photo Researchers, Inc. 28.20*b*: J. Robert Waaland/Biological Photo Service. 28.21*a*: Jeff Foott/Tom Stack & Assoc. 28.21*b*: J. N. A. Lott/Biological Photo Service. 28.23: James W. Richardson/Visuals Unlimited. 28.24*a*: Milton Rand/Tom Stack & Assoc. 28.24*b*: J. N. A. Lott/Biological Photo Service. 28.25*a*: Carolina Biological/Visuals Unlimited. 28.25*b*: Andrew J. Martinez/Photo Researchers, Inc. 28.28*a*: William Bourland/micro*scope. 28.28*b*: David Patterson & Aimlee Laderman/micro*scope. 28.30*a*: Andrew Syred/SPL/Photo Researchers, Inc. 28.30*b*: Eric Grave/SPL/Photo Researchers, Inc. 28.31*a*: Barbara J. Miller/Biological Photo Service. 28.31*b*: Carolina Biological/Visuals Unlimited. 28.32: Courtesy of R. Blanton & M. Grimson.

Chapter 29 *Opener*: Dr. Ray Clark & Mervyn De Calcina/SPL/Photo Researchers, Inc. 29.3*a*: Ron Dengler/Visuals Unlimited. 29.3*b*: Larry Mellichamp/Visuals Unlimited. 29.6: J. Robert Waaland/Biological Photo Service. 29.7*a*: Rod Planck/Dembinsky Photo Assoc. 29.7*b*: William Harlow/Photo Researchers, Inc. 29.7*c*: Science VU/Visuals Unlimited. 29.8: Daniel Vega/AGE Fotostock. 29.9*a*: Brian Enting/Photo Researchers, Inc. 29.9*b*: Courtesy of J. H. Troughton. 29.11: University of Michigan Exhibit Museum. 29.15*a*: Ed Reschke/Peter Arnold, Inc. 29.15*b*: Carolina Biological/Visuals Unlimited. 29.16*a*: J. N. A. Lott/Biological Photo Service. 29.16*b*: David Sieren/Visuals Unlimited. 29.17: W. Ormerod/Visuals Unlimited. 29.18*a*: Rod Planck/Dembinsky Photo Assoc. 29.18*b*: Nuridsany et Perennou/Photo Researchers, Inc. 29.18*c*: Dick Keen/Visuals Unlimited. 29.19: L. West/Photo Researchers, Inc.

Chapter 30 *Opener*: Warren Faidley/DRK Photo. 30.3: Phil Gates/Biological Photo Service. 30.4*a*: Roland Seitre/Peter Arnold, Inc.

30.4*b*: Bernd Wittich/Visuals Unlimited. 30.4*c*: M. Graybill & J. Hodder/Biological Photo Service. 30.4*d*: N. H. Cheatham/DRK Photo. 30.5*a left*: Michael P. Gadomski/Photo Researchers, Inc. 30.5*a right*: Stan W. Elems/Visuals Unlimited. 30.5*b left*: Gerald & Buff Corsi/Visuals Unlimited. 30.5*b right*: John D. Cunningham/Visuals Unlimited. 30.8*a*: Dick Poe/Visuals Unlimited. 30.8*b*: Richard Shiell. 30.8*c*: Richard Shiell/Dembinsky Photo Assoc. 30.9*a*: Sinclair Stammers/SPL/Photo Researchers, Inc. 30.9*b*: Rod Planck/Photo Researchers, Inc. 30.12*a*: Inga Spence/Tom Stack & Assoc. 30.12*b*: Holt Studios/Photo Researchers, Inc. 30.12*c*: Catherine M. Pringle/Biological Photo Service. 30.12*d*: Henry Beeker/AGE Fotostock. 30.14*a*: Courtesy of Stephen McCabe, U. California, Santa Cruz, & UCSC Arboretum. 30.14*b*: John Gerlach/Dembinsky Photo Assoc. 30.14*c*: Rob & Ann Simpson/Visuals Unlimited. 30.14*d*: R. C. Carpenter/Photo Researchers, Inc. 30.14*e*: Geoff Bryant/Photo Researchers, Inc. 30.14*f*: Chris Sharp/Photo Researchers, Inc. 30.15*a*: Ken Lucas/Visuals Unlimited. 30.15*b*: Ed Reschke/Peter Arnold, Inc. 30.15*c*: Adam Jones/Dembinsky Photo Assoc. 30.16*a*: Richard Shiell. 30.16*b*: Adam Jones/Dembinsky Photo Assoc. 30.16*c*: Alan & Linda Detrick/Photo Researchers, Inc.

Chapter 31 *Opener*: Dr. Gary Gaugler/Visuals Unlimited. 31.1*a*: Inga Spence/Tom Stack & Assoc. 31.1*b*: L. E. Gilbert/Biological Photo Service. 31.2: D. M. Phillips/Visuals Unlimited. 31.4: G. T. Cole/Biological Photo Service. 31.5: N. Allin & G. L. Barron/Biological Photo Service. 31.7: J. Robert Waaland/Biological Photo Service. 31.8: Biophoto Associates/Photo Researchers, Inc. 31.9: M. F. Brown/Visuals Unlimited. 31.10: John D. Cunningham/Visuals Unlimited. 31.11*a*: Richard Shiell/Dembinsky Photo Assoc. 31.11*b*: Matt Meadows/Peter Arnold, Inc. 31.12: Andrew Syred/SPL/Photo Researchers, Inc. 31.14*a*: Angelina Lax/Photo Researchers, Inc. 31.14*b*: Manfred Danegger/Photo Researchers, Inc. 31.14*c*: Stan Flegler/Visuals Unlimited. 31.15 *inset*: Biophoto Associates/Photo Researchers, Inc. 31.16*a*: R. L. Peterson/Biological Photo Service. 31.16*b*: Merton F. Brown/Visuals Unlimited. 31.17*a*: Ed Reschke/Peter Arnold, Inc. 31.17*b*: Gary Meszaros/Dembinsky Photo Assoc. 31.18*a*: J. N. A. Lott/Biological Photo Service.

Chapter 32 *Opener*: Doug Scott/AGE Fotopix. 32.5*a*: D. Fawcett/Visuals Unlimited. 32.5*b*: Paul Osmond/Painet Inc. 32.5*c*: Courtesy of Jean Vacelet. 32.6*a*: Larry Jon Friesen. 32.6*b*: Jett Britnell/DRK Photo. 32.6*c*: David J. Wrobel/Visuals Unlimited. 32.6*d*: Larry Jon Friesen. 32.7, 32.8: Adapted from F. M. Bayerand & H. B. Owre. 1968. *The Free-Living Lower Inverte-brates*, Macmillan Publishing Co. 32.9*a*: David Hall/Photo Researchers, Inc. 32.9*b*: Ed Robinson/Tom Stack & Assoc. 32.10, 32.11: Adapted from F. M. Bayerand & H. B. Owre. 1968. *The Free-Living Lower Invertebrates*, Macmillan Publishing Co. 32.12*b*: Kathie Atkinson/Oxford Scientific Films. 32.13: M. W. Martin, from *Science* 288: 841. 32.15*a*: Denise

Tackett/DRK Photo. 32.17*b*, 32.19*a*: Robert Brons/Biological Photo Service. 32.20: David J. Wrobel/Biological Photo Service. 32.21*b*: Oxford Scientific Films. 32.24*a*: Brian Parker/Tom Stack & Assoc. 32.24*b*: Roger K. Burnard/Biological Photo Service. 32.24*c*: Stanley Breeden/DRK Photo. 32.24*d*: Courtesy of R. R. Hessler, Scripps Institute of Oceanography. 32.26*a*: Jeff Foott/Nature Picture Library. 32.26*b*: Dave Fleetham/Tom Stack & Assoc. 32.26*c,d,e*: Larry Jon Friesen. 32.26*f*: Fred McConnaughey/Photo Researchers, Inc.

Chapter 33 *Opener*: Sharon Cummings/Dembinsky Photo Associates. 33.2: Courtesy of Jen Grenier & Sean Carroll, University of Wisconsin. 33.4: R. Calentine/Visuals Unlimited. 33.5*b,c*: James Solliday/Biological Photo Service. 33.6*a*: Michael Fogden/DRK Photo. 33.6*b*: Diane R. Nelson/Visuals Unlimited. 33.7: Ken Lucas/Visuals Unlimited. 33.9*a,b*: Larry Jon Friesen. 33.9*c*: Tom Branch/Photo Researchers, Inc. 33.9*d*: A. Flowers & L. Newman/Photo Researchers, Inc. 33.12: Oxford Scientific Films. 33.13*a*: Sinclair Stammers/Nature Picture Library. 33.13*b*: Larry Jon Friesen. 33.13*c*: David Maitland/Masterfile. 33.13*d*: Larry Jon Friesen. 33.13*e*: Colin Milkins/Oxford Scientific Films. 33.13*f*: Peter J. Bryant/Biological Photo Service. 33.13*g*: Courtesy of Scott Bauer/USDA ARS. 33.13*h*: L. West/Photo Researchers, Inc. 33.15*a*: Marty Cordano/DRK Photo. 33.15*b*: William Leonard/DRK Photo. 33.16*a*: Joel Simon. 33.16*b*: Fred Bruemmer/DRK Photo. 33.17*a,b*: Larry Jon Friesen. 33.12*c*: W. M. Beatty/Visuals Unlimited. 33.12*d*: Photo by Eric Erbe; colorization by Chris Pooley/USDA ARS.

Chapter 34 *Opener*: Michael Fogden/DRK Photo. 34.1: From Bengtson, S. 2000. Teasing fossils out of shales with cameras and computers. *Palaeontologia Electronica* 3(1). 34.4*a*: Hal Beral/Visuals Unlimited. 34.4*b*: Randy Morse/Tom Stack & Assoc. 34.4*c*: Mark J. Thomas/Dembinsky Photo Assoc. 34.4*d*: Randy Morse/Tom Stack & Assoc. 34.4*e*: Larry Jon Friesen. 34.5*a*: C. R. Wyttenbach/Biological Photo Service. 34.6*a*: Denise Tackett/DRK Photo. 34.6*b*: David Wrobel/Visuals Unlimited. 34.7*b*: Robert Brons/Biological Photo Service. 34.10*a*: Brian Parker/Tom Stack & Assoc. 34.10*b*: Gary Milburn/Tom Stack & Assoc. 34.12*a*: Dave Fleetham/Tom Stack & Assoc. 34.12*b*: Marty Snyderman/Masterfile. 34.12*c*: Dave Fleetham/Tom Stack & Assoc. 34.13*a*: Tobias Bernhard/Oxford Scientific Films. 34.13*b*: Fred Bavendam/Minden Pictures. 34.13*c*: Dave Fleetham/Visuals Unlimited. 34.13*d*: Dr. Paul A. Zahl/Photo Researchers, Inc. 34.14: Tom McHugh, Steinhart Aquarium/Photo Researchers, Inc. 34.15*a*: Ken Lucas/Biological Photo Service. 34.15*b*: Gary Meszaros/Dembinsky Photo Assoc. 34.15*c*: Nick Garbutt/Indri Images. 34.19*a*: Dave B. Fleetham/Tom Stack & Assoc. 34.19*b*: C. Alan Morgan/Peter Arnold, Inc. 34.19*c*: Gerry Ellis, DigitalVision/PictureQuest. 34.19*d*: Michael Fogden/DRK Photo. 34.19*e*: Mark J. Thomas/Dembinsky Photo Assoc. 34.20*a*: From Xu, X., et al., 2003.

Nature 421: 335. Macmillan Publishers Ltd. 34.20b: Tom & Therisa Stack/Tom Stack & Assoc. 34.20c: Fossil from the Natural History Museum of Basel, photographed by Severino Dahint. 34.21a: Joe McDonald/Tom Stack & Assoc. 34.21b: Skip Moody/Dembinsky Photo Assoc. 34.21c: John Shaw/Tom Stack & Assoc. 34.21d: Fred Bruemmer/DRK Photo. 34.22a: Ed Kanze/Dembinsky Photo Assoc. 34.22b: Dave Watts/Tom Stack & Assoc. 34.23a: Art Wolfe. 34.23b: Jany Sauvanet/Photo Researchers, Inc. 34.23c: Hans & Judy Beste/Animals Animals. 34.24a: Rod Planck/Dembinsky Photo Assoc. 34.24b: Joe McDonald/Tom Stack & Assoc. 34.24c: Michael S. Nolan/Tom Stack & Assoc. 34.24d: Erwin & Peggy Bauer/Tom Stack & Assoc. 34.26a: Martin Harvey/DRK Photo. 34.26b: Stanley Breeden/DRK Photo. 34.27a: Steve Kaufman/DRK Photo. 34.27b: John Bracegirdle/Masterfile. 34.28a: Stan Osolinsky/Dembinsky Photo Assoc. 34.28b: Anup Shah/Dembinsky Photo Assoc. 34.28c: Art Wolfe. 34.28d: Anup Shah/Dembinsky Photo Assoc. 34.30: Dembinsky Photo Assoc.

Chapter 35   *Opener*: D. Cavagnaro/Visuals Unlimited. 35.3a: Antonia Reeve/SPL/Photo Researchers, Inc. 35.3b: R. Calentine/Visuals Unlimited. 35.4a: Joyce Photographics/Photo Researchers, Inc. 35.4b: Renee Lynn/Photo Researchers, Inc. 35.4c: C. K. Lorenz/Photo Researchers, Inc. 35.7: Biophoto Associates/Photo Researchers, Inc. 35.9a: Biodisc/Visuals Unlimited. 35.9b: P. Gates /Biological Photo Service. 35.9c: Biophoto Associates/Photo Researchers, Inc. 35.9d: Jack M. Bostrack/Visuals Unlimited. 35.9e: John D. Cunningham/Visuals Unlimited. 35.9f: J. Robert Waaland/Biological Photo Service. 35.11, 35.13: J. Robert Waaland/Biological Photo Service. 35.16a: Jim Solliday/Biological Photo Service. 35.16b: Microfield Scientific LTD/Photo Researchers, Inc. 35.16c: Ray F. Evert, U. Wisconsin, Madison. 35.16d: John D. Cunningham/Visuals Unlimited. 35.18a,b left: Carolina Biological/Visuals Unlimited. 35.18a,b right: J. Robert Waaland/Biological Photo Service. 35.20: John N. A. Lott/Biological Photo Service. 35.21: Jim Solliday/Biological Photo Service. 35.22: P. Gates/Biological Photo Service. 35.23b: Courtesy of Thomas Eisner, Cornell U. 35.23c: C. G. Van Dyke/Visuals Unlimited.

Chapter 36   *Opener*: Bettman/Corbis. 36.6: From B. Bentwood & J. Cronshaw. 1978. *Planta* 140: 111. 36.7: Ed Reschke/Peter Arnold, Inc. 36.10: After M. A. Zwieniecki et al., 2001. *Science* 291: 1059–1061. 36.11a: D. M. Phillips/Visuals Unlimited. 36.13: M. H. Zimmermann.

Chapter 37   *Opener*: Courtesy of Emerson D. Nafziger. 37.3: Kathleen Blanchard/Visuals Unlimited. 37.5: Hugh Spencer/Photo Researchers, Inc. 37.7: E. H. Newcomb & S. R. Tandon/Biological Photo Service. 37.9a: J. H. Robinson/Photo Researchers, Inc. 37.9b: Milton Rand/Tom Stack & Assoc. 37.10: Gilbert S. Grant/Photo Researchers, Inc.

Chapter 38   *Opener*: Jeremy Woodhouse/DRK Photo. 38.2: Tom J. Ulrich/Visuals Unlimited. 38.3: John Eastcott, Yva Momatiuk/DRK Photo. 38.5: Courtesy of J. A. D. Zeevaart, Michigan State U. 38.11: Ed Reschke/Peter Arnold, Inc. 38.13: Biophoto Associates/Photo Researchers, Inc. 38.16: T. A. Wiewandt/DRK Photo. 38.19: Courtesy of Dr. Eva Huala, Carnegie Institution of Washington.

Chapter 39   *Opener*: RMF/Visuals Unlimited. 39.2: Stephen Dalton/Photo Researchers, Inc. 39.4: From Bowman, J. (ed.), 1994. *Arabiopsis: An Atlas of Morphology and Development*. Springer-Verlag, New York. Photo by S. Craig & A. Chaudhury, Plate 6.2. 39.8a: C. P. George/Visuals Unlimited. 39.8b: Tess & David Young/Tom Stack & Assoc. 39.14: After M. J. Yanovsky and S. A. Kay, 2002. *Nature* 419: 308–312. 39.16a: Nigel Cattlin, Holt Studios International/Photo Researchers, Inc. 39.16b: Jerome Wexler/Photo Researchers, Inc.

Chapter 40   *Opener*: Peter J. Bryant/Biological Photo Service. 40.2: D. Cavagnaro/Visuals Unlimited. 40.6: Courtesy of Thomas Eisner, Cornell U. 40.7: Adam Jones/Dembinsky Photo Assoc. 40.8: J. N. A. Lott/Biological Photo Service. 40.9: Richard Shiell. 40.10: Janine Pestel/Visuals Unlimited. 40.11: Chip Isenhart/Tom Stack & Assoc. 40.12: J. N. A. Lott/Biological Photo Service. 40.13: Robert & Linda Mitchell. 40.15: Budd Titlow/Visuals Unlimited.

Chapter 41   *Opener*: AP/Wide World Photos. 41.3a: CNRI/SPL/Photo Researchers, Inc. 41.3b: G. W. Willis/Visuals Unlimited. 41.9a: B. & C. Alexander/Photo Researchers, Inc. 41.9b: Ann & Steve Toon/NHPA. 41.11: Auscape (Parer-Cook)/Peter Arnold, Inc. 41.15: G. W. Willis/Biological Photo Service. 41.16a: Peter Chadwick/SPL/Photo Researchers, Inc. 41.16b: Jim Roetzel/Dembinsky Photo Assoc.

Chapter 42   *Opener*: Courtesy of R. D. Fernald, Stanford U. 42.6: Schwartzwald Lawrence/Corbis Sygma. 42.13b: Courtesy of Gerhard Heldmaier, Philipps University.

Chapter 43   *Opener*: Nik Wheeler. 43.1a: Biophoto Associates/Photo Researchers, Inc. 43.1b: Brian Parker/Tom Stack & Assoc. 43.2a: Patricia J. Wynne. 43.5: CNRI/SPL/Photo Researchers, Inc. 43.7a: Mitsuaki Iwago/Minden Pictures. 43.7b: Johnny Johnson/DRK Photo. 43.12: P. Bagavandoss/Photo Researchers, Inc. 43.17: Courtesy of The Institute for Reproductive Medicine and Science of Saint Barnabas, New Jersey.

Chapter 44   *Opener*: Dietmar Nill/Nature Picture Library. 44.3b: C. Raines/Visuals Unlimited.

Chapter 45   *Opener*: Courtesy of Grace Sours, ATF. 45.3 *left*: Courtesy of R. A. Steinbrecht. 45.3 *right*: G. I. Bernard/Animals Animals. 45.5,

45.10: P. Motta/Photo Researchers, Inc. 45.15a: Dennis Kunkel Microscopy, Inc. 45.18: Omikron/Science Source/Photo Researchers, Inc.

Chapter 46   *Opener*: From van Praag et al. 2002. *Nature* 415: 1030. Macmillan Publishers Ltd. 46.8: From Harlow, J. M., 1869. *Recovery from the passage of an iron bar through the head*. Boston: David Clapp & Son. 46.13a: David Joel Photography, Inc. 46.16: Wellcome Dept. of Cognitive Neurology/SPL/Photo Researchers, Inc.

Chapter 47   *Opener*: Mark Andersen/RubberBall Productions/PictureQuest. 47.1 *upper*: Innerspace Imaging/SPL/Photo Researchers, Inc. 47.1 *center*: SPL/Photo Researchers, Inc. 47.1 *lower*: Eric Grave/SPL/Photo Researchers, Inc. 47.3: Frank A. Pepe/Biological Photo Service. 47.8a: Courtesy of Jesper L. Andersen. 47.15: Robert Brons/Biological Photo Service. 47.19b *upper*: Ken Lucas/Visuals Unlimited. 47.19b *lower*: Fred McConnaughey/Photo Researchers, Inc.

Chapter 48   *Opener*: John Warden/Alaskan Express/PictureQuest. 48.1a: Larry Jon Friesen. 48.1b: Robert Brons/Biological Photo Service. 48.1c: Tom McHugh/Photo Researchers, Inc. 48.4b: Skip Moody/Dembinsky Photo Assoc. 48.4c: Courtesy of Thomas Eisner, Cornell U. 48.10a: SPL/Photo Researchers, Inc. 48.10c: P. Motta/Photo Researchers, Inc. 48.13: Fred Bruemmer/DRK Photo.

Chapter 49   *Opener*: Doc White/Nature Picture Library. 49.11: Dennis Kunkel Microscopy, Inc. 49.14a: Chuck Brown/Science Source/Photo Researchers, Inc. 49.14b: Biophoto Associates/Science Source/Photo Researchers, Inc. 49.15: After N. Campbell, 1990. *Biology*, 2nd Ed., Benjamin Cummings Publishing Co. 49.16b: CNRI/Photo Researchers, Inc.

Chapter 50   *Opener*: Marilyn "Angel" Wynn/Nativestock.com. 50.1a: Gerry Ellis, DigitalVision/PictureQuest. 50.1b: Tom Walker/Visuals Unlimited. 50.3: AP/Wide World Photos. 50.6: David Roberts/Nature's Images/Photo Researchers, Inc. 50.9: Dennis Kunkel Microscopy, Inc. 50.21: Jackson/Visuals Unlimited. 50.22: Katsutoshi Ito/Nature Productions.

Chapter 51   *Opener*: Michael Fogden/DRK Photo. 51.1a: Brian Kenney. 51.2b: Rod Planck/Photo Researchers, Inc. 51.8: From R. G. Kessel & R. H. Kardon. 1979. *Tissues and Organs*. W. H. Freeman, San Francisco. 51.11: Courtesy of Lise Bankir, INSERM Unit, Hôpital Necker, Paris.

Chapter 52   *Opener*: Frans de Waal, Emory U. 52.1: Bill Beatty/Visuals Unlimited. 52.3: Courtesy of Marc Chappell, U. California, Riverside. 52.5: Nina Leen/TimePix. 52.11: François Savigny/Animals Animals. 52.16: Fritz Pölking/Dembinsky Photo Assoc.

Chapter 53 *Opener*: Arturo M. Enríquez/ Painet, Inc. 53.1a: Francis Caldwell/Painet, Inc. 53.2: Anup Shah/DRK Photo. 53.6: Pete Oxford/DRK Photo. 53.8: Courtesy of John Alcock, Arizona State U. 53.9: Courtesy of Arild Johnsen, University of Oslo. 53.11: Art Wolfe. 53.12: David Houston/Bruce Coleman, Inc. 53.13: Courtesy of John Alcock, Arizona State U. 53.14: José Fuste Raga/AGE Fotostock.

Chapter 54 *Opener*: T. W. Davies/California Academy of Sciences. 54.3: Stephen G. Maka/ DRK Photo. 54.4: After P. A. Marquet, 2000. *Science* 289: 1487–1488. 54.5: Ed Reschke/Peter Arnold, Inc. 54.6: François Gohier/Photo Researchers, Inc. 54.9: From T. Coulson et al., 2002. *Science* 295: 2023–2024. 54.11: Tom Brakefield/DRK Photo. 54.13: T. W. Davies/ California Academy of Sciences. 54.15: Adam Jones/Photo Researchers, Inc. 54.19: Courtesy of John R. Hosking, NSW Agriculture, Australia.

Chapter 55 *Opener*: Bob Gibbons/Oxford Scientific Films. 55.1: Data from R. H. Whittaker, 1960. *Ecological Monographs* 30: 279–338. 55.10: Vitit Kantabutra/Painet, Inc. 55.11: Jim Zipp/Photo Researchers, Inc. 55.12: Tom Bean/DRK Photo. 55.15: After M. Begon, J. Harper, & C. Townsend. 1986. *Ecology*.

Blackwell Scientific Publications. 55.16a left: Bill Lea/Dembinsky Photo Associates. 55.16a right: Stephen J. Krasemann/Photo Researchers, Inc. 55.16b left: M. P. Kahl/DRK Photo. 55.16b center: Kenneth W. Fink/Photo Researchers, Inc. 55.16b right: Jeff Foott/DRK Photo.

Chapter 56 *Opener*: Peter Mead/Tom Stack & Assoc. 56.2 chameleon: Joe McDonald/Tom Stack & Assoc. 56.2 tenrec: Nigel J. Dennis/ Photo Researchers, Inc. 56.2 fossa: Pete Oxford/Nature Picture Library. 56.2 lemur: Wendy Dennis/Dembinsky Photo Associates. 56.2 baobab: Pete Oxford/Nature Picture Library. 56.2 Pachypodium: Michael Leach/ Oxford Scientific Films. Tundra, upper: Tom & Pat Leeson/DRK Photo. Tundra, lower: Elizabeth N. Orians. Boreal, upper: Carr Clifton/ Minden Pictures. Boreal, lower: James P. Rowan/DRK Photo. Temperate deciduous: Paul W. Johnson/Biological Photo Service. Temperate grasslands, upper: Robert & Jean Pollock/ Biological Photo Service. Temperate grasslands, lower: Elizabeth N. Orians. Cold desert, upper: Edward Ely/Biological Photo Service. Cold desert, lower: Art Wolfe. Hot desert, left: Terry Donnelly/Tom Stack & Assoc. Hot desert, right: Dave Watts/Tom Stack & Assoc. Chaparral: Elizabeth N. Orians. Thorn forest: Nick Garbutt/ Indri Images. Savanna: Tim Davis/Photo Re-

searchers, Inc. Tropical deciduous: Courtesy of Donald L. Stone. Tropical evergreen: Elizabeth N. Orians.

Chapter 57 *Opener*: Andrew J. Martinez/ Photo Researchers, Inc. 57.1a: Elizabeth N. Orians. 57.4: Richard Bierregaard, Courtesy of the Smithsonian Institution, Office of Environmental Awareness. 57.5: Adapted from S. H. Reichard and C. W. Hamilton, 1997. *Conservation Biology* 11: 193–203. 57.6a: Courtesy of Christopher Baisan and the Laboratory of Tree-Ring Research, U. Arizona, Tucson. 57.6b: Karen Wattenmaker/Painet, Inc. 57.7 photo: Stan Osolinsky/Oxford Scientific Films. 57.7b: After *BioScience*, 2002. 52: 809. 57.8a: R. B. Aronson et al., 2000. *Nature* 405: 36. 57.8b: Paul Osmond/Painet, Inc. 57.10: Tom Vezo/Nature Picture Library. 57.11: Julian Pender Hume/ NHMPL.

Chapter 58 *Opener*: John Wong/RubberBall Productions/PictureQuest. 58.1: Adapted from M. C. Jacobson et al., 2000. Introduction to *Earth System Science: From Biogeochemical Cycles to Global Climate Change*, Academic Press. 58.5a: Corbis Images/PictureQuest. 58.8: Adapted from B. A. Hungate et al., 1997. *Nature* 388: 576–578. 58.15: Courtesy of NASA.

Index

Numbers in **boldface** indicate a definition of the term; numbers in *italic* indicate there is information in a figure, figure caption, or table.

About the Book

Editor: Andrew D. Sinauer

Project Editor: Carol J. Wigg

Developmental Editor: James Funston

Review coordinator: Susan McGlew

Copy Editor: Norma Roche

Production Manager: Christopher Small

Book Layout and Production: Janice Holabird, Jefferson Johnson, and Joan Gemme

Art Editing and Illustration Program: Elizabeth Morales

Book and Cover Design: Jefferson Johnson

Photo Research: David McIntyre

Index: Acorn Indexing

Color Separations: Burt Russell Litho

Book and Cover Manufacture: Courier Companies, Inc.